IARD 2012: 8th Biennial Conference on Classical and Quantum Relativistic Dynamics of Particles and Fields

Journal of Physics: Conference Series Volume 437

Florence, Italy
29 May - 1 June 2012

ISBN: 978-1-62748-314-8
ISSN: 1742-6588

Printed from e-media with permission by:

Curran Associates, Inc.
57 Morehouse Lane
Red Hook, NY 12571

Some format issues inherent in the e-media version may also appear in this print version.

Printed by Curran Associates, Inc. (2013)

For permission requests, please contact the Institute of Physics at the address below.

Institute of Physics
Dirac House, Temple Back
Bristol BS1 6BE UK

Phone: 44 1 17 929 7481
Fax: 44 1 17 920 0979

techtracking@iop.org

Additional copies of this publication are available from:

Curran Associates, Inc.
57 Morehouse Lane
Red Hook, NY 12571 USA
Phone: 845-758-0400
Fax: 845-758-2634
Email: curran@proceedings.com
Web: www.proceedings.com

TABLE OF CONTENTS

Preface

Although the subject of relativistic dynamics has been explored, from both classical and quantum mechanical points of view, since the work of Einstein and Dirac, its most striking development has been in the framework of quantum field theory. The very accurate calculations of spectral and scattering properties, for example, of the anomalous magnetic moment of the electron and the Lamb shift in quantum electrodynamics, and many qualitative features of the strong and electroweak interactions, demonstrate the very great power of description achieved in this framework. Yet, many fundamental questions remain to be clarified, such as the structure of classical relativistic dynamical theories on the level of Hamilton and Lagrange in Minkowski space as well as on the curved manifolds of general relativity. There moreover remains the important question of the covariant classical description of systems at high energy for which particle production effects are not large, such as discussed in Synge's book, *The Relativistic Gas*, and in Balescu's book on relativistic statistical mechanics. In recent years, the study of high energy plasmas and heavy ion collisions has emphasized the importance of developing the techniques of relativistic mechanics.

The results of Lindner et al [Physical Review Letters **95**, 0040401 (2005)] as well as the more recent proposal of Palacios et al [Physical Review Letters **103**, 253001 (2009)] and others, have shown that there must be a quantum theory with coherence in time. Such a theory, manifestly covariant under the transformations of special relativity with an invariant evolution parameter, such as that of Stueckelberg [Helv. Phys. Acta **14**, 322, 588 (1941); **15**, 23 (1942); see also R.P. Feynman, Physical Review **80**, 4401 and J.S. Schwinger, Phys. Rev. **82**, 664, (1951)] could provide a suitable basis for the study of such questions, as well as many others for which the application of the standard methods of quantum field theory are difficult to manage, involving, in particular, local properties of spacetime structure.

The scope of this series of conferences is, however, much wider. There have been recent developments in the understanding of general relativity concerning questions associated with dark energy and the dark matter problem, the distribution of stars in galaxies, and the distribution of galaxies in the visible universe, as well as the internal structure of stars. There are, moreover fundamental questions in the applications of relativistic dynamics to physical problems, and in its mathematical and logical structure.

It was for this purpose, to bring together researchers from a wide variety of fields, such as particle physics, astrophysics, cosmology, heavy ion collisions, plasma research, and mathematical physics, with a common interest in relativistic dynamics, that this Association was founded.

The International Association for Relativistic Dynamics was organized at its first meeting as an informal session of seminars among researchers with common interest in February 1998 in Houston, Texas, with John R. Fanchi as president.

The second meeting took place, in 2000, at Bar Ilan University in Ramat Gan, Israel, the third, in 2002, at Howard University in Washington, D.C., and the fourth, on June 12-19, 2004, in Saas Fee, Switzerland. In 2006, the meeting took place at the University of Connecticut campus in Storrs, Connecticut, and the sixth meeting, in Thessaloniki, Greece. The seventh meeting took place at the National Dong Hwa University in Hualien, Taiwan from May 30 to June 1, 2010, and the eighth meeting, reported here, at the Galileo Galilei Institute for Theoretical Physics (GGI) in Florence, Italy, 29 May to 1 June 2012.

This meeting forms the basis for the Proceedings of IARD 2012, recorded in this issue of *Journal of Physics: Conference Series*. Along with the work of some of the founding members of the

Association, we were fortunate to have lecturers from application areas that provided strong challenges for further developments in cosmology and astrophysics, the geometry of spacetime, including the possible presence of extra dimensions, and in the dynamics of systems described in the framework of general relativity.

There have been recent developments as well in the foundations of relativity, and in the understanding of electrodynamics in the framework of relativistic quantum theory. There is a study of relativistic quantum mechanics in the rest frame instant form of dynamics, and an analysis of the Laplace-Runge-Lenz symmetry in relativistic dynamics reported here, as well as a discussion of the quantization of massless fields of any spin. Results are reported on the existence and definition of a covariant Berry phase associated with a perturbed covariant harmonic oscillator.

A generalization of Stueckelberg's original classical and quantum model for pair production is found to provide a simple framework for the phenomenon of neutrino oscillation, which, along with recently published work on the spin of a system of relativistic particles, appears to provide a simple mechanism for CP violation in the presence of nonabelian gauge fields that seems applicable to the K, B and D meson systems as well, from the point of view of their inner quark structure.

New functional methods applicable to both classical and quantum relativistic systems are reported here, and a deep mathematical and philosophical discussion is given on a unified view of nonlinear systems in many areas, including that of perception.

We thank the Scientific Advisory Committee for their invaluable guidance and advice:

Stephen Adler (Institute for Advanced Study)
Itzhak Bars (University of Southern California)
Gordon Baym (University of Illinois)
Jacob Bekenstein (Hebrew University)
Fred Cooper (Los Alamos National Laboratory)
Bei-Lok Hu (University of Maryland)
Werner Israel (University of Victoria)
E.V. Shuryak (Brookhaven National Laboratory)
L.S. Shulman (Clarkson University)
William Unruh (University of British Columbia)
Luca Lusanna (National Institute for Nuclear Physics, INFN)
Benoit Famaey (Observatoire Astronomique de Strasbourg, CNRS)

The organizers express their gratitude to the Galileo Galilei Institute for Theoretical Physics for its support and the use of its excellent facilities, and to INFN for its generous support.

Finally, we thank the participants who contributed through their lectures, personal discussions, and these papers, to the advancement of the subject and our understanding.

For the Editors and Organizing Committee,

L.P. Horwitz (Tel-Aviv University, Bar Ilan University), Editor-in-Chief
Luca Lusanna (INFN), Chairman of the Local Organizing committee
Tepper Gill (Howard University), IARD Treasurer
Martin Land (Hadassah College), IARD President
Paolo Salucci (SISSA)

Peer review statement

All papers published in this volume of *Journal of Physics: Conference Series* have been peer reviewed through processes administered by the proceedings Editors. Reviews were conducted by expert referees to the professional and scientific standards expected of a proceedings journal published by IOP Publishing.

Challenges for ΛCDM and MOND

Benoit Famaey
Observatoire Astronomique de Strasbourg, CNRS UMR 7550, France

Stacy McGaugh
Case Western Reserve University, USA

E-mail: benoit.famaey@astro.unistra.fr, ssm69@case.edu

Abstract. The Universe on large scales is well described by the ΛCDM cosmological model. There however remain some heavy clouds on our global understanding, especially on galaxy scales, which we review here. While some of these clouds might perhaps disappear through small compensatory adjustments of the model, such as changing the mass of the dark matter particles or accounting better for baryonic physics, others should rather be taken as strong indications that the physics of the dark sector is, at the very least, much richer and complex than currently assumed, and that our understanding of gravity and dynamics might also be at play. For instance, the empirically well-tested MOND phenomenology in galaxies, whatever its final explanation, should be understood in any model of galaxy formation and dynamics. Current alternatives to ΛCDM however bring with them many unsolved questions and challenges.

1. Introduction

Assuming General Relativity to be valid on all scales, data ranging from the Cosmic Microwave Background to individual galaxies point towards a Universe dominated by dark energy and dark matter, the nature of these being most certainly amongst the deepest problems of modern physics. While dark energy is well-represented by a cosmological constant Λ in Einstein's field equations, the currently preferred dark matter candidate is a collection of stable, neutral, elementary particles that condensed from the thermal bath of the early Universe, and which are known as 'cold dark matter' (CDM) particles (see, e.g., Frenk & White 2012, Strigari 2012 for recent reviews). On galaxy scales, however, predictions of this standard ΛCDM cosmological model, although plagued by the enormous complications of baryonic astrophysics, are difficult to reconcile with observations. We hereafter review these challenges for the ΛCDM model, and point out that some of them hint at a richer and more complex physics of the dark sector than currently assumed. In particular, many observed scaling relations involve the ubiquitous appearance of an acceleration constant $a_0 \approx \Lambda^{1/2} \sim 10^{-10}\mathrm{m/s}^2$, whose origin is a deep mystery in the standard context. Surprisingly, most of these scaling relations can be summarized by the empirical formula of Milgrom (1983). The success of this formula means that the gravitational field in galaxies mimics, for whatever reason, an effectively modified force law on galaxy scales, known as Modified Newtonian Dynamics (MOND). This formula however fails to account for dynamics and lensing of galaxy clusters (e.g., Clowe et al. 2006, Angus et al. 2007), meaning that if the formula would be reflecting a true modification of gravity as an alternative to galactic dark matter, it should rely at cluster scales on residual missing mass, which could be in baryonic or non-baryonic form (Milgrom 2008, Angus et al. 2007), or the formula should be extended (Zhao & Famaey 2012). It is also unclear how the angular power spectrum of the Cosmic Microwave Background (CMB) could be accounted for without resorting to some form of non-baryonic dark matter (e.g., Slosar et al. 2005, Angus 2009). Nevertheless, the main motivation for studying alternatives to ΛCDM based on the MOND phenomenology is *not* necessarily to get rid of any form of 'dark matter', but rather to explain why the observed gravitational field in galaxies is apparently mimicking a universal force law generated by the baryons alone. The simplest explanation is of course *a priori* not that dark matter particles arrange themselves (by a hitherto unknown physical mechanism) in order to mimic a fake force law, but rather that the force law itself is modified. However, at a fundamental level, covariant theories of modified gravity often have to include new fields in the dark sector to reproduce this effective force law (fields with an energy density nevertheless subdominant to the baryonic one, and a role completely different from that of CDM: they would mediate the force rather than sourcing it), or even introduce what could be called a 'dark matter medium' (with an energy density outweighing the baryonic one) exhibiting a kind of new fundamental interaction with baryons: this makes the confrontation between MOND and dark matter much less clear than often believed, since the former implies a more complex structure of the dark sector than the currently assumed CDM, but does not necessarily imply the absence of a dark sector. In MOND, the new fields responsible for the effects usually attributed to CDM would in fact be somewhat closer to dark energy fields than to CDM. It would of course be even more exciting if one would manage to find a physical connection between these putative new fields and the actual dark energy sector.

We herefater list a (non-exhaustive) series of problems for ΛCDM (Sect. 2), then show how theories based on the MOND phenomenology might address a lot of them (Sect. 3), before listing a (non-exhaustive) series of covariant theories currently proposed in this vein (Sect. 4), together with their own questionmarks and internal problems. The reading of this short proceeding can be completed by the reading of the exhaustive review paper recently published in Living Reviews in Relativity (Famaey & McGaugh 2012).

2. Current problems for our understanding of galaxy properties in ΛCDM

2.1. The nature and distribution of dwarf galaxies

2.1.a. **The missing satellites challenge.** ΛCDM simulations predict vast numbers of subhalos that are satellites to the main halo hosting a galaxy like the Milky Way. One would naively expect each of these subhalos to host their own miniature satellite galaxy. The Local Group looks nothing like this prediction, having only a small handful of dwarfs around each of the giant galaxies. Taking into account stellar feedback and heating processes (mainly at re-ionisation) in the galaxy formation process, the predicted number of faint satellites around a Milky Way-like galaxy is ~ 100 to 600. Since the majority of the 24 known satellites of the Milky Way have been largely discovered with the Sloan Digital Sky Survey (SDSS), and since this survey covered only one fifth of the sky, it has been argued that the problem was solved. However, models that successfully explain the properties of the Milky Way dwarf satellites predict unobserved dwarfs beyond the virial radius (Bovill & Ricotti 2011a). Moreover, the models tend to produce an overabundance of bright dwarf satellites ($L_V > 10^4 L_\odot$) with respect to observations (Bovill & Ricotti 2011b). A rather discomforting way-out is to simply state that the Milky Way must be a statistical outlier, but this is actually contradicted by the study of Strigari & Wechsler (2012) on the abundance of bright satellites around Milky Way-like galaxies in the SDSS survey. A somewhat related but nevertheless distinct problem is that simulations predict that the most massive subhalos of the Milky Way are too *dense* to host any of the observed bright satellite galaxies (Boylan-Kolchin et al. 2011, 2012). This is the 'too big to fail' aspect of the missing satellites challenge. A potential solution might be warm rather than cold dark matter.

2.1.b. **The satellites phase-space correlation challenge (disks of satellites).** The distribution of dark subhalos around Milky Way-sized halos is also predicted by ΛCDM to be roughly isotropic. However, the Milky Way satellites are currently observed to be highly correlated in phase-space: they lie within a seemingly rotationally supported, relatively thin disk (Fig. 1, see Kroupa et al. 2010). This differs from the prediction for subhalos in both coordinate and momentum space. Since the SDSS survey covered only one fifth of the sky, it will of course be most interesting to see whether ongoing surveys such as Pan-STARRS will confirm this state of affairs. Whether or not such a satellite phase-space correlation would be unique to the Milky Way should also be carefully checked: the Milky Way could be a statistical outlier, but if the ΛCDM model is a realistic description of nature, then the average satellite configurations in external galaxies should be only moderately flattened, and generally non-rotating. In this respect, the recent findings of Ibata et al. (2013) that half of the satellites of M31 define an extremely thin and extended rotating structure makes this phase-space correlation challenge even more severe (see Sect. 3 for a possible solution).

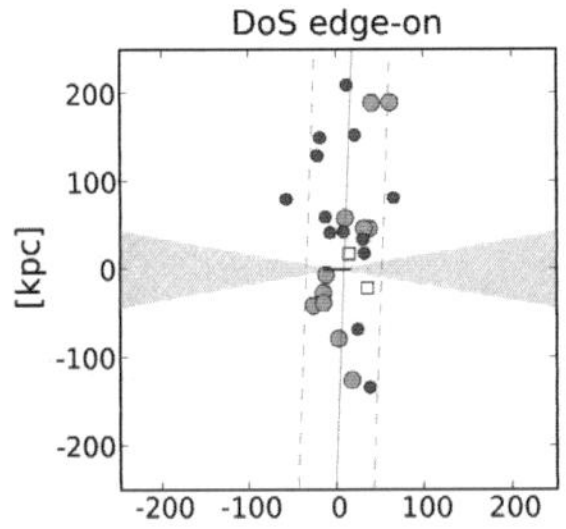

Figure 1. Spatial distribution of the Milky Way (MW) satellites (from Kroupa et al. 2010). Both axes in kpc. The 11 classical satellites are shown as large (yellow) circles, the 13 recently discovered satellites are represented by the smaller (green) dots, Pisces I and II being the two southern dots. The two open squares near the MW are Seg 1 and 2. The obscuration-region from the MW disk is given by the horizontal gray area.

2.1.c. **The density-morphology relation of dwarf ellipticals.** More dwarf elliptical galaxies are observed in denser environments (e.g. Kroupa et al. 2010). This relation, observed in the field, in galaxy groups and in galaxy clusters, is not yet understood.

2.2. The dynamical friction and stabilization from dark matter

2.2.a. **The angular momentum challenge.** Both the merger history of galaxy disks in a hierarchical formation scenario and the associated transfer of angular momentum from the baryonic disk to the dark halo cause the specific angular momentum of the baryons to end up being much too small in simulated disks. These simulated disks in turn end up much smaller than the observed ones, and elliptical systems end up too concentrated too. To address this challenge within the standard paradigm, one needs to form disks through late-time quiescent gas accretion from large-scale filaments, with less late-time mergers than currently predicted in ΛCDM.

2.2.b. **The pure disk challenge.** Related to the previous problem, large *bulgeless* thin disk galaxies are very difficult to produce in simulations. This is because major mergers typically create bulges. This is fundamental to the hierarchical nature of galaxy formation in ΛCDM, wherein galaxies are built by the merger of smaller galaxies. Bulgeless galaxies should therefore represent the rare, quiescent tail of a distribution of merger histories for galaxies of the Local Volume. However, these bulgeless disk galaxies are quite common, representing more than half of large galaxies (with $V_c > 150$ km/s) in the Local Volume.

2.2.c. **The stability challenge.** Quasi-spherical CDM halos stabilize very low surface density disks against the formation of bars and spirals, due to a lack of disk self-gravity. The observation (McGaugh et al. 1995) of Low Surface Brightness (LSB) disk galaxies with prominent bars and spirals is thus challenging in the absence of a significant disk component of dark matter. Moreover, in the absence of such a disky dark matter component, the lack of disk self-gravity prevents the creation of observed large razor thin LSB disks. This suggests more disk self-gravity than meets the eye (McGaugh & de Blok 1998). In the standard context, this would tend to point towards an additional disk-specific DM component, either a CDM-one linked to in-plane accretion of satellites, or a baryonic one, for instance in the form of molecular gas or due to a bottom-heavy IMF.

2.3. Structure formation not fast enough in ΛCDM?

2.3.a. **The bulk flow challenge.** In the ΛCDM model, peculiar velocities of galaxy clusters are predicted to be of order $\sim$ 200 km/s. These velocities can actually be measured through the fluctuations in the CMB generated by the scattering of the CMB photons by the hot X-ray-emitting gas inside galaxy clusters. These observations yield an observed coherent bulk flow of order $\sim$ 1000 km/s on scales out to at least 400 Mpc (Kashlinsky et al. 2012). This bulk flow challenge also appears in galaxy studies (Watkins et al. 2009). A related problem is the collision velocity of the merging bullet cluster 1E0657-56 at $z = 0.3$, which is larger than 3100 km/s, too high a value to be accounted for by ΛCDM. These observations seem to indicate that the attractive force is enhanced compared to what ΛCDM predicts.

2.3.b. **The high-z clusters challenge.** The observation of a single very massive cluster at high redshift can falsify ΛCDM. In this respect the existence of galaxy clusters like El Gordo with a mass of $\sim 2\times10^{15}\,M_\odot$ at $z = 0.87$ (Menanteau et al. 2012) and XMMU J2235.3-2557 with a mass of of $\sim 4\times10^{14}\,M_\odot$ at $z = 1.4$ is surprising. Though not sufficient to rule out the model, these clusters certainly push the envelope: there is only about a 50:50 chance that a cluster as massive as El Gordo exists at such high redshift. This has been the empirical experience: for every generation of surveys, structures are found further away, and appear sooner than expected.

2.3.c. **The Local Void challenge.** The 562 known galaxies at distances smaller than 8 Mpc from the center of the Local Group define the 'Local Volume'. Within this volume, the region known as the 'Local Void' hosts only 3 galaxies. This is much less than the expected $\sim$ 20 galaxies for a typical similar void in the ΛCDM model (Peebles & Nusser 2010). Moreover, in the Local Volume, large luminous galaxies are 6 times more abundant than predicted in the underdense regions. This could just mean that the Local Volume is a statistical outlier, but it could also point, in line with the two previous problems, towards more rapid structure formation in the early Universe, allowing sparse regions to form large galaxies cleaning their environment in a shorter time, thus making the voids emptier and the galaxies larger at early times.

2.4. Systematics in the dark-to-baryonic matter ratios and the ubiquitous a_0-scale

2.4.a. **The cusp-core challenge.** A long-standing problem of ΛCDM is the fact that the numerical simulations of the collapse of dark matter halos lead to a 'cuspy' density distribution that rises steeply as a decreasing function of radius, while rotation curves of external galaxies imply nearly constant density cores in the central parts. The state-of-the-art solution to this problem is to enforce strong supernovae outflows that move large amounts of low-angular-momentum gas from the central parts and that pull on the central dark matter concentration to create a core, but this is still a relatively fine-tuned process, which fails to account for cored profiles in the smallest galaxies, and fails to produce their observed baryon fractions

2.4.b. **The missing baryons challenge(s): baryon fraction defined by the acceleration in units of a_0.** Constraints from the CMB imply $\Omega_m = 0.27$ and $\Omega_b = 0.046$. However, our inventory of known baryons in the local Universe comes up short of the total. For example, Bell et al. (2003) estimate that the sum of cold gas and stars is only $\sim 5\%$ of Ω_b. While it now seems that many of the missing baryons are in the form of highly ionized gas in the warm-hot intergalactic medium (WHIM), we are still unable to give a confident account of where all the baryons reside. But there is another missing baryons challenge within each CDM halo: one would indeed naively expect each halo to have the same baryon fraction as the Universe, $f_b = \Omega_b/\Omega_m = 0.17$. On the scale of clusters of galaxies, this is approximately true (but still systematically low), but for galaxies, observations depart from this in a systematic way which remains unexplained in the standard context. In Fig. 2 (McGaugh et al. 2010), the ratio of the detected baryon fraction over the cosmological one, f_d, is plotted as a function of the potential well of the systems. There is a clear correlation: at large radii, the baryon fraction is actually directly equivalent to the acceleration in units of $a_0 = 10^{-10}\text{m/s}^2$. If we adopt a rough relation $M_{500} \simeq 1.5\times10^5\ M_\odot \times V_c^3(\text{km/s})^{-3}$, we get that the acceleration at R_{500}, and thus the baryon fraction, is $M_b/M_{500} = a_{500}/a_0 \simeq 4 \times 10^{-4} \times V_c(\text{km/s})^{-1}$. Divided by the cosmological baryon fraction, this explains the trend for $f_d = M_b/(0.17M_{500})$ with potential ($\Phi = V_c^2$) in Fig. 2. This missing baryons challenge is actually closely related to the baryonic Tully–Fisher relation.

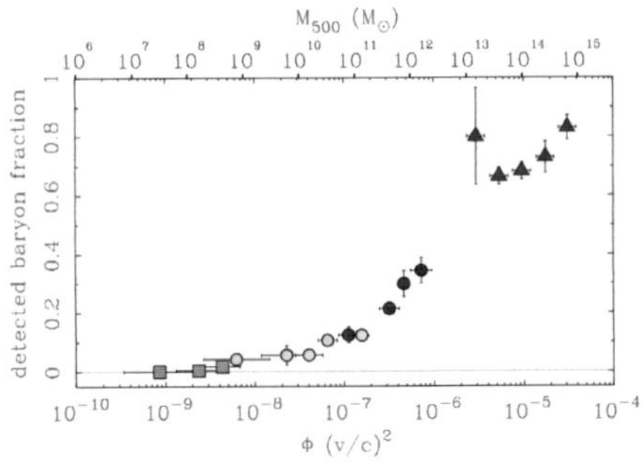

Figure 2. The fraction of detected over expected baryons $f_d = (M_b/M_{500})/f_b$ as a function of potential well and mass. Each point represents many objects. Orange squares are Local Group dwarf satellites, light blue circles gas-dominated disk galaxies, dark blue circles star-dominated spiral galaxies, grey triangles galaxy clusters.

2.4.c **The baryonic Tully–Fisher relation (with Ga_0 defining zero point).** The baryonic Tully–Fisher relation is an observed relation between the asymptotic circular velocity and baryonic mass in galaxies (Fig. 3). Recent work extending the relation to low mass, typically low surface brightness and gas rich galaxies, extends the dynamic range of the relation to five decades in baryonic mass. Over this range, the relation has remarkably little intrinsic scatter (consistent with zero given the observational errors) and is well described as a power law, or equivalently, as a straight line in log-log space $\log M_b = \alpha \log V_f - \log\beta$, with slope $\alpha = 4$. This slope is consistent with a constant acceleration scale $a_0 \approx V_f^4/(GM_b)$ such that the normalization constant $\beta = Ga_0$. All rotationally-supported galaxies are observed to follow the relation, irrespective of their size, surface brightness, or gas fraction.

2.4.d **Tidal dwarf galaxies.** Even young tidal dwarf galaxies (TDGs) formed in the collision of larger galaxies appear to obey the baryonic Tully–Fisher relation (Gentile et al. 2007). This is surprising because these galactic collisions should be very effective at segregating dark and baryonic matter: the rotating gas disks of galaxies that provide the fodder for tidal tails and the TDGs that form within them initially have nearly circular, coplanar orbits. In contrast, the dark matter particles are on predominantly radial orbits in a quasi-spherical distribution. This difference in phase space leads to tidal tails that themselves contain very little dark matter. When TDGs form from tidal debris, they should be largely devoid of dark matter. Nevertheless, TDGs do appear to contain enough dark matter to obey the same baryonic Tully-Fisher relation as galaxies that have not suffered this baryon-dark matter segregation process.

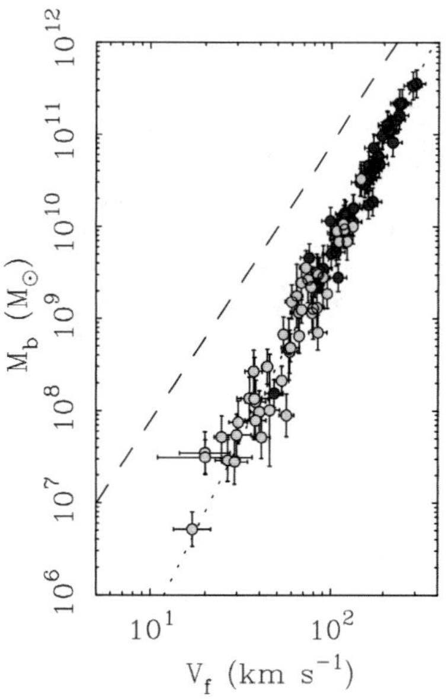

Figure 3. The Baryonic Tully–Fisher relation for galaxies with well-measured asymptotic velocities V_f. The dark blue points have $M_* > M_g$, while the light blue points have $M_* < M_g$ and are generally less precise in velocity, but more accurate in terms of mass. The dotted line has the observed slope 4, the dashed line has slope 3, the difference between these two lines being the origin of the variation in the detected baryon fraction in Fig. 2.

2.4.e. a_0 **as a transition acceleration.** The mass discrepancy in galaxies always appears (transition from baryon dominance to dark matter dominance) when $V_c^2/R \sim a_0$, yielding a clear mass-discrepancy acceleration relation (Fig. 4). This, again, is the case for every single rotationally supported system irrespective of its formation mechanism and history. For High Surface Brightness (HSB) galaxies, where there exists two distinct regions where $V_c^2/R > a_0$ in the inner parts and $V_c^2/R < a_0$ in the outer parts, locally measured mass-to-light ratios show no indication of hidden mass in the inner parts, but rise beyond the radius where $V_c^2/R \approx a_0$. It can never be emphasized enough that the role played by a_0 in the zero-point of the baryonic Tully-Fisher relation and this role as a transition acceleration have strictly no intrinsic link with each other, they are fully independent of each other within the CDM paradigm. There is nothing built in ΛCDM that stipulates that these two relations (the existence of a transition acceleration and the baryonic Tully-Fisher relation) should exist at all, and of course nothing that implies that these should exhibit an identical acceleration scale.

2.4.f. a_0/G **as a transition central surface density.** The acceleration a_0 defines the transition from LSB galaxies to HSB galaxies: in LSB galaxies whose *central* surface density is much smaller than some critical value of order $\Sigma_\dagger = a_0/G$, DM dominates everywhere, and the magnitude of the mass discrepancy is given by the inverse of the acceleration in units of a_0. On the other hand, baryons dominate in the inner parts of HSB galaxies whose central surface density is higher than $\Sigma_\dagger$. There is an anti-correlation between the baryonic surface density and the fractional contribution of DM to the rotation curve (Fig. 5). To explain this, there must be a strong fine-tuning between dark and baryonic surface densities, a sort of repulsion between them, a repulsion which is however contradicted by the correlations between baryonic and dark matter bumps and wiggles in rotation curves (known as 'Renzo's rule'). The shapes of rotation curves also depend on surface density: HSB galaxies have

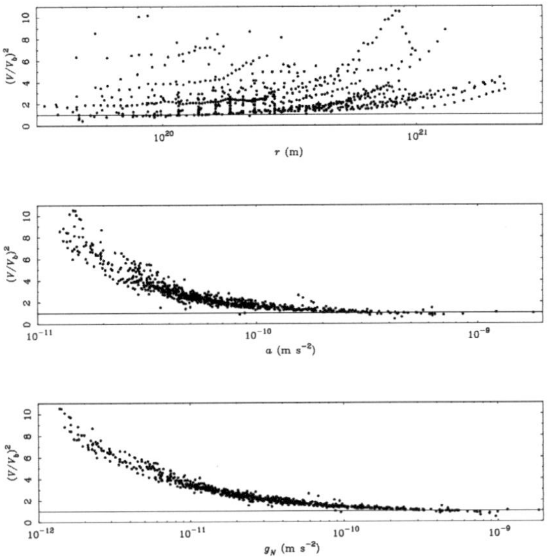

Figure 4. The mass discrepancy is defined as $(V/V_b)^2$ where V is the observed velocity and V_b is the velocity attributable to baryonic matter. Many hundreds of individual resolved measurements along the rotation curves of $\sim$ 100 spiral galaxies are plotted. The mass discrepancy is plotted as a function of radius (top panel), centripetal acceleration $a = V^2/r$ (middle panel), and Newtonian acceleration $g_N = V_b^2/r$ (bottom panel).

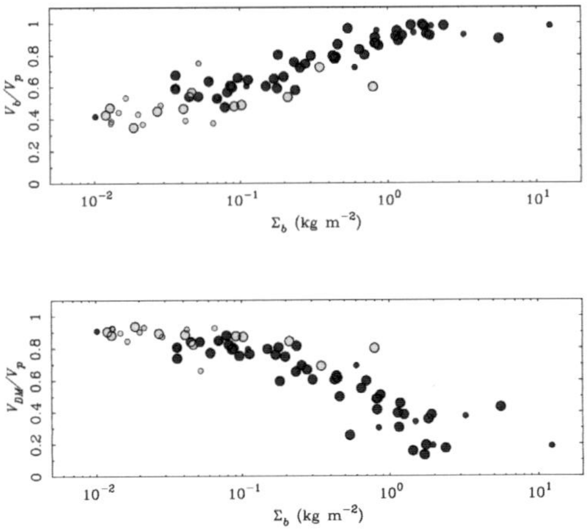

Figure 5. The fractional contribution of baryons (V_b/V_p, top) and dark matter (V_{DM}/V_p, bottom) to the rotation velocity V_p at the radius R_P where the contribution of the baryons peaks, plotted as a function of the characteristic surface density ($\Sigma_b = 0.75 M_b/R_p^2$). As the baryonic surface density increases, the fractional contribution of dark matter to the total gravitating mass decreases. This fine-tuning persists for any choice of stellar mass-to-light ratio.

rotation curves that rise steeply then become flat, or even fall somewhat to the not-yet-reached asymptotic flat velocity, while LSB galaxies have rotation curves that rise slowly to the asymptotic flat velocity. Finally the *total* (baryons+DM) acceleration declines with the mean *baryonic* surface density of galaxies, in the form $a \propto \Sigma_b^{1/2}$ (Fig. 6).

2.4.g. **Features in the baryonic distribution imply features in the rotation curve.** This is known as 'Renzo's rule' (Sancisi 2004, see also Swaters et al. 2012). While the effect of non-axisymmetric motions should be investigated in more detail, this is currently most easily interpreted by either a heavy 'dark baryonic' component (e.g., in the form of molecular H2) scaling with HI in galaxy disks, or by a theory where the baryons effectively act as the main source of gravity.

2.4.h. a_0/G **as a critical mean surface density for stability.** Disks with mean baryonic surface density $\langle\Sigma\rangle > \Sigma_\dagger = a_0/G$ are extremely rare (Fig. 7) and unstable. If stability is provided by a DM halo, there is no reason for this stability threshold to be related to the same acceleration scale that appears elsewhere (so this is again an independent occurence of the a_0-scale). Dense halos predicted by ΛCDM could (and should) in principle host higher surface brightness disks than are observed.

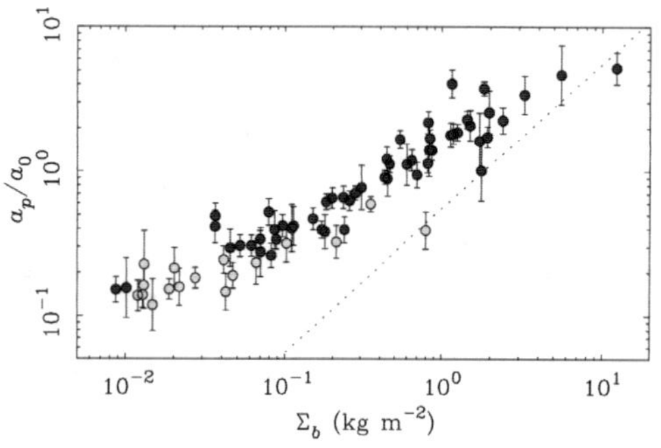

Figure 6. The characteristic centripetal acceleration $a_p = V_p^2/R_p$ in units of a_0 plotted against the characteristic baryonic surface density. The data do show a correlation ($a_p \propto \Sigma_b^{1/2}$), clearly indicating a dynamical role for the baryons, whilst CDM dominance should normally erase any such correlation.

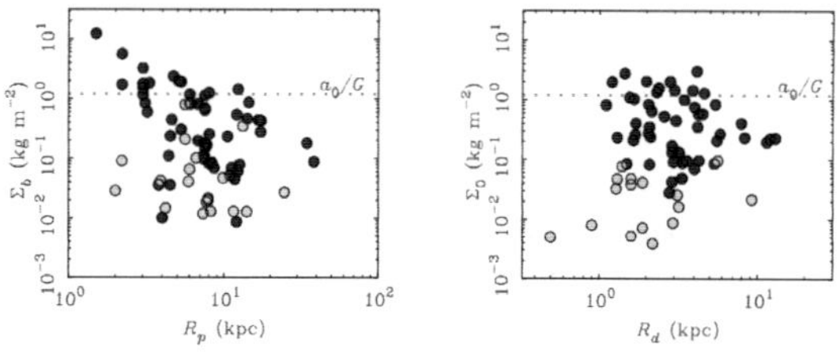

Figure 7. The characteristic surface density of baryons vs. their dynamical scale length R_p (Left), and disk-only central surface density vs. disk scale-length (Right). High characteristic surface densities at low R_p in the left panel are typical of bulge-dominated galaxies, and disks with characteristic surface density $> a_0/G$ are extremely rare.

3. Solving problems: Milgrom's law and MOND

As we have seen, the a_0 acceleration-scale is ubiquitous in galaxies. It defines the zero point of the Tully-Fisher relation and the amplitude of the mass-discrepancy in the weak-gravity limit, it appears as a transition-acceleration above which no DM is needed, and below which DM appears, and it defines the maximum surface density of pure disks. It would obviously be interesting if these observational occurrences could be summarized and empirically unified in some way: such a unification *is* actually feasible through the rather old idea of Milgrom (1983) that the total gravity $\mathbf{g}$ is related to the baryon-generated one $\mathbf{g_N}$ in a way analogous to Coulomb's law in a dielectric medium:

$$\mu\left(\frac{g}{a_0}\right)\mathbf{g} = \mathbf{g_N}, \tag{1}$$

where the interpolating function

$$\mu(x) \to 1 \text{ for } x \gg 1 \text{ and } \mu(x) \to x \text{ for } x \ll 1. \tag{2}$$

This law is at the basis of the Modified Newtonian Dynamics (MOND) paradigm. It does *not* simply make rotation curves flat, but predicted *all* the challenging observations of Sect. 2.4 in a quick and efficient way (see Sect. 5.2 of Famaey & McGaugh 2012). This formula was not constructed to explain already known observational facts, but predicted almost all of them *a priori*. Its salient properties can be elegantly summarized by simply stating that the underlying theory, whatever its deep nature, has to become scale-invariant under transformations $(t, \mathbf{r}) \to \lambda(t, \mathbf{r})$ in the regime of small acceleration $a \ll a_0$ (see, e.g., Milgrom 2012).

Addressing all the other challenges, from Sect. 2.1 to 2.3, would need a full-fledged theory predicting Milgrom's law and the associated MOND phenomenology. As we shall see in the next section, such a full-fledged theory does not exist at the moment. One can however speculate on how these problems could be addressed by such a theory. For instance, if the fields generating the MOND behavior on galaxy scales have an energy density subdominant to the baryonic one, effects such as dynamical friction with DM would disappear. Galaxy disks would then easily get more extended thanks to the absence of angular momentum transfer to the dark halo, but mergers would also take longer, and be subject to multiple passages (Tiret & Combes 2008). While MOND would naturally provide added stability compared to Newtonian dynamics without DM, it does not over-stabilize disks: features like bars and spiral arms are a natural result of disk self-gravity, and numerical simulations based on MOND appear to do a good job at reproducing the range of observed morphologies of spiral galaxies (Tiret & Combes 2007, 2008). Concerning structure formation, the MOND paradigm has *a priori* no cosmology, providing analogs for neither the Friedmann equation nor the Robertson-Walker metric. For these, one must appeal to specific hypotheses for the covariant parent theory of MOND (Sect. 4), which is far from unique. What is more, none of the candidates is theoretically satisfactory, as none emerges from first principles. It is thus not clear whether a compelling candidate MOND cosmology will ever emerge. One can however speculate on what would be a reasonable scenario, based on simulations using MOND in the weak-field regime at their basis (e.g., Sanders 1998, Angus & Diaferio 2011). In the absence of CDM, and with a MOND-like force driving structure formation, after initially lagging behind, structure formation would rapidly speed up compared to ΛCDM once the influence of radiation declines and perturbations begin to enter the MOND regime. Large galaxies could form by $z \approx 10$ (Sanders 2008) and clusters by $z \approx 2$, considerably earlier than in ΛCDM. By $z = 0$, the voids would become more empty than in ΛCDM, but otherwise MOND simulations (of collisionless particles, which is of course not the best representation of the baryon fluid) show the same qualitative features of the cosmic web (Llinares et al. 2008). The main difference is in the timing of when structures of a given mass would appear (assembling a large mass early being easier in MOND). This means that MOND is promising in addressing the high-z clusters challenge and Local Void challenge, as well as the bulk flow challenge and high collisional velocity of the bullet cluster (Llinares et al. 2009), due to the much larger than Newtonian MOND force in the structure formation context. What is more, it could allow large massive galaxies to form early ($z \approx 10$) from monolithic dissipationless collapse (Sanders 2008), with well-defined relationships between the mass, radius and velocity dispersion. Consequently, there would be less mergers than in ΛCDM at intermediate redshifts, but since these mergers would last longer, the number of interacting galaxies could remain unchanged. This could explain the observed abundance of large thin bulgeless disks unaffected by major mergers, and in those rare mergers between spirals, many tidal dwarf galaxies (TDGs) would be formed. This could lead to the possibility that many dwarf galaxies are not primordial but have been formed tidally in these encounters (Kroupa et al. 2010). These populations of satellite galaxies, associated with globular clusters that formed along with them, would naturally appear in closely related planes, thereby providing a natural solution to the satellites phase-space correlation problem (Ibata et al. 2013). To explain why these TDG dwarf satellites appear dark matter-dominated, while TDGs should not if DM is collisionless (see also problem 2.4.d hereabove), MONDian dynamics would be the only possible explanation. What is more, the density-morphology relation for dwarf ellipticals could also find a natural explanation (Dabringhausen & Kroupa 2012). However, all this would require a covariant parent theory of MOND, which will also have extreme challenges to overcome, and most notably the missing baryon problem that MOND suffers in rich clusters of galaxies and the high third peak in the acoustic power spectrum of the CMB.

4. MOND-inspired covariant theories: carrying their own problems

We provide hereafter a non-exhaustive list of the most popular covariant theories that have been proposed to reproduce the MOND phenomenology on galaxy scales. More details and references (as well as a more complete list including proposals based on non-local theories and modified inertia) can be found in Famaey & McGaugh (2012), as well as in Bruneton & Esposito-Farèse (2007) and Skordis & Zlosnik (2012). We also point out hereafter some of the main weaknesses of all these approaches. It should be highlighted again that these theories should *not* be confused with the MOND paradigm itself, which is not a theory *per se* (see e.g. Milgrom 2012).

4.1. Scalar-tensor k-essence (RAQUAL)

Main ingredients: An 'Einstein' metric $\tilde{g}_{\mu\nu}$ (with Einstein-Hilbert action), a k-essence scalar field ϕ, and normal matter from the standard model coupling to a combination of both.

Action: The matter couples to the 'physical metric':

$$g_{\mu\nu} \equiv e^{2\phi}\tilde{g}_{\mu\nu} \tag{3}$$

and the scalar field is given a so-called k-essence action, with no potential and a non-linear, aquadratic, kinetic term:

$$S_\phi \equiv -\frac{c^4}{2k^2l^2G}\int d^4x\sqrt{-\tilde{g}}\, f(X), \tag{4}$$

where k is a dimensionless constant, $l \sim c^2/a_0$ is a length-scale, and $X = kl^2\tilde{g}^{\mu\nu}\phi_{,\mu}\,\phi_{,\nu}$. The function $f(X)$ yields the MOND dynamics in the appropriate limit $X \ll 1$ (Bekenstein & Milgrom 1984).

Main weakness: Does not enhance gravitational lensing in galaxies because of the conformal relation between the Einstein and physical metrics, a priori enough to reject the theory.

4.2. Tensor-Vector-Scalar theory (TeVeS)

Main ingredients: An Einstein metric $\tilde{g}_{\mu\nu}$ (with Einstein-Hilbert action), a k-essence scalar field ϕ, a dynamical unit-norm vector field U_μ allowing a disformal relation between the Einstein and physical metrics (thus allowing to enhance gravitational lensing), and normal matter coupling to a combination of these fields.

Action: The matter couples to the physical metric:

$$g_{\mu\nu} \equiv e^{-2\phi}\tilde{g}_{\mu\nu} - 2\sinh(2\phi)U_\mu U_\nu, \tag{5}$$

the scalar field is given a k-essence action (Eq. 4), and the vector field (Bekenstein 2004, Skordis 2008):

$$S_U \equiv -\frac{c^4}{16\pi G}\int d^4x\sqrt{-\tilde{g}}\left[K^{\alpha\beta\mu\nu}U_{\beta;\alpha}U_{\nu;\mu} - \lambda(\tilde{g}^{\mu\nu}U_\mu U_\nu + 1)\right], \tag{6}$$

where

$$K^{\alpha\beta\mu\nu} = c_1\tilde{g}^{\alpha\mu}\tilde{g}^{\beta\nu} + c_2\tilde{g}^{\alpha\beta}\tilde{g}^{\mu\nu} + c_3\tilde{g}^{\alpha\nu}\tilde{g}^{\beta\mu} + c_4U^\alpha U^\mu\tilde{g}^{\beta\nu} \tag{7}$$

for a set of constants c_1, c_2, c_3, c_4, and λ is a Lagrange multiplier enforcing the unit-norm.

Main weaknesses: (i) Possible stability issues (especially inside matter), (ii) requires a fine-tuned form of $f(X)$ to evade strong-field observational constraints in the solar system and binary pulsars, (iii) cannot produce a high third peak in the angular power spectrum of the CMB without leading to an exaggerated integrated Sachs-Wolfe (ISW) effect, or without resorting to additional collisionless hot dark matter (appealing to which might overproduce superclusters and massive galaxy clusters as pointed out by Angus & Diaferio 2011), etc.

4.3. Generalized Einstein-Aether theories (GEA)

Main ingredients: Normal matter coupling to the physical metric $g_{\mu\nu}$ (with Einstein-Hilbert action), and a unit-norm dynamical vector field U_μ also coupling to the physical metric.

Action: Zlosnik et al. (2006) showed TeVeS to be expressible as a pure Tensor-Vector theory in the matter frame, with the *physical* metric at the same time satisfying the Einstein-Hilbert action and coupling minimally to the matter fields, just like in General Relativity (GR). The GEA approach makes use of this fact to devise a simpler Tensor-Vector action in the matter frame, where the Einstein-Hilbert and matter actions remain precisely as in GR, but with an additional unit-norm vector field (Zlosnik et al. 2007):

$$S_U \equiv -\frac{c^4}{16\pi G l^2}\int d^4x\sqrt{-g}\,\Big[f(X) - l^2\lambda(g^{\mu\nu}U_\mu U_\nu + 1)\Big], \tag{8}$$

where (see Eq. 7 and replacing $\tilde{g}^{\mu\nu}$ by $g^{\mu\nu}$)

$$X = l^2 K^{\alpha\beta\mu\nu}U_{\beta,\alpha}U_{\nu,\mu}, \tag{9}$$

and

$$l = \frac{(2-K)c^2}{3/2K^{3/2}a_0}. \tag{10}$$

In the static weak-field limit, the unit-norm constraint fixes the vector field in terms of the metric, and from there we have that, in the weak-field limit, $X \propto |\nabla\Phi|^2$.

Main weaknesses: Mostly the same as TeVeS for the CMB and ISW (Zuntz et al. 2010) and for the additional hot dark matter, *BUT*, contrary to TeVeS, it can be chosen to approach GR as fast as needed for high accelerations (with no fine tuning) so it avoids all the problems TeVeS has in binary pulsars, and the unacceptably large preferred-frame effects of TeVeS in the solar system.

4.4. Bimetric theories (BIMOND)

Main ingredients: Two metrics $g_{\mu\nu}$ and $\hat{g}_{\mu\nu}$, normal matter coupling to $g_{\mu\nu}$, and a new form of 'twin matter' coupling to the second metric $\hat{g}_{\mu\nu}$.

Action: The heart of this class of theories (Milgrom 2009) is an interaction term between the two metrics, through a function of scalars constructed from $C^\alpha_{\mu\nu}/a_0$ where $C^\alpha_{\mu\nu} = \Gamma^\alpha_{\mu\nu} - \hat{\Gamma}^\alpha_{\mu\nu}$. For instance, one can write:

$$S \equiv S_{\rm m}[\text{matter}, g_{\mu\nu}] + S_{\rm m}[\text{twin matter}, \hat{g}_{\mu\nu}] + \frac{c^4}{16\pi G}\int d^4x[\alpha\sqrt{-\hat{g}}\hat{R} + \beta\sqrt{-g}R - 2(g\hat{g})^{1/4}l^{-2}f(X)], \tag{11}$$

where $l = c^2/a_0$, and

$$X = l^2 g^{\mu\nu}(C^\alpha_{\mu\beta}C^\beta_{\nu\alpha} - C^\alpha_{\mu\nu}C^\beta_{\beta\alpha}). \tag{12}$$

Judiciously chosing the α and β parameters yields a whole class of BIMOND theories with various phenomenological behaviors. For instance, in matter-twin matter symmetric versions ($\alpha = \beta = 1$), and within a fully symmetric matter-twin matter system, a cosmological constant is given by the zero-point of the function f, naturally of the order of 1, thereby naturally leading to $\Lambda \sim a_0^2$ for the large-scale Universe. Matter and twin matter would not interact at all in the high-acceleration regime, and would repel each other in the MOND regime (i.e., when the acceleration difference of the two sectors is small compared to a_0), thereby possibly playing a crucial role in the Universe expansion and structure formation.

Main weaknesses: (i) Should still be checked against the existence of ghost modes, (ii) structure formation etc. to be studied in more details, (iii) not clear how these theories can explain the angular power spectrum of the CMB (even when appealing to twin matter).

4.5. Dipolar dark matter (DDM)

Main ingredients: Normal matter coupling to the GR physical metric $g_{\mu\nu}$, dark matter also coupling to the physical metric, but carrying a space-like[1] four-vector gravitational dipole moment ξ^μ.

Action: The dipolar dark matter medium (Blanchet 2007, Blanchet & Le Tiec 2009) is described as a fluid with mass current $J^\mu = \rho u^\mu$ endowed with the dipole moment vector ξ^μ (which will affect the total density in addition to the above mass density ρ), with the following action:

$$S_{\rm DM} \equiv \int d^4x\sqrt{-g}\,[c^2(J_\mu\dot{\xi}^\mu - \rho) - W(P)], \tag{13}$$

where P is the norm of the projection perpendicular to the four-velocity of the polarization field $P^\mu = \rho\xi^\mu$. The Poisson equation in the weak-field limit is then recovered as:

$$-\nabla.(\mathbf{g} - 4\pi\mathbf{P}) = 4\pi G(\rho_b + \rho). \tag{14}$$

From there, in order to reproduce the MOND phenomenology in galaxies, one appeals to a 'weak-clustering hypothesis', namely the fact that, in galaxies, the dark matter fluid does not cluster much ($\rho \ll \rho_b$) and is essentially at rest ($\mathbf{v} = 0$) because the internal force of the fluid precisely balances the gravitational force, in such a way that the polarization field $\mathbf{P}$ is precisely aligned with the gravitational one $\mathbf{g}$, and $g \propto -W'(P)$. At the cosmological level, the monopolar density of the dipolar atoms ρ will play the role of CDM, while the minimum of the potential $W(P)$ naturally adds a cosmological constant term $\Lambda \sim a_0^2$, thus explaining this numerical coincidence and making the theory precisely equivalent to the ΛCDM model at linear order for the expansion, for large scale structure formation, and for the CMB (naturally explaining the high third peak).

Main weaknesses: (i) Not clear that the weak-clustering hypothesis in galaxies would be the only natural outcome of structure formation within this model, (ii) the weak clustering hypothesis in itself might be problematic for explaining the residual missing mass in galaxy clusters, due to the fact that this residual mass should essentially be concentrated in the central parts of these objects, thus leaving the model to rely on baryonic dark matter or additional hot dark matter to explain galaxy clusters.

[1] This is to be contrasted with the time-like nature of TeVeS and GEA vector fields in the static weak-field limit

5. Conclusion

If one is familiar with cosmology and large scale structure, it must seem rather peculiar that anyone would seriously consider alternatives to ΛCDM based on the MOND phenomenology. But if one is more concerned with the observed phenomenology in a wide range of galaxy data, it seems just as odd to invoke non-baryonic and collisionless CDM together with fine-tuned feedback to explain the appearance of an effective force law that appears to act with only the observed baryons as a source in galaxies (see Sect 2.4 and 3, see also Famaey & McGaugh 2012 for an extensive review). Accepting this, and building a theory to account for this phenomenology, could perhaps help resolving (see Sect. 3) many of the other challenges (Sects. 2.1–2.3) currently faced by ΛCDM. Nevertheless, consistent covariant theories that have currently been proposed along these lines are at best effective, and bring with them their own challenges (see Sect. 4). Indeed, the most important aspect before one rejects any model is to have a 'simpler' model at hand, that still reproduces the successes of the earlier favored model but also naturally predicts the discrepant data. While MOND, as a paradigm, has *a priori* predicted a lot of observations which ΛCDM cannot explain in galaxies, it is absolutely fair to say that there is currently no alternative which does better *overall* than ΛCDM, and in favor of which Ockham's razor would be. It would however probably be a great mistake to persistently ignore the fine-tuning problems for ΛCDM and the related uncanny successes of the MOND paradigm on galaxy scales, as they could very plausibly point at a hypothetical better new theory of the dark sector. It would for instance be extremely exciting if one would manage to find a physical connection between the dark energy sector and the galactic MOND phenomenology in the weak-field limit through the $\Lambda \sim a_0^2$ coincidence. In this sense, ideas based on entropic gravity are an interesting line of thought (Verlinde 2011, Ho et al. 2010, Klinkhamer & Kopp 2011, and others, all somewhat inspired by Milgrom 1999), but one should nevertheless keep in mind the observational challenges that any such new theory of the dark sector could face, especially regarding the third peak of the acoustic power spectrum of the CMB. It is, on the other hand, conceivable that a fully successful alternative theory does not exist, and that the apparent MONDian behavior of galaxies will be explained through small compensatory adjustments of the current ΛCDM paradigm. But one has to realize that this is far from trivial, and one would have to demonstrate how this could possibly occur. One also has to realize that most currently proposed MOND-inspired effective theories (see Sect. 4) do include new fields, so that the MOND paradigm does not necessarily imply the absence of a dark sector. In any case, the existence of a characteristic acceleration $a_0 \sim \Lambda^{1/2}$ playing various roles in many apparently independent galactic scaling relations is by now an empirically established fact, and it is thus mandatory for *any* successful model of galaxy formation and evolution to explain it, together with all the other challenges mentioned in Sect. 2.

References

Angus, G.W., 2009, *Mon. Not. R. Astron. Soc.*, **394**, 527
Angus, G.W., Shan, H.Y., Zhao, H.S. and Famaey, B., 2007, *Astrophys. J.*, **654**, L13
Angus, G.W. and Diaferio, A., 2011, *Mon. Not. R. Astron. Soc.*, **417**, 941
Bekenstein, J., 2004, *Phys. Rev. D*, **70**, 083509
Bekenstein, J. and Milgrom, M., 1984, *Astrophys. J.*, **286**, 7
Bell, E.F., McIntosh, D.H., Katz, N. and Weinberg, M.D., 2003, *Astrophys. J.*, **585**, L117
Blanchet, L., 2007, *Class. Quantum Grav.*, **24**, 3541
Blanchet, L. and Le Tiec A., 2009, *Phys. Rev. D*, **80**, 023524
Bovill, M.S. and Ricotti, M., 2011, *Astrophys. J.*, **741**, 17
Bovill, M.S. and Ricotti, M., 2011, *Astrophys. J.*, **741**, 18
Boylan-Kolchin, M., Bullock, J. S. and Kaplinghat, M., 2011, *Mon. Not. R. Astron. Soc.*, **415**, L40
Boylan-Kolchin, M., Bullock, J. S. and Kaplinghat, M., 2012, *Mon. Not. R. Astron. Soc.*, **422**, 1203
Bruneton, J.-P. and Esposito-Farèse, G., 2007, *Phys. Rev. D*, **76**, 124012
Clowe, D., et al., 2006, *Astrophys. J.*, **648**, L109

Dabringhausen, J. and Kroupa, P., 2012, *Mon. Not. R. Astron. Soc.*, arXiv:1211.1382
Famaey, B. and McGaugh, S., 2012, *Living Reviews in Relativity*, **15**, 10
Frenk, C. and White, S., 2012, *Annalen der Physik*, **524**, 507
Gentile, G., Famaey, B., Combes, F., Kroupa, P., Zhao, H.S. and Tiret, O., 2007, *Astron. Astrophys.*, **472**, L25
Ho, C. M., Minic, D. and Jack Ng, Y., 2010, *Phys. Lett. B*, **693**, 567
Ibata, R., et al., 2013, *Nature*, **493**, 62
Kashlinsky, A., Atrio-Barandela, F. and Ebeling, H., 2012, *eprint* arXiv:1202.0717
Klinkhamer, F. R. and Kopp, M., 2011, *Modern Phys. Lett. A*, **26**, 2783
Kroupa, P. et al., 2010, *Astron. Astrophys.*, **523**, A32, (2010)
Llinares, C., Knebe, A. and Zhao, H.S., 2008, *Mon. Not. R. Astron. Soc.*, **391**, 1778
Llinares, C., Zhao, H.S. and Knebe, A., 2009, *Astrophys. J.*, **695**, L145
McGaugh, S.S., Schombert, J.M. and Bothun, G.D., 1995, *Astron. J.*, **109**, 2019
McGaugh, S.S. and de Blok, W.J.G, 1998, *Astrophys. J.*, **499**, 66
McGaugh, S.S., Schombert, J.M., de Blok, W.J.G. and Zagursky, M.J., 2010, *Astrophys. J.*, **708**, L14
Menanteau, F., et al., 2012, *Astrophys. J.*, **748**, 7
Milgrom, M., 1983, *Astrophys. J.*, **270**, 365
Milgrom, M., 1999, *Phys. Lett. A*, **253**, 273
Milgrom, M., 2008, *New Astron. Review*, **51**, 906
Milgrom, M., 2009, *Phys. Rev. D*, **80**, 123536
Milgrom, M., 2012, arXiv:1212.2568
Peebles, P.J.E. and Nusser, A., 2010, *Nature*, **465**, 565
Sancisi, R., 2004, *IAU Symposium*, **220**, 233
Sanders, R.H., 1998, *Mon. Not. R. Astron. Soc.*, **296**, 1009
Sanders, R.H., 2008, *Mon. Not. R. Astron. Soc.*, **386**, 1588
Skordis, C., 2008, *Phys. Rev. D*, **77**, 123502
Skordis, C. and Zlosnik, T.G., 2012, *Phys. Rev. D*, **85**, 044044
Slosar, A., Melchiorri, A. and Silk, J.I., 2005, *Phys. Rev. D*, **72**, 101301
Strigari, L.E. and Wechsler, R.H., 2012, *Astrophys. J.*, **749**, 75
Strigari, L.E., 2012, arXiv:1211.7090
Swaters, R., Sancisi, R., van der Hulst, J. and van Albada, T., 2012, *Mon. Not. R. Astron. Soc.*, **425**, 2299
Tiret, O. and Combes, F., 2007, *Astron. Astrophys.*, **464**, 517
Tiret, O. and Combes, F., 2008, *Astron. Astrophys.*, **483**, 719
Tiret, O. and Combes, F., 2008, *ASP Conference Series*, **396**, 259
Verlinde, E., 2011, *Journal of High Energy Physics*, **2011**, 29
Watkins, R., Feldman, H. A. and Hudson, M. J., 2009, *Mon. Not. Roy. Astron. Soc.*, **392**, 743
Zhao, H.S. and Famaey, B., 2012, *Phys. Rev. D*, **86**, 067301
Zlosnik, T.G., Ferreira, P.G. and Starkman, G.D., 2006, *Phys. Rev. D*, **74**, 044037
Zlosnik, T.G., Ferreira, P.G. and Starkman, G.D., 2007, *Phys. Rev. D*, **75**, 044017
Zuntz, J., Zlosnik, T. G., Bourliot, F., Ferreira, P. G. and Starkman, G. D., 2010, *Phys. Rev. D*, **75**, 104015

Galactic rotation curves in conformal gravity

Philip D Mannheim[1] **and James G O'Brien**[2]

[1]Department of Physics, University of Connecticut, Storrs, CT 06269, USA
[2]Department of Sciences, Wentworth Institute of Technology, Boston, MA 02115, USA

E-mail: [1]`philip.mannheim@uconn.edu`

E-mail: [2]`obrienj10@wit.edu`

Abstract. We review some recent work by Mannheim and O'Brien on the systematics of galactic rotation curves in the conformal gravity theory. In this work the conformal theory was applied to a comprehensive, high quality sample of spiral galaxies whose rotation curves extend well beyond the galactic optical disks. On galactic scales the conformal gravitational theory departs from the standard Newtonian theory in two distinct ways. One is a local way in which local matter sources within galaxies generate not just Newtonian potentials but linear potentials as well. The other is a global way in which two universal global potentials, one linear the other quadratic, are generated by the rest of the matter in the universe. The study involves a broad set of 138 spiral galaxies of differing luminosities and sizes, and is augmented here through the inclusion of an additional three tidal dwarf galaxies. With its linear and quadratic potentials the conformal theory can account for the systematics of an entire 141 galaxy sample without any need for galactic dark matter, doing so with only one free parameter per galaxy, namely the visible galactic mass to light ratio.

1. Introduction

Over the past three decades, the persistence of the missing mass or dark matter problem has generated an increasing interest in alternative gravitational theories. In the time since the classic studies (see e.g. [1]) of the missing mass problem in spiral galaxies galactic observational techniques have improved, and it has become possible to study and constrain the motions of luminous matter in galaxies with great precision. With rotational data for spiral galaxies now extending well beyond galactic optical disks, one finds that in essentially every case studied the measured rotational velocities do not conform with the familiar Newtonian gravity expectation associated with the observed visible material. It is important to note that the Newtonian expectation on galactic distance scales is derived by extrapolating standard gravity beyond solar system distance scales without any modification. To determine the expected Newtonian prediction one treats the galaxy as a collection of N^* individual sources each of typical mass $M_\odot$ and combines the Newtonian gravitational potentials generated by each of the individual sources. For a thin disk-shaped galaxy with a typical exponential surface brightness distribution $\Sigma(R) = \Sigma_0 e^{-R/R_0}$ with scale length R_0, the resulting circular velocity for a test particle at a radial distance R from the center of the disk is given by the Freeman formula (see e.g. [2])

$$v^2(R) = \frac{N^* M_\odot G R^2}{2R_0^3}\left[I_0\left(\frac{R}{2R_0}\right)K_0\left(\frac{R}{2R_0}\right) - I_1\left(\frac{R}{2R_0}\right)K_1\left(\frac{R}{2R_0}\right)\right]. \tag{1}$$

The failure of equation (1) using known visible matter sources alone can be alleviated via the ad hoc addition of dark, non-luminous, mass sources, as they can then bring the theory into agreement with data, and use of such dark matter sources has been extensively studied in the literature. However, the ability of such sources to fit data requires that the distribution of these sources be specifically chosen in each given galaxy on a case by case basis rather than in a way that is universal, with a substantial amount of dark matter always needing to be located in the outer regions of galaxies where there is little or no visible matter. Since so many galaxies over so extensive a range of sizes and luminous distributions have now been found to have a missing mass problem, some authors have considered the inability of luminous sources to fit the data on their own to be indicative of a failure of the Newtonian theory on galactic distance scales rather than of the presence of dark matter. Hence, these authors have suggested that is the extrapolation of solar system Newtonian gravity to galactic distance scales that is the cause of the problem (see e.g. [2] for a recent review). Various candidate alternative theories have been presented in the literature such as Milgrom's Modified Newtonian Dynamics (MOND) [3] and Moffat's Metric Skew Tensor Gravity (MSTG) [4]. It is the purpose of this review to highlight recent work done by Mannheim and O'Brien in applying another candidate alternative theory, namely the conformal gravity theory, to a large data sample containing some of the most well-studied spiral galaxies in the astronomical literature. Using the best galactic optical and radio data available, and a standardized, non-biased, treatment for selecting input galactic parameters, we find that the conformal theory is able to provide for a good accounting of the data without the need for any dark matter whatsoever. As with both MOND and MSTG, the conformal theory provides a universal prescription for departures from standard Newtonian gravity, to thus require no galaxy by galaxy dependent free parameters other than the standard galactic visible mass to light ratio common to all theories of galactic rotation curves.

The conformal theory of gravity was originally formulated by Weyl shortly after the advent of General Relativity. As such, it is a completely covariant metric theory of gravity, as endowed with an additional symmetry, namely local conformal invariance. This conformal symmetry forbids the presence of both any fundamental Einstein-Hilbert action term (with its dimensionful Newton constant) or any fundamental cosmological constant term in the gravitational action, with the gravitational action being uniquely prescribed to be of the form

$$I_{\rm W} = -\alpha_g \int d^4x\,(-g)^{1/2} C_{\lambda\mu\nu\kappa} C^{\lambda\mu\nu\kappa} \equiv -2\alpha_g \int d^4x\,(-g)^{1/2} \left[R_{\mu\kappa}R^{\mu\kappa} - \frac{1}{3}(R^{\alpha}{}_{\alpha})^2\right]. \qquad (2)$$

In equation (2) the the gravitational coupling constant α_g is dimensionless and the tensor $C_{\lambda\mu\nu\kappa}$ defined by

$$C_{\lambda\mu\nu\kappa} = R_{\lambda\mu\nu\kappa} - \frac{1}{2}\left(g_{\lambda\nu}R_{\mu\kappa} - g_{\lambda\kappa}R_{\mu\nu} - g_{\mu\nu}R_{\lambda\kappa} + g_{\mu\kappa}R_{\lambda\nu}\right) + \frac{1}{6}R^{\alpha}{}_{\alpha}\left(g_{\lambda\nu}g_{\mu\kappa} - g_{\lambda\kappa}g_{\mu\nu}\right) \qquad (3)$$

is the conformal Weyl tensor. The conformal action is invariant under local conformal transformations of the metric of the form $g_{\mu\nu}(x) \to e^{2\alpha(x)}g_{\mu\nu}(x)$ where $\alpha(x)$ is a local, spacetime-dependent phase. Because of its scale invariance, the conformal theory does not suffer from the cosmological problem present in the standard Einstein gravitational theory [5, 6].

Unlike the standard Einstein theory where the action is linear in the Riemann curvature tensor, the conformal action is quadratic in the Riemann tensor. Consequently, its equations of motion are fourth-order derivative functions of the metric, and in the presence of a matter source $T^{\mu\nu}$ take the form

$$4\alpha_g W^{\mu\nu} = 4\alpha_g \left[2C^{\mu\lambda\nu\kappa}{}_{;\lambda;\kappa} - C^{\mu\lambda\nu\kappa}R_{\lambda\kappa}\right] = 4\alpha_g\left[W^{\mu\nu}_{(2)} - \frac{1}{3}W^{\mu\nu}_{(1)}\right] = T^{\mu\nu}, \qquad (4)$$

where the tensors $W^{\mu\nu}_{(1)}$ and $W^{\mu\nu}_{(2)}$ are given by

$$W^{\mu\nu}_{(1)} = 2g^{\mu\nu}(R^{\alpha}{}_{\alpha})^{;\beta}{}_{;\beta} - 2(R^{\alpha}{}_{\alpha})^{;\mu;\nu} - 2R^{\alpha}{}_{\alpha}R^{\mu\nu} + \frac{1}{2}g^{\mu\nu}(R^{\alpha}{}_{\alpha})^2,$$
$$W^{\mu\nu}_{(2)} = \frac{1}{2}g^{\mu\nu}(R^{\alpha}{}_{\alpha})^{;\beta}{}_{;\beta} + R^{\mu\nu;\beta}{}_{;\beta} - R^{\mu\beta;\nu}{}_{;\beta} - R^{\nu\beta;\mu}{}_{;\beta} - 2R^{\mu\beta}R^{\nu}{}_{\beta} + \frac{1}{2}g^{\mu\nu}R_{\alpha\beta}R^{\alpha\beta}. \quad (5)$$

Despite its somewhat formidable appearance, Mannheim and Kazanas [7, 8] were able to find a closed form analytic solution to equation (4) in the case of a standard static spherically symmetric geometry with line element

$$ds^2 = B(r)c^2dt^2 - A(r)dr^2 - r^2d\Omega_2, \quad (6)$$

where $d\Omega_2 = d\theta^2 + \sin^2\theta d\phi^2$. Specifically, using the conformal symmetry they showed that one could transform this line element into one that was conformal to a line element with $A(r) = 1/B(r)$, with equation (4) then reducing (up to an irrelevant conformal transformation) to the remarkably simple

$$\nabla^4 B(r) = B'''' + \frac{4B'''}{r} = \frac{3}{4\alpha_g B(r)}(T^0{}_0 - T^r{}_r) \equiv f(r), \quad (7)$$

a relation that is exact without approximation. The general solution to equation (7) is given by

$$\begin{aligned} B(r) \;=\; & -\frac{r}{2}\int_0^r dr'\, r'^2 f(r') - \frac{1}{6r}\int_0^r dr'\, r'^4 f(r') \\ & -\frac{1}{2}\int_r^\infty dr'\, r'^3 f(r') - \frac{r^2}{6}\int_r^\infty dr'\, r' f(r') + \hat{B}(r), \end{aligned} \quad (8)$$

where $\hat{B}(r)$ obeys $\nabla^4\hat{B}(r) = 0$. Solutions to equation (7) will depend on the matter distribution $f(r)$, and unlike the standard second-order theory where only local matter inside of a source generates a gravitational force, in the conformal theory both local matter inside a source (the first two integrals in equation (8)) and the global matter exterior to it (the third and fourth integrals in equation (8)) can make contributions [9]. Thus to determine motions within galaxies one needs to consider local contributions coming from the visible material inside the galaxy and global contributions coming from material outside the galaxy (viz. the rest of the universe). As we will see below, the combined effect of these local and global contributions will enable us to provide an accounting of galactic rotation curve data without the need to invoke dark matter.

Given equation (8), a star of radius r_0 and source function $f^*(r)$ will generate a gravitational potential of the form

$$V^*(r > r_0) = -\frac{\beta^* c^2}{r} + \frac{\gamma^* c^2 r}{2} \quad (9)$$

(as normalized to unit solar mass units), where

$$\gamma^* = -\frac{1}{2}\int_0^{r_0} dr'\, r'^2 f^*(r'), \qquad 2\beta^* = \frac{1}{6}\int_0^{r_0} dr'\, r'^4 f^*(r'). \quad (10)$$

A star such as the sun thus puts out both a Newtonian potential $-\beta^* c^2/r$ and a linear potential $\gamma^* c^2 r/2$. As can be seen, on identifying $\beta^* = M_\odot G/c^2$ the potential given in equation (9) reduces to the standard Newtonian potential in the small r limit where the $\gamma^* r$ term becomes negligible. However, for large enough r this same term would become competitive with the Newtonian term, to thereby precisely provide for our desired departure from standard Newtonian gravity on large distance scales.

For applications to galaxies, we need to integrate $V^*(r)$ over the local matter distribution in the galaxy. On taking the galaxy to be a thin disk with N^* stars distributed with surface brightness $\Sigma(R) = \Sigma_0 e^{-R/R_0}$ and luminosity $L = 2\pi R_0^2\Sigma_0$, the net contribution to circular velocities due the local material within galaxies is found to be given by [2]

$$v_{\rm LOC}^2({\rm disk}) = \frac{N^*\beta^*c^2R^2}{2R_0^3}\left[I_0\left(\frac{R}{2R_0}\right)K_0\left(\frac{R}{2R_0}\right) - I_1\left(\frac{R}{2R_0}\right)K_1\left(\frac{R}{2R_0}\right)\right] + \frac{N^*\gamma^*c^2R^2}{2R_0}I_1\left(\frac{R}{2R_0}\right)K_1\left(\frac{R}{2R_0}\right). \quad (11)$$

The ratio $N^*M_\odot/L$ is recognized as the galactic mass to light ratio M/L. Equation (11) generalizes the Newtonian expectation given in equation (1) and gives the net conformal gravity contribution to circular velocities due to the local material within a galaxy.

As regards the global effect due to the rest of the material in the universe, we note that cosmologically there are two main global components, a cosmological background that is homogeneous and isotropic, and inhomogeneities in it that arise due to fluctuations around that background. In the conformal theory both of these components make a contribution to motions within galaxies, but they do so in quite different ways. Since the cosmological background is both homogeneous and isotropic, it can be described by a Robertson-Walker (RW) geometry with isotropic coordinate line element

$$ds^2 = c^2d\tau^2 - \frac{R^2(\tau)}{[1+K\rho^2/4]^2}\left(d\rho^2 + \rho^2 d\Omega_2\right). \quad (12)$$

With an RW geometry being conformal to flat, in it both the Weyl tensor $C^{\lambda\mu\nu\kappa}$ and the $W^{\mu\nu}$ tensor introduced in equation (4) vanish identically. In consequence the background matter energy-momentum tensor must vanish also (something it both can do and does do non-trivially in the conformal case [2]), with the cosmic background thus only contributing to the $\hat{B}(r)$ term in equation (8). Since the presence of inhomogeneities breaks the conformal to flat structure of the background geometry, inhomogeneities then lead to both a non-vanishing $W^{\mu\nu}$ and a non-vanishing $f(r)$, and hence contribute to the global third and fourth integrals in equation (8). Moreover, no matter in what specific way the cosmological background and the inhomogeneities in it might contribute to equation (8), they must do so in a way that is independent of any given galaxy, i.e. they must contribute universally via an effect that is to act in exactly the same way on each and every galaxy, without any regard to their morphologies, sizes, luminosities, or masses. We thus anticipate, and will in fact find, that there must be some universality in the galactic rotation curve data.

To determine the specific contribution of the cosmological background to galactic motions we need to transform the comoving cosmological RW geometry to the Schwarzschild coordinate rest frame coordinate system given in (6). To this end we note [7] that the coordinate transformation

$$\rho = \frac{4r}{[2(1+\gamma_0 r)^{1/2} + 2 + \gamma_0 r]}, \qquad t = \int \frac{d\tau}{R(\tau)} \quad (13)$$

transforms the RW line element into

$$ds^2 = e^{-2\alpha(\tau,\rho)}\left[(1+\gamma_0 r)c^2dt^2 - \frac{dr^2}{(1+\gamma_0 r)} - r^2 d\Omega_2\right], \quad (14)$$

where

$$\gamma_0 = 2(-K)^{1/2}, \qquad e^{-\alpha(\tau,\rho)} = R(\tau)\frac{(1-\gamma_0\rho/4)}{(1+\gamma_0\rho/4)}. \quad (15)$$

With a conformal transformation of the form $g_{\mu\nu} \rightarrow g_{\mu\nu}e^{2\alpha(\tau,\rho)}$ bringing the line element in equation (14) to the Schwarzschild coordinate system, we see that in the rest frame of a galaxy that is comoving with the Hubble flow, conformal cosmology acts like a universal linear potential. With conformal cosmology naturally leading [2] to a spatially open RW geometry with negative spatial curvature K, the coefficient γ_0 of the universal linear potential will be real, just as is needed to enable $\gamma_0 c^2 r/2$ is to serve as a gravitational potential. Since the measured rotational velocities in galaxies are non-relativistic we can incorporate the universal linear potential using weak gravity, with (11) then being augmented [9] to

$$v_{\rm TOT}^2(R) = v_{\rm LOC}^2({\rm disk}) + \frac{\gamma_0 c^2 R}{2}. \tag{16}$$

To determine the specific contribution due to inhomogeneities in the cosmological background, we return to the third and fourth integrals in equation (8) to which such cosmological inhomogeneities will contribute. With our interest being in applying equation (8) on galactic distance scales, we note that since such scales are much smaller than any typical cluster of galaxies scale $r_{\rm clus}$ that would be associated with cosmological inhomogeneities, for the purposes of evaluating $v_{\rm TOT}^2(R \ll r_{\rm clus})$ we can replace the lower limits in each of those two integrals by $r_{\rm clus}$. Thus, on setting

$$\kappa = \frac{1}{6}\int_{r_{\rm clus}}^{\infty} dr'\, r' f(r'). \tag{17}$$

we augment equation (16) to [10, 11]

$$v_{\rm TOT}^2(R) = v_{\rm LOC}^2({\rm disk}) + \frac{\gamma_0 c^2 R}{2} - \kappa c^2 R^2, \tag{18}$$

with associated asymptotic limit

$$v_{\rm TOT}^2(R) \rightarrow \frac{N^*\beta^* c^2}{R} + \frac{N^*\gamma^* c^2 R}{2} + \frac{\gamma_0 c^2 R}{2} - \kappa c^2 R^2. \tag{19}$$

With $v_{\rm LOC}^2({\rm disk})$ readily generalizing (see equation (24) below) to include more than one visible component (as well as an optical disk of stars galaxies can possess spherical stellar bulges and a disk of atomic gas), equation (18) thus gives the expectation of conformal gravity for galactic rotational velocities. With the mass of the atomic gas being measurable, even with its inclusion, $v_{\rm TOT}^2(R)$ would still only contain one free parameter per galaxy, namely the total number of stars N^*, and is thus highly constrained. In the original study of [9] equation (16) was used to fit an 11 galaxy sample, and good fitting was found with parameters

$$\gamma^* = 5.42 \times 10^{-41}\ {\rm cm}^{-1}, \qquad \gamma_0 = 3.06 \times 10^{-30}\ {\rm cm}^{-1}. \tag{20}$$

At the time the fitting presented in [9] was made, it was not appreciated that equation (16) would need to be augmented by the quadratic $-\kappa c^2 R^2$ term. Basically, the data that were available at the time did not go out far enough in distance for any possible large distance behavior other than a linear growth in $v_{\rm TOT}^2(R)$ to matter. However as more data accumulated it became clear that continuing linear growth was not being found in the data. When a large data sample of 111 galaxies was explored [10, 11] a set of 21 of them was identified [10] in which the expectation of equations (16) and (20) completely overshot the data at large distances. Remarkably, it was found in [10, 11] that with the addition of just the one universal $-\kappa c^2 R^2$ term with a repulsive overall sign and a magnitude

$$\kappa = 9.54 \times 10^{-54}\ {\rm cm}^{-2}, \tag{21}$$

and with the continuing use of the values for γ^* and γ_0 given in (20), fitting to the entire 21 galaxies could then be brought into accord with data, with fitting to none of the other 90 galaxies in the 111 galaxy sample being found to be adversely affected. The realization that the κ term was significant on the largest available galactic distance scales underscores the value in having a data sample large enough that one could uncover unanticipated regularities. Subsequently, equations (18), (20) and (21) were successfully applied [12] to a further set of 27 galaxies, and below we describe the fitting to the full 138 galaxy set in detail. In addition, we present an application of our theory to three tidal dwarf galaxies. With only one free parameter per galaxy the successful fitting to a total of now 141 galaxies is very encouraging for the conformal theory.

Self-consistently, we note that the value we obtain for γ^* ensures that the linear potential of the sun is indeed negligible for the solar system, with the standard gravitational solar system phenomenology being left intact, just as desired. Additionally, we note that γ^* is so small that one needs to go to a galactic system (N^* of order 10^{10} or so, and R of order a galactic distance scale) in order for the $N^*\gamma^*c^2R/2$ term to build up into something that could compete with Newtonian gravity. Moreover, the magnitude of the γ_0 parameter shows it to indeed be cosmological in scale, and with $\gamma_0/\gamma^* \sim 5\times10^{10}$, the γ_0 term also first manifests itself on galactic scales. Finally, we note that with $\kappa^{-1/2} \sim 3\times10^{26}$ cm the parameter κ is indeed a cosmological inhomogeneity distance scale, again just as required. Beyond being able to fit data at all with the very compact and highly constrained equation (18), we note that the specific values that we obtain for the γ^*, γ_0 and κ parameters are completely compatible with the motivation that led us to consider them in the first place.

2. Details of the Fitting Formalism

In applying the conformal theory to galactic rotation curves we need to specify the local contribution to the rotational velocity due to each of the individual visible matter components within each galaxy. These components typically consist of a stellar disk, a stellar bulge, and atomic gas. Both of the stellar components are located within the optically visible disk, with any stellar bulge being located in the galactic center. For a stellar disk with scale length R_0 the optical disk typically becomes faint at a distance R of order $4R_0$ or so. The atomic gas is distributed both within and beyond the optical disk. Since the Newtonian contribution of the stellar material (disk and bulge combined when applicable) would lead to a Keplerian fall off in rotational velocities in the $R > 4R_0$ region, study of rotation curves via HI atomic gas radio observations provides the primary probe of this expectation in the $R > 4R_0$ region. With the measured circular velocities showing no sign of any Keplerian fall off beyond $R = 4R_0$, and with the Newtonian contribution of the atomic gas itself not being substantial enough to affect the circular velocities in any significant way, the $R > 4R_0$ HI radio data provide the clearest evidence that there is a missing mass problem in galaxies. To test whether the conformal theory can account for the detected departures from the luminous Newtonian expectation without any need to introduce dark matter, we have therefore chosen to fit the HI rotation curve data of a large and very comprehensive set of 138 galaxies (we fit the HI data over the full available R range and not just over the $R > 4R_0$ range of course), with the data that we use being the best HI data available in the astronomical literature.

Of the galaxies we study, either the atomic gas mass is much less than the stellar mass (the case in high surface brightness (HSB) spirals) or the atomic gas mass and the stellar mass are both small (typically the case in low surface brightness (LSB) spirals or dwarf (DWF) spirals). With the atomic gas being distributed in a thin disk, then just like the stellar disk we shall model the gas profile as an exponential disk with an effective surface mass density $\Sigma_{\rm gas}(R) = \Sigma_0({\rm gas})e^{-R/R_0({\rm gas})}$. (Even though gas profiles are typically not as close to a single exponential disk as the stellar distributions, since the gas is not a substantial contributor to the total rotational velocity, our fits are not sensitive to our modeling of the gas profile.) Since the

HI gas extends well beyond the optical disk, the effective scale length of the gas in any galaxy must be quite a bit larger than that of the disk stars in the same galaxy, and after modeling a few measured gas surface densities we found an effective gas scale length that was typically four times that of the accompanying stellar R_0. Since in every case we study there is little sensitivity to the HI gas content, for each galaxy in our sample we shall use a gas scale length that is four times that of the stellar scale length of that galaxy. (For some of our chosen galaxies we had to use a model profile anyway since an actual gas profile had not been reported in the literature.) Since the atomic gas in a galaxy consists of both hydrogen and helium, we multiply the measured HI mass by 1.4 to account for primordial helium, to then obtain a total atomic gas mass $N_{\rm gas}M_\odot$ as expressed in solar mass units. With such a local gas distribution the contribution of the gas to rotation velocities duplicates equation (11) and is given by

$$\begin{aligned} v^2_{\rm LOC}({\rm gas}) &= \frac{N_{\rm gas}\beta^* c^2 R^2}{2R_0^3({\rm gas})}\left[I_0\left(\frac{R}{2R_0({\rm gas})}\right)K_0\left(\frac{R}{2R_0({\rm gas})}\right)-I_1\left(\frac{R}{2R_0({\rm gas})}\right)K_1\left(\frac{R}{2R_0({\rm gas})}\right)\right] \\ &+ \frac{N_{\rm gas}\gamma^* c^2 R^2}{2R_0({\rm gas})}I_1\left(\frac{R}{2R_0({\rm gas})}\right)K_1\left(\frac{R}{2R_0({\rm gas})}\right). \end{aligned} \tag{22}$$

Some of the galaxies in our sample have a spherical bulge. With a spherical mass density $\sigma(r)$ per unit solar mass and a total mass $N_{\rm bulge}M_\odot$ where $N_{\rm bulge} = 4\pi\int dr'\, r'^2\sigma(r')$ the net gravitational potential due to a spherical bulge can then be obtained by integrating equation (9) over the $\sigma(r)$ distribution, and is found to yield circular velocities of the form [2]

$$\begin{aligned} v^2_{\rm LOC}({\rm bulge}) &= \frac{4\pi\beta^* c^2}{r}\int_0^r dr'\,\sigma(r')r'^2 \\ &+ \frac{2\pi\gamma^* c^2}{3r}\int_0^r dr'\,\sigma(r')(3r^2r'^2-r'^4)+\frac{4\pi\gamma^* c^2 r^2}{3}\int_r^\infty dr'\,\sigma(r')r'. \end{aligned} \tag{23}$$

Finally, on including the gas and bulge contributions we augment equation (18) to

$$v^2_{\rm TOT}(R) = v^2_{\rm LOC}({\rm disk}) + v^2_{\rm LOC}({\rm gas}) + v^2_{\rm LOC}({\rm bulge}) + \frac{\gamma_0 c^2 R}{2} - \kappa c^2 R^2, \tag{24}$$

with associated asymptotic limit

$$v^2_{\rm TOT}(R) \to \frac{N_{\rm TOT}\beta^* c^2}{R} + \frac{N_{\rm TOT}\gamma^* c^2 R}{2} + \frac{\gamma_0 c^2 R}{2} - \kappa c^2 R^2, \tag{25}$$

where

$$N_{\rm TOT} = N^* + N_{\rm gas} + N_{\rm bulge}. \tag{26}$$

Equation (24) is our main result, and we proceed below to apply it to galactic rotation curve data. As regards its use, we note that with the mass of the atomic gas $N_{\rm gas}M_\odot$ being measured directly in the radio observations, and with the parameters γ^*, γ_0 and κ being universal (they can be determined by just a few rotation curve data points in just a single one of the 138 galaxies in our sample), in equation (24) there is only one parameter that can vary from one galaxy to the next, namely the total stellar mass $(N^* + N_{\rm bulge})M_\odot$ in each galaxy. The galactic stellar mass is not measured directly since in photometric studies of the stars in the optical disk one instead measures the galactic luminosity. However, with the conformal gravity contributions to $v^2_{\rm TOT}(R)$ growing with distance, in the inner regions of galaxies the Newtonian contribution is the most important, to thereby permit a determination of $(N^* + N_{\rm bulge})M_\odot$ from inner region rotational velocity data alone. Encouragingly for our fits we find that when the values for $(N^* + N_{\rm bulge})M_\odot$ that we obtain this way are divided by the measured luminosities we get ratios that are close to the mass to light ratio measured in the local solar neighborhood (i.e. within a factor of the mass to light ratio $M_\odot/L_\odot$ of the sun). Then, with $N^* + N_{\rm bulge}$ being determined by inner region data alone, the use of equation (24) to fit the outer regions is effectively parameter free. The conformal gravity theory outer region predictions are thus highly constrained, and yet as we show, the theory is fully capable of accounting for the rotation curve systematics that are observed right across our 138 galaxy sample.

3. Input Parameters for the Fits

In order to apply equation (24) to galaxies we need to supply input values for stellar scale lengths R_0 and distances D to galaxies. In order to be able to test the conformal theory in as unbiased and standardized a way as possible, we have established a common baseline for selecting these parameters from data. Specifically, while there is often a range of stellar disk scale lengths reported in the literature (a spread we took advantage of in a few cases), by and large we have used the stellar scale length values as measured in the longest wavelength band available for each galaxy (most often the K band). As regards distances to galaxies, we have used distances to galaxies as given by the NASA Extragalactic Database (NED). Determining such distances D is a key requirement for our theory not just because such distances are needed to convert angles on the sky to absolute scales for R, for R_0, and for the HI gas profile and total mass $M_{\rm HI}$, but with the parameters γ^*, γ_0 and κ being absolute quantities, any variation in R would affect the absolute normalization of the contributions of the linear and quadratic terms in equation (24). Our theory is thus very sensitive to distance determinations.

In the NED two procedures for determining distances are presented, one based on visual observations of nearby galaxies (usually Cepheid or Tully-Fisher determinations), and the other based on redshift measurements for more distant galaxies. The visually determined distance values are listed as a world mean value and its one standard deviation, while the redshift-based determinations depend on cosmological models and are listed with a spread in values depending on the model chosen. We have systematically used visually determined mean values and galactocentric cosmological model redshift-based distance determinations. With such an assumption we found [10, 11, 12] that our fits were immediately falling on the data in all but 15 of the 138 cases, and in those 15 cases the fits were close enough that we could then bring them into agreement with data by allowing for the above noted uncertainties in the NED distance determinations. The fact that our fits work so well at the NED distances is thus a noteworthy achievement for our theory.

Also we note that there can be another uncertainty in the data, namely there can be some uncertainty in the inclination angle of a galaxy. Although this has no effect on the shape of the measured rotation curve, it would serve to change the normalization of the curve, since a change in the inclination angle would result in a changed determination of the actual values of the rotational velocities. Specifically, with the inclination angle being the angle between the normal to the plane of the galactic disk and our line of sight, a measured Doppler shift $v(\text{meas})$ is actually a measurement of the projection $v(i)\sin i$ at inclination angle i along our line of sight (so that for edge-on galaxies $v(90^\circ) = v(\text{meas})$). At two different assumed inclinations the inferred velocities would thus be related as $v(\text{meas}) = v(i_1)\sin i_1 = v(i_2)\sin i_2$, with a decrease or increase in assumed inclination leading to an increase or decrease in inferred rotation velocity. For five of the galaxies in our 138 galaxy sample we found it advantageous to allow for inclination uncertainties. Since we use single exponential models for both the stellar disk and the atomic gas profile (equivalent to a face-on 0° inclination angle), we have no need to take into consideration any inclination effects that would be needed if we were to use the actually measured stellar or gas profiles themselves.

4. The Conformal Gravity Fits

In making a selection of which galactic rotation curves to study, for each galaxy we required that there be both good optical photometry (as needed to determine the explicit contribution due to the visible material in each galaxy), and reliable, well-accepted HI rotation curve data. For the HI data we explicitly sought data that extended the furthest in radial distance, so that we could explore the kinematic region that is the most sensitive to the linear and quadratic terms in equation (24). Our sample contains galaxies that range in luminosity by as much as four orders of magnitude, ranging from very bright HSB galaxies (as bright as $2 \times 10^{11} L_{\odot}^{\rm B}$), to LSB galaxies and to DWF galaxies (as dim as $3 \times 10^{7} L_{\odot}^{\rm B}$). In our sample there are 21 galaxies that go out to a last radial data point $R_{\rm last}$ that is more than 31 kpc, with the largest being Malin 1, a galaxy for which $R_{\rm last}$ is a mammoth 98 kpc.

Given these requirements, we have identified a sample consisting of 138 galaxies in total. The sample is comprised of 18 galaxies from The HI Nearby Galactic Survey (THINGS), 30 galaxies from an Ursa Major study, 20 galaxies from an LSB galaxy study, 21 from a second LSB study, 22 miscellaneous galaxies, 20 galaxies from a dwarf galaxy study, 2 additional LSB galaxies, and 5 new THINGS dwarf galaxies. The miscellaneous 22 galaxy set contains many of the classic rotation curves that first established that there actually was a galactic missing mass problem. The galaxies we fit have frequently been studied in the literature for dark matter fitting, as well for as alternative gravity fitting based on theories such as MOND [13, 14] or MSTG [15].

We reported conformal gravity fitting to the first 111 galaxies of our sample in [11] (with the largest 21 also being reported in [10]), and reported fitting to the remaining 27 in [12]. To illustrate how our theory fares, for our purposes here we present a few typical fits from each of the components of our sample. From the THINGS sample we present fits to NGC 3198, NGC 3521 and NGC 5055 (all HSB). From the Ursa Major sample we present fits to NGC 3893, NGC 4183 (both HSB) and UGC 6923 (LSB). From the 20 galaxy LSB survey we present fits to F563-1, NGC 959 and UGC 128. From the 21 galaxy LSB survey we present fits to F579-V1, UGC 6614, and UGC 11557. From the 22 miscellaneous galaxy set we present fits to NGC 247 (LSB) , NGC 2683 (HSB) and Malin 1 (HSB). From the 20 DWF galaxy sample we present fits to UGC 731 and UGC 11707, and from the 5 THINGS DWF galaxies we present a fit to UGC 5423. In Table 1 we present the input and output data associated with these selected galaxies and provide the relevant data sources for rotational velocities (column labelled v), luminosities (L), stellar disk scale lengths (R_0), and HI gas masses (HI). Of the 18 galaxies we present here two have spherical bulges, NGC 5055 and Malin 1. For theses two galaxies we find respective bulge masses $0.73 \times 10^{10} M_{\odot}$ and $9.46 \times 10^{10} M_{\odot}$. In Table 1 the mass to light ratios listed for these two galaxies are ratios of the total stellar mass (disk plus bulge) to total blue luminosity.

In Figure 1 we present the conformal gravity fitting to the rotational velocities (in km sec^{-1}) together with their quoted errors as plotted as a function of radial distance (in kpc). For each of the 18 selected galaxies we have exhibited the contribution due to the luminous Newtonian term alone (dashed curve), the contribution from the two linear terms alone (dot-dashed curve), the contribution from the two linear terms and the quadratic terms combined (dotted curve), with the full curve showing the total contribution. As we see, the tightly constrained equation (24) captures the essence of the data, and does so without needing any dark matter whatsoever. It is a notable achievement for the theory that not only do we fit the data with only one free parameter per galaxy, viz. the galactic mass to light ratio, we are able to do so with fitted mass to light ratios that are typical of the mass to light ratio found in the local solar neighborhood.

As we can see from the fitting to the largest galaxies in Figure 1, were it not for the negative quadratic potential term the contribution of the linear terms would totally overshoot the data at the largest radial distances. Moreover, despite the fact that the quadratic term is universal and has no dependence at all on the galaxy-dependent $N_{\rm TOT}$ factor in (25), nonetheless the quadratic term is able to arrest the linear rise in each and every case. So dramatic is this effect, that if we were to go to even larger radial distances, we would expect the rotation velocities to begin to decline and eventually be cut-off at some large radial distance of order $R \sim (\gamma_0 + N_{\rm TOT}\gamma^*)/2\kappa$ where $v^2_{\rm TOT}(R)$ would otherwise become negative. In the conformal theory galaxies thus have some natural limiting size that is fixed by an interplay of the local galaxy with the global cosmologically generated γ_0 and κ terms. This eventual decline in the rotation curve provides a falsifiable test of the conformal gravity theory. To illustrate this effect, in Figure 2 we provide extended distance predictions for some six of the largest galaxies in our sample (NGC 3198, NGC 3521, NGC 5055, UGC 128, NGC 2683, and Malin 1). With the last few data points in UGC 128 and NGC 2683 perhaps already suggesting the beginnings of a fall off, testing the prediction of an eventual fall off in rotation curves could be within observational reach.

In regard to the issue of a possible limiting size to galaxies, we take note of the study of [16], in which it was noted that there would be constraints on circular orbits if the potential is treated as a one-body potential. Specifically, if we introduce a point particle Lagrangian $L = \dot{r}^2/2 + r^2\dot{\phi}^2/2 - V(r)$ for motion in a plane with central potential $V(r)$, we obtain circular orbits that obey $\ddot{r} = r\dot{\phi}^2 - V'(r) = 0$ and $r^2\dot{\phi} = J$, and thus $J^2/r^3 = V'(r)$. If we define an effective potential of the form $V_{\rm eff}(r) = V(r) + J^2/2r^2$ (so that $\ddot{r} = -V'_{\rm eff}(r)$), then the radii of circular orbits have to obey $V'(r) \geq 0$ if $r\dot{\phi}$ is to be real, and have to obey $V''_{\rm eff} = V''(r) + 3V'(r)/r \geq 0$ if the orbit is to be stable. For a potential of the form $V(r) = \gamma c^2 r/2 - \kappa c^2 r^2/2$, $V'(r)$ vanishes at $r = \gamma/2\kappa$, while the quantity $V''(r) + 3V'(r)/r$ vanishes at $r = 3\gamma/8\kappa$. Thus only orbits with $r \leq 3\gamma/8\kappa$ would be stable, and those with $3\gamma/8\kappa \leq r \leq \gamma/2\kappa$ would not be. Even with this stability condition rotation velocities would still undergo a fall off, but the radius of the largest stable orbit would be smaller than that associated with the vanishing of $v^2_{\rm TOT}(R)$.

While the above stability analysis would apply if the problem were a one-body problem, in the conformal theory the potential itself arises through an interplay between each local galaxy and the global cosmology, and is thus actually many body in nature. Moreover, the very existence of galaxies at all requires that inhomogeneities form as perturbations in an otherwise uniform cosmological background. With these perturbations generating the κ term (c.f. equation (17)), galaxies generated this way will already be stable as long as the cosmological perturbations themselves are stable. At the present time a theory for the growth of conformal gravity fluctuations is still being developed [17], and a cosmological stability analysis has yet to be made. It could thus be very instructive to explore both theoretically and observationally where (and of course whether) galaxies might actually terminate.

In our study we have found that the simple universal formula given in equation (24) accounts for a large body of data. The reason that this formula is able to do so is because there is a lot of universality in the data themselves. Specifically, across the entire 138 galaxies in our sample, and as specifically noted in Table 1 for our 18 selected galaxies, the value of the centripetal acceleration $(v^2/c^2R)_{\rm last}$ at the last data point is found to be close in magnitude to the numerical value we found for the universal γ_0 parameter. It is this universality that drives the fits, with such universality naturally being present in the conformal theory because $\gamma_0 = 2(-K)^{1/2}$ has a natural origin in the spatial curvature of the universe. At the present time there is no indication of the emergence of any such universality for $(v^2/c^2R)_{\rm last}$ in dark matter theory, and to account for the 138 galaxies using dark matter no less than 276 additional adjustable free parameters are required (two free numerical parameters per dark matter halo). Since our equation (24) does account for the data, if it is to be dark matter theory that is to be the correct explanation of the missing mass problem, then one would need to be able to derive equation (24) in it. Given the phenomenological success of equation (24), and given the fact that it is derived from a fundamental gravitational theory, the conformal gravity theory poses a quite considerable challenge to the standard dark matter paradigm.

5. Tidal Galaxies

We conclude this review with an application of the conformal gravity theory to tidal dwarf galaxies (TDGs). These dwarf galaxies are formed in the tidal tails of collisions of disk galaxies, and are thought [18] to be predominantly composed of material expelled from the galactic disk of a parent galaxy. With any dark matter present in the parent galaxies expected to predominantly be in spherical haloes, tidal galaxies should thus have a very low dark matter content, and thus should not themselves be expected to possess the substantial spherical dark matter haloes that are ordinarily required to accompany and stabilize disk galaxies in standard gravity. In consequence, in the standard dark matter picture TDG rotation curves should not be expected to display any substantial mass discrepancies. However, in any explanation of the missing mass problem that involves a change in Newton's law of gravity on galactic distance scales as opposed to a change in the matter content of galaxies, one should expect mass discrepancies. Tidal dwarf galaxies thus provide a quite unusual laboratory for exploring the missing mass problem. Rotation curve data have become available for three TDGs associated with the parent galaxy NGC 5291 [19] (see especially the accompanying supporting online material), and it has been shown [19] that there are in fact mass discrepancies, and that a good accounting of the data can be provided by MOND [20]. Here we show that conformal gravity can also provide a good accounting of the data.

Because of their locations the three TDGs in NGC 5291 are designated NGC 5291N, NGC 5291S, and NGC 5291SW. HI rotation curves are available for all three, and for NGC 5291N some inner region optical rotation curve data are available as well [21]. For the TDGs there is little reported photometry in the literature and for only one, NGC 5291N, is there a reported luminosity. However, there are good HI gas profiles for all three TDGs, and so we can extract effective gas scale lengths. From them we can then infer effective stellar scale lengths by dividing the gas scale lengths by a factor of four. With the HI gas masses being known, we shall estimate the luminosities of NGC 5291S and NGC 5291SW by assuming that the luminosity scales uniformly with the total gas mass for all three TDGs. With the luminosity of NGC 5291N being reported in the V band, to convert to a blue luminosity we shall use the $B - V = 0.3$ modulus that was found [22] to be characteristic of the debris around NGC 5291. Also, for the total gas mass $M_{\rm gas}^{\rm tot}$ in each TDG we use the combined atomic hydrogen, atomic helium and molecular hydrogen total mass values reported in [19]. With there only being redshift determinations of the distance D to NGC 5291N and nothing at all reported in the NED for NGC 5291S or NGC 5291SW, we shall follow [19] and [20] and for all three TDGs use the mean redshift determined distance of 62 Mpc for NGC 5291N as given in the NED. Finally, again following [19] we initially take the inclination angle i of each of the three TDGs to be 45°. In our fitting we shall follow [11] and [12] and require the mass to light ratio of each galaxy to be at least 0.2.

Using equation (24) we obtain the plots given in Figure 3. As can be seen, the conformal theory captures the essence of the TDG data. Since there is a reported uncertainty in inclination angle for NGC 5291N [19, 20], for it we can use inclination angles up to 55°. At $i = 55°$ we obtain the improved fit to NGC 5291N given in Figure 4. We list all of the input and output parameters for our fits in Table 2.

As we see the conformal theory nicely accounts for the tidal dwarf galaxy data. That it is able to do so is because, as can be seen in Table 2, just as with our earlier 138 galaxies, for all three TDGs the value of the centripetal acceleration at the last data point is again found to be very close to the numerical value obtained for γ_0. In conclusion therefore, we note that given its success in fitting no less than 141 galactic rotation curves with only one free parameter per galaxy, and given in addition its theoretical underpinnings, conformal gravity can realistically be advanced as a candidate alternate theory of gravity.

Acknowledgments

The authors wish to thank Dr. J. R. Brownstein, Dr. W. J. G. de Blok, Dr. J. W. Moffat, Dr. S. S. McGaugh, Dr. R. A. Swaters, Dr. S.-H. Oh, Dr. B. Famaey, and Dr. F. Bournaud for helpful communications, and especially for providing their galactic data bases. This research has made use of the NASA/IPAC Extragalactic Database (NED) which is operated by the Jet Propulsion Laboratory, California Institute of Technology, under contract with the National Aeronautics and Space Administration. This review is based in part on a talk given by one of of us (JGO) at the 8th Biennial Conference on Classical and Quantum Relativistic Dynamics of Particles and Fields held in Florence, Italy in May 2012. JGO would like to thank the International Association of Relativistic Dynamics for the kind hospitality of the conference and would like to thank the Istituto Nazionale di Fisica Nucleare for related financial support.

Table 1. Properties of the Selected 18 Galaxy Sample

Galaxy	D (Mpc)	L_B ($10^{10}L_\odot^B$)	$(R_0)_{disk}$ (kpc)	R_{last} (kpc)	M_{HI} ($10^{10}M_\odot$)	M_{disk} ($10^{10}M_\odot$)	$(M/L)_{stars}$ ($M_\odot/L_\odot^B$)	$(v^2/c^2R)_{last}$ (10^{-30} cm^{-1})	Data Sources v	L	R_0	HI
NGC 3198	14.1	3.24	4.0	38.6	1.06	3.64	1.12	2.09	[23]	[24]	[25]	[24]
NGC 3521	12.2	4.77	3.3	35.3	1.03	9.25	1.94	4.21	[23]	[24]	[26]	[24]
NGC 5055	9.2	3.62	2.9	44.4	0.76	6.04	1.87	2.36	[23]	[24]	[26]	[24]
NGC 3893	18.1	2.93	2.4	20.5	0.59	5.00	1.71	3.85	[27]	[28]	[29]	[28]
NGC 4183	16.7	1.04	2.9	19.5	0.30	1.43	1.38	2.36	[27]	[28]	[29]	[28]
UGC 6923	18.0	0.30	1.5	5.3	0.08	0.35	1.18	4.43	[27]	[28]	[30]	[28]
F563-1	46.8	0.14	2.9	18.2	0.29	1.35	9.65	2.44	[31]	[32]	[32]	[33]
NGC 959	13.5	0.33	1.3	2.9	0.05	0.37	1.11	7.43	[34]	[35]	[36]	[35]
UGC 128	64.6	0.60	6.9	54.8	0.73	2.75	4.60	1.03	[37]	[32]	[38]	[38]
F579-V1	86.9	0.56	5.2	14.7	0.21	3.33	5.98	3.18	[31]	[33]	[32]	[33]
UGC 6614	86.2	2.11	8.2	62.7	2.07	9.70	4.60	2.39	[31]	[39]	[38]	[38]
UGC 11557	23.7	1.81	3.0	6.7	0.25	0.37	0.20	3.49	[31]	[39]	[40]	[40]
NGC 247	3.6	0.51	4.2	14.3	0.16	1.25	2.43	2.94	[41]	[42]	[42]	[41]
NGC 2683	10.2	1.88	2.4	36.0	0.15	6.03	3.20	2.28	[43]	[43]	[44]	[45]
Malin 1	338.5	7.91	84.2	98.0	5.40	1.00	1.32	1.77	[46]	[46]	[47]	[46]
UGC 731	11.8	0.07	2.4	10.3	0.16	0.32	4.63	1.91	[48]	[49]	[14]	[40]
UGC 11707	21.5	0.11	5.8	20.3	0.68	0.99	8.76	1.77	[48]	[49]	[14]	[40]
UGC 5423	7.1	0.01	0.6	2.0	0.01	0.03	2.01	1.82	[50]	[24]	[50]	[24]

Table 2. Properties of the 3 Tidal Dwarf Galaxies

Galaxy	D (Mpc)	L_B ($10^8L_\odot^B$)	i °	$(R_0)_{gas}$ (kpc)	R_{last} (kpc)	M_{gas}^{tot} ($10^8M_\odot$)	M_{disk} ($10^8M_\odot$)	$(M/L)_{disk}$ ($M_\odot/L_\odot^B$)	$(v^2/c^2R)_{last}$ (10^{-30}cm^{-1})
NGC 5291N	62.0	9.5	45	0.8	4.7	7.7	3.4	0.35	5.7
NGC 5291S	62.0	10.7	45	0.8	5.2	8.6	2.1	0.20	1.9
NGC 5291SW	62.0	5.7	45	0.8	2.7	4.6	1.1	0.20	3.4
NGC 5291N	62.0	9.5	55	0.8	4.7	7.7	1.9	0.20	4.3

Journal of Physics: Conference Series **437** (2013) 012002 doi:10.1088/1742-6596/437/1/012002

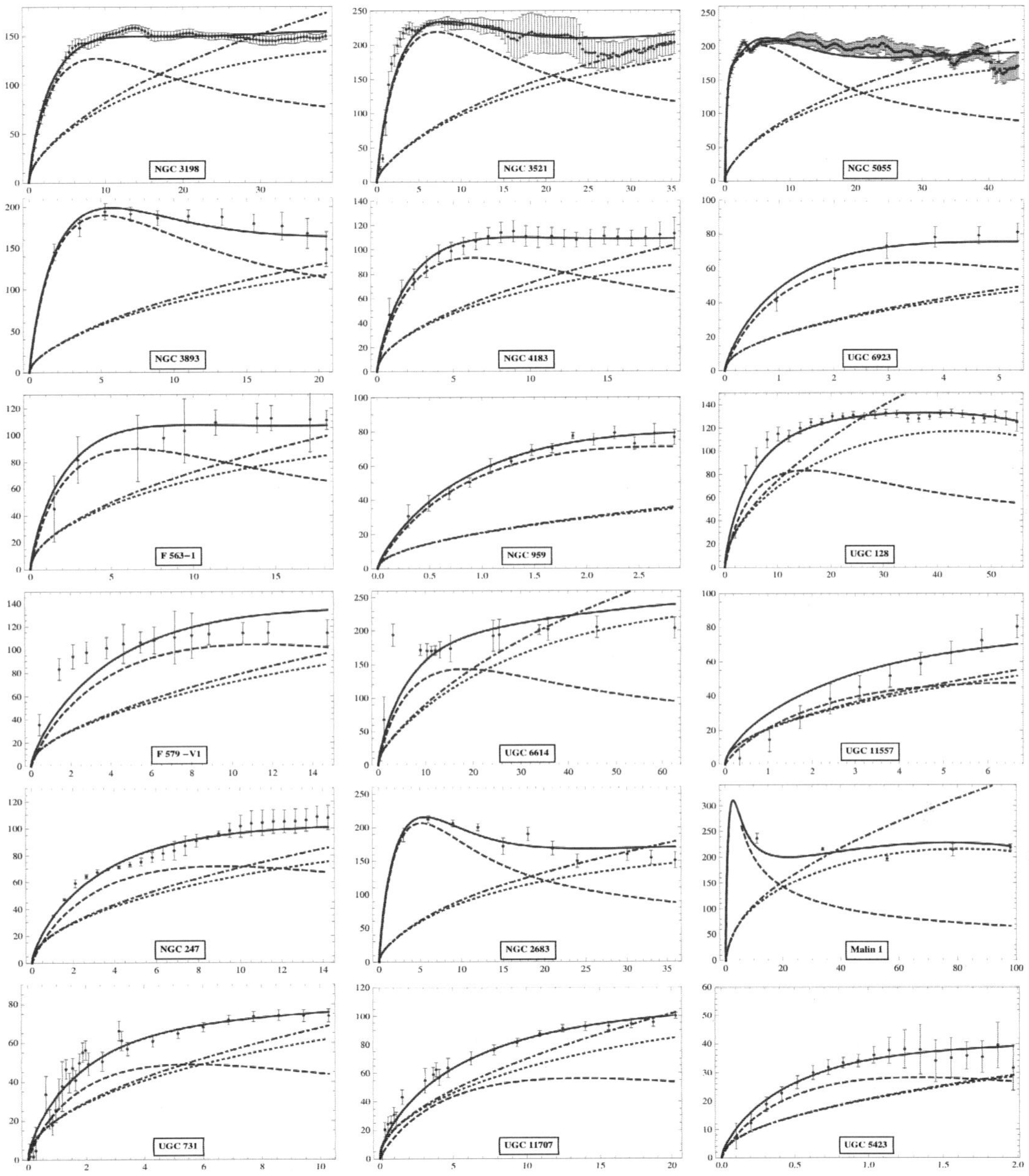

Figure 1. Conformal gravity fitting to the rotational velocities (in km $\sec^{-1}$) of the selected 18 galaxy sample with their quoted errors as plotted as a function of radial distance (in kpc). For each galaxy we have exhibited the contribution due to the luminous Newtonian term alone (dashed curve), the contribution from the two linear terms alone (dot-dashed curve), the contribution from the two linear terms and the quadratic terms combined (dotted curve), with the full curve showing the total contribution. No dark matter is assumed.

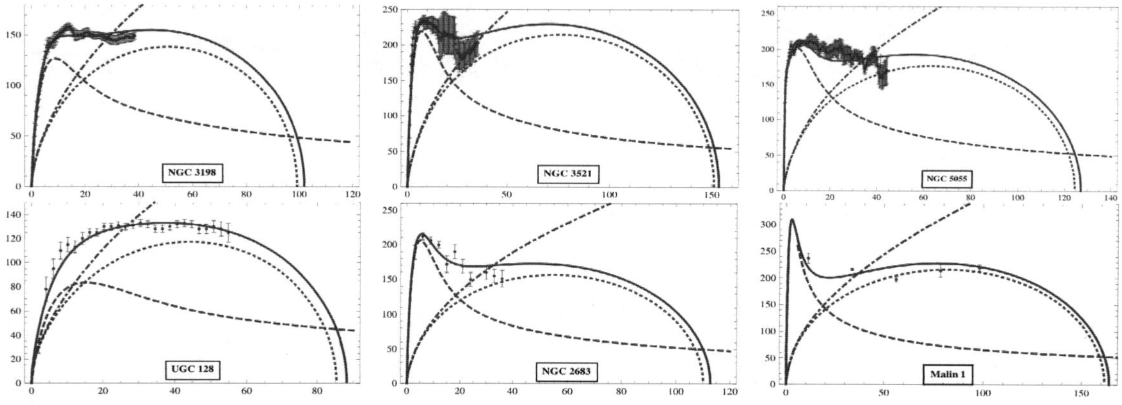

Figure 2. Extended distance predictions for NGC 3198, NGC 3521, NGC 5055, UGC 128, NGC 2683 and Malin 1. The curves are the same as in Figure 1.

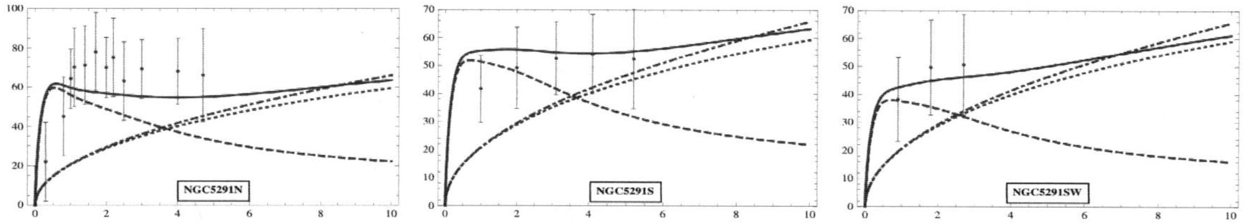

Figure 3. Rotation curves for the tidal dwarf galaxies NGC 5291N, NGC 5291S, and NGC 5291SW, all with inclination 45°. The curves are the same as in Figure 1.

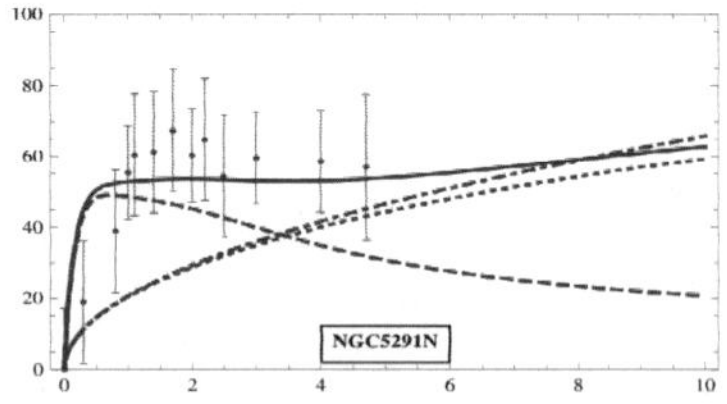

Figure 4. Rotation curve for the tidal dwarf galaxy NGC 5291N with inclination 55°. The curves are the same as in Figure 1.

References

[1] Rubin V C 2000 *Pub. Ast. Soc. Pac.* **112** 747
[2] Mannheim P D 2006 *Prog. Part. Nucl. Phys.* **56** 340
[3] Milgrom M 1983 *ApJ* **270** 365, 371, 384
[4] Moffat J W 2005 *J. Cos. Ast. Phys.* **05** 003; 2006 *J. Cos. Ast. Phys.* **03** 004
[5] Mannheim P D 2011 *Gen. Rev. Gravit.* **43** 703
[6] Mannheim P D 2012 *Found. Phys.* **42** 388
[7] Mannheim P D and Kazanas D 1989 *ApJ* **342** 635
[8] Mannheim P D and Kazanas D 1994 *Gen. Rel. Gravit.* **26** 337
[9] Mannheim P D 1997 *ApJ* **479** 659
[10] Mannheim P D and O'Brien J G 2011 *Phys. Rev. Lett.* **106** 121101
[11] Mannheim P D and O'Brien J G 2012 *Phys. Rev.* D **85** 124020
[12] O'Brien J G and Mannheim P D 2012 *MNRAS* **421** 1273
[13] Sanders R H and McGaugh S S 2002 *Ann. Rev. Astron. Ap.* **40** 263
[14] Swaters R A, Sanders R H and McGaugh S S 2010 *ApJ* **718** 380
[15] Brownstein J R and Moffat J W 2006 *ApJ* **636** 721
[16] Nandi K K and Bhadra A 2012 *Phys. Rev. Lett.* **109** 079001
[17] Mannheim P D 2012 *Phys. Rev.* D **85** 124008
[18] Barnes J E and Hernquist L 1992 *Nature* **360** 715
[19] Bournaud F, Duc P-A, Brinks E, Boquien M, Amram P, Lisenfeld U, Koribalski B S, Walter F and Charmandaris V 2007 *Science* **316** 1166
[20] Gentile G, Famaey B, Combes F, Kroupa P, Zhao H S and Tiret O 2007 *A&A* **472** L25
[21] Bournaud F, Duc P-A, Amram P, Combes F and Gach J-L 2004 *A&A* **425** 813
[22] Duc P-A and Mirabel I F 1998 *A&A* **333** 813
[23] de Blok W J G, Walter F, Brinks E, Trachternach C, Oh S-H and Kennicutt R C 2008 *AJ* **136** 2648
[24] Walter F, Brinks E, de Blok W J G, Bigiel F, Kennicutt R C, Thornley M D and Leroy A 2008 *AJ* **136** 2563
[25] Wevers B M H R, van der Kruit P C and Allen R J 1986 *A&A Suppl. Ser.* **66** 505
[26] Leroy A K, Walter F, Brinks E, Bigiel F, de Blok W J G, Madore B and Thornley M D 2008 *AJ* **136** 2782
[27] Verheijen M A W and Sancisi R 2001 *A&A* **370** 765
[28] Sanders R H and Verheijen M A W 1998 *ApJ* **503** 97
[29] Tully R B, Verheijen M A W, Pierce M J, Huang J-S and Wainscoat R J 1996 *AJ* **112** 2471
[30] Tully R B and Verheijen M A W 1997 *ApJ* **484** 145
[31] McGaugh S S, Rubin V C and de Blok W J G 2001 *AJ* **122** 2381
[32] de Blok W J G and McGaugh S S 1997 *MNRAS* **290** 533
[33] de Blok W J G, McGaugh S S and van der Hulst J M 1996 *MNRAS* **283** 18
[34] Kuzio de Naray R, McGaugh S S and de Blok W J G 2008 *ApJ* **676** 920
[35] Fisher J R and Tully R B 1981 *ApJ Suppl. Ser.* **47** 139
[36] Esipov V F, Kyazumov G A and Dzhafarov A R 1991 *Sov. Astr. (tr. Astron. Zh.)* **35** 452
[37] Verheijen M A W and de Blok W J G 1999 *As. Sp. Sc.* **269-270** 673
[38] van der Hulst J M, Skillman E D, Smith T R, Bothun G D, McGaugh S S and de Blok W J G 1993 **AJ** 106, 548
[39] de Blok W J G, McGaugh S S and Rubin V C 2001 *AJ* **122** 2396
[40] Swaters R A, van Albada T S, van der Hulst J M and Sancisi R 2002 *A&A* **390** 829
[41] Carignan C and Puche D 1990 *AJ* **100** 641
[42] Carignan C 1985 *ApJ Suppl. Ser.* **58** 107
[43] Casertano S and van Gorkom J H 1991 *AJ* **101** 1231
[44] Kent S M 1985 *ApJ Suppl. Ser.* **59** 115
[45] Sanders R H 1996 *ApJ* **473** 117
[46] Lelli F, Fraternali F and Sancisi R 2010 *A&A* **516** A11
[47] Pickering T E, Impey C D, van Gorkom J H and Bothun G D 1997 *AJ* **114** 1858
[48] Swaters R A, Madore B F, van den Bosch F C and Balcells M 2003 *ApJ* **583** 732; Swaters R A, Sancisi R, van Albada T S and van der Hulst J M 2009 *A&A* **493** 871
[49] Swaters R A and Balcells M 2002 *A&A* **390** 863
[50] Oh S-H, de Blok W J G, Brinks E, Walter F and Kennicutt R C 2011*AJ* **141** 193

IARD2012 IOP Publishing
Journal of Physics: Conference Series **437** (2013) 012003 doi:10.1088/1742-6596/437/1/012003

Coframe geometry, gravity and electromagnetism

Yakov Itin

Institute of Mathematics, Hebrew University of Jerusalem and
Jerusalem College of Technology, Jerusalem 91904, Israel

E-mail: itin@math.huji.ac.il

Abstract. The extensions of GR for description of fermions on a curved space, for supergravity, and for the loop quantum gravity ordinary use a set of 16 independent variables instead of 10 components of metric. These variables can be assembled in a coframe field, i.e., a set of four linearly independent 1-forms. In this presentation we review a geometrical structure based on the coframe field. We construct a complete class of the coframe connections which are linear in the first order derivatives of the coframe field on an n dimensional manifolds with and without a metric. The subclasses of the torsion-free, metric-compatible and flat connections are derived. We also study the behavior of the geometrical structures under local transformations of the coframe. The remarkable fact is an existence of a subclass of connections which are invariant when the infinitesimal transformations satisfy the Maxwell-like system of equations.

1. Introduction. Why do we have to go beyond Riemannian geometry?

General relativity (GR) is, probably, the best of the known theories of gravity. From mathematical and aesthetic points of view, it can be used as a standard of what a physical theory has to be. Einstein's theory is in a very good agreement with all up-to-date observation data. The central idea of Einstein's GR is that the physical properties of the gravitational field are in one-to-one correspondence with the geometry of the base manifold. The standard GR is based on a (pseudo) Riemannian geometry with a unique metric tensor and a unique Levi-Civita connection constructed from this tensor.

After classical works of Weyl, Cartan and others, it is known that Riemannian construction is not a unique possible geometry. A most general geometric framework involves independent metric and independent connection. Gravity field models based on such general geometry (metric-affine gravity) was studied intensively, see [1]—[17] and the references given therein. Probably a main problem of these constructions is a huge number of geometrical fields which do not find their physical co-partners.

In this paper we study a much more economical construction based on a unique geometrical object — coframe field. Absolute (teleparallel) frame/coframe variables (repèr, vierbein, ...) were introduced in physics by Einstein in 1928 with an aim of a unification of gravitational and electromagnetic fields (for classical references, see [18]). The physical models for gravity based on the coframe variable are well studied, see [19]—[43]. In some aspects such models are even preferable to the standard GR. In particular, they involve a meaningful definition of the gravitational energy, which is in a proper correspondence with the Noether procedure [33], [36]. Moreover some problems inside and beyond Einstein's gravity, such as: i) Hamiltonian formulation of GR[44],[45]; (ii) positive energy proofs [46]; (iii) fermions on a curved manifold

[47],[48]; (iv) supergravity [49]; (v) loop quantum gravity [50], require a richer set of 16 independent variables of the coframe.

In this paper, we present a geometric structure that can be constructed from the vierbein (frame/coframe) variable.

The organization of the paper is as follows:

In the first section, we construct a geometrical structure based on a coframe variable as unique building block. In an addition to the coframe volume element and metric, we present a most general coframe connection. The Levi-Civita and flat connections are special cases of it. The torsion and nonmetricity tensors of the general coframe connection are calculated. We identify the subclasses of symmetric (torsion-free) connections and of metric-compatible connections. The unique symmetric metric-compatible connection is of Levi-Civita. In section 2, we study the transformation properties of the coframe field and identify a subclass of connections which are invariant under restricted coframe transformations. In section 3, we study the invariant conditions on the coframe transformations. In particular, we present a quite remarkable fact that invariance conditions are approximated by a Maxwell-type system. It gives a possibility to model the electromagnetic field by the elements of the transformation group. Exact spherical symmetric solution for our model is presented. It is bounded near the Schwarzschild radius. Further off, the solution is close to the Coulomb field.

In the last section, some proposals of possible developments of a geometrical coframe construction and its applications to gravity and electromagnetism are presented.

2. Coframe geometry

In this section, we define a geometry based on a coframe field. It is instead of the of the standard Riemannian geometry based on a metric tensor field.

2.1. Coframe manifold. Definitions and notations

Coframe field. Let an $(n + 1)$-dimensional differential manifold M endowed with a smooth nondegenerate coframe field ϑ^α be given. The coframe comes together with its dual — the frame field e_α. In an arbitrary chart of local coordinates $\{x^i\}$, these fields are expressed as

$$\vartheta^\alpha = \vartheta^\alpha{}_i dx^i\,, \qquad e_\alpha = e_\alpha{}^i\partial_i\,, \tag{2.1}$$

i.e., by two nondegenerate matrices $\vartheta^\alpha{}_i$ and $e_\alpha{}^i$ which are the inverse to each-other. In other words, we are dealing with a set of n^2 independent smooth functions on M. The coframe indices change in the same range $\alpha, \beta, \ldots = 0, \ldots, n$ as the coordinate indices $i, j, \ldots$. However they must be strictly distinguished. In particular, the indices in $\vartheta^\alpha{}_i$ or $e_\alpha{}^i$ cannot be contracted.

Coframe transformation. For most physical models based on the coframe field, this field is defined only up to global transformations. It is natural to consider a wider class of coframe fields related by local pointwise transformations

$$\vartheta^\alpha \to L^\alpha{}_\beta(x)\vartheta^\beta\,, \qquad e_\alpha \to L_\alpha{}^\beta(x)e_\beta\,. \tag{2.2}$$

Here $L^\alpha{}_\beta(x)$ and $L_\alpha{}^\beta(x)$ are inverse to each-other at arbitrary point x. Denote the group of matrices $L^\alpha{}_\beta(x)$ by G. Note two specially important cases: (i) G is a group of global transformations with a constant matrix $L^\alpha{}_\beta$; (ii) G is a group of arbitrary local transformations such that the entries of $L^\alpha{}_\beta$ are arbitrary functions of a point.

Consequently we involve an additional element of the coframe structure — *the coframe transformations group*

$$G = \Big\{L^\alpha{}_\beta(x) \in GL(n+1, \mathbb{R});\ \text{for every}\, x \in M\Big\}\,. \tag{2.3}$$

On this stage, we only require the matrices $L^\alpha{}_\beta(x)$ to be invertible at an arbitrary point $x \in M$. The successive specializations of the coframe transformation matrix will be considered in sequel.

Coframe field volume element. We assume the coframe field to be non-degenerate at an arbitrary point $x \in M$. Consequently, a special $n+1$-form, *the coframe field volume element*, is defined and nonzero. Define

$$\mathrm{vol}(\vartheta^\alpha) = \frac{1}{n!}\varepsilon_{\alpha_o\cdots\alpha_n}\vartheta^{\alpha_o}\wedge\cdots\wedge\vartheta^{\alpha_n}\,, \tag{2.4}$$

where $\varepsilon_{\alpha_o\cdots\alpha_n}$ is the Levi-Civita permutation symbol normalized by $\varepsilon_{01\cdots n} = 1$. Treating the coframe volume element as one of the basic elements of the coframe geometric structure, we obtain the following invariance condition.

Volume element invariance condition: Volume element $\mathrm{vol}(\vartheta^\alpha)$ is assumed to be invariant under pointwise transformations of the coframe field

$$\mathrm{vol}\,(\vartheta^\alpha) = \mathrm{vol}\left(L^\alpha{}_\beta\vartheta^\beta\right)\,. \tag{2.5}$$

This condition is satisfied by matrices with unit determinant. Consequently, the coframe transformation group (2.3) is restricted to

$$G = \left\{L^\alpha{}_\beta(x) \in SL(n+1,\mathbb{R});\ \text{for every}\, x \in M\right\}. \tag{2.6}$$

Metric tensor. For a meaningful physical field model, it is necessary to have a metric structure on M. Moreover, the metric tensor has to be of the Lorentzian signature. In a coordinate basis, a generic metric tensor is written as

$$g = g_{ij}dx^i \otimes dx^j\,, \tag{2.7}$$

where the components g_{ij} and g_{ab} are smooth functions of a point $x \in M$.

On a coframe manifold, a metric tensor is not an independent quantity. Instead, we are looking for a metric explicitly constructed from a given coframe field, $g = g(\vartheta^\alpha)$. We assume the metric tensor to be quadratic in the coframe field components and independent of its derivatives. Moreover, it should be reducible at a point to the Lorentzian metric $\eta_{\alpha\beta} = \mathrm{diag}(-1, 1, \cdots, 1)$. With these restrictions, we come to a definition of the *coframe field metric tensor*

$$g = \eta_{\alpha\beta}\vartheta^\alpha \otimes \vartheta^\beta\,, \qquad g_{ij} = \eta_{\alpha\beta}\vartheta^\alpha{}_i\vartheta^\beta{}_j\,. \tag{2.8}$$

Note that the equations (2.8) often appear as a definition of a (non unique) orthonormal basis of reference for a given metric. Another interpretation treats (2.8) as an expression of a given metric in a special orthonormal basis of reference, as in (2.7). In our approach, (2.8) has a principle different meaning. It is a definition of the metric tensor field via the coframe field. Certainly the form of the metric $\eta_{\alpha\beta}$ in the tangential vector space T_xM is an additional axiom of our construction. With an aim to define an invariant coframe geometric structure we require:

Metric tensor invariance condition: Metric tensor is assumed to be invariant under pointwise transformations of the coframe field, i.e.,

$$g\,(\vartheta^\alpha) = g\left(L^\alpha{}_\beta\vartheta^\beta\right)\,. \tag{2.9}$$

This condition is satisfied by pseudo-orthonormal matrices,

$$\eta_{\mu\nu}L^\mu{}_\alpha L^\nu_\beta = \eta_{\alpha\beta}. \tag{2.10}$$

Consequently, the invariance of the coframe metric restricts the coframe transformation group to

$$G=\Big\{L^\alpha{}_\beta(x)\in O(1,n,\mathbb{R});\ \text{for every}\, x\in M\Big\}\,. \tag{2.11}$$

Volume&metric invariance condition: In order to have simultaneously a metric and a volume element structures both constructed from the coframe field, we have to assume a successive restriction of the coframe transformation group:

$$G=\Big\{L^\alpha{}_\beta(x)\in SO(1,n,\mathbb{R});\ \text{for every}\, x\in M\Big\}\,. \tag{2.12}$$

Topological restrictions. A global smooth coframe field may be defined only on a parallelizable manifold, i.e., on a topological manifold of a zero second Whitney class. This topological restriction is equivalent to existence of a spinorial structure on M. We restrict ourselves to a local consideration, thus the global definiteness problems will be neglected. Moreover, we assume the coframe field to be smooth and nonsingular only in a "weak" sense. Namely, the components $\vartheta^\alpha{}_i$ and $e_\alpha{}^i$ are required to be differentiable and linearly independent at almost all points of M, except of a zero measure set. So, in general, the coframe field can degenerate at singular points, on singular lines (strings), or even on singular submanifolds (p-branes). This assumption leaves a room for the standard singular solutions of the physics field equations such as the Coulomb field, the Schwarzschild metric, the Kerr metric etc..

2.2. *Coframe connections*

A differential manifold endowed with a coframe field is a rather poor structure. In particular, we can not determine if two vectors attached at distance points are parallel to each-other or not. In order to have a meaningful geometry and, consequently, meaningful geometrical field models, we have to consider a reacher structure. In this section we define a coframe manifold with a linear coframe connection. The connection 1-form $\Gamma_a{}^b$ will not be an independent variable, as in the Cartan geometry or in MAG [5]. Alternatively, in our construction, the connection will be explicitly constructed from the coframe field and its first order derivatives. Thus we are dealing with a category of *coframe manifolds with a linear coframe connection:*

$$\Big\{M\,,\vartheta^\alpha,G\,,\Gamma_a{}^b(\vartheta^\alpha)\Big\}\,. \tag{2.13}$$

We start with a coframe manifold without an addition metric structure. Metric contributions to the connection will be considered in sequel.

Affine connection. Recall the main properties of a generic linear affine connection on an $(n+1)$ dimensional differential manifold. Relative to a local coordinate chart x^i, a connection is represented by a set of $(n+1)^3$ independent functions $\Gamma^k{}_{ij}(x)$ — *the coefficients of the connection.* The only condition these functions have to satisfy is to transform, under a change of coordinates $x^i\mapsto y^i(x)$, by an inhomogeneous linear rule:

$$\Gamma^i{}_{jk}\mapsto\left(\Gamma^l{}_{mn}\frac{\partial y^m}{\partial x^j}\frac{\partial y^n}{\partial x^k}+\frac{\partial^2 y^l}{\partial x^j\partial x^k}\right)\frac{\partial x^i}{\partial y^l}\,. \tag{2.14}$$

When instead of the coordinate basis $\{dx^i,\partial/\partial x^i\}$, an arbitrary reference basis $\{\theta^a\,,f_b\}$ is involved, the coefficients of the connection are arranged in a $GL(n,\mathbb{R})$-valued *connection 1-form*, which is defined as [51]

$$\Gamma_a{}^b=f_a{}^k\left(\theta^b{}_i\Gamma^i{}_{jk}-\theta^b{}_{k,j}\right)dx^j\,. \tag{2.15}$$

To return to the holonomic coordinate basis, we can simply use the identities ${\theta^a}_i = \delta^a_i$ and ${f_a}^i = \delta^i_a$. Consequently, in a coordinate basis, the connection 1-form is

$$\varGamma_j{}^i = \Gamma^i{}_{jk}\, dx^k\,. \tag{2.16}$$

Due to (2.14), this quantity transforms under the coordinate transformations as

$$\varGamma_j{}^i \to \left[\Gamma_m{}^l \frac{\partial y^m}{\partial x^j} + d\left(\frac{\partial y^l}{\partial x^j}\right)\right] \frac{\partial x^i}{\partial y^l}\,. \tag{2.17}$$

Alternatively, the connection 1-form (2.15) is invariant under smooth transformations of coordinates. The inhomogeneous linear behavior is shifted here to the transformations of $\varGamma_a{}^b$ under a linear local map of the reference basis ($\theta^a \to A^a{}_b\theta^b$, $f_a \to A_a{}^b f_b$)

$$\varGamma_a{}^b \mapsto \left(\Gamma_c{}^d A_a{}^c + dA_a{}^d\right) A^b{}_d\,. \tag{2.18}$$

On a manifold with a given coframe field ϑ^α, the connection 1-form (2.15), can also be referred to this field. We denote this quantity by $\varGamma_\alpha{}^\beta$. Due to (2.15) it reads:

$$\varGamma_\alpha{}^\beta = e_\alpha{}^k \left(\vartheta^\beta{}_i \Gamma^i{}_{jk} - \vartheta^\beta{}_{k,j}\right) dx^j\,. \tag{2.19}$$

This quantity can be treated as an expression of a generic connection (2.15) in a special basis. An essential difference between two very similar equations (2.15) and (2.19) is vizualized when we use in both formulas the connection constructed explicitly from the derivatives of the coframe field itself $\Gamma^i{}_{jk}(\vartheta^\alpha)$.

Linear coframe connections. We restrict ourselves to the quasi-linear $\Gamma^i{}_{jk}(\vartheta^\alpha)$, i.e., we consider a connection constructed as a linear combination of the first order derivatives of the coframe field. The coefficients in this linear expression may depend on the frame/coframe components. In other words, we are looking for a coframe analog of an ordinary Levi-Civita connection.

Let us assist ourselves with a similar construction from the Riemannian geometry. In this case, we are looking for a most general connection that can be constructed from the metric tensor components. Consider a general linear combination of the first order derivatives of the metric tensor:

$$g^{mi}(\alpha_1 g_{mj,k} + \alpha_2 g_{mk,j} + \alpha_3 g_{jk,m})\,. \tag{2.20}$$

Although this expression has the same index content as $\Gamma^i{}_{jk}$, it is a connection only for some special values of the parameters $\alpha_1, \alpha_2.\alpha_3$. Indeed, any two connections are differ by a tensor. Thus an arbitrary connection can be expressed as a certain special connection plus a tensor

$$\Gamma^i{}_{jk} = \overset{*}{\Gamma}{}^i{}_{jk} + K^i{}_{jk}\,. \tag{2.21}$$

Use for $\overset{*}{\Gamma}{}^i{}_{jk}$ the Levi-Civita connection

$$\overset{*}{\Gamma}{}^i{}_{jk} = \frac{1}{2} g^{im}(g_{mj,k} + g_{mk,j} - g_{jk,m})\,. \tag{2.22}$$

However in Riemannian geometry, does not exist a tensor constructed from the first order derivatives of the metric. Therefore $K^i{}_{jk} = 0$, thus the Levi-Civita connection is a unique connection that can be constructed from the first order derivatives of the metric tensor. It is evidently symmetric and metric compatible.

In an analogy to this construction, we will look for a most general coframe connection of the form

$$\Gamma^i{}_{jk}(\vartheta^\alpha) = \overset{\circ}{\Gamma}{}^i{}_{jk}(\vartheta^\alpha) + K^i{}_{jk}(\vartheta^\alpha)\,. \tag{2.23}$$

Here $\overset{\circ}{\Gamma}{}^i{}_{jk}$ is a certain special connection, while $K^i{}_{jk}$ is a tensor. To start with, we need a certain analog of the Levi-Civita connection, i.e., a special connection constructed from the coframe field. Note that alternatively to the metric geometry, non-trivial tensor $K^i{}_{jk}(\vartheta^\alpha)$ do exist.

The flat Weitzenböck connection. On a bare differentiable manifold M, without any additional structure, the notion of parallelism of two vectors attached to distance points depends on a curve joint the points. Oppositely, on a coframe manifold $\{M\,,\vartheta^\alpha\}$, a certain type of the parallelism of distance vectors may be defined in an absolute (curve independent) sense [53]. Namely, two vectors $u(x_1)$ and $v(x_2)$ may be declared parallel to each other, if, being referred to the local elements of the coframe field $u(x_1) = u_\alpha(x_1)\vartheta^\alpha(x_1)$ and $v(x_2) = v_\alpha(x_2)\vartheta^\alpha(x_2)$, they have the proportional components $u_\alpha(x_1) = Cv_\alpha(x_2)$. This definition is independent on the coordinates used on the manifold and on the nonholonomic frame of reference. It do depends on the coframe field. Since, by local transformations, the coframes at distance points change differently, only rigid linear coframe transformations preserve such type of a parallelism.

This geometric picture may be reformulated in term of a special connection. The elements of the coframe field attached to distinct points have to be assumed parallel to each other. It means that a special connection $\overset{\circ}{\Gamma}{}^i{}_{jk}$ exists such that the corresponding covariant derivative of the coframe field components is zero:

$$\vartheta^\alpha{}_{j;k} = \vartheta^\alpha{}_{j,k} - \overset{\circ}{\Gamma}{}^i{}_{jk}\vartheta^\alpha{}_i = 0\,. \tag{2.24}$$

Multiplying by $e_\alpha{}^i$, we have an explicit expression

$$\overset{\circ}{\Gamma}{}^i{}_{jk} = e_\alpha{}^i\vartheta^\alpha{}_{k,j}\,. \tag{2.25}$$

Under a smooth transform of coordinates, this expression is transformed in accordance with the inhomogeneous linear rule (2.14). Consequently, (2.25) indeed gives the coefficients of a special connection which is referred to as the *Weitzenböck flat connection*. This connection is unique for a class of coframes related by rigid linear transformations.

In an arbitrary nonholonomic reference basis (θ^a, f_a), we have correspondingly a unique Weitzenböck's connection 1-form which is constructed by (2.15) from (2.25)

$$\Gamma_a{}^b = f_a{}^k\left(\theta^b{}_i\,\overset{\circ}{\Gamma}{}^i{}_{jk} - \theta^b{}_{k,j}\right)dx^j\,. \tag{2.26}$$

Substituting the coframe field ϑ^α instead of the nonholonomic basis θ^a we have

$$\overset{\circ}{\Gamma}{}_\alpha{}^\beta = \left(-\vartheta^\beta{}_{i,j} + \vartheta^\beta{}_k e_\alpha{}^k\vartheta^\alpha{}_{i,j}\right)e_\alpha{}^i\,dx^j = 0\,. \tag{2.27}$$

Thus the Weitzenböck connection 1-form is zero, when it is referred to the coframe field $(\vartheta^\alpha, e_\alpha)$ itself. Certainly, this property is only a basis related fact. It yields, however, vanishing of the curvature of the Weitzenböck connection, which is a basis independent property.

General coframe connections. Recall that we are looking for a general coframe connection constructed from the first order derivatives of the coframe field components. In the Riemannian geometry, the analogous construction yields an unique connection of Levi-Civita. In the coframe geometry, however, the situation is different [37].

Proposition 3: The general linear connection constructed from the first order derivatives of the coframe field is given by a 3-parametric family:

$$\Gamma^i{}_{jk} = \overset{\circ}{\Gamma}{}^i{}_{jk} + \alpha_1 C^i{}_{jk} + \alpha_2 C_j \delta^i_k + \alpha_3 C_k \delta^i_j \,, \tag{2.28}$$

where

$$C^i{}_{jk} = \frac{1}{2} e_\alpha{}^i \left(\vartheta^\alpha{}_{k,j} - \vartheta^\alpha{}_{j,k}\right) \,, \qquad C_i = C^m{}_{mi} \,. \tag{2.29}$$

By (2.15), the connection 1-form corresponded to the coefficients (2.28), being referred to a nonholonomic basis, takes the form

$$\Gamma_a{}^b = f_a{}^k \left(-\theta^b{}_{k,m} + \theta^b{}_l \overset{\circ}{\Gamma}{}^l{}_{mk} + K^l{}_{mk} \theta^b{}_l\right) dx^m \,. \tag{2.30}$$

When this quantity is referred to the coframe field itself, it can be expressed by the exterior derivative of the coframe:

$$\Gamma_\alpha{}^\beta = \left(\alpha_1 C^\beta{}_{\gamma\alpha} + \alpha_2 C_\gamma \delta^\beta_\alpha + \alpha_3 C_\alpha \delta^\beta_\gamma\right) \vartheta^\gamma \,. \tag{2.31}$$

Metric-coframe connection. Consider a manifold endowed with the coframe metric tensor (2.8). Again, we are looking for a most general coframe connection that can be constructed from the first order derivatives of the coframe field. We will refer to it as the *metric-coframe connection.* Thus we are deal with a category of *coframe manifolds with a coframe metric and a linear coframe connection:*

$$\left\{M\,, \vartheta^\alpha, G\,, g(\vartheta^a), \Gamma_a{}^b(\vartheta^\alpha)\right\}. \tag{2.32}$$

Now the connection expression will involve some additional terms which depend on the metric tensor (2.8). To describe all possible combinations of the metric tensor components and frame/coframe components it is useful to pull down all the indices. Define:

$$\Gamma_{ijk} = g_{im} \Gamma^m{}_{jk} \,, \qquad C_{ijk} = g_{im} C^m{}_{jk} \,. \tag{2.33}$$

Proposition 4: The most general metric-coframe connection constructed from the first order derivatives of the coframe field is represented by a 6-parametric family:

$$\Gamma_{ijk} = \overset{\circ}{\Gamma}_{ijk} + \alpha_1 C_{ijk} + \alpha_2 g_{ik} C_j + \alpha_3 g_{ij} C_k + \beta_1 g_{jk} C_i + \beta_2 C_{jki} + \beta_3 C_{kij} \,. \tag{2.34}$$

The expression (2.34) can be rewritten in a

$$\Gamma^i{}_{jk} = \overset{\circ}{\Gamma}{}^i{}_{jk} + \alpha_1 C^i{}_{jk} + \alpha_2 \delta^i_k C_j + \alpha_3 \delta^i_j C_k + \beta_1 g^{il} g_{jk} C_l + \beta_2 g^{il} C_{jkl} + \beta_3 g^{il} C_{klj} \,. \tag{2.35}$$

Here we can identify: (i) The terms with the coefficient α_i that do not depend on the metric; (ii) The terms with the coefficient β_i that can be constructed only by use of the metric tensor.

With respect to a nonholonomic basis (f_a, θ^a), the coefficients of a connection (2.35) correspond to a connection 1-form (2.15)

$$\Gamma_a{}^b = \overset{\circ}{\Gamma}_a{}^b + K^i{}_{jk} f_a{}^k \theta^b{}_i dx^j \,. \tag{2.36}$$

When (2.36) is referred to the coframe field itself, it can be expressed by the exterior derivative of the coframe. We have

$$\Gamma_\alpha{}^\beta = \left(\alpha_1 C^\beta{}_{\gamma\alpha} + \alpha_2 C_\gamma \delta^\beta_\alpha + \alpha_3 C_\alpha \delta^\beta_\gamma + \beta_1 C^\beta \eta_{\alpha\gamma} + \beta_2 C_{\gamma\alpha\nu}\eta^{\beta\nu} + \beta_3 C_{\alpha\nu\gamma}\eta^{\beta\nu}\right)\vartheta^\gamma . \qquad (2.37)$$

2.3. Torsion of the coframe connection

Torsion tensor and torsion 2-form. Definitions. Consider a connection 1-form $\Gamma_b{}^a$ referred to an arbitrary basis (θ^a, f_a). For a tensor valued p-form of a representation type $\rho\left(A_a{}^b\right)$, the *covariant exterior derivative* operator $D : \Omega^p(\mathcal{M}) \to \Omega^{p+1}(\mathcal{M})$ is defined as [24], [5]

$$D = d + \Gamma_b{}^a \rho\left(A_a{}^b\right) \wedge . \qquad (2.38)$$

In particular, the covariant exterior derivative of a scalar-valued form ϕ is $D\phi = d\phi$. For a vector-valued form ϕ^a, it is given by $D\phi^a = d\phi^a + \Gamma_b{}^a \wedge \phi^b$, etc.

For a connection 1-form $\Gamma_a{}^b$ written with respect to a nonholonomic basis, the *torsion 2-form* $\mathcal{T}^a$ is defined as

$$\mathcal{T}^a = D\theta^a = d\theta^a + \Gamma_b{}^a \wedge \theta^b . \qquad (2.39)$$

Substituting (2.15) into (2.39), we observe that the coframe derivative term $d\vartheta^a$ cancels out. Hence,

$$\mathcal{T}^a = \Gamma^i{}_{jk}\theta^a{}_i dx^j \wedge dx^k = \Gamma^i{}_{[jk]}\theta^a{}_i dx^j \wedge dx^k . \qquad (2.40)$$

In a coordinate coframe, this expression is simplified to

$$\mathcal{T}^i = \Gamma^i{}_{[jk]} dx^j \wedge dx^k . \qquad (2.41)$$

Consequently, the torsion 2-form $\mathcal{T}^a$ is completely determined by an antisymmetric combination of the coefficients of the connection. Observe that such combination is a tensor. Thus, the torsion 2-form is completely equivalent to a $(1,2)$-rank *torsion tensor* which is defined as

$$T^i{}_{jk} = 2\Gamma^i{}_{[jk]} . \qquad (2.42)$$

In a holonomic and a nonholonomic bases, the torsion 2-form is expressed respectively as

$$\mathcal{T}^i = \frac{1}{2} T^i{}_{jk} dx^j \wedge dx^k , \qquad \mathcal{T}^a = \frac{1}{2} T^i{}_{jk}\theta^a{}_i dx^j \wedge dx^k . \qquad (2.43)$$

It is useful to define also a quantity

$$\mathcal{T}^\alpha = \frac{1}{2} T^i{}_{jk}\vartheta^\alpha{}_i dx^j \wedge dx^k . \qquad (2.44)$$

With respect to the coframe field, the torsion 2-form of the Weitzenböck connection (2.44) reads

$$\overset{\circ}{\mathcal{T}}{}^\alpha = d\vartheta^\alpha . \qquad (2.45)$$

Torsion of the metric-coframe connection. For the metric-coframe connection (2.34), the covariant components $T_{ijk} = 2g_{im}\Gamma^m{}_{[jk]}$ of the torsion tensor take the form

$$T_{ijk} = 2(1+\alpha_1)C_{ijk} + (\alpha_2 - \alpha_3)(g_{ik}C_j - g_{ij}C_k) + (\beta_2 + \beta_3)(C_{jki} + C_{kij}) . \qquad (2.46)$$

The corresponded torsion 2-form is expressed in the coordinate basis as

$$T^i = \Big[(1+\alpha_1)C^i{}_{jk} + (\alpha_2-\alpha_3)C_j\delta^i_k + (\beta_2+\beta_3)g^{im}C_{jkm}\Big]dx^j\wedge dx^k\,. \qquad (2.47)$$

Torsion-free metric-coframe connection. Let us look for which values of the parameters the torsion of the metric-coframe connection is identically zero. The corresponded connection is called *the symmetric or torsion-free connection.* It is clear from (2.46) that the metric-coframe connection is symmetric if

$$\alpha_1 = -1\,, \qquad \alpha_2 = \alpha_3\,, \qquad \beta_2 = -\beta_3\,. \qquad (2.48)$$

The necessity of this condition can be derived from the irreducible decomposition [37].

Thus on a manifold of the dimension $D \geq 3$ there exists a 3-parametric family of the symmetric (torsion-free) connections:

$$\Gamma^i{}_{jk} = \overset{\circ}{\Gamma}{}^i{}_{jk} - C^i{}_{jk} + \alpha_2\left(\delta^i_k C_j + \delta^i_j C_k\right) + \beta_1 g_{jk}g^{im}C_m + \beta_2 g^{im}\left(C_{jkm} - C_{kmj}\right)\,. \qquad (2.49)$$

2.4. Nonmetricity of the metric-coframe connection

Nonmetricity tensor and nonmetricity 2-form. Definition. When Cartan's manifold is endowed with a metric tensor, the connection generates an additional tensor field called *the nonmetricity tensor.* It is expressed as a covariant derivative of the metric tensor components. For a metric given in a local system of coordinates as $g = g_{ij}dx^i \otimes dx^j$, the nonmetricity tensor is defined as

$$Q_{kij} = -\nabla_k g_{ij} = -g_{ij,k} + \Gamma^m{}_{ik}g_{mj} + \Gamma^m{}_{jk}g_{im}\,, \qquad (2.50)$$

or,

$$Q_{kij} = -g_{ij,k} + \Gamma_{jik} + \Gamma_{ijk}\,. \qquad (2.51)$$

Evidently, this tensor is symmetric in the last pair of indices $Q_{kij} = Q_{kji}$. Hence, on a D dimensional manifold, the nonmetricity tensor has $D(D^2+D)/2$ independent components.

For the exterior form representation, it is useful to define *the nonmetricity 1-form.* In a coordinate basis, it is given by

$$Q_{ij} = Q_{kij}dx^k = -dg_{ij} + \Gamma_{ij} + \Gamma_{ji}\,. \qquad (2.52)$$

In an arbitrary reference basis $(f_a\,,\theta^a)$, the metric tensor is expressed as $g = g_{ab}\theta^a\otimes\theta^b$. Correspondingly, the nonmetricity 1-form reads

$$Q_{ab} = -dg_{ab} + \Gamma_{ab} + \Gamma_{ba}\,. \qquad (2.53)$$

With respect to the coframe field ϑ^α, the components of the metric are constants $\eta_{\alpha\beta}$, thus the nonmetricity is merely the symmetric combination of the connection 1-form components

$$Q_{\alpha\beta} = \Gamma_{\alpha\beta} + \Gamma_{\beta\alpha}\,. \qquad (2.54)$$

Note, that this expression is not a usual tensorial quantity. In fact, it is an expression of a tensor-valued 1-form of nonmetricity with respect to a special class of bases. Its relation to a proper tensorial valued 1-form (2.52) is, however, very simple. By a substitution of (2.19) into (2.54) we have

$$Q_{ij} = Q_{\alpha\beta}\vartheta^\alpha{}_i\vartheta^\beta{}_j\,. \qquad (2.55)$$

The following generalization of the Levi-Civita theorem from the Riemannian geometry provides a decomposition of an arbitrary affine connection [54].

Proposition 5: Let a metric g on a manifold M be fixed and two tensors T_{ijk} and Q_{ijk} with the symmetries

$$T_{ijk} = -T_{ikj}\,, \qquad Q_{kij} = Q_{kji}\,. \tag{2.56}$$

be given. A unique connection Γ_{ijk} exists on M such that T_{ijk} is its torsion and Q_{ijk} is its nonmetricity. Explicitly,

$$\Gamma_{ijk} = \overset{*}{\Gamma}_{ijk} - \frac{1}{2}\big(Q_{ijk} - Q_{jki} - Q_{kij}\big) + \frac{1}{2}\big(T_{ijk} + T_{jki} - T_{kij}\big)\,, \tag{2.57}$$

where

$$\overset{*}{\Gamma}_{ijk} = \frac{1}{2}\big(g_{ij,k} + g_{ik,j} - g_{jk,i}\big) \tag{2.58}$$

are the components of the Levi-Civita connection.

Nonmetricity of the metric-coframe connection. We calculate now the nonmetricity tensor of the metric-coframe connection (2.34)

$$\begin{aligned} Q_{kij} \;=\; & \Big(-g_{ij,k} + \overset{\circ}{\Gamma}_{ijk} + \overset{\circ}{\Gamma}_{jik}\Big) + (\alpha_1 - \beta_2)(C_{ijk} - C_{jki}) + \\ & (\alpha_2 + \beta_1)(g_{ik}C_j + g_{jk}C_i) + 2\alpha_3 g_{ij}C_k\,. \end{aligned} \tag{2.59}$$

The first parenthesis represent the nonmetricity tensor of the Weitzenböck connection. This expression vanishes identically, i.e., the Weitzenböck connection is metric-compatible. Indeed, we have

$$g_{ij,k} = \eta_{\alpha\beta}(\vartheta^\alpha{}_{i,k}\vartheta^\beta_j + \vartheta^\alpha{}_i\vartheta^\beta_{j,k}) = \overset{\circ}{\Gamma}_{ijk} + \overset{\circ}{\Gamma}_{jik}\,. \tag{2.60}$$

Consequently, (2.59) is simplified to

$$Q_{kij} = (\alpha_1 - \beta_2)(C_{ijk} + C_{jki}) + (\alpha_2 + \beta_1)(g_{ik}C_j + g_{jk}C_i) + 2\alpha_3 g_{ij}C_k\,. \tag{2.61}$$

Metric compatible metric-coframe connection. Let us look for which values of the coefficients the connection is *metric-compatible*, i.e., has an identically zero non-metricity tensor. Recall that both quantities, the metric tensor and the connection, are constructed from the same building block — the coframe field ϑ^α. It is clear from (2.61) that the metric-coframe connection is metric-compatible if

$$\alpha_1 = \beta_2\,, \qquad \alpha_2 = -\beta_1\,, \qquad \alpha_3 = 0\,. \tag{2.62}$$

The necessity of this condition can be derived from the irreducible decomposition of the nonmetricity tensor [37].

Metric compatible and torsion-free metric-coframe connection. Let us look now for a general coframe connection of a zero torsion and zero non-metricity, i.e., for a symmetric metric compatible connection constructed from the coframe field. The system of conditions (2.48) and (2.62) has a unique solution

$$\alpha_1 = \beta_2 = -\beta_3 = -1\,, \qquad \beta_1 = \alpha_2 = \alpha_3 = 0\,. \tag{2.63}$$

Consequently, a metric-compatible symmetric connection is unique. This is in a correspondence to the original Levi-Civita theorem, and the unique connection is of Levi-Civita. Moreover, substituting (2.63) into (2.34) we can express now the standard Levi-Civita connection $\overset{*}{\Gamma}{}^i{}_{jk}$ via the flat connection of Weitzenböck $\overset{\circ}{\Gamma}{}^i{}_{jk}$

$$\overset{*}{\Gamma}_{ijk} = \overset{\circ}{\Gamma}_{i(jk)} + C_{kij} - C_{jki}\,. \tag{2.64}$$

3. Gauge transformations

Local transformations of the coframe field. The geometrical structure considered above is well defined for a fixed coframe field e_α. Moreover, it is invariant under rigid coframe transformations. The gauge paradigm suggests now to look for a localization of such transformations:

$$\vartheta^\alpha \mapsto L^\alpha{}_\beta\,\vartheta^\beta\,, \qquad e_\alpha \mapsto L_\alpha{}^\beta\,e_\beta\,, \tag{3.1}$$

or, in the components,

$$\vartheta^\alpha{}_i \mapsto L^\alpha{}_\beta\,\vartheta^\beta{}_i\,, \qquad e_\alpha{}^i \mapsto L_\alpha{}^\beta\,e_\beta{}^i\,. \tag{3.2}$$

Here the matrix $L^\alpha{}_\beta$ and its inverse $L_\alpha{}^\beta$ are functions of a point $x \in M$. We require the volume element (2.4) and the metric tensor (2.8) both to be invariant under the pointwise transformations (3.1). Consequently, $L^\alpha{}_\beta$ is assumed to be a pseudo-orthonormal matrix whit enters are smooth functions of a point. We will also use an infinitesimal version of the transformation (3.2) with $L^\alpha{}_\beta = \delta^\alpha_\beta + X^\alpha{}_\beta$. In the components, it takes the form

$$\vartheta^\alpha{}_i \mapsto \vartheta^\alpha{}_i + X^\alpha{}_\beta\,\vartheta^\beta{}_i\,, \qquad e_\alpha{}^i \mapsto e_\alpha{}^i - X^\beta{}_\alpha\,e_\beta{}^i\,. \tag{3.3}$$

As the elements of the algebra $so(1,n)$, the matrix $X_{\alpha\beta} = \eta_{\alpha\mu}X^\mu{}_\beta$ is antisymmetric. We define a corresponded antisymmetric tensor

$$F_{ij} = \vartheta^\alpha{}_i\vartheta^\beta{}_j X_{\alpha\beta}\,. \tag{3.4}$$

We have already postulated the invariance of the volume element and of the metric tensor under the coframe transformations. It is natural to involve now an additional invariance requirement concerning the affine connection.
Connection invariance condition. Affine coframe connection is assumed to be invariant under pointwise transformations of the coframe field

$$\Delta\Gamma^i{}_{jk} := \Gamma^i{}_{jk}\left(L^\alpha{}_\beta\vartheta^\beta\right) - \Gamma^i{}_{jk}\left(\vartheta^\alpha\right) = 0\,. \tag{3.5}$$

Since the coframe connection is constructed from the first order derivatives of the coframe field, (3.5) is a first order PDE for the elements of the group G and for the components of the coframe field.

Weitzenböck connection transformation. Since the Weitzenböck connection is a basis tool of our construction, it is useful to calculate the change of this quantity under the coframe transformations (3.1). We have

$$\Delta\overset{\circ}{\Gamma}{}^i{}_{jk} = e_\alpha{}^i\vartheta^\beta{}_k Y^\alpha{}_{\beta j}\,, \qquad \text{where} \qquad Y^\alpha{}_{\beta j} = L^\alpha{}_\gamma L^\gamma{}_{\beta,j}\,. \tag{3.6}$$

All matrices involved here are nonsingular, consequently the Weitzenböck connection is preserved only under the rigid transformations of the coframe field with $L^\gamma{}_{\beta,j} = 0$.

Let us rewrite (3.6) in alternative forms. Since the metric tensor is invariant under the transformations (3.1) we have

$$\Delta\overset{\circ}{\Gamma}{}_{ijk} = \Delta\left(g_{im}\overset{\circ}{\Gamma}{}^m{}_{jk}\right) = g_{im}\Delta\overset{\circ}{\Gamma}{}^m{}_{jk}\,. \tag{3.7}$$

Consequently

$$\Delta\overset{\circ}{\Gamma}{}_{ijk} = \vartheta^\alpha{}_i\vartheta^\beta{}_k Y_{\alpha\beta j}\,, \qquad \text{where} \qquad Y_{\alpha\beta j} = \eta_{\alpha\mu}Y^\mu{}_{\beta j}\,. \tag{3.8}$$

In the infinitesimal approximation, (3.6) takes the form

$$\Delta \overset{o}{\Gamma}{}^i{}_{jk} = e_\alpha{}^i \vartheta^\beta{}_k X^\alpha{}_{\beta,j} \,. \tag{3.9}$$

while (3.7) with $X_{\alpha\beta} = \eta_{\alpha\mu} X^\mu{}_\beta$ reads

$$\Delta \overset{o}{\Gamma}_{ijk} = \vartheta^\alpha{}_i \vartheta^\beta{}_k X_{\alpha\beta,j} \,. \tag{3.10}$$

Note that since $X_{\alpha\beta}$ is antisymmetric, we have in this approximation

$$\Delta \overset{o}{\Gamma}_{ijk} = -\Delta \overset{o}{\Gamma}_{kji} \,. \tag{3.11}$$

We will also consider an additional physical meaningful approximation when the derivatives of the coframe is considered to be small relative to the derivatives of the transformation matrix. In this case, (3.9) and (3.10) read

$$\Delta \overset{o}{\Gamma}{}^i{}_{jk} = F^i{}_{k,j} \,, \qquad \text{where} \qquad F^i{}_k = e_\alpha{}^i \vartheta^\beta{}_k X^\alpha{}_\beta \,, \tag{3.12}$$

and

$$\Delta \overset{o}{\Gamma}_{ijk} = F_{ik,j} \,, \qquad \text{where} \qquad F_{ij} = \vartheta^\alpha{}_i \vartheta^\beta{}_j X_{\alpha\beta} \,. \tag{3.13}$$

Transformations preserved the geometric structure. Since the coframe field appears in the coframe geometrical structure only implicitly, (3.1) is a type of a gauge transformation. Invariance of the metric tensor and of the volume element restricts $L^\alpha{}_\beta$ to a pseudo-orthonormal matrix $G = SO(1, n)$. Let us ask now, under what conditions the general coframe connection (2.34) is invariant under the coframe transformations (3.1). First we rewrite (2.34) via the Levi-Civita connection. Using (2.63) we have

$$\overset{o}{\Gamma}_{ijk} = \overset{*}{\Gamma}_{ijk} + C_{ijk} - C_{kij} + C_{jki} \,. \tag{3.14}$$

Thus (2.34) takes the form

$$\Gamma_{ijk} = \overset{*}{\Gamma}_{ijk} + (\alpha_1 + 1)C_{ijk} + \alpha_2 g_{ik} C_j + \alpha_3 g_{ij} C_k + \beta_1 g_{jk} C_i + (\beta_2 + 1)C_{jki} + (\beta_3 - 1)C_{kij} \,. \tag{3.15}$$

Since the Levi-Civita connection $\overset{*}{\Gamma}_{ijk}$ is invariant under the transformations (3.1), the equation $\Delta\Gamma_{ijk} = 0$ takes the form

$$(\alpha_1 + 1)\Delta C_{ijk} + \alpha_2 g_{ik} \Delta C_j + \alpha_3 g_{ij} \Delta C_k + \beta_1 g_{jk} \Delta C_i + (\beta_2 + 1)\Delta C_{jki} + (\beta_3 - 1)\Delta C_{kij} = 0 \,. \tag{3.16}$$

Hence in order to have an invariant coframe connection, we have to look for possible solutions of equation (3.16).

Trivial solutions of the invariance equation. Consider first two trivial solutions of (3.16) which turn out to be non-dynamical.

(i) *Arbitrary transformations — Levi-Civita connection.*
The equation (3.16) is evidently satisfied when all the numerical coefficients mutually equal to zero. It is easy to check that these six relations are equivalent to (2.63). Thus the corresponded connection is of Levi-Civita. In this case, the elements of the matrix $L^\alpha{}_\beta$ are arbitrary functions of a point. Thus we come to a trivial fact that the Levi-Civita connection is a unique coframe connection which is invariant under arbitrary local $SO(1, n)$ transformations of the coframe field.

(ii) *Rigid transformations.*
Another trivial solution of the system (3.16) emerges when we require $\Delta C_{ijk} = 0$. All permutations and traces of this tensor are also equal to zero so (3.16) is trivially valid. Due to (3.8), it means that the matrix of transformations is independent on a point. In this case, an arbitrary coframe connection, in particular the Weitzenböck connection, remains unchanged. Thus we come to another trivial fact that the coframe connection is invariant under rigid transformations of the coframe field.

Dynamical solution. We will look now for nontrivial solutions of the system (3.16). Three traces of this system yield the equations $\Delta C_i = 0$. Thus we obtain the first condition

$$\Delta C_i = 0\,. \tag{3.17}$$

The system (3.16) remains now in the form

$$(\alpha_1 + 1)\Delta C_{ijk} + (\beta_2 + 1)\Delta C_{jki} + (\beta_3 - 1)\Delta C_{kij} = 0\,. \tag{3.18}$$

Applying the complete antisymetrization in three indices we derive the second equation

$$\Delta C_{[ijk]} = 0\,. \tag{3.19}$$

The equation (3.18) remains now in the form

$$(\beta_2 - \alpha_1)\Delta C_{jki} + (\beta_3 - \alpha_1 - 2)\Delta C_{jki} = 0\,. \tag{3.20}$$

We have to restrict now the coefficients, otherwise we obtain $\Delta C_{ijk} = 0$, i.e., only the rigid transformations. Consequently we require

$$\beta_2 = \alpha_1\,, \qquad \beta_3 = \alpha_1 + 2\,. \tag{3.21}$$

Thus we have proved

Proposition 6: The coframe connection

$$\Gamma_{ijk} = \overset{*}{\Gamma}_{ijk} + (\alpha_1 + 1)C_{[ijk]} + \alpha_2 g_{ik}C_j + \alpha_3 g_{ij}C_k + \beta_1 g_{jk}C_i\,. \tag{3.22}$$

is invariant under the coframe transformations satisfied the equations

$$\Delta C_i = 0\,. \qquad \Delta C_{[ijk]} = 0\,. \tag{3.23}$$

Observe that this family includes the Levi-Civita connection, which is invariant under a wider group of arbitrary transformations of the coframe field.

The torsion tensor of the connection (3.22) is expressed as

$$T_{ijk} = (\alpha_1 + 1)C_{[ijk]} + (\alpha_2 - \alpha_3)(g_{ik}C_j - g_{ij}C_k)\,. \tag{3.24}$$

Thus a torsion-free subfamily of (3.22) is given by

$$\Gamma_{ijk} = \overset{*}{\Gamma}_{i(jk)} + \alpha_2(g_{ik}C_j + g_{ij}C_k) + \beta_1 g_{jk}C_i\,. \tag{3.25}$$

The nonmetricity tensor of the connection (3.22) reads

$$Q_{kij} = (\alpha_2 + \beta_1)(g_{ik}C_j + g_{jk}C_i) + 2\alpha_3 g_{jj}C_k\,. \tag{3.26}$$

Thus a metric compatible subfamily of (3.22) is given by

$$\Gamma_{ijk} = \overset{*}{\Gamma}_{i(jk)} + (\alpha_1 + 1)C_{[ijk]} + \alpha_2(g_{ik}C_j - g_{jk}C_i)\,. \tag{3.27}$$

From (3.24) and (3.26) we derive an interesting conclusions:

$$\Delta Q_{kij} = 0 \quad \Longleftrightarrow \quad \Delta C_i = 0\,. \tag{3.28}$$

and, together with this relation,

$$\Delta T_{ijk} = 0 \quad \Longleftrightarrow \quad \Delta C_{[ijk]} = 0\,. \tag{3.29}$$

Thus the relations (3.23) obtain a geometric meaning, they correspond to invariance of the torsion and nonmetricity tensors under coframe transformations.

4. Maxwell-type system

Let us examine now what physical meaning can be given to the invariance conditions [42]

$$\Delta C_{[ijk]} = 0\,, \qquad \Delta C_i = 0\,. \tag{4.1}$$

Denote $K_{ijk} = \Delta C_{ijk}$. Thus (4.1) takes the form

$$K_{[ijk]} = 0\,, \qquad K^m{}_{im} = 0\,. \tag{4.2}$$

The tensor K_{ijk} depends on the derivatives of the Lorentz parameters $X_{\alpha\beta}$ and on the components of the coframe field

$$K_{ijk} = \frac{1}{2}\vartheta^\alpha{}_k\Big(X_{\alpha\beta,j}\vartheta^\beta{}_i - X_{\alpha\beta,i}\vartheta^\beta{}_j\Big)\,. \tag{4.3}$$

Thus, in fact, we have in (4.2), two first order partial differential equations for the entries of an antisymmetric matrix $X_{\alpha\beta}$. We can translate this matrix into an antisymmetric tensor F_{ij}

$$F_{ij} = X_{\mu\nu}\vartheta^\mu{}_i\vartheta^\nu{}_j\,, \qquad X_{\mu\nu} = F_{ij}e_\mu{}^i e_\nu{}^j\,. \tag{4.4}$$

Substituting into (4.3), we derive

$$\begin{aligned} K_{ijk} &= F_{k[i,j]} - \frac{1}{2}X_{\alpha\beta}\Big[(\vartheta^\alpha{}_k\vartheta^\beta{}_i)_{,j} - (\vartheta^\alpha{}_k\vartheta^\beta{}_j)_{,i}\Big] \\ &= F_{k[i,j]} - F_{km}C^m{}_{ij} - \frac{1}{2}\Big(F_{mi}\,\overset{\circ}{\Gamma}{}^m{}_{kj} - F_{mj}\,\overset{\circ}{\Gamma}{}^m{}_{ki}\Big)\,. \end{aligned} \tag{4.5}$$

Consequently, the first equation from (4.2) takes the form

$$F_{[ij,k]} = \frac{2}{3}(C^m{}_{ij}F_{km} + C^m{}_{jk}F_{im} + C^m{}_{ki}F_{jm})\,, \tag{4.6}$$

while the second equation from (4.2) is rewritten as

$$F^i{}_{j,i} = -2F^i{}_mC^m{}_{ij} + F_{kj}g^{ki}{}_{,i} + F_{mj}g^{ki}\,\overset{\circ}{\Gamma}{}^m{}_{ki} - F_{mi}g^{ki}\,\overset{\circ}{\Gamma}{}^m{}_{kj}\,. \tag{4.7}$$

Observe first a significant approximation to (4.6—4.7). If the right hand sides in both equations are neglected, the equations take the form of the ordinary Maxwell equations for the electromagnetic field in vacuum —

$$F_{[ij,k]} = 0\,, \qquad F^i{}_{j,i} = 0\,. \tag{4.8}$$

In the coframe models, the gravity is modeled by a variable coframe field, i.e., by nonzero values of the quantities $\overset{\circ}{\Gamma}{}_{ij}{}^{k}$. Consequently, the right hand sides of (4.6—4.7) can be viewed as curved space additions, i.e., as the gravitational corrections to the electromagnetic field equations. In the flat spacetime, when a suitable coordinate system is chosen, these corrections are identically equal to zero. Consequently, in the flat spacetime, the invariance conditions (4.2) take the form of the vacuum Maxwell system.

On a curved manifold, the standard Maxwell equations are formulated in a covariant form. Let us show that our system (4.6—4.7) is already covariant. We rewrite (4.5) as

$$K_{ijk} = \frac{1}{2}(F_{ki,j} - F_{km}\,\overset{\circ}{\Gamma}{}^{m}{}_{ij} - F_{mi}\,\overset{\circ}{\Gamma}{}^{m}{}_{kj}) - \frac{1}{2}(\,i \longleftrightarrow j\,)\,. \tag{4.9}$$

Consequently,

$$K_{ijk} = F_{k[i;j]}\,, \tag{4.10}$$

where the covariant derivative (denoted by the semicolon) is taken relative to the Weitzenböck connection. Consequently, the system (4.6—4.7) takes the covariant form

$$F_{[ij;k]} = 0\,, \qquad F^{i}{}_{j;i} = 0\,. \tag{4.11}$$

These equations are literally the same as the electromagnetic sector field equations of the Maxwell-Einstein system. The crucial difference is encoded in the type of the covariant derivative. In the Maxwell-Einstein system, the covariant derivative is taken relative to the Levi-Civita connection, while, in our case, the corresponding connection is of Weitzenböck. Observe that, due to our approach, the Weitzenböck connection is rather natural in (4.11). Indeed, since the electromagnetic-type field describes the local change of the coframe field, it should itself be referred only to the global changes of the coframe. As we have shown, such global transformations correspond precisely to the teleparallel geometry with the Weitzenböck connections.

Let us show how these modifications alternate the electromagnetic field solutions. On Schwarzschild space-time the spherically symmetric solution of (4.11) is given [42] by

$$F_{0}{}^{\hat{i}} = x^{\hat{i}}\frac{Q}{\rho^3}\frac{1 - m/2\rho}{(1 + m/2\rho)^3}\,. \tag{4.12}$$

The Coulomb-type force acted on a test charge q (of a small mass) takes the form

$$\mathcal{F} = \frac{Qq}{\rho^2}\frac{1 - m/2\rho}{(1 + m/2\rho)^3}\,. \tag{4.13}$$

Here the isotropic coordinates are used. The ordinary Cartesian radius r is related to the isotropic radius ρ as

$$r = \rho\left(1 + \frac{m}{2\rho}\right)^2\,. \tag{4.14}$$

Hence

$$\rho = \frac{r - m + \sqrt{r^2 - 2mr}}{2} \approx r\left(1 - \frac{m}{r}\right)\,. \tag{4.15}$$

Observe that the isotropic coordinates are defined only for $r > 2m$. Consequently, the modernized Coulomb force takes the form

$$F = \frac{Qq}{r^2}\left(1 - \frac{m^2}{4\rho^2}\right)\,. \tag{4.16}$$

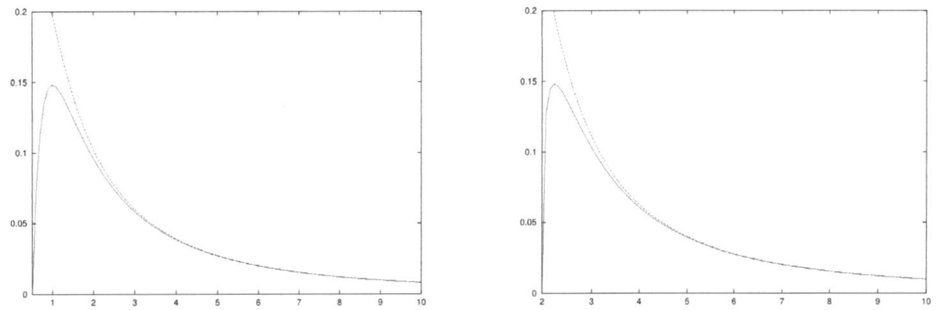

Figure 1. The graphs represent the dependence of the force F on ρ/m and r/m correspondingly relative to the Coulomb force (the top lines). In both cases F is given in the units of Qq/m^2.

Finally, in the Cartesian coordinates, the mass-correction terms take the form

$$\begin{aligned} F &= Qq\frac{r-2m+\sqrt{r^2-2mr}}{4(r+\sqrt{r^2-2mr})^3} \\ &= \frac{Qq}{r^2}\left(1-\frac{m^2}{4r^2}-\frac{m^3}{2r^3}+\cdots\right). \end{aligned} \tag{4.17}$$

We depict the dependence of the force F on the distances ρ and r on Fig. 2. The graphs start from $\rho = m/2$ and $r = 2m$ which are the minimal possible value. The deviation from the Coulomb values appears only for small distances $\rho, r \sim m$. The maximal value of the force between two charged particles predicts as

$$F \approx 0.15\, Qq/m^2\,. \tag{4.18}$$

For two electrons, it gives $F \approx 0.76 \cdot 10^{86} N$.

5. Conclusion

GR is a well-posed classical field theory for 10 independent variables — the components of the metric tensor. Although, this theory is completely satisfactory in the pure gravity sector, its possible extensions to other physics phenomena is rather problematic. In particular, the description of fermions on a curved space and the supergravity constructions require a richer set of 16 independent variables. These variables can be assembled in a coframe field, i.e., a local set of four linearly independent 1-forms. Moreover, in supergravity, it is necessary to involve a special flat connection constructed from the derivatives of the coframe field. These facts justify the study of the field models based on a coframe variable alone.

The classical field construction of the coframe gravity is based on a Yang-Mills-type Lagrangian which is a linear combination of quadratic terms with dimensionless coefficients. Such model turns to be satisfactory in the gravity sector and has the viable Schwarzschild solutions even being alternative to the standard GR. Moreover, the coframe model treating of the gravity energy makes it even preferable than the ordinary GR where the gravity energy cannot be defined at all. A principle problem that the coframe gravity construction does not have any connection to a specific geometry even being constructed from the geometrical meaningful objects. A geometrization of the coframe gravity is an aim of this presentation.

We construct a general family of coframe connections which involves as the special cases the Levi-Civita connection of GR and the flat Weitzenböck connection. Every specific connection generates a geometry of a specific type. We identify the subclasses of metric-compatible and torsion-free connections. Moreover we study the local linear transformations of the coframe fields and identify a class of connections which are invariant under restricted coframe transformations. Quite remarkable that the restriction conditions are necessary approximated by a Maxwell-type system of equations.

Acknowledgments

I thank Roman Jackiw (MIT), Friedrich Hehl (Cologne/Missouri), and Yuri Obukhov (Moscow) for fruitful discussions. My acknowledgments to the organizers of IARD 2012 conference, especially Lawrence Horwitz (Tel-Aviv), Martin Land(Jerusalem), and Luca Lusanna (Florence), for their encouragement and assistance. This work was partially supported by the German-Israeli Foundation for Scientific Research and Development (GIF), Research Grant No. 1078-107.14/2009.

References

[1] F. W. Hehl, P. Von Der Heyde, G. D. Kerlick and J. M. Nester, Rev. Mod. Phys. *48*, 393 (1976).
[2] S. Hojman, M. Rosenbaum, M. P. Ryan and L. C. Shepley, Phys. Rev. D *17*, 3141 (1978).
[3] W. Kopczynski, Acta Phys. Polon. B *10* (1979) 365.
[4] J. D. McCrea, Class. Quant. Grav. *9*, 553 (1992).
[5] F. W. Hehl, J. D. McCrea, E. W. Mielke and Y. Neeman, Phys. Rept. *258*, 1 (1995);
[6] F. Gronwald and F. W. Hehl, arXiv:gr-qc/9602013.
[7] G. Giachetta and G. Sardanashvily, Class. Quant. Grav. *13*, L67 (1996)
[8] Yu. N. Obukhov, E. J. Vlachynsky, W. Esser and F. W. Hehl, Phys. Rev. D *56*, 7769 (1997).
[9] J. Socorro, C. Lammerzahl, A. Macias and E. W. Mielke, Phys. Lett. A *244*, 317 (1998)
[10] A. Garcia, F. W. Hehl, C. Laemmerzahl, A. Macias and J. Socorro, Class. Quant. Grav. *15*, 1793 (1998)
[11] F. Gronwald, Int. J. Mod. Phys. D *6*, 263 (1997)
[12] F. W. Hehl and A. Macias, Int. J. Mod. Phys. D *8*, 399 (1999)
[13] D. Puetzfeld, Class. Quant. Grav. *19*, 3263 (2002)
[14] M. Godina, P. Matteucci and J. A. Vickers, J. Geom. Phys. *39*, 265 (2001)
[15] D. Vassiliev, Annalen Phys. *14*, 231 (2005)
[16] V. Pasic and D. Vassiliev, Class. Quant. Grav. *22*, 3961 (2005)
[17] Y. N. Obukhov, Phys. Rev. D *73*, 024025 (2006)
[18] H. F. M. Goenner, Living Rev. Rel. *7*, 2 (2004).
[19] K. Hayashi and T. Shirafuji, Phys. Rev. D *19*, 3524 (1979)
[20] E. Sezgin and P. van Nieuwenhuizen, Phys. Rev. D *21*, 3269 (1980).
[21] J. Nitsch and F. W. Hehl, Phys. Lett. B *90*, 98 (1980);
[22] F. Mueller-Hoissen and J. Nitsch, Phys. Rev. D *28*, 718 (1983);
[23] R. Kuhfuss and J. Nitsch, Gen. Rel. Grav. *18*, 1207 (1986).
[24] E. W. Mielke, Annals Phys. *219*, 78 (1992);
[25] U. Muench, F. Gronwald and F. W. Hehl, Gen. Rel. Grav. *30*, 933 (1998);
[26] R. S. Tung and J. M. Nester, Phys. Rev. D *60*, 021501 (1999);
[27] Y. Itin and S. Kaniel, J. Math. Phys. *41*, 6318 (2000)
[28] M. Blagojevic and M. Vasilic, Class. Quant. Grav. *17*, 3785 (2000);
[29] Y. Itin, Int. J. Mod. Phys. D *10*, 547 (2001)
[30] M. Blagojevic and I. A. Nikolic, Phys. Rev. D *62*, 024021 (2000)
[31] I. L. Shapiro, Phys. Rept. *357*, 113 (2001)
[32] R. T. Hammond, Rept. Prog. Phys. *65*, 599 (2002).
[33] Y. Itin, Class. Quant. Grav. *19*, 173 (2002);
[34] Y. Itin, Gen. Rel. Grav. *34*, 1819 (2002);
[35] Y. N. Obukhov and J. G. Pereira, Phys. Rev. D *67*, 044016 (2003)
[36] Y. Itin, J. Phys. A *36*, 8867 (2003)
[37] Y. Itin, in *Classical and Quantum Gravity Research Progress*, ed. M. N. Christiansen and T. K. Rasmussen, (Hauppauge, NY: Nova Science Publishers) 2008 (arXiv:0711.4209)
[38] M. Leclerc, Phys. Rev. D *71*, 027503 (2005)

[39] Y. N. Obukhov, G. F. Rubilar and J. G. Pereira, Phys. Rev. D *74*, 104007 (2006)
[40] Y. Itin, J. Math. Phys. *46* 12501 (2005).
[41] F. B. Estabrook, Class. Quant. Grav. *23*, 2841 (2006)
[42] Y. Itin, Class. Quant. Grav. *23* (2006) 3361.
[43] G. G. L. Nashed and T. Shirafuji, Int. J. Mod. Phys. D *16*, 65 (2007)
[44] A. Ashtekar, Phys. Rev. D *36*, 1587 (1987).
[45] S. Deser and C. J. Isham, Phys. Rev. D *14*, 2505 (1976).
[46] J. M. Nester and R. S. Tung, Phys. Rev. D *49*, 3958 (1994)
[47] S. Deser and P. van Nieuwenhuizen, Phys. Rev. D *10*, 411 (1974).
[48] P. G. Bergmann, V. de Sabbata, G. T. Gillies and P. I. Pronin, *International School of Cosmology and Gravitation: 15th Course: Spin in Gravity: Is it Possible to Give an Experimental Basis to Torsion?*, Erice, Italy, 13-20 May 1997.
[49] P. Van Nieuwenhuizen, Phys. Rept. *68*, 189 (1981).
[50] A. Perez and C. Rovelli, Phys. Rev. D *73*, 044013 (2006)
[51] S. Kobayashi and K. Nomizu, *Foundations of Differential Geometry*, vol. 1 and 2, Interscience Tracts in Pure and Applied Mathematics, Interscience Publ., New-York, 1969.
[52] S. Kobayashi, *Transformation Groups in Differential Geometry*, Springer-Verlag, 1972.
[53] T.Y. Thomas: *The differential invariants on generalized spaces*, Cambridge, The University Press, 1934.
[54] J.A. Schouten, *Ricci-Calculus, An Introduction to Tensor Analysis and its Geometrical Applications* (2nd ed., Springer-Verlag, New York, 1954).
[55] V. Iyer, R.M. Wald *Phys.Rev.* , **D50**, (1994), 846-864.
[56] F.W. Hehl and Yu.N. Obukhov, Lecture Notes in Physics Vol. *562* (Springer: Berlin, 2001) pp. 479-504.
[57] J. W. York, Phys. Rev. Lett. *28*, 1082 (1972).
[58] G. W. Gibbons and S. W. Hawking, Phys. Rev. D *15*, 2752 (1977).

Bulk renormalization and particle spectrum in codimension-two brane worlds

Alberto Salvio

Scuola Normale Superiore and INFN, Piazza dei Cavalieri 7, 56126 Pisa, Italy

Departamento de Física Teórica, Universidad Autónoma de Madrid and
Instituto de Física Teórica IFT-UAM/CSIC, Cantoblanco, 28049 Madrid, Spain

E-mail: `alberto.salvio@sns.it`

Abstract. We study the Casimir energy due to bulk loops of matter fields in codimension-two brane worlds and discuss how effective field theory methods allow us to use this result to renormalize the bulk and brane operators. In the calculation we explicitly sum over the Kaluza-Klein (KK) states with a new convenient method, which is based on a combined use of zeta function and dimensional regularization. Among the general class of models we consider we include a supersymmetric example, 6D gauged chiral supergravity. Although much of our discussion is more general, we treat in some detail a class of compactifications, where the extra dimensions parametrize a rugby ball shaped space with size stabilized by a bulk magnetic flux. The rugby ball geometry requires two branes, which can host the Standard Model fields and carry both tension and magnetic flux (of the bulk gauge field), the leading terms in a derivative expansion. The brane properties have an impact on the KK spectrum and therefore on the Casimir energy as well as on the renormalization of the brane operators. A very interesting feature is that when the two branes carry exactly the same amount of flux, one half of the bulk supersymmetries survives after the compactification, even if the brane tensions are large. We also discuss the implications of these calculations for the natural value of the cosmological constant when the bulk has two large extra dimensions and the bulk supersymmetry is partially preserved (or completely broken).

1. Introduction

Higher dimensional theories have a variety of motivations in the physics of fundamental interactions. The original one, in the pioneering works of Kaluza and Klein (KK), was the unification of the known forces. This idea survived as the decades passed by through superstring theory, which provides a framework where gravity can be unified with the other interactions and strongly indicates that the space-time contains a number of extra dimensions. Other advantages of higher dimensional setups include the possibility to understand the structure of fermion masses and couplings [1] in the Standard Model.

Remarkably, extra dimensions also provide mechanisms to understand technical naturalness problems. For example, the gauge hierarchy problem, which queries why the Fermi scale is so much smaller than the Planck mass, can be addressed in the large extra dimension (LED) scenario or by so called warped geometries. It is then natural to ask whether extra dimensions can help us understand the most serious fine tuning issue we know, the cosmological constant problem [2].

To understand the origin of such problem it is useful to look at the usual Einstein-Hilbert gravity[1] coupled to matter

$$\frac{\mathcal{L}}{\sqrt{-g}} = -\frac{R}{2\kappa_4^2} - \Lambda_0 + \frac{\mathcal{L}_{matter}}{\sqrt{-g}}, \tag{1}$$

where $\mathcal{L}_{matter}$ is the part which depends on the matter fields. This represents the leading gravitational tensor theory in the large wavelength limit. Here κ_4 is the 4D Planck scale and Λ_0 the tree level value of the cosmological constant. From this equation it is clear that the quantum vacuum energy $\langle\rho\rangle$ produced by the gravitational field $g_{\mu\nu}$ and any matter particle contributes to the cosmological constant $\Lambda = \Lambda_0 + \langle\rho\rangle$. The cosmological constant problem consists in the mismatch between the observed value $\Lambda \sim (10^{-3}\text{eV})^4$ and the (relatively) enormous contribution to $\langle\rho\rangle$ due to the known forms of matter: any Standard Model particle with mass m adds to Λ a quantity of order m^4 and therefore a huge fine tuning of Λ_0 is required to obtain the observed value.

Any solution of such problem must involve a modification of gravity at an energy scale roughly of order 10^{-3}eV. Indeed we know all non-gravitational forces up to the $\sim$ 10 TeV scale and the only possible modification should therefore occur in the gravity sector, which is much less constrained by the experiments; if such modification emerges only at some intermediate energy scale E_{int}, between 10^{-3}eV and 10 TeV, then the low energy effective theory at energies much smaller than E_{int} would be again the Standard Model plus Einstein's gravitational theory and we would be back to the original problem.

The required modification occurs in any model with two large extra dimensions. Let us see why. The LED scenario predicts the following ratio between the 4D Planck mass, $M_{\rm Pl} \equiv 1/(\sqrt{2}\kappa_4)$, and its higher dimensional counterpart, $M \equiv 1/(\sqrt{2}\kappa)^{2/(D-2)}$, [4]

$$\frac{M_{\rm Pl}^2}{M^2} = (Mr)^{D-4}, \tag{2}$$

where D is the full space-time dimension and r is the linear size of the $D-4$ dimensional space volume (that is the length scale of the extra dimensions). In this framework one addresses the gauge hierarchy problem by choosing roughly $M \sim$ TeV, which for $D = 6$ gives us a KK scale of order $1/r \sim 10^{-3}$eV (above which gravity is modified). It is important to notice that a true solution of the gauge hierarchy problem requires a mechanism to dynamically stabilize r to this large value; one way to do so is to consider a compactification on a topologically non-trivial space with a magnetic flux (flux stabilization). Also, notice that in order for the extra dimensions to be so large, only gravity (among the known interactions) can propagate in the bulk while the Standard Model fields should be confined on a 3+1 dimensional brane (3-brane henceforth), which in the case of interest here, $D = 6$, has to be of codimension-two.

Of course, the fact that gravity is modified at the required energy scale is not by itself a solution of the cosmological constant problem. We need a further mechanism protecting Λ from large quantum corrections. Supersymmetry is a possible candidate because the fermion and boson contributions to the vacuum energy cancel exactly if supersymmetry is unbroken, or, if supersymmetry is broken at an energy scale m_S, produce a net vacuum energy of order m_S^4. A crucial observation now is that the supersymmetry breaking scale in the bulk does not have to be of the same order of that on the brane, which is required to be larger than the TeV scale

[1] We adopt the mostly plus signature and the curvature conventions of Ref. [3].

by the LHC experiments. A supersymmetry breaking brane Lagrangian δL_b, which, including quantum corrections, is expected to be of order TeV gives rise to a supersymmetry breaking splitting δm^2_{KK} in the KK spectrum of order $\kappa^2 \delta L_b / r^2$. Here the factor κ^2 reminds us that the brane physics is communicated to the bulk through gravitational interactions while $1/r^2$ is there for dimensional reasons. To the extent that m^2_S is given by δm^2_{KK} we therefore obtain a cosmological constant of the correct order of magnitude. To summarize

$$m_S^2 \overset{?}{\sim} \delta m^2_{KK} \sim \frac{\kappa^2 \delta L_b}{r^2} \overset{LED}{\sim} \frac{1}{r^2},$$

where the question mark reminds us our expectation. To confirm it explicit calculations in a concrete model are therefore needed.

The idea that supersymmetric large extra dimensions with codimension-two branes can address the gauge hierarchy and the cosmological constant problem was originally proposed in [5]. We refer to these works for an extended discussion. But the fact that extra dimensions in general can help with the cosmological constant problem is older (see for example Refs. [6])

Since any issue of technical naturalness is a quantum problem we believe it is important to have a sistematic approach to compute quantum corrections in codimension-two brane worlds. Here we discuss quantum corrections due to bulk (massive) states, focusing on the one-loop approximation. The main subject of this article is the Casimir energy produced by integrating out a (massive) bulk field, and how to obtain it from its KK mass spectrum. Much of our discussions is based on Refs. [7, 8, 9, 10, 11], but here we also obtain some original results. The main one is a convenient way to compute Casimir energies from KK spectra, which is based on a combined use of dimensional regularization (for the ultraviolet divergences) and zeta function regularization (for the KK sums). The Casimir energy can then be used to calculate the renormalization group equations (RGEs) of both bulk and brane coefficients in the quantum action.

Although some results can be applied to any codimension-two model, we study in some detail a concrete class of compactifications, where the extra dimensions form a rugby ball shaped space and the flux of a gauge field stabilizes their size, r. These solutions are supported by two 3-branes having both tension [12] and magnetic flux (the leading terms in a derivative expansion of the brane action) [13]. We discuss a class of models having these solutions, including one with bulk supersymmetry, 6D gauged chiral supergravity [14, 15, 16], which we refer to throughout this paper as our supersymmetric example. As shown in Ref. [11], when the two 3-branes have identical localized fluxes one half of the bulk supersymetries are preserved (if the bulk is supersymmetric), implying, interestingly, that the 4D vacuum energy vanishes in this limit.

For rugby ball compactifications, the KK towers of many bulk fields are known from previous calculations [7, 8, 9, 10, 11] and we make use of these results to compute explicitly the Casimir energy and the renormalization of bulk and brane coefficients as a function of the bulk mass m of the field we integrate out. Our computational method is very convenient and confirms the results of Ref. [10]. The final expression for the Casimir energy is a polynomial function in mr of degree 6, where the coefficients depend on the brane tension and the bulk and brane localized fluxes. In the supersymmetric case, all these coefficients vanish for identical localized fluxes or, more generically, they are suppressed by the difference between these fluxes. As shown in [11] the 4D cosmological constant inherits this suppression and can be of the observed size.

Let us give the outline of this article. In section 2 we introduce our family of models, including both the bulk and brane actions, and the rugby ball solutions; as a concrete supersymmetric

Journal of Physics: Conference Series **437** (2013) 012004 doi:10.1088/1742-6596/437/1/012004

example we define 6D gauged chiral supergravity and reexamine the technical naturalness of the cosmological constant in this specific case. We then review the KK spectra of scalars, fermions and vectors for rugby ball compactifications in section 3. The Casimir energy calculation and the bulk renormalization is performed in section 4, treating in some detail the codimension-two case and, more specifically, the rugby ball solutions. In section 5 we finally discuss the form of the Casimir energy due to loops of specific matter fields: the simplest case of a real scalar and (massive) matter multiplets of 6D supergravity.

2. The class of models

We focus on a class of models which include, in addition to the metric tensor g_{MN}, a set of gauge fields A^a_M, scalars ϕ^i and fermions ψ^r. The bosonic part of the Lagrangian for these fields is

$$\frac{\mathcal{L}_B}{\sqrt{-g}} = -\frac{1}{2\kappa^2}R - \frac{1}{2}\mathcal{G}_{ij}(\phi)D_M\phi^i\, D^M\phi^j - \frac{1}{4}\mathcal{H}_{ab}(\phi)\, F^a_{MN}F^{MNb} - V(\phi)\,, \tag{3}$$

where D_M is the gauge-covariant derivatives for the scalars, F^a_{MN} is the field strength of A^a_M, and $\mathcal{G}_{ij}(\phi)$, $\mathcal{H}_{ab}(\phi)$ and $V(\phi)$ are generic functions of the scalars.

This is general enough to describe the linearized dynamics of matter supermultiplets in explicit 6D supergravities. The supersymmetric example we shall refer to throughout this paper is 6D gauged chiral supergravity. However, the general class of models we consider also includes non-supersymmetric theories and the results we shall discuss in the following sections hold for them as well (unless otherwise stated). The supersymmetric field content is given by a supergravity-tensor multiplet (g_{MN}, B_{MN}, ϕ, ψ_M, χ) - with metric tensor, anti-symmetric Kalb-Ramond field (with field strength G_{MNP}), dilaton, gravitino and dilatino - coupled to some gauge multiplets (A^a_M, λ) - with gauge fields and gauginos - and some hypermultiplets (Φ^I, Ψ) - each of them having hyperscalars and their chiral hyperini. The fermions are all Weyl spinors and satisfy $\Gamma_7\psi_M = \psi_M$, $\Gamma_7\chi = -\chi$, $\Gamma_7\lambda = \lambda$ and $\Gamma_7\Psi = -\Psi$. We will consider a matter content with gauge group of the form $\mathcal{G} = \tilde{\mathcal{G}} \times U(1)_R$, where $U(1)_R$ is an Abelian R-symmetry, and $\tilde{\mathcal{G}}$ is a generic product of simple groups. The bosonic Lagrangian of 6D gauged chiral supergravity is given by

$$\frac{\mathcal{L}_B}{\sqrt{-g}} = -\frac{1}{2\kappa^2}(R+\partial_M\phi\,\partial^M\phi) - \frac{e^{-\phi}}{4g_a^2}\, F^a_{MN}F^{MN}_a - \frac{1}{2}G_{IJ}(\Phi)\, g^{MN}D_M\Phi^I D_N\Phi^J - \frac{2g^2}{\kappa^4}\, e^{\phi}\, U(\Phi)\,, \tag{4}$$

where g is the $U(1)_R$ gauge coupling and we have set $G_{MNP} = 0$ for simplicity.

As we have already mentioned a realistic realization of the large extra dimensions idea requires the presence of 3-branes, one of which supports the Standard Model fields. We take the brane action to be [13].

$$\begin{aligned} S_b &= -\int d^4x\,\sqrt{-g}\,L_b \\ \text{with}\quad L_b &= T_b - \frac{\mathcal{A}_b}{2g_b^2}\,\epsilon^{mn}F_{mn} + \frac{\mathcal{B}_b}{\kappa}\,R + \frac{\mathcal{C}_b}{\kappa}(\partial\phi)^2 + \cdots\,, \end{aligned} \tag{5}$$

where g_b is the gauge coupling of the gauge field corresponding to the second term in (5) (which has to be Abelian in order for that term to be gauge invariant). The ellipses denote other terms involving two or more derivatives (including in principle the Standard Model fields), and T_b, $\mathcal{A}_b$, $\mathcal{B}_b$, $\mathcal{C}_b$ and so on could depend on the scalars ϕ^i.

One can now ask again whether the observed value of the cosmological constant emerges naturally in this framework. While no positive answer is found in non-supersymmetric cases, the supersymmetric model has been proven to have very interesting features [5, 17]. One can compute the cosmological constant in three steps: first, the brane fields are integrated out at the quantum level; second, the classical integration in the bulk is performed; third, the quantum corrections of the bulk integration are computed. The first and second steps always give a vanishing cosmological constant if the dilaton does not couple directly to the branes [5, 18]. Ref. [17] has recently shown that the third step is not dangerous either, at least for representations of the supersymmetry algebra which are massless in the 6D sense. The essence of the argument is that $e^{2\phi}$ acts as a loop counting quantity (as it can be proved by going to the frame defined by $\hat{g}_{MN} = e^{\phi} g_{MN}$) and is very small, of order $1/(r^4 \mathrm{TeV}^4)$. The last property can be understood by noticing that the bulk Lagrangian enjoys a classical scale invariance under which $e^{\phi} g_{MN}$ does not change; in the large extra dimension setup the value $e^{\phi} r^2$ is therefore expected to be fixed by the field equations to a value of order $1/\mathrm{TeV}^2$. The contents of the following sections are useful tools to address the same question in the presence of massive 6D supermultiplets (see also [19] for a study of bulk UV sensitivity for Ricci-flat geometries, including the gravity sector, but without branes.).

Let us conclude this section by discussing the solutions of the models we have considered. The simplest solution preserving 4D Poincaré invariance and providing a flux stabilization is a $(Minkowski)_4 \times S^2$ compactification, where the metric of the two extra dimensions is that of a sphere with radius r,

$$ds^2 = r^2 (d\theta^2 + \sin^2\theta \, d\varphi^2) \, . \tag{6}$$

($0 \leq \theta \leq \pi$ and $0 \leq \varphi < 2\pi$). Also, the scalars ϕ^i are constant and the non-vanishing components of the field strength are $F_{mn} = f \epsilon_{mn}$, where f is a constant. If there are some matter fields having a non-trivial charge[2], $q\tilde{g}$, under this background gauge field, then its field strength F_{mn} has to satisfy the following quantization condition [20]

$$2\pi N = q \int_{S_2} F = 4\pi r^2 q f \, , \quad \text{(without brane sources)} \tag{7}$$

where $N = 0, \pm 1, ...$ is an arbitrary integer. Then f must satisfy

$$f = \frac{N}{2qr^2} \, . \quad \text{(without brane sources)} \tag{8}$$

In the supersymmetric model, the field equations also imply

$$e^{\phi} = \frac{\kappa^2}{4g^2 r^2} \quad \text{and} \quad f = \pm \frac{\tilde{g}}{2\, g\, r^2} \, . \quad \text{(supersymmetric case)} \tag{9}$$

Since all the parameters of the bulk theory are expected to be of order TeV (to the appropriate power) in the large extra dimensions setup, we notice that the first of these two conditions confirms our general expectation that $e^{\phi} r^2$ is of order $1/\mathrm{TeV}^2$. The second one, combined with Eq. (8), gives us $N = \pm q\tilde{g}/g$. When the background gauge field is along the $U(1)_R$, and so $\tilde{g} = g$, $N = \pm 1$ and this solution preserves half of the supersymmetries of the bulk theory [16]. For other embeddings of the background gauge field supersymmetry is instead completely broken.

[2] We denote with $\tilde{g}$ the gauge coupling of the background gauge field, while q is the charge of a given field, which we assume to be integer.

A simple way to introduce 3-branes in this case is starting from the sphere metric and demanding φ to have a modified period, $\varphi \sim \varphi + 2\pi\alpha$, where α is a positive constant [12]. This procedure introduces two conical singularities, one at the north pole and the other one at the south pole, with identical defect angles, $\delta = 2\pi(1-\alpha)$. The resulting internal space is called the *rugby ball.*

The brane action in (5) can support these conical singularities and the near-brane boundary conditions imply [13] that the defect angle is

$$\delta_b = \kappa^2 L_b\,. \tag{10}$$

In order for this angle to be positive we will assume that $L_b \geq 0$ at the background solution. The brane sources also changes the flux quantization condition. This arises because the branes themselves can support a localized flux:

$$\Phi_b = \frac{q\mathcal{A}_b}{2\pi}\,. \tag{11}$$

If the branes are identical the total flux localized in this way is $\Phi := \sum_b \Phi_b$, in terms of which the flux-quantization condition is (when the background gauge field has some localized flux)

$$2\pi N = q\left(\sum_b \mathcal{A}_b + \int_{S_2(\alpha)} F\right) = 2\pi\,\Phi + 4\pi\alpha\, r^2 q f\,, \tag{12}$$

where $S_2(\alpha)$ is the rugby ball. The normalization constant f is then given by

$$f = \frac{\mathcal{N}}{2qr^2}\,, \quad \text{(with brane sources)} \tag{13}$$

where $\mathcal{N} := \omega(N - \Phi)$ and $\omega := 1/\alpha$.

In the supersymmetric case, the above-mentioned classical scale invariance can be broken by the boundary localized fluxes, the second term in Eq. (5), which gives mass to the corresponding flat direction [13]; another reason why these fluxes are useful is that they allow us to recover supersymmetry, when the background gauge field is along the $U(1)_R$, in a continuous limit [13, 17]. To understand this let us take the rugby ball solution, $\tilde{g} = g$ and $\Phi_b = 0$, then $\mathcal{N} = \omega N$ and the second condition in Eq. (9) together with (13) implies $\omega N = \pm q$, which does not allow you to approach the supersymmetric value $\omega = 1$ continuously. In the absence of Φ_b instead we do not find this obstruction: using the second condition of (9) in (13) this time we obtain

$$\frac{\Phi}{q} = \frac{N}{q} \mp \alpha\frac{\tilde{g}}{g}\,, \quad \text{(supersymmetric case)} \tag{14}$$

which, in the case $\tilde{g} = g$ and $N/q = \pm 1$, gives

$$\frac{\Phi}{q} = \pm(1-\alpha)\,. \quad \text{(supersymmetric case)} \tag{15}$$

In the presence of boundary localized fluxes supersymmetry can be broken by an arbitrarily small amount. For this reason in the supersymmetric model we will always take the boundary localized flux in the $U(1)_R$ direction, $g_b = g$.

Let us conclude this section by mentioning that, for branes carrying both tension and flux, there is, remarkably, a case in which one half of the 6D supersymmetries is preserved; this occurs when the background gauge field is along the $U(1)_R$ generator, for which there is exactly the same amount of localized flux on the two branes [11] (i.e. $\Phi_b = \Phi/2$ both at the north and south branes). Such property has important implications regarding the smallness of the cosmological constant.

3. Spectrum in codimension-two brane worlds

We now move to the analysis of the linear perturbations around the 3-brane solutions we have considered. Since these configurations preserve 4D Poncaré symmetry, such analysis is equivalent to computing the 4D particle spectrum (defined in the usual sense). This will provide us with additional physical information; it is an important computation to study the stability of the background solutions [8, 9] and, of special relevance for this article, it is an intermediate step to determine the quantum corrections to the 4D vacuum energy. Indeed in dimensional regularization the one-loop contribution to the quantum potential (the Casimir energy) due to a generic field with bulk mass m is

$$V = \frac{1}{2}\mu^{4-d}(-1)^F \sum_{\underline{n}} \int \frac{d^d p}{(2\pi)^d} \ln\left(\frac{p^2 + m_{\underline{n}}^2 + m^2}{\mu^2}\right), \tag{16}$$

where the collective index $\underline{n}$ includes all KK numbers and $m_{\underline{n}}$ represents the full set of KK masses. Also $(-1)^F$ is 1 for bosons and -1 for fermions. The 4D particle spectrum is therefore an important ingredient to compute V and we will use Eq. (16) to compute the renormalization of bulk and brane couplings in sections 4 and 5.

In the presence of two extra dimensions we have two KK numbers, $\underline{n} = j, n$; in the rest of this section we give the form of $m_{jn}^2 := \lambda_{jn}/r^2$ for (minimally coupled) scalars, fermions and gauge fields on top of the rugby ball geometry[3] sourced by branes with both tension T_b and flux Φ_b (the leading terms in a derivative expansion). In the absence of localized fluxes, these m_{jn}^2 were computed[4] in Refs. [7, 25, 8]. Subsequently, their form in the presence of Φ_b was derived in [10].

Let us consider first the simple case of a minimally coupled real scalar, satisfying the equation $(D_M D^M + m^2)\phi = 0$, that is coupled to the background gauge field through a monopole number N and boundary localized fluxes Φ_b. In this case the scalar spectrum (in the north patch of the gauge potential) is

$$\lambda_{jn}^{\rm s} = \left(j + \frac{\omega}{2}|n - \Phi_+| + \frac{\omega}{2}|n - N + \Phi_-| + \frac{1}{2}\right)^2 - \frac{1 + \mathcal{N}^2}{4}, \tag{17}$$

where

$$j = 0, 1, 2, \dots, \qquad n = 0, \pm 1, \pm 2$$

and Φ_+ (Φ_-) is the flux localized on the north (south) pole of the rugby ball, where $\cos\theta = +1$ (-1).

Moving to fermions, the KK spectrum for a field, satisfying $(\not{D} + m)\psi = 0$, that is charged under the $U(1)$ is (using again the north patch of the gauge potential)

$$\lambda_{jn}^{{\rm f}\sigma} = \left(j + \frac{\omega}{2}\left|n_{1/2} - \Phi_+ - \frac{\sigma}{2\omega}\right| + \frac{\omega}{2}\left|n_{1/2} - N + \Phi_- + \frac{\sigma}{2\omega}\right| + \frac{1}{2}\right)^2 - \frac{\mathcal{N}^2}{4}, \tag{18}$$

where $n_{1/2} = n - \sigma/2$ and $\sigma \in \{\pm 1\}$ corresponds to the 4D helicity of the spinor, of which there are 2 (4) each for a 6D Weyl (Dirac) spinor.

[3] Another analysis of the spectrum in codimension-two brane worlds, but for other background solutions can be found for example in [21] (see also [22] for preliminary technical aspects).

[4] Those works considered a more general compactification [23] than the rugby ball one, which involves generically non-factorizable geometries. Here, however, we focus on the rugby ball case for the sake of simplicity (see [24] for the explicit unwarped limit).

Finally let us give the KK spectrum arising from a gauge field, which assume *not* to be in the Lie algebra direction where the background gauge field lies[5]. There a two cases: the gauge field can either be massless or massive. We begin with the massless case, when the field satisfy the equation $g^{MN}D_M F_{NP} = 0$. In an appropriate gauge (e.g. light-cone gauge [26, 27, 28, 8, 9]) the 6D gauge field can be decomposed into four components, each with a spectrum (once again in the north patch of the gauge potential) given by

$$\lambda_{jn}^{\mathrm{gf}\xi} = \left(j + \frac{\omega}{2}\left|n - \Phi_+ + \frac{\xi}{\omega}\right| + \frac{\omega}{2}\left|n - N + \Phi_- - \frac{\xi}{\omega}\right| + \frac{1}{2}\right)^2 - \frac{(1+\mathcal{N}^2)}{4}, \tag{19}$$

where $\xi \in \{0, 0, +1, -1\}$ for each of the four components and we assume $N = 0, \pm 1$ to ensure stability [29, 8]. We observe that two modes have exactly the same spectrum as scalars (i.e. those with $\xi = 0$), while the other two (with $\xi = \pm 1$) have almost the same spectrum. For massive gauge fields we have a more complicated situation because we need a scalar field Φ that is charged under the gauge field we are studying in order to give mass through the Higgs mechanism. In order to interpret $\langle \Phi \rangle \neq 0$ as a 6D spontaneous symmetry breaking, we require $\langle \Phi \rangle$ to be constant and to be at the minimum of U. Then, in order to solve the background scalar equation, $D_M D^M \Phi = 0$, we also demand that $\langle \Phi \rangle \neq 0$ does not break the $U(1)$ where the background gauge field lies: otherwise it would not be possible to have $\langle \Phi \rangle$ constant, at least in the sphere compactification of interest in this paper [30, 31, 25]. If we choose again the light-cone gauge, it is possible to show that a massive gauge field leads to the 4D spectrum of a massless gauge field, Eq. (19), plus that of a scalar, Eq. (17).

4. Casimir energy and bulk renormalization

The purpose of this section is to compute the Casimir energy due to bulk loops and to show how, consequently, the bulk and brane couplings renormalize. This is an important step to address any problem of technical naturalness, such as the gauge hierarchy and the cosmological constant problem. We will consider in some detail the case in which the branes are of codimension-two, but some of our results will be valid in more general brane worlds. The renormalized couplings will depend as usual on a renormalization energy μ and we will compute explicitly their RGEs.

4.1. A general technique to compute the Casimir energy

The starting point of this calculation is the formula for the one-loop contribution of a single field to the quantum potential in Eq. (16). Notice that, modulo terms which are independent of X, we have $\ln X = -\int_0^\infty (ds/s)\exp(-sX)$ and therefore

$$V = -\frac{1}{2}\mu^{4-d}(-1)^F \sum_{\underline{n}} \int \frac{d^d p}{(2\pi)^d} \int_0^\infty \frac{ds}{s} \exp\left[-s\left(p^2 + m_{\underline{n}}^2 + m^2\right)\right]. \tag{20}$$

Performing the integral in $d^d p$ and rescaling the variable s we obtain

$$V = -\frac{\mu^4}{2(2\pi)^d}(-1)^F \sum_{\underline{n}} \int_0^\infty \frac{dt}{t^{1+d/2}} e^{-\pi t\left(m_{\underline{n}}^2 + m^2\right)/\mu^2}. \tag{21}$$

[5] This assumption simplifies our calculation because the gauge field perturbations along the same Lie algebra direction as the background gauge field mix with the metric perturbations. See, however, Ref. [9], for a discussion of this more complicated sector, but in the absence of boundary localized fluxes.

Consider first the case $\mathrm{Re}(d) < 0$ and rewrite the integral in Eq. (21) as

$$\int_0^\infty \frac{dt}{t^{1+d/2}} e^{-\pi t\,(m_{\underline{n}}^2+m^2)/\mu^2} = \left(\frac{\pi m_{\underline{n}}^2 + \pi m^2}{\mu^2}\right)^{d/2} \Gamma\left(-d/2\right) . \tag{22}$$

We now extend this integral function by analytic continuation to all complex d except the non-positive integers (where the function has simple poles). By using

$$\Gamma\left(-d/2\right) = -\frac{1}{d-4} + \frac{3}{4} - \frac{\gamma}{2} + \mathcal{O}\left(d-4\right) , \tag{23}$$

where γ is Euler's constant, one then obtains

$$V = -\frac{1}{2\,(4\pi)^2\, r^4}(-1)^F \sum_{\underline{n}} (r^2 m_{\underline{n}}^2 + r^2 m^2)^{d/2} \left(-\frac{1}{d-4} + \ln(r\mu) + ...\right) , \tag{24}$$

where the dots represent finite r-independent terms.

One possible renormalization scheme (which we will adopt from now on) is to subtract the divergent part in the brackets of Eq. (24). The renormalized potential V_r (the Casimir energy) can then be written as

$$V_r = \frac{C}{(4\pi)^2 r^4} \log\left(\frac{r}{r_0}\right) , \tag{25}$$

where r_0^{-1} has to be identified as a ultraviolet (UV) scale which can be computed once the UV completion is known and

$$C := -\frac{1}{2}(-1)^F \sum_{\underline{n}} (r^2 m_{\underline{n}}^2 + r^2 m^2)^{d/2} . \tag{26}$$

It is important to notice that all we need in order to compute this coefficient is the divergent part of V, as it is clear from Eq. (24).

The quantity on the right hand side of Eq. (26) is divergent and has to be regularized. Notice that the exponent $d/2$ can effectively act as a regulator for the sum over $\underline{n}$ (zeta function regularization [32]): for $d = 4$ the sum is divergent, but one can (and we will) compute (26) for those d such that the sum is convergent and then consider the analytic continuation of the final result at $d = 4$.

4.2. The Casimir energy for rugby ball compactifications

A generic form of the spectrum which covers all the cases encountered here, Eqs. (17), (18) and (19), is

$$m_{jn}^2 := \frac{1}{r^2}\lambda_{jn} , \qquad \text{where} \quad \lambda_{jn} = \left(j + \frac{\omega}{2}\left|n + b_+\right| + \frac{\omega}{2}\left|n - b_-\right| + a\right)^2 - \tau , \tag{27}$$

and $b_\pm$, a and τ are real parameters; the only assumption we make is that $b_\pm$ and a are independent of j and τ is independent of both j and n.

Then the contribution to the C parameter from a single field is

$$C = -\frac{(-1)^F}{2} \sum_{n,j} \left\{ \left(j + \frac{\omega}{2}\left|n + b_+\right| + \frac{\omega}{2}\left|n - b_-\right| + a\right)^2 - \tau + (mr)^2 \right\}^{d/2} . \tag{28}$$

where $(-1)^F = 1$ for a boson and $(-1)^F = -1$ for a fermion. We can now expand $\{...\}^{d/2}$ by using the binomial series to obtain

$$C = -\frac{(-1)^F}{2}\sum_{n,j}\sum_{k=0}^{\infty}\frac{\Gamma(k-d/2)}{k!\Gamma(-d/2)}\left[j+\frac{\omega}{2}|n+b_+|+\frac{\omega}{2}|n-b_-|+a\right]^{d-2k}\left[\tau-(mr)^2\right]^k. \quad (29)$$

The sum over j can be performed by means of the following representation of the Hurwitz zeta function $\zeta(s,c)$ (valid for $\mathrm{Re}(s) < 0$ and $\mathrm{Re}(c) > 0$)

$$\zeta(-s,c) := \sum_{j=0}^{\infty}(j+c)^s = \frac{1}{\Gamma[-s]}\int_0^{\infty} dy\, \frac{y^{-1-s}\, e^{-cy}}{1-e^{-y}}. \quad (30)$$

Indeed setting $s = d-2k$ and $c = c_n = (|n+b_+|+|n-b_-|)\omega/2 + a$ in (30) we have

$$C = -\frac{(-1)^F}{2}\sum_{k=0}^{\infty}\frac{\Gamma(k-d/2)\left[\tau-(mr)^2\right]^k}{k!\Gamma(-d/2)\Gamma(2k-d)}\int_0^{\infty} dy\, \frac{y^{-1+2k-d}}{1-e^{-y}}\sum_n e^{-c_n y}. \quad (31)$$

Observing that integrals of the form $\int_0^{\infty} dy \frac{y^h e^{-y}}{1-e^{-y}}$ can be computed explicitly for $\mathrm{Re}(h) > 0$,

$$\int_0^{\infty} dy \frac{y^h e^{-y}}{1-e^{-y}} = \Gamma[1+h]\mathrm{Li}_{1+h}(1), \qquad \mathrm{Li}_{1+h}(1) := \sum_{k=1}^{\infty}\frac{1}{k^{1+h}}, \quad (32)$$

we Taylor-expand $ye^y\sum_n e^{-c_n y}$ around $y = 0$,

$$ye^y\sum_n e^{-c_n y} = \sum_{k'=0}^{\infty}\frac{\epsilon_{k'}(\omega)}{k'!}y^{k'}, \quad (33)$$

and obtain

$$C = -\frac{(-1)^F}{2}\sum_{k=0}^{\infty}\frac{\Gamma(k-d/2)\left[\tau-(mr)^2\right]^k}{k!\Gamma(-d/2)\Gamma(2k-d)}\sum_{k'=0}^{\infty}\frac{\epsilon_{k'}(\omega)}{k'!}\Gamma(-1-d+2k+k')\mathrm{Li}_{-1-d+2k+k'}(1). \quad (34)$$

We now need to take the limit $d \to 4$. The previous expression turns out to be well defined in this limit. Also, because of the $\Gamma(-d/2)$ in the denominator, only a finite number of k and k' contributes to the sum: $k = 0, 1, 2, 3$ and $k' = 0, 1, 2, 3, 4, 5, 6$. This implies two things: (i) C is computable once we know $\sum_n e^{-c_n y}$, (ii) C has the generic structure

$$C = \frac{s_{-1}}{6}(mr)^6 - \frac{s_0}{2}(mr)^4 + s_1(mr)^2 - s_2\,. \quad (35)$$

Since the coefficient C can be computed from the mere knowledge of the UV divergent part of the Casimir energy, we should expect that this formula can be applied to codimension-two compactifications which are more general than the rugby ball one. This is because UV divergences are related to the local structure of the space-time and are therefore insensitive to global properties such as its topology.

4.3. Bulk and brane counterterms and renormalized couplings

The models we have considered are non-renormalizable and so not all divergences can be reabsorbed in counterterms of the same form as the terms in the classical action. However, one can show in very general terms that the number of counterterms needed is always finite at a given order in perturbation theory. In this subsection we review [10] the renormalization due to bulk loops in codimension-two braneworlds (and in particular for rugby ball compactifications) at the one-loop level. At the end we will therefore obtain a finite number of counterterms and renormalized couplings. To perform this calculation we will use effective field theory methods (see for example [33]).

Although only bulk loops[6] are computed here, both bulk ad brane counterterms are needed. There is an important difference between them. The bulk counterterms, unlike the brane ones, do not depend on the brane properties and so they can be computed in the sphere limit, $\alpha \to 1$.

4.3.1. Renormalization of the bulk interactions. To capture all the terms needed to reabsorb the UV divergences we write down the most general local Lagrangian with the chosen field content and set of symmetries, which we organize in a derivative expansion $\mathcal{L}_B^{\rm ct} = \mathcal{L}_{B0}^{\rm ct} + \mathcal{L}_{B2}^{\rm ct} + \mathcal{L}_{B4}^{\rm ct}$. After renormalization this generates a corresponding series of renormalized interactions $\mathcal{L}_B^r = \mathcal{L}_{B0}^r + \mathcal{L}_{B2}^r + \mathcal{L}_{B4}^r$. Focusing on the fields that are non-zero in the background we have

$$\begin{aligned}
\mathcal{L}_{B0}^r &= -\sqrt{-g}\,\lambda\,, \\
\mathcal{L}_{B2}^r &= -\sqrt{-g}\left[\frac{\zeta_R}{2\kappa^2}\,R + \frac{\zeta_A}{4\tilde{g}^2}\,F_{MN}F^{MN}\right], && (36)\\
\mathcal{L}_{B4}^r &= -\sqrt{-g}\left[\frac{\kappa\zeta_{AR}}{8\tilde{g}^2}\,R\,F_{MN}F^{MN} + \frac{\zeta_{R^2}}{\kappa}\,\bar{R}^2\right], && (37)\\
\mathcal{L}_{B6}^r &= -\sqrt{-g}\left[\zeta_{R^3}\,\bar{R}^3 + ...\right], && (38)\\
&\cdots\,, && (39)
\end{aligned}$$

where $\bar{R}^2$ ($\bar{R}^3$) is a generic linear combination of terms which are quadratic (cubic) in the curvature, that is

$$\bar{R}^2 = a_R\,R^2 + 2b_R\,R_{MN}R^{MN} + c_R\,R_{MNPQ}R^{MNPQ} \tag{40}$$

where $a_R + b_R + c_R = 1$ so that $\bar{R}^2 = R^2$ when specialized to the sphere geometry (for which $R_{mnpq}R^{mnpq} = 2R_{mn}R^{mn} = R^2 = 4/r^4$). A similar expression is used for $\bar{R}^3$. Calculations on a sphere can only provide the overall couplings ζ_{R^2}, ζ_{R^3} and not the separate parameters such as a_R, b_R and c_R (see however [10] and references therein to know the latter quantities).

Evaluating the renormalized action at the background sphere solution and integrating over the extra dimensions gives

$$\begin{aligned}
V_B^r = -\int d^2x\,\mathcal{L}_B^r &= \left(4\pi r^2\right)\left\{\lambda - \frac{\zeta_R}{\kappa^2 r^2} + \frac{f^2}{2\tilde{g}^2}\left[\zeta_A - \frac{\kappa\zeta_{AR}}{r^2}\right] + \frac{4\zeta_{R^2}}{\kappa\,r^4} - \frac{8\zeta_{R^3}}{r^6} + ...\right\}\\
&= \left(4\pi r^2\right)\left\{\lambda - \frac{\zeta_R}{\kappa^2 r^2} + \frac{\mathcal{N}^2}{8\,q^2\tilde{g}^2 r^4}\left[\zeta_A - \frac{\kappa\zeta_{AR}}{r^2}\right] + \frac{4\zeta_{R^2}}{\kappa\,r^4} - \frac{8\zeta_{R^3}}{r^6} + ...\right\}. && (41)
\end{aligned}$$

Therefore λ, ζ_R, ζ_{R^2}, and ζ_{R^3} can be read off respectively from the r^2, r^0, r^{-2}, and r^{-4} terms in V_r (see Eqs. (25) and (35)), while the ζ_A and ζ_{AR} coefficients are identified as the $\mathcal{N}^2/r^2$

[6] Loops of brane localized fields can be calculated and the corresponding divergences can be reabsorbed with the usual 4D field theory methods.

and $\mathcal{N}^2/r^4$ terms respectively. This implies that integrating a bulk field with mass m gives the following contribution to the RGEs

$$\mu\,\frac{\partial\lambda}{\partial\mu} = \frac{m^6}{6(4\pi)^3}\,s_{-1}^{\text{sph},0}\,, \qquad \mu\,\frac{\partial}{\partial\mu}\left(\frac{\zeta_R}{\kappa^2}\right) = \frac{m^4}{2(4\pi)^3}\,s_0^{\text{sph},0}\,, \tag{42}$$

$$\mu\,\frac{\partial}{\partial\mu}\left(\frac{\zeta_{R^2}}{\kappa}\right) = \frac{m^2}{4(4\pi)^3}\,s_1^{\text{sph},0}\,, \qquad \mu\,\frac{\partial\zeta_{R^3}}{\partial\mu} = \frac{1}{8(4\pi)^3}\,s_2^{\text{sph},0}\,. \tag{43}$$

$$\mu\,\frac{\partial}{\partial\mu}\left(\frac{\zeta_A}{\tilde{g}^2}\right) = \frac{2\,m^2}{(4\pi)^3 r^4 f^2}\,s_1^{\text{sph},2} = \frac{8\,q^2 m^2}{(4\pi)^3\mathcal{N}^2}\,s_1^{\text{sph},2} \tag{44}$$

and

$$\mu\,\frac{\partial}{\partial\mu}\left(\frac{\kappa\zeta_{AR}}{\tilde{g}^2}\right) = \frac{2}{(4\pi)^3 r^4 f^2}\,s_2^{\text{sph},2} = \frac{8\,q^2}{(4\pi)^3\mathcal{N}^2}\,s_2^{\text{sph},2} \tag{45}$$

and so on. The "sph" in $s_i^{\text{sph},k}$ emphasizes that these quantities are evaluated on the sphere, while the superscript 'k' denotes terms involving k powers of $\mathcal{N}$. The renormalization of the gauge-field terms, ζ_A and ζ_{AR}, has been done by looking at the $\mathcal{N}$-dependent divergences produced when a particle with charge $q\tilde{g}$ runs in the loop.

4.3.2. Renormalization of the brane interactions. In this case we have a dependence on the boundary conditions used near the brane but the result should be independent of the boundary conditions on distant branes.

To compute the brane contributions we first subtract the (boundary condition independent) bulk contributions found above. Noticing that the bulk counterterms should be integrated over the volume of the rugby ball, which is $4\pi\alpha\,r^2$, we define

$$\delta s_i^{\text{tot}} = s_i - \alpha\, s_i^{\text{sph}}\,, \tag{46}$$

and use $\delta s_i^{\text{tot}} = \sum_b \delta s_{i(b)}$ to extract how the interactions on each individual brane renormalize. This can be done as before, by distinguishing the interactions that depend on the gauge field which is non-zero on the background from those that do not. In the following we understand the label (b) in $\delta s_{i(b)}$ to have a simpler notation.

Writing the most general local brane Lagrangian organized in a derivative expansion, $\mathcal{L}_b^r = \mathcal{L}_{b0}^r + \mathcal{L}_{b1}^r + \mathcal{L}_{b2}^r + \mathcal{L}_{b3}^{\text{ct}} + ...$, and dropping terms that vanish at the background, we have

$$\mathcal{L}_{b0}^r = -\sqrt{-\gamma}\;T_b\,, \tag{47}$$

$$\mathcal{L}_{b1}^r = \sqrt{-\gamma}\left[\frac{\zeta_{\tilde{A}b}}{2\tilde{g}^2}\,\epsilon^{mn}F_{mn}\right], \tag{48}$$

$$\mathcal{L}_{b2}^r = -\sqrt{-\gamma}\left[\frac{\zeta_{Rb}}{\kappa}\,R + \frac{\kappa\zeta_{Ab}}{4\tilde{g}^2}\,F_{MN}F^{MN}\right], \tag{49}$$

$$\mathcal{L}_{b3}^r = \sqrt{-\gamma}\left[\frac{\kappa\zeta_{\tilde{A}Rb}}{2\tilde{g}^2}\,R\,\epsilon^{mn}F_{mn}\right], \tag{50}$$

$$\mathcal{L}_{b4}^r = -\sqrt{-\gamma}\left[\zeta_{R^2b}\,R^2 + \frac{\kappa^2\,\zeta_{ARb}}{8\tilde{g}^2}\,R\,F_{MN}F^{MN}\right], \tag{51}$$

and so on, where $\gamma_{\mu\nu} := g_{MN}\partial_\mu x^M\partial_\nu x^N$ (with the right-hand side computed at the brane position) is the induced metric on the brane.

Evaluating these at the background solution gives the following contribution to the Casimir energy

$$\begin{aligned} V_b^\tau &= T_b - \frac{\zeta_{\tilde{A}b} f}{\tilde{g}^2} - \frac{2\zeta_{Rb}}{\kappa\, r^2} + \frac{\kappa \zeta_{Ab} f^2}{2\tilde{g}^2} + \frac{2\,\kappa \zeta_{\tilde{A}R\,b} f}{\tilde{g}^2 r^2} + \frac{4\zeta_{R^2 b}}{r^4} - \frac{\kappa^2 \zeta_{AR\,b} f^2}{2\,\tilde{g}^2 r^2} + \dots \\ &= T_b - \frac{\zeta_{\tilde{A}b}\mathcal{N}}{2\,q\tilde{g}^2\,r^2} - \frac{2\zeta_{Rb}}{\kappa\, r^2} + \frac{\kappa\zeta_{Ab}\mathcal{N}^2}{8\,q^2\tilde{g}^2 r^4} + \frac{\kappa\zeta_{\tilde{A}R\,b}\mathcal{N}}{q\tilde{g}^2 r^4} + \frac{4\zeta_{R^2 b}}{r^4} - \frac{\kappa^2\zeta_{AR\,b}\mathcal{N}^2}{8\,q^2\tilde{g}^2 r^6} + \dots . \end{aligned} \tag{52}$$

Using Eqs. (25) and (35) then gives the following RGEs

$$\begin{aligned} \mu\,\frac{\partial\, T_b}{\partial\mu} &= -\frac{m^4}{2(4\pi)^2}\,\delta s_0^0\,, & \mu\,\frac{\partial}{\partial\mu}\left(\frac{\zeta_{\tilde{A}b}}{\tilde{g}^2}\right) &= -\frac{2\,qm^2}{(4\pi)^2\mathcal{N}}\,\delta s_1^1\,, \\ \mu\,\frac{\partial}{\partial\mu}\left(\frac{\zeta_{Rb}}{\kappa}\right) &= -\frac{m^2}{2(4\pi)^2}\,\delta s_1^0\,, & \mu\,\frac{\partial}{\partial\mu}\left(\frac{\kappa\zeta_{\tilde{A}R\,b}}{\tilde{g}^2}\right) &= -\frac{q}{(4\pi)^2\mathcal{N}}\,\delta s_2^1\,, \\ \mu\,\frac{\partial\zeta_{R^2 b}}{\partial\mu} &= -\frac{1}{4(4\pi)^2}\,\delta s_2^0\,, & \mu\,\frac{\partial}{\partial\mu}\left(\frac{\kappa\zeta_{Ab}}{\tilde{g}^2}\right) &= -\frac{8\,q^2}{(4\pi)^2\mathcal{N}^2}\,\delta s_2^2\,, \end{aligned} \tag{53}$$

where δs_2^k are terms with k powers of $\mathcal{N}$.

5. Casimir energy for rugby balls in explicit cases

In this section we apply the method of sections 4.1 and 4.2 to compute the Casimir energy, i.e. the coefficients s_i, produced by specific matter fields, for rugby ball compactifications sourced by branes with tension and flux. Indeed the s_i is all we need to obtain the renormalization of the bulk ad brane couplings, as it is clear from Eqs. (42)-(45) and (53). Moreover, as we shall comment later on, the s_i can be used to extract the 4D cosmological constant [10, 11].

5.1. A single real scalar.

We observe that the method of section 4.2 can be applied in the case of real scalars because the spectrum in (17) has the form (27) with

$$b_+ = -\Phi_+\,, \quad b_- = N - \Phi_-\,, \quad a = \frac{1}{2}\,, \quad \tau = \frac{1+\mathcal{N}^2}{4}\,. \tag{54}$$

The (renormalized) Casimir energy produced by a real scalar is given by Eq. (25) with C given in Eq. (35) and one obtains the following s_i coefficients:

$$s^{\rm s}_{-1} = \frac{1}{\omega}\,, \tag{55}$$

$$s^{\rm s}_0(\omega, N, \Phi_b) = \frac{1}{\omega}\left[\frac{1}{6} + \frac{\omega^2}{6}(1-3F)\right], \tag{56}$$

$$s^{\rm s}_1(\omega, N, \Phi_b) = \frac{1}{\omega}\left[\frac{1}{180} - \frac{\mathcal{N}^2}{24} + \frac{\omega^2}{18}(1-3F) - \frac{\omega^3\mathcal{N}}{12}\sum_b \Phi_b\, G_b + \frac{\omega^4}{180}(1-15F^{(2)})\right], \tag{57}$$

$$\begin{aligned} s^{\rm s}_2(\omega, N, \Phi_b) = \frac{1}{\omega}\Bigg[&-\frac{1}{504} - \frac{11\,\mathcal{N}^2}{720} + \left(\frac{1}{90} - \frac{\mathcal{N}^2}{144}\right)(1-3F)\omega^2 - \frac{\omega^3\mathcal{N}}{24}\sum_b \Phi_b\, G_b \\ &+ \frac{\omega^4(1-\mathcal{N}^2)}{360}(1-15F^{(2)}) - \frac{\omega^5\mathcal{N}}{120}\sum_b \Phi_b\, G_b(1+3F_b) \end{aligned}$$

$$+\left(\frac{1}{1260}-\frac{F^{(2)}}{120}-\frac{F^{(3)}}{60}\right)\omega^6\Bigg], \tag{58}$$

where we introduced the notation

$$F_b := |\Phi_b|\,(1-|\Phi_b|)\,, \quad G_b := (1-|\Phi_b|)\,(1-2|\Phi_b|)\,, \quad F^{(n)} := \sum_b F_b^n\,. \quad F^{(1)} := F\,. \tag{59}$$

The method of sections 4.1 and 4.2 can also be applied to fermions and gauge fields as their spectra, Eqs. (18) and (19), are both of the form given in (27). The explicit expressions for the s_i coefficients for fermions and (massive) gauge fields can be found in [10].

5.2. *Supermultiplets*

Let us now consider supermultiplets of 6D gauged chiral supergravity focusing on the hypermultiplets and the gauge multiplets (for which we further restrict to the case in which the gauge field is zero in the background). The main reason is that the cancellation of gauge and gravitational anomalies typically require hundreds of such supermultiplets [34, 35] and so their contribution is expected to dominate the Casimir energy. For example, the first anomaly free theory of this sort that has been found has a large $(E_6 \times E_7 \times U(1)_R)$ gauge symmetry with many (456) hypermultiplets [34, 36].

A massless hypermultiplet consists of four massless scalars (called hyperscalars) and one 6D Weyl fermion, the hyperino. A massless gauge multiplets is made of one gauge field and a 6D Weyl fermion, the gaugino. By contrast, a massive 6D matter multiplet consists of a massive gauge field, a massive Dirac fermion and three scalars, a total of eight bosonic and eight fermionic states. Since this is also the number of degrees of freedom of a gauge plus a hypermultiplet, one expects to form a massive supermultiplet by having the gauge boson from a gauge multiplet 'eat' one of the hyperscalars through the Higgs mechanism.

5.2.1. Non-supersymmetric embedding of the background gauge field. We first consider the case in which the background gauge field is *not* embedded in the $U(1)_R$. In this case supersymmetry is broken both by the branes and the bulk solution. Since the gauge field whose flux is localized on the branes is the $U(1)_R$ gauge field, the spectrum and the Casimir energy as well as the renormalization will not depend on Φ_b for this choice of the gauge field embedding.

The contribution of a hypermultiplet to the s_i coefficients is obtained by summing the result for a 6D Weyl fermion to that produced by four hyperscalars[7]. We obtain

$$s_{-1}^{\rm hm}(\omega,N) = 0\,, \tag{60}$$

$$s_0^{\rm hm}(\omega,N) = \frac{1}{\omega}(-1+\omega^2)\,, \tag{61}$$

$$s_1^{\rm hm}(\omega,N) = \frac{1}{\omega}\left[\frac{5}{24}-\frac{\omega^2}{12}+\frac{\omega^4}{24}-\frac{\omega^2N^2}{2}\right], \tag{62}$$

$$s_2^{\rm hm}(\omega,N) = \frac{1}{\omega}\left[-\frac{23}{1440}+\frac{31\,\omega^2}{1440}+\frac{7\,\omega^4}{1440}+\frac{\omega^6}{160}+\omega^2N^2\left(-\frac{1}{48}-\frac{\omega^2}{24}-\frac{\omega^4}{48}\right)\right],$$

[7] The hyerscalar contribution is given by the scalar result after substituting $m^2 \to m^2 + 1/(2r^2)$ [11].

where here N is the common monopole number of the hyperscalars and hyperino. For a massless ($m = 0$) hypermultiplet the only coefficient which matters is $s_2^{\rm hm}$ and the corresponding Casimir energy is $-s_2^{\rm hm}\ln(r/r_0)/(4\pi r^2)^2$.

As far as the gauge multiplet is concerned, we should sum the contribution of a gauge field to that of a 6D Weyl fermion, obtaining for the s_i coefficients

$$
\begin{aligned}
s_{-1}^{\rm gm}(\omega,N) &= 0\,, &(63)\\
s_0^{\rm gm}(\omega,N) &= \frac{1}{\omega}\left(1-2\,\omega+\omega^2\right), &(64)\\
s_1^{\rm gm}(\omega,N) &= \frac{1}{\omega}\left[\frac{1}{24}+\frac{\omega^2}{4}+\frac{\omega^4}{24}+\frac{\omega^2N^2}{2}\right], &(65)\\
s_2^{\rm gm}(\omega,N) &= \frac{1}{\omega}\left[-\frac{7}{1440}+\frac{71\,\omega^2}{1440}+\frac{23\,\omega^4}{1440}+\frac{\omega^6}{160}+\omega^2N^2\left(-\frac{5}{48}-\frac{\omega}{4}-\frac{\omega^2}{24}-\frac{\omega^4}{48}\right)\right], &(66)
\end{aligned}
$$

where now N is the common monopole number of the gauge field and gaugino. For a massless gauge multiplet the only important coefficient is $s_2^{\rm gm}$ and the corresponding Casimir energy is $-s_2^{\rm gm}\ln(r/r_0)/(4\pi r^2)^2$.

Therefore, if one has massless supermultiplets only, the total contribution to the Casimir energy is approximately given by $-(s_2^{\rm hm}+s_2^{\rm gm})\ln(r/r_0)/(4\pi r^2)^2$. The overall sign depends on the particular anomaly free model [34] that one chooses.

For a massive multiplet made of a hypermultiplet and a gauge multiplet we have that the s_i coefficients are $s_i^{\rm mm}=s_i^{\rm hm}+s_i^{\rm gm}$. Therefore, by using the explicit expressions for $s_i^{\rm hm}$ and $s_i^{\rm gm}$ given before, we obtain

$$
\begin{aligned}
s_{-1}^{\rm mm}(\omega,N) &= 0\,, &(67)\\
s_0^{\rm mm}(\omega,N) &= 2(\omega-1)\,, &(68)\\
s_1^{\rm mm}(\omega,N) &= \frac{1}{\omega}\left(\frac{1}{4}+\frac{\omega^2}{6}+\frac{\omega^4}{12}\right), &(69)\\
s_2^{\rm mm}(\omega,N) &= \frac{1}{\omega}\left[-\frac{1}{48}+\frac{17\,\omega^2}{240}+\frac{\omega^4}{48}+\frac{\omega^6}{80}+\omega^2N^2\left(-\frac{1}{8}-\frac{\omega}{4}-\frac{\omega^2}{12}-\frac{\omega^4}{24}\right)\right]. &(70)
\end{aligned}
$$

From this result, and from Eqs. (35) and (25), we note that for mr at most of order 1 the obtained Casimir energy is $\sim\ln(r/r_0)/(4\pi r^2)^2$. However, for $mr\gg 1$, integrating out a massive multiplet gives a dangerously large contribution. We will see that an extra suppression can be obtained when the gauge background is along the R-symmetry generator.

5.2.2. Supersymmetric embedding of the background gauge field. Let us now turn to the case in which the background gauge field is along the $U(1)_R$. Hyper and gauge multiplet contributions to the Casimir energy have been computed in [11] and are quite involved. We therefore refer to this work for explicit expressions associated with hyper, gauge and massive multiplets.

One important point we want to emphasize here is that $s_i^{\rm sph}=0$ for this background gauge field embedding and therefore the renormalization group equation of the bulk Casimir energy vanishes, $\mu\partial_\mu V_B^r=0$. The reason is that, as we mentioned, one half of bulk supersymmetry is not broken in this case.

In the particular case of identical boundary localized fluxes ($\Phi_b = \Phi/2$) also the branes preserve one half of the 6D supersymmetries (see the discussion at the end of section 2). This implies that for balanced fluxes on the two branes also the brane Casimir energy does not run, $\mu\partial_\mu V_b^r = 0$.

When the boundary localized fluxes are unbalanced a non-trivial result arises. However, by continuity the result should be suppressed by the difference of the two localized fluxes, $\Delta\Phi$. Also, as a remnant of 6D supersymmetry the coefficient s_{-1} for massive supermultiplets vanishes. When the bulk mass is such that mr is at most of order one the Casimir energy is $\sim \ln(r/r_0)/(4\pi r^2)^2$. When $mr \gg 1$ the contribution of a massive supermultiplet to the Casimir energy is of order $\Delta\Phi(mr)^4 \ln(r/r_0)/(4\pi r^2)^2$ and can be again as small as $\sim \ln(r/r_0)/(4\pi r^2)^2$ for $\Delta\Phi$ appropriately small. It is very interesting to notice that the localized flux difference does not receive quantum corrections from loops involving brane localized fields only (when they are not charged under the corresponding gauge symmetry) and therefore taking this difference to be very small does not need to have the usual fine tuning.

Let us conclude this section by mentioning that the Casimir energy is not exactly equal to the 4D cosmological constant, but is rather equal to the quantum action evaluated at the classical solution. The backreaction of the brane on the bulk is important in this case and leads to a sizeble correction of the solution. This point is extensively discussed in [11]. After taking into account this effect, however, the prediction for the (most UV sensitive part of the) 4D cosmological constant can remain as small as $\ln(r/r_0)/(4\pi r^2)^2$.

6. Conclusions and outlook

We discussed the Casimir energy as well as the renormalization of bulk and brane coefficients in the quantum action produced by integrating out (massive) bulk matter in codimension-two brane worlds. In the calculation we focused on the one-loop approximation and explicitly summed over the KK towers. Much of what we presented here is a review of [7, 8, 9, 10, 11], but we also provided some new results, in particular in the technique to perform the KK sums.

Regarding the motivations, as discussed extensively in the introduction, codimension-two brane worlds may provide a framework to solve the gauge hierarchy and, when the bulk is supersymmetric, the cosmological constant problem. These are technical naturalness problems and as such they require sistematic ways of computing quantum corrections.

We considered in some detail the rugby ball compactifications in which the size r of the extra dimensions is stabilized by the flux of a bulk field. A class of model having these configurations as solutions has been presented, including a concrete supersymmetric theory, 6D gauged chiral supergravity. The two 3-branes required to support these solutions can not only carry tension, but also a localized flux of the same gauge field which stabilizes the extra dimensions (tensions and localized fluxes are the leading terms in a derivative expansion of the brane Lagrangians, Eq. (5)).

As discussed in [11], when the localized fluxes on the two 3-branes are identical one half of the bulk supersymmetries is unbroken, which implies that the Casimir energy and, remarkably, the 4D cosmological constant vanish.

For the rugby ball compactifications the KK spectra of many types of bulk fields are known [7, 8, 9, 10, 11] and we reviewed their structure for scalars, fermions and gauge fields (see Eqs. (17), (18) and (19) rescpectively). The explicit form of the spectra allows us to have explicit expressions for the Casimir energy and, consequently, for the renormalization group equations, which depend on the flux and brane tensions and fluxes. The calculation of the Casimir energy starting from the KK spectrum exploits a novel efficient technique (see sections 4.1 and 4.2) which allows us to confirm the results of [10].

The explicit form for the Casimir energy we obtain is a polynomial function of mr of degree six, see eqs. (25) and (35). The coefficients s_i of the polynomial are computed explicitly as a function of the brane tensions and bulk and brane fluxes; for example, in the simple case of a bulk scalar they are given in Eqs. (55)-(58).

When the bulk particle that is integrated out is massless, or m is at most of order $1/r$, the 4D cosmological constant has the desired order of magnitude regardless of the fact that there is bulk supersymmetry. We show that for large m the final result can be appropriately suppressed if we select the supersymmetric model, the bulk solution preserves one of the supersymetries and localized fluxes have very similar values, such that we are close to a supersymmetric setup.

Let us mention some outlook of the results presented here.

A possible extension of our work is the inclusion of warped geometries, which represent the most general solutions with 4D maximal symmetry. A first step towards this goal is the codimension-one case, which would be interesting by itself as the Randall-Sundrum model can address the gauge hierarchy problem [37] (bulk fields in the Randall-Sundrum model have been considered in [38]). Through the AdS/CFT correspondence quantum loops in the bulk would correspond to $1/N_c$ corrections in the CFT side, where N_c counts the number of "colors".

Regarding again holography, we notice that the formalism of Refs. [7, 8, 9, 10, 11] to compute the spectrum for codimension-two brane worlds that we reviewed here can also be used to analyze spectral properties of holographic models in the confined phase [39]: these are obtained from models with one extra dimension (the holographic coordinate) by an additional compactified dimension and scalar, vector and fermion fields in the bulk can have a variety of uses ranging from condensed matter[8] [41] to quantum chromodynamics [42].

Finally, an interesting property of supersymmetry with two large extra dimensions is that it could provide a link between the observed value of the cosmological constant and the scale at which modifications of gravity should occur, $1/r$ [43]. It would be interesting to know how gravity gets modified in the concrete supersymmetric model we discussed. In the absence of localized fluxes graviton contributions have been computed in [44], but the role of these fluxes, which are important for the dilaton stabilization, remains an interesting target for future research.

Acknowledgements

We would like to thank Cliff Burgess, Leo van Nierop, Susha Parameswaran and Matt Williams for collaborations, Hyun-Min Lee for much help trying to diagonalize the supergravity sector in early stages of this work and Riccardo Barbieri, Oriol Pujolàs, Seifallah Randjbar-Daemi and George Thompson for useful discussions. This work was partly supported by the EU ITN "Unification in the LHC Era", contract PITN-GA-2009-237920 (UNILHC) and by MIUR under contract 2006022501.

[8] See, however, [40] for an alternative to compactification in this context.

References

[1] Grossman Y and Neubert M 2000 *Phys. Lett. B* **474** 361 (*Preprint* hep-ph/9912408). Huber S J and Shafi Q 2001 *Phys. Lett. B* **498** 256 (*Preprint* hep-ph/0010195). Salvio A and Shaposhnikov M 2007 *JHEP* **0711** 037 (*Preprint* arXiv:0707.2455 [hep-th]).
[2] Weinberg S 1989 *Rev. Mod. Phys.* **61** 1.
[3] Weinberg S 1972 *Gravitation and cosmology: principles and applications of the general theory of relativity* (New York London Sydney Toronto: John Wiley & Sons).
[4] Arkani-Hamed N, Dimopoulos S and Dvali G R 1998 *Phys. Lett. B* **429** 263 (*Preprint* hep-ph/9803315). Antoniadis I, Arkani-Hamed N, Dimopoulos S and Dvali G R 1998 *Phys. Lett. B* **436** 257 (*Preprint* hep-ph/9804398). Arkani-Hamed N, Dimopoulos S and Dvali G R 1999 *Phys. Rev. D* **59** 086004 (*Preprint* hep-ph/9807344]).
[5] Aghababaie Y, Burgess C P, Parameswaran S L and Quevedo F 2004 *Nucl. Phys. B* **680** 389 (*Preprint* hep-th/0304256). Burgess C P 2004 *Annals Phys.* **313** 283 (*Preprint* hep-th/0402200). Burgess C P 2005 *AIP Conf. Proc.* **743** 417 (*Preprint* hep-th/0411140).
[6] Rubakov V A and Shaposhnikov M E 1983 *Phys. Lett. B* **125** 139. Randjbar-Daemi S and Wetterich C 1986 "Kaluza-Klein Solutions With Noncompact Internal Spaces," *Phys. Lett. B* **166** 65. Gherghetta T and Shaposhnikov M E 2000 "Localizing gravity on a string - like defect in six-dimensions," *Phys. Rev. Lett.* **85**, 240 (*Preprint* [hep-th/0004014]). Cline J. M., Descheneau J, Giovannini M and Vinet J 2003 *JHEP* **0306** 048 (*Preprint* [hep-th/0304147]).
[7] Parameswaran S L, Randjbar-Daemi S and Salvio A 2007 *Nucl. Phys. B* **767** 54 (*Preprint* hep-th/0608074).
[8] Parameswaran S L, Randjbar-Daemi S and Salvio A 2007 *JHEP* **0801** 051 (*Preprint* arXiv:0706.1893 [hep-th]).
[9] Parameswaran S L, Randjbar-Daemi S and Salvio A 2009 *JHEP* **0903** 136 (*Preprint* arXiv:0902.0375 [hep-th]).
[10] Williams M, Burgess C P, van Nierop L and Salvio A 2012 *Preprint* arXiv:1210.3753 [hep-th].
[11] Burgess C P, van Nierop L, Parameswaran S, Salvio A and Williams M *Preprint* arXiv:1210.5405 [hep-th].
[12] Carroll S M and Guica M M 2003 *Preprint* hep-th/0302067. Navarro I 2003 *JCAP* **0309** 004 (*Preprint* hep-th/0302129). Navarro I 2003 *Class. Quant. Grav.* **20** 3603 (*Preprint* hep-th/0305014). Nilles H P, Papazoglou A and Tasinato G 2004 *Nucl. Phys. B* **677** 405 (*Preprint* hep-th/0309042). Bostock P, Gregory R, Navarro I and Santiago J 2004 *Phys. Rev. Lett.* **92** 221601 (*Preprint* hep-th/0311074). Graesser M L, Kile J E and Wang P 2004 *Phys. Rev. D* **70** 024008 (*Preprint* hep-th/0403074). Vinet J and Cline J M 2004 *Phys. Rev. D* **70** 083514 (*Preprint* hep-th/0406141). Navarro I and Santiago J 2005 *JHEP* **0502** 007 (*Preprint* hep-th/0411250). Kofinas G *Preprint* hep-th/0506035.
[13] Burgess C P and van Nierop L 2011 *JHEP* **1102** 094 (*Preprint* arXiv:1012.2638 [hep-th]). Burgess C P and van Nierop L 2011 *JHEP* **1104** 078 (*Preprint* arXiv:1101.0152 [hep-th]).
[14] Marcus N and Schwarz J H 1982 *Phys. Lett.* **115B** 111.
[15] Nishino H and Sezgin E 1984 *Phys. Lett.* **144B** 187.
[16] Salam A and Sezgin E 1984 *Phys. Lett.* **147B** 47.
[17] Burgess C P and van Nierop L 2011 (*Preprint* arXiv:1108.0345 [hep-th]).
[18] Aghababaie Y, Burgess C P, Cline J M, Firouzjahi H, Parameswaran S L, Quevedo F, Tasinato G and Zavala I 2003 *JHEP* **0309** 037 (*Preprint* [hep-th/0308064]).
[19] Burgess C P and Hoover D 2007 *Nucl. Phys. B* **772** 175 (*Preprint* [hep-th/0504004]). Hoover D and Burgess C P 2006 *JHEP* **0601** 058 (*Preprint* [hep-th/0507293]). Ghilencea D M, Hoover D, Burgess C P and Quevedo F 2005 *JHEP* **0509** 050 (*Preprint* [hep-th/0506164]).
[20] Randjbar-Daemi S, Salam A and Strathdee J A 1983 *Nucl. Phys. B* **214** 491.
[21] Giovannini M, Meyer H and Shaposhnikov M E 2001 *Nucl. Phys. B* **619** 615 (*Preprint* [hep-th/0104118]). Giovannini M 2002 *Phys. Rev. D* **66** 044016 (*Preprint* [hep-th/0205139]). Giovannini M, Le Be J V and Riederer S 2002 *Class. Quant. Grav.* **19** 3357 (*Preprint* [hep-th/0205222]).
[22] Giovannini M 2001 *Phys. Rev. D* **64** 064023 (*Preprint* [hep-th/0106041]).
[23] Gibbons G W, Gueven R and Pope C N 2004 *Phys. Lett. B* **595** 498 (*Preprint* hep-th/0307238).
[24] Parameswaran S L, Randjbar-Daemi S and Salvio A 2010 *Preprint* arXiv:1001.3271 [hep-th].
[25] Salvio A 2007 author's Ph.D. thesis *Preprint* hep-th/0701020.
[26] Randjbar-Daemi S, Salam A and Strathdee J A 1984 *Nuovo Cim. B* **84** 167.
[27] Randjbar-Daemi S and Sarmadi M H 1985 *Phys. Lett. B* **151** 343.
[28] Randjbar-Daemi S and Shaposhnikov M 2002 *Nucl. Phys. B* **645** 188 (*Preprint* hep-th/0206016).
[29] Randjbar-Daemi S, Salam A and Strathdee J A 1983 *Phys. Lett. B* **124**, 345 [Erratum-ibid. B **144**, 455 (1984)]. Dvali G R, Randjbar-Daemi S and Tabbash R 2002 *Phys. Rev. D* **65** 064021 (*Preprint* [arXiv:hep-ph/0102307]).
[30] Randjbar-Daemi S, Salvio A and Shaposhnikov M 2006 *Nucl. Phys. B* **741** 236 (*Preprint* hep-th/0601066);
[31] Salvio A 2006 *AIP Conf. Proc.* **881** 58 (*Preprint* hep-th/0609050).

[32] Hawking S W 1977 *Commun. Math. Phys.* **55** 133.
[33] Burgess C P 2004 *Living Rev. Rel.* **7** 5 (*Preprint* [gr-qc/0311082]). Burgess CP 2007 *Ann. Rev. Nucl. Part. Sci.* **57** 329 (*Preprint* [hep-th/0701053]).
[34] Randjbar-Daemi S, Salam A, Sezgin E and Strathdee J A 1985 *Phys. Lett. B* **151** 351.
[35] Avramis S D, Kehagias A and Randjbar-Daemi S 2005 *JHEP* **0505** 057 (*Preprint* [hep-th/0504033]). Avramis S D, Kehagias A 2005 *JHEP* **0510** 052 (*Preprint* [hep-th/0508172]).
[36] Randjbar-Daemi S and Sezgin E 2004 *Nucl. Phys. B* **692** 346 (*Preprint* [hep-th/0402217]).
[37] Randall L and Sundrum R 1999 *Phys. Rev. Lett.* **83** 3370 (*Preprint* [hep-ph/9905221]).
[38] Goldberger W D and Wise M B 1999 *Phys. Rev. D* **60** 107505 (*Preprint* [arXiv:hep-ph/9907218]). Davoudiasl H, Hewett J L and Rizzo T G 2000 *Phys. Lett. B* **473** 43 (*Preprint* [arXiv:hep-ph/9911262]). Gherghetta T and Pomarol A 2000 *Nucl. Phys. B* **586** 141 (*Preprint* [arXiv:hep-ph/0003129].
[39] Witten E 1998 *Adv. Theor. Math. Phys.* **2** 505-532. (*Preprint* [hep-th/9803131]). Horowitz G T, Myers R C 1998 *Phys. Rev.* **D59** 026005. (*Preprint* [hep-th/9808079]).
[40] Salvio A 2012 *JHEP* **1209** 134 (*Preprint* [arXiv:1207.3800 [hep-th]]).
[41] Nishioka T, Ryu S and Takayanagi T 2010 *JHEP* **1003** 131 (*Preprint* [arXiv:0911.0962 [hep-th]]). Montull M, Pujolas O, Salvio A and Silva P J 2011 *Phys. Rev. Lett.* **107** 181601 (*Preprint* [arXiv:1105.5392 [hep-th]]). Montull M, Pujolas O, Salvio A and Silva P J 2012 *JHEP* **1204** 135 (*Preprint* [arXiv:1202.0006 [hep-th]]). Bhattacharya J, Ogawa N, Takayanagi T and Ugajin T 2012 *JHEP* **1202** 137 (*Preprint* [arXiv:1201.0764 [hep-th]]).
[42] Basu P, Nogueira F, Rozali M, Stang J B and Van Raamsdonk M 2011 *New J. Phys.* **13** 055001 (*Preprint* [arXiv:1101.4042 [hep-th]]).
[43] Callin P and Burgess C P 2006 *Nucl. Phys.* B **752** 60 (*Preprint* [hep-ph/0511216]).
[44] Salvio A 2009 *Phys. Lett. B* **681** 166 (*Preprint* [arXiv:0909.0023 [hep-th]]).

IARD2012 IOP Publishing
Journal of Physics: Conference Series **437** (2013) 012005 doi:10.1088/1742-6596/437/1/012005

On the conformal version of Schwarzschild - de Sitter spacetime

Hristu Culetu
Ovidius University, Dept.of Physics,
B-dul Mamaia 124, 900527 Constanta, Romania
E-mail: hculetu@yahoo.com

Abstract. An imperfect cosmic fluid with energy flux is analyzed. Even though its energy density ρ is positive, the pressure $p = -\rho$ due to the fact that the metric is asymptotically de Sitter. The kinematical quantities for a nongeodesic congruence are computed. The scalar expansion is time independent but divergent at the singularity $r = 2m$. Far from the central mass m and for a cosmic time $\bar{t} << H^{-1}$, the heat flux q does not depend on Newton's constant G.

1. Introduction

The exterior Schwarzschild solution of Einstein's field equations describes the stationary gravitational field around a spherically symmetric star or black hole (BH). Such asymptotically flat solutions apply to a region outside the central object, far from other compact sources.

A 2nd set of solutions are the cosmological ones that model the behaviour of the universe at its largest scale [1] as, for example, the homogeneous and isotropic Friedmann - Robertson - Walker (FRW) metric on which the standard model is based. The purpose is to combine both classes of solutions and to find exact solutions for the gravitational field of a compact object embedded in a cosmological background [1]. One means the solutions describe BHs which are asymptotically flat and, in addition, dynamical [2, 3, 4, 5, 6]. A complication arises when a BH is "immersed" in a cosmological background (excepting the static de Sitter case): the phenomenon of accretion will be present. Namely, a radial flow of energy of the cosmic fluid tends to infall toward the central object [1, 2, 7, 8].

McVittie [9], in his well known paper, has taken into account the effect of the cosmic expansion on local systems. To forbid the accretion of the cosmic fluid into the central object he imposed a special relation between the (time dependent) mass of the object and the scale factor. Since in this case the stress tensor has no non-diagonal components, there is no radiative heat flow. However, the physical meaning of the McVittie spacetime is not completely clear and is still under debate.

Another exact solution that models a spherical object in a cosmological context was proposed by Sultana and Dyer [10]. Their metric is conformally equivalent to the exterior Schwarzschild metric and the cosmic matter is composed of two non - interacting perfect fluids, one being pressureless dust and the other - a null fluid. In addition, the Sultana - Dyer spacetime does not belong to the class of McVittie models (see [1]).

Throughout the paper we use the geometrical units $G = c = 1$.

2. Time dependent conformal Schwarzschild metric

Let us begin with a general geometry

$$ds^2 = a^2(t)\left[-(1-\frac{2m}{r})dt^2 + (1-\frac{2m}{r})^{-1}dr^2 + r^2 d\Omega^2\right], \tag{1}$$

where $a(t)$ is a positive function of the time t, m is the BH mass (or the mass of a spherically symmetric compact object) and $r > 2m$. When $m = 0$, Eq. (1) gives us a conformally flat spacetime. In contrast, when $a(t) = 1$ we obtain the exterior Schwarzschild metric around the isolated mass m.

With the help of the GrTensor II package and software package Maple, the following expressions for the components of the Einstein's tensor $G^a{}_b$ are obtained :

$$G^0{}_0 = -\frac{3r\dot{a}^2}{a^4(r-2m)},\quad G^0{}_1 = -\frac{2m\dot{a}}{a^3(r-2m)^2},\quad G^1{}_0 = \frac{2m\dot{a}}{a^3 r^2},\quad G^1{}_1 = G^2{}_2 = G^3{}_3 = \frac{r(\dot{a}^2 - 2a\ddot{a})}{a^4(r-2m)}, \tag{2}$$

where $\dot{a} = da/dt$, the Latin indices run from 0 to 3 and the coordinates are in order (t, r, θ, ϕ). We observe that in (2) there are nonzero non - diagonal components. That means a flux of radial energy will be present in the system. Therefore, the cosmic fluid must be imperfect [3] and the corresponding energy - momentum tensor given by the Einstein equations $G_{\mu\nu} = 8\pi T_{\mu\nu}$ should have terms dependent on the heat flux vector. It is worth noting that the scalar curvature

$$R^a{}_a = \frac{6\ddot{a}}{a^3(1-\frac{2m}{r})} \tag{3}$$

is singular at $r = 2m$, a property valid also for the McVittie spacetime [3], where the mass m is time dependent. In addition, the metric (1) has another (expected) singularity at $r = 0$, where the Kretschmann scalar $R^{abcd}R_{abcd}$ diverges. As far as the Weyl tensor is concerned, all its components are of the form $a^2(t)\ C^{Schw}_{abcd}$, where C^{Schw}_{abcd} corresponds to the Schwarzschild metric. The Weyl tensor is, of course, vanishing when $m = 0$ because the metric is conformally flat in that case.

3. Imperfect fluid stress tensor

Let us choose now the following congruence of observers

$$u^a = \left(\frac{1}{a(t)\sqrt{1-\frac{2m}{r}}}, 0, 0, 0\right) \tag{4}$$

with u^a the velocity vector field associated to "static" observer ($dr/d\tau = 0, u^a u_a = -1$, τ being here the proper time). We assume the stress tensor is given by the following expression

$$T_{ab} = (p+\rho)u_a u_b + p g_{ab} + u_a q_b + u_b q_a, \tag{5}$$

where $u^a q_a = 0$. The energy density ρ, the pressure p and the heat flux q^a acquire the form

$$\rho = T_{ab}u^a u^b,\quad p = T^r_r = T^\theta_\theta = T^\phi_\phi,\quad q^a = -T^a_b u^b - \rho u^a. \tag{6}$$

Using now Eqs. (2), one obtains

$$\rho = \frac{3\dot{a}^2}{8\pi a^4\left(1-\frac{2m}{r}\right)},\quad p = \frac{\dot{a}^2 - 2a\ddot{a}}{8\pi a^4\left(1-\frac{2m}{r}\right)},\quad q^a = \left(0,\ -\frac{m\dot{a}}{4\pi a^4 r^2\sqrt{1-\frac{2m}{r}}},\ 0,\ 0\right). \tag{7}$$

Let us consider the case of an expanding universe ($\dot{a} > 0$). Hence, $q^1 \equiv q^r < 0$, namely a radial inflow will be present. In addition, the parameters ρ, p and q^r are divergent at $r = 2m$, a consequence of the singularity of the curvature invariant scalars there.

It is worth to compute the total radial energy flowing across a $r = const.$ hypersurface Σ [11] for some time interval $\Delta t = t_2 - t_1$

$$W = \int T^a_{\ b} u^b n_a \sqrt{-h}\ dt\ d\theta\ d\phi \tag{8}$$

where $n^a = (0, \sqrt{1 - \frac{2m}{r}}/a, 0, 0)$ is the unit spacelike normal vector to Σ, $h_{ab} = g_{ab} - n_a n_b$ is the induced metric on Σ and $h = det(h_{ab})$. By means of $T^1_{\ 0} = m\dot{a}/4\pi a^3 r^2$ from (2) and u^a from (4), we get

$$W = \frac{m}{\sqrt{1 - \frac{2m}{r}}} \int_{t_2}^{t_1} \dot{a}(t) dt = \frac{m\Delta a(t)}{\sqrt{1 - \frac{2m}{r}}} \tag{9}$$

where $\Delta a(t) = a(t_2) - a(t_1)$. For an observer located at $r >> 2m$, we have $W \cong m\Delta a(t)$, i. e. the gain in the physical mass of the compact object. In other words, all the accreted energy contributes to the increase of the object mass. We of course have neglected here the back reaction effect, keeping $m = const.$ in the metric (1). Moreover, W grows indefinitely when $r \to 2m$. If $\Delta t = t_2 - t_1$ is small, so is Δa and, therefore, W is also small excepting near the horizon $r = 2m$.

The congruence (4) is not geodesic as could be seen from the expression of the acceleration vector

$$a^b = \left(0,\ \frac{m}{a^2 r^2},\ 0,\ 0\right), \tag{10}$$

with $\sqrt{a^b a_b} = m/ar^2\sqrt{1 - \frac{2m}{r}}$. As far as the other kinematical quantities is concerned, the shear and the vorticity tensors of the congruence are vanishing (due to the spherical symmetry) while the expansion scalar is given by

$$\Theta \equiv \nabla_a u^a = \frac{3\dot{a}}{a^2\sqrt{1 - \frac{2m}{r}}} \tag{11}$$

If we formally compute the surface gravity on the horizon $r = 2m$, one obtains

$$\sqrt{a^b a_b}\ \sqrt{-g_{00}}|_{r=2m} = \frac{m}{(2m)^2} = \frac{1}{4m}, \tag{12}$$

as for a static BH. It is, however, problematic whether one might define a surface gravity in our nonstatic spacetime and, in addition, at $r = 2m$ the metric is singular.

4. Time dependent Schwarzschild - de Sitter geometry

Our next task is to take into consideration a particular expression of $a(t)$ and analyze that case in detail. A situation not studied till now, to our knowledge, is the Schwarzschild BH embedded in a de Sitter universe, written in conformally flat coordinates. We prefer a spacetime conformal to the Schwarzschild one in order to preserve the null trajectories and the causal horizon. Let us firstly consider the de Sitter line element written in comoving coordinates

$$ds^2 = -d\bar{t}^2 + e^{2H\bar{t}}(dr^2 + r^2 d\Omega^2), \tag{13}$$

where $H = 1/\bar{R}$ is the Hubble constant, $\bar{R}$ is the radius of the de Sitter universe (from the static form of the metric, related to the cosmological constant Λ, with $1/\bar{R}^2 = \Lambda/3$), $\bar{t}$ is the comoving time and $d\Omega^2 = d\theta^2 + sin^2\theta d\phi^2$ represents the metric on the unit two-sphere. By means of the transformation [12]

$$\eta = -\frac{1}{H}e^{-H\bar{t}}, \quad (\bar{t} > 0, \quad -\frac{1}{H} < \eta < 0), \tag{14}$$

the metric (13) may be put in a conformally flat form,

$$ds^2 = \frac{1}{H^2\eta^2}(-d\eta^2 + dr^2 + r^2 d\Omega^2), \tag{15}$$

η being the conformal time. We intend to study a BH embedded in the above spacetime, namely in the geometry

$$ds^2 = \frac{1}{H^2\eta^2}\left[-(1-\frac{2m}{r})d\eta^2 + \frac{dr^2}{1-\frac{2m}{r}} + r^2 d\Omega^2\right], \tag{16}$$

for to investigate the impact of an expanding universe on local physics. We have now $a(\eta) = 1/H|\eta|$ which leads to

$$\dot{a} = \frac{da}{d\eta} = \frac{1}{H\eta^2}, \quad \frac{\dot{a}}{a} = \frac{1}{|\eta|}, \quad \ddot{a} = -\frac{2}{H\eta^3} > 0, \quad 2a\ddot{a} - \dot{a}^2 = \frac{3}{H^2\eta^4} \tag{17}$$

We notice that $\ddot{a} > 0$, that is the spacetime is accelerating, as it is confirmed for our universe. For the above particular value for $a(\eta)$ we have from (2) for the components of the stress tensor

$$8\pi T^0_{\ 1} = G^0_{\ 1} = \frac{2mH^2\eta}{r^2(1-\frac{2m}{r})^2} = -\frac{8\pi T^1_{\ 0}}{(1-\frac{2m}{r})^2}, \quad 8\pi\rho = -8\pi T^0_{\ 0} = -8\pi p = \frac{3H^2}{1-\frac{2m}{r}} \tag{18}$$

We see that the equation of state of the fluid is $p = -\rho$, irrespective of the value of m. Moreover, they have no a time dependence. For $r >> 2m$ (far from the central mass), ρ acquires the well-known expression $3H^2/8\pi$ for a spatially flat FRW universe. Even though the energy density of the fluid is positive and the weak energy condition holds, the strong energy condition is not satisfied. We have, indeed

$$\left(T_{ab} - \frac{1}{2}g_{ab}T^c_c\right)u^a u^b = -\rho < 0, \tag{19}$$

a basic property of the dark energy (DE), due to the negative pressure.

The kinematical quantities a^b, Θ and q^a become now

$$a^b = \left(0, \frac{mH^2\eta^2}{r^2}, 0, 0\right), \quad \sqrt{a^b a_b} = \frac{mH|\eta|}{r^2\sqrt{1-\frac{2m}{r}}}, \quad \Theta = \frac{3H}{\sqrt{1-\frac{2m}{r}}}, \quad q^a = \left(0, -\frac{mH^3\eta^2}{4\pi r^2\sqrt{1-\frac{2m}{r}}}, 0, 0\right) \tag{20}$$

In addition, (20) yields

$$\dot{\Theta} \equiv u^a\nabla_a\Theta = 0, \quad q \equiv \sqrt{q^a q_a} = \frac{mH^2|\eta|}{4\pi r^2\left(1-\frac{2m}{r}\right)} \tag{21}$$

We see that, in spite of the time dependance of the metric (16), Θ is a function on the radial coordinate only. One means that for an observer sitting at $r = const.$ the congruence has a constant expansion. It becomes, however, divergent when $r \to 2m$ because of the singularity. An interesting property of the spatial vector field q^a that represents the current density of heat is its independence of the Newton constant G. Far from the constant mass and for the cosmic time $\bar{t} << H^{-1}$ (or from (14), $H|\eta| \approx 1$), the expression of the heat flow becomes (we introduce here the fundamental constants G and c)

$$q = \frac{c^4}{G}\frac{Gm}{c^2}\frac{H}{4\pi r^2} = \frac{mc^2 H}{4\pi r^2}. \tag{22}$$

Let us find, as an example, the ingoing heat flux on the Sun surface, using the above approximation. With $M_S \approx 2\text{x}10^{33}g$, $R_S \approx 7\text{x}10^{10}cm$ and $H \approx 2.3\text{x}10^{-18}s^{-1}$, we get $q_S \approx 6.2\text{x}10^{13}ergs/cm^2s = 6.88\text{x}10^{-8}g/cm^2s$, or a power of $1.4\text{x}10^{44}$ergs/year.

5. Misner - Sharp mass

For to interpret the spacetime (16) as a model for an inhomogeneity in a FRW universe (spatially flat in our situation) it is useful to compute the quasi-local Misner - Sharp (MS) mass M [13, 14, 15, 16], with its Weyl and Ricci parts, namely M_W and M_R, respectively. It will help to detect localized sources of gravity . The mass M is obtained from

$$1 - \frac{2M(t,r)}{R} = g^{ab}R_{,a}R_{,b}, \tag{23}$$

where $R_{,a} \equiv \partial R/\partial x_a$ is the areal (physical) radius which in our case is given by $R(t,r) = ra(t)$ (see Eq. (1)). Eq. (23) yields

$$M(t,r) = ma + \frac{\dot{a}^2 r^3}{2a(1-\frac{2m}{r})} \tag{24}$$

We identify the 1st term from the r.h.s. with the Weyl mass M_W and the 2nd term with the Ricci mass M_R that gives the contribution of the fluid to the MS energy. We note that M_W is the gravitational energy and $ma(t)$ represents the physical mass.

Let us now apply the formula (24) for the spacetime (16). One obtains

$$M(\eta,r) = \frac{m}{H|\eta|} + \frac{r^3}{2H|\eta|\eta^2\left(1-\frac{2m}{r}\right)} \tag{25}$$

The fact that $M_W = m/H|\eta|$ grows from m to infinity when η varies from $-1/H$ to zero is interpreted as saying that energy is accreted from the ambient matter onto the central object, as one sees from the expression of q. M_W comes from the Weyl curvature and may be identified with the gravitational mass of the central object. The Ricci part of the Misner - Sharp mass may be expressed in terms of the density ρ from (18) and the areal radius, to give $M_R = \rho\, 4\pi R^3/3$, as expected (see, for instance, [1]). We wish now to evaluate the MS mass at $\bar{t} = 0$ or $|\eta| = 1/H$, for $r >> 2m$. While $M_W \approx m$, the Ricci mass appears as

$$M_R \approx \frac{H^2 r^3}{2G} \tag{26}$$

where the gravitational constant G has been introduced, to have correct units. Taking again the Sun as the central object and using the previous values of M_S, H and R_S, one obtains $M_R \approx 2.8$ Kg. We see that by far the basic contribution to $M(\eta, r)$ comes from the Weyl mass. However, the situation changes drastically near $r = 2m$ or for $\eta \to 0$, when the 2nd term from (25) dominates. It is clear that "to be near the singularity $r = 2m$" or "to wait for a long time" have similar effects.

6. Conclusions

We studied in this paper a BH (or a compact object with spherical symmetry) immersed in a de Sitter space written in a conformally flat form. To be a solution of Einstein's equations, we need an imperfect fluid stress tensor on its RHS. The proposed model leads to the presence of an inward energy flux that diverges at $r = 2m$. The kinematical quantities associated to a congruence of non-geodesic observers are calculated. One finds the shear and vorticity tensors are vanishing but the scalar expansion does not depend on time and is asymptotically constant. The cosmic fluid stress tensor does not observe the strong energy condition (as for dark energy) due to the negative pressure $p = -\rho$. We stress the very simple form of the formula (22) for the energy flux (q asymptotically equals the star energy times the Hubble constant divided by its surface area). In addition, it is independent of the gravitational constant.

Acknowledgments

I would like to thank the Organizers of the IARD 2012 Conference, held at Florence, Italy, May 29 - June 1, for inviting me to attend it and for financial support.

References

[1] Carrera M and Giulini D 2010 *Phys. Rev.* D**81**: 043521
[2] Faraoni V, 2009 *Phys. Rev.* D**80**: 044013; Faraoni V, Moreno A Z and Nandra R, 2012 Making sense of the bizarre behaviour of horizons in the McVittie spacetime (*Preprint* 1202.0719 [gr-qc])
[3] Faraoni V and Jacques A, 2007 *Phys. Rev.* D**76**, 063510; Faraoni V, Gao C, Chen X and Shen Y 2009 *Phys. Lett.* **B**671 7
[4] Saida H, Harada T and Maeda H 2007 *Class. Quant. Grav.* **24**: 4711; Sun C 2011 *Comm. Theor. Phys.* **55**, 597 (2011)
[5] Said J L and Adami K Z Cosmological Effects on Black Hole Formation (*Preprint* 1201.0750 [gr-qc])
[6] Anderson M 2011 J. Phys.: Conf. Series **283**, 012001
[7] Herrera L, di Prisco A and Ibanez J, 2011 *Phys. Rev.* D **84**, 064036
[8] Nolan B C 1999 *Class. Quan. Grav.* **16**, 1227
[9] McVittie G C 1933 *Mon. Not. R. Astr. Soc.* **93**, 325
[10] Sultana L and Dyer C C 2005 *Gen. Relat. Grav.* **37**, 1349
[11] Guariento D C and Horvath J E 2012 *Gen. Relat. Grav.* **44**, 985
[12] Birrell N D and Davies P C W 1982 *Quantum fields in curved spaces*, (Cambridge: Cambridge University Press) p. 130
[13] Herrera L, Di Prisco A, Fuenmayor E and Troconis O 2008 *Int. J. Mod. Phys.* D**18**: 129
[14] Nielsen A and Yeom D 2009 *Int. J. Mod. Phys.* A**24**: 5261
[15] Giambo' R, Giannoni F, Magli G and Piccione P 2003 *Commun. Math. Phys.* **235**, 545
[16] Firouzjaee J T, Parsi Mood M and Mansouri R 2010 Do we know the mass of a black hole? Mass of some cosmological black hole models *Preprint* 1010.3971 [gr-qc].

Quantum field theories in spaces with neutral signatures

Matej Pavšič
Jožef Stefan Institute, Jamova 39, SI-1000, Ljubljana, Slovenia
E-mail: matej.pavsic@ijs.si

Abstract. We point out that quantum field theories based on the concept of Clifford space and Clifford algebra valued-fields involve both positive and negative energies. This is a consequence of the indefinite signature (p, q) of the Clifford space. When the signature is neutral, $p = q$, then vacuum energy vanishes and there is no cosmological constant problem. A question of the stability of such theories in the presence of interactions arises. We investigate a toy model of the harmonic oscillator in the space $M_{1,1}$. We have found that in the presence of certain interactions the amplitude of oscillations can remain finite. In general this is not the case and the amplitude grows to infinity, but only when the two frequencies are exactly the same. When they are even slightly different, the amplitude remains finite and the system is stable. We show how such oscillator comes from the Stueckelberg action in curved space, and how it can be generalized to field theories.

1. Introduction

Research in quantum gravity leads us to higher derivative theories that imply negative energies. They have turned out to be a stumbling block against further progress, because they imply vacuum instability. Negative energies also occur in spaces with non Lorentzian signature, $M_{p,q}$. Therefore, such spaces are commonly considered as unsuitable for physics. In the literature it is usually stated that negative signatures imply negative probabilities or ghosts. In ref. ([1]–[5]) it has been pointed out that by an appropriate redefinition of vacuum, instead of negative probabilities and positive energies, one has positive probabilities and negative energies.

In this paper we investigate the issue of negative energies in some detail. We consider a classical interacting oscillator in the space $M_{1,1}$. Solutions of its equations of motion can be obtained numerically. We have found that the solutions are not always runaway, unstable ones. In many cases they have oscillatory behavior, and do not escape into infinity. The same is true for a truncated quantum oscillator.

Then we consider field theories in field spaces with neutral signature, an example of which is the Clifford space, discussed in Sec. 2. The vacuum energy of such fields is zero, which resolves the notorious cosmological constant problem. But there remain the problem of "instantaneous" vacuum decay of an interacting field. We show in Sec. 6 how this problem can be circumvented, and the "explosion" of the vacuum stabilized.

2. Clifford space: An extension of spacetime

An extended object, O, can be be sampled by a finite set of parameters, for instance, by the center of mass coordinates, and by the orientation of its axes of symmetries. Higher multipole deformations, such as the dipole and the quadrupole ones, can also be taken into account. For practical reasons, only a finite number of mulipoles can be taken into account. Instead of the infinite number of degrees of freedom, we consider only a finite number of degrees of freedom. We thus perform a mapping from an infinite dimensional configuration space, associated with the object O, to a finite dimensional subspace.

Extended objects of particular interest for theoretical physics are strings and branes. They can be described by coordinate functions $X^\mu(\xi^a)$, $\mu = 0,1,2,...,N-1$, $a = 0,1,2,...,n-1$, where $n \leq N$. Such a description is infinite dimensional. In refs. [6] it was pointed out how one can employ a finite description in terms of a quenched mini superspace.

The idea has been further developed [7]–[11] by means of Clifford algebras, a very useful tool for description of geometry [12]. Here we are interested in description of spacetime, M_N, and the objects embedded in M_N.Therefore, let us start by considering the line element in M_N:

$$Q = \mathrm{d}s^2 = g_{\mu\nu}\mathrm{d}x^\mu \mathrm{d}x^\nu \ , \qquad \mu,\nu = 0,1,2,...N-1. \tag{1}$$

If we take the square root, $\sqrt{Q}$, we have the following possibilities:

i) $\sqrt{Q} = \sqrt{g_{\mu\nu}\mathrm{d}x^\mu \mathrm{d}x^\nu}$ scalar

ii) $\sqrt{Q} = \gamma_\mu \mathrm{d}x^\mu$ vector

Here γ_μ are generators of the Clifford algebra $Cl(p,q)$, $p+q=N$, satisfying

$$\gamma_\mu \cdot \gamma_\nu \equiv \frac{1}{2}(\gamma_\mu\gamma_\nu + \gamma_\nu\gamma_\mu) = g_{\mu\nu}, \tag{2}$$

where $g_{\mu\nu}$ is the metric of M_N.

The generators γ_μ have the role of basis vectors of the spacetime M_N. The symmetric product $\gamma_\mu \cdot \gamma_\nu$ represents *the inner product.* The antisymmetric (wedge) product of two basis vectors gives a unit bivector:

$$\gamma_\mu \wedge \gamma_\nu \equiv \frac{1}{2}(\gamma_\mu\gamma_\nu - \gamma_\nu\gamma_\mu) \tag{3}$$

and has thus the role of outer product. In analogous way we obtain 3-vectors, 4-vectors, etc..

We assume that the signature of an N-dimensional spacetime is $(1, N-1)$, i.e., $(+---...)$. In the case of the 4-dimensional spacetime we thus have the signature $(1,3)$, i.e., $(+---)$. The corresponding Clifford algebra is $Cl(1,3)$.

The basis of $Cl(1, N-1)$ is

$$\{1, \gamma_\mu, \gamma_{\mu_1} \wedge \gamma_{\mu_2}, ..., \gamma_{\mu_1} \wedge \gamma_{\mu_2} \wedge ... \wedge \gamma_{\mu_N}\}. \tag{4}$$

A generic element, $X \in Cl(1, N-1)$, is a superposition

$$X = \sum_{r=0}^{N} \frac{1}{r!} X^{\mu_1\mu_2...\mu_r} \gamma_{\mu_1} \wedge \gamma_{\mu_2} \wedge ... \wedge \gamma_{\mu_r} \equiv X^M \gamma_M, \tag{5}$$

called a *Clifford aggregate* or *polyvector.*

In refs. [10, 11, 13] it has been demonstrated that r-vectors $X^{\mu_1\mu_2...\mu_r}$ can be associated with closed instantonic $(r-1)$-branes or open instantonic r-branes. A generic polyvector, $X = X^M\gamma_M$, can be associated with a conglomerate of (instantonic) r-branes for various values of $r = 0,1,2,...,N$.

Our objects are instantonic r-branes, which means that they are localized in spacetime[1]. They generalize the concept of 'event', a spacetime point, x^μ, $\mu = 0, 1, 2, 3$. Instead of an event, we have now an extended event, $\mathcal{E}$, described by coordinates $X^{\mu_1\mu_2...\mu_r}$, $r = 0, 1, 2, 3, 4$. The space of extended events is called *Clifford space*, C. It is a manifold whose tangent space at any of its points is a Clifford algebra $Cl(1,3)$. If C is a flat space, then it is isomorphic to the Clifford algebra $Cl(1,3)$ with elements $X = \sum_{r=0}^{4} \frac{1}{4!} X^{\mu_1\mu_2...\mu_r}\gamma_{\mu_1\mu_2...\mu_r} \equiv X^M \gamma_M$.

In flat C-space, the basis vectors are equal to the wedge product

$$\gamma_M = \gamma_{\mu_1} \wedge \gamma_{\mu_2} \wedge ... \wedge \gamma_{\mu_r} \tag{6}$$

at every point $\mathcal{E} \in C$. This not true in curved C-space: if we (parallel) transport a polyvector $A = A^M \gamma_M$ from a point $\mathcal{E} \in C$ along a closed path back to the original point, $\mathcal{E}$, then the orientation of the polyvector A after such transport will not coincide with the initial orientation of A. After the transport along a closed path we will obtain a new polyvector $A' = A'^M \gamma_M$. If, in particular, the initial polyvector is one of the Clifford algebra basis elements, $A = \gamma_M$, i.e., an object with definite grade, then the final polyvector will be $A' = A'^M \gamma_M$, which is an object with mixed grade. A consequence is that in curved Clifford space C, basis vectors cannot have definite grade at all points of C.

The situation in a curved Clifford space, C, is analogous to that in a usual curved space, where after the (parallel) transport along a closed path, a vector changes its orientation. In Clifford space, a change of orientation in general implies a change of a polyvector's grade, so that, e.g., a definite grade polyvector changes into a mixed grade polyvector.

However, if we impose a condition that, under parallel transport, the grade of a polyvector does not change, then one has a very special kind of curved Clifford space [14]. In such a space, afer a parallel transport along a closed path, the vector part $\langle A \rangle_1 = a^\mu \gamma_\mu$ changes into $\langle A' \rangle_1 = a'^\mu \gamma_\mu$, the bivector part $\langle A \rangle_2 = a^{\mu\nu} \gamma_\mu \wedge \gamma_\nu$ changes into $\langle A' \rangle_2 = a'^{\mu\nu} \gamma_\mu \wedge \gamma_\nu$, etc., but one grade does not change into another grade. Such special Clifford space, in which the consequences of curvature manifest themselves *within* each of the subspaces with definite grade separately, but not *between* those subspaces, is very complicated. We will not consider such special Clifford spaces, because they are analogous to the usual curved spaces of the product form $M = M_1 \times M_2 \times ...M_n$, where $M_i \subset M$ is a curved lower dimensional subspace of M, and where only those (parallel) transports are allowed that bring tangent vectors of M_i into another tangent vectors of the same subspace M_i.

The squared line element in Clifford space, C, is

$$\mathrm{d}S^2 = G_{MN} \mathrm{d}x^M \mathrm{d}x^N = \text{X}^\ddagger * \mathrm{d}X = \langle \mathrm{d}X^\ddagger \mathrm{d}X \rangle. \tag{7}$$

Here $\mathrm{d}X = \mathrm{d}x^M \gamma_M$, and $\mathrm{d}X^\ddagger = \mathrm{d}x^M \gamma_M^\ddagger$, where $\ddagger$ denotes the operation of inversion: $(\gamma_{\mu_1}\gamma_{\mu_2}...\gamma_{\mu_r})^\ddagger = \gamma_{\mu_r}\gamma_{\mu_{r-1}}...\gamma_{\mu_r}$. The metric of C is

$$G_{MN} = \gamma_M^\ddagger * \gamma_N = \langle \gamma^\ddagger \gamma_N \rangle_0, \tag{8}$$

[1] The usual p-branes are localized in space, but they are infinitely extended into a time-like direction, so that they are $(p+1)$-dimensional worldsheets in spacetime.

where $\langle\ \rangle_0$ means the scalar part. A Clifford space with such a metric has signature [11] $(8, 8)$, i.e., $(++++++++--------)$. This is ultrahyperbolic space with neutral signature.

In the following sections we will show that, contrary to the wide spread belief, the physics in spaces with signature (n, n) makes sense.

3. A toy model: Harmonic oscillator in the pseudo Euclidean space $M_{r,s}$

3.1. Case $M_{1,1}$

In ref. [4] we considered the harmonic oscillator described by the Lagrangian

$$L = \frac{1}{2}(\dot{x}^2 - \dot{y}^2) - \frac{1}{2}\omega^2(x^2 - y^2). \tag{9}$$

The change of sign in front of the y-terms has no influence on the equations of motion. A difference occurs when we calculate the canonical momenta

$$p_x = \frac{\partial L}{\partial \dot{x}} = \dot{x}\,, \qquad p_y = \frac{\partial L}{\partial \dot{y}} = -\dot{y} \tag{10}$$

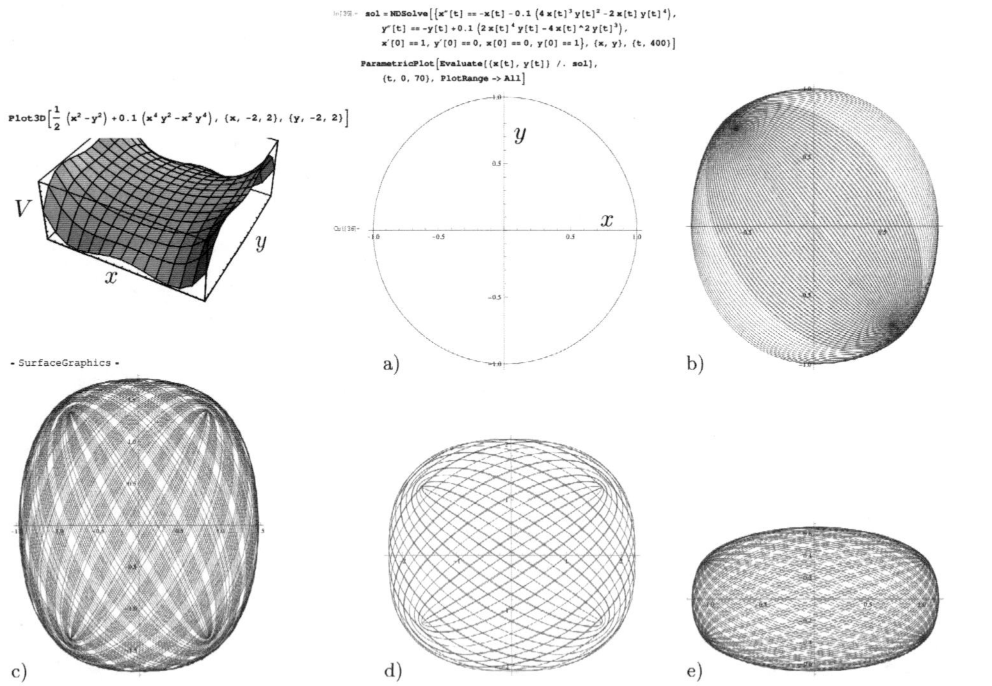

Figure 1. Examples of solutions, obtained by Mathematica, for various initial conditions, of the interactive pseudo-Euclidean oscillator for the potential $V = (1/2)(x^2 - y^2) + V_1$, with $V_1 = 0.1(x^4y^2 - x^2y^4)$. a) $\dot{x}(0) = 1$, $\dot{y}(0) = 0$, $x(0) = 0$, $y(0) = 1$; b) $\dot{x}(0) = 0.9$, $\dot{y}(0) = 0$, $x(0) = 0$, $y(0) = 1$; c) $\dot{x}(0) = 0.9$, $\dot{y}(0) = 0.2$, $x(0) = 0.6$, $y(0) = 1.5$; d) $\dot{x}(0) = 2$, $\dot{y}(0) = 1$, $x(0) = 0.3$, $y(0) = 1$; e) $\dot{x}(0) = 0.2$, $\dot{y}(0) = 0.2$, $x(0) = 0.3$, $y(0) = 1$.

The Hamiltonian

$$H = p_x\dot{x} + p_y\dot{y} - L = \frac{1}{2}(p_x^2 - p_y^2) + \frac{\omega^2}{2}(x^2 - y^2) \tag{11}$$

can have positive or negative values, but this does not mean that the system is unstable. Namely, the equations of motion

$$\ddot{x} = -\frac{\partial V}{\partial x}\,, \qquad \ddot{y} = \frac{\partial V}{\partial y}\,, \quad V = \frac{\omega^2}{2}(x^2 - y^2) \tag{12}$$

have different signs in front of the force terms. The criterion for stability of the y-degree of freedom is that the potential has to have a *maximum* in the (y, V)-plane [5].

Stability could be destroyed if we include an extra interaction term into the potential V. But for

$$V = \frac{1}{2}(x^2 - y^2) + \lambda(x^4 y^2 - x^2 y^4) \tag{13}$$

we find, by solving numerically the equations of motion, that the system is stable (Fig. 1). We find the same phenomenon also for other potentials (Fig. 2). Those figures show that the trajectories in the (x, y) space remain confined, and do not escape into infinity, which reveals that the interacting pseudo-Euclidean oscillator is not automatically unstable. Later we will study this in more detail.

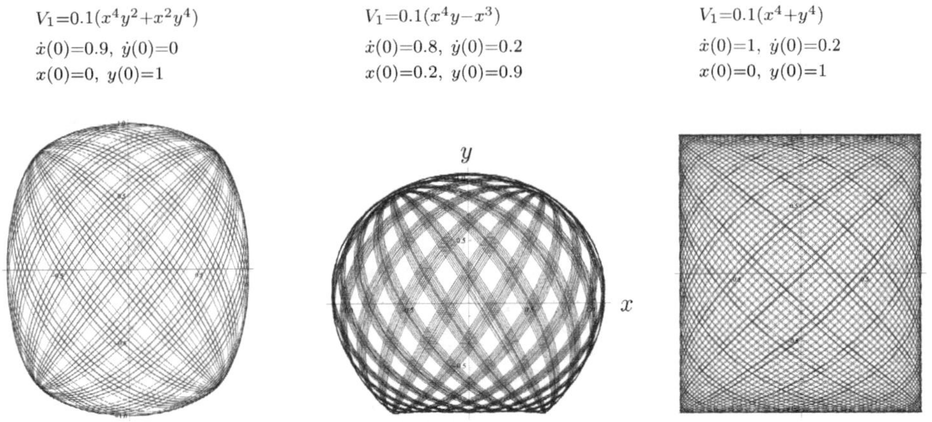

Figure 2. Examples of solutions, for various initial conditions and potentials, of the interactive pseudo-Euclidean oscillator

In refs. [4, 5] we also studied the quantized non interacting system (9) by replacing the coordinates and momenta with the operators, satisfying $[x, p_x] = i$, $[y, p_y] = i$, $[x, y] = [p_x, p_y] = 0$, and by introducing the operators

$$c_x = \frac{1}{\sqrt{2}}(\sqrt{\omega}\,x + \frac{i}{\sqrt{\omega}}\,p_x)\,, \quad c_x^\dagger = \frac{1}{\sqrt{2}}(\sqrt{\omega}\,x - \frac{i}{\sqrt{\omega}}\,p_x) \tag{14}$$

$$c_y = \frac{1}{\sqrt{2}}(\sqrt{\omega}\,y + \frac{i}{\sqrt{\omega}}\,p_y)\,, \quad c_y^\dagger = \frac{1}{\sqrt{2}}(\sqrt{\omega}\,y - \frac{i}{\sqrt{\omega}}\,p_y) \tag{15}$$

satisfying the commutation relations

$$[c_x, c_x^\dagger] = 1\,, \qquad [c_y, c_y^\dagger] = 1, \tag{16}$$

$$[c_x, c_y] = [c_x^\dagger, c_y^\dagger] = 0 \tag{17}$$

The Hamilton operator then reads

$$H = \omega\,(c_x^\dagger c_x + c_x c_x^\dagger - c_y^\dagger c_y - c_y c_y^\dagger). \tag{18}$$

Let us define vacuum so that it is annihilated by c_x and c_y:

$$c_x|0\rangle = 0\ , \qquad c_y|0\rangle = 0. \tag{19}$$

If we arrange the terms in the Hamiltonian so that the creation operators are on the left, we obtain

$$H = \omega\,(c_x^\dagger c_x - c_y^\dagger c_y) \tag{20}$$

which has vanishing vacuum expectation value,

$$\langle 0|H|0\rangle = 0. \tag{21}$$

The excited states $c_x^\dagger|0\rangle$, $c_x^\dagger c_x^\dagger|0\rangle$,..., have positive energies, whereas the states $c_y^\dagger|0\rangle$, $c_y^\dagger c_y^\dagger|0\rangle$,..., have negative energies. But all those states have positive norms, because the commutators (16) are positive.

In the Schrödinger representation, $x, y \in \mathbb{R}$, $p_x = -i\partial/\partial x$, $p_y = -i\partial/\partial y$, and $\langle x, y|0\rangle \equiv \psi_0(x, y)$. The relations (19) become

$$\frac{1}{2}\left(\sqrt{\omega}x + \frac{1}{\sqrt{\omega}}\frac{\partial}{\partial x}\right)\psi_0(x, y) = 0 \tag{22}$$

$$\frac{1}{2}\left(\sqrt{\omega}y + \frac{1}{\sqrt{\omega}}\frac{\partial}{\partial y}\right)\psi_0(x, y) = 0 \tag{23}$$

and their solution is the following vacuum wave function:

$$\psi_0 = A e^{\frac{1}{2}\omega(x^2+y^2)}. \tag{24}$$

It is normalized according to $\int \psi_0^2\,\mathrm{d}x\mathrm{d}y = 1$ if $A = \sqrt{\omega/\pi}$, and it satisfies the Schrödinger equation $H\psi = E\psi$, with $E = 0$.

4. Generalization to $M_{r,s}$

The system (9) can be generalized [4] to a space of arbitrary dimension and signature, $M_{r,s}$. The Lagrangian is then

$$L = \frac{1}{2}\dot{x}^a\dot{x}_a - \frac{1}{2}\omega^2 x^a x_a. \tag{25}$$

The corresponding Hamiltonian is

$$\frac{1}{2}p^a p_a + \frac{1}{2}\omega^2 x^a x_a, \tag{26}$$

where $p_a = \partial L/\partial\dot{x}^a = \dot{x}_a = \eta_{ab}\dot{x}^b$ are the canonical momenta, and $\eta_{ab} = \mathrm{diag}(1, 1, 1, ..., -1, -1, -1)$ the metric tensor with signature (r, s). Upon quantization, the classical variables x^a, p_a are replaced by the operators, satisfying

$$[x^a, p_b] = i\delta_b^a \quad \text{or} \quad [x^a, p^b] = i\eta^{ab} \tag{27}$$

Because the momenta p_a are canonically conjugated to x^a, a natural choice of creation and annihilation operators is

$$c^a = \frac{1}{\sqrt{2}}\left(\sqrt{\omega}x^a + \frac{i}{\sqrt{\omega}}p_a\right) \tag{28}$$

$$c^{a\dagger} = \frac{1}{\sqrt{2}}\left(\sqrt{\omega}x^a - \frac{i}{\sqrt{\omega}}p_a\right) \tag{29}$$

the commutation relations being

$$[c^a, c^{b\dagger}] = \delta^{ab}\,, \qquad [c^a, c^b] = [c^{a\dagger}, c^{b\dagger}] = 0. \tag{30}$$

The Hamiltonian then becomes

$$H = \frac{1}{2}\,\omega\,(c_a^\dagger c^a + c^a c_a^\dagger). \tag{31}$$

Defining vacuum according to

$$c^a|0\rangle = 0, \tag{32}$$

and using $c^a c_a^\dagger = \eta_{ab} c^a c^{b\dagger} = \eta_{ab}(c^{b\dagger}c^a + \delta^{ab}) = c^{a\dagger}c_a + r - s$, the Hamiltonian (31) can be rewritten as

$$H = \omega\,(c_a^\dagger c^a + \frac{r}{2} - \frac{s}{2}). \tag{33}$$

Vacuum expectation of the latter Hamiltonian is zero if the signature is neutral, i.e., if $r = s$.

In eqs. (28),(29) we composed the creation and annihilation operators in terms of the *contravariant* components x^a and the *covariant* components p_a. But in the literature, the following operators are used:

$$a^a = \frac{1}{2}\left(\sqrt{\omega}\,x^a + \frac{i}{\sqrt{\omega}}\,p^a\right) \tag{34}$$

$$a^{a\dagger} = \frac{1}{2}\left(\sqrt{\omega}\,x^a - \frac{i}{\sqrt{\omega}}\,p^a\right) \tag{35}$$

They are composed in terms of x^a and $p^a = \eta^{ab}p_b$, and satisfy

$$[a^a, a_b^\dagger] = \delta^a{}_b\,, \qquad [a^a, a^{b\dagger}] = \eta^{ab}. \tag{36}$$

The Hamiltonian is

$$H = \frac{1}{2}\omega\,(a^{a\dagger}a_a + a_a a^{a\dagger}). \tag{37}$$

There are two possible definitions of the vacuum:

Possibility I

$$a^a|0\rangle = 0. \tag{38}$$

This is the usual definition. Rewriting the Hamiltonian in the form

$$H = \omega\,(a^{a\dagger}a_a + \frac{r}{2} + \frac{s}{2}), \tag{39}$$

we see that the vacuum energy, $\langle 0|H|0\rangle = (r+s)/2$, is positive. The eigenvalues of H are all positive. But there exist negative norm states, called ghosts.

Possibility II

$$a^{\bar{a}}|0\rangle = 0 \ , \qquad a^{\underline{a}\dagger}|0\rangle = 0, \tag{40}$$

where we have split the index a according to

$$\bar{a} = 1, 2, ..., r \ ; \qquad \underline{a} = r+1, r+2, ..., r+s. \tag{41}$$

Such definition of vacuum (and its consequences) has been used in refs. [3, 1, 4]. Writing the Hamiltonian (37) in the form

$$H = \omega \ (a^{\bar{a}\dagger} a_{\bar{a}} + a_{\underline{a}} a^{\underline{a}\dagger} + \frac{r}{2} - \frac{s}{2}), \tag{42}$$

we see that the energy of the vacuum (40) is $E = \omega(r-s)/2$.. If $r = s$, then the vacuum energy vanishes. The excitation states have positive or negative energies, $E = \omega(m - n + r/2 - s/2)$, depending on the mode (m, n) of excitation. There are no negative norm (ghost) states.

The vacuum definition (40) is equivalent to the definition (32), because the operators a^a, $a^{a\dagger}$ are just rewritten operators c^a and $c^{a\dagger}$ From the definition (28), (29) of the latter operators it follows that the vacuum defined according to (40) is covariant under the transformations of the group SO(r,s).

5. Non interacting quantum field theory

5.1. The scalar fields

In refs. [4] we considered the scalar field theory described by the action

$$I[\varphi^a] = \frac{1}{2} \int \mathrm{d}^4 x \, \sqrt{-g} (g^{\mu\nu} \partial_\mu \varphi^a \, \partial_\nu \varphi^b - m^2 \varphi^a \varphi^b) \gamma_{ab}, \tag{43}$$

where γ_{ab} is a metric in the space of fields $\varphi^a(x)$ at a fixed point $x \equiv x^\mu$ of spacetime. We will consider the case $\gamma_{ab} = \mathrm{diag}(1, 1, 1, ..., -1, -1, -1) = \gamma^{ab}$ and $g_{\mu\nu} = \mathrm{diag}(1, -1, -1, -1) = g^{\mu\nu}$.

The canonical momenta are

$$\pi_a = \frac{\partial L}{\partial \partial_0 \varphi^a} = \partial^0 \varphi_a = \partial_0 \varphi_a \equiv \dot{\varphi}_a, \tag{44}$$

and the Hamiltonian is

$$H = \frac{1}{2} \int \mathrm{d}^3 x \, (\dot{\varphi}^a \dot{\varphi}^b - \partial_i \varphi^a \partial^i \varphi^b + m^2 \varphi^a \varphi^b) \gamma_{ab}. \tag{45}$$

In the quantized theory, $\varphi^a(x)$ and $\pi_a(x)$ are operators satisfying

$$[\varphi^a(\mathbf{x}), \pi_b(\mathbf{x}')] = i\delta^3(\mathbf{x} - \mathbf{x}')\delta^a{}_b. \tag{46}$$

Expanding the field according to

$$\varphi^a = \int \frac{\mathrm{d}^3\mathbf{k}}{(2\pi)^3} \frac{1}{2\omega_{\mathbf{k}}} (a^a(\mathbf{k}) \mathrm{e}^{-ikx} + a^{a\dagger}(\mathbf{k}) \mathrm{e}^{ikx}), \tag{47}$$

where $\omega_{\mathbf{k}} = \sqrt{m^2 + \mathbf{k}^2}$, and

$$[a^a(\mathbf{k}), a_b^\dagger(\mathbf{k}')] = (2\pi)^3 \, 2\omega_{\mathbf{k}} \, \delta^3(\mathbf{k} - \mathbf{k}')\delta^a{}_b, \tag{48}$$

i.e.,

$$[a^a(\mathbf{k}), a^{b\dagger}(\mathbf{k}')] = (2\pi)^3\, 2\omega_{\mathbf{k}}\, \delta^3(\mathbf{k}-\mathbf{k}')\gamma^{ab}, \tag{49}$$

the Hamiltonian (45) becomes

$$H = \frac{1}{2}\int \frac{\mathrm{d}^3\mathbf{k}}{(2\pi)^3}\frac{\omega_{\mathbf{k}}}{2\omega_{\mathbf{k}}}(a^{a\dagger}(\mathbf{k})a^b(\mathbf{k}) + a^a(\mathbf{k})a^{b\dagger}(\mathbf{k}))\gamma_{ab}. \tag{50}$$

Analogously to eq. (40), we define the vacuum as

$$a^{\bar{a}}(\mathbf{k})|0\rangle = 0\,, \qquad a^{\underline{a}\dagger}(\mathbf{k})|0\rangle = 0, \tag{51}$$

where we split the index a into the part $\bar{a}$, runing over the positive, and the part $\underline{a}$, runing over the negative signature components.

Using (49) we can rewrite the Hamiltonian (50) into the form

$$H = \int \frac{\mathrm{d}^3\mathbf{k}}{(2\pi)^3}\frac{\omega_{\mathbf{k}}}{2\omega_{\mathbf{k}}}(a^{\bar{a}\dagger}(\mathbf{k})a_{\bar{a}}(\mathbf{k}) + a(\mathbf{k})a^{\dagger}) + \frac{1}{2}\int \mathrm{d}^3\mathbf{k}\,\omega_{\mathbf{k}}\delta^3(0)(r-s). \tag{52}$$

If the signature has equal number of plus and minus signs, i.e., if $r = s$, then the zero point energies cancel out from the Hamiltonian.

5.2. Generalization to Clifford space

5.2.1. The generalized Klein-Gordon equation In ref. [8, 15, 16, 17] it has been proposed to consider the Clifford algebra valued field in C-space:

$$\Phi(X) = \phi^A(X)\gamma_A, \tag{53}$$

where $\gamma_A \equiv (1/r!)\gamma_{a_1}\wedge\gamma_{a_2}\wedge...\wedge\gamma_{a_p}$, $r = 0,1,2,...,n$, is a basis of the Clifford algebra $Cl(1,n-1)$ of the spacetime $M_{1,n-1}$.

In general, the field depends on coordinates $X^M \equiv X^{\mu_1...\mu_r}$ of C-space. In this paper we will consider the case in which the field depends on spacetime coordinates only. The action is then

$$I = \frac{1}{2}\int \mathrm{d}^n x\,\sqrt{-g}(g^{\mu\nu}\partial_\mu\varphi^A\partial_\nu\varphi^B - m^2\varphi^A\varphi^B)G_{AB}, \tag{54}$$

where $\mu = 0,1,2,...,n-1$. The metric is defined as $G_{AB} = \gamma_A^{\ddagger} * \gamma_B \equiv \langle\gamma_A^{\ddagger}\gamma_B\rangle_0$, where $\langle\ \rangle_0$ means the scalar part of the expression, and the operation $\ddagger$ reverses the order of vectors. It turns out [5] that such metric has signature (R,S) with $R = S$, where $R + S = 2^n$ is the dimension of $Cl(1,n-1)$. Because the signature is neutral, the action (54) is just like the action (43). Following the same procedure as before, we obtain that the vacuum energy vanishes. Therefore, in such theory there is no cosmological constant problem. Recall that in Einstein's equations there is a term with the cosmological constant, $\lambda g_{\mu\nu}$, and the term with the stress-energy tensor, $8\pi G T_{\mu\nu}$. The vacuum of a quantum field gives $T_{\mu\nu} = \rho g_{\mu\nu}$, where ρ is the energy density. The two contributions sum to an effective cosmological constant $\Lambda = \lambda + 8\pi G\rho$. Here λ is a free parameter that can have in principle any values. Usual field theoretic calculations give infinite vacuum energy density ρ. When taking into account the cutoff at the Planck scale, it turns out that ρ is 120 orders of magnitude bigger than expected from observations. Therefore, the effective cosmological constant, Λ, is too big as well. This is the notorious cosmological constant problem. Here we have found a way, how ρ can vanish. The cosmological constant problem is thus resolved, because what remains is $\Lambda = \lambda$, and there is no longer the annoying huge contribution coming from quantum vacuum. How λ can be theoretically calculated remains, of course, a problem. But this is another sort of problem from the one, why the ρ is so big. The small observed cosmological constant could be a residual effect of something else, e.g., of a spacetime filling brane [18, 19].

5.2.2. Generalized Dirac equation (Dirac-Kähler equation) Another possibility is to assume that the Clifford algebra valued field (53) satisfies the Dirac-Kähler equation [20]

$$(i\gamma^\mu \partial_\mu - m)\Phi = 0. \tag{55}$$

Instead of expanding Φ in terms of $\gamma_A = (1, \gamma_{a_1}, \gamma_{a_1} \wedge \gamma_{a_2}, ..., \gamma_{a_1} \wedge ... \wedge \gamma_{a_n})$, we can expand it in terms of the spinor basis, $\xi_{\tilde{A}}$ of $Cl(1, n-1)$, considered in refs. [21, 17, 22]. Then

$$\Phi = \psi^{\tilde{A}} \xi_{\tilde{A}} \equiv \psi^{\alpha i} \xi_{\alpha i}, \tag{56}$$

where α is the spinor index of a left minimal ideal of $Cl(1, n-1)$, whereas i runs over $2^{n/2}$ left ideal of $Cl(1, n-1)$. From now on we will consider the case $n = 4$.

The γ^μ are abstract objects, the Clifford numbers, i.e., the elements of $Cl(1, 3)$. They can be represented as matrices according to

$$\langle (\xi^{\tilde{A}})^\ddagger \gamma^\mu \xi_{\tilde{B}} \rangle_S \equiv (\gamma^\mu)^{\tilde{A}}{}_{\tilde{B}}, \tag{57}$$

where the subscript S means the normalized scalar part of the expression (see [16]). Multiplying eq. (55) from the left by $(\xi^{\tilde{A}})^\ddagger$ and taking the scalar part, we obtain

$$\left(i\, (\gamma^\mu)^{\tilde{A}}{}_{\tilde{B}}\, \partial_\mu - m \delta^{\tilde{A}}{}_{\tilde{B}} \right) \psi^{\tilde{B}} = 0. \tag{58}$$

The 16×16 matrices, $(\gamma^\mu)^{\tilde{A}}{}_{\tilde{B}}$, can be factorized as $(\gamma^\mu)^{\tilde{A}}{}_{\tilde{B}} = (\gamma^\mu)^\alpha{}_\beta \delta^i{}_j$, and eq. (58) written in the form

$$\left(i\, (\gamma^\mu)^\alpha{}_\beta \partial_\mu - m \delta^\alpha{}_\beta \right) \psi^{\beta i} = 0, \tag{59}$$

or shortly,

$$(i\, \gamma^\mu \partial_\mu - m)\psi^i = 0. \tag{60}$$

In the last equation, the spinor index, α, has been omitted, and kept only the index $i = 1, 2, 3, 4$ denoting four left ideals. Bear in mind that the i as an index (superscript or subscript) has different a role than the i as a factor in front of γ^μ, in which case it is the imaginary unit, $i^2 = -1$.

The action is

$$I = \int \mathrm{d}^4 x\, \bar{\psi}^i (i\, \gamma^\mu \partial_\mu - m) \psi^j z_{ij}, \tag{61}$$

where z_{ij} is the metric in the space of ideals. It comes[2] from the metric $z_{\tilde{A}\tilde{B}}$ of the 16-dimensional spinor space of $Cl(1,3)$:

$$(\xi_{\tilde{A}})^\ddagger * \xi_{\tilde{B}} = z_{\tilde{A}\tilde{B}} = z_{(\alpha i)(\beta j)} = z_{\alpha\beta} z_{ij}, \tag{62}$$

where

$$z_{ij} = \begin{pmatrix} 1 & 0 & 0 & 0 \\ 0 & 1 & 0 & 0 \\ 0 & 0 & -1 & 0 \\ 0 & 0 & 0 & -1 \end{pmatrix}, \qquad z_{\alpha\beta} = \begin{pmatrix} 1 & 0 & 0 & 0 \\ 0 & 1 & 0 & 0 \\ 0 & 0 & -1 & 0 \\ 0 & 0 & 0 & -1 \end{pmatrix}. \tag{63}$$

The fields $\psi^i \equiv \psi^{\alpha i}(x)$ and their conjugate momenta[3] $\pi_{\alpha i}(x) = i\psi^{*\alpha}{}_i(x)$ satisfy the equal time anticommutation relations

$$\{\psi^{\alpha i}(t,\mathbf{x}), \pi_{\beta j}(t,\mathbf{x}')\}, = i\delta^\alpha{}_\beta \delta^i{}_j \delta^3(\mathbf{x}-\mathbf{x}') \tag{64}$$

$$\{\psi^{\alpha i}(t,\mathbf{x}), \psi^{\beta j}(t,\mathbf{x}')\} = 0\,, \qquad \{\pi^\dagger_{\alpha i}(t,\mathbf{x}), \pi^\dagger_{\beta j}(t,\mathbf{x}')\} = 0. \tag{65}$$

The Hamiltonian, belonging to the action (61) is

$$H = \int \mathrm{d}^3x\, \bar{\psi}^i(-i\gamma^r\partial_r + m)\psi^j z_{ij}. \tag{66}$$

Expanding the ψ^i in terms of the annihilation and creation operators, we obtain

$$H = \sum_{n=1}^{2} \frac{\mathrm{d}^3\mathbf{p}}{(2\pi)^3}\, m \left(b_s^{i\dagger}(\mathbf{p})\, b_s^j(\mathbf{p}) - d_s^i(\mathbf{p})\, d_s^{j\dagger}(\mathbf{p})\right) z_{ij}. \tag{67}$$

The index $i = 1,2,3,4$ distinguishes the spinors of different left ideals of $Cl(1,3)$. The index $s = 1,2$ is the usual one that distinguishes 'spin up' and 'spin down' states. We have the following anticommutation relations:

$$\{b_s{}^i(\mathbf{p}), b_{s'}{}^{j\dagger}(\mathbf{p}') = (2\pi)^3 \frac{E}{M}\,\delta^3(\mathbf{p}-\mathbf{p}')\delta_{ss'} z^{ij}, \tag{68}$$

$$\{d_s{}^i(\mathbf{p}), d_{s'}{}^{j\dagger}(\mathbf{p}')\} = (2\pi)^3 \frac{E}{M}\,\delta^3(\mathbf{p}-\mathbf{p}')\delta_{ss'} z^{ij}, \tag{69}$$

where $E = |\sqrt{\mathbf{p}^2 + m^2}$.

Let us now split the index according to $i = (\bar{i}, \underline{i})$, where $\bar{i} = 1,2$ and $\underline{i} = 3,4$, and define vacuum as follows:

$$\begin{array}{ll} b_s^{\bar{i}}|0\rangle = 0\,, & d_s^{\bar{i}}|0\rangle = 0 \\ b_s{}^{\underline{i}}|0\rangle, = 0\,, & d_s{}^{\underline{i}}|0\rangle = 0. \end{array} \tag{70}$$

The terms in the Hamiltonian (67) can be arranged so that we obtain

$$H = \sum_{s=1}^{2} \frac{\mathrm{d}^3p}{(2\pi)^3}\, m \left\{ \left[b_s{}^{\bar{i}\dagger}(\mathbf{p}) b_s{}^{\bar{j}}(\mathbf{p}) + d_s{}^{\bar{i}\dagger}(\mathbf{p}) d_s{}^{\bar{j}}(\mathbf{p}) \right] z_{\bar{i}\bar{j}} \right.$$

[2] For more details see ref. [16].

[3] They come from the action (61) written in the explicit form $I = \int \mathrm{d}^4x\, \psi^*_{\tilde{A}} \left((\gamma^\mu)^{\tilde{A}}{}_{\tilde{B}}\, \partial_\mu - m\right) \psi^{\tilde{B}}$, from which we obtain $\pi_{\tilde{C}} = \partial\mathcal{L}/\partial\dot{\psi}^{\tilde{c}} = i\psi^*_{\tilde{A}}(\gamma^0)^{\tilde{A}}{}_{\tilde{C}} = i\psi^*_{\alpha i}(\gamma^0)^\alpha{}_\delta \delta^i{}_j = i\psi^*_{\alpha j}(\gamma^0)^\alpha{}_\delta = i\psi^{*\beta i} z_{\alpha\beta} z_{ij} (\gamma^0)^\alpha{}_\delta$. Because $z_{\alpha\beta}(\gamma^0)^\alpha{}_\delta = \delta_{\alpha\beta}$ (see ref. [16]), we have $\pi_{\delta j} = i\psi^{*\delta}{}_j$.

$$+ \left[b_s{}^{\underline{i}\dagger}(\mathbf{p}) b_s{}^{\underline{j}}(\mathbf{p}) + d_s{}^{\underline{i}\dagger}(\mathbf{p}) d_s{}^{\bar{j}}(\mathbf{p}) \right] z_{\underline{i}\,\underline{j}} \Big\}$$

$$+ \sum_{s=1}^{2} \mathrm{d}^3\mathbf{p}\, E\delta(0)(-\delta^{\bar{i}\,\bar{j}} z_{\bar{i}\,\bar{j}} - \delta^{\underline{i}\,\underline{j}} z_{\underline{i}\,\underline{j}}) \tag{71}$$

The term with $z_{\bar{i}\bar{j}}$ gives positive. whereas the term with $z_{\underline{i}\underline{j}}$ gives negative expectation values. In the term with $\delta(0)$, the positive and negative contributions cancel out, because $z_{\bar{i}\bar{j}} = \delta_{\bar{i}\bar{j}}$, and $z_{\underline{i}\underline{j}} = -\delta_{\underline{i}\underline{j}}$. Therefore, the vacuum expectation value of this Hamiltonian is zero.

Each fermion ψ^i couples to the corresponding gauge field. The Casimir force between two metallic plates, consisting of ψ^i, $i = 1$, is not expected to vanish in this theory [4]. The vacuum expectation value $\langle H \rangle = \langle T^{00} \rangle$ is the source of the gravitational field. Because $\langle 0|H|0 \rangle = 0$, the cosmological constant vanishes. There is no problem of the huge cosmological constant.

6. Presence of interactions

6.1. Classical oscillator

We will now add an interaction term to the Lagrangian for an oscillator in the pseudo Euclidean space $M_{1,1}$:

$$L = \frac{1}{2}(\dot{x}^2 - \dot{y}^2) - V\,, \qquad V = \frac{\omega}{2}(x^2 - y^2) + V_1 \tag{72}$$

Equations of motions are

$$\ddot{x} + \omega^2 x + \frac{\partial V_1}{\partial x} = 0, \tag{73}$$

$$\ddot{y} + \omega^2 y - \frac{\partial V_1}{\partial y} = 0, \tag{74}$$

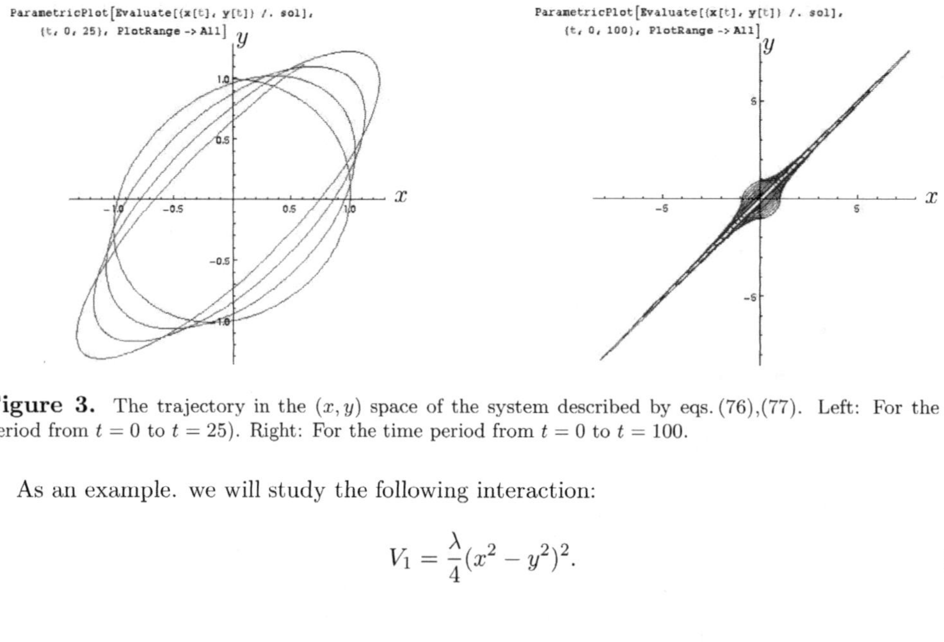

Figure 3. The trajectory in the (x, y) space of the system described by eqs. (76),(77). Left: For the time period from $t = 0$ to $t = 25$). Right: For the time period from $t = 0$ to $t = 100$.

As an example. we will study the following interaction:

$$V_1 = \frac{\lambda}{4}(x^2 - y^2)^2. \tag{75}$$

So we have

$$\ddot{x} + \omega^2 x + \lambda\, x(x^2 - y^2) = 0, \tag{76}$$

$$\ddot{y} + \omega^2 y + \lambda\, y(x^2 - y^2) = 0. \tag{77}$$

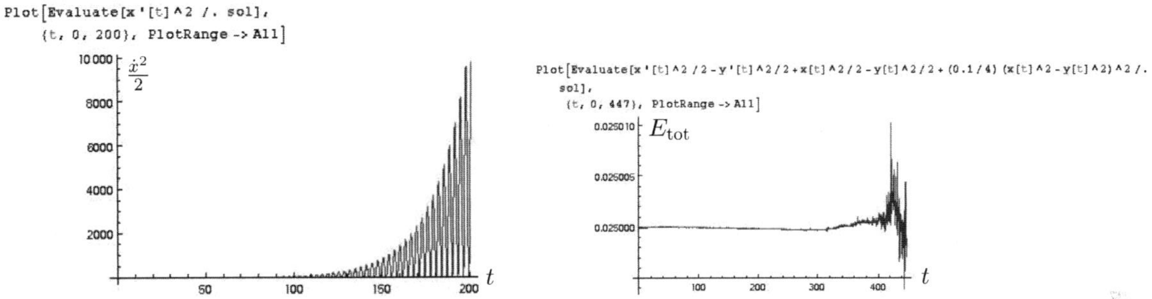

Figure 4. Left: The kinetic energy $\dot{x}^2/2$ as function of time. The envelop of oscillations grows to infinity. Right: The total energy, $E_{\rm tor}$, remains constant within the numerical error.

These equations can be solved numerically by the program Mathematica. In Figs. 3,4 we show the result for $\omega = 1$, $\lambda = 0.1$ and the initial conditions $\dot{x}(0) = 1$, $\dot{y}(0) = 0$, $x(0) = 0$, $y(0) = 1$. We see that the system is unstable in the sense that the amplitudes of $x(t)$, $\dot{x}(t)$ grow to infinity. The total energy, $E_{\rm tot} = (\dot{x}^2 - \dot{y}^2)/2 + V(x, y)$, remains constant (Fig.4).

In Fig. 5 we have the solution for the initial conditions $\dot{x}(0) = 1$, $\dot{y}(0) = -1.2$, $x(0) = 0$, $y(0) = 0.5$. Here also the system is unstable.

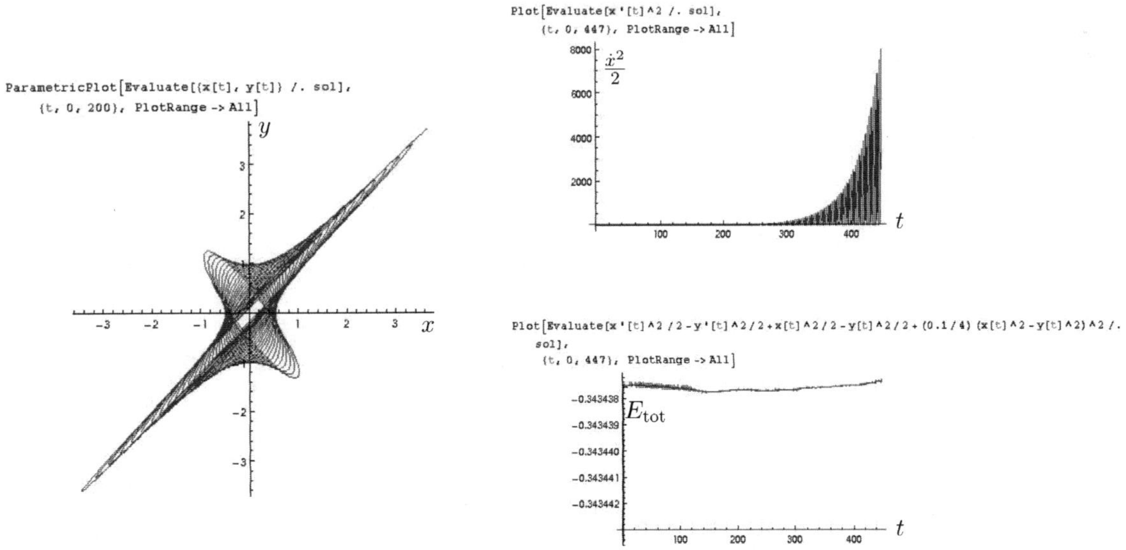

Figure 5. The solution of the system described by eqs. (76),(77), for the initial conditions $\dot{x}(0) = 1$, $\dot{y}(0) = -1.2$, $x(0) = 0$, $y(0) = 0.5$.

Something fascinating happens, if instead of eqs. (76),(77), we take slightly different equations,

$$\ddot{x} + \mu[\omega^2 x + \lambda\, x(x^2 - y^2)] = 0, \tag{78}$$

$$\ddot{y} + \nu[\omega^2 y + \lambda\, y(x^2 - y^2)] = 0. \tag{79}$$

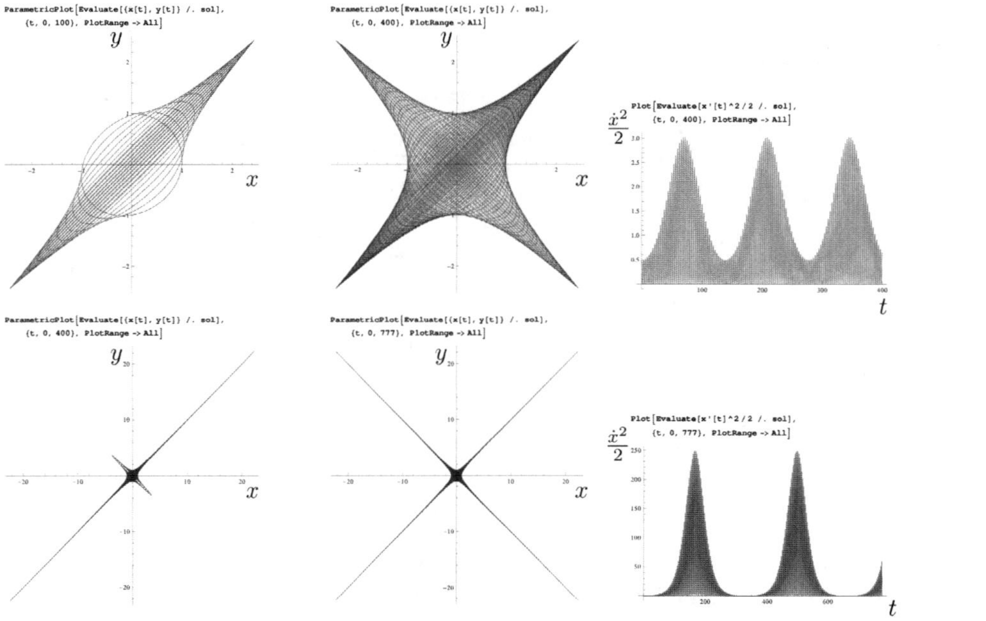

Figure 6. Up left and middle: The (x, y) plot of the solution to eqs. (78)(79) for constants $\mu = 1.01$ and $\nu = 1$. Up right: The kinetic energy $\dot{x}^2/2$ as function of time. Low left and middle: The (x, y) plot of the solution to eqs. (78)(79) for constants $\mu = 1.0001$ and $\nu = 1$. Low right: The kinetic energy $\dot{x}^2/2$ as function of time.

The solution for $\mu = 1.01$, $\nu = 1$, and the same initial condition as in Figs. 3,4 is shown in Fig. 6. We see that the trajectory in the (x, y) space does not escape into infinity. Instead, a second arm is formed, and the trajectory remains confined within a star-like envelope. The kinetic energy of one component, $\dot{x}^2/2$, now does not grow to infinity. Instead, the amplitude of the kinetic energy oscillations is modulated by a slowly oscillating envelope.

In Fig. 6 we also repeat the calculation for $\mu = 1.0001$ and $\nu = 1$. This is now very close to $\mu = 1$, $\nu = 1$, the case of eqs. (76),(77). Now the trajectory in the (x, y) space goes much farther from the origin than in the case $\mu = 1.01$. But again it does not go to infinity; instead, after some time, the trajectory start to move within a second arm. The envelope of the kinetic energy oscillations now consists of separated peaks. If μ approaches 1, the height of the peaks becomes higher and higher, and their separation increases. In the limit $\mu \to 1$, the height of the first peak is infinite, and there is no second or other peaks (because their positions recedes to infinity).

We have found a very interesting result that $\mu = 1$, $\nu = 1$ is a singular case, in which the system is unstable. But if μ slightly differs from ν, then the system is stable; its trajectory

Journal of Physics: Conference Series **437** (2013) 012006 doi:10.1088/1742-6596/437/1/012006

remains confined within a finite region of the (x, y) space, and the kinetic energy remains finite as well.

We will now show that the stable system (78),(79) is a special case of a more general system, described by the action

$$I = \frac{1}{2} \int \mathrm{d}\tau \, g_{\mu\nu} \, \dot{X}^\mu \dot{X}^\nu \, , \qquad \mu, \nu = 0, 1, 2, ..., D-1. \tag{80}$$

This is the Stueckelberg action [23] for a point particle in a D-dimensional curved space with a metric $g_{\mu\nu}$ [8] (see also [24]). Introducing

$$\gamma_{ab} = g_{ab} - \frac{g_{0a} g_{0b}}{g_{00}}, \tag{81}$$

we can split[4] the quadratic form according to

$$g_{\mu\nu} \dot{X}^\mu \dot{X}^\nu = \gamma_{ab} \dot{X}^a \dot{X}^b + \frac{\dot{X}_0^2}{g_{00}}. \tag{82}$$

Here $a, b = 1, 2, ..., D-1$. In general, the signature is (r, s), with $r + s = D - 1$. We wil take $r = s$. Using (81),(82), the action (80) then becomes

$$I = \frac{1}{2} \int \mathrm{d}\tau \left(\gamma_{ab} \dot{X}^a \dot{X}^b + \frac{\dot{X}_0^2}{g_{00}} \right). \tag{83}$$

If $g_{\mu\nu} = 0$, then $\dot{X}_0$ is a constant of motion[5]. We will assume that this is the case.

Assuming that $\gamma_{ab,c} \equiv (\partial/\partial X^c)\gamma_{ab} = 0$, the equations of motion derived from (83) are

$$\ddot{X}^a + \frac{1}{2} \frac{C}{g_{00}^2} \, g_{00,b} \, \gamma^{ab} = 0, \tag{84}$$

where $C = \dot{X}_0^2$ is a constant.

Introducing

$$V = -\frac{1}{2} \frac{\dot{X}_0^2}{g_{00}} = -\frac{1}{2} \frac{C}{g_{00}}, \tag{85}$$

we have

$$\ddot{X}^a + \, V_{,b} \, \gamma^{ab} = 0 \tag{86}$$

The latter equations correspond to eqs. (78),(79), if $\gamma^{ab} = \mathrm{diag}(\gamma^{11}, \gamma^{22}, ..., \gamma^{D-1,D-1})$, i.e., a diagonal metric. Then eqs. (86) become

$$\begin{aligned} \ddot{X}^1 + \, V_{,1} \, \gamma^{11} &= 0, \\ \ddot{X}^2 + \, V_{,2} \, \gamma^{22} &= 0, \\ &\vdots \end{aligned} \tag{87}$$

[4] An analogous splitting is used in Kaluza-Klein theories.

[5] Analogously, in Kaluza-Klein theories, $\dot{X}_5$ is a constant of motion, if the $5D$ metric does not depend on the fifth dimension.

If we consider the case $D-1=2$, identify $X^1 \equiv x$, $X^2 \equiv y$, $\gamma^{11} \equiv \mu$, $\gamma^{22} \equiv \nu$, and take $V = \frac{\omega^2}{2}(x^2-y^2) + \frac{\lambda}{4}(x^2-y^2)^2$, then the system (87) becomes the system (78),(79).

The $g_{\mu\nu}(x)$ in the action (80) need not be a fixed background metric field. It can be a dynamical field. Then one has to include into the action (80) a kinetic term for $g_{\mu\nu}$, e.g., the Einstein-Hilbert term with the curvature scalar. The dynamics of such a system is involved, and, in view of the results given in Fig. 6, one cannot a priori claim that the system is unstable.

6.2. Collision of the oscillator with a free particle

Let us assume that in the surroundings of the oscillator, O, described by the Lagrangian (72), there is free particle, P. Depending on the initial conditions, it may happen that the oscillator hits the particle. Such combined system of O and P can be modeled by the Lagrangian

$$L = \frac{1}{2}(\dot{x}^2-\dot{y}^2) - \frac{1}{2}(x^2-y^2) - \frac{\lambda}{2}(x^2-y^2)^2 + \frac{1}{2}(\dot{u}^2+\dot{v}^2) - \frac{\alpha/5}{[(u-x)^2+(v-y)^2+a]^5}, \tag{88}$$

where u, v are particle's coordinates. The equations of motions are

$$\begin{aligned} &\ddot{x} + x + \lambda x(x^2-y^2) + \frac{\alpha(u-x)}{[(u-x)^2+(v-y)^2+a]^{5/2}} = 0 \\ &\ddot{y} + y + \lambda y(x^2-y^2) + \frac{\alpha(v-y)}{[(u-x)^2+(v-y)^2+a]^{5/2}} = 0 \\ &\ddot{u} - \frac{\alpha(u-x)}{(u-x)^2+(v-y)^2+a} = 0 \\ &\ddot{v} - \frac{\alpha(v-y)}{[(u-x)^2+(v-y)^2+a]^{5/2}} = 0 \end{aligned} \tag{89}$$

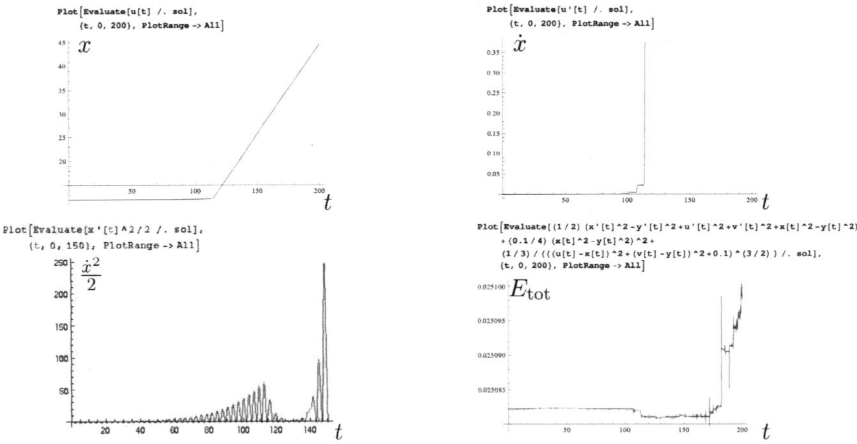

Figure 7. Collision of the oscillator with a particle. Up: Particle's position and velocity as functions of time. Low left: Oscillator's kinetic energy $\dot{x}^2/2$ as function of time. Low left: The total energy of the the oscillator and the particle as function of time.

The numerical solution for the constants $\Lambda = 0.1$, $\alpha = 1$, and the initial conditions $\dot{x}(0) = 1$, $\dot{y}(0) = 0$, $\dot{u}(0) = 0$, $\dot{v}(0) = 0$, $x(0) = 0$, $y(0) = 1$, $u(0) = 12$, $v(0) = 11.5$ is shown in Fig. 7. We see that the solution properly reproduces the fact that the particle P

is initially at rest, and after the interaction with the oscillator O, its moves with a constant velocity.

The positive component of the oscillator's kinetic energy starts to increase, but at time around $t = 112$ it drops down to zero. After a while, the positive kinetic energy "recovers" and start to increase again. A further collision with some other particle would again drop down $\dot{x}^2/2$. Analogous holds for $\dot{y}^2/2$. We see that, according to this numerical solution, the surrounding particles stabilize the oscillator and prevent it to escape into the infinity. We expect that many such oscillators, immersed into a bath of particles would increase their average $\dot{u}^2/2$ and $\dot{v}^2/2$, i.e., the temperature of the bath. Further investigation of this interesting topics is beyond the scope of the present paper.

6.3. Quantum oscillator

The quantum oscillator is described by the wave function $\psi(t,x,y)$, satisfying the Schrödinger equation

$$i\frac{\partial\psi}{\partial t} = H\psi, \tag{90}$$

with

$$H = \frac{1}{2}\left(-\frac{\partial}{\partial x^2} + \frac{\partial}{\partial y^2}\right) + V(x,y). \tag{91}$$

Let us expand the wave function in terms of the basis functions of the $2D$ harmonic pseudo Euclidean oscillator, which are the same as those of the Euclidean oscillator:

$$\psi = \sum_{m,n=0}^{\infty} c_{mn}(t)\,\psi_{mn}. \tag{92}$$

Here $\psi_{mn} = \frac{1}{\pi^{1/4}\sqrt{2^n n!}}\frac{1}{\pi^{1/4}\sqrt{2^m m!}}H_n(x)H_m(y)\mathrm{e}^{-(x^2+y^2)/2}$ are orthonormal eigenfunctions (with positive or negative energies), of the Hamiltonian

$$H_0 = \frac{1}{2}\left(-\frac{\partial}{\partial x^2} + \frac{\partial}{\partial y^2}\right) + \frac{1}{2}(x^2 - y^2). \tag{93}$$

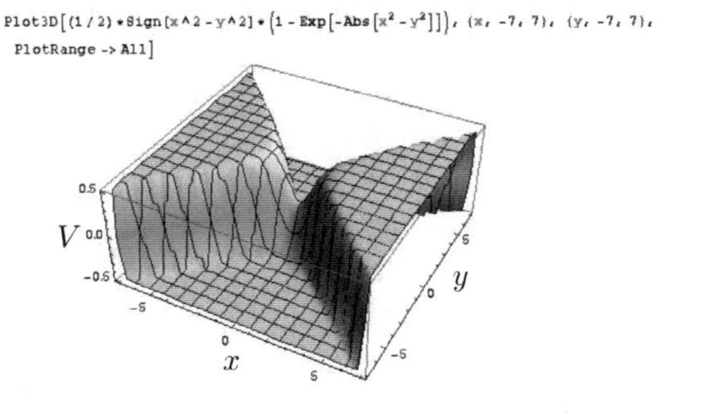

Figure 8. The potential used in the calculations of the quantum pseudo-Euclidean harmonic oscillator described by eqs. (90),(91).

Using the matrix elements

$$H_{mn;rs} = \int \mathrm{d}x\,\mathrm{d}y\,\psi^*_{mn} H \psi_{rs}, \tag{94}$$

the Schrödinger equation (90) can be rewritten as

$$i\dot{c}_{mn} = \sum_{rs} H_{mn;rs}\, c_{rs}. \tag{95}$$

We will investigate the case of the potential (see Fig. 8)

$$V(x,y) = \frac{1}{2}\varepsilon\left(1 - e^{-\varepsilon(x^2-y^2)}\right), \tag{96}$$

where $\epsilon = \text{sign}\,(x^2 - y^2)$, i.e., $\epsilon = 1$, if $x^2 - y^2 > 0$, $\epsilon = -1$, if $x^2 - y^2 < 0$, and $\epsilon = 0$, if $x^2 - y^2 = 0$. Such a potential is a 2D model of a potential that could eventually occur in a three or higher dimensional world, where potentials running into infinity are unrealistic.

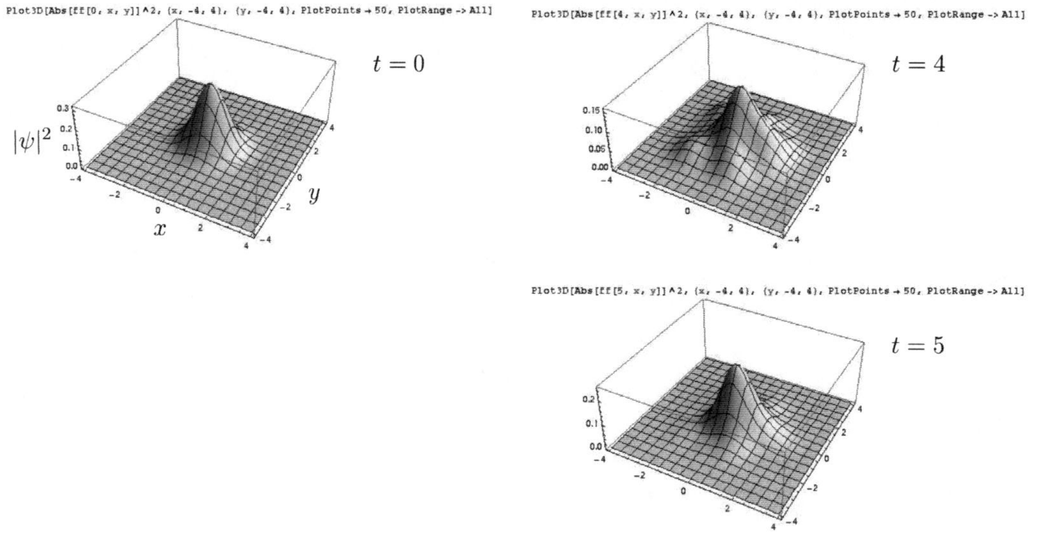

Figure 9. The plot of $|\psi(t,x,y)|^2$, calculated for the initial conditions $c_{00}(0) = 1$, $c_{01}(0) = C_{10}(0) = c_{11}(0) = ... = 0$, at different values of the time t.

Instead of the system (95) of infinite number of first order differential equations, let us consider a finite system with $m, n = 0, 1, 2, ..., N$. Then we can numerically solve the system and calculate the coefficients $C_{mn}(t)$ for given initial conditions. So we obtain the wave functions

$$\psi(t,x,y) = \sum_{m,n=0}^{N} c_{mn}(t)\,\psi_{mn}, \tag{97}$$

and its absolute square, $|\psi(t,x,y)|^2$.

In Fig. 9 we show the result for $N = 4$, and the initial condition $c_{00} = 1$, with the remaining coefficients being equal to zero. We see that with increasing time t, the vacuum $\psi(0) = \psi_{00} =$ gradually decays, and, after a while (at $t \approx 5$) occurs again. The wave function thus oscillates between the vacuum and a decayed vacuum.

In Fig. 10 we show the calculation of $|\psi(t,x,y)|^2$ for the initial conditions $c_{01}(0) = 1/\sqrt{2}$, $c_{10}(0) = 1/\sqrt{2}$, $c_{00}(0) = 0$, $c_{11}(0) = 0$, $c_{12}(0) = 0, ..., c_{44}(0) = 0$. Initially, we have two

Journal of Physics: Conference Series **437** (2013) 012006 doi:10.1088/1742-6596/437/1/012006

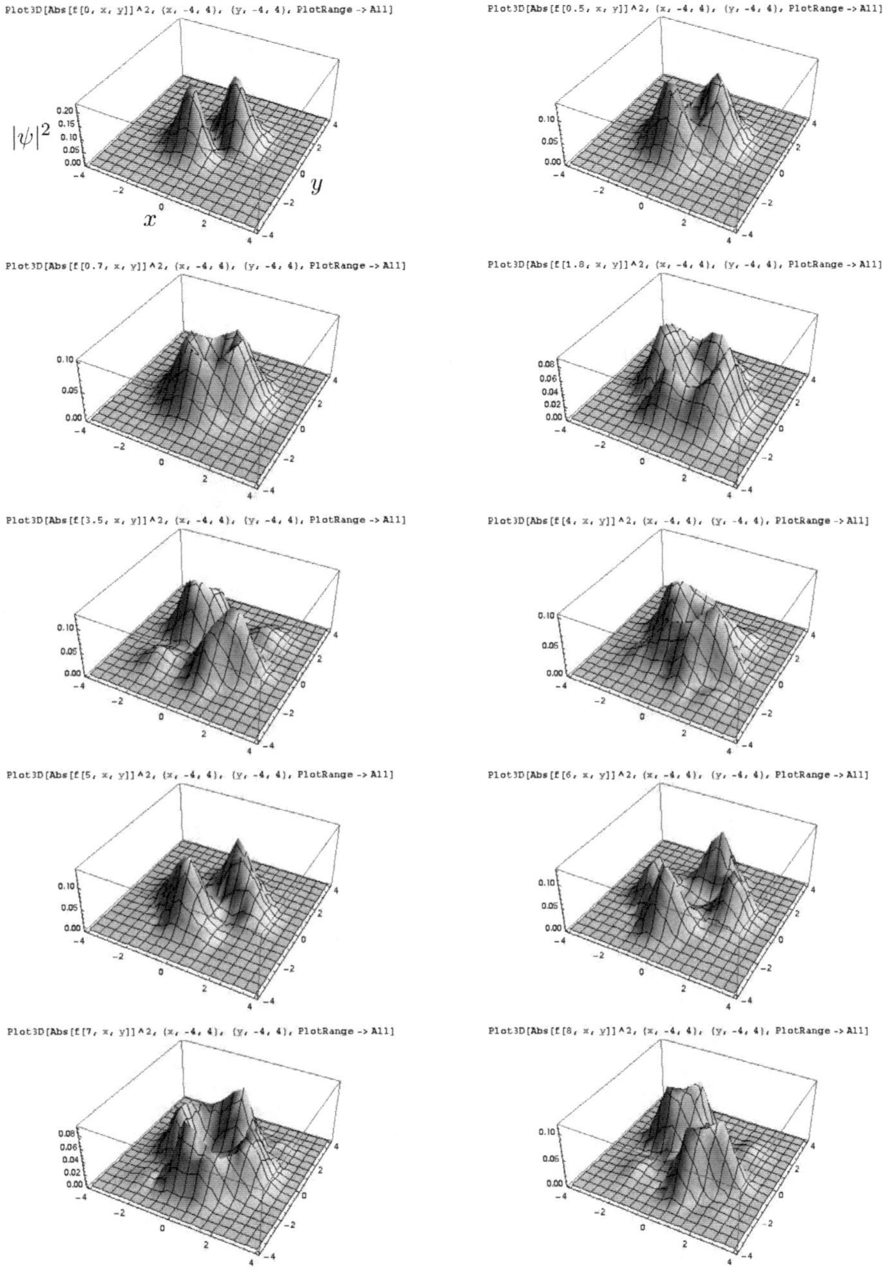

Figure 10. The plot of $|\psi(t,x,y)|^2$, calculated for the initial conditions $c_{01}(0) = C_{10}(0) = 1/\sqrt{2}$, $c_{00}(0) = c_{11}(0) = c_{12}(0) = ... = 0$, at different values of the time t.

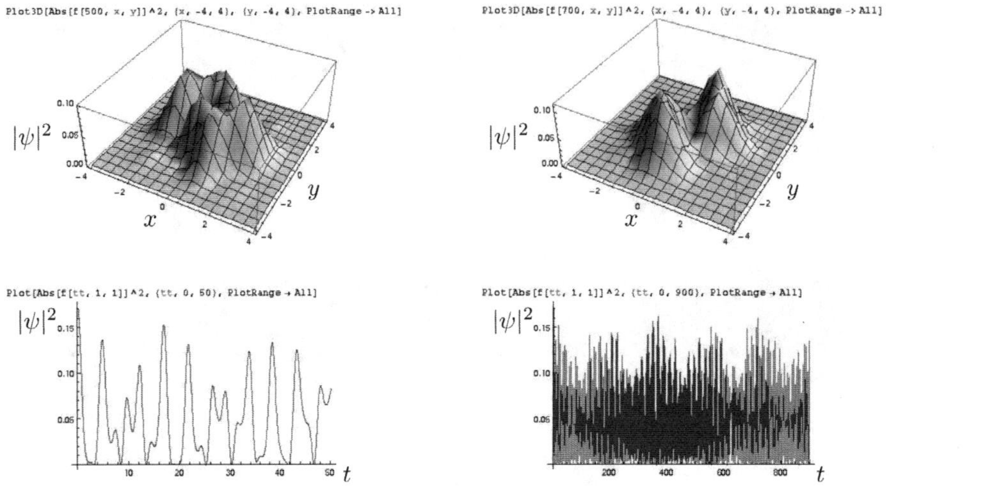

Figure 11. Up: The plot of $|\psi(t,x,y)|^2$, calculated for the initial conditions $c_{01}(0) = C_{10}(0) = 1/\sqrt{2}$, $c_{00}(0) = c_{11}(0) = c_{12}(0) = ... = 0$, at two very large values the time, $t = 500$ and $t = 700$. Low: The plot of $|\psi(t, x = 0, y = 50)|^2$, as a function of t, calculated for the same initial conditions as above.

peaks, one for the positive energy excitation, $c_{10}(0)$, and the other one for the negative energy excitations, $c_{01}(0)$. The system then oscillates as shown in Fig. 10 and Fig. 11.

Such a wave function (98), of course, is not a solution of the Schrödinger equation (90) for H given in eqs. (91), (96). Since $c_{mn}(t)$, $m, n = 0, 1, 2, 3, 4$, are solution of the truncated system (95), also $\psi(t, x, y)$ of eq. (97) is a solution of a truncated system, presumably of the Schrödinge equation on the lattice. A realistic lattice is a discrete set of closely separated points corresponding to the system, not of the 5×5 equations (95), but of the $N \times N$ equations, with N very large. The N is related to the maximum absolute energy of the oscillator excitations that, in turn, is related to the minimum distance (cutoff). Regardles of how large is N, and how small is the cutoff distance, the examples of Figs. 9–11 indicate that there are oscillations, and that the system is thus stable. Only in the limit of infinitely small cutoff distance, implying infinite quantum numbers m, n of poisitive and negative energy excitations, would the system be unstable. Then the peak in Fig. 9 would spread to infinity. However, according to common belief, the Planck length is the cutoff distance under which it is no longer possible to probe spacetime distances. Our spacetime is then like a huge lattice. Then also the corresponding Clifford space, C, is like a lattice. The space $M_{1,1}$ is a subspace of C. Instead of the infinite system of differential equations (95), we then have a finite, though very big, system of differential equations.

6.4. Interacting quantum fields

6.4.1. Scalar fields As an example let us consider the scalar fields described by the action

$$I = \frac{1}{2} \int \mathrm{d}^4 x [g^{\mu\nu} \partial_\mu \varphi^a \partial_\nu \varphi^b \gamma_{ab} + V(\varphi)], \tag{98}$$

which is similar to the action (43), except that now $V(\varphi)$ is a more general potential with an interaction term, e.g., $V(\varphi) = m^2\varphi^a\varphi^b\gamma_{ab} + \varphi^a\varphi^b\varphi^c\varphi^d\lambda_{abcd}$.

Upon quantization, the system is described by a state vector,

$$|\Psi\rangle = \sum |P\rangle \langle P|\Psi\rangle, \tag{99}$$

that can be expanded in terms of the Fock space basis vectors $|P\rangle \equiv |p_1 p_2 ... p_n\rangle$ that are eigenvectors of the free field Hamiltonian H_0 (without the interaction quartic term $\varphi^a\varphi^b\varphi^c\varphi^d\lambda_{abcd}$). Also the vacuum, $|0\rangle$, is an eigenstate H_0. If the metric γ_{ab} has signature (r,s), the states $|P\rangle$ can have positive or negative energies.

The time evolution of the state vectors is governed by the Hamilton operator H, corresponding to the field action (98):

$$|\Psi(t)\rangle = e^{-iH(t-t_0)}|\Psi(t_0)\rangle. \tag{100}$$

From (100) we have

$$\langle P|\Psi(t)\rangle = \sum_{P'} \langle P|e^{-iH(t-t_0)}|P'\rangle \langle P'|\Psi(t_0)\rangle. \tag{101}$$

If the initial state is the vacuum, $|\psi(t_0)\rangle = |0\rangle$, an eigenstate of the unperturbed Hamiltonian, H_0, then we have

$$\langle P|\Psi(t)\rangle = \langle P|e^{-iH(t-t_0)}|0\rangle, \tag{102}$$

Such transition is possible, because a state $\langle P|$ contains particles with positive and negative energies. Therefore, the total energy in such transition is conserved and remains equal to the energy of the initial state $|0\rangle$, which is zero.

The vacuum thus decays into a superposition of many particle states:

$$|\Psi(t)\rangle = \sum_{n=0}^{\infty} |p_1 p_2 ... p_n\rangle \langle p_1 p_2 ... p_n|\Psi(t)\rangle. \tag{103}$$

Here $\langle p_1 p_2 ... p_n|\Psi(t)\rangle$ is the amplitude that we will measure the multiparticle state $|p_1 p_2 ... p_n\rangle$. The probabilities that the vacuum decays into any of the states $|p_1\rangle$, $|p_1 p_2\rangle$, $|p_1 p_2 ... p_n\rangle$, ..., are not drastically different from each other, and they sum to 1:

$$\sum_{p_1} |\langle p_1|\Psi\rangle|^2 + \sum_{p_1,p_2} |\langle p_1,p_2|\Psi\rangle|^2 + \sum_{p_1,p_2,...,p_n} |\langle p_1,p_2,...,p_n|\Psi\rangle|^2 + ... = 1. \tag{104}$$

We see that the probability of vacuum decay into 2,4,6,8, or any finite number of particles, is negligible in comparison to the probability of the decay into infinite number of particles, because such configuration spaces occupy the vast majority of the terms in the sum (104). According to such reasoning, the vacuum instantly decays into infinitely many particles. It is then argued that because such instantaneous vacuum decay makes no sense in physics, ultrahyperbolic spaces are unphysical. In my opinion such conclusion is not unavoidable, for the reasons described below and in the remaining parts of the paper.

Let us consider a generalization of the field action (98). We can rewrite it in a more compact notation:

$$I = \frac{1}{2}\partial_\mu\varphi^{a(x)}\partial_\nu\varphi^{b(x')}\gamma^{\mu\nu}{}_{a(x)b(x')} - U[\varphi] \tag{105}$$

Here $\varphi^{a(x)} \equiv \varphi^a(x)$, where (x) is the continuous index, denoting components of an infinite dimensional vector. In addition, for every (x), the components are also denoted by a discrete index a. Altogether, vector components are denoted by the double index $a(x)$. Alternative notation, often used in the litarature, is $\varphi^a(x) \equiv \varphi^{ax}$ or $\varphi^a(x) \equiv \varphi^{(ax)}$.

The action (105) may be obtained from a higher dimensional action

$$I_\phi = \frac{1}{2}\partial_\mu \phi^{A(x)} \partial_\nu \phi^{B(x')} \, G^{\mu\nu}{}_{A(x)B(x')}, \tag{106}$$

where $A = (a, \bar{A})$, and $\phi^{A(x)} = (\phi^{a(x)}, \phi^{\bar{A}(x)})$. The higher dimensional metric is a functional of $\phi^{A(x)}$. Performing the Kaluza-Klein split,

$$G^{\mu\nu}{}_{AB} = \begin{pmatrix} \gamma^{\mu\nu}{}_{ab} + A_a{}^{\bar{A}} A_b{}^{\bar{B}} \bar{G}^{\mu\nu}_{\bar{A}\bar{B}}, & A_a{}^{\bar{B}} \bar{G}^{\mu\nu}_{\bar{A}\bar{B}} \\ A_b{}^{\bar{B}} \bar{G}^{\mu\nu}_{\bar{A}\bar{B}}, & \bar{G}^{\mu\nu}_{\bar{A}\bar{B}} \end{pmatrix}, \tag{107}$$

where for simplicity we have omitted the index (x), we find,

$$I_\phi = \frac{1}{2}\partial_\mu \phi^{a(x)} \partial_\nu \phi^{b(x')} \gamma^{\mu\nu}{}_{a(x)b(x')} + \frac{1}{2}\partial_\mu \phi_{\bar{A}(x)} \, \partial_\nu \phi_{\bar{B}(x)} \, \bar{G}^{\mu\nu \bar{A}(x)\bar{B}(x)}. \tag{108}$$

Identifying $\phi^{a(x)} \equiv \varphi^{a(x)}$, and denoting $\frac{1}{2}\partial_\mu \phi_{\bar{A}(x)} \, \partial_\nu \phi_{\bar{B}(x)} \, \bar{G}^{\mu\nu \bar{A}(x)\bar{B}(x)} = -U[\varphi]$, we obtain the action (105).

Thus, the field action (98) is embedded in a higher dimensional action with a metric $G^{\mu\nu}{}_{A(x)B(x)}$ in field space. A question arises as to which field space metric to choose. The lesson from general relativity tells us that the metric itself is dynamical. Let us therefore assume that this is so in the case of field theory[8] as well. Then (106) must be completed by a kinetic term, I_G, for the field space metric [8]. The total action is then

$$I[\phi, G] = I_\phi + I_G. \tag{109}$$

According to such dynamical principle, not only the field ϕ, but also the metric $G^{\mu\nu}{}_{A(x)B(x)}$ changes with the evolution of the system. This implies that also the potential $U[\varphi]$ of eq. (105) changes with evolution, and so does the potential $V(\varphi)$, occurring in eq. (98).

Let us assume that the action (109) describes the whole universe[6]. Then $|\psi(t)\rangle$ of eq. (103) contains everything in such universe, including observers. There is no external observer, $\mathcal{O}_{\text{ext}}$, according to whom the coefficients $\langle p_1...p_n|\psi(t)\rangle$ in eq. (103) could be related to the probability densities $|\langle p_1...p_n|\psi(t)\rangle|^2$ of finding the system in an n-particle states with momenta p_1, ..., p_n, $n = 0, 1, 2, 3, ..., \infty$. There are only inside observers, $\mathcal{O}$, incorporated within appropriate multiparticles states $|p_1...p_n\rangle$, $n = 0, 1, 2, 3, ..., \infty$, of the "universal" state $|\psi(t)\rangle$. According to the Everett interpretation of quantum mechanics, all states, $|0\rangle$, $|p_1 p_2\rangle$,..., $|p_1...p_n\rangle$, ..., $n = 0, 1, 2, ..., \infty$, in the superposition (103) actually occur, each in a different world. The Everett interpretation is now getting increasing support among cosmologists (see, e.g., Ref. [25]). For such an inside observer, $\mathcal{O}$, there is no instantaneous vacuum decay into infinitely many

[6] Of course, such a model universe is not realistic, because our universe contains fermions and accompanying gauge fields as well.

particles. For $\mathcal{O}$, at a given time t, there exists a configuration $|P\rangle$ of n-particles (that includes $\mathcal{O}$ himself), and a certain field potential $V(\varphi)$ (coming from $G^{\mu\nu}{}_{A(x)B(x)}$). At some later time, $t+\Delta t$, there exists a slightly different configuration $|P'\rangle$ and potential $V'(\varphi)$, etc. Because $\mathcal{O}$ nearly continuously measures the state $|\psi(t)\rangle$ of his universe, the evolution of the system is being "altered", due to the notorious "watchdog effect" or "Zeno effect" [26] of quantum mechanics (besides being altered by the evolution of the potential $V(\varphi)$). The peculiar behavior of a quantum system between two measurements has also been investigated in Refs. [27].

6.4.2. The generalized Dirac field In Sec. 4.2.2 we considered the generalized Dirac field described by the Dirac-Kähler equation (55). In the spinor basis of $Cl(1,3)$, the field can be represented by a 4×4 matrix

$$\psi = \begin{pmatrix} \psi^{11} & \psi^{12} & \psi^{13} & \psi^{14} \\ \psi^{21} & \psi^{22} & \psi^{23} & \psi^{24} \\ \psi^{31} & \psi^{32} & \psi^{33} & \psi^{34} \\ \psi^{41} & \psi^{42} & \psi^{43} & \psi^{44} \end{pmatrix}, \tag{110}$$

where each column represents a Dirac spinor[7]. The components $\psi^{\alpha i}$ have either positive or negative energies, according to the following scheme:

$$\text{Energy} = \begin{pmatrix} + & + & - & - \\ + & + & - & - \\ - & - & + & + \\ - & - & + & + \end{pmatrix}, \tag{111}$$

which is a consequence of the metric (62),(63).

If we add an interaction term to the action (61), then transitions between positive and negative energy states become possible. On the other hand, positive and energy states of the usual Dirac spinors do not mix in our Universe. Even if once they did mix, the evolution of the Universe must have led to the current situation with no mixing. This was not so clear when Dirac proposed his theory. The occurrence of negative energy states was puzzling at that time, and Fermi [28] wrote:

"It is well known that the most serious difficulty in Dirac's relativistic wave equations lies in the fact that it yields besides the normal positive states also negative ones, which have no physical significance. This would do no harm if no transition between positive and negative state were possible (as are, e.g., transitions between states with symmetrical and antisymmetrical wave function). But this is unfortunately not the case: Klein has shown by a very simple example that eleectrons impinging against a very high potential barrier have a finite probability of going over in a negative state."

Thus, Dirac's relativistic wave equation could have been put aside and ignored. Fortunately, this did not happen. The problem was resolved by the Dirac sea of negative enery states.

Now the situation is analogous as in 1932. Within the Clifford algebra framework, besides the negative energy states of the first left ideal of $Cl(1,3)$, we have also negative energy states of the second, third and forth ideal. The existence of such states also should not be considered a priori as problematic. A deeper investigation is necessary before making any definite conclusion.

[7] A column may represent the Weyl and Majorana spinors as well

6.4.3. Clifford algebra description of fermionic fields It is known that spinors can be represented as Fock space like objects [21, 17, 22] embedded in a Clifford algebra. An analogous procedure can be carried out for spinor fields [8, 22]. A vector Ψ in an infinite dimensional space, $\mathcal{S}$, can be represented as [8, 22]

$$\Psi = \psi^{r(x)} h_{r(x)} \,, \quad r = 1,2; \quad x \in \mathbb{R}^3 \text{ or } x \in \mathbb{R}^{1,3}. \tag{112}$$

Here $\psi^{r(x)}$ are the vector components (see Sec. 6.4.1), and $h_{r(x)}$ are basis vectors, represented as generators of the infinite dimensional Clifford Algebra, $Cl(\infty)$, satisfying

$$h_{r(x)} \cdot h_{r'(x')} \equiv \frac{1}{2}(h_{r(x)} h_{r'(x')} + h_{r'(x')} h_{r(x)}) = \rho_{r(x)r'(x')}, \tag{113}$$

where $\rho_{r(x)r'(x')}$ is the metric of $\mathcal{S}$. In particular, it can be $\rho_{r(x)'(x')} = \delta_{rr'}\delta(x - x')$. The 4-component spinor indices, $\alpha = 1,2,3,4$, and the index $i = 1,2,3,4$, denoting the ideals, is implicit in $\psi^{r(x)} \equiv \psi^{\alpha i r(x)}$, and $h_{r(x)} \equiv h_{\alpha i r(x)}$. In a new basis, called the *Witt basis,*

$$h_{\ (x)} = \frac{1}{\sqrt{2}}(h_{1(x)} + i\, h_{2(x)}) \tag{114}$$

$$h_{*(x)} = \frac{1}{\sqrt{2}}(h_{1(x)} - i\, h_{2(x)}), \tag{115}$$

instead of (113), we have[8]

$$h_{\ (x)} \cdot h_{*(x')} = \rho_{\ (x)*(x')} \tag{116}$$

$$h_{\ (x)} \cdot h_{\ (x')} = h_{*(x)} \cdot h_{*(x')} = 0, \tag{117}$$

which are *fermionic anticommutation relations.* A vector Ψ can be expanded as

$$\Psi = \psi^{\ (x)} h_{\ (x)} + \psi^{*(x)} h_{*(x)}. \tag{118}$$

The scalar product is

$$\langle \Psi\, \Psi \rangle_S = \psi^{\ (x)} \rho_{\ (x)*(x')} \psi^{*(x')} + \psi^{*(x)} \rho_{*(x)\ (x')} \psi^{\ (x')}. \tag{119}$$

Let us introduce the object

$$\Omega = \prod_x h_{*(x)}, \tag{120}$$

satisfying

$$h_{*(x)} \Omega = 0. \tag{121}$$

We see that Ω has the role of vacuum, and $h_{*(x)}$ are *annihilation operators.* The object $h_{\ (x)}$ gives a state $h_{\ (x)}\Omega \neq 0$, with a hole in vacuum. Therefore, $h_{\ (x)}$ are *creation operators* (of holes in Ω).

If we act with a vector Ψ on the vacuum Ω, we obtain

$$\Psi\, \Omega = \psi^{\ (x)}\, h_{\ (x)}\, \Omega. \tag{122}$$

[8] Instead of the indices 1, 2, denoting real and imaginary componet, we now use the space, , and the star, $*$.

This is a superposition of one particle (or better, one hole) states.

A generalization of eq. (122) is

$$(\psi_0 + \psi^{(x)} h_{(x)} + \psi^{(x)(x')} h_{(x)} h_{(x')} + \ldots)\Omega. \tag{123}$$

This is a superposition of zero, one two,...,many,..., particle states. Such a state is the infinite dimensional analog [22] of the spinor as an element of a left ideal[9] of Clifford algebra [21, 22].

Besides (121), there are other possible vacuums, e.g.,

$$\Omega = \prod_x h_{(x)}, \quad , \quad h_{(x)}\Omega = 0 \tag{124}$$

$$\Omega = \left(\prod_{x\in R_1} h_{*(x)}\right)\left(\prod_{x\in R_2} h_{(x)}\right), \quad h_{*(x)}\Omega = 0, \ \text{if } x \in R_1$$

$$h_{(x)}\Omega = 0, \ \text{if } x \in R_2 \tag{125}$$

etc.

where[10] $R_1 \in \mathbb{R}^3$, $R_2 \in \mathbb{R}^3$, $R_1 \cup R_2 = \mathbb{R}^3$.

An analogous situation also holds in momentum representation, where the operators are $c_{(p)}$, $c_{*(p)}$, with $p = (p^0, \mathbf{p})$. The vacuum can then be defined as, e.g.,

$$\Omega = \left(\prod_{p^0>0,\mathbf{p}} c_{*(p^0,\mathbf{p})}\right)\left(\prod_{p^0<0,\mathbf{p}} c_{(p^0,\mathbf{p})}\right). \tag{126}$$

Instead of the above notation, adapted to the Clifford algebraic description, we will now again use our notation of sec. 5.2.2, which is a straightforward generalization of the usual notation. We have the following operators in momentum space:

$$b_s^{\bar{i}}(\mathbf{p}), \quad d_s^{\bar{i}}(\mathbf{p}) \tag{127}$$

$$b_s^{\underline{i}}(\mathbf{p}), \quad d_s^{\underline{i}}(\mathbf{p}) \tag{128}$$

which annihilate the vacuum

$$\Omega = \left(\prod_{s,\mathbf{p}} b_s^{\bar{i}}(\mathbf{p})\right)\left(\prod_{s,\mathbf{p}} d_s^{\bar{i}}(\mathbf{p})\right)\left(\prod_{s,\mathbf{p}} b_s^{\underline{i}}(\mathbf{p})\right)\left(\prod_{s,\mathbf{p}} d_s^{\underline{i}}(\mathbf{p})\right). \tag{129}$$

The Fock space states are

$$\underbrace{b_s^{\bar{i}\dagger}(\mathbf{p})\,\Omega\,, \quad d_s^{\bar{i}\dagger}(\mathbf{p})\,\Omega}_{\text{positive energies}}\,, \quad \underbrace{b_s^{\underline{i}\dagger}\Omega(\mathbf{p})\,, \quad d_s^{\underline{i}\dagger}\Omega(\mathbf{p})}_{\text{negative energies}}\,, \ldots, \ \text{and all many particle states} \tag{130}$$

We see that, though considered nowadays as obsolete, the concept of the Dirac sea finds its revival with the Clifford algebra approach to field theory. Besides (129), other vacua can also be constructed by taking various combinations of the operators $b_s^{i\dagger}$, b_s^i, $d_s^{i\dagger}$, d_s^i (see ref. [22]).

[9] In the case of a finite dimensional Clifford algebra $Cl(p,q)$, a left ideal is a subspace of $Cl(p,q)$ that is invariant under the left multiplication by any element of $Cl(p,q)$.

[10] If we consider the Stueckelberg theory [23], then instead of $\mathbb{R}^3$ we have $\mathbb{R}^{1,3}$.

In the absence of interactions, the vacuum (129) has vanishing energy, i.e., the expectation value of the Hamiltonian (67) is $\langle \Omega^\dagger H \Omega \rangle_S = 0$. In the presence of interactions that mix the positive and negative energy states, the vacuum Ω decays into a superposition of positive and negative energy states. The finite state with infinitely many positive and negative energy particles, $|p_1 p_2 ... p_\infty\rangle$, is the state in which all the operators were removed from the vacuum Ω:

$$\Psi(t) = b_{s_1}^{\bar{i}_1 \dagger}(\mathbf{p}_1) b_{s_2}^{\bar{i}_2 \dagger}(\mathbf{p}_2) \ldots d_{s_1}^{\bar{i}_1 \dagger}(\mathbf{p}_1) d_{s_2}^{\bar{i}_2 \dagger}(\mathbf{p}_2) \ldots d_{s_1}^{i_1 \dagger}(p_1) d_{s_2}^{i_2 \dagger}(\mathbf{p}_2) \ldots \Omega = 1 \tag{131}$$

The expectation value of the free Hamiltonian (67) in the latter state is also zero. The state (131) also is "unstable" and can evolve into another state that is a superposition of the basis states

$$b_s^{i\dagger}(\mathbf{p}) , \quad d_s^{i\dagger}(\mathbf{p}) , \quad b_s^{i}(\mathbf{p}), \quad d_s^{i}(\mathbf{p}) , \quad \text{and all many operator states} \tag{132}$$

In a finite (closed) universe, with a minimal (e.g., Planck) distance, there would be a maximal absolute value of the particle's energy, and there would be a finite number of discrete momenta values. The $\mathbf{p}$ in eqs. (129) would be discrete. The unstable vacuum Ω would then decay, in a finite time, into a finite number of positive and negative energy states. The latter state would then "decay" back into Ω (just as in the simple model of the oscillator illustrated in Fig. 9). So there would be oscillations between $\Psi = \Omega$ and $\Psi = 1$. Such would be the situation for an external observer, $\mathcal{O}_{\text{ext}}$, who does not disturb the state $\Psi(t)$. But for an observer who is a part of the universe described by the state $\Psi(t)$, and who "nearly" continuously watches (at least a part of) his universe, the evolution of $\Psi(t)$ is "altered" according to the watchdog effect of QM; it is no longer an undisturbed evolution. The discussion at the end of Sec. 6.3.1 is valid for fermionic fields as well.

7. Discussion

Clifford algebras and Clifford spaces appear as a promising framework for the unification of particles and (gauge) fields. Such spaces do not have Lorentzian signature, $(+----...)$, but a more general signature $(+++...---...)$. The Clifford space, C, whose tangent space at any of its point is the Clifford algebra $Cl(1,3)$, has the neutral signature $(8,8)$. We have investigated the stability of the classical and quantum harmonic oscillator in the space $M_{1,1}$, which is a subspace of C. It is known [1, 4] that the negative energies occurring in a space with neutral signature pose no problem, if there are no interactions. We have shown that even the presence of an interaction causing transitions between positive and negative energy states, need not be problematic, and that such a system may have stable solutions.

In the case of *the classical oscillator*, we have found that certain interactions prevent the runaway solutions, and make the system stable. Collisions of an otherwise unstable pseudo Euclidean oscillator with surrounding particles also stabilize the oscillator. After such collisions, the particles gain kinetic energy. A material made of such oscillators would thus increase the temperature of the surrounding medium, after being immersed into it. This is very hypothetical, but the history of physics teaches us that we can never be sure about what surprises are waiting ahead of us. Recall that quasicrystals, with crystallographically forbidden symmetries[11], cannot

[11] Schechtman's discovery of quasicrystals [29] was ridiculed, because, in view of the established crystalographic

be explained in terms of local interactions in three dimensions. They can be explained as regular crystals in 6-dimensions, projected in 3-dimensions. The Clifford space, associated with objects in spacetime, has sixteen dimensions, and one can envisage that there exist in C the crystals that, from the 3D point of view, appear as quasicrystals. If such an explanation of quasicrystals eventually turns out to be correct, then the next step would be to investigate whether the pseudo Euclidean oscillators, allowed by the physics in Clifford space, actually exist in nature.

In the case of *the quantum oscillator* in $M_{1,1}$, we have calculated numerical solutions for a truncated system. Because the truncated system is finite, the solution is oscillating, and does not run away into infinity. For practical reasons we considered the energy modes up to $m, n = 4$, so that all higher energies were cut off. We can envisage a cutoff at much higher energies. This corresponds to a small minimum cutoff distance[12] in the space $M_{1,1}$. Then, instead of continuous space $M_{1,1}$, we have a finite periodic lattice. An analogous situation occurs if we consider an oscillator in 16D Clifford space.

After having investigated how the theory works on the examples of the classical and quantum pseudo Euclidean oscillator, we considered quantum field theories. If the metric of field space is neutral, then there occur positive and negative energy states. An interaction causes transitions between those states. A vacuum therefore decays into a superposition of states with positive and negative energies. Because of the vast phase space of infinitely many particles, such a vacuum decay is instantaneous—for an external observer. But we, as a part of our universe, are not external observers. We are internal observers entangled with the "wave function" of the rest of our universe. According to the Everett interpretation of quantum mechanics, we find ourselves in one of the branches of the universal wave function. In the scenario with decaying vacuum, our branch can consist of a finite number of particles. Once being in such a branch, it is improbable that at the next moment we will find ourselves in a branch with infinitely many particles. For us, because of the "watchdog effect" of quantum mechanics, the evolution of the universal wave function relative to us is frozen to the extent that instantaneous vacuum decay is not possible. The Everett interpretation is gaining increasing support among quantum cosmologists [25].

We have also pointed out that the field potential $V(\varphi)$ need not be constant during the evolution of the universe. It may change, and thus alter a system's dynamics from an unstable to a stable regime.

If there exists a minimal distance (e.g., at the Planck scale), and if the universe is closed, then it contains a finite, albeit very large, number of points; it is a sort of lattice. Such a finite system cannot run into infinity, because there is neither infinite energy or infinitely small distance, neither infinite number of particles. Therefore, such a system is stable even in the presence of negative energies. It oscillates with large, but finite, amplitudes. In Sec. 6.3 we showed how does oscillate a simple quantum finite oscillator. An analogous situation holds for quantum fields on a finite lattice.

In a field theory with neutral signature, there is an outburst of positive and negative energy

theory, it was considered as impossible. Fortunately, that was a simple experiment and it was not difficult for other labs to repeat it.

[12] It is generally believed that the Planck distance is the minimal distance, therefore the idea that space or Clifford space is in fact a lattice makes sense.

particles, like an explosion, that is eventually stabilized. This is reminiscent of the Big Bang. Our universe indeed emerged in an explosion. But in our universe we do not "see" equal number of positive and negative energy particles. Can then such a quasi unstable vacuum be an explanation for Big Bang?

Description of our universe requires fermions and accompanying gauge fields, including gravitation. According to the Clifford algebra generalized Dirac equation—Dirac-Kähler equation—there are four sorts of the 4-component spinors, with energy signs a shown in eq. (111). The vacuum of such field has vanishing energy and evolves into a superposition of positive and negative energy fermions, so that the total energy is conserved. A possible scenario is that the branch of the superposition in which we find ourselves, has the sea of negative energy states of the first and the second, and the sea positive energy states of the third and forth minimal left ideal of $Cl(1,3)$. According to ref. [17], the former states are associated with the familiar, weakly interacting particles, whereas the latter states are associated with mirror particles, coupled to mirror gauge fields, and thus invisible to us. According to the field theory based on the Dirac-Kähler equation, the unstable vacuum could be an explanation for Big Bang.

Introduction of spaces with neutral signature that imply the existence of negative energies, is potentially significant for further important progress of theoretical physics. The usual arguments—mostly related to stability—why such spaces are not suitable for physics, can be circumvented along the lines indicated in this paper.

Acknowledgments

This research was supported by the Slovenian Research Agency.

References

[1] Woodard R 2007 *Lect. Notes Phys.* **720** 403
[2] Pauli W 1943 *Rev. Mod. Phys.* **15** 175
[3] Cangemi D Jackiw R and Zwiebach B 1996 *Annals of Physics* **245** 408
Benedict E, Jackiw R and Lee H J 1996 *Phys. Rev. D* **54** 6213
[4] Pavšič M 1999 *Phys. Lett. A* **254** 119 [hep-th/9812123].
[5] Pavšič M 2005 *Found. Phys.* **35** 1617 [hep-th/0501222].
[6] Ansoldi S, Aurilia A, Castro C and Spallucci E 2001 *Phys. Rev. D* **64** 026003 [arXiv:hep-th/0105027]
Aurilia A, Ansoldi S and Spallucci E 2002 *Class. Quant. Grav.* **19** 3207 [arXiv:hep-th/0205028].
[7] Castro C 1999 *Chaos, Solitons and Fractals* **10** 295
Chaos, Solitons and Fractals 12 (2001) 1585; Castro C and Pavšič M 2002 *Phys. Lett. B* **539** 133 [arXiv:hep-th/0110079]
Castro C and Pavšič M 2005 *Prog. Phys.* **1** 31
[8] Pavšič M 2001 *The Landscape of Theoretical Physics: A Global View; From Point Particles to the Brane World and Beyond, in Search of a Unifying Principle* (Dordrecht: Kluwer)
[9] Pavšič M 2001 *Found. Phys.* **31** 1185 [hep-th/0011216].
[10] Pavšič M 2003 *Found. Phys.* **33** 1277 [gr-qc/0211085].
[11] Pavšič M 2007 *Found. Phys.* **37** 1197 [hep-th/0605126].
[12] Hestenes D 1966 *Space-Time Algebra* (New York:Gordon and Breach)
Hestenes D and Sobcyk G 1984 *Clifford Algebra to Geometric Calculus* (Dordrecht: Reidel).
[13] Pavšič M 2012 *Localized Propagating Tachyons in Extended Relativity Theories* arXiv:1201.5755 [hep-th].
[14] Castro C 2012 *Int. J. Theor. Phys.* DOI 10.1007/s10773-012-1295-3
Castro C 2012 *Adv. Appl. Clifford Algebras* DOI 10.1007/s00006-012-0370-4
[15] Pavšič M 2005 *Phys. Lett. B* **614** 85 [hep-th/0412255].
[16] Pavšič M 2006 *Int. J. Mod. Phys. A* **21** 5905 [gr-qc/0507053].
[17] Pavšič M 2010 *Phys. Lett. B* **692** 212 [arXiv:1005.1500 [hep-th]].

[18] Bandos I A 2010 *Spacetime Brout-Englert-Higgs effect in General Relativity interacting with p-brane matter J. Phys. Conf. Ser.* 229 012017, 1001.4388 [hep-th].
[19] Pavšič M 2011 *Phys. Lett. B* **703** 614 [arXiv:1106.2017 [gr-qc]].
[20] Kähler E 1962 *Rendiconti di Matematica* **21** 425
Kruglov S I 2002 *Int. J. Theor. Phys.* **41** 653 (and references therein)
Spehled D and Marques G C 2009 *Eur. Phys. J.* **61** 75
[21] Budinich P 1986 *Phys. Rep.* **137** 35
Budinich P and Trautman A 1986 *Lett. Math. Phys.* **11** 315
Giler G, Kosinski P, Rembielinski J and Maslanka P 1987 *Acta Phys. Pol. B* **18** 713
M. Budinich M 2012 *Adv. Appl. Clifford Algebras* **22** 283296
[22] Pavšič M 2012 *Adv. Appl. Clifford Algebras* **22** (2012) 449 [arXiv:1104.2266 [math-ph]].
[23] Fock V 1937 *Phys. Z. Sowj.* **12** 404
Stueckelberg E C G 1941 *Helv. Phys. Acta*
Stueckelberg E C G 1942 **15** 23
Horwitz L P and C. Piron C 1973 *Helv. Phys. Acta* **46** 316
Horwitz L P and Rohrlich F 1981 *Physical Review D* **24** 1528
Horwitz L P, Arshansky R I and Elitzur A C 1988 *Found. Phys* **18** 1159
Feynman R P 1951 *Phys. Rev* **84** 108
Fanchi J R 1993 *Found. Phys.* **23** 287, and many references therein
Fanchi J R 1993 *Parametrized Relativistic Quantum Theory* (Dordrecht: Kluwer) Pavšič M 1991 *Found. Phys.* **21** 1005
[24] L. Horwitz, Gershon A and Schiffer M 2011 *Found Phys* **41** 141
[25] Tegmark M 1998 *Fortschritte der Physik* **46** 855 [quant-ph/9709032]
[26] Sudarshan E C G and Misra B 1977 *J. Math. Phys.* **18** 763
[27] Albert Z, Aharonov Y and D'Amato S *Phys. Rev. Lett.* **54** 5;
Aharonov Y, Massar S, Popescu S, Tollaksen J and Vaidman L 1996 *Phys. Rev. Lett.* **77** 983
[28] Fermi E 1932 *Rev. Mod. Phys.* **4** 87
[29] Shechtman D, Blech I, Gratias D and . Cahn J W 1984 *Phys. Rev. Let.* **53** 1951

Constants of the motion, universal time and the Hamilton-Jacobi function in general relativity

Paul O'Hara

Dept. of Mathematics, Northeastern Illinois University, 5500 North St. Louis Avenue, Chicago, Illinois 60625-4699

E-mail: pohara@neiu.edu

Abstract.
In most text books of mechanics, Newton's laws or Hamilton's equations of motion are first written down and then solved based on initial conditions to determine the constants of the motions and to describe the trajectories of the particles. In this essay, we take a different starting point. We begin with the metrics of general relativity and show how they can be used to construct by inspection constants of motion, which can then be used to write down the equations of the trajectories. This will be achieved by deriving a Hamiltonian-Jacobi function from the metric and showing that its existence requires all of the above mentioned properties. The article concludes by showing that a consistent theory of such functions also requires the need for a universal measure of time which can be identified with the "worldtime" parameter, first introduced by Steuckelberg and later developed by Horwitz and Piron.

Max Born in his book "Natural Philosophy of Cause and Chance" gives a derivation of Newton's laws of gravity from Kepler's laws of planetary motion noting that it "is the basis on which [his] whole conception of causality in physics rests"([1]). In the spirit of that insight, this essay will explore the metrics of general relativity and show how it is possible to use them to derive both the constants of the motion and the particle trajectory in a gravitational field. We will also show that a consistent development of the above mentioned approach requires the use of a universal time parameter which can be identified with the "worldtime" defined by Steuckelberg ([5])and further developed by Horwitz et al.([2]).

The key to this development will be rewriting the metric as an exact differential associated with the Hamiltonian-Jacobi function. Such exact differentials can always be constructed by noting that the inner product of any gradient vector $\nabla\psi(s)$ with a tangent vector to a curve $\mathbf{ds}$ is always exact. More precisely, using spinor notation, a local tetrad can be constructed at any point on the curve ([4]), with one-form $\tilde{ds} = \gamma^a dx_a$ and its dual $\tilde{\partial}_s\psi = \gamma^a \frac{\partial\psi}{\partial x_a}$ such that

$$\frac{\tilde{ds}}{d\tau}.\tilde{\partial}_s\psi = \frac{1}{2}\left\{\frac{\tilde{ds}}{d\tau},\tilde{\partial}_s\psi\right\} + \frac{1}{2}\left[\frac{\tilde{ds}}{d\tau},\tilde{\partial}_s\psi\right] \tag{1}$$

$$= \frac{d\vec{\psi}}{d\tau} + \frac{\mathbf{ds}}{d\tau}\wedge\frac{\partial\vec{\Psi}}{\partial s}. \tag{2}$$

Equations (1) and (2) can be identified by noting that the anti-commutator and commutator relationships in $\frac{\tilde{ds}}{d\tau}.\tilde{\partial}_s\psi$ associated with the spinor tetrad, define a dot product and a cross product respectively. Historically, equation (2) was first discovered by Grossman in 1841 and then independently rediscovered by Clifford.

We say that $\psi(\tau) \equiv W$ is a Hamilton-Jacobi function if (1) is true and $\left[\frac{\tilde{ds}}{d\tau}, \tilde{\partial}_s W\right] = 0$, or equivalently W is a Hamilton Jacobi function whenever $dW = p_a^* dx^a$ is an exact differential, where $p_a^* = \frac{\partial W(s)}{\partial x^a} = W' p_a$, and $p_a = \frac{\partial s}{\partial x}$. Usually, p_o is denoted by $-H$ and $dW = p_1^* dx^1 + p_2^* dx^2 + p_3^* dx^3 - H^* dt$. In particular, if s is a parameter denoting the length of a smooth curve then [4]

$$dW = p_1^* dx^1 + p_2^* dx^2 + p_3^* dx^3 - H^* dt \ \text{ iff } \ ds = p_1 dx^1 + p_2 dx^2 + p_3 dx^3 - H dt. \tag{3}$$

It follows that Hamilton-Jacobi functions can be constructed at will starting from the metric. Consider

$$ds^2 = g_{ij} dx^i dx^j \tag{4}$$

this is equivalent to

$$\frac{ds}{d\tau} ds = g_{ij} \frac{dx^i}{d\tau} dx^j \qquad \tau \text{ a parameter.} \tag{5}$$

Now choose W such that $dW = \frac{ds}{d\tau} ds$ then W will be a Hamilton-Jacobi function with

$$p_j^* \equiv \frac{\partial W(s)}{\partial x^j} = g_{ij} \frac{dx^i}{d\tau}, \qquad \tau \text{ a parameter.} \tag{6}$$

It is easy to check that this is consistent, and that for all i, j

$$p_i^* = p_j'^* \frac{\partial x'^j}{\partial x^i} \qquad \text{and} \qquad \frac{\partial p_j^*}{\partial x^i} = \frac{\partial p_i^*}{\partial x^j}. \tag{7}$$

In the case of a specified space (for example defined by a Schwarzschild or a Robertson-Walker metric), the parameter τ can in principle be conveniently chosen to be the proper time associated with a unit rest mass along a specified geodesic passing through the origin and to which all other time parameters can then be referred, although it does not have to be so. It can be identified with the "worldtime" defined by Horwitz et al.in [2]. Indeed, in the case of a particle moving with constant speed along a geodesic there is an affine relationship between it and the proper time along the curve. However, in the case of non-geodesic motion no such affine connection will exist in general between the proper time and the worldtime. In practice, the choice of this universal time is not so clear. A particle in motion within the Schwarzschild space of the earth, is in turn also moving within the Schwarzschild metric of the sun which in turn is to a first approximation moving within a solar system that is part of a specific galaxy moving along a geodesic associated with the Robertson-Walker metric. The best that one can hope to do is invoke the principle of equivalence and establish a universal time parameter with respect to some standard "laboratory" frame defined with respect to the fixed stars. We shall postpone further discussion until the next section.

1. Curves and constants of the motion

For the purpose of this article we shall confine ourselves to working with a four dimensional pseudo-Riemannian manifold and/or Minkowski space, although many of the observations could easily be extended to higher dimension spaces. Also, for what follows, we will continue to work with an arbitrary parameter τ but will simplify notation by using p for p^*. Our interest is to classify Hamilton-Jacobi functions $W = W(s)$ associated with specific forms of the metric $ds = p_1 dx^1 + p_2 dx^2 + p_3 dx^3 - H dt$, where x^i are generalized coordinates. Three specific cases arise:

(1) All of the p_i are constant.
(2) Some of the p_i are constant
(3) None of the p_i are constant.
All three cases apply to all metrics. However, in practise the specific solutions chosen and the coordinate system used usually reflect the problem under investigation. In either case, the simplest possible solutions occur when all the p_i are constants, although these are not usually the only solutions and identifying them as constants depends on the coordinate system used. For example, the differential

$$\begin{aligned} dW(s) &= \gamma(s)(2xydx + x^2dy) && (8) \\ &= p_x dx + p_y dy && (9) \end{aligned}$$

is an exact differential for all parameterizations $x = x(s)$, $y = y(s)$,where $\gamma(s) = \gamma(x^2y)$ is smooth. Also, if we require that they are both constant, they pick out a very specific family of curves (one for each k) associated with $x^2 = 2kxy$. On the other hand, rewriting the above metric in polar coordinates gives

$$\begin{aligned} dW(s) &= \gamma(s)(2xydx + x^2dy) && (10) \\ &= \gamma(s)[(3r^2\cos^2\theta\sin\theta dr + (r^3\cos^3\theta - 2r^3\cos\theta\sin^2\theta)d\theta] && (11) \\ &= p_r dr + p_\theta d\theta && (12) \end{aligned}$$

Once again, it is easy to check that $\frac{\partial p_r}{\partial\theta} = \frac{\partial p_\theta}{\partial r}$. However, the previous requirement that

$$p_x = 2\gamma xy = k, \qquad \text{a constant,}$$

now precludes that

$$p_r = 3\gamma r^2\cos^2\theta\sin\theta = \frac{3}{2}k\cos\theta$$

is constant, except in the trivial case when r and θ are both constants and vice versa. So the choice of coordinate systems is important.

Secondly, it is worth recalling that if $\dot{p}_i = 0$ for all i then this implies motion along a geodesic only if the coordinate system (x^i) forms a tetrad. In other cases, motion will be along some other type of curve. This leads to the following definitions and lemma [6](p44).

Definition 1 *If $s = s(\tau)$ where τ is a parameter and*

$$2f_i = \frac{d}{d\tau}\left(\frac{\partial\dot{s}^2}{\partial\dot{x}^i}\right) - \frac{\partial\dot{s}^2}{\partial x^i}$$

then $f_i = f_i(\tau)$ is called the acceleration tensor with respect to τ.

Lemma 1 *Let $ds^2 = g_{ij}dx^i dx^j$, with ds an exact differential and $p_i(\tau) = g_{ij}\dot{x}^j = \frac{\partial s}{\partial x^i}\dot{s}$. If in phase space $\frac{\partial\dot{s}^2}{\partial x^i} = 0$ then $\dot{p}_i(\tau) = 0$ iff $f_i = 0$.*

Proof:

$$2f_i \equiv \frac{d}{d\tau}\left(\frac{\partial\dot{s}^2}{\partial\dot{x}^i}\right) - \frac{\partial\dot{s}^2}{\partial x^i}. \qquad (13)$$

But $\frac{\partial\dot{s}^2}{\partial x^i} = 0$ implies $\frac{d}{d\tau}\left(\frac{\partial\dot{s}^2}{\partial\dot{x}^i}\right) = 2\dot{p}_i$ from the definition of p_i. Therefore, $\dot{p}_i = 0$ iff $f_i = 0$. The result follows.□

In practice, as we shall see below, the lemma above permits one to solve for those curves and determine those potentials for which momentum is conserved. It also allows us in special cases to read off the constants of the motion by inspection. Also, in the event that not all p_i are constant, then as noted above, the requirement that s be a Hamilton-Jacobi curve in a given coordinate system allows one to determine all possible motions not involving spin or vortex motion.

Example 1: As an application of the above theory, we begin by considering planar motion in Minkowski space with metric

$$ds^2 = dr^2 + r^2 d\theta^2 - c^2 dt^2. \tag{14}$$

This can be written with respect to a parameter τ by

$$\frac{ds}{d\tau} ds = \frac{dr}{d\tau} dr + r^2 \frac{d\theta}{d\tau} d\theta - c^2 \frac{dt}{d\tau} dt. \tag{15}$$

The requirement that $f_\theta = 0$ and $f_t = 0$ (see Def. 1) implies from the lemma that $p_\theta = r^2\dot{\theta}$ and $p_t = -c^2\dot{t}$ are constants of the motion along the curve $s(r,\theta,t) = h(r) + k_1\theta + k_2 t$ with respect to the parameter τ, where $s(\tau)$ is a Hamilton-Jacobi function, h a function and $\dot{p}_i(\tau) = \frac{\partial s}{\partial x^i}\dot{s}$.

Two cases are of particular interest:
(1) $\tau = s$ and $h(r) = kr$ (equivalent to $\dot{r} = constant \neq 0$)
(2) $\tau = s$ and $h(r) = k$ (equivalent to $\dot{r} = 0$).
In the first case

$$f_r(s) = -r\dot{\theta}^2 = -(r^2\dot{\theta})^2/r^3 = -k^2/r^3$$

defines an inverse cube law of motion. Moreover, both $\dot{r}$ and $r^2\dot{\theta}$ are non-zero constants and consequently are proportional to each other. It follows that in general the trajectory of a particle subjected to this force in Minkowski space obeys the equation $\dot{r} = \epsilon r^2\dot{\theta}$ which is equivalent to $k_3 r - \epsilon r\theta = 1$. In the second case $r = constant$. This defines circular motion.

Finally, if in addition to $f_\theta = f_t = 0$, we restrict ourselves to purely geodesic motion given by taking $f_r = 0$, then it is easy to check that the only possible trajectory is given by $\theta = constant$ which is equivalent to straight line motion in $\Re^2$.

It should be clear that the form of the metric is key to associating specific curves with specific constants of the motion. In the above example, we derived the general form of the trajectories associated with the conservation of angular momentum $\dot{p}_\theta = r^2\dot{\theta}$. However, one might seek curves in the same metric space associated with other constants. For example, if we define $\dot{\phi} = r\dot{\theta}$ then the metric takes the form

$$ds^2 = dr^2 + d\phi^2 - c^2 dt^2 \tag{16}$$

on the Minkowski manifold $M(r,\phi,t)$. The geodesic equations are given by $f_r = f_\phi = f_t = 0$ or equivalently $\dot{r} = k\dot{\phi}$. Substituting $r\dot{\theta}$ for $\dot{\phi}$ and solving for $r \neq constant$, gives the trajectory $r = Ae^{k\theta}$ on the manifold $M(r,\theta,t) \setminus \{(0,0)\}$, while in the case $\dot{r} = 0$ we obtain the circle $r = r_o$, r_o a constant. In other words, the requirement that $r\dot{\theta}$ be a constant of the motion (and not $r^2\dot{\theta}$) determines the family of trajectories

$$\mathcal{F} = \{r = Ae^{k\theta} | A, k \text{ are constant}\} \bigcup \{r = r_o | r_o \text{ a constant}\}.$$

Example 2: As a second example consider the metric

$$\dot{s}ds = \frac{l\dot{r}}{\sin\theta} dr + r^2\dot{\theta}d\theta - c^2\dot{t}dt, \tag{17}$$

derivable from Kepler's first and second laws of planetary motion which state that planets move on ellipses given by $l/r = 1 + \epsilon\cos\theta$, with constant angular momentum. The Ricci curvature tensor $R_{ij} \neq 0$ and consequently the space is not flat. For the purpose of this essay, let us begin with the metric and require that $\dot{s}$ be an exact differential. Note that $p_t = -c^2\frac{\partial s}{\partial t}$ in terms of phase space is independent of r and θ. It follows from the lemma that $\dot{t}$ is a constant of the motion if $f_t = 0$. However, the lemma does not apply to both $p_r = \frac{l\dot{r}}{\sin\theta}$ and $p_\theta = r^2\dot{\theta}$, except in the case of $\dot{\theta} = 0$ (a geodesic) and therefore precludes these from being constants of the motion along other geodesics. However, there are other trajectories for which they are constants of the motion, determined by the equation $p_r = \epsilon p_\theta$ or equivalently that $\frac{l\dot{r}}{\sin\theta} = \epsilon r^2\dot{\theta}$. Integrating out gives the equation of a conic for an inverse square law of motion, which is Kepler's first law of motion.

Example 3: The same techniques can also be used to identifying the constants of the motion associated with all metrics in which the equations of motion obey the Hamilton-Jacobi equation. Consider the Schwarzschild space with a metric of the form

$$ds^2 = B(r)dt^2 - A(r)dr^2 - r^2d\theta^2 - r^2\sin^2\theta d\phi^2. \tag{18}$$

For $f_t = 0$ and $f_\phi = 0$, it follows from the lemma that in phase space

$$B(r)\dot{t} = k_1, r^2\sin^2\theta\dot{\phi} = k_2.$$

Also,

$$\frac{\partial r^2\dot{\theta}}{\partial\phi} = \frac{\partial r^2\sin^2\theta\dot{\phi}}{\partial\theta} = 0$$

implies $\theta = k_3$. Taking $k_3 = \frac{\pi}{2}$, the Hamilton-Jacobi function takes the form $s(r,\theta,\phi,t) \equiv k_1t + k_2 + k_3\phi + h(r)$ with p_t, p_θ, p_ϕ being constants of the motion, and h a function such that $\frac{\partial h}{\partial r} = -A(r)$.

The three constants are well known and can be easily shown to be associated with geodesic motion. To fully obtain a geodesic, we would also require that $f_r = 0$, in addition to f_t, f_θ and f_ϕ already given above. Also by noting that any two constants can be related by a constant of proportionality ϵ, it follows that if $f_t = f_\phi = 0$ then any trajectory for such a motion must obey the equation $k_1 = \epsilon k_3$ or equivalently $\frac{B(r)}{r^2} = \epsilon\frac{d\phi}{dt}$.

Example 4: Similarly in the case of the Robertson-Walker metric

$$ds^2 = dt^2 - R^2(t)\left\{\frac{dr^2}{1-kr^2} + r^2d\theta^2 + r^2\sin^2\theta d\phi^2\right\} \tag{19}$$

there exists trajectories for which $\dot{t}, R^2(t)\frac{\dot{r}}{1-kr^2}, R^2(t)r^2\dot{\phi}$ and θ are constants of the motion, and in this case a generalized first law of Kepler would require that galaxies move on trajectories given by $\frac{\dot{r}}{r^2(1-kr^2)} = \epsilon\dot{\phi}$. This can be integrated out, using partial fractions, to give the family of curves

$$-\epsilon r\phi + k_5r + \frac{r\sqrt{k}}{2}\ln\left(\frac{1+\sqrt{k}r}{1-\sqrt{k}r}\right) = 1. \tag{20}$$

These curves are not geodesics, since only $R^2(t)r^2\dot{\phi} = constant$ and $\theta = constant$ satisfy the geodesic equation $f_t(s) = 0$.

2. Universal time

As previously noted, the requirement that $dW(\tau) \equiv \frac{ds}{d\tau}ds$ be a Hamilton-Jacobi function such that

$$p_j(\tau) = \frac{\partial W(\tau)}{\partial x^j} = g_{ij}\frac{dx^i}{d\tau}, \qquad \tau \text{ a parameter}$$

begs the question as to how we should determine the parameter τ. On the one hand it does not matter in that one can easily pass from one parameter τ to another by means of the transformation

$$p(\tau') = \frac{d\tau}{d\tau'}p(\tau). \tag{21}$$

Note the similarity between this and coordinate transformations of the form

$$p_i^* = p_j'^* \frac{\partial x'^j}{\partial x^i}.$$

This latter case applies when we always use the same proper time s/c. Indeed, if we were confined to a single curve then the proper time would be the most convenient. However, things are more complicated as soon as we pass over into higher dimensions and try to seek a consistent parametrization that applies to all curves. In this case, we will be guided by equation (21). Either way, as we shall see below, it will be necessary to define a universal time scale and basis within the space in order to have a common and consistent parametrization. We now explore this more in depth.

Consider for example the difficulty in trying to coherently compare the two straight lines,

$$ds^2 = c^2dt^2 \text{ and } ds^2 = c^2dt^2 - dx^2 = c^2dt^2(1 - \tanh^2\phi) = c^2dt^2\sec^2\theta$$

defined over the two dimensional Minkowski space $M(t, x)$. Using the proper time s/c as a parameter and $\phi = i\theta$ a constant, the first can be parameterized in terms of s by $t = s/c$ and defines the proper time for the curve, while the second line will have a parametrization given by $t = \sec(\theta)s/c$. Unless $\theta = 0$, a clear contradiction arises on equating s or t in each equation. To avoid this paradox there are two choices to make: either we rewrite the equations of the two curves as $s = tc$ and $s' = c\sec(\theta)t$ ($s \neq s'$) or as $s = tc$ and $s = c\sec(\theta)t'$ ($t \neq t'$). Both choices are valid and should never be confused.

At the core of the distinction is the difference between having two different lines or parameterizations with respect to the same reference frame versus the same line defined with respect to two different references frames. Equivalently it is the difference between parameterizing the two curves with respect to the same local time (common reference frame) or parameterizing with respect to the same proper time. The advantage of the latter case is that one can easily pass from one curve to another by means of a Lorentz transformation, if they are of the same type. A simple rotation of the axes will transform one curve into another. This is Weinberg's approach. It reflects the fact that one is examining the same phenomena from two different references frames ([7]).

However, in the event that two particles are moving along the same curve (geodesic) but with different speeds (not to mention accelerations) this second perspective is inadequate. Indeed, the requirement that $ds = ds'$ would be equivalent to saying that both particles have proper speed c by definition (albeit defined with respect to two different references frames) although they are clearly moving at different speeds from the perspective of the same reference frames. From this perspective, the equations $ds^2 = c^2dt^2$ and $ds'^2 = c^2\sec^2\theta dt^2$ can be viewed as two different parameterizations along the same curve reflecting the different speeds of the particles with respect to the laboratory frame. The problem becomes even more pronounced when we try to compare the velocities and accelerations between two particles moving on totally different curves. In such cases, unless the problem is formulated with respect to the same basis $\{cdt, dx, dy, dz\}$ it risks being ill-posed.

In terms of the general theory, consider two curves defined with respect to a basis $\{x^i\}$ and a parameter τ such that

$$\frac{ds}{d\tau}ds = p_i(s)dx^i, \qquad \frac{ds'}{d\tau}ds' = p'_i(s')dx^i$$

on a pseudo-Riemannian manifold. Since both s and s' are considered to be Hamilton-Jacobi functions the existence of p_i and p'_i are guaranteed by equation (6). Let us further assume that the first curve is both a geodesic and a unit speed curve when parameterized with respect to s, while the second curve parameterized with respect to s is an arbitrary non-unit speed curve. In other words, motion along the non-unit curve is defined in terms of a tetrad $\{x^i\}$ which can be Fermi transported along the unit geodesic. Transposed to the laboratory rest frame these can be expressed by the equations

$$ds^2 = c^2 d\tau^2, \qquad ds'^2 = p'_\tau d\tau^2,$$

with $p'_t = c\frac{ds'}{ds}$. By construction these curves cannot be transformed one into another by means of a Lorentz transformation. Their corresponding proper times are different. However, it is possible to define one proper time as a non-trivial function of the other. To do this in an effective and consistent way it is necessary to establish a universal measure of time, a standard time to which all others can be compared. In theory, by the Principle of Equivalence it is sufficient to parameterize any curve in terms of a local unit speed geodesic proper time and use this as a universal time parameter. In practice, the only way one can know if this has been successful is by comparing the time calibrations of the standard laboratory clock with a universally pre-established time standard. It can be identified with the worldtime defined by Horwitz et al. in [2].

3. Mass and momentum

Thus far the role of mass in our equations has been ignored and the above discussion related to different parameterizations of a curve suggests that the concept of rest mass will also be affected by such changes. Returning to equation (1), we see that for the curve $x^i = x^i(s)$ defined with respect to a local tetrad, the function $W(s) = m_o s$ is a Hamilton-Jacobi function with $\frac{\partial s}{\partial x^i} = \frac{dx^i}{ds}$, where s/c is the proper time along the curve. It follows that $p_i = \frac{\partial W}{\partial x^i} = m_o c \frac{dx}{ds}$. However, if the parameter is changed to s' then

$$p^i = m_o c \frac{dx^i}{ds} = m(s') c \frac{dx^i}{ds'}, \qquad \text{with} \qquad m(s') \equiv m_o \frac{ds'}{ds} \tag{22}$$

which means the momentum remains invariant under a change of parameter.

In particular, in the case of motion defined with respect to two different parameters s and s' respectively, the four momentum $m_o c$ in the rest frame associated with s will will transform into

$$m_o c = m_o c \frac{ds}{ds'}\frac{ds'}{ds} = m'_o c \frac{ds}{ds'} \tag{23}$$

in the frame associated with s'. Equivalently, the rest mass defined with respect to two different parameterizations of a curve may be different. For example, if $s' = t$ the local time then the change in mass will correspond to the usual mass change under Lorentz transformations. If $s' = 2.2s$ then a mass measurement in kilos in the s frame would correspond to the same measurement in pounds in the s' frame. It also follows from equation (23) that if s'/c is the worldtime parameter defined in [3] then

$$ds^2 = \frac{m^2}{M^2} ds'^2, \qquad m_o = m, \; M = m(s')$$

corresponds to equation (14) in [3] and $m^2 = M^2$ defines the so called "mass shell" condition. Indeed, the mass will be the same under any change of basis that does not affect the curve parameter. If $ds = ds'$ then $m(s') \equiv c\frac{dx^i}{ds} = m_o$. In other words, the rest mass along any geodesic measured with respect to a standard clock and length will always be the same. Real differences will only occur in a gravitational field.

On a final note, if we return to equation (22), then the conservation of momentum requires

$$\frac{ds}{m(s)} = \frac{ds'}{m(s')},$$

which implies that $\frac{ds^2}{m^2(s)}$ is invariant both under general coordinate transformations given by equation (7) and under changes of parameter given by equation (21) and it unifies the two different perspectives as outlined in the previous section. For example, it allows one to pass from the worldtime parameter to a proper time parameter in one step.

4. Conclusion

In the spirit of Born's observation, there is something special about Hamilton-Jacobi functions. Not only can they be used to derive Hamilton's equations but they allow us to identify equations and constants of motion as well as a new relativistic invariant $\frac{ds}{m(s)}$ associated with the motion, with the rest mass m_o being meaningfully defined only if a "worldtime" parameter is introduced. They also determine the trajectories in general for natural motions.

In that regard, it should be recalled that if $s(\tau)$ is a Hamilton-Jacobi function then so also are smooth functions $W(s)$ and more sophisticated motions will require their use. Indeed, in the context of the overall field of mechanics the Hamiltonian-Jacobi functions with gradient ∇W serve as gauge terms for the more general motion which can be written (see equation (2)) as

$$\frac{\tilde{ds}}{d\tau}.\tilde{\partial}_s \psi = \frac{dW}{d\tau} + \vec{A}.\frac{ds}{d\tau} + \vec{A} \wedge \frac{\vec{ds}}{d\tau}. \tag{24}$$

But this is a discussion for another day.

[1] Born, Max. *Natural Philosophy of Cause and Chance.* Crarendon Press (Oxford),1949, 129.
[2] Horwitz, L.P. and C. Piron, Helv. Phys. Acta 46 (1973) 316
[3] Horwitz, L.P., Foun. Phys. Vol 22, No. 3, 1992.
[4] O'Hara, Paul. J. of Phys. (Conf.) Vol 330, 2011.
[5] Steuckelberg,E.C.G. Helv. Phys. Acta 14 (1941) 322.
[6] Synge, J.L. and A. Schild. *Tensor Calculus.* Dover Publ.(New York), 1969.
[7] Weinberg, S. *Gravitation and Cosmology: Principles and Applications of the General Theory of Relativity.* Wiley & Sons, 1972, 29.

Is General Relativity a simplified theory?

Martín Rivas

Theoretical Physics Department, University of the Basque Country,
Apdo. 644, 48080 Bilbao, Spain

E-mail: martin.rivas@ehu.es

Abstract. Gravity is understood as a geometrization of spacetime. But spacetime is also the manifold of the boundary values of the spinless point particle in a variational approach. Since all known matter, baryons, leptons and gauge bosons are spinning objects, it means that the manifold, which we call the kinematical space, where we play the game of the variational formalism of an elementary particle is greater than spacetime. This manifold for any mechanical system is a Finsler metric space such that the variational formalism can always be interpreted as a geodesic problem on this space. This manifold is just the flat Minkowski space for the free spinless particle. Any interaction modifies its flat Finsler metric as gravitation does. The same thing happens for the spinning objects but now the Finsler metric space has more dimensions and its metric is modified by any interaction, so that to reduce gravity to the modification only of the spacetime metric is to make a simpler theory, the gravitational theory of spinless matter. Even the usual assumption that the modification of the metric only involves dependence on the metric coefficients of the spacetime variables is also a restriction because in general these coefficients are dependent on the velocities. In the spirit of unification of all forces, gravity cannot produce, in principle, a different and simpler geometrization than any other interaction.

1. Introduction

Things should be made simple, but not simpler. From this sentence attributed to Albert Einstein is where we take the title of this work to show that if the spin concept of elementary particles had been known to physics before General Relativity was born most probably the geometrization of spacetime proposed by its creator should be changed by the geometrization of a different manifold, larger than spacetime, so that today's General Relativity would be considered as a theory of gravitation of simpler and spinless matter.

The variational approach of classical mechanics can always be interpreted as a geodesic statement on the space X of the boundary variables of the variational formalism [1]. But this metric manifold X, is not a pseudo-Riemannian space but rather a Finsler space [2], [3], where the symmetric metric $g_{ij}(x,\dot{x})$ is not only a function of the point $x \in X$, but also of its velocity $\dot{x}$, where the overdot means derivative with respect to some arbitrary evolution parameter. For the relativistic spinless or point particle this manifold X is just the spacetime $\mathbb{R}^4$ and in the free case the metric is Minkowski's metric $\eta_{\mu\nu}$. But if the particle has spin it would have more degrees of freedom, so that the variational approach will be described as a geodesic problem in a larger manifold than spacetime. Interactions and gravitation would modify the metric of this larger manifold, so that to restrict ourselves to the geometrization of the spacetime submanifold is to simplify the problem, or in physical terms, to reduce the gravitational behaviour of real spinning matter to that of spinless and unexistent matter.

In the next section 2, I will make a summary of the variational approach of classical mechanics, which shows how it can be interpreted as a geodesic statement and the way to obtain the metric from the Lagrangian. If the Lagrangian of a free elementary particle is modified by any interaction, the metric of the boundary X manifold is modified. Any interaction modifies the geometry of the X space.

In section 3 we analyze some examples to show the Finsler metric structure of the spacetime of a charged point particle under some external interactions, which include a uniform magnetic field and a constant gravitational field of a point mass M. In section 4 we also consider the example of the point particle under a static Newtonian gravitational potential. In all cases the modification of the metric coefficients involve dependence on the velocities of the point particle. A Riemannian approximation to the metric can be obtained in the low velocity limit. All these analysis are done in a special relativity framework.

In section 5 after a short introduction to the concept of classical elementary particle I will describe the most general X manifold of a relativistic spinning particle which satisfies Dirac's equation when quantized. Since it seems that there are no spinless elementary particles in nature, it is on this larger manifold that the plausible generalization of Einstein's gravitational formalism has to be worked out.

As a general conclusion, General Relativity is a restricted theory of gravitation and therefore a simplified theory in two aspects. One, that the manifold whose geometry is changed by any interaction is larger than spacetime because real elementary particles are spinning particles. The second is that the modification of the metric coefficients should involve dependence on the velocities, i.e., the metric should be a Finsler metric instead of a Riemannian metric.

2. The geodesic interpretation of the variational formalism

Let us consider any mechanical system of n degrees of freedom q_i, $i = 1, \ldots, n$, described by a Lagrangian, $L(t, q_i, dq_i/dt)$, where t is the time. The variational approach is stated in such a way that the path followed by the system makes stationary the action functional

$$\mathcal{A}[q(t)] = \int_{t_1}^{t_2} L(t, q_i, dq_i/dt)dt,$$

between the initial state $x_1 \equiv (t_1, q_i(t_1))$ and final state $x_2 \equiv (t_2, q_i(t_2))$ on the X manifold, which is the $(n+1)$-th dimensional manifold spanned by the time t and the n degrees of freedom q_i. If instead of describing the evolution in terms of time we express the evolution in parametric form $\{t(\tau),\ q_i(\tau)\}$ in terms of some arbitrary evolution parameter τ, then $dq_i/dt = \dot{q}_i/\dot{t}$, where now the overdot means τ-derivative. The variational approach will be written as

$$\int_{\tau_1}^{\tau_2} L(t, q_i, \dot{q}_i/\dot{t})\ \dot{t}d\tau = \int_{\tau_1}^{\tau_2} \widetilde{L}(x, \dot{x})d\tau, \quad \widetilde{L} = L\dot{t},$$

with the same boundary values on the X manifold as before x_1 and x_2. But now the Lagrangian $\widetilde{L}$ is independent on the evolution parameter τ and it is a homogeneous function of first degree of the derivatives $\dot{x}$ [3]. In fact in L each time derivative dq_i/dt has been replaced by a quotient $\dot{q}_i/\dot{t}$ and therefore it is a homogeneous function of zero-th degree of $\dot{t}, \dot{q}_i$. But it is the last term $dt = \dot{t}d\tau$, which makes $\widetilde{L} = L\dot{t}$, homogeneous of first degree of $\dot{x} \equiv \{\dot{t}, \dot{q}_i\}$.

This means that $\widetilde{L}^2$ is a positive definite homogeneous function of second degree of the derivatives $\dot{x}$, so that Euler's theorem on homogeneous functions allows us to write

$$\widetilde{L}^2 = g_{ij}(x, \dot{x})\dot{x}^i\dot{x}^j, \quad i, j = 0, 1, \ldots n$$

where index 0 corresponds to the time variable and the g_{ij} are computed from $\widetilde{L}^2$ by

$$g_{ij}(x,\dot{x}) = \frac{1}{2}\frac{\partial^2 \widetilde{L}^2}{\partial \dot{x}^i \partial \dot{x}^j} = g_{ji}, \quad i,j = 0,\ldots,n. \tag{1}$$

Between the allowed boundary states x_1 and x_2, since $\widetilde{L}^2 > 0$, the metric $g_{ij}(x,\dot{x})$ represents a definite positive metric which transforms as a second rank covariant tensor under transformations which leave $\widetilde{L}$ invariant [2, 3]. The variational problem can be stated as

$$\int_{\tau_1}^{\tau_2} \widetilde{L}(x,\dot{x})d\tau = \int_{\tau_1}^{\tau_2} \sqrt{\widetilde{L}^2(x,\dot{x})}d\tau = \int_{\tau_1}^{\tau_2} \sqrt{g_{ij}(x,\dot{x})\dot{x}^i\dot{x}^j}d\tau =$$

$$= \int_{x_1}^{x_2} \sqrt{g_{ij}(x,\dot{x})dx^i dx^j} = \int_{x_1}^{x_2} ds,$$

where ds is the arc length on the X manifold with respect to the metric g_{ij}. The variational statement has been transformed into a geodesic problem with a Finsler metric. As shown in [1] this is valid even for Lagrangian systems depending on higher order derivatives $L(t, q_i, q_i^{(1)}, \ldots, q_i^{(k)})$, $q_i^{(k)} = d^k q_i/dt^k$. In this case the manifold of the boundary variables X, which will be called the kinematical space from now on, is spanned by the time t, the n degrees of freedom q_i and their corresponding time derivatives up to order $k-1$.

Since $\widetilde{L}$ is homogeneous of first degree in terms of the derivatives $\dot{x}$ can also be decomposed as a sum of terms with dimensions of action if the arbitrary evolution parameter is taken dimensionless,

$$\widetilde{L} = \frac{\partial \widetilde{L}}{\partial \dot{x}^i}\dot{x}^i = F_i(x,\dot{x})\dot{x}^i,$$

where the $F_i(x,\dot{x})$ are homogeneous functions of zero-th degree of the $\dot{x}^i$, so that they involve time derivatives of the different degrees of freedom. The metric coefficients can be expressed as

$$g_{ij} = F_i F_j + \widetilde{L}\frac{\partial^2 \widetilde{L}}{\partial \dot{x}^i \partial \dot{x}^j} = F_i F_j + \widetilde{L}\frac{\partial F_i}{\partial \dot{x}^j} = g_{ji} \tag{2}$$

and are also homogeneous functions of zero-th degree of the $\dot{x}^i$.

As an example, the relativistic point particle of mass m and spin 0 has a kinematical space spanned by time t and the position of the point $\boldsymbol{r}$, so that the free Lagrangian $\widetilde{L}_0 = -mc\sqrt{c^2\dot{t}^2 - \dot{\boldsymbol{r}}^2}$, is clearly a homogeneous function of first degree of the derivatives $\dot{t}$ and $\dot{\boldsymbol{r}}$. With $x^0 \equiv ct$, the Finsler metric becomes

$$\widetilde{L}_0^2 = m^2c^2(\dot{x}_0^2 - \dot{\boldsymbol{r}}^2), \quad g_{\mu\nu} = \frac{1}{2}\frac{\partial^2 \widetilde{L}_0^2}{\partial \dot{x}^\mu \partial \dot{x}^\nu} = m^2c^2\eta_{\mu\nu},$$

where $\eta_{\mu\nu}$ is diag$(1,-1,-1,-1)$. The interaction with some external electromagnetic field is described by the new Lagrangian $\widetilde{L} = \widetilde{L}_0 + \widetilde{L}_I$, with $\widetilde{L}_I = -eA_\mu(x)\dot{x}^\mu$, so that the variational problem is transformed into a geodesic problem with a new metric on X space, given by

$$\widetilde{L}^2 = \left(\widetilde{L}_0 + \widetilde{L}_I\right)^2, \quad F_\mu = \frac{\partial \widetilde{L}}{\partial \dot{x}^\mu} = -p_\mu - eA_\mu, \quad p_\mu = \frac{mc\dot{x}_\mu}{\sqrt{\dot{x}_\nu \dot{x}^\nu}}$$

$$g_{\mu\nu}(x,\dot{x}) = m^2c^2\eta_{\mu\nu} + e^2A_\mu A_\nu + e(p_\mu A_\nu + p_\nu A_\mu) + eA_\sigma\dot{x}^\sigma\frac{\partial p_\mu}{\partial \dot{x}^\nu}. \tag{3}$$

The modification of the metric vanishes when $e \to 0$. Because p_μ is not explicitely dependent on the variables x, the dependence of the metric on the spacetime coordinates is coming only from the external fields $A_\mu(x)$. But it depends on the $\dot{x}$ variables through its dependence on the p_μ and its derivatives.

Journal of Physics: Conference Series **437** (2013) 012008 doi:10.1088/1742-6596/437/1/012008

3. Examples of Finsler spaces

In figure 1 we show three possible motions of a charged point particle in its kinematical space, which reduces in this case to the spacetime. The three trajectories are geodesics of spacetime but with respect to three different metrics. In part (a) the motion is free, the trajectory is a straight line. In (b) the particle is under the action of an external uniform magnetic field, and the trajectory has curvature and torsion. In this case the Finsler metric of spacetime is different than the metric in the free case. The external magnetic field modifies the metric. Finally in (c) it is the same free trajectory but as seen by an accelerated observer. According to the equivalence principle this is equivalent to the analysis under a global and constant gravitational field. Also in this case the metric has been modified.

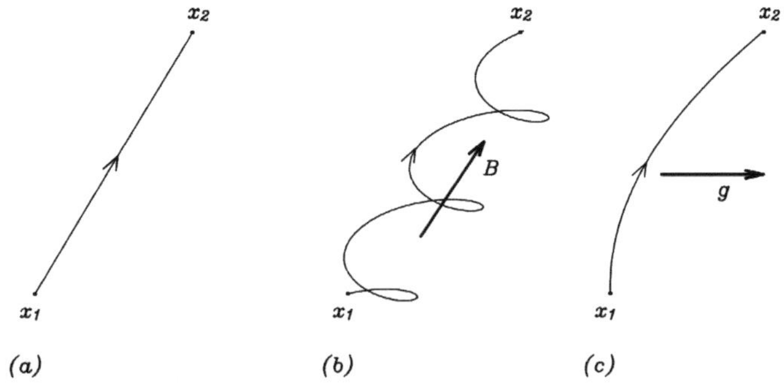

Figure 1. Three motions of a point particle in its kinematical space between the boundary points x_1 and x_2. (a) in the free case, (b) under a uniform magnetic field $\boldsymbol{B}$, and (c) under a uniform gravitational field $\boldsymbol{g}$. In the three cases the kinematical space is the same, the spacetime, the trajectories are geodesics but with respect to three different Finslerian metrics. The spatial part of the trajectories is in the case (a) a straight line with no curvature and no torsion, in (b) with curvature and torsion and in (c) a flat trajectory with curvature.

In case (a) the metric is $g_{\mu\nu} = m^2c^2\eta_{\mu\nu}$, where $\eta_{\mu\nu}$ is diag$(1, -1, -1, -1)$. It is the constant Minkowski metric.

In the case (b), let us assume a uniform magnetic field along OZ axis of intensity B. We can take as the potential vector $\boldsymbol{A} = (0, Bx, 0)$ and scalar potential $A_0 = 0$. The Lagrangian of the point particle under this field is

$$\widetilde{L}_B = -mc\sqrt{\dot{x}_0^2 - \dot{\boldsymbol{r}}^2} + eBx\dot{y}. \tag{4}$$

This Lagrangian leads to the Lorentz force dynamical equation

$$\frac{d\boldsymbol{p}}{dt} = e\,\boldsymbol{u} \times \boldsymbol{B}, \quad \boldsymbol{u} = \frac{d\boldsymbol{r}}{dt}. \tag{5}$$

According to (1) and (3) and calling $K = eBmc$, the Finsler metric coefficients of spacetime are,

$$g_{00} = m^2c^2 + \frac{Kxu^2u_y}{(c^2-u^2)^{3/2}}, \quad g_{11} = -m^2c^2 + \frac{Kxu_y}{(c^2-u^2)^{3/2}}\left(c^2 - u_y^2 - u_z^2\right),$$

$$g_{22} = -m^2c^2 + e^2B^2x^2 + \frac{Kxu_y}{(c^2-u^2)^{3/2}}\left(3c^2 - 3u_x^2 - 2u_y^2 - 3u_z^2\right),$$

Journal of Physics: Conference Series **437** (2013) 012008 doi:10.1088/1742-6596/437/1/012008

$$g_{33} = -m^2c^2 + \frac{Kxu_y}{(c^2-u^2)^{3/2}}\left(c^2 - u_x^2 - u_y^2\right),$$

$$g_{01} = -\frac{Kxcu_xu_y}{(c^2-u^2)^{3/2}}, \quad g_{02} = -\frac{Kxc}{(c^2-u^2)^{3/2}}(c^2 - u_x^2 - u_z^2),$$

$$g_{03} = -\frac{Kxcu_yu_z}{(c^2-u^2)^{3/2}}, \quad g_{12} = \frac{Kxu_x}{(c^2-u^2)^{3/2}}\left(c^2 - u_x^2 - u_z^2\right),$$

$$g_{13} = \frac{Kx}{(c^2-u^2)^{3/2}}u_xu_yu_z, \quad g_{23} = \frac{Kxu_z}{(c^2-u^2)^{3/2}}\left(c^2 - u_x^2 - u_z^2\right),$$

As we see, the metric coefficients are functions of the point, i.e., of the variable x, but they are functions of the three components of the velocity of the particle u_x, u_y, u_z, and therefore the metric depends on both x and $\dot{x}$. The dependence on $\dot{x}$ of the $g_{\mu\nu}$ is a homogeneous function of zero-th degree, and thus it depends on the time derivatives $dr_i/dt = u_i$. If the velocity is negligible with respect to c, the metric coefficients become

$$g_{00} = m^2c^2, \quad g_{02} = -eBmcx, \quad g_{11} = -m^2c^2, \quad g_{22} = -m^2c^2 + e^2B^2x^2, \quad g_{33} = -m^2c^2,$$

vanishing the remaining ones, and since the dependence on the velocity has disappeared the metric has been transformed into a Riemannian metric. Spacetime metric is Riemannian in the low velocity limit.

These metric coefficients give rise to a restricted Lagrangian $\widetilde{L}_R$,

$$\widetilde{L}_R^2 = m^2c^2(c^2\dot{t}^2 - \dot{\boldsymbol{r}}^2) + e^2B^2x^2\dot{y}^2 - 2eBmc^2x\dot{t}\dot{y}, \tag{6}$$

such that when compared with (4) we have an additional term

$$\widetilde{L}_B^2 = \widetilde{L}_R^2 - 2emcBx\dot{y}\left(\sqrt{c^2\dot{t}^2 - \dot{\boldsymbol{r}}^2} - c\dot{t}\right).$$

The Lorentz force dynamical equations (5) are

$$\frac{du_x}{dt} = \frac{eB}{m\gamma(u)}u_y = \frac{1}{\gamma(u)}kcu_y, \quad \frac{du_y}{dt} = -\frac{eB}{m\gamma(u)}u_x = -\frac{1}{\gamma(u)}kcu_x, \quad \frac{du_z}{dt} = 0,$$

with $k = eB/mc$, which gives rise to $u_xdu_x/dt + u_ydu_y/dt + u_zdu_z/dt = \boldsymbol{u}\cdot d\boldsymbol{u}/dt = 0$, so that the motion is at a velocity of constant modulus u, the factor $\gamma(u)$ is constant and the point particle moves at a constant velocity along OZ axis and rotates on the plane XOY with angular velocity $\omega = eB/\gamma(u)m = k\sqrt{c^2-u^2}$. However, the geodesic dynamical equations obtained from the restricted metric (6) are

$$\frac{du_x}{dt} = kcu_y(1 - kxu_y/c), \quad \frac{du_y}{dt} = -kcu_x(1 - kxu_y/c), \quad \frac{du_z}{dt} = 0,$$

which reduce to the above equations when $u/c \to 0$. From the restricted Lagrangian (6) the force acting on the point particle is not longer the Lorentz force.

In the case (c) in a uniform gravitational field $\boldsymbol{g}$, the dynamical equations

$$d\boldsymbol{p}/dt = m\boldsymbol{g}, \tag{7}$$

independent of the mass of the particle, come from the Lagrangian

$$\widetilde{L}_g = \widetilde{L}_0 + m\boldsymbol{g}\cdot\boldsymbol{r}\dot{t}. \tag{8}$$

From the geodesic point of view it corresponds to an evolution in a spacetime with the Finslerian metric given by:

$$g_{00} = m^2c^2 + m^2(\boldsymbol{g}\cdot\boldsymbol{r})^2/c^2 - \frac{m^2c(\boldsymbol{g}\cdot\boldsymbol{r})}{(c^2-u^2)^{3/2}}(2c^2-3u^2),$$

$$g_{11} = -m^2c^2 + \frac{m^2c(\boldsymbol{g}\cdot\boldsymbol{r})}{(c^2-u^2)^{3/2}}(c^2-u_y^2-u_z^2),$$

$$g_{22} = -m^2c^2 + \frac{m^2c(\boldsymbol{g}\cdot\boldsymbol{r})}{(c^2-u^2)^{3/2}}(c^2-u_x^2-u_z^2),$$

$$g_{33} = -m^2c^2 + \frac{m^2c(\boldsymbol{g}\cdot\boldsymbol{r})}{(c^2-u^2)^{3/2}}(c^2-u_x^2-u_y^2),$$

$$g_{01} = -\frac{m^2u^2(\boldsymbol{g}\cdot\boldsymbol{r})}{(c^2-u^2)^{3/2}}u_x, \quad g_{02} = -\frac{m^2u^2(\boldsymbol{g}\cdot\boldsymbol{r})}{(c^2-u^2)^{3/2}}u_y, \quad g_{03} = -\frac{m^2u^2(\boldsymbol{g}\cdot\boldsymbol{r})}{(c^2-u^2)^{3/2}}u_z,$$

$$g_{12} = \frac{m^2c(\boldsymbol{g}\cdot\boldsymbol{r})}{(c^2-u^2)^{3/2}}u_xu_y, \quad g_{23} = \frac{m^2c(\boldsymbol{g}\cdot\boldsymbol{r})}{(c^2-u^2)^{3/2}}u_yu_z, \quad g_{13} = \frac{m^2c(\boldsymbol{g}\cdot\boldsymbol{r})}{(c^2-u^2)^{3/2}}u_xu_z.$$

If again, the velocity is negligible with respect to c, the nonvanishing coefficients are

$$g_{00} = m^2c^2 + m^2(\boldsymbol{g}\cdot\boldsymbol{r})^2/c^2 - 2m^2(\boldsymbol{g}\cdot\boldsymbol{r}), \quad g_{11} = -m^2c^2 + m^2(\boldsymbol{g}\cdot\boldsymbol{r}),$$

$$g_{22} = -m^2c^2 + m^2(\boldsymbol{g}\cdot\boldsymbol{r}), \quad g_{33} = -m^2c^2 + m^2(\boldsymbol{g}\cdot\boldsymbol{r}).$$

i.e.,

$$g_{00} = m^2c^2\left(1-\frac{\boldsymbol{g}\cdot\boldsymbol{r}}{c^2}\right)^2, \quad g_{ii} = -m^2c^2\left(1-\frac{\boldsymbol{g}\cdot\boldsymbol{r}}{c^2}\right), \quad i=1,2,3,$$

where the g_{00} component is the same as the corresponding component of the Rindler metric corresponding to a noninertial accelerated observer or to the presence of a global uniform gravitational field, in General Relativity.

4. Example: Point particle in a Newtonian potential

A final example is the relativistic point particle in the Newtonian potential of a point mass M. The dynamical equations

$$\frac{d\boldsymbol{p}}{dt} = -\frac{GmM}{r^3}\boldsymbol{r}, \tag{9}$$

are independent of the mass of the particle and come from the Lagrangian

$$\widetilde{L}_N = \widetilde{L}_0 + \frac{GmM}{cr}c\dot{t}. \tag{10}$$

If we take into account (1) the metric coefficients are

$$g_{00} = m^2c^2 + \frac{G^2m^2M^2}{c^2r^2} - \frac{Gm^2Mc}{r(c^2-u^2)^{3/2}}(2c^2-3u^2),$$

$$g_{11} = -m^2c^2 + \frac{Gm^2Mc^3}{r(c^2-u^2)^{3/2}} - \frac{Gm^2Mc(u_y^2+u_z^2)}{r(c^2-u^2)^{3/2}},$$

$$g_{22} = -m^2c^2 + \frac{Gm^2Mc^3}{r(c^2-u^2)^{3/2}} - \frac{Gm^2Mc(u_x^2+u_z^2)}{r(c^2-u^2)^{3/2}},$$

$$g_{33} = -m^2c^2 + \frac{Gm^2Mc^3}{r(c^2-u^2)^{3/2}} - \frac{Gm^2Mc(u_x^2+u_y^2)}{r(c^2-u^2)^{3/2}},$$

$$g_{01} = -\frac{Gm^2Mu^2u_x}{r(c^2-u^2)^{3/2}}, \quad g_{02} = -\frac{Gm^2Mu^2u_y}{r(c^2-u^2)^{3/2}}, \quad g_{03} = -\frac{Gm^2Mu^2u_z}{r(c^2-u^2)^{3/2}},$$

$$g_{12} = \frac{Gm^2Mcu_xu_y}{r(c^2-u^2)^{3/2}}, \quad g_{23} = \frac{Gm^2Mcu_yu_z}{r(c^2-u^2)^{3/2}}, \quad g_{31} = \frac{Gm^2Mcu_zu_x}{r(c^2-u^2)^{3/2}},$$

It is a Finsler metric, which in the case of low velocity only the diagonal components survive

$$g_{00} = m^2c^2\left(1 - \frac{2GM}{c^2r} + \frac{G^2M^2}{c^4r^2}\right) = m^2c^2\left(1 - \frac{GM}{c^2r}\right)^2.$$

and the

$$g_{ii} = -m^2c^2\left(1 - \frac{GM}{c^2r}\right), \quad i = 1,2,3.$$

This corresponds to a static and spherically symmetric Riemannian metric which in spherical coordinates and supressing the constant factor m^2c^2, becomes

$$\left(1 - \frac{GM}{c^2r}\right)^2 c^2dt^2 - \left(1 - \frac{GM}{c^2r}\right)(dr^2 + r^2d\Omega^2),$$

where $d\Omega^2 = d\theta^2 + \sin^2\theta d\phi^2$.

This metric is not a vacuum solution of Einstein's equations, so that it cannot be transformed into the Schwarzschild metric in isotropic coordinates.

In all the examples, the free Lagrangian $\widetilde{L}_0$ of the spinless particle, has been transformed by the interactions in the way

$$\widetilde{L}_0^2 = m^2c^2\eta_{\mu\nu}\dot{x}^\mu\dot{x}^\nu \quad \Rightarrow \quad \widetilde{L}^2 = g_{\mu\nu}(x,\dot{x})\dot{x}^\mu\dot{x}^\nu, \tag{11}$$

where the new metric $g_{\mu\nu}(x,\dot{x})$ is a Finslerian metric. The low velocity limit of the above metrics produce a Riemannian spacetime approximation which does not give rise to the usual (and expected) dynamical equations. All these examples have been worked out in a special relativity context.

However, General Relativity states that gravity modifies the metric of spacetime producing a new (pseudo-)Riemannian metric $g_{\mu\nu}(x)$, which is related through Einstein's equations to the energy momentum distribution $T^{\mu\nu}$ of all forms of matter and energy. The motion of a test point particle in this gravitational background is a geodesic on spacetime, and therefore can be treated as a Lagrangian dynamical problem with a Lagrangian

$$\widetilde{L}_g^2 = g_{\mu\nu}(x)\dot{x}^\mu\dot{x}^\nu. \tag{12}$$

Because electromagnetism produces a Finsler metric of spacetime, and in the spirit of unification of all interactions, one is tempted to extend the formulation of gravity (12) to the more general *ansatz* (11) by allowing to the metric coefficients produced by gravity to be also a function of the derivatives $\dot{x}$. Otherwise, to assume only a Riemannian metric is to consider that gravity, as far as dynamical equations are concerned, produces a different geometrization than any other interaction. In a region where the gravitational field can be considered uniform, or is given by the Newtonian potential the Lagrangian dynamics of the point particle is equivalent to a geodesic problem in that region where the metric is necessarily a Finsler metric, as seen in the above examples. Any theory of gravitation when restricted to a region of uniform gravitational field or to the sourroundings of a point mass M should reproduce the above dynamical equations (7) and (9), respectively. We expect that, when restricted to inertial observers, general relativity should reproduce the analysis under a special relativity context. But these dynamical equations are derived from Finsler metrics, so that a Riemannian metric $g_{\mu\nu}(x)$ cannot reproduce them. The elimination of the presence of the velocities in the metric coefficients could be interpreted as a low velocity limit of a more general gravitational theory.

5. The spin structure of elementary particles

In the mentioned reference [1] and previous works cited in, an elementary particle is defined as a mechanical system whose kinematical space X (the boundary space of its Lagrangian description) is necessarily a homogeneous space of the Poincaré group $\mathcal{P}$. The idea is that an elementary particle cannot be divided and, if not annihilated with its antiparticle, cannot be deformed so that any state is just a kinematical modification of any one of them [4]. When the initial state x_1 is modified by the dynamics, the subsequent states $x(\tau)$ can always be obtained from it by some change $g \in \mathcal{P}$ of inertial observer $x = gx_1$ and also $x_2 = gx_1$, so that given any two points $x_1, x_2 \in X$ we can always find some $g \in \mathcal{P}$ which links them. It is clear that the point particle manifold, the spacetime, is a homogeneous space of $\mathcal{P}$ and thus fulfills this requirement. Nevertheless, it describes a spinless object and there are no spinless elementary particles in nature. To describe spin we have to enlarge this kinematical space with the above constraint to obtain the largest homogeneous space of $\mathcal{P}$ to describe the elementary particle with the more complex structure. The classical system that when quantized satisfies Dirac's equation corresponds to a kinematical space spanned by the following variables $x \equiv (t, \boldsymbol{r}, \boldsymbol{u}, \boldsymbol{\alpha})$, which are interpreted as the time t, position of the center of charge $\boldsymbol{r}$, velocity of the center of charge $\boldsymbol{u} = d\boldsymbol{r}/dt$ at the speed of light $u = c$, and the orientation $\boldsymbol{\alpha}$ of a cartesian system located at point $\boldsymbol{r}$ [5]. It is a nine-dimensional kinematical space described by four noncompact variables $(t, \boldsymbol{r})$ and five compact ones $(\theta, \phi, \boldsymbol{\alpha})$, being θ, ϕ the orientation of the velocity vector $\boldsymbol{u}$ and the orientation of the particle local frame $\boldsymbol{\alpha}$. The particle has a center of mass $\boldsymbol{q}$ which is expressed in terms of $\boldsymbol{r}$ and its time derivatives. Elementary spinning particles have two distinguished points the center of mass and the center of charge which are different points. The cartesian frame can be taken as the Frenet-Serret triad, so that the angular velocity of the particle can also be expressed in terms of the derivatives of the position of the point $\boldsymbol{r}$.

The free motion of the center of charge $\boldsymbol{r}$ corresponds to a helix of constant curvature and torsion when expressed in terms of the Frenet-Serret triad, and at a velocity of constant absolute value c.

This classical model of elementary particle can be applied for leptons and quarks if, as assumed, they satisfy in the quantum formalism Dirac's equation.

If we want to include gravitation we have to admit arbitrary changes of spacetime coordinates, not only those given by the Poincaré group. This will produce a modification of the metric of the spacetime submanifold, but also the modification of the remaining components of the metric on the whole kinematical X-space. Because all known baryonic and leptonic matter and the gauge bosons are spinning objects we cannot start the geometrization of matter by assuming that it is only the metric of the kinematical space of the point particle which is modified, because there are no spinless objects in nature. We have to geometrize the complete kinematical space of the spinning particle accordingly. We cannot make things simpler.

6. Conclusions

We consider that General Relativity is a constrained, and therefore a simpler formalism for describing gravity for two reasons: One is that the geometrization of spacetime has to be enlarged to consider Finsler metrics instead of pseudo-Riemannian metrics, and another that the manifold which describes the boundary states of spinning matter is larger than spacetime.

The manifold X of the boundary variables of any Lagrangian dynamical system is always a Finsler metric space, so that any variational approach is equivalent to a geodesic statement on this metric manifold. This metric, which is in general a function of the variables $x \in X$ and their derivatives $\dot{x}$, depends on the interactions, and to assume that gravitation only produces a modification of the metric which is only a function of the point x, is a restriction of a more general formalism which allows for this modification, in the spirit of unification of all interactions.

The second constraint of a gravitational theory is that it has to be applied to a manifold larger than spacetime, because spacetime is the boundary manifold of the spinless point particle and spinless elementary particles seem not to exist in nature.

Without any assumption about general covariance, or any assumption about the dynamical behaviour of spacetime, but in a Lagrangian framework, we have analysed several examples of the Finsler space structure of spacetime under different interactions. In all of them the new metrics are true Finsler metrics which in the case of low velocity limit, and therefore a metric independent on the velocities, resemble the metrics obtained in a general relativity formalism but they are not vacuum solutions of Einstein's equations.

It is possible that the spin structure of matter plays no role in the gravitational analysis of the solar system and in a cosmological background, so that the usual treatment in terms of only spacetime variables is sufficient to describe planetary motions. But in cosmological models, when the velocity of particles is not negligible, redshifts of order 6 and higher have been quoted for several galaxies which correspond to velocities of $0.9c$, a metric dependent on the velocities would produce a different analysis than a Riemannian one.

If we need to take into account the spin content of matter, may be in a neutron star where the magnetic moments of the particles are aligned, or in a gravitational colapse with a huge density of matter where gravitational effects associated to the spin structure are expected, it is unavoidable to enlarge the dynamical formalism to include the spin description of matter, in which the space X is larger than spacetime and the metric should depend also on the velocities. It is on this larger manifold that gravity has to be worked out. The physical restrictions have to applied in the analysis of the particular cases, not at the very begining of the formalism.

Acknowledgments

This work has been partially supported by Universidad del País Vasco/Euskal Herriko Unibertsitatea grant 9/UPV00172.310-14456/2002.

References

[1] Rivas M 2001 *Kinematical theory of spinning particles*, (Dordrecht: Kluwer), Chapter 6.
[2] Asanov GS 1985 *Finsler geometry, Relativity and Gauge theories*, (Dordrecht: Reidel Pub. Co)
[3] Rund H 1973 *The Hamilton-Jacobi theory in the calculus of variations*, (N.Y: Krieger Pub. Co.)
[4] Rivas M 2008 The atomic hypothesis: Physical consequences, *J. Phys.* A **41** 304022
[5] Rivas M 1994 Quantization of generalized spinning particles: New derivation of Dirac's equation, *J. Math. Phys.* **35** 3380
Some other works related to this formalism can be obtained through the webpage <http://tp.lc.ehu.es/martin.htm>

Covariant Uniform Acceleration

Yaakov Friedman and Tzvi Scarr

Jerusalem College of Technology
Departments of Mathematics and Physics
P.O.B. 16031 Jerusalem 91160, Israel

E-mail: friedman@jct.ac.il, tzviscarr@gmail.com

Abstract. We derive a 4D *covariant* Relativistic Dynamics Equation. This equation canonically extends the 3D relativistic dynamics equation $\mathbf{F} = \frac{d\mathbf{p}}{dt}$, where $\mathbf{F}$ is the 3D force and $\mathbf{p} = m_0\gamma\mathbf{v}$ is the 3D relativistic momentum. The *standard* 4D equation $F = \frac{dp}{d\tau}$ is only partially covariant. To achieve full Lorentz covariance, we replace the four-force F by a rank 2 antisymmetric *tensor* acting on the four-velocity. By taking this tensor to be constant, we obtain a covariant definition of *uniformly accelerated motion*. This solves a problem of Einstein and Planck.

We compute *explicit* solutions for uniformly accelerated motion. The solutions are divided into four *Lorentz-invariant* types: null, linear, rotational, and general. For *null* acceleration, the worldline is *cubic* in the time. *Linear* acceleration covariantly extends 1D hyperbolic motion, while *rotational* acceleration covariantly extends pure rotational motion.

We use Generalized Fermi-Walker transport to construct a *uniformly accelerated family* of inertial frames which are instantaneously comoving to a uniformly accelerated observer. We explain the connection between our approach and that of Mashhoon. We show that our solutions of uniformly accelerated motion have constant acceleration in the comoving frame. Assuming the Weak Hypothesis of Locality, we obtain local spacetime transformations from a uniformly accelerated frame K' to an inertial frame K. The spacetime transformations between two uniformly accelerated frames with the same acceleration are *Lorentz*. We compute the metric at an arbitrary point of a uniformly accelerated frame.

We obtain velocity and acceleration transformations from a uniformly accelerated system K' to an inertial frame K. We introduce the *4D velocity*, an adaptation of Horwitz and Piron's notion of "off-shell." We derive the general formula for the time dilation between accelerated clocks. We obtain a formula for the angular velocity of a uniformly accelerated object. Every rest point of K' is uniformly accelerated, and its acceleration is a function of the observer's acceleration and its position. We obtain an interpretation of the Lorentz-Abraham-Dirac equation as an acceleration transformation from K' to K.

1. Introduction

Newton's Second Law $\mathbf{F} = m\mathbf{a}$ can be written as $\mathbf{F} = \frac{d\mathbf{p}}{dt}$, where $\mathbf{p} = m\mathbf{v}$ is the classical momentum. In special relativity, the classical momentum is replaced by the relativistic momentum $\mathbf{p} = m_0\gamma\mathbf{v}$, and Newton's Second Law is replaced by the standard 3D relativistic dynamics equation [1, 2, 3]

$$\mathbf{F} = m_0\frac{d(\gamma\mathbf{v})}{dt}. \tag{1}$$

When the 3D force **F** is *constant*, the solutions to (1) are traditionally called *uniformly accelerated motion.* Equation (1), however, is covariant only with respect to the *little Lorentz group* and *not* covariant with respect to the full Lorentz group.

As a 4D extension of (1), we have

$$F = \frac{dp}{d\tau}, \tag{2}$$

where F is the four-force, p is the four-momentum, and τ is proper time (see [1]). Unfortunately, equation (2) is also covariant only with respect to the little Lorentz group. Moreover, when F is a constant, as in a homogeneous gravitational field, equation (2) has no solution! This follows from the fact that the four-velocity and the four-acceleration are perpendicular. This was noticed by Planck, who wrote to Einstein about it. This, in turn, prompted Einstein to submit a "correction" [4] to [5]. In the correction, he states that the "concept 'uniformly accelerated' needs further clarification." This was a call for a fully Lorentz covariant relativistic dynamics equation and for a better definition of "uniform acceleration."

It was clear, even in 1908, that the *physical* definition of "uniformly accelerated motion" is motion whose acceleration is *constant in the comoving frame.* This definition is found widely in the literature, as early as [6] and [7], again in [8], and as recently as [9] and [10]. This definition is natural, since the acceleration in the comoving frame is "precisely the push we feel when sitting in an accelerating rocket" or automobile. Similarly, "by the equivalence principle, the gravitational field in our terrestrial lab is the negative of our proper acceleration, our instantaneous rest frame being an imagined Einstein cabin falling with acceleration g" ([2], page 71). Indeed, we will, at times, invoke the Equivalence Principle and interpret accelerations as the effect of a gravitational field of a massive object.

If the acceleration is constant in the comoving frame, then the length of the four-acceleration a is constant:

$$a^{\mu}a_{\mu} = \text{constant}. \tag{3}$$

Equation (3) is a good candidate to replace (2). It's even *fully* Lorentz covariant. However, as in the case of equations (1) and (2), existing techniques have produced only *1D hyperbolic motion* as solutions to (3). There are clearly some missing solutions, since equation (3) is covariant, while the class of 1D hyperbolic motions is *not.*

We are thus faced with two problems:

(1) Can $\mathbf{F} = \frac{d\mathbf{p}}{dt}$ be extended to a 4D Lorentz covariant version?

(2) What is the "right" equation for uniformly accelerated motion?

In this paper, we derive a 4D Lorentz covariant Relativistic Dynamics Equation:

$$c\frac{du^{\mu}}{d\tau} = A^{\mu}_{\nu}u^{\nu}, \tag{4}$$

where u is the four-velocity, τ is proper time, and $A_{\mu\nu}$ is a rank 2 antisymmetric tensor, or, equivalently, A^{μ}_{ν} is skew adjoint with respect to the Minkowski inner product $\eta_{\mu\nu} = \text{diag}(1, -1, -1, -1)$. As will be shown here, equation (4) has the following advantages:

- It canonically extends the Relativistic Dynamics Equation (1) and is covariant with respect to the *full Lorentz group*
- By redefining *uniformly accelerated motion* as the solutions to (4) when A is *constant*, we obtain the clarification that Einstein was looking for
- It admits four Lorentz-invariant classes of solutions: null acceleration, linear acceleration, rotational acceleration, and general acceleration. The null, rotational, and general classes were previously unknown. The linear class is a *covariant extension* of 1D hyperbolic motion and contains the motion of an object in a *homogeneous gravitational field*

- It can be extended in a straightforward manner to obtain a covariant definition of the "comoving frame" of a uniformly accelerated observer. In this comoving frame, all of the solutions of (4) have constant acceleration.
- It can be modified to accommodate a universal maximal acceleration. Thus, this paper is an important step in the study of evidence for and implications of the existence of a universal maximal acceleration (see [11, 12, 13]).

The plan of the paper is as follows. In section 2, we derive a 4D Lorentz covariant *Relativistic Dynamics Equation.* We also show that equations (1) and (2) are *not* Lorentz covariant. They are, however, canonically embedded in (4). By taking the tensor A to be constant, we obtain a covariant definition of *uniformly accelerated motion.* In section 3, we obtain *explicit* solutions to equation (4) in the case of a constant force. Our solutions are divided into four Lorentz-invariant types: null acceleration, linear acceleration, rotational acceleration and general acceleration. The linear acceleration is a covariant extension of 1D hyperbolic motion. The rotational class is a covariant extension of rotations. By integrating our solutions, we obtain trajectories of a uniformly accelerated observer. We conclude this section by computing the nonrelativistic limits of our solutions.

In section 4, we attach to our observer a *uniformly accelerated frame*, which is defined covariantly by extending equation (4). Our technique is essentially that of Generalized Fermi-Walker transport and provides a covariant definition of the *comoving frame of a uniformly accelerated observer.* We then show that all of our solutions to (4) have *constant acceleration in the comoving frame.* We also show that our definition of the comoving frame is equivalent to that of Mashhoon [14]. We also have the surprising result that if two uniformly accelerated frames have a common acceleration tensor A, then the spacetime transformations between them are *Lorentz*, despite the fact that neither frame is inertial. In section 5, we use the Weak Hypothesis of Locality to compute the spacetime transformations from a uniformly accelerated frame to an inertial frame. We show that these transformations extend the Lorentz transformations. Section 6 is devoted to examples of these transformations.

In section 7, we adapt Horwitz and Piron's notion of "off-shell" [15] to the four-velocity. We call the new notion the *4D velocity* and use it to derive velocity transformations from K' to K. We show that when K' is inertial, our velocity transformations reduce to the usual Einstein velocity addition. Using the ratio between the four-velocity and the 4D velocity, we obtain, in section 8, the general formula for the time dilation between clocks located at different positions in K'. We also derive a formula for the *angular velocity of a uniformly accelerated body.* In section 9, we consider an observer located at an arbitrary point (not necessarily the origin, as has been the case until now) of a uniformly accelerated frame and show that this observer is also uniformly accelerated. We then derive the transformation formulas for the components $\mathbf{g}$ and $\boldsymbol{\omega}$ of the acceleration tensor A. Section 10 is devoted to acceleration transformations from K' to K. We use our acceleration transformations to explain the Lorentz-Abraham-Dirac equation. In section 11, we discuss the continuation of this research.

Accelerated reference frames were also studied in [16] and [17].

2. Covariant Relativistic Dynamics Equation

In this section, we derive a 4D Lorentz covariant *Relativistic Dynamics Equation* (equation (22) below). We also show that equations (1) and (2) are *not* Lorentz covariant. They are, however, canonically embedded in (22). First, we will review some basic notions and establish our notation.

2.1. Basic Notions

In flat Minkowski space, the spacetime coordinates of an event are denoted by x^μ ($\mu = 0, 1, 2, 3$), with $x^0 = ct$. The *Minkowski inner product* is

$$x \cdot y = \eta_{\mu\nu} x^\mu y^\nu, \tag{5}$$

where η is the Minkowski metric $\eta_{\mu\nu} = \text{diag}(1, -1, -1, -1)$. The worldline of a particle is $x(t) = (ct, \mathbf{x}(t))$. The particle's 3D velocity is $\mathbf{v} = \frac{d\mathbf{x}}{dt}$. Then $\frac{dx}{dt} = (c, \mathbf{v})$, and the dimensionless scalar γ is defined by

$$\gamma = \gamma(\mathbf{v}) = \frac{1}{\left|\frac{dx}{cdt}\right|} = \frac{1}{\sqrt{1 - \frac{\mathbf{v}^2}{c^2}}}. \tag{6}$$

The particle's *proper time*, denoted by τ, is defined by

$$\gamma d\tau = dt. \tag{7}$$

Since $cd\tau = ds$, where ds is the differential of arc length along the particle's worldline, the proper time is a Lorentz invariant quantity. The particle's dimensionless *four-velocity* u^μ is defined, as usual, by

$$u^\mu = \frac{dx^\mu}{ds} = \frac{1}{c}\frac{dx^\mu}{d\tau}, \quad (u^0, u^1, u^2, u^3) = \gamma\left(1, \frac{\mathbf{v}}{c}\right), \tag{8}$$

and its *proper velocity* $\mathbf{u}$ is c times the spatial part of the four-velocity:

$$\mathbf{u} = \gamma(\mathbf{v})\mathbf{v}. \tag{9}$$

A straightforward calculation shows that we can write γ as a function of the proper velocity:

$$\gamma = \sqrt{1 + \frac{\mathbf{u}^2}{c^2}}. \tag{10}$$

The four-velocity always has "length" 1 in the Minkowski metric:

$$|u| = \sqrt{\eta_{\mu\nu} u^\mu u^\nu} = 1. \tag{11}$$

The particle's *four-acceleration* a^μ is defined by

$$a^\mu = c\frac{du^\mu}{d\tau} \tag{12}$$

and has units of acceleration. Differentiating $u \cdot u = 1$, we see that the four-acceleration and the four-velocity are always perpendicular:

$$u \cdot a = \eta_{\mu\nu} u^\mu a^\nu = 0. \tag{13}$$

This implies that the four-acceleration is spacelike.

The *rest-mass* of an object is denoted by m_0, and we let $m = m(\mathbf{v}) = m_0\gamma = m_0\gamma(\mathbf{v})$. The *3D momentum* is $\mathbf{p} = m\mathbf{v} = m_0\gamma\mathbf{v} = m_0\mathbf{u}$, and the *four-momentum* is $p = m_0 c u = (m_0\gamma c, m_0\gamma\mathbf{v}) = (mc, \mathbf{p})$.

2.2. Embedding $\mathbf{F} = \frac{d\mathbf{p}}{dt}$ *in Four Dimensions*

The standard *Relativistic Dynamics Equation* is the 3D equation

$$\mathbf{F} = \frac{d\mathbf{p}}{dt}. \tag{14}$$

In special relativity, however, we require a 4D version of this equation. Since the 3D vector $\mathbf{p}$ is part of the four-momentum p, we seek an appropriate expression for $\frac{dp}{d\tau}$ (or $\frac{du}{d\tau}$).

It is natural to consider the 4D equation

$$F = \frac{dp}{d\tau}. \tag{15}$$

This equation, however, *has no solution* when F is a *constant* four-vector. To see this, suppose F is constant. Then, since $F \sim a$, equation (13) implies that a is both lightlike and perpendicular to the timelike vector u, which is impossible. This implies that the four-acceleration *cannot* be constant in an inertial frame. Hence, equation (15) cannot be used to model constant-force motion and is inappropriate as a dynamics equation.

Next, we show that $\mathbf{F} = \frac{d\mathbf{p}}{dt} = m_0\frac{d\mathbf{u}}{dt}$ can be written in the form $c\frac{du}{d\tau} = Au$, where A is an antisymmetric tensor. Since $u = (\gamma, \mathbf{u}/c)$, we have

$$\frac{du}{d\tau} = \left(\frac{d\gamma}{d\tau}, \frac{d\mathbf{u}}{cd\tau}\right). \tag{16}$$

Using (10) and then (7), we have

$$\frac{d\gamma}{d\tau} = \frac{d\gamma}{d\mathbf{u}}\frac{d\mathbf{u}}{d\tau} = \frac{\mathbf{u}/c^2}{\gamma}\cdot\gamma\frac{d\mathbf{u}}{dt} = \frac{\mathbf{u}}{c^2}\cdot\frac{d\mathbf{u}}{dt} = \frac{1}{m_0c^2}\mathbf{u}\cdot\mathbf{F}, \tag{17}$$

and

$$\frac{d\mathbf{u}}{d\tau} = \gamma\frac{d\mathbf{u}}{dt} = \frac{\gamma}{m_0}\mathbf{F}. \tag{18}$$

Combining (17) and (18), we have

$$c\frac{du}{d\tau} = \frac{1}{m_0}\begin{pmatrix} 0 & \mathbf{F}^T \\ \mathbf{F} & 0 \end{pmatrix} u, \tag{19}$$

where the superscript T denotes matrix transposition. This shows that the 3D Relativistic Dynamics Equation (14) is equivalent to

$$c\frac{du}{d\tau} = Au, \quad \text{with} \quad A = \frac{1}{m_0}\begin{pmatrix} 0 & \mathbf{F}^T \\ \mathbf{F} & 0 \end{pmatrix}. \tag{20}$$

Note that A is an antisymmetric tensor of the particular form $\begin{pmatrix} 0 & \mathbf{g}^T \\ \mathbf{g} & 0 \end{pmatrix}$, where $\mathbf{g} = \frac{1}{m_0}\mathbf{F}$. As an operator, $A = A^\mu_\nu$ has mixed indices, one upper and one lower. If we lower the upper index using the Minkowski metric, $A_{\mu\nu}$ is antisymmetric.

2.3. Achieving Lorentz Covariance

A tensor of the form $\begin{pmatrix} 0 & \mathbf{g}^T \\ \mathbf{g} & 0 \end{pmatrix}$ is not Lorentz covariant. In fact, a Lorentz transformation of such a tensor will, in general, produce *any* antisymmetric tensor. Therefore, in order to achieve Lorentz covariance, we must allow A to be *any* antisymmetric tensor. In fact, if $F = Au$, where is a tensor, then A *must* be antisymmetric. To see this, first note that since, from (15), $F \sim a$, we have, from (13), that $u \cdot F = 0$. Substituting $F = Au$, we obtain

$$0 = \eta_{\mu\nu} u^\mu F^\nu = \eta_{\mu\nu} u^\mu A^\nu_\alpha u^\alpha = u^\mu A_{\mu\alpha} u^\alpha,$$

and so

$$A_{\alpha\beta} = -A_{\beta\alpha}. \tag{21}$$

The need for antisymmetry can be understood as follows. The Lorentz transformation produces a *rotation* in Minkowski spacetime. Similarly, *acceleration* can be interpreted as a rotation of the four-velocity, since the four-acceleration is perpendicular to the four-velocity. It is known that a rotation in 3D Euclidean space is given by the exponent of an antisymmetric tensor. The antisymmetry of A is the 4D extension of this fact.

We thus arrive at a *4D Lorentz covariant Relativistic Dynamics Equation*

$$\boxed{c\frac{du^\mu}{d\tau} = A^\mu_\nu u^\nu} \tag{22}$$

where $A_{\mu\nu}$ is an antisymmetric rank 2 tensor, or, equivalently, A^μ_ν is skew adjoint with respect to the inner product (5). The components of A have units of acceleration and may be *functions* of the position x and the four-velocity u. We refer to A as the *acceleration tensor* associated with the given motion. Equation (22) solves the first problem mentioned in the introduction.

In the $1+3$ decomposition, the tensors $A_{\mu\nu}$ and A^μ_ν take the form

$$A_{\mu\nu}(\mathbf{g},\boldsymbol{\omega}) = \begin{pmatrix} 0 & \mathbf{g}^T \\ -\mathbf{g} & -c\pi(\boldsymbol{\omega}) \end{pmatrix}, \quad A^\mu_\nu(\mathbf{g},\boldsymbol{\omega}) = \begin{pmatrix} 0 & \mathbf{g}^T \\ \mathbf{g} & c\pi(\boldsymbol{\omega}) \end{pmatrix}, \tag{23}$$

where $\mathbf{g}$ is a 3D vector with units of acceleration, $\boldsymbol{\omega}$ is a 3D vector with units of 1/time, and, for any 3D vector $\boldsymbol{\omega} = (\omega^1, \omega^2, \omega^3)$,

$$\pi(\boldsymbol{\omega}) = \varepsilon_{ijk}\omega^k,$$

where ε_{ijk} is the Levi-Civita tensor. The factor c in A provides the necessary units of acceleration. The 3D vectors $\mathbf{g}$ and $\boldsymbol{\omega}$ are related to the linear, or translational, acceleration and the angular velocity, respectively, of the motion. We will obtain a more precise explanation of the physical meaning of these vectors in sections 4 and 8. We will also show there that if a uniformly accelerated system was at rest at time $t = 0$, then $\mathbf{F}/m_0$ is the constant acceleration in the comoving frame.

By (20), $\mathbf{F} = \frac{d\mathbf{p}}{dt}$ is equivalent to a 4D equation

$$c\frac{du}{d\tau} = \begin{pmatrix} 0 & \mathbf{g}^T \\ \mathbf{g} & c\pi(\boldsymbol{\omega}) \end{pmatrix} u, \tag{24}$$

where $\boldsymbol{\omega} = 0$. Hence, $\mathbf{F} = \frac{d\mathbf{p}}{dt}$ is covariant only with respect to transformations which preserve the condition $\boldsymbol{\omega} = 0$. A straightforward calculation shows that the only Lorentz transformations which preserve the condition $\boldsymbol{\omega} = 0$ are boosts in the direction of $\mathbf{F}$ and spatial rotations about the axis of $\mathbf{F}$. The little Lorentz group, as defined in [18], is the stabilizer of the spatial axis in a given direction, which we may choose to be the direction of the force $\mathbf{F}$. Thus, $\mathbf{F} = \frac{d\mathbf{p}}{dt}$ is covariant only with respect to this little Lorentz group and *not* to the full Lorentz group. It follows immediately that *1D hyperbolic motion* is also covariant only with respect to this little Lorentz group.

For an additional proof, note that the spatial part of (24) is

$$\frac{d\mathbf{u}}{d\tau} = \gamma\mathbf{g} + \mathbf{u} \times \boldsymbol{\omega}, \quad \text{or} \quad \frac{d\mathbf{p}}{dt} = \mathbf{F} + \gamma^{-1}\mathbf{p} \times \boldsymbol{\omega}. \tag{25}$$

Hence, $\mathbf{F} = \frac{d\mathbf{p}}{dt}$ if and only if $\boldsymbol{\omega} = 0$.

By taking the tensor A to have *constant* components, we obtain a covariant definition of uniformly accelerated motion, thus solving the second problem mentioned in the introduction. We define *uniformly accelerated motion* as motion whose four-velocity $u(\tau)$ is a solution to the initial value problem

$$c\frac{du^\mu}{d\tau} = A^\mu_\nu u^\nu \quad , \quad u(0) = u_0, \tag{26}$$

where $A_{\mu\nu}$ is an antisymmetric rank 2 tensor with *constant* components, or, equivalently, A^μ_ν is constant and skew adjoint with respect to the inner product (5). In the next section, we compute *explicit* solutions to (26).

3. Explicit Trajectories for Uniformly Accelerated Motion

In this section, we obtain explicit trajectories for uniformly accelerated motion, that is, we obtain the explicit solutions $u(\tau)$ of (26). It is known (see [19], page **1**-65) that for a given initial condition $u(0) = u_0$, (26) has the unique solution

$$u(\tau) = \exp(A\tau/c)u_0 = \left(\sum_{n=0}^{\infty} \frac{A^n}{n!c^n}\tau^n\right) u_0\,. \tag{27}$$

The worldline $\widehat{x}(\tau)$ of a uniformly accelerated observer may then be obtained by integrating $u(\tau)$.

Since A is antisymmetric, all solutions of the form (27) are Lorentz transformations of the initial velocity u_0, with an angle that is linear in τ.

3.1. Lorentz-Invariant Classification of Solutions

Fix $A = A^\mu_\nu(\mathbf{g}, \boldsymbol{\omega})$ as in (23). It can be shown by direct calculation that

$$\det(A - aI) = a^4 - l_1 a^2 - c^2(l_2)^2, \tag{28}$$

where $l_1 = \mathbf{g}^2 - c^2\boldsymbol{\omega}^2$ and $l_2 = \mathbf{g} \cdot \boldsymbol{\omega}$ are Lorentz invariants, similar to the two known Lorentz invariants associated with the electromagnetic field. Hence, the matrix $A^\mu_\nu(\mathbf{g}, \boldsymbol{\omega})$ (one upper index, one lower index) has the eigenvalues $\pm\alpha$ and $\pm i\beta$, where

$$\alpha = \sqrt{\frac{\sqrt{(l_1)^2 + 4(cl_2)^2} + l_1}{2}} \quad \text{and} \quad \beta = \sqrt{\frac{\sqrt{(l_1)^2 + 4(cl_2)^2} - l_1}{2}}.$$

We classify our solutions for uniformly accelerated motion into four types, depending on the values of the Lorentz invariants and the eigenvalues of $A^\mu_\nu(\mathbf{g},\boldsymbol{\omega})$:

Type	Lorentz invariants	Eigenvalues
Null	$l_1 = l_2 = 0$	$\alpha = 0, \beta = 0$
Linear	$l_1 > 0, l_2 = 0$	$\alpha = \sqrt{l_1} > 0, \beta = 0$
Rotational	$l_1 < 0, l_2 = 0$	$\alpha = 0, \beta = \sqrt{-l_1} > 0$
General	$l_2 \neq 0$	$\alpha > 0, \beta > 0$

Note that each type is a *Lorentz-invariant subset.*

For each of the four types of uniformly accelerated motion, we now obtain the explicit solutions $u(\tau)$ of (26).

3.2. Null Acceleration $(\alpha = 0, \beta = 0)$

In this case, $|\mathbf{g}| = c|\boldsymbol{\omega}|$ and $\mathbf{g} \perp \boldsymbol{\omega}$. Direct calculation shows that $A^3 = 0$. Thus, from (27), we have

$$\boxed{u(\tau) = u(0) + Au(0)\tau/c + \frac{1}{2}A^2u(0)\tau^2/c^2} \tag{29}$$

The four-acceleration is

$$a(\tau) = c\frac{du}{d\tau} = Au(0) + A^2u(0)\tau/c. \tag{30}$$

Despite the apparent dependence of the four-acceleration on τ, we will show in section 6 that the length of $a(\tau)$ is, in fact, constant.

The worldline $\widehat{x}(\tau)$ of a uniformly accelerated observer is, then, *cubic* in τ and may be obtained by integrating $u(\tau)$:

$$\widehat{x}(\tau) = \widehat{x}(0) + c\int_0^\tau u(s)ds = \widehat{x}(0) + u(0)c\tau + \frac{1}{2}Au(0)\tau^2 + \frac{1}{6}A^2u(0)\tau^3/c. \tag{31}$$

A similar cubic equation was obtained in [20] and page 83 of [21].

Note that, in general, $A^2 \neq 0$, as can be seen from the example

$$A = \begin{pmatrix} 0 & 1 & 0 & 0 \\ 1 & 0 & 0 & -1 \\ 0 & 0 & 0 & 0 \\ 0 & 1 & 0 & 0 \end{pmatrix}, \quad A^2 = \begin{pmatrix} 1 & 0 & 0 & -1 \\ 0 & 0 & 0 & 0 \\ 0 & 0 & 0 & 0 \\ 1 & 0 & 0 & -1 \end{pmatrix}.$$

3.3. Linear, Rotational and General Acceleration

We will obtain the solutions for the remaining three types using the eigenvalues and eigenvectors of A. The following two claims use the *linear* extension η^* of the Minkowski inner product (5) to complex Minkowski space (see [22], p. 12).

Claim 1 Let A be a skew adjoint matrix with respect to η^*. Let a be a non-zero eigenvalue of A. If v is an eigenvector of A corresponding to a, then v is lightlike.

To see this, note that

$$av^2 = v \cdot av = v \cdot Av = -Av \cdot v = -av \cdot v = -av^2.$$

Since $a \neq 0$, we must have $v^2 = 0$. This proves the claim.

Claim 2 Let A be a skew adjoint matrix with respect to η^*. If v_1 is an eigenvector corresponding to the eigenvalue a, and v_2 is an eigenvector corresponding to the eigenvalue μ, where $a \neq -\mu$, then v_1 and v_2 are orthogonal.

To see this, note that

$$a(v_1 \cdot v_2) = av_1 \cdot v_2 = Av_1 \cdot v_2 = v_1 \cdot -Av_2 = v_1 \cdot -\mu v_2 = -\mu(v_1 \cdot v_2).$$

Since $a \neq -\mu$, we must have $v_1 \cdot v_2 = 0$. This proves the claim.

We will also need the following notation. Let r be a real four-vector, and let $\{E_0, E_1, E_2, E_3\}$ be linearly independent (possibly complex) four-vectors. Let c_0, c_1, c_2, c_3 be the unique (possibly complex) numbers such that $r = \sum_{k=0}^{3} c_k E_k$. Let $\Im(z)$ denote the imaginary part of a complex scalar or vector z. Define

$$\begin{aligned} D_0(r) &= c_0 E_0 + c_1 E_1, \quad D_1(r) = c_0 E_0 - c_1 E_1, \\ D_2(r) &= c_2 E_2 + c_3 E_3, \quad D_3(r) = -2\Im(c_2 E_2). \end{aligned} \tag{32}$$

3.4. Linear Acceleration $(\alpha > 0, \beta = 0)$

In this case, we have $\alpha = \sqrt{\mathbf{g}^2 - c^2\boldsymbol{\omega}^2}$ and $\mathbf{g} \perp \boldsymbol{\omega}$. Let E_0 and E_1 be eigenvectors corresponding to the eigenvalues α and $-\alpha$, respectively, and let E_2 and E_3 be linearly independent eigenvectors corresponding to the eigenvalue 0. Note that we may choose all of the eigenvectors to be real. Then

$$u(\tau) = c_0 E_0 e^{\alpha\tau/c} + c_1 E_1 e^{-\alpha\tau/c} + c_2 E_2 + c_3 E_3,$$

where $u(0) = \sum_{k=0}^{3} c_k E_k$. Using (32), with $D_\mu = D_\mu(u(0))$, we have

$$\boxed{u(\tau) = D_0 \cosh(\alpha\tau/c) + D_1 \sinh(\alpha\tau/c) + D_2} \tag{33}$$

We claim that D_0, D_1, D_2 are mutually orthogonal. By Claim 2, $D_2 \in \text{Span}(E_2, E_3)$ is orthogonal to both D_0 and D_1, which belong to $\text{Span}(E_0, E_1)$. To show that $D_0 \cdot D_1 = 0$, note that $E_0 = (1/2c_0)(D_0 + D_1)$ and $E_1 = (1/2c_1)(D_0 - D_1)$ are lightlike, by Claim 1. Hence, $(D_0)^2 \pm 2D_0 \cdot D_1 + (D_1)^2 = 0$, implying that $D_0 \cdot D_1 = 0$. This also implies that $(D_1)^2 = -(D_0)^2$.

From (12) and (33), we obtain

$$a(\tau) = \alpha D_0 \sinh(\alpha\tau/c) + \alpha D_1 \cosh(\alpha\tau/c). \tag{34}$$

The length of $a(\tau)$ is constant:

$$a^2(\tau) = -\alpha^2 (D_0)^2, \tag{35}$$

and thus satisfies (3). Since a is spacelike, D_0 is timelike.

The interpretation of the solutions (33) are as follows. In the plane generated by D_0 and D_1, there is 1D hyperbolic motion, and this plane is moving in a normal direction with four-velocity D_2. These solutions form a *covariant extension of 1D hyperbolic motion.*

3.5. Rotational Acceleration $(\alpha = 0, \beta > 0)$

In this case, we have $\beta = \sqrt{c^2\boldsymbol{\omega}^2 - \mathbf{g}^2}$ and $\mathbf{g} \perp \boldsymbol{\omega}$. Let E_2 and E_3 be eigenvectors corresponding to the eigenvalues $i\beta$ and $-i\beta$, respectively, and let E_0 and E_1 be linearly independent eigenvectors corresponding to the eigenvalue 0. Since the two complex eigenvalues are complex conjugates of each other, we may choose $E_3 = \overline{E}_2$. Then

$$u(\tau) = c_0 E_0 + c_1 E_1 + c_2 E_2 e^{i\beta\tau/c} + c_3 \overline{E}_2 e^{-i\beta\tau/c},$$

where $u(0) = \sum_{k=0}^{3} c_k E_k$. Since $u(\tau)$ is real, we must have $c_3 = \overline{c_2}$. Using (32), with $D_\mu = D_\mu(u(0))$, we have

$$\boxed{u(\tau) = D_0 + D_2 \cos(\beta\tau/c) + D_3 \sin(\beta\tau/c)} \tag{36}$$

We claim that D_0, D_2, D_3 are mutually orthogonal. By Claim 2, $D_0 \in \text{Span}(E_0, E_1)$ is orthogonal to both D_2 and D_3, which belong to $\text{Span}(E_2, E_3)$. To show that $D_2 \cdot D_3 = 0$, note that $E_2 = 1/2ic_2(iD_2 + D_3), E_3 = 1/2ic_3(iD_2 - D_3)$ are lightlike, by Claim 1. Hence, $-(D_2)^2 \pm 2iD_2 \cdot D_3 + (D_3)^2 = 0$, implying that $D_2 \cdot D_3 = 0$. This also implies that $(D_2)^2 = (D_3)^2$.

From (12) and (36), we obtain

$$a(\tau) = -\beta D_2 \sin(\beta\tau/c) + \beta D_3 \cos(\beta\tau/c). \tag{37}$$

The length of $a(\tau)$ is constant:

$$a^2(\tau) = \beta^2 (D_2)^2, \tag{38}$$

and thus satisfies (3). Since a is spacelike, D_2 is also spacelike. Substituting $\tau = 0$ into (36), we obtain that D_0 is timelike.

The interpretation of the solutions (36) are as follows. In the plane generated by D_2 and D_3, there is pure rotational motion, and this plane is moving in a normal direction with four-velocity D_0. The solutions (36) form a *covariant extension of pure rotational motion.*

3.6. General Acceleration $(\alpha > 0, \beta > 0)$

Since the four eigenvalues $\pm\alpha, \pm i\beta$ are distinct, there are linearly independent eigenvectors E_0, E_1, E_2, E_3 of $\alpha, -\alpha, i\beta, -i\beta$, respectively. Since the two complex eigenvalues are complex conjugates of each other, we may choose $E_3 = \overline{E}_2$. Then

$$u(\tau) = c_0 E_0 e^{\alpha\tau/c} + c_1 E_1 e^{-\alpha\tau/c} + c_2 E_2 e^{i\beta\tau/c} + c_3 \overline{E}_2 e^{-i\beta\tau/c},$$

where $u(0) = \sum_{k=0}^{3} c_k E_k$. Since $u(\tau)$ is real, we must have $c_3 = \overline{c_2}$. Using (32), with $D_\mu = D_\mu(u(0))$, we have

$$\boxed{u(\tau) = D_0 \cosh(\alpha\tau/c) + D_1 \sinh(\alpha\tau/c) + D_2 \cos(\beta\tau/c) + D_3 \sin(\beta\tau/c)} \tag{39}$$

As in the previous cases, the vectors D_0, D_1, D_2, D_3 are mutually orthogonal, $(D_1)^2 = -(D_0)^2$, and $(D_2)^2 = (D_3)^2$.

From (12) and (39), we obtain

$$a(\tau) = \alpha D_0 \sinh(\alpha\tau/c) + \alpha D_1 \cosh(\alpha\tau/c) - \beta D_2 \sin(\beta\tau/c) + \beta D_3 \cos(\beta\tau/c). \tag{40}$$

The length of $a(\tau)$ is constant:

$$a^2(\tau) = -\alpha^2 D_0^2 + \beta^2 D_2^2, \tag{41}$$

and thus satisfies (3).

Since, in this case, $\mathbf{g}$ and $\boldsymbol{\omega}$ are not perpendicular, there exists a basis in which they are parallel (see [8]). Here, we have in fact obtained the explicit form of this basis, namely, $\{D_0, D_1, D_2, D_3\}$.

In the plane generated by D_2 and D_3, there is pure rotational motion, and this plane is uniformly accelerated in a normal direction.

This completes all of the cases. The general solution to (26) is

$$u(\tau) = \left\{ \begin{array}{l} u(0) + Au(0)\tau/c + \frac{1}{2}A^2u(0)\tau^2/c^2 \text{ , if } \alpha = 0, \beta = 0 \text{ (null acceleration)} \\ \\ D_0\cosh(\alpha\tau/c) + D_1\sinh(\alpha\tau/c) + D_2 \text{ , if } \alpha > 0, \beta = 0 \text{ (linear acceleration)} \\ \\ D_0 + D_2\cos(\beta\tau/c) + D_3\sin(\beta\tau/c) \text{ , if } \alpha = 0, \beta > 0 \text{ (rotational acceleration)} \\ \\ D_0\cosh(\alpha\tau/c) + D_1\sinh(\alpha\tau/c) \\ +D_2\cos(\beta\tau/c) + D_3\sin(\beta\tau/c) \text{ , if } \alpha > 0, \beta > 0 \text{ (general acceleration)} \end{array} \right\}. \tag{42}$$

3.7. Nonrelativistic Limit

Here we compute the nonrelativistic limit ($c \to \infty$) for uniformly accelerated motion. Notice from (23) that c appears in the definition of the acceleration tensor A. However, this is only to provide the correct units. Thus, when taking the limit $c \to \infty$, we may consider $c\pi(\boldsymbol{\omega})$ as a *constant* $\pi(\boldsymbol{\omega}')$. We refer to this method as "holding the tensor A constant". This is equivalent to holding the eigenvalues α and β constant. Alternatively, when taking the limit $c \to \infty$, we may "hold the *components* $\mathbf{g}$ and $\boldsymbol{\omega}$ constant" and let the c of $c\pi(\boldsymbol{\omega})$ also go to infinity.

For null acceleration, we must hold the tensor A constant since letting the c of $c\pi(\boldsymbol{\omega})$ go to infinity breaks the condition $\mathbf{g}^2 = c^2\boldsymbol{\omega}^2$. Note that the spatial part of cu is the velocity $\mathbf{v}$ in the nonrelativistic limit, and, in particular, the spatial part of $cu(0)$ is $\mathbf{v}(0)$ in this limit. Thus, the nonrelativistic limit of (29) is

$$\mathbf{v}(t) = \lim_{c\to\infty} \left(\mathbf{v}(0) + A\mathbf{v}(0)\tau/c + O(c^{-2})\right) = \mathbf{v}(0), \tag{43}$$

implying that the velocity in such motion is constant. Thus, the nonrelativistic limit of null acceleration is *zero* acceleration. This justifies the name *null* acceleration.

Next, we compute the nonrelativistic limits for linear, rotational, and general acceleration when we hold the tensor A constant. First, we show that the 3D acceleration $\mathbf{a} = \frac{d\mathbf{v}}{dt}$ is the nonrelativistic limit of the spatial part of $a(\tau)$. To see this, note that the spatial part of $a = c\frac{du}{d\tau}$ is $\frac{d\mathbf{u}}{d\tau}$. Using (17), we have

$$\frac{d\mathbf{u}}{d\tau} = \frac{d}{d\tau}(\gamma\mathbf{v}) = \mathbf{v}\frac{d\gamma}{d\tau} + \gamma\frac{d\mathbf{v}}{d\tau} = \left(\frac{\mathbf{u}}{c^2}\cdot\frac{d\mathbf{u}}{dt}\right)\mathbf{v} + \gamma^2\frac{d\mathbf{v}}{dt}, \tag{44}$$

which tends to $\frac{d\mathbf{v}}{dt}$ in the nonrelativistic limit.

In the linear case, from (34), we have

$$\lim_{c\to\infty} a(\tau) = \alpha D_1. \tag{45}$$

Similarly, for the rotational case, from (37) we get $\lim_{c\to\infty} a(\tau) = \beta D_3$, and for general acceleration, from (40) we get $\lim_{c\to\infty} a(\tau) = \alpha D_1 + \beta D_3$. Since D_1 and D_3 do not depend on c, the 3D acceleration $\mathbf{a}$ is constant in all of these cases in the nonrelativistic limit.

To identify the acceleration, we take the nonrelativistic limit of the first equation of (25) and substitute $t = 0$. This gives $\mathbf{a} = \mathbf{g} + \mathbf{v}(0) \times \boldsymbol{\omega}$. Thus, the 3D velocity in the nonrelativistic limit is

$$\mathbf{v}(t) = \mathbf{v}(0) + (\mathbf{g} + \mathbf{v}(0) \times \boldsymbol{\omega})t. \tag{46}$$

The nonrelativistic limit is constant *linear* acceleration. For $\boldsymbol{\omega} = 0$, we obtain the usual nonrelativistic result. A nonzero value of $\boldsymbol{\omega}$, however, will change both the magnitude and the direction of the acceleration.

Finally, we compute the nonrelativistic limit by holding the *components* $\mathbf{g}$ and $\boldsymbol{\omega}$ constant. For linear acceleration, we have $\mathbf{g}^2 > c^2\boldsymbol{\omega}^2$, and so we cannot let $c \to \infty$. Thus, we will consider only rotational and general acceleration. Since, in this case, α, β and the D_i all vary with c, it is easier to pass to the nonrelativistic limit of the first equation of (25), which is

$$\frac{d\mathbf{v}}{dt} = \mathbf{g} + \mathbf{v} \times \boldsymbol{\omega}. \tag{47}$$

Equation (47) describes nonrelativistic motion under a Lorentz-type force. If $\mathbf{g} = 0$, we obtain rotation with uniform angular velocity $\boldsymbol{\omega}$. If $\boldsymbol{\omega} = 0$, we obtain linear acceleration.

4. Covariant Comoving Frame

In this section, we use Generalized Fermi-Walker transport to define the notion of the *comoving frame of a uniformly accelerated observer*. We then show that in this comoving frame, all of our solutions to equation (22) have *constant acceleration*. We show that our definition of the comoving frame is equivalent to that of Mashhoon [14]. We also show that if two uniformly accelerated frames have a common acceleration tensor A, then the spacetime transformations between them are *Lorentz*, despite the fact that neither frame is inertial.

Gupta and Padmanabhan [23] applied Fermi-Walker transport to an accelerated charge and obtained the Lorentz-Abraham-Dirac equation.

4.1. Uniformly Accelerated Family of Inertial Frames

First, we define the notion of a one-parameter family of inertial frames which are instantaneously comoving to a uniformly accelerated observer. We call such a family a *uniformly accelerated family*. The coordinates in this family of comoving frames will be used as a bridge between the observer's coordinates and the coordinates in an inertial frame K, which, without loss of generality, we may take to be the initial frame K_0 of the observer himself. The family of frames is constructed by Generalized Fermi-Walker transport of the initial frame K_0 along the worldline of the observer. In the case of 1D hyperbolic motion, this construction reduces to Fermi-Walker transport [24, 25].

In fact, Fermi-Walker transport may *only* be used in the case of 1D hyperbolic motion. This is because Fermi-Walker transport uses only a part of the Lorentz group - the boosts. This subset of the group, however, is not a *subgroup*, since the combination of two boosts entails a rotation. Generalized Fermi-Walker transport, on the other hand, uses the full homogeneous Lorentz group, and can be used for all four types of uniform acceleration: null, linear, rotational, and general.

The construction of the uniformly accelerated family $\{K_\tau : \tau \geq 0\}$ is according to the following definition.

Definition 1. *Let $\widehat{x}(\tau)$ be the worldline of a uniformly accelerated observer whose motion is determined by the acceleration tensor A, the initial four-velocity $u(0)$, and the initial position $\widehat{x}(0)$.*

To specify the initial frame K_0, *we take the origin at time $\tau = 0$ to be $\widehat{x}(0)$. For the basis of K_0, choose any orthonormal basis $\widehat{\lambda} = \{u(0), \widehat{\lambda}_{(1)}, \widehat{\lambda}_{(2)}, \widehat{\lambda}_{(3)}\}$.*

For each $\tau > 0$, define K_τ as follows. The origin of K_τ at time τ is set as $\widehat{x}(\tau)$. The basis of K_τ is defined to be the unique solution $\lambda(\tau) = \{\lambda_{(\kappa)}(\tau) : \kappa = 0, 1, 2, 3\}$, to the initial value *problem*

$$c\frac{d\lambda^{\mu}_{(\kappa)}}{d\tau} = A^{\mu}_{\nu}\lambda^{\nu}_{(\kappa)} \quad , \quad \lambda_{(\kappa)}(0) = \widehat{\lambda}_{(\kappa)}. \tag{48}$$

We make the following observations about definition 1:

(1) The choice of the initial four-velocity $u(0)$ for $\widehat{\lambda}_{(0)}$ is deliberate and required by Generalized Fermi-Walker transport.

(2) For all τ, we have $\lambda_{(0)}(\tau) = u(\tau)$. This follows immediately from (26).

(3) From (27), the unique solution to (48) is

$$\lambda_{(\kappa)}(\tau) = \exp(A\tau/c)\widehat{\lambda}_{(\kappa)}. \tag{49}$$

Using matrix multiplication, we combine the solutions for $\kappa = 0, 1, 2, 3$ into one equation

$$\lambda(\tau) = \exp(A\tau/c)\widehat{\lambda}. \tag{50}$$

(4) Since A is antisymmetric, $\exp(A\tau/c)$ is an *isometry*. Thus, $\lambda(\tau)$ is an orthonormal basis.

(5) Analogously to (42), the general solution to (48) is

$$\lambda_{(\kappa)}(\tau) = \left\{ \begin{array}{l} \widehat{\lambda}_{(\kappa)} + A\widehat{\lambda}_{(\kappa)}\tau/c + \frac{1}{2}A^2\widehat{\lambda}_{(\kappa)}\tau^2/c^2 \text{ , if } \alpha = 0, \beta = 0 \text{ (null acceleration)} \\[2ex] D_0(\widehat{\lambda}_{(\kappa)})\cosh(\alpha\tau/c) + D_1(\widehat{\lambda}_{(\kappa)})\sinh(\alpha\tau/c) + D_2(\widehat{\lambda}_{(\kappa)}), \\ \qquad \text{if } \alpha > 0, \beta = 0 \text{ (linear acceleration)} \\[2ex] D_0(\widehat{\lambda}_{(\kappa)}) + D_2(\widehat{\lambda}_{(\kappa)})\cos(\beta\tau/c) + D_3(\widehat{\lambda}_{(\kappa)})\sin(\beta\tau/c), \\ \qquad \text{if } \alpha = 0, \beta > 0 \text{ (rotational acceleration)} \\[2ex] D_0(\widehat{\lambda}_{(\kappa)})\cosh(\alpha\tau/c) + D_1(\widehat{\lambda}_{(\kappa)})\sinh(\alpha\tau/c) \\ +D_2(\widehat{\lambda}_{(\kappa)})\cos(\beta\tau/c) + D_3(\widehat{\lambda}_{(\kappa)})\sin(\beta\tau/c) \text{ , if } \alpha > 0, \beta > 0 \text{ (general acceleration)} \end{array} \right\}. \tag{51}$$

(6) For a given A, all four solutions $\lambda_{(\kappa)}(\tau), \kappa = 0, 1, 2, 3$ are of the same type (null, linear, rotational, or general). The comoving frame of a rotating observer "rotates" along with him, and he feels that in *this* frame, the acceleration is constant.

(7) Along the worldline, the components of the acceleration tensor A remain constant. To see this, let A denote the tensor as computed in the lab frame K, and let $A(\tau)$ denote the tensor as computed in the comoving frame K_τ. Then, since $\lambda(\tau)$ is the change of matrix basis from K to K_τ, we have

$$A(\tau) = \lambda(\tau)^{-1}A\lambda(\tau) = (\exp(A\tau/c)\widehat{\lambda})^{-1}A\exp(A\tau/c)\widehat{\lambda} = \widehat{\lambda}^{-1}A\widehat{\lambda} = A(0). \tag{52}$$

(8) For all τ, we have $\lambda(\tau)A(\tau) = A\lambda(\tau)$.

4.2. Uniformly Accelerated Frame

Two frames are said to be *comoving* at time τ if at this time, the origins of the two frames coincide, their respective axes are parallel, and they have the same four-velocity.

We now define the notion of a *uniformly accelerated frame.*

Definition 2. *A frame K' is* uniformly accelerated *if there exists a uniformly accelerated family $\{K_\tau(A, \widehat{x}(0), \widehat{\lambda})\}$ such that at every time τ, the frame K_τ is comoving to K'.*

In light of this definition, we may regard our uniformly accelerated observer as positioned at the spatial origin of a uniformly accelerated frame. This approach is motivated by the following statement of Brillouin [26]: a frame of reference is a "heavy laboratory, built on a rigid body of tremendous mass, as compared to the masses in motion."

Our construction of a uniformly accelerated family should be contrasted with Mashhoon's approach [14], which is well suited to curved spacetime, or a manifold setting. There, the orthonormal basis is defined by

$$c\frac{d\lambda^{\mu}_{(\kappa)}}{d\tau} = \widetilde{A}^{(\nu)}_{(\kappa)}\lambda^{\mu}_{(\nu)}, \tag{53}$$

where $\widetilde{A} = \widetilde{A}_{\mu\nu}$ is an antisymmetric tensor. Notice that the derivative of each of Mashhoon's basis vectors depends on *all* of the basis vectors, whereas the derivative of each of our basis vectors depends only on its own components. In particular, Mashhoon's observer's four-acceleration depends on both his four-velocity $\lambda_{(0)}$ *and* on the spatial vectors of his basis, while our observers's four-acceleration depends *only* on his four-velocity. This seems to be the more natural physical model: is there any *a priori* reason why the four-acceleration of the observer should depend on his *spatial* basis? We show now, however, that the two approaches are, in fact, equivalent.

The two approaches are equivalent if we identify Mashhoon's tensor $\widetilde{A}$ as our tensor A computed along the worldline: $\widetilde{A} = A(\tau) = A(0)$. Then, by equation (48) and observation (8) above, we have

$$c\frac{d\lambda^{\mu}_{(\kappa)}}{d\tau} = A^{\mu}_{\nu}\lambda^{\nu}_{(\kappa)} = \lambda^{\mu}_{(\nu)}\widetilde{A}^{(\nu)}_{(\kappa)},$$

which is (53).

Despite the equivalence of the two approaches, the four-acceleration of Mashhoon's observer *does* depend on his spatial basis. This is because the components of Mashhoon's tensor $\widetilde{A}$ are computed in the comoving frame. In *this* frame, the observer's spatial basis must be perpendicular to his four-velocity. In the inertial lab frame K, on the other hand, there is no such restriction.

Unless specifically mentioned otherwise, we will always choose the lab frame K to be the initial comoving frame K_0. This implies that $\widehat{\lambda} = I$. Moreover, *we will always use the acceleration tensor as computed in the initial comoving frame* K_0 *and denote it by* A *instead of by* $\widetilde{A}$.

We now show that all of our solutions of equation (26) have constant acceleration in the comoving frame. Let A be as in (23). Since $u(0) = (1,0,0,0)^T$, $u(\tau) = \exp(A\tau/c)u(0)$, and $A\lambda = \lambda A$, we have

$$a(\tau) = Au(\tau) = A\exp(A\tau/c)u(0) = A\lambda(\tau)u(0) = \lambda(\tau)Au(0) = \lambda_{(i)}(\tau)\mathbf{g}^{(i)}. \tag{54}$$

Thus, the acceleration of the observer in the comoving frame is constant and equals $\mathbf{g}$.

We end this section by showing that if K' and K'' are two uniformly accelerated frames with a common acceleration tensor A, then the spacetime transformations between K' and K'' are *Lorentz*, despite the fact that neither K' nor K'' is inertial:

$$\exp(\widehat{\lambda}A\widehat{\lambda}^{-1}\tau/c)\widehat{\lambda} = \widehat{\lambda}\exp(A\tau/c)\widehat{\lambda}^{-1}\widehat{\lambda} = \widehat{\lambda}\exp(A\tau/c). \tag{55}$$

This implies, in particular, that there is a Lorentz transformation from a lab frame on Earth to an airplane flying at constant velocity, since we are both subject to the same gravitational field.

5. Spacetime Transformations from a K' to K

In this section, we construct the spacetime transformations from a uniformly accelerated frame K' to the inertial frame $K = K_0$. This will be done in two steps.

Step 1: From K_τ to K

First, we will derive the spacetime transformations from K_τ to K. The idea here is as follows. Let $\widehat{x}(\tau)$ be the worldline of a uniformly accelerated observer. Fix an event X. Find the time τ for which $\widehat{x}(\tau)$ is simultaneous to X in the comoving frame K_τ. Define the 0-coordinate in K_τ to be $y^{(0)} = c\tau$. Use the basis $\lambda(\tau)$ of K_τ to write the relative spatial displacement of the event X with respect to the observer as $y^{(i)}\lambda_{(i)}(\tau), i = 1,2,3$. The spacetime transformation from K_τ to K is then defined to be

$$x^\mu = \widehat{x}^\mu(\tau) + y^{(i)}\lambda^\mu_{(i)}(\tau). \tag{56}$$

Transformations of the form (56) have a natural physical interpretation: the vector sum of the *observer* positioned at the origin of the comoving frame K_τ and the *spatial coordinates* of the event as measured in K_τ. These transformations were also used in [14]. A similar construction can be found in [27], in which the authors use *radar 4-coordinates*, and in [28], but in the less general setting of (non-rotating) Fermi-Walker transport.

The above construction relies on the splitting of spacetime into *locally* disjoint 3D spatial hyperplanes X_τ. This is indeed possible. Since X_τ is *perpendicular* to $u(\tau)$, there exist a neighborhood of τ and a spatial neighborhood of the observer in which the X_τ are pairwise disjoint. This insures that, at least locally, the same event does not occur at two different times. Hence, the observer may uniquely define coordinates for any event within the locality restriction. Thus, at least *locally*, the spacetime transformations from K_τ to K are given by (56).

Note then when K' is inertial ($A = 0$), one may split spacetime into *globally* pairwise disjoint 3D spatial hyperplanes. In this case, there are *no* locality restrictions, and the transformations (56) are defined *everywhere*. We show now, in fact, that the transformations (56) extend the *Lorentz transformations*.

The Lorentz transformations are normally written

$$x^\mu = \Lambda^\mu_\nu y^\nu, \quad x^\mu = (x^0 = ct, x^1, x^2, x^3), \tag{57}$$

where the 4×4 matrix Λ has *constant* entries. Note, however, that the *columns* $\lambda_{(0)}, \lambda_{(1)}, \lambda_{(2)}, \lambda_{(3)}$ of Λ form an orthonormal basis of Minkowski space and that $\lambda_{(0)} = u$, the four-velocity of the observer. Rewriting (57) in terms of this basis, we obtain

$$x^\mu = y^{(\nu)}\lambda^\mu_{(\nu)}. \tag{58}$$

Separating out the time component, we arrive at

$$x = y^{(0)}\lambda_{(0)} + y^{(i)}\lambda_{(i)}, \tag{59}$$

showing that the coordinates of an event in the lab frame K are the usual vector sum of the *observer* positioned at the origin of the comoving frame K_τ and the *spatial coordinates* of the event as measured in K_τ.

Step 2: From K' to K_τ

At this point, we invoke a weaker form of the Hypothesis of Locality introduced by Mashhoon [29, 30]. This *Weak Hypothesis of Locality* is an extension of the Clock Hypothesis.

The Weak Hypothesis of Locality *Let K' be a uniformly accelerated frame, with an accelerated observer with worldline $\widehat{x}(\tau)$. For any time τ_0, the rates of the clock of the accelerated* *observer and the clock at the origin of the comoving frame K_{τ_0} are the same, and, for events simultaneous to $\widehat{x}(\tau_0)$ in the comoving frame K_{τ_0}, the comoving and the accelerated observers measure the same spatial components.*

Consider an event X. By step 1, the coordinates of X in K are $x = \widehat{x}(\tau_0) + y^{(i)}\lambda_{(i)}(\tau_0)$, where τ_0 is the unique value of τ such that X and $\widehat{x}(\tau_0)$ are simultaneous in the comoving frame, and $(y^{(0)} = c\tau_0, y^{(i)})$ are the coordinates of X in K_{τ_0}. Since x and $\widehat{x}(\tau_0)$ are simultaneous in K_{τ_0}, the Weak Hypothesis of Locality implies that the spatial coordinates $y^{(i)}$ coincide with the spatial coordinates in K'. Therefore, *the spacetime transformations from K' to K* are

$$x = \widehat{x}(\tau) + y^{(i)}\lambda_{(i)}(\tau), \quad \text{with } \tau = y^{(0)}/c. \tag{60}$$

We end this section by calculating the metric at the point y of K'. First, we calculate the differential of the transformation (60). Differentiating (60), we have

$$dx = \lambda_{(0)}(\tau)dy^{(0)} + \lambda_{(i)}(\tau)dy^{(i)} + y^{(i)}\frac{1}{c}\frac{d\lambda_{(i)}}{d\tau}dy^{(0)}.$$

Define $\bar{y} = (0, \mathbf{y})$. Using (53) (but writing A for $\widetilde{A}$, as is our convention), this becomes

$$dx = \lambda_{(0)}(\tau)dy^{(0)} + \lambda_{(i)}(\tau)dy^{(i)} + c^{-2}(A\bar{y})^{(\nu)}\lambda_{(\nu)}(\tau)dy^{(0)}. \tag{61}$$

Finally, since

$$A\bar{y} = (\mathbf{g}\cdot\mathbf{y}, \mathbf{y}\times c\boldsymbol{\omega}), \tag{62}$$

we obtain

$$dx = \left(\left(1 + \frac{\mathbf{g}\cdot\mathbf{y}}{c^2}\right)\lambda_{(0)} + c^{-1}(\mathbf{y}\times\boldsymbol{\omega})^{(i)}\lambda_{(i)}\right)dy^{(0)} + \lambda_{(j)}dy^{(j)}. \tag{63}$$

Therefore, the metric at the point $\bar{y}$ is

$$s^2 = dx^2 = \left(\left(1 + \frac{\mathbf{g}\cdot\mathbf{y}}{c^2}\right)^2 - c^{-2}(\mathbf{y}\times\boldsymbol{\omega})^2\right)(dy^{(0)})^2 + \frac{2}{c}(\mathbf{y}\times\boldsymbol{\omega})_{(i)}dy^{(0)}dy^{(i)} + \delta_{jk}dy^{(j)}dy^{(k)}. \tag{64}$$

This formula was also obtained by Mashhoon [14]. We point out that the metric is dependent only on the *position* in the accelerated frame and not on *time*.

6. Examples of Uniformly Accelerated frames and Spacetime Transformations

In this section, we consider examples of uniformly accelerated frames and the corresponding spacetime transformations.

6.1. Null Acceleration $(\alpha = 0, \beta = 0)$

Since, in this case, $|\mathbf{g}| = |c\boldsymbol{\omega}|$ and $\mathbf{g}\cdot\boldsymbol{\omega} = 0$, we may choose $\mathbf{g} = (g, 0, 0)$ and $c\boldsymbol{\omega} = (0, 0, g)$. From (23), we have

$$A^{\mu}_{\nu} = \begin{pmatrix} 0 & g & 0 & 0 \\ g & 0 & g & 0 \\ 0 & -g & 0 & 0 \\ 0 & 0 & 0 & 0 \end{pmatrix}. \tag{65}$$

Then

$$A^2 = \begin{pmatrix} g^2 & 0 & g^2 & 0 \\ 0 & 0 & 0 & 0 \\ -g^2 & 0 & -g^2 & 0 \\ 0 & 0 & 0 & 0 \end{pmatrix}. \tag{66}$$

Journal of Physics: Conference Series **437** (2013) 012009 doi:10.1088/1742-6596/437/1/012009

Thus, from (51), we have

$$\lambda(\tau) = I + A\tau/c + \frac{1}{2}A^2\tau^2/c^2 = \begin{pmatrix} 1+\frac{g^2\tau^2}{2c^2} & g\tau/c & \frac{g^2\tau^2}{2c^2} & 0 \\ g\tau/c & 1 & g\tau/c & 0 \\ -\frac{g^2\tau^2}{2c^2} & -g\tau/c & 1-\frac{g^2\tau^2}{2c^2} & 0 \\ 0 & 0 & 0 & 1 \end{pmatrix}. \tag{67}$$

The observer's four-velocity is, therefore,

$$u(\tau) = \lambda_{(0)}(\tau) = \left(1+\frac{g^2\tau^2}{2c^2}, g\tau/c, -\frac{g^2\tau^2}{2c^2}, 0\right). \tag{68}$$

His four-acceleration is

$$a(\tau) = \left(\frac{g^2\tau}{c}, g, -\frac{g^2\tau}{c}, 0\right) = g\lambda_{(1)}(\tau), \tag{69}$$

which shows that the acceleration is constant in the comoving frame.

Integrating (68), we have

$$\widehat{x}(\tau) = \left(c\tau + \frac{g^2\tau^3}{6c}, \frac{g\tau^2}{2}, -\frac{g^2\tau^3}{6c}, 0\right).$$

Using (67) and $y^{(0)} = c\tau$, the spacetime transformations (60) are

$$\begin{pmatrix} x^0 \\ x^1 \\ x^2 \\ x^3 \end{pmatrix} = \begin{pmatrix} c\tau + \frac{g^2\tau^3}{6c} + y^{(1)}g\tau/c + y^{(2)}\frac{g^2\tau^2}{2c^2} \\ \frac{g\tau^2}{2} + y^{(1)} + y^{(2)}g\tau/c \\ -\frac{g^2\tau^3}{6c} - y^{(1)}g\tau/c + y^{(2)} - y^{(2)}\frac{g^2\tau^2}{2c^2} \\ y^{(3)} \end{pmatrix}. \tag{70}$$

6.2. Linear Acceleration $(\alpha > 0, \beta = 0)$

Without loss of generality, we may choose

$$A^\mu_\nu = \begin{pmatrix} 0 & g & 0 & 0 \\ g & 0 & c\omega & 0 \\ 0 & -c\omega & 0 & 0 \\ 0 & 0 & 0 & 0 \end{pmatrix}, \tag{71}$$

where $g > c\omega > 0$. In order to simplify the calculation of the exponent of A, we perform a Lorentz boost

$$B = \begin{pmatrix} g/\alpha & 0 & -c\omega/\alpha & 0 \\ 0 & 1 & 0 & 0 \\ -c\omega/\alpha & 0 & g/\alpha & 0 \\ 0 & 0 & 0 & 1 \end{pmatrix} \tag{72}$$

to the drift frame corresponding to the velocity

$$\mathbf{v} = (c^2/g, 0, 0) \times (0, 0, \omega). \tag{73}$$

Since $g > c\omega > 0$, we have $|\mathbf{v}| \leq c$.

In the drift frame, the acceleration tensor A becomes

$$A_{dr} = B^{-1}AB = \begin{pmatrix} 0 & \alpha & 0 & 0 \\ \alpha & 0 & 0 & 0 \\ 0 & 0 & 0 & 0 \\ 0 & 0 & 0 & 0 \end{pmatrix},$$

and leads to 1D hyperbolic motion. Hence,

$$\lambda(\tau) = \exp(A\tau/c) = B\exp(A_{dr}\tau/c)B^{-1}$$

$$= \begin{pmatrix} \frac{g^2}{\alpha^2}(\cosh(\alpha\tau/c)-1)+1 & \frac{g}{\alpha}\sinh(\alpha\tau/c) & \frac{gc\omega}{\alpha^2}(\cosh(\alpha\tau/c)-1) & 0 \\ \frac{g}{\alpha}\sinh(\alpha\tau/c) & \cosh(\alpha\tau/c) & \frac{c\omega}{\alpha}\sinh(\alpha\tau/c) & 0 \\ \frac{-gc\omega}{\alpha^2}(\cosh(\alpha\tau/c)-1) & \frac{-c\omega}{\alpha}\sinh(\alpha\tau/c) & \frac{-c^2\omega^2}{\alpha^2}(\cosh(\alpha\tau/c)-1)+1 & 0 \\ 0 & 0 & 0 & 1 \end{pmatrix}. \quad (74)$$

If $\omega = 0$, we recover the usual hyperbolic motion of a frame. Thus, the previous formula is a covariant extension of hyperbolic motion.

From the first column of (74), the observer's four-velocity is

$$u(\tau) = (\frac{g^2}{\alpha^2}(\cosh(\alpha\tau/c)-1)+1, \frac{g}{\alpha}\sinh(\alpha\tau/c), \frac{-gc\omega}{\alpha^2}(\cosh(\alpha\tau/c)-1), 0). \quad (75)$$

Hence, the observer's four-acceleration is

$$a(\tau) = \left(\frac{g^2}{\alpha}\sinh(\alpha\tau/c), g\cosh(\alpha\tau/c), \frac{-gc\omega}{\alpha}\sinh(\alpha\tau/c), 0\right) = g\lambda_{(1)}(\tau), \quad (76)$$

which shows that the acceleration is constant in the comoving frame.

To see how formula (75) extends 1D hyperbolic motion, we compute the proper velocity $\mathbf{u}$. Using (75), we have

$$\mathbf{u} = \left(\frac{cg}{\alpha}\sinh(\alpha\tau/c), \frac{-gc^2\omega}{\alpha^2}(\cosh(\alpha\tau/c)-1), 0\right).$$

Since $\frac{d\tau}{dt} = \gamma^{-1}$, and γ is the zero component of $u(\tau)$, we have

$$\frac{d\mathbf{u}}{dt} = \frac{d\mathbf{u}}{d\tau}\frac{d\tau}{dt} = \frac{\left(g\cosh(\alpha\tau/c), \frac{-gc\omega}{\alpha}\sinh(\alpha\tau/c), 0\right)}{\frac{g^2}{\alpha^2}(\cosh(\alpha\tau/c)-1)+1}. \quad (77)$$

If $\omega = 0$, then (77) reduces to the 1D hyperbolic motion $\frac{d\mathbf{u}}{dt} = \mathbf{g}$, where $\mathbf{g} = (g, 0, 0)$. If $\omega \neq 0$, then $\frac{d\mathbf{u}}{dt}$ will also depend on ω. Thus, we have here an explicit example of the fact that $\mathbf{F} = \frac{d\mathbf{p}}{dt}$ if and only if $\boldsymbol{\omega} = 0$ (see the end of section 2).

Integrating (75), we have

$$\widehat{x}(\tau) = \left(\frac{c^2}{\alpha^2}\left(\frac{g^2}{\alpha}\sinh(\alpha\tau/c)+c\omega\tau\right), \frac{c^2 g}{\alpha^2}(\cosh(\alpha\tau/c)-1), \frac{-c^2 g\omega}{\alpha^2}\left(\frac{c}{\alpha}\sinh(\alpha\tau/c)-\tau\right), 0\right).$$

Using (74) and $y^{(0)} = c\tau$, the spacetime transformations (60) are

$$\begin{pmatrix} x^0 \\ x^1 \\ x^2 \\ x^3 \end{pmatrix} = \begin{pmatrix} \frac{c^2}{\alpha^2}\left(\frac{g^2}{\alpha}\sinh(\alpha\tau/c)+c\omega\tau\right) + y^{(1)}\frac{g}{\alpha}\sinh(\alpha\tau/c) + y^{(2)}\frac{cg\omega}{\alpha^2}(\cosh(\alpha\tau/c)-1) \\ \frac{c^2 g}{\alpha^2}(\cosh(\alpha\tau/c)-1) + y^{(1)}\cosh(\alpha\tau/c) + y^{(2)}\frac{c\omega}{\alpha}\sinh(\alpha\tau/c) \\ \frac{-c^2 g\omega}{\alpha^2}\left(\frac{c}{\alpha}\sinh(\alpha\tau/c)-\tau\right) - y^{(1)}\frac{c\omega}{\alpha}\sinh(\alpha\tau/c) - y^{(2)}\frac{c^2\omega^2}{\alpha^2}(\cosh(\alpha\tau/c)-1) + y^{(2)} \\ y^{(3)} \end{pmatrix}. \quad (78)$$

6.3. Rotational Acceleration $(\alpha = 0, \beta > 0)$
Without loss of generality, we may choose

$$A^{\mu}_{\nu} = \begin{pmatrix} 0 & g & 0 & 0 \\ g & 0 & c\omega & 0 \\ 0 & -c\omega & 0 & 0 \\ 0 & 0 & 0 & 0 \end{pmatrix}, \tag{79}$$

where $c\omega > g > 0$. In order to simplify the calculation of the exponent of A, we perform a Lorentz boost

$$B = \begin{pmatrix} c\omega/\beta & 0 & -g/\beta & 0 \\ 0 & 1 & 0 & 0 \\ -g/\beta & 0 & c\omega/\beta & 0 \\ 0 & 0 & 0 & 1 \end{pmatrix}$$

to the drift frame corresponding to the velocity

$$\mathbf{v} = (g, 0, 0) \times (0, 0, 1/\omega). \tag{80}$$

Since $c\omega > g > 0$, we have $|\mathbf{v}| \le c$.

In the drift frame, the acceleration tensor A becomes

$$A_{dr} = B^{-1}AB = \begin{pmatrix} 0 & 0 & 0 & 0 \\ 0 & 0 & \beta & 0 \\ 0 & -\beta & 0 & 0 \\ 0 & 0 & 0 & 0 \end{pmatrix},$$

and leads to pure rotational motion. Hence,

$$\lambda(\tau) = \exp(A\tau/c) = B\exp(A_{dr}\tau/c)B^{-1}$$

$$= \begin{pmatrix} \frac{g^2}{\beta^2}(1-\cos(\beta\tau/c))+1 & \frac{g}{\beta}\sin(\beta\tau/c) & \frac{gc\omega}{\beta^2}(1-\cos(\beta\tau/c)) & 0 \\ \frac{g}{\beta}\sin(\beta\tau/c) & \cos(\beta\tau/c) & \frac{c\omega}{\beta}\sin(\beta\tau/c) & 0 \\ \frac{-gc\omega}{\beta^2}(1-\cos(\beta\tau/c)) & -\frac{c\omega}{\beta}\sin(\beta\tau/c) & \frac{-c^2\omega^2}{\beta^2}(1-\cos(\beta\tau/c))+1 & 0 \\ 0 & 0 & 0 & 1 \end{pmatrix}. \tag{81}$$

If $g = 0$, we recover the usual rotation of the basis about the z axis. Thus, the previous formula is a covariant extension of rotational motion.

From the first column of (81), the observer's four-velocity is

$$u(\tau) = \left(\frac{g^2}{\beta^2}(1-\cos(\beta\tau/c))+1, \frac{g}{\beta}\sin(\beta\tau/c), \frac{-gc\omega}{\beta^2}(1-\cos(\beta\tau/c)), 0\right). \tag{82}$$

Hence, the observer's four-acceleration is

$$a(\tau) = \left(\frac{g^2}{\beta}\sin(\beta\tau/c), g\cos(\beta\tau/c), \frac{-gc\omega}{\beta}\sin(\beta\tau/c), 0\right) = g\lambda_{(1)}(\tau), \tag{83}$$

which shows that the acceleration is constant in the comoving frame.

Integrating (82), we have

$$\widehat{x}(\tau) = \left(\frac{-c^2g^2}{\beta^3}\sin(\beta\tau/c) + \left(\frac{g^2}{\beta^2}+1\right)c\tau, \frac{c^2g}{\beta^2}\left(\cos(\beta\tau/c)-1\right), \frac{c^3g\omega}{\beta^3}\sin(\beta\tau/c) - \frac{c^2g\omega}{\beta^2}\tau, 0\right).$$

Using (81) and $y^{(0)} = c\tau$, the spacetime transformations (60) are

$$\begin{pmatrix} x^0 \\ x^1 \\ x^2 \\ x^3 \end{pmatrix} = \begin{pmatrix} \frac{-c^2 g^2}{\beta^3}\sin(\beta\tau/c) + \left(\frac{g^2}{\beta^2}+1\right)c\tau - \frac{g y^{(1)}}{\beta}\sin(\beta\tau/c) + \frac{y^{(2)}cg\omega}{\beta^2}(1-\cos(\beta\tau/c)) \\ \frac{c^2 g}{\beta^2}(\cos(\beta\tau/c)-1) + y^{(1)}\cos(\beta\tau/c) - \frac{y^{(2)}c\omega}{\beta}\sin(\beta\tau/c) \\ \frac{c^3 g\omega}{\beta^3}\sin(\beta\tau/c) - \frac{c^2 g\omega}{\beta^2}\tau + \frac{y^{(1)}c\omega}{\beta}\sin(\beta\tau/c) - \frac{y^{(2)}c^2\omega^2}{\beta^2}(1-\cos(\beta\tau/c)) + y^{(2)} \\ y^{(3)} \end{pmatrix}. \tag{84}$$

7. Velocity Transformations

In this section, we obtain the transformation of a particle's velocity in a uniformly accelerated frame K' to its four-velocity in the initial comoving inertial frame $K = K_0$.

A particle's four-velocity in K is, by definition, $\frac{dx^\mu}{d\tau_p}$, where $x(\tau_p)$ is the particle's worldline, and τ_p is the particle's proper time. However, from Special Relativity, it is known that the proper time of a particle depends on its velocity. In addition, it is known that the rate of a clock in an accelerated system also depends on its position, as occurs, for example, for linearly accelerated systems, due to gravitational time dilation. As a result, the quantity $d\tau_p$ depends on both the position and the velocity of the particle, that is, on the state of the particle.

Since we do not yet know the particle's proper time, it is not clear how to calculate the particle's four-velocity in K directly from its velocity in K'. To get around this problem, we will differentiate the particle's worldline by a parameter $\tilde{\tau}$ instead of τ_p. For convenience, we will choose $\tilde{\tau}$ to be a constant multiple of the time. For example, we often choose $\tilde{\tau} = c\tau$. The same technique was used by Horwitz and Piron [15], using the four-momentum instead of the four-velocity, thereby introducing the area known as"off-shell" electrodynamics.

We now introduce the following definition.

Definition 3. *Let x^μ be the worldline of a particle. The particle's* 4D velocity with respect to the parameter $\tilde{\tau}$ *is denoted by $\tilde{u}$ and is defined by*

$$\tilde{u}^\mu = \frac{dx^\mu}{d\tilde{\tau}}. \tag{85}$$

Note that the 4D velocity has the same direction as the four-velocity. In fact, the particle's four-velocity is

$$u = \frac{\tilde{u}}{|\tilde{u}|}, \tag{86}$$

the normalization of $\tilde{u}$. In the particular case $\tilde{\tau} = ct$, where t is the time in an *inertial* frame, then

$$|\tilde{u}| = \left|\frac{dx^\mu}{cdt}\right| = \left|\left(1, \frac{\mathbf{v}}{c}\right)\right| = \sqrt{1 - \frac{\mathbf{v}^2}{c^2}} = \gamma^{-1}. \tag{87}$$

Consider now a moving particle in K'. Let $\tilde{w}^{(\nu)} = \frac{dy^{(\nu)}}{dy^{(0)}}$ denote the particle's 4D velocity in K' with respect to $y^{(0)} = c\tau$. We will calculate the particle's 4D velocity in K at the point $\mathbf{y}$ of K', with respect to $c\tau = c\gamma^{-1}t$, where γ corresponds to the observer's velocity in K. Recall from formula (61) that the differential of the spacetime transformations

$$x = \widehat{x}(\tau) + y^{(i)}\lambda_{(i)} \tag{88}$$

is

$$dx = \lambda_{(0)}(\tau)dy^{(0)} + \lambda_{(i)}(\tau)dy^{(i)} + c^{-2}(A\bar{y})^{(\nu)}\lambda_{(\nu)}(\tau)dy^{(0)}\,. \tag{89}$$

Thus, the particle's 4D velocity in K at the point y is

$$\tilde{u} = \frac{1}{c}\frac{dx}{d\tau} = \frac{dx}{dy^{(0)}} = \frac{dy^{(\nu)}}{dy^{(0)}}\lambda_{(\nu)}(\tau) + c^{-2}(A\bar{y})^{(\nu)}\lambda_{(\nu)}(\tau) = (\tilde{w}^{(\nu)} + c^{-2}(A\bar{y})^{(\nu)})\lambda_{(\nu)}(\tau)\,, \tag{90}$$

where $\bar{y} = (0, \mathbf{y})$. Thus, the 4D velocity $\tilde{u}$ is the sum of the particle's 4D velocity $\tilde{w}$ within K' and an additional *4D velocity due to the acceleration of* K' with respect to the inertial frame K, which we denote

$$\tilde{u}_a = c^{-2}(A\bar{y})^{(\nu)}\lambda_{(\nu)}(\tau). \tag{91}$$

Finally, from (86), the four-velocity of the particle in K is

$$u = \frac{\tilde{u}}{|\tilde{u}|} = \frac{\tilde{w} + \tilde{u}_a}{|\tilde{w} + \tilde{u}_a|} = \frac{(\tilde{w}^{(\nu)} + c^{-2}(A\bar{y})^{(\nu)})\lambda_{(\nu)}(\tau)}{|\tilde{w} + c^{-2}A\bar{y}|}. \tag{92}$$

Writing $\tilde{w}^{(\nu)} = (1, \mathbf{w}/c)$, the four-velocity in the $1+3$ decomposition becomes

$$u = \frac{(1 + \frac{\mathbf{g}\cdot\mathbf{y}}{c^2}, c^{-1}(\mathbf{w} + \mathbf{y}\times\boldsymbol{\omega}))}{\sqrt{(1 + \frac{\mathbf{g}\cdot\mathbf{y}}{c^2})^2 - (\frac{\mathbf{w}+\mathbf{y}\times\boldsymbol{\omega}}{c})^2}}. \tag{93}$$

By the Equivalence Principle, we may interpret the zero component of the vector in the numerator as the gravitational time dilation. This factor depends only on the translational acceleration, which is the only cause of the change in time. The spatial part is the correction due to the particle's velocity in K' and the rotational velocity of K' with respect to K_τ.

We now show that substituting $A = 0$ in formula (90) yields the Einstein velocity addition formula for inertial systems. Assume that K' is inertial, say with uniform 3D velocity $\mathbf{v} = (v, 0, 0)$ with respect to the inertial frame K. Suppose a particle has 3D velocity $\mathbf{w}$ in K'. We wish to compute $\mathbf{v} \oplus_E \mathbf{w}$, defined to be the particle's 3D velocity in K. The particle's 4D velocity in K' with respect to ct', where t' is the time in K', is $\tilde{w}^{(\nu)} = (1, \mathbf{w}/c)$. The comoving frame of K' in this case is

$$\lambda_{(0)} = \gamma(1, v/c, 0, 0), \lambda_{(1)} = \gamma(v/c, 1, 0, 0), \lambda_{(2)} = (0, 0, 1, 0), \lambda_{(3)} = (0, 0, 0, 1). \tag{94}$$

From (90), we get

$$\tilde{u} = \tilde{w}^{(\nu)}\lambda_{(\nu)} = \left(\gamma\left(1 + \frac{w^1 v}{c^2}\right), \gamma\left(\frac{v + w^1}{c}\right), w^2/c, w^3/c\right). \tag{95}$$

Dividing c times the spatial part of $\tilde{u}$ by the time component, we obtain the 3D velocity of the particle in K as

$$\mathbf{v} \oplus_E \mathbf{w} = \frac{(v + w^1, \gamma^{-1}w^2, \gamma^{-1}w^3)}{1 + \frac{w^1 v}{c^2}} = \frac{\mathbf{v} + P_{\mathbf{v}}\mathbf{w} + \gamma^{-1}(I - P_{\mathbf{v}})\mathbf{w}}{1 + \frac{w^1 v}{c^2}}, \tag{96}$$

where $P_{\mathbf{v}}\mathbf{w}$ denotes the projection of $\mathbf{w}$ onto $\mathbf{v}$. This is the well-known Einstein velocity addition formula (see [2], formula (3.7)).

8. Time Dilation

We turn now to time dilation. We will compute the time dilation between the clock of a uniformly accelerated observer located at the origin of K' and the clocks at other positions in K'. Let τ_p be the proper time of a particle positioned at $\mathbf{y}$ in K', with 4D velocity $\tilde{w}$ with respect to $y^{(0)}$. Since the four-velocity (in K) of the particle is

$$u = \frac{1}{c}\frac{dx}{d\tau_p} = \frac{1}{c}\frac{dx}{d\tau}\frac{d\tau}{d\tau_p} = \tilde{u}\frac{d\tau}{d\tau_p}, \tag{97}$$

equation (86) implies that

$$d\tau = \frac{1}{|\tilde{u}|}d\tau_p = \tilde{\gamma}d\tau_p, \tag{98}$$

where $\tilde{\gamma} := \frac{1}{|\tilde{u}|}$ is a function of $\tilde{w}$ and $\mathbf{y}$, or, in short, of the state of the particle. The definition of $\tilde{\gamma}$ for accelerated systems is analogous to the definition of γ for inertial systems. In fact, we will see below that if $A = 0$, then $\tilde{\gamma} = \gamma$. The factor $\tilde{\gamma}$ expresses the time dilation between the particle at $\mathbf{y}$ and the observer at the origin of K'. To obtain the time dilation of the particle with respect to the inertial frame K, one must also multiply by the time dilation of the observer with respect to K, which is the zero component of the observer's four-velocity, explicitly obtained in section 3.

We now express the time dilation (98) in the $1+3$ decomposition. If a particle has velocity $\tilde{w}^{(\nu)} = \frac{dy^{(\nu)}}{dy^{(0)}} = (1, \mathbf{w}/c)$ in K', then the time dilation between the particle and the observer is given by

$$d\tau_p = \sqrt{\left(1+\frac{\mathbf{g}\cdot\mathbf{y}}{c^2}\right)^2 - \left(\frac{\mathbf{w}+\mathbf{y}\times\boldsymbol{\omega}}{c}\right)^2}\, d\tau\,, \quad \tilde{\gamma} = \frac{1}{\sqrt{\left(1+\frac{\mathbf{g}\cdot\mathbf{y}}{c^2}\right)^2 - \left(\frac{\mathbf{w}+\mathbf{y}\times\boldsymbol{\omega}}{c}\right)^2}}. \tag{99}$$

Thus, the time dilation between the particle and the observer in K' combines the gravitational time dilation (via the Equivalence Principle) and an additional time dilation due to the velocity of the particle together with the rotational velocity of the system. The same formula was obtained in [31]. Note that the expression underneath the square root must be nonnegative. This limits the admissible values for $\mathbf{y}$ and is a manifestation of the locality of the spacetime transformations from section 5. The same limitation was obtained by Mashhoon [14].

If $A = 0$, then

$$\tilde{\gamma} = \frac{1}{\sqrt{1-\frac{\mathbf{w}^2}{c^2}}} = \gamma(\mathbf{w}),$$

expressing the time dilation due to the velocity of the particle in K', which is an inertial system in this case.

For a clock at rest in K', the particular case $\boldsymbol{\omega} = 0$ gives a time dilation of $1 + \frac{\mathbf{g}\cdot\mathbf{y}}{c^2}$, which is the known formula for gravitational time dilation. If $\mathbf{g} = 0$, the time dilation is $\sqrt{1-\left(\frac{\mathbf{y}\times\boldsymbol{\omega}}{c}\right)^2}$, which is the time dilation due to the rotational velocity of a rotating system.

The lower order terms of the expansion of the time dilation of (99) are

$$1+\frac{\mathbf{g}\cdot\mathbf{y}}{c^2} - \frac{1}{2}\left(\frac{\mathbf{w}+\mathbf{y}\times\boldsymbol{\omega}}{c}\right)^2 = 1+\frac{\mathbf{g}\cdot\mathbf{y}}{c^2} - \frac{1}{2}\frac{(\mathbf{y}\times\boldsymbol{\omega})^2}{c^2} - \frac{1}{2}\frac{\mathbf{w}^2}{c^2} - \frac{\mathbf{w}\cdot(\mathbf{y}\times\boldsymbol{\omega})}{c^2}. \tag{100}$$

The second term represents the gravitational time dilation. The third and fourth terms are the transversal Doppler shifts due to the rotation of the system and the velocity of the particle, respectively. The last term is new in the setting of flat Minkowski space but was also obtained recently by Grøn and Braeck ([32], equation (29)) in Schwarzschild spacetime.

We now obtain the physical meaning of $\boldsymbol{\omega}$ in the acceleration matrix A. From (93), the four-velocity of a rest point $\mathbf{y}$ in the comoving frame is

$$u = \frac{\left(1 + \frac{\mathbf{g}\cdot\mathbf{y}}{c^2}, c^{-1}(\mathbf{y}\times\boldsymbol{\omega})\right)}{\sqrt{\left(1+\frac{\mathbf{g}\cdot\mathbf{y}}{c^2}\right)^2 - \left(\frac{\mathbf{y}\times\boldsymbol{\omega}}{c}\right)^2}}, \tag{101}$$

and the corresponding 3D velocity is

$$\mathbf{v} = \frac{d\mathbf{x}}{dt} = \frac{\mathbf{y}\times\boldsymbol{\omega}}{1+\frac{\mathbf{g}\cdot\mathbf{y}}{c^2}}. \tag{102}$$

This formula defines the angular velocity of a uniformly accelerated body. Note that for rest points on the axis of rotation, we have $\frac{d\mathbf{x}}{dt} = 0$. Also, if $\mathbf{y}$ belongs to the plane perpendicular to $\mathbf{g}$, then $\frac{d\mathbf{x}}{dt} = \mathbf{y}\times\boldsymbol{\omega}$, the classical angular velocity. Multiplying both sides of (102) by γ, we obtain

$$\mathbf{u} = \frac{d\mathbf{x}}{d\tau} = \tilde{\gamma}(\mathbf{y}\times\boldsymbol{\omega}). \tag{103}$$

The explanation of this formula is as follows. Measure the angular velocity of each point with respect to a common clock, in this case, the clock at the origin. Then, since each point of the rotating object must have the same period, the classical angular velocity must be multiplied by the time dilation between the clock at the origin and the clock at the point in question.

9. The Acceleration of Rest Points in a Uniformly Accelerated Frame

We will show here that every rest point in a uniformly accelerated frame is also uniformly accelerated, but the acceleration differs from point to point. Until now, our uniformly accelerated observer has been located at the origin of a uniformly accelerated system K'. In this section, we consider an observer at rest at an arbitrary point $\mathbf{y}$ of K'.

First, we show that this observer is also uniformly accelerated. To see this, note that the coefficients of $\lambda_{(\nu)}(\tau)$ in the four-velocity (92) do not depend on τ. Hence,

$$c\frac{du}{d\tau_p} = c\frac{du}{d\tau}\frac{d\tau}{d\tau_p} = c\tilde{\gamma}\frac{du}{d\tau} = c\tilde{\gamma}u^{(\nu)}\frac{d\lambda_{(\nu)}(\tau)}{d\tau}$$

$$= \tilde{\gamma}u^{(\nu)}A\lambda_{(\nu)}(\tau) = \tilde{\gamma}Au. \tag{104}$$

This shows that the observer at $\mathbf{y}$ is also uniformly accelerated, with acceleration tensor $\tilde{\gamma}A$. In the particular case $\boldsymbol{\omega} = 0$, the comoving frame at $\mathbf{y}$ coincides with the comoving frame of the observer at the origin, and $\tilde{\gamma} = \frac{1}{1+\frac{\mathbf{g}\cdot\mathbf{y}}{c^2}}$. Thus, in this case, $\mathbf{g}(\mathbf{y}) = \tilde{\gamma}\mathbf{g}$. The same formula was obtained by Franklin [33].

In the general case ($\boldsymbol{\omega} \neq 0$), we now find explicit expressions for the components $\mathbf{g}(\mathbf{y})$ and $\boldsymbol{\omega}(\mathbf{y})$ of the acceleration matrix $A(\mathbf{y})$ of a rest point $\mathbf{y}$. Assume, without loss of generality, that $\boldsymbol{\omega}$ is in the direction of the positive z axis and that $\mathbf{y}$ has spatial coordinates $(r, 0, z)$ in the comoving frame.

In order to compute the acceleration matrix $A(\mathbf{y})$ for the observer at $\mathbf{y}$, we need to calculate the comoving frame at $\mathbf{y}$. By (101), the 3D velocity $\mathbf{v}$ of the point $\mathbf{y}$ in the comoving frame $\lambda(\tau)$ is

$$\mathbf{v} = \frac{\omega r(0,-1,0)}{1+\frac{\mathbf{g}\cdot\mathbf{y}}{c^2}}, \quad \gamma = \frac{1+\frac{\mathbf{g}\cdot\mathbf{y}}{c^2}}{\sqrt{\left(1+\frac{\mathbf{g}\cdot\mathbf{y}}{c^2}\right)^2 - \left(\frac{\omega r}{c}\right)^2}} = \tilde{\gamma}\left(1+\frac{\mathbf{g}\cdot\mathbf{y}}{c^2}\right). \tag{105}$$

Hence, the initial comoving frame at the point $\mathbf{y}$ is B_v, where B_v is a Lorentz boost in the direction of $\mathbf{v}$, with $v = |\mathbf{v}| = \frac{\omega r}{1+\frac{\mathbf{g}\cdot\mathbf{y}}{c^2}}$. Since the acceleration matrix for this observer is

$$A(\mathbf{y}) = \tilde{\gamma} B_v A B_v^{-1}, \tag{106}$$

the comoving frame at $\mathbf{y}$ is

$$\lambda'(\tau') = \exp(A(\mathbf{y})\tau'/c)B_v = \exp(\tilde{\gamma} B_v A B_v^{-1}\tau'/c)B_v = B_v \exp(\tilde{\gamma} A\tau'/c)$$
$$= B_v \exp(A\tau/c) = B_v\lambda(\tau).$$

Hence, from (106), we obtain

$$A(\mathbf{y}) = \tilde{\gamma}\begin{pmatrix} 0 & \gamma g^1 - \tilde{\gamma}\omega^2 r & g^2 & g^3\gamma \\ \gamma g^1 - \tilde{\gamma}\omega^2 r & 0 & \gamma c\omega - \tilde{\gamma} g^1 \omega r/c & 0 \\ g^2 & -\gamma c\omega + \tilde{\gamma} g^1\omega r/c & 0 & \tilde{\gamma} g^3 \omega r/c \\ g^3\gamma & 0 & -\tilde{\gamma} g^3\omega r/c & 0 \end{pmatrix}. \tag{107}$$

For a pure rotation ($\mathbf{g} = 0$), we have $\mathbf{g}(\mathbf{y}) = (-\gamma^2\omega^2 r, 0, 0)$ and $\boldsymbol{\omega}(\mathbf{y}) = (0, 0, \gamma^2\omega)$.

10. Acceleration Transformations in a Uniformly Accelerated Frame

Our next goal is to obtain the transformation of a particle's acceleration in K' to its four-acceleration in K. First, however, we will calculate the 4D acceleration $\tilde{a} = c\frac{d\tilde{u}}{d\tau}$ in K. Let $\tilde{b} = c\frac{d\tilde{w}}{d\tau}$. Using (90), we have

$$\tilde{a} = c\frac{d\tilde{u}}{d\tau} = c\frac{d\tilde{w}^{(\nu)}}{d\tau}\lambda_{(\nu)}(\tau) + A\frac{d\bar{y}^{(\nu)}}{dy^{(0)}}\lambda_{(\nu)}(\tau) + c(\tilde{w}^{(\nu)} + c^{-2}(A\bar{y})^{(\nu)})\frac{d\lambda_{(\nu)}(\tau)}{d\tau} = \tilde{b} + A\bar{w} + A\tilde{u}. \tag{108}$$

The quantity $\tilde{d} := \tilde{b} + A\bar{w}$ is the acceleration of the particle with respect to the comoving frame, since this is the part of $\tilde{a}$ which treats $\lambda(\tau)$ as constant. We now write equation (108) in the $1+3$ decomposition. Write $\tilde{b}^{(\nu)} = (0, \mathbf{a}_p)$, where $\mathbf{a}_p$ is the 3D acceleration of the particle in K'. Since, for any four-vector $r = (r^0, \mathbf{r})$, we have

$$Ar = (\mathbf{g}\cdot\mathbf{r}, r^0\mathbf{g} + \mathbf{r}\times c\boldsymbol{\omega}), \tag{109}$$

we obtain

$$\tilde{a}^{(\nu)} = \left(2\mathbf{g}\cdot\mathbf{w} + \frac{1}{c}\mathbf{g}\cdot(\mathbf{y}\times\boldsymbol{\omega}), \mathbf{a}_p + \left(1 + \frac{\mathbf{g}\cdot\mathbf{y}}{c^2}\right)\mathbf{g} + 2\mathbf{w}\times c\boldsymbol{\omega} + (\mathbf{y}\times\boldsymbol{\omega})\times\boldsymbol{\omega}\right). \tag{110}$$

For a rest particle, we have $\tilde{w}^{(\nu)} = (1, 0, 0, 0)$, and, in this case, formula (110) becomes

$$\tilde{a}^{(\nu)} = \left(\frac{1}{c}\mathbf{g}\cdot(\mathbf{y}\times\boldsymbol{\omega}), \left(1 + \frac{\mathbf{g}\cdot\mathbf{y}}{c^2}\right)\mathbf{g} + (\mathbf{y}\times\boldsymbol{\omega})\times\boldsymbol{\omega}\right). \tag{111}$$

Now we obtain the particle's four-acceleration in K:

$$a = c\frac{du}{d\tau_p} = c\frac{du}{d\tau}\frac{d\tau}{d\tau_p} = c\tilde{\gamma}\frac{d}{d\tau}(\tilde{\gamma}\tilde{u}) = \tilde{\gamma}^2\tilde{a} - \tilde{\gamma}^2(\tilde{a}\cdot u)u. \tag{112}$$

Notice that

$$A\tilde{u}\cdot u = A\tilde{u}\cdot\tilde{\gamma}\tilde{u} = 0,$$

since A is antisymmetric. Now, substituting $\tilde{a} = \tilde{d} + A\tilde{u}$ into (112), we have

$$a = \tilde{\gamma}^2 A\tilde{u} + \tilde{\gamma}^2 \tilde{d} - \tilde{\gamma}^2 (\tilde{d} \cdot u)u = \tilde{\gamma} \left(Au + \tilde{\gamma}(\tilde{d} - (\tilde{d} \cdot u)u) \right). \tag{113}$$

Let $P_u \tilde{d}$ be the projection of $\tilde{d}$ onto u, and let $\tilde{d}_\perp = (I - P_u)\tilde{d}$. Then we can write the four-acceleration as

$$a = \tilde{\gamma}^2 (A\tilde{u} + \tilde{d}_\perp) = \tilde{\gamma} Au + \tilde{\gamma}^2 \tilde{d}_\perp. \tag{114}$$

The explanation of formula (114) is as follows. The acceleration of a rest point in the comoving frame is Au, where u is the point's four-velocity in K. The factor $\tilde{\gamma}$ is the time dilation factor between the observer's clock and the proper time of the particle and arises here because we differentiated the four-velocity by τ instead of τ_p. The term $\tilde{\gamma} Au$ accounts only for the acceleration of K' with respect to K. Thus, we must add the term $\tilde{\gamma}^2 \tilde{d}_\perp$ to account for the acceleration of the particle inside K'. In this term, the factor $\tilde{\gamma}^2$ appears because we differentiated *twice* by τ instead of τ_p. Since the four-acceleration is always perpendicular to the four-velocity, and Au is perpendicular to u, the four-acceleration can contain only the component of $\tilde{d}$ which is perpendicular to u. This completes the explanation of formula (114).

Consider the motion of a charged particle in a constant electromagnetic field F. We decompose its motion into motion under a constant Lorentz force and acceleration produced by the self-force due to the radiation. We consider the particle to be at the origin ($\tilde{\gamma} = 1$) of a uniformly accelerated system K', with acceleration tensor $A = \frac{e}{m} F$. The acceleration due to the radiation will be considered as motion with respect to K'. The self-force generates an acceleration, which is known to be $\tilde{d} = \tau_0 A^2 u$, where τ_0 is a universal constant. Thus, in this case, formula (113) coincides with the Lorentz-Abraham-Dirac equation ([9], equation S-10, page 259)

$$\frac{du}{d\tau} = Au + \tau_0 \left(A^2 u - (A^2 u \cdot u)u \right), \tag{115}$$

which Rohrlich calls the *correct* equation of motion of a classical point charge.

If the acceleration due to the radiation is in the direction of the four-velocity, then $\tilde{d}_\perp = 0$. In this case, the motion of a uniformly accelerated charge also satisfies equation (115). This occurs, for example, in the particular cases of 1D hyperbolic motion ($\boldsymbol{\omega} = 0$) and pure rotation ($\mathbf{g} = 0$).

11. Summary and Discussion

The fully Lorentz covariant *Relativistic Dynamics Equation* (22) extends the 3D relativistic dynamic equation $\mathbf{F} = \frac{d\mathbf{p}}{dt}$. We have shown that the *standard* 4D equation $F = \frac{dp}{d\tau}$ is only partially covariant. To achieve full Lorentz covariance, we replaced the four-force F by a rank 2 antisymmetric *tensor* $A_{\mu\nu}$ acting on the four-velocity.

In section 3, we obtained *explicit* solutions to (22) in the case of *constant force*. We call the solutions *uniformly accelerated motion*. We have shown that uniformly accelerated motions are divided into four *Lorentz-invariant* types: null, linear, rotational, and general. For *null* acceleration, the worldline (31) is *cubic* in the time. *Linear* acceleration (33) covariantly extends 1D hyperbolic motion, while *rotational* acceleration (36) covariantly extends pure rotational motion. We have shown that if we keep the tensor A constant, the nonrelativistic limit (46) of our uniformly accelerated motion is motion with constant linear acceleration. A different nonrelativistic limit is obtained for rotational and general uniform acceleration by keeping the components $\mathbf{g}$ and $\boldsymbol{\omega}$ of the tensor A constant. This limit (47) describes motion under a Lorentz-type force, which includes uniform rotation.

In [34], it is shown that a *photon* is uniformly accelerated. To which of the four types of uniform acceleration does the photon belong? The two Lorentz invariants of the electromagnetic field of a photon are both 0, as in a constant, uniform electromagnetic field in the absence of sources. Since motion in such a field is an example of *null* acceleration, we conjecture that the motion of a photon is described by null acceleration.

Our results here are restricted to the particular case of *constant force*. In [35], the first author considers the one-dimensional *non-constant force* case $F = -kx$.

In [11]and [34], the first author proved that the spacetime transformations between two frames which are linearly uniformly accelerated with respect to each other must be one of two types. Type I assumes Mashhoon's Weak Hypothesis of Locality. The transformations of section 6 are explicit examples of Type I. If the Weak Hypothesis of Locality fails, then there exists a universal maximal acceleration a_{max}, and the resulting transformations are Type II. It still remains to compute the explicit Type II transformations corresponding to those of section 6. Since accelerations are bounded by a_m, the set of admissible accelerations form a *bounded symmetric domain* known as a JC^*-triple. Thus, one could use the machinery of chapter 4 of [21] to compute Type II transformations.

We have some preliminary results on the *rigidity* of a uniformly accelerated frame, including a possible resolution of the Ehrenfest paradox. This paradox concerns a rotating disk. Let r be the radius of the disk, measured in the lab frame, when the disk is at rest. When the disk rotates, the radius is always perpendicular to the disk's motion. Thus, an observer in the lab frame will measure the radius of the rotating disk to be r. However, the circumference should appear to be Lorentz-contracted by a factor of $\gamma(\omega r)$, where ω is the angular velocity of the disk. This implies that the radius of the disk is $r\sqrt{1-\frac{(\omega r)^2}{c^2}}$, which is less than r. We plan to present our resolution of this paradox in an upcoming paper.

We are also studying ways in which to incorporate *radiation* into our model.

Acknowledgments

This research is supported in part by the German-Israel Foundation for Scientific Research and Development: GIF No. 1078-107.14/2009

References

[1] Goldstein H, Poole C and Safko J 2002 *Classical Mechanics, 3rd edition* (New Jersey: Pearson Educational International) p. 299
[2] Rindler W 2001 *Relativity: Special, General, and Cosmological* (New York: Oxford University Press)
[3] Moller C 1972 *The Theory of Relativity* (Oxford: Clarendon Press)
[4] Einstein A 1908 *Jahrb. Rad. Elektr.* **5** p. 98, translated in 1989 *The Collected Papers of Albert Einstein* (Princeton University Press) vol 2 p. 316
[5] Einstein A 1907 Über das relativitäts prinzip und die aus demselben gezogenen folgerungen *Jahrbuch der Radioaktivität und Elektronik* **4** p. 411 - 62
[6] Minkowski H 1915 *Ann. Phys. Lpz.* **47** p. 927
[7] Born M 1909 *Ann. Phys. Lpz.* **30** p. 1
[8] Landau L and Lifshitz E 1971 *The Classical Theory of Fields* Course of Theoretical Physics vol. 2 (Pergamon, Addison-Wesley)
[9] Rohrlich F 2007 *Classical Charged Particles, 3rd edition* (London: World Scientific)
[10] Lyle S 2008 *Uniformly Accelerating Charged Particles* (Berlin: Springer-Verlag)
[11] Friedman Y and Gofman Y 2010 *Phys. Scr.* **82** 015004
[12] Friedman Y 2011 *Ann. Phys.* **523** p. 408
[13] Friedman Y and Resin E 2012 Dynamics of hydrogen-like atom bounded by maximal acceleration *Phys. Scr.* **86** 015002.
[14] Mashhoon B and Muench U 2002 Length measurement in accelerated systems *Ann. Phys.* **11** p. 532-547
[15] Horwitz L P and Piron C 1973 *Helv. Phys. Acta* **46** p. 316
[16] Nelson R A 1987 Generalized Lorentz transformation for an accelerated, rotating frame of reference *J. Math. Phys.* **28**, p. 2379

[17] Turyshev S, Minazzoli O and Toth V 2012 Accelerating relativistic reference frames in Minkowski space-time *J. Math. Phys.* **53** 032501
[18] Wigner E 1939 On the unitary representations of the inhomogeneous Lorentz group *Ann. Math.* **40** p. 149-204
[19] Condon E U and Odishaw H 1958 *Handbook of Physics* (New York: McGraw-Hill)
[20] Friedman Y and Semon M 2005 Relativistic acceleration of charged particles in uniform and mutually perpendicular electric and magnetic fields as viewed in the laboratory frame *Phys. Rev.* E **72** 026603-1–10
[21] Friedman Y 2004 *Homogeneous Balls and their Physical Applications* (Boston: Birkhäuser)
[22] Barut A 1980 *Electrodynamics and Classical Theory of Fields and Particles* (New York: Dover)
[23] Gupta A and Padmanabhan T 1998 Radiation from a charged particle and radiation reaction - revisited *Phys. Rev. D* **57** p. 7241-7250
[24] Hehl F W and Obukhov Y 2003 *Foundations of Classical Electrodynamics: Charge, Flux, and Metric* (Boston: Birkhäuser)
[25] Hehl F W, Lemke J and Mielke E W 1991 Two lectures on fermions and gravity *Geometry and Theoretical Physics* Debrus J and Hirshfeld A C (eds.) (Berlin:Springer) p. 56140
[26] Brillouin L 1970 *Relativity Reexamined* (New York: Academic Press)
[27] Alba D and Lusanna L 2007 Generalized radar 4-coordinates and equal-time Cauchy surfaces for arbitrary accelerated observers *Int J. Mod. Phys.* **D16** p. 1149
[28] Misner C W, Thorne K S and Wheeler J A 1973 *Gravitation* (San Francisco: W.H. Freeman and Company)
[29] Mashhoon B 1990 Limitations of spacetime measurements *Phys. Lett.* A **143** p. 176-82
[30] Mashhoon B 1990 The hypothesis of locality in relativistic physics *Phys. Lett.* A **145** p. 147-53
[31] Nikolić H 2000 Relativistic contraction and related effects in noninertial frames *Phys. Rev.* A **61** 032109
[32] Grøn Ø and Braeck S 2011 The twin paradox in a cosmological context *Eur. Phys. J. Plus* **126** p. 79
[33] Franklin J 2010 Lorentz contraction, Bell's spaceships and rigid body motion in special relativity *Eur. J. Phys.* **31** p. 291-98
[34] Friedman Y Extending the Relativity of Time "Proceedings of IARD 2012, to appear 2013"
[35] Friedman Y Quantum-like behavior of the harmonic oscillator in extended relativity, submitted to *Phys. Scr.*

IARD2012 IOP Publishing
Journal of Physics: Conference Series **437** (2013) 012010 doi:10.1088/1742-6596/437/1/012010

The principle of finiteness – a guideline for physical laws

Abraham Sternlieb
Ariel University Center for Samaria, ISRAEL
E-mail: asternlieb@ariel.ac.il

Abstract. I propose a new principle in physics–the principle of finiteness (FP).It stems from the definition of physics as a science that deals with measurable dimensional physical quantities. Since measurement results including their errors, are always finite, FP postulates that the mathematical formulation of legitimate laws in physics should prevent exactly zero or infinite solutions. I propose finiteness as a postulate, as opposed to a statement whose validity has to be corroborated by, or derived theoretically or experimentally from other facts, theories or principles.

Some consequences of FP are discussed, first in general, and then more specifically in the fields of special relativity, quantum mechanics, and quantum gravity. The corrected Lorentz transformations include an additional translation term depending on the minimum length epsilon. The relativistic gamma is replaced by a corrected gamma, that is finite for v=c. To comply with FP, physical laws should include the relevant extremum finite values in their mathematical formulation.

An important prediction of FP is that there is a maximum attainable relativistic mass/energy which is the same for all subatomic particles, meaning that there is a maximum theoretical value for cosmic rays energy. The Generalized Uncertainty Principle required by Quantum Gravity is actually a necessary consequence of FP at Planck's scale. Therefore, FP may possibly contribute to the axiomatic foundation of Quantum Gravity.

1. Introduction: the principle of "finiteness"

The notion of finiteness of physical entities has appeared in the scientific literature for quite some time, as opposed to either the notion of absolute zero or to the notion of infinity. The Born and Infeld model [1], trying to deal with the infinite self-energy of a point charge, was founded on a "principle of finiteness" that says (in their words) that a satisfactory theory should avoid physical quantities becoming infinite (this was before the arrival of renormalization). Albert Einstein [2], referring to the meaning of R_s, the Schwarzschild radius, and the black-hole singularity, wrote that physical models, cannot exhibit such a singularity, including the need to modify his own general relativity in the future, when new relevant evidence will be available.

A. Vilenkin [3] has proposed a cosmological model that is intended to avoid the Big-Bang singularity.

T. D. Lee [4] mentions the necessity of time discretization because of the finite number of measurements that can be performed in a finite time.

Gambini and Pullin [5] discuss loss of coherence and unitarity in the measurement process and consequently obtain an expression for a minimum possible clock accuracy:

$$\delta t \propto t_{pl}^{2/3} \bullet t^{1/3} \qquad (t_{pl}\text{= Planck time}).$$

Refs. [6] and [7] mention that infinities generally arise because of the point-like definition of elementary particles, and that string theory introduces finiteness (finite-sized strings) as a possible remedy. Finally, to conclude this literature sampling, in Ref. [8] mention is made of the (until now unknown) contributions of Ettore Majorana to the ideas of elementary (finite) length and time scales. The central theme evolving from the above-mentioned references, is that there is no physical reality besides measureable entities. In other words, there is no place in physical models for zero-size (point-like) entities or for infinite (singular) entities.

It is clear that the measurement process and its outcome are of outmost importance in defining the "legitimate" content of physics, as we shall see in the following. We begin our discussion by mentioning summarily the axiomatic basis of special relativity.

Albert Einstein developed his special relativity theory on the basis of two principles (postulates): a) the constancy of the speed of light, and b) the principle of special relativity, that requires invariance of the mathematical expressions of "legitimate" physical laws (under appropriate transformations) when passing from one inertial frame of reference to another which is moving at a constant velocity with respect to the first. The two postulates were chosen because of being based on experimental evidence and on its logical consequences, as available at that time.

By adhering to his postulates and studying carefully the measurement process, Einstein derived the Lorentz Transformations (which had been invented before him, ad-hoc, by Lorentz), which explained naturally the length contraction and time dilation. As a consequence, the immovable ether frame of reference became superfluous.

New principles (postulates) in physics do not appear frequently. They should be based on well-established experimental facts and/or on reasonable logical thinking. Also, they should be as general as possible, in order to be relevant to most of physics (Special and General Relativity, Quantum Mechanics, etc.)

Accordingly, it seems desirable to derive such a postulate from the very definition of physics. This would be a fundamental postulate, to which the laws of physics should obey.

I choose to (ad-hoc) define physics as follows: physics is the science that deals (not exclusively) with well-defined physical quantities, including theories about their functional relationships. It also deals with measurement methodologies relevant to the measurable quantities. Concerning the complex relationship between theory and measurement (experiment), one should consult books on the philosophy of science [9, 10], which is outside the scope of this paper.

In measurement methodology there is always a need to include a measurement error ("accuracy", or "error bar") when stating an experimental result R:

$$\mathrm{R} = x \pm \Delta x \tag{1}$$

where x is the expected value (single or average) of a measurement result of any physical measurable dimensional quantity/entity, and Δx (>0) is the measurement error (accuracy).

It is not within the scope of this paper to discuss the theory of measurement error. At this point I would like to mention that errors can be improved (reduced) by increasing the number of measurements, or, by improving the measurement methods. However, even if measurement errors (accuracy) of measurable quantities do improve in the course of time, they have never become exactly zero, because of finiteness of resolution of measurement devices and because of finiteness of possible number of measurements.

It would be safe to assume that errors will never become exactly zero (for example, as saying that "c"— the speed of light in vacuum—is never going to change). This is basically because non-zero accuracy is a consequence of the finiteness of human capabilities, and this is a rather permanent and fundamental limitation.

The summary of the discussion above is that accuracy (error) of any physical measurable dimensional quantity is always finite (non-zero):

$$\Delta x \geq \varepsilon > 0 \tag{2}$$

ε is the lower limit value of error for the physical entity called x. In this section (section 1), ε is positive.

Usually there is a finite ratio between a measurement result and its error Δx. It follows that measurement results of any physical measurable dimensional quantity cannot be exactly zero, i.e. they are finite. Even if $x = 0$ (which might occur), for the net results R, $|R| = \Delta x \geq \varepsilon > 0$. Therefore we assert that the result R can never be $R = 0 \pm 0$. Therefore:

$$|R| \geq \varepsilon > 0 \tag{3}$$

According to the definition of physics, as stated above, we conclude that values of exactly zero should not be allowed as acceptable solutions of mathematical expressions describing legitimate physical laws, because exactly zero experimental results (0 ± 0) do not occur. A corollary would be that physical laws should include ε in their formulation.

Next, we discuss very shortly the infinity problem. For reasons <u>similar</u> to those stated above, measurement results are <u>always</u> finite, <u>this</u> time in the sense of being non-infinite:

$$|R| \leq E < \infty \tag{4}$$

E is the upper limit value for the physical entity called x. To summarize our discussion till now, measurement results are <u>always</u> finite, namely <u>non-zero and non-infinite</u>:

$$0 < \varepsilon \leq |R| \leq E < \infty \tag{5}$$

Physical laws are formulated by means of mathematical expressions. As a consequence of the previous discussion, I propose that a new basic principle (postulate) of physics should be formulated, as follows:

"A legitimate law of physics is one whose mathematical expression does not allow exactly zero or infinite solutions as possible values for the measurable dimensional physical quantities present in the expression."

I call this the "Principle of Finiteness". By logical terminology, this is established as a necessary condition for the "legitimacy" of physical laws.

In the following section, we explore some implications of the finiteness principle when applied to several well-known laws of physics. Deliberately, I do this in a mostly deductive way, independently of any other physical theory or principle, besides, perhaps, the "simplicity" principle, as we shall see later.

2. Some consequences of the finiteness principle

2.1. General consequences

There are some immediate general consequences that can be derived from the finiteness principle:

a) Every measurable dimensional physical entity (quantity) has a minimum finite (non-zero) value (ε) and a maximum finite (non-infinite) value (E).

The minimum finite (non-zero)—or—the maximum finite (non-infinite) values of any measurable dimensional physical quantity are by definition impenetrable (irreducible) limits, namely no lower, or respectively no higher, values, are allowed.

b) Mathematical expressions of physical laws, which do not comply with the finiteness principle, should be changed accordingly in the simplest possible way (the "simplicity" principle), desirably without violating existing well-established principles.

2.2. Consequences of the finiteness principle in special relativity

I follow Einstein's derivation of the Lorentz Transformation ([11], 1920). We use two coordinate systems, k and k', whose x-axes permanently coincide. We consider only events on the x-axis. An event is represented in system k by distance x (from k-origin) and time t, and in system k' by distance x' (from k'-origin) and time t'.

The origin of k' is moving relative to k with velocity v, along the x-axis. We assume that at $t = t' = 0$, the origins of k and k' coincide, and a light-signal is transmitted from the origin along the positive x-axis. The choice of t=0 is, of course, arbitrary. We can set $t=t_0$, where t_0 is non-zero, and translate to zero by a variable change.

We follow faithfully Einstein's derivation, with the difference that I include in my derivation a minimum finite length error ε in his initial equations, as required by the Finiteness Principle (ε is the "Gedanken" minimum length error ([11], 1905)).

In this section (2.2), we assume that ε is a real invariant number (same in k and k').

Einstein's original equations for the light signal are:

$$\left.\begin{array}{r} x - ct = 0 \\ x' - ct' = 0 \end{array}\right\}$$

for the positive x-axis.

My corrected equations become:

$$\left.\begin{array}{r} x - ct = \varepsilon \\ x' - ct' = \varepsilon \end{array}\right\} \tag{6}$$

Similar relations are obtained for the negative x-axis.

After some calculations we get the corrected Lorentz Transformations:

$$\left\{\begin{array}{l} x' = \gamma\, x - v\gamma\, t + \varepsilon(1-\gamma) \\ t' = -\frac{v\gamma}{c^2}x + \gamma\, t + \frac{\varepsilon v \gamma}{c^2} \end{array}\right\} \tag{7}$$

$$\gamma = \frac{1}{\sqrt{1 - v^2/c^2}}. \tag{8}$$

It is important to notice that the length measurement error ε is an integral part of the Lorentz Transformations.

The corrected relativistic length contraction factor Γ_{CL} becomes:

$$\Gamma_{CL} \equiv \frac{L_C}{L_0} = \frac{1}{\gamma}\left[1 - \frac{\varepsilon(1-\gamma)}{L_0}\right], \tag{9}$$

where $\frac{1}{\gamma}$ is the usual contraction factor ($\varepsilon = 0$),L_0 is length measured at rest, and L_C is contracted length (should be $\geq \varepsilon$ according to the Finiteness Principle, where ε is the minimum length).

Two remarks are in order:

a) The maximal possible contraction of <u>any</u> distance L_0 (from (9)) is (ε/L_0), as it should be, as required by the finiteness principle.

b) If $L_0 = \varepsilon$, $\Gamma_{CL} = 1$ (as it should be).

Next, the corrected relativistic time-dilation factor is:

$$\Gamma_{CT} \equiv \frac{\tau_D}{\tau_0} = \gamma\left[1 - \frac{v\varepsilon}{\tau_0 c^2}\right] \tag{10}$$

τ_0 = time duration measured at rest

τ_D = dilated time duration

γ = usual time dilation ($\varepsilon = 0$)

Two remarks are in order:

a) one can test (10) by trying to detect changes in lifetimes of ultra-high velocity ($v \sim c$) and/or very short-lived subatomic particles. In this case we may use:

$$\frac{\tau_D}{\tau_0} \cong \gamma\left[1 - \frac{\varepsilon}{\tau_0 c}\right] \tag{11}$$

to detect departure from γ which is larger than γ's measurement error.

b) We notice that Γ_{CT} goes to infinity when $v \to c$. In this case and in general, the finiteness principle requires that γ should be replaced by a corrected function γ_c that satisfies the following requirements:

- **i.** γ_c should be finite for v=c (consequently we should get finite extremum values for all physical quantities).
- **ii.** γ_c should be identically equal to 1 (one) for all velocities when rest-frame (proper) measured values of physical quantities are equal to their extremum values.
- **iii.** Otherwise, γ_c should be identical to γ (of course, this should include all known physics and obviously much beyond).

It is easy to show that a simple function satisfying these requirements is:

$$\gamma_C \equiv \gamma \left[\frac{1-e^{-\frac{\alpha}{\gamma}}}{1-e^{-\alpha}}\right] \tag{12}$$

where α is a non-negative dimensionless number. We do not know yet how to derive eq.(12) **ab initio**, and we do not know if this equation is simplest or unique.

We require the replacement of γ by $\gamma_c$in Lorentz transformations and in the Special Relativity Lagrangian. Lorentz invariance and the Lorentz transformation group property are not observed for cases (i) and (ii) above, but are observed for case (iii) which includes all known physics.

In general, α is a function of the rest-frame values and extremum values of physical quantities. Usually α is extremely large, besides case (ii) above, where α is zero.

We will estimate later in the paper the value of α for subatomic particles

In this case γ_C is practically identical to γ, unless v is extremely close to c (ultra-high velocities). In this case equation (10) becomes:

$$\frac{\tau_D}{\tau_0} = \gamma \left[\frac{1-e^{-\frac{\alpha}{\gamma}}}{1-e^{-\alpha}}\right]\left[1-\frac{v\varepsilon}{\tau_0 c^2}\right] \tag{13}$$

When $\gamma >> \alpha$, we obtain (ultra-high velocities regime):

$$\frac{\tau_D{}^{\max}}{\tau_0} = \left[\frac{\alpha}{1-e^{-\alpha}}\right]\left[1-\frac{\varepsilon}{\tau_0 c}\right] \tag{14}$$

Since α is very large:

$$\frac{\tau_D{}^{\max}}{\tau_0} \approx \alpha \left[1-\frac{\varepsilon}{\tau_0 c}\right] \tag{15}$$

$\tau_D{}^{\max}$ = maximum allowable time dilation (!)

Assuming$\tau_0 >> \frac{\varepsilon}{c}$, we get:

$$\tau_D{}^{\max} \approx \alpha\tau_0$$

This and eq. (15) may be tested under appropriate conditions.

We now use (12) to correct Einstein's mass/energy equation such as to obey the Finiteness Principle.

Einstein's equation:

$$m = m_0\gamma \tag{16}$$

m_0 = rest mass

The corrected equation becomes:

$$m = m_0\gamma \left[\frac{1-e^{-\frac{\alpha}{\gamma}}}{1-e^{-\alpha}}\right] \tag{17}$$

Then, for $\gamma >> \alpha$, we obtain:

$$m_{\max} \approx m_0\alpha \tag{18}$$

$m_{\max}$ =maximum attainable relativistic mass (energy) (!)

In Section 2.3 we estimate α for sub-atomic particles.

As we shall see, $m_{\max}$is a fixed value for all sub-atomic particles, irrespective of m_0. In this case, the maximum possible relativistic mass/energy for all sub-atomic particles is a constant:

$$E_{\max} = m_{\max} c^2 \tag{19}$$

We conclude that length contraction and time dilation depend on ε, γ, and also on rest values of length and time interval being measured. Finally it is easy to show that the relativistic velocity addition law and $E = mc^2$ do not change, because the new translational terms in (7) are time-independent.

2.3. Consequences in quantum mechanics

I choose to apply the Finiteness Principle to the uncertainty principle, because of its central significance in Quantum Mechanics.:

$$\Delta x \cdot \Delta p \sim h \tag{20}$$

$$\therefore \Delta x \sim \frac{h}{\Delta p} \tag{21}$$

h is Planck's constant

In this discussion, Δx and Δp are assumed positive.

According to the Finiteness Principle, we have to correct the above equation such that Δx cannot become less than ε, the minimum length error (same ε as in Section 2.2) , for any finite value of Δp.

Note: in this section, (2.3), $\varepsilon > 0$.

Equation (21) represents Δx as a decreasing function ofΔp. According to the Finiteness Principle, Δx should become a corrected function of Δp, which has a minimum finite value $\varepsilon > 0$, for a finite value of Δp.

The simplest (increasing) function to ensure this and to be added to (21) is: $a\Delta p$, where a is a positive number, to be estimated in the following.

Therefore the necessary corrected equation is (see paragraph 2.1.b above) :

$$\Delta x \sim \frac{h}{\Delta p} + a\Delta p \tag{22}$$

To find a, we set the derivative of (22) to zero, which gives the minimum:

$$-\frac{h}{(\Delta p)^2_{\min}} + a = 0 \tag{23}$$

So:

$$(\Delta p)_{\min} = \sqrt{\frac{h}{a}} \tag{24}$$

$(\Delta p)_{\min}$ is the Δp value for which Δx is a minimum ($\Delta x = \varepsilon > 0$).

We insert (24) in (22), and remember that: $(\Delta x)_{\text{min}} = \varepsilon > 0$. We get:

$$\varepsilon = \frac{h}{\sqrt{\frac{h}{a}}} + a\sqrt{\frac{h}{a}} \tag{25}$$

Then we get:

$$a = \frac{\varepsilon^2}{4h} \tag{26}$$

Therefore the simplest corrected form of the uncertainty principle that obeys the Finiteness Principle is:

$$\Delta x \sim \frac{h}{\Delta p} + \frac{\varepsilon^2}{4h}\Delta p \tag{27}$$

We summarize the minimum coordinates of (27):

$$\left.\begin{array}{l} (\Delta x)_{\text{min}} = \varepsilon \\ (\Delta p)_{\text{min}} = \frac{2h}{\varepsilon} \end{array}\right\} \tag{28}$$

Now we estimate α (see (12)) for particles obeying the corrected uncertainty principle. We assume that$(\Delta p)_{\text{min}}$is also the maximum value of Δp if we want equation (27) to be a single-valued monotonous function.

We obtain:

$$\frac{2h}{\varepsilon} = \max(\gamma_C\ m_0 v) \tag{29}$$

(since γ_C replaces γ for $v \sim c$, see equation (12)); therefore:

$$\max(\gamma_C\ m_0 v) = m_0 \max(\gamma_C\ v) = m_0 \max(\gamma_C\ c) = m_0 c \max(\gamma_C) = m_0 c\alpha$$

and we have:

$$\frac{2h}{\varepsilon} = \alpha\ m_0 c$$

$$\therefore \alpha = \left(\frac{2h}{\varepsilon\ c}\right)\frac{1}{m_0} \tag{30}$$

Then equation (18) becomes:

$$m_{\text{max}} = m_0\alpha = \frac{2h}{\varepsilon c} \tag{31}$$

and equation (19) becomes:

$$E_{\text{max}} = \alpha\ m_0 c^2 = \frac{2h}{\varepsilon\ c}c^2 = \frac{2hc}{\varepsilon} \tag{32}$$

Therefore we conclude that all sub-atomic particles have the same maximum possible relativistic mass/energy, irrespective of their rest mass/energy. They obtain this value at $v \approx c$ (ultra-high velocities).

3. Discussion

Before continuing, we remark parenthetically that contradictions may arise between the consequences of the finiteness principle and the consequences of other well-established principles. In this case, the discrepancies might occur because different principles may deal with different regimes of the parameters (e.g., different velocity ranges).

Returning to the discussion on the results obtained in this paper, we first note that equation (27), with $\varepsilon = l_p$ (Planck's length) is required in quantum gravity as a replacement for the usual uncertainty principle, and it is referred to as the "generalized uncertainty principle" [12]. In this context a generalized quantum-mechanical Hamiltonian is constructed, on the basis of the generalized uncertainty principle, which becomes important at the Planck scale. In Ref. [13] a generalized uncertainty principle is derived from micro-black hole Gedanken experiment, using only the uncertainty principle and R_s (Schwarzschild radius).

Historically, a generalized uncertainty principle was first derived in [14] and then in [15] with what corresponds to our $\varepsilon = l_p$, and $l_p^2 \approx \alpha'$ (string tension).

From the results obtained in our paper, it is clear that the Finiteness Principle explains in a fundamental way the additional term in the generalized uncertainty principle ($\varepsilon = l_p$ is a special case, representing the inevitable gravitational interference uncertainty). It is indeed very encouraging to find that quantum gravity is consistent with the Finiteness Principle.

It is also important to mention that the T-Duality in string theory requires a minimum observable length $R_{\text{min}} \sim \alpha'^2$ and brings about a generalized uncertainty principle [16, 17]. This means that T-Duality (and maybe other aspects of string theory) is intimately connected with the Finiteness Principle.

Next, we note that neutrino measurements data are very important in testing the consequences of the Finiteness Principle, because of their $m_0 \neq 0$, and $v \sim c$ [see eqs. (17), (18), (19), (31), (32) with $\varepsilon = l_p$ (Planck length) and $m_{\text{max}} = m_p$ (Planck mass)]. It is especially significant to have experiments giving each at least two of the three following parameters: energy, rest-mass, speed.

However, the usefulness of the existing neutrino data [18, 19, 20, 21, 22, 23] is limited, because of lack of ultra-high energy measurements, lack of absolute mass measurements, and finally, lack of sufficient accuracy of speed measurement at $v \sim c$.

Finally, because of the importance of the fundamental constants to the Planck scale limits, astronomical measurements are being made and evaluated extensively, in order to determine the accuracy and constancy of various parameters [24].

4. Conclusions

I propose a new postulate in physics, the Principle of Finiteness. I believe it is a very basic postulate, to which any "legitimate" physical law should obey. I explore the implications of the finiteness principle for some simple relations of special relativity and quantum mechanics.

The corrected relativistic length contraction and time dilation factors depend on: ε, γ, τ_0, L_0, and m_0 (through α).

An important prediction is that the maximum relativistic mass/energy attainable is the same for all sub-atomic particles (independent of m_0).

It also seems that the generalized uncertainty principle required by quantum gravity (and T-duality) as a replacement for the usual Heisenberg uncertainty principle, is actually a necessary consequence of the Finiteness Principle. It follows that quantum gravity is consistent with the Finiteness Principle. The Finiteness Principle apparently explains in a fundamental way the need of quantum gravity to replace the usual uncertainty principle with a corrected ("generalized") version, at the minimum length scale.

Therefore, the Finiteness Principle may contribute to the axiomatic foundation of quantum gravity, in addition to many other consequences.

Acknowledgments

It is my pleasure to acknowledge useful remarks by: Prof. J. Bekenstein, Prof. L. Dorman, Prof. N. Fisch, Prof. S. Cohen, Prof. A. Yahalom, Prof. L. Horwitz, Dr. A. Yarom, and Prof. A. Polyakov.

I would like to thank J. A. Baumgaertel and L. Peterson for carefully preparing the manuscript for publication.

References

[1] Born M 1934 *Proc. R. Soc. A.* **143** 410
Born M and Infeld L 1934 *Proc. R. Soc. A.* **144** 425
[2] Einstein A 1939 *Annals of Mathematics* **40** (4) 922-936
[3] Vilenkin A 1982 *Phys. Lett. B* **117** (1-2) 25-28
[4] Lee T D 1983 *Phys. Lett. B* **122** 217
Lee T D 1987 *J. Stat. Phys.* **46** 843
[5] Gambini R and Pullin J. 2007 *Found. Phys.* **37** (7) 1074-1092
[6] Nicolai H 2009*Physics* **2** 70
[7] Bern Z, Carrasco J J, Dixon LJ, Johansson H and Roiban R 2009 *Phys. Rev. Lett.* **103** 081301
[8] Esposito S and Salesi G 2010 *Ann. Phys.* Berlin 1-11
[9] Popper KR 1968 *The Logic of Scientific Discovery* Hutchinson, London
[10] Margenau H 1950 *The Nature of Physical Reality* McGraw-Hill, New-York
[11] Einstein A 1920 *Relativity: The Special and General Theory* Henry Holt, New York
Einstein A 1905 On the Electrodynamics of Moving Bodies *Annalen der Physik,* **17** footnote No.3. Footnote text: "We shall not here discuss the inexactitude which lurks in the concept of simultaneity of two events at approximately the same place, which can only be removed by an abstraction." If the Principle of Finiteness is accepted then this abstraction is prohibited and the inexactitude remains.
[12] Das S and Vagenas EC 2009 *Can. J. Phys.* **87** 233-240
[13] Scardigli F 1999 *Phys. Lett. B* **452** (1-2) 39-44
[14] Veneziano G 1986*Euro Phys. Lett.* **2** (3) 199
[15] Gross DJ and Mende P F 1987*Phys. Lett. B*, **197** 129
Gross DJ and Mende P F 1988 *Nucl. Phys. B* **303** 407
[16] Witten E, April 1996 *Phys. Today* 24
[17] Ne'eman Y *Ann. Phys.* Leipzig, **8** (1) 3-17
[18] Spolyar D, Buckley M, Freese K, Hooper D and Murayama H 2009, High Energy Neutrinos as a Test of Leptophilic Dark Matter, *arxiv: 0905.4764 v1 Astro-ph.co*
[19] *Rev. of Particle Physics* http://pdg.lbl.gov
[20] Witten E 2002 *Nature* **415** 969
[21] Bahcall JN 1989 *Neutrino Astrophysics* Cambridge Univ. Press, ISBN 052137975x
[22] Dorman LI 2004 *Cosmic Rays in the Earth's Atmosphere and Underground* (Kluwer Academic Publishers (Astroph. And Space Sci. Library)) 52-74.

[23] Pagliaroli G, Rossi-Torres F and Vissani F 2010 *Astroparticle Phys.* **33** 287-291
[24] Garcia-Berro E , Isern J and Kubyshin YA 2007 *Astron. Astroph. Rev.* **14** 113-170

Offshell quantum electrodynamics

Martin Land[1] **and Lawrence P. Horwitz**[2,3]

[1] Department of Computer Science, Hadassah College, 37 HaNeviim Street, Jerusalem
[2] School of Physics, Tel Aviv University, Ramat Aviv 69978, Israel
[3] Department of Physics, Bar Ilan University, Ramat Gan 52900, Israel

E-mail: [1]`martin@hadassah.ac.il`

Abstract.
In this paper, we develop the quantum field theory of off-shell electromagnetism, and use it to calculate the Møller scattering cross-section. This calculation leads to *qualitative* deviations from the usual scattering cross-sections, which are, however, small effects, but may be visible at small angles near the forward direction.

1. Introduction

Covariant Quantum Mechanics

In 1941, Stueckelberg [1] (see also Fock [2]) proposed a Poincaré invariant Hamiltonian mechanics in which particle worldlines are traced out by the evolution of events according to an invariant parameter τ. Stueckelberg's purpose was to describe pair annihilation as the evolution of a single worldline, first forward and then backward in time. Since the Einstein time coordinate $x^0 = t$ does not increase monotonically as the system evolves in such a model (the basis of the Feynman-Stueckelberg interpretation of anti-particles [3]), it was necessary to replace x^0 with a new chronological parameter. The Poincaré invariant parameter τ is formally similar to the Galilean invariant time in Newtonian theory, and its introduction permits the adaptation of many techniques from nonrelativistic classical and quantum mechanics. In the symplectic mechanics which follows from this formulation, one writes classical Hamilton equations in the form

$$\frac{dx^\mu}{d\tau} = \frac{\partial K}{\partial p_\mu} \qquad \frac{dp^\mu}{d\tau} = -\frac{\partial K}{\partial x_\mu} \tag{1}$$

where

$$\eta_{\mu\nu} = \text{diag}(-1,1,1,1) \qquad \text{and} \qquad \mu,\nu = 0,\cdots,3 \tag{2}$$

and K is the analog of the Hamiltonian which generates system evolution according to τ. As in the nonrelativistic case, one makes the transition from classical to quantum mechanics by regarding the Hamiltonian as the Hermitian generator of unitary τ evolution and writing

$$i\frac{\partial}{\partial\tau}\psi_\tau(x) = K\psi_\tau(x) \tag{3}$$

as a covariant Schrödinger equation with first order τ evolution. The electromagnetic system

$$i\partial_\tau\psi(x,\tau) = \frac{1}{2m}\left[p^\mu - eA^\mu(x)\right]\left[p_\mu - eA_\mu(x)\right]\psi(x,\tau) \tag{4}$$

enjoys U(1) local gauge symmetry

$$\psi(x,\tau) \longrightarrow \exp\left[ie\Lambda(x)\right]\ \psi(x,\tau) \qquad\qquad A_\mu \longrightarrow A_\mu + \partial_\mu\Lambda(x) \tag{5}$$

and global gauge symmetry leading to the conserved current

$$\partial_\mu j^\mu + \partial_\tau\rho = 0 \tag{6}$$

where

$$\rho(x,\tau) = \left|\psi(x,\tau)\right|^2 \qquad j^\mu = -\frac{i}{2M}\left\{\psi^*(\partial^\mu - ieA^\mu)\psi - \psi(\partial^\mu + ieA^\mu)\psi^*\right\} \tag{7}$$

with an interpretation of $\rho(x,\tau)$ as the probability density at τ of finding the event at the spacetime point x. The non-zero divergence of the four-vector current $j^\mu(x,\tau)$ prevents its identification as the source of the field $A^\mu(x)$, but Stueckelberg observed that

$$\rho(x,\tau) \xrightarrow[\tau\to\pm\infty]{} 0 \qquad \Rightarrow \qquad \partial_\mu J^\mu(x) = \partial_\mu \int_{-\infty}^{\infty} d\tau\ j^\mu(x,\tau) = 0\ . \tag{8}$$

However, there is no guarantee that the particle worldline associated with the *a posteriori* current $J^\mu(x)$ will induce the field $A^\mu(x)$ that interacts with the particle instantaneously at τ.

In order to obtain a well-posed theory, Sa'ad, Horwitz, and Arshansky [4] generalized (4) by introducing τ-dependent potentials and a fifth gauge field. Writing $x^5 = \tau$ and adopting the index convention

$$\alpha,\beta,\gamma = 0,1,2,3,5 \qquad\qquad \lambda,\mu,\nu = 0,1,2,3 \qquad\qquad i,j,k = 1,2,3 \tag{9}$$

the Stueckelberg-Schrodinger equation

$$\left[i\partial_\tau + e_0 a_5(x,\tau)\right]\psi(x,\tau) = \frac{1}{2M}\left[p^\mu - e_0 a^\mu(x,\tau)\right]\left[p_\mu - e_0 a_\mu(x,\tau)\right]\psi(x,\tau) \tag{10}$$

is locally gauge invariant under τ-dependent gauge transformations

$$\psi \to e^{ie_0\Lambda(x,\tau)}\psi \qquad\qquad a_\alpha \to a_\alpha + \partial_\alpha\Lambda(x,\tau) \tag{11}$$

and leads to the conserved current

$$\partial_\mu j^\mu + \partial_\tau j^5 = 0 \tag{12}$$

where

$$j^5 \equiv \rho = \left|\psi(x,\tau)\right|^2 \qquad j^\mu = \frac{-i}{2M}\left\{\psi^*(\partial^\mu - ie_0 a^\mu)\psi - \psi(\partial^\mu + ie_0 a^\mu)\psi^*\right\}. \tag{13}$$

Equation (10) may be derived by variation of the action

$$S = \int d^4x d\tau \left\{\psi^*(i\partial_\tau + e_0 a_5)\psi - \frac{1}{2M}\psi^*(p_\mu - e_0 a_\mu)(p^\mu - e_0 a^\mu)\psi - \frac{\lambda}{4} f_{\alpha\beta} f^{\alpha\beta}\right\} \tag{14}$$

to which Sa'ad, et. al. added a kinetic term for the fields generalizing the Maxwell term from four to five dimensions. In raising the 5-index one chooses a value for η^{55} in

$$f_{\alpha\beta}f^{\alpha\beta} = f_{\mu\nu}f^{\mu\nu} + 2f_{5\mu}f^{5\mu} = f_{\mu\nu}f^{\mu\nu} + 2\eta^{55}f_{5\mu}f_5{}^{\mu} \tag{15}$$

corresponding to a formal metric

$$\eta^{\alpha\beta} = \mathrm{diag}(-1,1,1,1,\sigma) \ . \tag{16}$$

It has been shown [5] that choosing $\sigma = -1$ leads to a consistent description of low energy classical Coulomb scattering. Varying the action (14) with respect to the gauge fields, the equations of motion are found to be

$$\partial_\beta f^{\alpha\beta} = \frac{e_0}{\lambda} j^\alpha = e j^\alpha \qquad\qquad \epsilon^{\alpha\beta\gamma\delta\epsilon}\partial_\alpha f_{\beta\gamma} = 0 \tag{17}$$

where j^α is given by (13). The fields satisfy the wave equation

$$\partial_\alpha\partial^\alpha f^{\beta\gamma} = (\partial_\mu\partial^\mu + \partial_\tau\partial^\tau)f^{\beta\gamma} = (\partial_\mu\partial^\mu + \sigma\,\partial_\tau^2)f^{\beta\gamma} = -e(\partial^\alpha j^\beta - \partial^\beta j^\alpha) \tag{18}$$

depending on the signature σ of ∂_τ. Under the boundary conditions

$$\tau \to \pm\infty \qquad \Rightarrow \qquad a^5(x,\tau) \to 0 \qquad\qquad j^5(x,\tau) \to 0 \tag{19}$$

the standard Maxwell theory is extracted as the equilibrium limit of (18) by integration over the worldline

$$\left.\begin{array}{c} \partial_\beta f^{\alpha\beta}(x,\tau) = e j^\alpha(x,\tau) \\ \\ \partial_{[\alpha}f_{\beta\gamma]} = 0 \end{array}\right\} \xrightarrow[\int d\tau]{} \left\{\begin{array}{c} \partial_\nu F^{\mu\nu}(x) = e J^\mu(x) \\ \\ \partial_{[\mu}F_{\nu\rho]} = 0 \end{array}\right. \tag{20}$$

where

$$F^{\mu\nu}(x) = \int_{-\infty}^{\infty} d\tau\ f^{\mu\nu}(x,\tau) \qquad\qquad A^\mu(x) = \int_{-\infty}^{\infty} d\tau\ a^\mu(x,\tau) \ . \tag{21}$$

This integration has been called concatenation [6] and as in (8) links the event current $j^\mu(x,\tau)$ with the particle current $J^\mu(x)$ defined on the entire particle worldline. It is seen from (10) and (21) that e_0 and λ must have dimensions of time, so that the dimensionless ratio $e = e_0/\lambda$ can be identified as the Maxwell charge. The microscopic τ-dependent fields have been called pre-Maxwell fields. For further background on the classical theory underlying off-shell QED, see [7] and references contained therein.

Off-Shell Quantum Electrodynamics

In this paper, we develop the methods required to apply off-shell quantum electrodynamics as a scattering theory, and calculate certain elementary processes. In Section 2, we perform a canonical quantization of the interacting off-shell theory. As a parameterized evolution theory on an unconstrained phase space, it is natural to apply Jackiw's method of quantizing an action that can be made linear in time derivatives, without introducing auxiliary fields [8]. In conventional quantum theories, the momentum is related to a t-derivative, so putting the action into first order form may break manifest Lorentz covariance. For the off-shell theory, the classical velocity is a τ-derivative, and so Jackiw's approach remains manifestly covariant throughout. A path integral quantization following this method was presented in [9]. The results obtained are in complete agreement with those of Shnerb and Horwitz who used a bosonic gauge fixing method [10]. Having identified the essential degrees of freedom, we perform the detailed Fourier expansion of the operators of the free fields, required in perturbation theory. In Section 3, we carry out a canonical quantization of the free off-shell matter field, and provide an expansion of the field operators in terms of annihilation operators in momentum space. We use these expansions to show that the vacuum expectation value of τ-ordered fields is precisely the Stueckelberg-Schrödinger equation Green's function, with Schwinger-Feynman boundary conditions. In Section 4, we carry out the canonical quantization of the five dimensional electromagnetic field and the expansion of field operators in momentum space. This field has three independent polarizations, one of which decouples in the Maxwell (zero mode) limit. We show that vacuum expectation value of products of two τ-ordered fields is the Feynman Green's function for the wave equation, and that it preserves the gauge condition. In Section 5, we develop the perturbation theory for the S-matrix in the interaction picture. In order to provide the link with the usual formulas for the perturbation expansion of Green's functions, we adapt the LSZ reduction formulas to the off-shell theory. We then derive the Feynman rules. In Section 6, we derive the scattering cross-section in terms of the transition amplitude, and demonstrate the relationship of this expression to the usual cross-section. In Section 7, we analyze the Møller scattering amplitude in detail and examine the qualitative deviation from the corresponding amplitude calculated in conventional QED. In Section 8, we consider the problem of renormalization of the off-theory. We derive the Ward identity associated with current conservation and show that it connects the matter field propagators with both the 3-particle and the 4-particle vertex functions. We conclude that the off-shell theory is renormalizable when a cut-off is placed on the mass of the off-shell photons. We relate this cut-off to a modification of the electromagnetic action required in classical off-shell electrodynamics.

2. Canonical quantization of the interacting theory

Following [4] we take the action for off-shell electromagnetism to be

$$S = \int d^4x d\tau \left\{ \psi^*(i\partial_\tau + e_0 a_5)\psi - \frac{1}{2M}\psi^*(-i\partial_\mu - e_0 a_\mu)(-i\partial^\mu - e_0 a^\mu)\psi - \frac{\lambda}{4} f_{\alpha\beta} f^{\alpha\beta} \right\} \tag{22}$$

using the index conventions (9)and (16). Because of their support properties, 4-divergences as vanishing under spacetime integration while τ-derivatives vanish *independently* under τ-integration because of Riemann-Lebesgue oscillations (the asymptotic mass dependence of the fields $exp\{-i(m^2/2M)\,\tau\}$ varies arbitrarily rapidly for large τ). Thus, for the τ-derivative terms in (22),

$$\frac{1}{2}\left[\psi^*(i\partial_\tau + e_0 a_5)\psi + \text{h.c.}\right] = i\psi^*\dot{\psi} + e_0 a_5\, \psi^*\psi + \partial_\tau(\psi^*\psi) \tag{23}$$

where $\dot{\psi} = \partial\psi/\partial\tau$. We will henceforth regard (22) as hermitian.

In Dirac's method [11] of quantization for gauge theories, we would form the Hamiltonian from (22), including a momentum π_5 conjugate to a^5 and a Lagrange multiplier to enforce the primary constraint $\pi_5 = 0$ (in [10], this is accomplished through the gauge fixing term). The secondary constraint (that the primary constraint commute with the Hamiltonian) would lead to the Gauss Law for the off-shell theory, which is the $\alpha = 5$ term of the first of (17).

We follow a quantization scheme advocated by Jackiw [8], in which one first eliminates the constraint from the Lagrangian, and then constructs the Hamiltonian from the unconstrained degrees of freedom. In order to perform the transformation which eliminates the constrained degrees of freedom, it is necessary that the action be made linear in the time derivatives. Because the scalar τ plays the role of time in the off-shell theory, Lorentz covariance is maintained throughout, in a manner consistent with the original spirit of the proper time method.

In Jackiw's method, a choice of gauge is made implicitly by solving the constraints and introducing a decomposition of the fields consistent with that solution. The elimination of the longitudinal polarizations is carried out as an application of the Darboux theorem, diagonalizing the Hamiltonian. This procedure is evidently related to the quantization method proposed by Fermi in 1932 [12]. Applying Fermi's method to (22), we would perform a gauge transformation which guarantees the condition $\partial_\mu a^\mu = 0$, and leads to the Gauss law for the component a_5. The presence of sources in the action would require the decomposition of the fields into longitudinal terms induced by the sources and transverse propagating terms with vanishing our-divergence. We could then argue that the longitudinal fields do not satisfy wave equations and need not be quantized; this would leave the unconstrained transverse fields in the action, and we would obtain a consistent canonical quantization. Elements of such a procedure may be recognized in the method presented below.

The kinetic term for the matter field is linear in ∂_τ by construction. In order to put the kinetic term for the gauge field into explicitly canonical form, we rewrite $f^{5\mu} f_{5\mu}$ as $f^{5\mu}(\partial_\tau a_\mu - \partial_\mu a_5)$ and take the quantity $f^{5\mu}$ to be independent of the fields a_α (this is a variant of the first order Lagrangian form for the usual electromagnetic field). Expanding the electromagnetic term,

$$f_{\alpha\beta} f^{\alpha\beta} = f_{\mu\nu} f^{\mu\nu} + 2\sigma f^{5\mu} f^5{}_\mu = f_{\mu\nu} f^{\mu\nu} + 2\sigma \left[2(\sigma\partial_\tau a^\mu - \partial^\mu a^5) f^5{}_\mu - f^{5\mu} f^5{}_\mu \right] \tag{24}$$

when we vary the action with respect to $f^{5\mu}$, we recover its relationship to the a^α. Integrating by parts, we obtain

$$(\sigma\partial_\tau a^\mu - \partial^\mu a^5) f^5{}_\mu = \sigma(\partial_\tau a^\mu) f^5{}_\mu + a^5 \partial^\mu f^5{}_\mu - \text{divergence} \tag{25}$$

from which we drop the divergence and introduce the notation

$$\epsilon^\mu = f^{5\mu} \, . \tag{26}$$

Using (25) and (26), the action becomes

$$\begin{aligned} S \;=\; & \int d^4x d\tau \Big[i\psi^* \dot\psi - \lambda \epsilon_\mu \dot a^\mu - \frac{1}{2M} \psi^* (-i\partial_\mu - e_0 a_\mu)(-i\partial^\mu - e_0 a^\mu)\psi \\ & \quad - \frac{\lambda}{4} f_{\mu\nu} f^{\mu\nu} + \frac{\lambda\sigma}{2} \epsilon^\mu \epsilon_\mu + a_5 (e_0 \psi^* \psi - \lambda \partial^\mu \epsilon_\mu) \Big] \end{aligned} \tag{27}$$

where we have collected terms in a_5 and written $\partial_\tau \psi = \dot\psi$. Following Jackiw, we regard (27) as an action for the conjugate pairs $\{i\psi^*, \psi\}$ and $\{\epsilon_\mu, a^\mu\}$, with a_5 playing the role of a Lagrange multiplier for the constraint

$$e_0 \psi^* \psi - \lambda \partial^\mu \epsilon_\mu = 0 \quad \Rightarrow \quad \partial^\mu \epsilon_\mu = \frac{e_0}{\lambda} \psi^* \psi = e\rho, \tag{28}$$

which is just the Gauss law for the off-shell theory. The constraint equation (28) can be solved through the decomposition

$$\epsilon^\mu = (\epsilon_\perp)^\mu + e\partial^\mu [G\rho] \tag{29}$$

where

$$\partial_\mu(\epsilon_\perp)^\mu = 0 \tag{30}$$

and where $G\rho$ is a shorthand for the functional

$$[G\rho](x,\tau) = \int d^4y\, G(x-y)\, \rho(y,\tau) \tag{31}$$

in which we specify the Maxwell Green's function

$$G(x-y) = \delta\left[(x-y)^2\right] \quad \Rightarrow \quad \Box G = 1. \tag{32}$$

Performing a similar decomposition of a^μ,

$$a^\mu = (a_\perp)^\mu + \partial^\mu[G\Lambda] \qquad \partial_\mu(a_\perp)^\mu = 0 \tag{33}$$

(by which we implicitly choose the gauge condition $\partial_\mu a^\mu = \Lambda$) the remaining terms of the theory are expressed as

$$\dot{a}^\mu = (\dot{a}_\perp)^\mu + \partial^\mu[G\dot{\Lambda}] \tag{34}$$

$$f^{\mu\nu} = \partial^\mu(a_\perp)^\nu - \partial^\nu(a_\perp)^\mu = (f_\perp)^{\mu\nu} \tag{35}$$

$$-i\partial^\mu - e_0 a^\mu = -i\partial^\mu - e_0(a_\perp)^\mu - e_0\partial^\mu[G\dot{\Lambda}] \tag{36}$$

$$\epsilon^\mu\epsilon_\mu = [\,(\epsilon_\perp)^\mu + e\partial^\mu[G\rho]\,][\,(\epsilon_\perp)_\mu + e\partial_\mu[G\rho]\,] = (\epsilon_\perp)^\mu(\epsilon_\perp)_\mu - e^2\rho[G\rho] + \text{divergence} \tag{37}$$

$$\epsilon_\mu\dot{a}^\mu = (\epsilon_\perp)_\mu(\dot{a}_\perp)^\mu - e\rho[G\dot{\Lambda}]\,. \tag{38}$$

In terms of this decomposition, the action becomes

$$\begin{aligned} S \;=\; \int d^4x d\tau \Big\{ & i\psi^*\dot{\psi} - \lambda(\epsilon_\perp)_\mu(\dot{a}_\perp)^\mu + \lambda e\rho[G\dot{\Lambda}] \\ & - \frac{1}{2M}\psi^*(-i\partial_\mu - e_0(a_\perp)_\mu - e_0\partial_\mu[G\dot{\Lambda}])(-i\partial^\mu - e_0(a_\perp)^\mu - e_0\partial^\mu[G\dot{\Lambda}])\psi \\ & - \frac{\lambda}{4}(f_\perp)_{\mu\nu}(f_\perp)^{\mu\nu} + \frac{\lambda\sigma}{2}(\epsilon_\perp)^\mu(\epsilon_\perp)_\mu - \frac{\lambda\sigma}{2}e^2\rho[G\rho]\Big\}\,. \end{aligned} \tag{39}$$

We now perform the gauge transformation

$$\psi \longrightarrow e^{ie_0[G\Lambda]}\psi \tag{40}$$

which entails

$$\dot{\psi} \longrightarrow e^{ie_0[G\Lambda]}[\dot{\psi} + ie_0[G\dot{\Lambda}]\psi] \qquad\qquad i\psi^*\dot{\psi} \longrightarrow i\psi^*\dot{\psi} - e_0\rho[G\dot{\Lambda}] \tag{41}$$

and

$$\begin{aligned} (-i\partial^\mu - e_0(a_\perp)^\mu - e_0\partial^\mu[G\dot{\Lambda}])\psi \;&\longrightarrow\; (-i\partial^\mu - e_0(a_\perp)^\mu - e_0\partial^\mu[G\dot{\Lambda}])e^{ie_0[G\Lambda]}\psi \\ &=\; e^{ie_0[G\Lambda]}(-i\partial^\mu - e_0(a_\perp)^\mu - e_0\partial^\mu[G\dot{\Lambda}] + e_0\partial^\mu[G\dot{\Lambda}])\psi \\ &=\; e^{ie_0[G\Lambda]}(-i\partial^\mu - e_0(a_\perp)^\mu)\psi\,. \end{aligned} \tag{42}$$

This transforms the action to the form

$$\begin{aligned} S \;=\; \int d^4x d\tau \Big\{ & i\psi^*\dot{\psi} - \lambda(\epsilon_\perp)_\mu(\dot{a}_\perp)^\mu - \frac{1}{2M}\psi^*(-i\partial_\mu - e_0(a_\perp)_\mu)(-i\partial^\mu - e_0(a_\perp)^\mu)\psi \\ & - \frac{\lambda}{4}(f_\perp)_{\mu\nu}(f_\perp)^{\mu\nu} + \frac{\lambda\sigma}{2}(\epsilon_\perp)^\mu(\epsilon_\perp)_\mu - \frac{\lambda\sigma}{2}e^2\rho[G\rho]\Big\} \end{aligned} \tag{43}$$

which is an unconstrained functional (only the unconstrained field degrees of freedom are present) and so may be canonically quantized. Henceforth, we may drop the subscript $_\perp$ from the quantized gauge field variables and assume the transversality conditions (30) and (33) for the field operators.

The conjugate momenta are found from (43) to be

$$\pi_\psi = \frac{\partial \mathcal{L}}{\partial \dot{\psi}} = i\psi^* \qquad\qquad \pi_{a_\mu} = \frac{\partial \mathcal{L}}{\partial \dot{a}_\mu} = -\lambda \epsilon^\mu \tag{44}$$

from which we compute the Hamiltonian

$$\begin{aligned} \mathcal{K} &= \pi_\psi \dot{\psi} + \pi_{a_\mu} \dot{a}_\mu - \mathcal{L} \\ &= \frac{1}{2M}\left[(i\partial_\mu - e_0 a_\mu)\psi^*\right]\left[(-i\partial^\mu - e_0 a^\mu)\psi\right] + \frac{\lambda}{4} f_{\mu\nu} f^{\mu\nu} - \frac{\lambda\sigma}{2}\epsilon^\mu \epsilon_\mu + \frac{\lambda\sigma}{2} e^2 \rho [G\rho] \end{aligned} \tag{45}$$

which may be decomposed as

$$\mathcal{K} = \mathcal{K}_{\text{photon}} + \mathcal{K}_{\text{matter}} + \mathcal{K}_{\text{interaction}} \tag{46}$$

where

$$\mathcal{K}_{\text{photon}} = \frac{\lambda}{4} f_{\mu\nu} f^{\mu\nu} - \frac{\lambda\sigma}{2}\epsilon^\mu \epsilon_\mu \tag{47}$$

$$\mathcal{K}_{\text{matter}} = \frac{1}{2M}\left(\partial_\mu \psi^*\right)\left(\partial^\mu \psi\right) \tag{48}$$

$$\mathcal{K}_{\text{interaction}} = \frac{ie_0}{2M} a_\mu (\psi^* \partial^\mu \psi - \psi \partial^\mu \psi^*) + \frac{e_0^2}{2M} a_\mu a^\mu |\psi|^2 + \frac{\lambda\sigma}{2} e^2 \rho [G\rho] \tag{49}$$

The last term in (49) has the form of a c-number energy density which represents the mass-energy equivalent required to assemble the matter field (see also [10]). The coupling $e_0\, e = \lambda\, e^2$ is characteristic of the classical Lorentz force due to a charge distribution. We demonstrate below that the remaining terms in (49), which lead to 3-particle and 4-particle interactions in the Feynman diagrams, are connected by the Ward identity for the conserved current (12).

3. Canonical quantization of the free spinless matter field

From the Hamiltonian density (48) the Hamiltonian operator is

$$\mathrm{K} = \int d^4x \mathcal{K}_{\text{matter}} = \frac{1}{2M}\int d^4x \left(\partial_\mu \psi^*\right)\left(\partial^\mu \psi\right) . \tag{50}$$

The fields must satisfy the canonical equal-τ commutation relations

$$[\psi(x,\tau), \pi_\psi(x',\tau)] = i\delta^4(x - x') \tag{51}$$

which together with the first of (44) leads to

$$[\psi(x,\tau), \psi^*(x',\tau)] = \delta^4(x - x'). \tag{52}$$

The field evolves dynamically according to the Heisenberg equation

$$i\partial_\tau \psi = [\psi, \mathrm{K}] \tag{53}$$

and using equations (48), (50), (51), and (53), we find

$$\begin{aligned}
i\partial_\tau \psi(x,\tau) &= \left[\psi(x,\tau), \frac{1}{2M}\int d^4x' \left(\partial'_\mu \psi^*(x',\tau)\right)\left(\partial'^\mu \psi(x',\tau)\right)\right] \\
&= -\frac{1}{2M}\int d^4x' \left[\psi(x,\tau), \psi^*(x',\tau)\partial'_\mu \partial'^\mu \psi(x',\tau)\right] \\
&= -\frac{1}{2M}\int d^4x' \left[\psi(x,\tau), \psi^*(x',\tau)\right] \partial'_\mu \partial'^\mu \psi(x',\tau) \\
&= -\frac{1}{2M}\int d^4x' \delta^4(x-x')\partial'_\mu \partial'^\mu \psi(x',\tau) \\
&= -\frac{1}{2M}\partial_\mu \partial^\mu \psi(x,\tau)
\end{aligned} \tag{54}$$

which is the Schrödinger equation (4) for the field operator $\psi(x,\tau)$ in the absence of interaction. Because it is satisfied, we may perform the Fourier expansion

$$\psi(x,\tau) = \int \frac{d^4k}{(2\pi)^4} b(k) e^{i(k\cdot x - \kappa\tau)} \qquad \psi^*(x,\tau) = \int \frac{d^4k}{(2\pi)^4} b^*(k) e^{-i(k\cdot x - \kappa\tau)} \tag{55}$$

where from (54) we see that $\kappa = k^2/2M$. From (51), we find that

$$\left[b(k), b^*(k')\right] = (2\pi)^4 \delta^4(k-k') \qquad \left[b(k), b(k')\right] = \left[b^*(k), b^*(k')\right] = 0. \tag{56}$$

In terms of the Fourier expansion, the Hamiltonian is given by

$$\begin{aligned}
\mathrm{K} &= \int d^4x \frac{1}{2M}\left[\partial_\mu \psi^*\right]\left[\partial^\mu \psi\right] \\
&= \frac{1}{2M}\int d^4x \int \frac{d^4k}{(2\pi)^4}(-ik_\mu) b^*(k) e^{-i(k\cdot x - \kappa\tau)} \int \frac{d^4k'}{(2\pi)^4}(ik'_\mu) b(k') e^{i(k'\cdot x - \kappa'\tau)} \\
&= \frac{1}{2M}\int \frac{d^4k}{(2\pi)^4}\frac{d^4k'}{(2\pi)^4} b^*(k) b(k') (2\pi)^4 \delta^4(k-k') e^{i\kappa\tau} e^{-i\kappa'\tau}(k\cdot k') \\
&= \int \frac{d^4k}{(2\pi)^4}\frac{k^2}{2M} b^*(k) b(k) \\
&= \int \frac{d^4k}{(2\pi)^4}\, \kappa\, b^*(k) b(k)
\end{aligned} \tag{57}$$

The ground state mass for this Hamiltonian vanishes even without normal ordering.

We write the Green's function $G_i(x,\tau)$ for the Schrödinger equation (54) in the integral form,

$$G(x,\tau) = \frac{1}{(2\pi)^5}\int_{C_i} d^4k d\kappa \frac{e^{i(k\cdot x - \kappa\tau)}}{\frac{1}{2M}k^2 - \kappa} \tag{58}$$

where the contour of κ integration C_i determines the boundary conditions. If C_i includes the interval $(-\infty, \infty)$, then $G_i(x,\tau)$ will satisfy the inhomogeneous equation

$$(i\partial_\tau + \frac{1}{2M}\partial^\mu \partial_\mu) G_i(x,\tau) = -\delta^4(x)\delta(\tau). \tag{59}$$

Journal of Physics: Conference Series **437** (2013) 012011 doi:10.1088/1742-6596/437/1/012011

We take the Green's function for the matter field to be

$$G(x,\tau) = \frac{1}{(2\pi)^5}\int d^4k d\kappa \frac{e^{i(k\cdot x-\kappa\tau)}}{\frac{1}{2M}k^2-\kappa-i\epsilon} \tag{60}$$

displacing the pole at $\kappa = k^2/2M - i\epsilon$ into the lower half plane. The κ-integration vanishes for $\tau < 0$ (closing the contour in the upper half plane) and for $\tau > 0$ (closing in the lower half plane), we obtain

$$\int d\kappa \frac{e^{-i\kappa\tau}}{\frac{1}{2M}k^2-\kappa-i\epsilon} = 2\pi i \,\text{Res}\frac{e^{-i\kappa\tau}}{\frac{1}{2M}k^2-\kappa-i\epsilon} \times \text{index of contour} = 2\pi i e^{-i(\frac{k^2}{2M}-i\epsilon)\tau}\theta(\tau). \tag{61}$$

Thus, we find

$$G(x,\tau) = \frac{i}{(2\pi)^4}\int d^4k e^{i(k\cdot x-\frac{k^2}{2M}\tau+i\epsilon\tau)}\theta(\tau) \tag{62}$$

which exhibits retarded causality through $\theta(\tau)$. We may also consider the Green's function for the matter field as expressed through the vacuum expectation value of the (τ-ordered) operator products. Using the momentum expansions (55), we find that

$$\begin{aligned} &\langle 0|T\psi(x_1,\tau_1)\psi^*(x_2,\tau_2)|0\rangle = \\ &\quad = \theta(\tau_1-\tau_2)\langle 0|\psi(x_1,\tau_1)\psi^*(x_2,\tau_2)|0\rangle + \theta(\tau_2-\tau_1)\langle 0|\psi^*(x_2,\tau_2)\psi(x_1,\tau_1)|0\rangle \\ &\quad = \theta(\tau_1-\tau_2)\int \frac{d^4k}{(2\pi)^4}\frac{d^4k'}{(2\pi)^4} e^{-i(k\cdot x_2-\frac{k^2}{2M}\tau_2)}e^{i(k'\cdot x_1-\frac{k'^2}{2M}\tau_1)}\langle 0|b(k')b^*(k)|0\rangle \\ &\quad = \theta(\tau_1-\tau_2)\int \frac{d^4k}{(2\pi)^4} e^{i[k\cdot(x_1-x_2)-\frac{k^2}{2M}(\tau_1-\tau_2)]} \\ &\quad = -iG(x_1-x_2,\tau_1-\tau_2) \end{aligned} \tag{63}$$

where we have used $b(k)|0\rangle = 0$ and (56). We thus verify that the Green's function that solves the Schrödinger equation is equal to the vacuum expectation value of the τ-ordered product of the operators, as required for the application of Wick's theorem in perturbation theory.

In his 1951 calculation of vacuum polarization, Schwinger transformed the Dirac problem into a dynamical theory by representing the Green's function as a parametric integral [13]. Applying this method to the Klein-Gordon equation, we represent the Green's function operator as

$$G = \frac{1}{(p-eA)^2+m^2-i\epsilon} \tag{64}$$

with spacetime representation

$$G(x,x') = \langle x|G|x'\rangle = i\int_0^\infty ds e^{-i(m^2-i\epsilon)s}\langle x|e^{-i(p-eA)^2 s}|x'\rangle\,. \tag{65}$$

The function

$$G(x,x';s) = \langle x(s)|x'(0)\rangle = \langle x|e^{-i(p-eA)^2 s}|x'\rangle \tag{66}$$

satisfies the Schrödinger equation (4) with the boundary condition

$$\lim_{s\to 0}\langle x(s)|x'(0)\rangle = \delta^4(x-x') \tag{67}$$

and following DeWitt [14] we regard (65) as defining the Green's function for the field $\psi(x,s)$ that solves (4). In deriving the path integral for the Klein-Gordon equation, Feynman [15] a similar Schrödinger equation. He regarded the integration of the Green's function with the weight e^{-im^2s} in (65) as the requirement (see also Nambu [16]) that asymptotic solutions of the be stationary eigenstates of the mass operator $i\partial_s$. From this point of view, one picks the mass eigenvalue by extending the lower limit of integration in (65) from 0 to $-\infty$ and adding the requirement that $G(x,x';s)=0$ for $s<0$.

Returning to the off-shell Green's function (61), we easily verify that the τ-integral

$$\begin{aligned}\int_{-\infty}^{\infty} d\tau e^{-i(m^2/2M)\tau} G(x,\tau) &= \frac{i}{(2\pi)^4}\int_0^{\infty} d\tau e^{-i(m^2/2M)\tau}\int d^4k e^{i(k\cdot x-\frac{k^2}{2M}\tau+i\epsilon\tau)} \\ &= i\int_0^{\infty} d\tau \int \frac{d^4k}{(2\pi)^4} e^{-i(m^2/2M+\frac{k^2}{2M}-i\epsilon)\tau} e^{ik\cdot x} \\ &= \int \frac{d^4k}{(2\pi)^4}\frac{e^{ik\cdot x}}{\frac{1}{2M}(k^2+m^2)-i\epsilon} \\ &= 2M\,\Delta_F(x)\end{aligned} \tag{68}$$

goes over to the Feynman propagator for a particle of mass m with an overall factor of $2M$. Thus, as noted by Feynman [15], the Feynman causality employed in standard QED emerges naturally from retarded causality in τ.

4. Quantization of the free gauge field

We begin with the Hamiltonian obtained from (47) as

$$\mathrm{K} = \int d^4x \mathcal{K}_{\text{photon}} = \int d^4x \left[\frac{\lambda}{4} f_{\mu\nu}f^{\mu\nu} - \frac{\lambda\sigma}{2}\epsilon^{\mu}\epsilon_{\mu}\right] . \tag{69}$$

In order to evaluate the canonical commutation relations for the transverse fields, we first consider the projection operator

$$\delta_{\perp}^{\mu\nu}(x-y) = \int \frac{d^4k}{(2\pi)^4}\left[g^{\mu\nu} - \frac{k^{\mu}k^{\nu}}{k^2}\right] e^{ik\cdot(x-y)} = \int \frac{d^4k}{(2\pi)^4}\mathcal{P}^{\mu\nu}(k) e^{ik\cdot(x-y)} \tag{70}$$

which projects onto the transverse part of a vector function. For a function $f_{\nu}(x,\tau)$ with Fourier transform $\tilde{f}_{\nu}(k,\tau)$,

$$\left[\delta_{\perp}^{\mu\nu} f_{\nu}\right](x,\tau) = \int d^4y\, \delta_{\perp}^{\mu\nu}(x-y) f_{\nu}(y,\tau) = \int d^4k \left[g^{\mu\nu} - \frac{k^{\mu}k^{\nu}}{k^2}\right] f_{\nu}(k,\tau) e^{ik\cdot x} \tag{71}$$

which clearly satisfies $\partial^{\mu} f_{\mu}(x,\tau)=0$. We represent the transverse fields as

$$a_{\perp}^{\mu}(x,\tau) = \left[\delta_{\perp}^{\mu\rho}a_{\rho}\right](x,\tau) \qquad \epsilon_{\perp}^{\mu}(x,\tau) = \left[\delta_{\perp}^{\mu\rho}\epsilon_{\rho}\right](x,\tau). \tag{72}$$

Canonical quantization requires the equal-τ commutation relations for the field components

$$\left[a^{\mu}(x,\tau), \pi_{a_{\nu}}(y,\tau)\right] = i g^{\mu\nu}\delta^4(x-y) \tag{73}$$

which becomes

$$[a^\mu(x,\tau), \epsilon^\nu(y,\tau)] = -\frac{i}{\lambda} g^{\mu\nu}\delta^4(x-y) \tag{74}$$

using (44) for π_{a_ν}. We find for the transverse field components

$$\begin{aligned}[a_\perp^\mu(x,\tau), \epsilon_\perp^\nu(y,\tau)] &= [\,[\delta_\perp^{\mu\rho} a_\rho](x,\tau),\ [\delta_\perp^{\nu\sigma}\epsilon_\sigma(y,\tau)]\,] = \frac{-i}{\lambda} g_{\rho\sigma}\left[\delta_\perp^{\mu\rho}\delta_\perp^{\nu\sigma}\delta^4\right](x-y) \\ &= \frac{-i}{\lambda}\delta_\perp^{\mu\nu}(x-y)\end{aligned} \tag{75}$$

and so drop the subscript $_\perp$ from the field variables, using (75) as the canonical commutation relation. This relation insures that the Gauss Law constraint (28) commutes with the other variables of the theory, and so, with the Hamiltonian (69).

The Heisenberg equations for the fields a^μ are

$$\begin{aligned} i\partial_\tau a^\mu(x,\tau) &= [a^\mu(x,\tau), \int d^4y\ \frac{\lambda}{4} f_{\rho\nu}f^{\rho\nu} - \frac{\lambda\sigma}{2}\epsilon^\rho\epsilon_\rho] \\ &= -\frac{\lambda\sigma}{2}\int d^4y\{[a^\mu(x,\tau),\epsilon^\rho(y,\tau)]\epsilon_\rho(y,\tau) + \epsilon^\rho(y,\tau)[a^\mu(x,\tau),\epsilon_\rho(y,\tau)]\} \\ &= -\frac{\lambda\sigma}{2}[-2\frac{i}{\lambda}\epsilon^\mu(x,\tau)],\end{aligned}$$

leading to

$$\partial_\tau a^\mu(x,\tau) = \sigma\epsilon^\mu(x,\tau) = \sigma f^{5\mu} = f_5{}^\mu \tag{76}$$

in agreement with the classical definition of $f^{5\mu}$ in the gauge $a_5 = 0$. The Heisenberg equation for the fields ϵ^μ are

$$\begin{aligned} i\partial_\tau \epsilon^\mu(x,\tau) &= [\epsilon^\mu(x,\tau), \int d^4y\ \frac{\lambda}{4} f_{\mu\nu}f^{\mu\nu}] \\ &= \frac{\lambda}{4}\int d^4y\ [\epsilon^\mu(x,\tau), (\partial_\lambda a_\nu - \partial_\nu a_\lambda)]f^{\lambda\nu} + f^{\lambda\nu}[\epsilon^\mu(x,\tau), (\partial_\lambda a_\nu - \partial_\nu a_\lambda)] \\ &= \frac{\lambda}{4}\int d^4y\ \partial_\lambda[\epsilon^\mu, a_\nu]f^{\lambda\nu} - \partial_\nu[\epsilon^\mu, a_\lambda]f^{\lambda\nu} + f^{\lambda\nu}\partial_\lambda[\epsilon^\mu, a_\nu] - f^{\lambda\nu}\partial_\nu[\epsilon^\mu, a_\lambda] \\ &= -i\partial_\lambda f^{\lambda\mu}(x,\tau).\end{aligned} \tag{77}$$

Since $\epsilon^\mu = f^{5\mu}$ (77) can be written

$$\partial_\lambda f^{\lambda\mu} + \partial_\tau f^{5\mu} = 0 \tag{78}$$

and combined with (30) to write

$$\partial_\alpha f^{\alpha\beta} = 0 \tag{79}$$

for $\alpha, \beta = 0, \cdots, 3, 5$. These are half of the pre-Maxwell field equations; the other pre-Maxwell equations follow from the definition of $f^{\mu\nu}$ and (76). By virtue of (33), (78) may be written in the form

$$0 = \partial_\tau f^{5\mu} + \partial_\lambda(\partial^\lambda a^\mu - \partial^\mu a^\lambda) = \partial_\tau f^{5\mu} + \partial_\lambda\partial^\lambda a^\mu - \partial^\mu\partial_\lambda a^\lambda = \partial_\tau f^{5\mu} + \partial_\lambda\partial^\lambda a^\mu. \tag{80}$$

Combining (80) with (76), we obtain the wave equation

$$0 = \partial_\lambda\partial^\lambda a^\mu + \partial_\tau\epsilon^\mu = \partial_\lambda\partial^\lambda a^\mu + \partial_\tau(\sigma\partial_\tau a^\mu) = (\partial_\lambda\partial^\lambda + \sigma\partial_\tau^2)a^\mu. \tag{81}$$

Therefore, we may perform a Fourier expansion of the field operator,

$$a^\mu(x,\tau) = \sum_{s=\text{polarizations}} \int \frac{d^4k}{2\kappa} \left[\varepsilon_s^\mu a(k,s) e^{i(k\cdot x+\sigma\kappa\tau)} + \varepsilon_s^{\mu *} a^*(k,s) e^{-i(k\cdot x+\sigma\kappa\tau)} \right] \tag{82}$$

where the five-dimensional mass shell condition is $\kappa = \sqrt{-\sigma k^2}$. For free fields, the condition $-\sigma k^2 > 0$ is required to prevent divergence at $\tau \to \pm\infty$, however this condition need not apply for virtual photons. Using (76), ϵ^μ is given by

$$\epsilon^\mu(x,\tau) = i \sum_{s=\text{polarizations}} \int \frac{d^4k}{2\kappa} \kappa \left[\varepsilon_s^\mu a(k,s) e^{i(k\cdot x+\sigma\kappa\tau)} - \varepsilon_s^{\mu *} a^*(k,s) e^{-i(k\cdot x+\sigma\kappa\tau)} \right]. \tag{83}$$

In momentum space (33) becomes $k_\mu a^\mu = 0$, so that there will be three independent polarizations (see also [17]). To choose the polarizations in a covariant way, we begin by choosing an arbitrary timelike vector n^μ ($n^2 = -1$). There are then two orthonormal spacelike vectors, ε_1 and ε_2, which satisfy the conditions

$$n \cdot \varepsilon_1 = n \cdot \varepsilon_2 = k \cdot \varepsilon_1 = k \cdot \varepsilon_2 = 0 \tag{84}$$

and these may be constructed in the following way. Since $n^2 = -1$,

$$(\vec{n})^2 - (n^0)^2 = -1 \quad \Rightarrow \quad n^0 = \pm\sqrt{(\vec{n})^2 + 1} \tag{85}$$

so that taking

$$\varepsilon_{1,2}^0 = \frac{\vec{n} \cdot \vec{\varepsilon}_{1,2}}{\sqrt{(\vec{n})^2 + 1}} \tag{86}$$

guarantees that $n \cdot \varepsilon_{1,2} = 0$. We now require that

$$0 = k \cdot \varepsilon_{1,2} = \vec{k} \cdot \vec{\varepsilon}_{1,2} - (k^0)\varepsilon_{1,2}^0 = \left[\vec{k} - \frac{k^0}{\sqrt{(\vec{n})^2 + 1}} \vec{n} \right] \cdot \vec{\varepsilon}_{1,2}. \tag{87}$$

Since two orthogonal 3-vectors can be found which satisfy (87), this establishes that (84) can also be satisfied. We normalize $\varepsilon_{1,2}$ so that

$$\varepsilon_1 \cdot \varepsilon_2 = 0 \qquad\qquad (\varepsilon_1)^2 = (\varepsilon_2)^2 = 1 \tag{88}$$

and form the third polarization, which we will denote ε_5, by taking a linear combination of n and k to guarantee linear independence from $\varepsilon_{1,2}$. Writing

$$\varepsilon_5 = A\, k + B\, n \tag{89}$$

we apply (33) and find that

$$0 = k \cdot \varepsilon_5 = k \cdot (A\, k + B\, n) = Ak^2 + B(k \cdot n) \tag{90}$$

We may take

$$A = -B\, \frac{k \cdot n}{k^2} = B\sigma \frac{k \cdot n}{\kappa^2} \tag{91}$$

so that

$$\varepsilon_5 = B\left[n + \sigma\frac{k \cdot n}{\kappa^2}k\right] \tag{92}$$

and

$$(\varepsilon_5)^2 = B^2\left[n^2 + 2\sigma\frac{(k \cdot n)^2}{\kappa^2} + \frac{k^2}{\kappa^2}\right] = B^2\frac{\sigma}{\kappa^2}\left[k^2 + (k \cdot n)^2\right] = B^2\frac{\sigma}{\kappa^2}\left[k + n(k \cdot n)\right]^2 \tag{93}$$

The expression $k + n(k \cdot n)$ is just $k_\perp$ and its square is positive definite (expanding as a quadratic in k^0, one sees that there are no roots). We thus find that choosing

$$B = \frac{\kappa}{|k + n(k \cdot n)|} \tag{94}$$

(93) becomes

$$(\varepsilon_5)^2 = \sigma. \tag{95}$$

In the simple case that $n = (1,0,0,0)$, we may take as ε_1 and ε_2 the vectors

$$\varepsilon_{1,2} = \begin{bmatrix} 0 \\ \vec{\varepsilon}_{1,2} \end{bmatrix} \tag{96}$$

where

$$\vec{k} \cdot \vec{\varepsilon}_{1,2} = 0 \qquad \vec{\varepsilon}_1 \cdot \vec{\varepsilon}_2 = 0 \qquad \vec{\varepsilon}_1 \cdot \vec{\varepsilon}_1 = \vec{\varepsilon}_2 \cdot \vec{\varepsilon}_2 = 1 \tag{97}$$

Thus, the set $\{\vec{\varepsilon}_1, \vec{\varepsilon}_2, \hat{k} = \vec{k}/|\vec{k}|\}$ forms an orthonormal basis for the 3-space in this frame. We find for the third polarization ε_5,

$$\varepsilon_5 = \frac{1}{\kappa}\begin{bmatrix} |\vec{k}| \\ k^0\hat{k} \end{bmatrix}. \tag{98}$$

Notice that for $k^2 \to 0$, $\kappa \to 0$ and $\varepsilon_5 \cdot \varepsilon_5 \to 0/0$, so we will treat the lightlike case as the limit of the massive case, in which we take $\kappa \to 0$ at the end of all other computations.

In order to evaluate the commutation relations among the operators $a(k,s)$ and $a^*(k,s)$, we must first invert the Fourier expansion of $a^\mu(x,\tau)$. To do this, we write the orthogonality relations among the polarization vectors in the form

$$\varepsilon_s \cdot \varepsilon_{s'} = g_{ss'} = g(s)\,\delta_{ss'} \qquad \text{for } s,s' = 1,2,5 \tag{99}$$

where

$$g(s) = \begin{cases} 1, & \text{if } s = s' = 1,2 \\ \sigma, & \text{if } s = s' = 5 \end{cases} \tag{100}$$

so that we have from (82)

$$\begin{aligned}\int d^4x e^{-i(k\cdot x+\sigma\kappa\tau)}\varepsilon_s \cdot a(x,\tau) &= \\ &= \sum_{s'}\int d^4x\frac{d^4k'}{2\kappa'}\varepsilon_s \cdot \varepsilon_{s'}\left[a(k',s')e^{i(x\cdot(k'-k)+\sigma(\kappa'-\kappa)\tau)} + a^*(k',s')e^{-i(x\cdot(k+k')+\sigma(\kappa+\kappa')\tau)}\right] \\ &= \frac{(2\pi)^4}{2\kappa}g(s)\left[a(k,s) + a^*(k,s)e^{-2i\sigma\kappa\tau}\right]\end{aligned} \tag{101}$$

Similarly,

$$\begin{aligned}
&\int d^4x e^{-i(k\cdot x+\sigma\kappa\tau)}\varepsilon_s\cdot\dot{a}(x,\tau) = \\
&= \sum_{s'}\int d^4x\frac{d^4k'}{2\kappa'}\varepsilon_s\cdot\varepsilon_{s'}(i\sigma\kappa') \\
&\qquad \left[a(k',s')e^{i(x\cdot(k'-k)+\sigma(\kappa'-\kappa)\tau)} - a^*(k',s')e^{-i(x\cdot(k+k')+\sigma(\kappa+\kappa')\tau)}\right] \\
&= \frac{i\sigma}{2}(2\pi)^4 g(s)\left[a(k,s) - a^*(k,s)e^{-2i\sigma\kappa\tau}\right]
\end{aligned} \tag{102}$$

Combining (101) and (102), we obtain

$$\begin{aligned}
i\sigma g(s)(2\pi)^4 a(k,s) &= \int d^4x e^{-i(k\cdot x+\sigma\kappa\tau)}\varepsilon_s\cdot[\dot{a}(x,\tau)+i\sigma\kappa a(x,\tau)] \\
&= \int d^4x e^{-i(k\cdot x+\sigma\kappa\tau)}\varepsilon_s\cdot[\overrightarrow{\partial_\tau}\, a(x,\tau) - \overleftarrow{\partial_\tau}\, a(x,\tau)] \\
&= \int d^4x e^{-i(k\cdot x+\sigma\kappa\tau)}\varepsilon_s\cdot[\overleftrightarrow{\partial_\tau}\, a(x,\tau)],
\end{aligned} \tag{103}$$

so that

$$a(k,s) = \frac{-i\sigma g(s)}{(2\pi)^4}\int d^4x e^{-i(k\cdot x+\sigma\kappa\tau)}\varepsilon_s\cdot[\overleftrightarrow{\partial_\tau}\, a(x,\tau)] \tag{104}$$

and

$$a^*(k,s) = \frac{i\sigma g(s)}{(2\pi)^4}\int d^4x e^{i(k\cdot x+\sigma\kappa\tau)}\varepsilon_s\cdot[\overleftrightarrow{\partial_\tau}\, a(x,\tau)] \tag{105}$$

where we have used $1/g(s) = g(s)$.

Expressions (104) and (105) permit us to evaluate the commutators

$$\begin{aligned}
[a(k,s),a^*(k',s')] &= \frac{g(s)g(s')}{(2\pi)^8}\int d^4x\int d^4x' e^{-i(k\cdot x+\sigma\kappa\tau)}e^{i(k'\cdot x'+\sigma\kappa'\tau')} \\
&\qquad \left[\varepsilon_s(k)\cdot[\overleftrightarrow{\partial_\tau}\, a(x,\tau)],\varepsilon_{s'}(k')\cdot[\overleftrightarrow{\partial_{\tau'}}\, a(x',\tau')]\right],
\end{aligned} \tag{106}$$

where

$$\begin{aligned}
\left[\overleftrightarrow{\partial_\tau}\, a^\mu(x,\tau),\overleftrightarrow{\partial_{\tau'}}\, a^\nu(x',\tau')\right] &= \left[\left(\overrightarrow{\partial_\tau}-\overleftarrow{\partial_\tau}\right)a^\mu(x,\tau),\left(\overrightarrow{\partial_{\tau'}}-\overleftarrow{\partial_{\tau'}}\right)a^\nu(x',\tau')\right] \\
&= \left[\overrightarrow{\partial_\tau}\, a^\mu(x,\tau),\overrightarrow{\partial_{\tau'}}\, a^\nu(x',\tau')\right] + \overleftarrow{\partial_\tau}\overleftarrow{\partial_{\tau'}}\left[a^\mu(x,\tau),a^\nu(x',\tau')\right]
\end{aligned} \tag{107}$$

so

$$\begin{aligned}
[a(k,s),a^*(k',s')] &= \frac{g(s)g(s')}{(2\pi)^8}\varepsilon_{s\mu}(k)\varepsilon_{s'\nu}(k')\int d^4x\int d^4x' e^{-i(k\cdot x+\sigma\kappa\tau)}e^{i(k'\cdot x'+\sigma\kappa'\tau')} \\
&\qquad \left\{\left[\dot{a}^\mu(x,\tau),\dot{a}^\nu(x',\tau')\right]+\kappa\kappa'\left[a^\mu(x,\tau),a^\nu(x',\tau')\right]\right. \\
&\qquad \left.-i\sigma\kappa'\left[\dot{a}^\mu(x,\tau),a^\nu(x',\tau')\right]+i\sigma\kappa\left[a^\mu(x,\tau),\dot{a}^\nu(x',\tau')\right]\right\}.
\end{aligned} \tag{108}$$

Evaluating the commutators at equal-τ and using (74) and (76) to establish

$$\left[\dot{a}^\mu(x,\tau),a^\nu(x',\tau)\right] = \frac{i\sigma}{\lambda}\delta_\perp^{\mu\nu}(x-x'), \tag{109}$$

one obtains

$$\begin{aligned}[a(k,s),a^*(k',s')] &= \frac{g(s)g(s')}{(2\pi)^8}\varepsilon_{s\mu}(k)\varepsilon_{s'\nu}(k')\int d^4x\int d^4x' e^{-i(k\cdot x+\sigma\kappa\tau)}e^{i(k'\cdot x'+\sigma\kappa'\tau')} \\ &\quad \left\{-i\sigma\kappa'\left(\frac{i\sigma}{\lambda}\delta_\perp^{\mu\nu}(x-x')\right)+i\sigma\kappa\left(\frac{-i\sigma}{\lambda}\delta_\perp^{\mu\nu}(x-x')\right)\right\} \\ &= \frac{g(s)g(s')}{(2\pi)^8}\varepsilon_{s\mu}(k)\varepsilon_{s'\mu}(k')\frac{\kappa+\kappa'}{\lambda}\int d^4x e^{i[x\cdot(k'-k)+\sigma\tau(\kappa'-\kappa)]} \\ &= \frac{2\kappa}{\lambda}\frac{g(s)}{(2\pi)^4}\delta_{ss'}\delta^4(k-k') \end{aligned} \tag{110}$$

where we have used $\sigma^2 = 1$, $g(s)^2 = 1$, (84), and (99).

Given the commutation relations for the operators $a(k,s)$ and $a^*(k,s)$, we may evaluate the photon propagator as the vacuum expectation value of the τ-ordered product of the fields. Thus,

$$\begin{aligned}\langle 0|\mathrm{T}a_\mu(x,\tau)a_\nu(x',\tau')|0\rangle = \theta(\tau-\tau')\langle 0|a_\mu(x,\tau)a_\nu(x',\tau')|0\rangle \\ +\theta(\tau'-\tau)\langle 0|a_\nu(x',\tau')a_\mu(x,\tau)|0\rangle\end{aligned} \tag{111}$$

where we use (82) to expand

$$\begin{aligned}\langle 0|a^\mu(x,\tau)a^\nu(x',\tau')|0\rangle &= \sum_{s,s'}\int\frac{d^4k}{2\kappa}\frac{d^4k'}{2\kappa'}\varepsilon_s^\mu\varepsilon_{s'}^\nu \\ &\quad \langle 0|\left[a(k,s)e^{i(k\cdot x+\sigma\kappa\tau)}+a^*(k,s)e^{-i(k\cdot x+\sigma\kappa\tau)}\right] \\ &\quad \left[a(k',s')e^{i(k'\cdot x'+\sigma\kappa'\tau')}+a^*(k',s')e^{-i(k'\cdot x'+\sigma\kappa'\tau')}\right]|0\rangle \\ &= \sum_{s,s'}\int\frac{d^4k}{2\kappa}\frac{d^4k'}{2\kappa'}\varepsilon_s^\mu\varepsilon_{s'}^\nu e^{i(k\cdot x+\sigma\kappa\tau)}e^{-i(k'\cdot x'+\sigma\kappa'\tau')}\langle 0|a(k,s)a^*(k',s')|0\rangle \\ &= \sum_{s,s'}\int\frac{d^4k}{2\kappa}\frac{d^4k'}{2\kappa'}\varepsilon_s^\mu\varepsilon_{s'}^\nu e^{i(k\cdot x+\sigma\kappa\tau)}e^{-i(k'\cdot x'+\sigma\kappa'\tau')}\frac{2\kappa}{\lambda}\frac{g(s)}{(2\pi)^4}\delta_{ss'}\delta^4(x-x') \\ &= \sum_s\frac{g(s)}{(2\pi)^4\lambda}\int\frac{d^4k}{2\kappa}\varepsilon_s^\mu\varepsilon_s^\nu e^{i[k\cdot(x-x')+\sigma\kappa(\tau-\tau')]}\end{aligned} \tag{112}$$

so that

$$\begin{aligned}\langle 0|\mathrm{T}a^\mu(x,\tau)a^\nu(x',\tau')|0\rangle &= \sum_s\frac{g(s)}{(2\pi)^4\lambda}\int\frac{d^4k}{2\kappa}\varepsilon_s^\mu\varepsilon_s^\nu e^{i[k\cdot(x-x')]} \\ &\quad \left[\theta(\tau-\tau')e^{i\sigma\kappa(\tau-\tau')}+\theta(\tau'-\tau)e^{i\sigma\kappa(\tau'-\tau)}\right].\end{aligned} \tag{113}$$

We recognize the

$$\frac{1}{2\kappa}\left[\theta(\tau-\tau')e^{i\sigma\kappa(\tau-\tau')}+\theta(\tau'-\tau)e^{i\sigma\kappa(\tau'-\tau)}\right] = -\frac{i}{2\pi}\int dK\frac{e^{i\sigma K(\tau-\tau')}}{k^2+\sigma K^2-i\epsilon} \tag{114}$$

which we may use to put the Green's function (the vacuum expectation value of the τ-ordered product of the fields) in the form of the Feynman propagator for the five-dimensional field:

$$\begin{aligned}d^{\mu\nu}(x-x',\tau-\tau') &= i\langle 0|\mathrm{T}a^\mu(x,\tau)a^\nu(x',\tau')|0\rangle \\ &= \int\frac{d^4k d\kappa}{(2\pi)^5}\sum_s\frac{g(s)}{\lambda}\varepsilon_s^\mu\varepsilon_s^\nu\frac{e^{i[k\cdot(x-x')+\sigma\kappa(\tau-\tau')]}}{k^2+\sigma\kappa^2-i\epsilon}.\end{aligned} \tag{115}$$

In order to evaluate the sum over polarizations in (115), we must consider the cases $\sigma = \pm 1$ separately. For the case that $\sigma = 1$, we must satisfy $k^2 = -\kappa^2 < 0$, and we may take

$$k = \lim_{\alpha \to 0} (\sqrt{\kappa^2 + \alpha^2}, 0, 0, \alpha) = (\kappa, 0, 0, 0). \tag{116}$$

Choosing

$$n = (1,0,0,0) \qquad \varepsilon_1 = (0,1,0,0) \qquad \varepsilon_2 = (0,0,1,0) \tag{117}$$

we find from (98) and (116) that

$$\varepsilon_5 = (|\vec{k}|, k^0 \hat{k}) \frac{1}{\sqrt{-k^2}} = \lim_{\alpha \to 0} \frac{1}{\sqrt{\kappa^2 + \alpha^2}} (\alpha, 0, 0, \sqrt{\kappa^2 + \alpha^2}) = (0,0,0,1). \tag{118}$$

The completeness relation may then be written in the form

$$g^{\mu\nu} = -\frac{1}{\kappa^2} k^\mu k^\nu + (\varepsilon_1)^\mu (\varepsilon_1)^\nu + (\varepsilon_2)^\mu (\varepsilon_2)^\nu + (\varepsilon_5)^\mu (\varepsilon_5)^\nu \tag{119}$$

which we may rearrange to write

$$\sum_{s=1,2,5} g(s) (\varepsilon_s)^\mu (\varepsilon_s)^\nu = \sum_{s=1,2,5} (\varepsilon_s)^\mu (\varepsilon_s)^\nu = g^{\mu\nu} + \frac{1}{\kappa^2} k^\mu k^\nu = g^{\mu\nu} - \frac{1}{k^2} k^\mu k^\nu \tag{120}$$

where we have used $g(1,2) = 1$ and $g(5) = \sigma = 1$.

For the case that $\sigma = -1$, we have $k^2 = \kappa^2 > 0$ so we may take

$$k = (0,0,0,\kappa). \tag{121}$$

and using (98) we find that

$$\varepsilon_5 = (|\vec{k}|, k^0 \hat{k}) \frac{1}{\sqrt{-k^2}} = (1,0,0,0). \tag{122}$$

In this case the completeness relation is

$$g^{\mu\nu} = -(\varepsilon_5)^\mu (\varepsilon_5)^\nu + (\varepsilon_1)^\mu (\varepsilon_1)^\nu + (\varepsilon_2)^\mu (\varepsilon_2)^\nu + \frac{1}{\kappa^2} k^\mu k^\nu \tag{123}$$

and it may be rearranged as

$$\sum_{s=1,2,5} g(s) (\varepsilon_s)^\mu (\varepsilon_s)^\nu = \sum_{s=1,2} (\varepsilon_s)^\mu (\varepsilon_s)^\nu - (\varepsilon_5)^\mu (\varepsilon_5)^\nu = g^{\mu\nu} - \frac{1}{\kappa^2} k^\mu k^\nu = g^{\mu\nu} - \frac{1}{k^2} k^\mu k^\nu \tag{124}$$

where we have used $g(1,2) = 1$ and $g(5) = \sigma = -1$. Notice that (124) and (120) are identical and are consistent with the transversality requirements on the polarization states. Using this expression and observing that the sum over polarization states gives the projection operator $\mathcal{P}^{\mu\nu}(k)$ defined in (70), equation (75) may be verified explicitly.

To evaluate the Hamiltonian in the momentum representation,

$$\frac{\lambda}{4} f_{\mu\nu} f^{\mu\nu} = \frac{\lambda}{4} (\partial_\mu a_\nu - \partial_\nu a_\mu)(\partial^\mu a^\nu - \partial^\nu a^\mu) = -\frac{\lambda}{2} a_\nu \Box a^\nu + \text{divergence} \tag{125}$$

and

$$-\frac{\lambda\sigma}{2}\epsilon^\mu\epsilon_\mu = -\frac{\lambda\sigma}{2}(\sigma\partial_\tau a_\mu)(\sigma\partial_\tau a^\mu) = -\frac{\lambda\sigma}{2}[\partial_\tau(a^\mu\partial_\tau a_\mu) - a_\mu\partial_\tau^2 a^\mu] = \frac{\lambda\sigma}{2}a_\mu\partial_\tau^2 a^\mu \quad (126)$$

where we drop divergences and total τ-derivatives. The Hamiltonian becomes

$$\mathrm{K} = \frac{\lambda}{2}\int d^4x a_\mu[-\Box + \sigma\partial_\tau^2]a^\mu = \lambda\sigma\int d^4x \; a_\mu\partial_\tau^2 a^\mu \quad (127)$$

where we have used the wave equation (81). Using the momentum expansion (82) and the commutation relations (110) we find for the normal ordered Hamiltonian

$$:\mathrm{K}: = \lambda\sigma\int d^4x \sum_{s,s'}\int \frac{d^4k d^4k'}{4\kappa\kappa'}(\varepsilon_s)^\mu(\varepsilon_{s'})_\mu(i\sigma\kappa')^2 \quad (128)$$

$$: \Big\{a(k,s)a(k',s')e^{i[(k+k')\cdot x+\sigma(\kappa+\kappa')\tau]} + a^*(k,s)a^*(k',s')e^{-i[(k+k')\cdot x+\sigma(\kappa+\kappa')\tau]}$$
$$+a^*(k,s)a(k',s')e^{-i[(k-k')\cdot x+\sigma(\kappa-\kappa')\tau]} + a^*(k',s')a(k,s)e^{i[(k-k')\cdot x+\sigma(\kappa-\kappa')\tau]}\Big\}:$$
$$= -\lambda\sigma(2\pi)^4\sum_{s,s'}\int\frac{d^4k}{4\kappa^2}\kappa^2(\varepsilon_s)^\mu(\varepsilon_{s'})_\mu$$
$$: \Big\{a(k,s)a(-k,s')e^{2i\sigma\kappa\tau} + a^*(k,s)a^*(-k,s')e^{-2i\sigma\kappa\tau}$$
$$+a^*(k,s)a(k,s') + a^*(k,s')a(k,s)\} : \quad (129)$$

By replacing $k \to -k$ in (82), we see that we may identify

$$a(-k,s)e^{i\sigma\kappa\tau} = a^*(k,s)e^{-i\sigma\kappa\tau} \qquad a^*(-k,s)e^{-i\sigma\kappa\tau} = a(k,s)e^{i\sigma\kappa\tau}. \quad (130)$$

Using (130) in (129), we obtain

$$\begin{aligned} :\mathrm{K}: &= -\lambda\sigma(2\pi)^4\sum_{s,s'}\int\frac{d^4k}{4\kappa^2}\kappa^2(\varepsilon_s)^\mu(\varepsilon_{s'})_\mu 2 : \{a^*(k,s)a(k,s') + a^*(k,s')a(k,s)\} : \\ &= -\lambda\sigma(2\pi)^4\sum_{s,s'}\int\frac{d^4k}{4\kappa^2}2\kappa^2[g(s)\delta_{ss'}]\{a^*(k,s)a(k,s') + a^*(k,s')a(k,s)\} \\ &= -\lambda\sigma(2\pi)^4\sum_s g(s)\int d^4k\, a^*(k,s)a(k,s) \end{aligned} \quad (131)$$

where we have used (99).

5. Perturbation theory

Defining Heisenberg states $|\psi\rangle$ as τ-independent and the corresponding Schrödinger states through

$$|\psi(\tau)\rangle = U(\tau)|\psi(0)\rangle = U(\tau)|\psi\rangle \quad (132)$$

the familiar t-dependent perturbation theory becomes a manifestly covariant τ-dependent theory. The time independence of $\langle\psi(\tau)|\psi(\tau)\rangle$ imposes unitarity on $U(\tau)$ which is easily shown to be generated by the Hamiltonian operator through $U = e^{-i\mathrm{K}\tau}$, such that $|\psi(\tau)\rangle$

satisfies the Schrödinger equation. Writing $\mathrm{K} = \mathrm{K}_0 + \mathrm{K}_I$ where K_0 is the free Hamiltonian and K_I represents the interaction, we are led to the interaction picture [18] in which

$$i\partial_\tau |\phi(\tau)\rangle_I = e^{i\mathrm{K}_0\tau}\mathrm{K}_I e^{-i\mathrm{K}_0\tau}\, |\phi(\tau)\rangle_I = \mathrm{K}_I(\tau)\, |\phi(\tau)\rangle_I. \tag{133}$$

Equation (133) has the formal solution

$$|\phi(\tau)\rangle_I = V(\tau,\tau')|\phi(\tau')\rangle_I = \mathrm{T}e^{-i\int_{\tau'}^{\tau} d\tau''\, \mathrm{K}_I(\tau'')}|\phi(\tau')\rangle_I \tag{134}$$

in which the symbol T refers to τ-ordering. The S-matrix is given by

$$S = \lim_{\substack{\tau'\to-\infty \\ \tau\to\infty}} \mathrm{T}e^{-i\int_{\tau'}^{\tau} d\tau''\, \mathrm{K}_I(\tau'')} = \mathrm{T}e^{-i\int d^4x d\tau\, \mathcal{K}_I} \tag{135}$$

Scattering amplitudes may be computed from the interaction picture states as

$$W_{i\to f} = \langle f|S|i\rangle. \tag{136}$$

Since the initial and final states are defined when there is no interaction, the asymptotic interaction picture states are identical with the asymptotic Heisenberg states. Moreover, since the Heisenberg picture Hamiltonian is formed of bilinear combinations of the fields without τ-derivatives, the interacting Hamiltonian in (135) has the same form as the Heisenberg picture Hamiltonian.

Reduction Formulas

In order to use Wick's theorem for perturbation expansions of the S-matrix, [3], we must adapt the LSZ reduction formulas [3] to the present theory. Consider the amplitude

$$\langle k' \text{ out}|k \text{ in}\rangle = \langle k' \text{ out}|b^*_{in}(k)|0\rangle \tag{137}$$

From (55) we find that

$$b^*(k) = \int d^4x e^{i(k\cdot x - \kappa\tau)}\, \psi^*(x,\tau) \qquad b(k) = \int d^4x e^{i(k\cdot x - \kappa\tau)}\, \psi(x,\tau) \tag{138}$$

which we may insert in (137) to obtain

$$\langle k' \text{ out}|k \text{ in}\rangle = \int d^4x \langle k' \text{ out}|\psi^*_{\text{in}}(x,\tau)|0\rangle e^{i(k\cdot x - \kappa\tau)} \tag{139}$$

which is valid for arbitrary τ_{in}. Now, consider the integral

$$\begin{aligned}\int_{\tau_i}^{\tau_f} d\tau \quad \int d^4x\, \partial_\tau \left\{ \langle k'\text{out}|\psi^*(x,\tau)|0\rangle e^{i(k\cdot x-\kappa\tau)} \right\} &= \\ &= \int d^4x e^{i(k\cdot x - \kappa\tau)} \langle k'\text{out}|\psi^*(x,\tau_f) - \psi^*(x,\tau_i)|0\rangle \\ &= \langle k' \text{ out}|b^*_{out}(k)|0\rangle - \int d^4x e^{i(k\cdot x-\kappa\tau)} \langle k' \text{ out}|\psi^*(x,\tau_i)|0\rangle,\end{aligned} \tag{140}$$

which we may rearrange as

$$\begin{aligned}
\langle k' \text{ out}|k \text{ in}\rangle &= -\int d^4x d\tau \partial_\tau \left\{ e^{i(k\cdot x-\kappa\tau)} \langle k' \text{ out}|\psi^*(x,\tau)|0\rangle \right\} + \langle k' \text{ out}|b^*_{out}(k)|0\rangle \\
&= -\int d^4x d\tau e^{i(k\cdot x-\kappa\tau)} \{-i\kappa + \partial_\tau\} \langle k' \text{ out}|\psi^*(x,\tau)|0\rangle + \langle k'\text{out}|b^*_{out}(k)|0\rangle \\
&= -\int d^4x d\tau e^{i(k\cdot x-\kappa\tau)} \left\{ i\frac{k^2}{2M} + \partial_\tau \right\} \langle k' \text{ out}|\psi^*(x,\tau)|0\rangle + \langle k' \text{ out}|b^*_{out}(k)|0\rangle \\
&= -\int d^4x d\tau e^{i(k\cdot x-\kappa\tau)} \left\{ \frac{i}{2M}\Box + \partial_\tau \right\} \langle k' \text{ out}|\psi^*(x,\tau)|0\rangle + \langle k' \text{ out}|b^*_{out}(k)|0\rangle \\
&= i\int d^4x d\tau e^{i(k\cdot x-\kappa\tau)} \left\{ i\partial_\tau - \frac{1}{2M}\Box \right\} \langle k' \text{ out}|\psi^*(x,\tau)|0\rangle + \langle k' \text{ out}|b^*_{out}(k)|0\rangle
\end{aligned} \tag{141}$$

where we performed two integrations by parts in the second to last step. The last term is a "disconnected term"; it corresponds to the non-scattering path and makes no contribution to the scattering amplitude. Now consider the integral

$$\begin{aligned}
\int_{\tau_i}^{\tau_f} d\tau' \int d^4x'\, \partial_{\tau'} \left[\langle 0|T\psi(x',\tau')\psi^*(x,\tau)|0\rangle e^{-i(k'\cdot x' - \kappa'\tau')} \right] &= \\
&= \int d^4x' e^{-i(k'\cdot x'-\kappa'\tau')} \langle 0|T\psi(x',\tau')\psi^*(x,\tau)|0\rangle \,|_{\tau_i}^{\tau_f} \\
&= \langle 0|b_{out}(k')\psi^*(x,\tau)|0\rangle - \langle 0|\psi^*(x,\tau)b_{in}(k')|0\rangle
\end{aligned} \tag{142}$$

in which the second term in (142) vanishes. For the second term in (141), we use (142) to similarly expand as

$$\begin{aligned}
\langle k' \text{ out}|\psi^*(x,\tau)|0\rangle &= \langle 0|b_{out}(k')\psi^*(x,\tau)|0\rangle \\
&= \int d^4x' e^{i(k'\cdot x'-\kappa'\tau')} \langle 0|\psi_{\text{out}}(x',\tau')\psi^*(x,\tau)|0\rangle \\
&= \int d^4x' d\tau' \partial_{\tau'} \left\{ e^{-i(k'\cdot x'-\kappa'\tau')} \langle 0|T\psi(x',\tau')\psi^*(x,\tau)|0\rangle \right\} + \\
&\quad \langle 0|\psi^*(x,\tau)b_{in}(k')|0\rangle
\end{aligned} \tag{143}$$

and again the second term vanishes. Finally, we arrive at

$$\begin{aligned}
\langle k' \text{ out}|k \text{ in}\rangle &= i^2 \int d^4x d\tau d^4x' d\tau' e^{i(k\cdot x-\kappa\tau)} e^{-i(k'\cdot x'-\kappa'\tau')} \\
&\quad [i\partial_{\tau'} + \frac{1}{2M}\Box'][i\partial_\tau - \frac{1}{2M}\Box] \langle 0|T\psi(x',\tau')\psi^*(x,\tau)|0\rangle
\end{aligned} \tag{144}$$

By continuing this process, we can write

$$\begin{aligned}
\langle k'_1 \cdots k'_n \text{ out}|k_1 \cdots k_m \text{ in}\rangle &= i^{n+m} \int d^5x_1 \cdots d^5x_m d^5x'_1 \cdots d^5x'_n \\
& e^{i(k_1\cdot x_1 + \cdots + k_m\cdot x_m - \kappa_1\tau_1 - \cdots - \kappa_m\tau_m)} e^{-i(k'_1\cdot x'_1 + \cdots + k'_n\cdot x'_n - \kappa'_1\tau'_1 - \cdots - \kappa'_n\tau'_n)} \\
& [i\partial_{\tau'_1} + \frac{1}{2M}\Box'_1] \cdots [i\partial_{\tau'_n} + \frac{1}{2M}\Box'_n][i\partial_{\tau_1} - \frac{1}{2M}\Box_1] \cdots [i\partial_{\tau_m} - \frac{1}{2M}\Box_m] \\
& \langle 0|T\psi(x'_1,\tau'_1) \cdots \psi(x'_n,\tau'_n)\psi^*(x_1,\tau_1) \cdots \psi^*(x_m,\tau_m)|0\rangle
\end{aligned} \tag{145}$$

where we denote $d^5x = d^4x d\tau$.

For the gauge field, the reduction formula follows closely the development for the usual Maxwell case. We begin with

$$\begin{aligned}\langle\beta,k,s\text{ out}|\alpha\text{ in}\rangle &= \langle\beta\text{ out}|a(k,s)|\alpha\text{ in}\rangle \\ &= \frac{-i\sigma g(s)}{(2\pi)^4}\int d^4x e^{-i(k\cdot x+\sigma\kappa\tau)}\langle\beta\text{ out}|\varepsilon_s\cdot\overleftrightarrow{\partial}_\tau a(x,\tau)|\alpha\text{ in}\rangle\end{aligned}\tag{146}$$

where α and β are the other quantum numbers for the states. We may again use

$$\int_{\tau_i}^{\tau_f} d\tau\int d^4x\partial_\tau g(x,\tau) = \int d^4x[g(x,\tau_f)-g(x,\tau_i)]\tag{147}$$

to write

$$\begin{aligned}&\int_{\tau_i}^{\tau_f} d\tau\int d^4x\partial_\tau\frac{-i\sigma g(s)}{(2\pi)^4}e^{-i(k\cdot x+\sigma\kappa\tau)}\langle\beta\text{ out}|\varepsilon_s\cdot\overleftrightarrow{\partial}_\tau a(x,\tau)|\alpha\text{ in}\rangle = \\ &= \frac{-i\sigma g(s)}{(2\pi)^4}\int d^4x e^{-i(k\cdot x+\sigma\kappa\tau)}\Big\{\langle\beta\text{ out}|\varepsilon_s\cdot\overleftrightarrow{\partial}_\tau a(x,\tau_f)|\alpha\text{ in}\rangle \\ &\qquad -\langle\beta\text{ out}|\varepsilon_s\cdot\overleftrightarrow{\partial}_\tau a(x,\tau_i)|\alpha\text{ in}\rangle\Big\} \\ &= \frac{-i\sigma g(s)}{(2\pi)^4}\int d^4x e^{-i(k\cdot x+\sigma\kappa\tau)}\langle\beta\text{ out}|\varepsilon_s\cdot\overleftrightarrow{\partial}_\tau a(x,\tau_f)|\alpha\text{ in}\rangle\end{aligned}\tag{148}$$

where we have used the fact that $a(x,\tau_i)|0\text{ in}\rangle = 0$. By expanding $\overleftrightarrow{\partial}_\tau$, using the wave equation (81), and performing two integrations by parts, we arrive at

$$\langle\beta,k,s\text{ out}|\alpha\text{ in}\rangle = \frac{-ig(s)}{(2\pi)^4}\int d^5x e^{-i(k\cdot x+\sigma\kappa\tau)}[\Box+\sigma\partial_\tau^2]\langle\beta\text{ out}|\varepsilon_s\cdot a(x,\tau)|\alpha\text{ in}\rangle\tag{149}$$

A second application of this procedure enables us to write

$$\begin{aligned}\langle\beta,k,s\text{ out}|\alpha,k',s'\text{ in}\rangle &= \frac{(-i)^2 g(s)g(s')}{(2\pi)^8}\int d^5x d^5x' e^{-i(k\cdot x+\sigma\kappa\tau)}e^{i(k'\cdot x'+\sigma\kappa'\tau')} \\ &\quad [\Box+\sigma\partial_\tau^2][\Box'+\sigma\partial_{\tau'}^2]\langle\beta\text{ out}|\varepsilon_s\cdot a(x,\tau)\varepsilon_{s'}\cdot a(x',\tau')|\alpha\text{ in}\rangle\end{aligned}\tag{150}$$

Feynman Rules

In order to write the Feynman rules for the interacting off-shell theory, may use the general expression for Green's functions, which may be derived from the path integral [19] or by operator methods [3],

$$G^{(n)}(x_1,\tau_1,\cdots,x_n,\tau_n) = \langle 0|\mathrm{T}\phi(x_1,\tau_1),\cdots,\phi(x_n,\tau_n)e^{i\int d^4yd\tau\mathcal{L}_{\text{int}}}|0\rangle\tag{151}$$

in which vacuum-vacuum diagrams are excluded from (151), and ϕ represents the non-interacting free fields of the theory, which permits us to use Wick's theorem and the free propagators (63) and (115). In (151), we adopt the conventional notation for Green's functions in field theory; in relation to the non-interaction propagators calculated above, these are

$$\begin{aligned}G^{(2)}(x,\tau) &= \langle 0|\mathrm{T}\psi(x,\tau)\psi^*(0)|0\rangle_{\text{tree}} = -iG(x,\tau) \\ d_{\mu\nu}^{(2)}(x,\tau) &= \langle 0|\mathrm{T}a_\mu(x,\tau)a_\nu(0)|0\rangle_{\text{tree}} = -id_{\mu\nu}(x,\tau)\end{aligned}\tag{152}$$

where $G(x,\tau)$ and $d_{\mu\nu}(x,\tau)$ are defined in (60) and (115).

Using (49) for the interacting Hamiltonian, we have

$$\mathcal{L}_{\rm int} = -\mathcal{K}_{\rm interaction} = -\frac{ie_0}{2M}a_\mu(\psi^*\partial^\mu\psi - (\partial^\mu\psi^*)\psi) - \frac{e_0^2}{2M}a_\mu a^\mu|\psi|^2. \qquad (153)$$

The two terms in (153) correspond to the two basic diagrams

and we may calculate the vertex factors separately. For the first diagram, we have

$$G^{(3)}_\mu(x_1,\tau_1,x_2,\tau_2,x_3,\tau_3) = \langle 0|T\psi^*(x_1,\tau_1)\psi(x_2,\tau_2)a_\mu(x_3,\tau_3)e^{i\int d^4y d\tau e_0 a_\mu(y,\tau)j^\mu(y,\tau)}|0\rangle_{\rm tree} \qquad (154)$$

where only the tree-level diagram is considered in (154) and j^μ is the vector part of the free matter field current (13),

$$j^\mu = \frac{-i}{2M}(\psi^*\partial^\mu\psi - (\partial^\mu\psi^*)\psi) \qquad (155)$$

Expanding the exponential in (154), we find that the tree-level term is given by

$$\begin{aligned}
G^{(3)}_\mu&(x_1,\tau_1,x_2,\tau_2,x_3,\tau_3) = \\
&= \langle 0|T\psi^*(x_1,\tau_1)\psi(x_2,\tau_2)a_\mu(x_3,\tau_3)\left\{ie_0\int d^4y d\tau a_\mu(y,\tau)j^\mu(y,\tau)\right\}|0\rangle \\
&= \frac{e_0}{2M}\langle 0|T\psi^*(x_1,\tau_1)\psi(x_2,\tau_2)a_\mu(x_3,\tau_3)\int d^4y d\tau a^\nu(y,\tau)\psi^*(y,\tau)\overleftrightarrow{\partial}_{y^\nu}\psi(y,\tau)|0\rangle \\
&= \frac{e_0}{2M}\int d^4y d\tau\langle 0|Ta_\mu(x_3,\tau_3)a^\nu(y,\tau)|0\rangle\langle 0|T\psi^*(x_1,\tau_1)\psi(x_2,\tau_2)\psi^*(y,\tau)\overleftrightarrow{\partial}_{y^\nu}\psi(y,\tau)|0\rangle \\
&= \frac{e_0}{2M}\int d^4y d\tau\langle 0|Ta_\mu(x_3,\tau_3)a^\nu(y,\tau)|0\rangle \\
&\qquad [\langle 0|T\psi(x_2,\tau_2)\psi^*(y,\tau)|0\rangle\overleftrightarrow{\partial}_{y^\nu}\langle 0|T\psi^*(x_1,\tau_1)\psi(y,\tau)|0\rangle] \\
&= \frac{e_0}{2M}\int d^4y d\tau[-id^\nu_\mu(x_3-y,\tau-\tau_3)] \\
&\qquad [(-iG(x_2-y,\tau_2-\tau))\overleftrightarrow{\partial}_{y^\nu}(-iG(y-x_1,\tau-\tau_1))]
\end{aligned} \qquad (156)$$

We now use the Fourier expansions for the free propagators to write

$$\begin{aligned}
G^{(3)}_\mu(x_1,\tau_1,x_2,\tau_2,x_3,\tau_3) &= \frac{ie_0}{2M}\int d^4y d\tau\frac{d^5k d^5p d^5p'}{(2\pi)^{15}\lambda}\mathcal{P}^\nu_\mu(k)\frac{e^{i[k\cdot(x_3-y)+\sigma\kappa(\tau_3-\tau)]}}{k^2+\sigma\kappa^2-i\epsilon} \\
&\qquad\left[\frac{e^{i(p'\cdot(x_2-y)-P'(\tau_2-\tau))}}{\frac{1}{2M}(p')^2-P'-i\epsilon}\overleftrightarrow{\partial}_{y^\nu}\frac{e^{i(p\cdot(y-x_1)-P(\tau-\tau_1))}}{\frac{1}{2M}p^2-P-i\epsilon}\right] \\
&= \frac{ie_0}{2M}\int d^4y d\tau\frac{d^5k d^5p d^5p'}{(2\pi)^{15}\lambda}\mathcal{P}^\nu_\mu(k)\, i(p+p')^\nu\, e^{iy\cdot[p-p'-k]-i\tau[P-P'+\sigma\kappa]} \\
&\qquad\frac{e^{i(k\cdot x_3+\sigma\kappa\tau_3)}}{k^2+\sigma\kappa^2-i\epsilon}\frac{e^{i(p'\cdot x_2)-P'\tau_2)}}{\frac{1}{2M}(p')^2-P'-i\epsilon}\frac{e^{-i(p\cdot x_1)-P\tau_1)}}{\frac{1}{2M}p^2-P-i\epsilon}
\end{aligned}$$

$$
\begin{aligned}
&= \frac{-e_0}{2M}\int \frac{d^5k d^5p d^5p'}{(2\pi)^{10}\lambda}\mathcal{P}^{\nu}_{\mu}(k)\;(p+p')^{\nu}\;\delta^4(p-p'-k)\delta(P-P'+\sigma\kappa) \\
&\qquad \frac{e^{i(k\cdot x_3+\sigma\kappa\tau_3)}}{k^2+\sigma\kappa^2-i\epsilon}\frac{e^{i(p'\cdot x_2)-P'\tau_2)}}{\frac{1}{2M}(p')^2-P'-i\epsilon}\frac{e^{-i(p\cdot x_1)-P\tau_1)}}{\frac{1}{2M}p^2-P-i\epsilon}
\end{aligned}
\tag{157}
$$

Transforming to the momentum space Green's function, with diagram

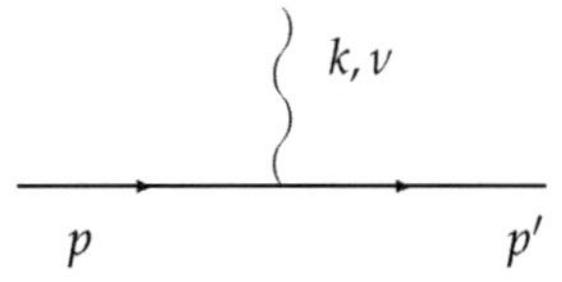

we find

$$
\begin{aligned}
G^{(3)}_{\mu}(p,p',k) &= \frac{e_0}{2M}\mathcal{P}^{\nu}_{\mu}(k)\;i(p+p')_{\nu}\;(2\pi)^5\delta^4(p-p'-k)\delta(P-P'+\sigma\kappa) \\
&\qquad \frac{1}{\lambda}\frac{-i}{k^2+\sigma\kappa^2-i\epsilon}\frac{-i}{\frac{1}{2M}(p')^2-P'-i\epsilon}\frac{-i}{\frac{1}{2M}p^2-P-i\epsilon}
\end{aligned}
\tag{158}
$$

which we identify as the product of the three propagators and the vertex factor for the interaction

$$
\frac{e_0}{2M}i(p+p')^{\nu}\;(2\pi)^5\delta^4(p-p'-k)\delta(P-P'+\sigma\kappa)
\tag{159}
$$

in which

$$
\kappa=\sqrt{-\sigma k^2} \qquad P=\frac{p^2}{2M} \qquad P'=\frac{p'^2}{2M}
\tag{160}
$$

For the second diagram, we calculate

$$
\begin{aligned}
&G^{(4)}_{\mu\nu}(x_1,\tau_1,x_2,\tau_2,x_3,\tau_3,x_4,\tau_4) \\
&\quad= \langle 0|\mathrm{T}\psi^*(x_1,\tau_1)\psi(x_2,\tau_2)a_\mu(x_3,\tau_3)a_\nu(x_4,\tau_4)e^{-i\int d^4y d\tau \frac{e_0^2}{2M}a_\rho a^\rho|\psi|^2}|0\rangle_{\text{tree}} \\
&\quad= \frac{-ie_0^2}{2M}\int d^4y d\tau\langle 0|\mathrm{T}\psi^*(x_1,\tau_1)\psi(x_2,\tau_2)a_\mu(x_3,\tau_3)a_\nu(x_4,\tau_4) \\
&\qquad a_\rho(y,\tau)a^\rho(y,\tau)\psi^*(y,\tau)\psi(y,\tau)|0\rangle \\
&\quad= \frac{-ie_0^2}{2M}\int d^4y d\tau\langle 0|\mathrm{T}\psi^*(x_1,\tau_1)\psi(x_2,\tau_2)\psi^*(y,\tau)\psi(y,\tau)|0\rangle \\
&\qquad\qquad \langle 0|\mathrm{T}a_\mu(x_3,\tau_3)a_\nu(x_4,\tau_4)a_\rho(y,\tau)a^\rho(y,\tau)|0\rangle \\
&\quad= \frac{-ie_0^2}{2M}\int d^4y d\tau\langle 0|\mathrm{T}\psi^*(x_1,\tau_1)\psi(y,\tau)|0\rangle\langle 0|\mathrm{T}\psi(x_2,\tau_2)\psi^*(y,\tau)|0\rangle \\
&\qquad\qquad 2\left\{\langle 0|\mathrm{T}a_\mu(x_3,\tau_3)a_\rho(y,\tau)|0\rangle\langle 0|\mathrm{T}a_\nu(x_4,\tau_4)a^\rho(y,\tau)|0\rangle\right\} \\
&\quad= \frac{-ie_0^2}{M}\int d^4y d\tau\frac{d^5k d^5k' d^5p d^5p'}{(2\pi)^{20}\lambda^2}\mathcal{P}^{\rho}_{\mu}(k)\frac{e^{i[k\cdot(y-x_3)+\sigma\kappa(\tau-\tau_3)]}}{k^2+\sigma\kappa^2-i\epsilon}\mathcal{P}_{\rho\nu}(k')\times \\
&\qquad \times\frac{e^{i[k'\cdot(x_4-y)+\sigma\kappa'(\tau_4-\tau)]}}{(k')^2+\sigma(\kappa')^2-i\epsilon}\frac{e^{i[p'\cdot(x_2-y)-P'(\tau_2-\tau)]}}{\frac{1}{2M}(p')^2-P'-i\epsilon}\frac{e^{i[p\cdot(y-x_1)-P(\tau-\tau_1)]}}{\frac{1}{2M}p^2-P-i\epsilon}
\end{aligned}
$$

$$= \frac{-ie_0^2}{M\lambda^2} \int \frac{d^5k d^5k' d^5p d^5p'}{(2\pi)^{20}} (2\pi)^5 \delta^4(k - k' - p' + p) \delta(\sigma\kappa - \sigma\kappa' + P' - P) \mathcal{P}_\mu^\rho(k)$$
$$\mathcal{P}_{\rho\nu}(k') e^{-i[k\cdot x_3 + \sigma\kappa\tau_3]} e^{i[k'\cdot x_4 + \sigma\kappa'\tau_4]} e^{i[p'\cdot x_2 - P'\tau_2]} e^{-i[p\cdot x_1 - P\tau_1]}$$
$$\frac{-i}{k^2 + \sigma\kappa^2 - i\epsilon} \frac{-i}{(k')^2 + \sigma(\kappa')^2 - i\epsilon} \frac{-i}{\frac{1}{2M}(p')^2 - P' - i\epsilon} \frac{-i}{\frac{1}{2M}p^2 - P - i\epsilon} \tag{161}$$

Notice that a factor of 2 appears in the fourth line, because there are two ways to contract the photon operators. Transforming to the momentum space Green's function, with diagram

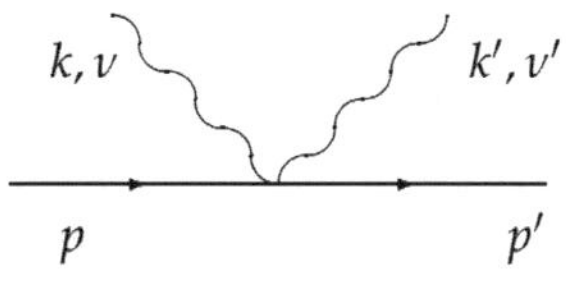

we find

$$G_{\mu\nu}^{(4)}(p, p', k, k') = \frac{-ie_0^2}{M\lambda^2} (2\pi)^5 \delta^4(k - k' - p' + p) \delta(\sigma\kappa - \sigma\kappa' + P' - P) \mathcal{P}_\mu^\rho(k) \mathcal{P}_{\rho\nu}(k')$$
$$\frac{-i}{k^2 + \sigma\kappa^2 - i\epsilon} \frac{-i}{(k')^2 + \sigma(\kappa')^2 - i\epsilon} \frac{-i}{\frac{1}{2M}(p')^2 - P' - i\epsilon} \frac{-i}{\frac{1}{2M}p^2 - P - i\epsilon} \tag{162}$$

which we recognize as the product of the four propagators and the vertex factor

$$\frac{-ie_0^2}{M} (2\pi)^5 g_{\mu\nu} \delta^4(k - k' - p' + p) \delta(\sigma\kappa - \sigma\kappa' + P' - P) \tag{163}$$

We may summarize the Feynman rules for the momentum space Green's functions as follows:

(i) For each matter field propagator, draw a directed line associated with the factor

$$\frac{1}{(2\pi)^5} \frac{-i}{\frac{1}{2M}p^2 - P - i\epsilon}$$

(ii) For each photon propagator, draw a photon line associated with the factor

$$\frac{1}{\lambda} \mathcal{P}^{\mu\nu} \frac{-i}{k^2 + \sigma\kappa^2 - i\epsilon}$$

(iii) For the three-particle interaction, write the vertex factor

$$\frac{e_0}{2M} i(p + p')^\nu \ (2\pi)^5 \delta^4(p - p' - k) \delta(P - P' + \sigma\kappa)$$

(iv) For the four-particle interaction, write the vertex factor

$$\frac{-ie_0^2}{M} (2\pi)^5 g_{\mu\nu} \delta^4(k - k' - p' + p) \delta(\sigma\kappa - \sigma\kappa' + P' - P)$$

To obtain the Feynman rules for the S-matrix elements, we use the reduction formulas together with the free propagators for the incoming and outgoing particles. From (145) for the matter fields, we see that inserting a propagator will precisely cancel the derivative operator, so that in calculating S-matrix elements, the incoming and outgoing propagators are replaced by 1.

Using (149) for the photons, we see that the derivative operator will cancel the factor $-i/(k^2 + \sigma\kappa^2 - i\epsilon)$, but not the factor $\frac{1}{\lambda}$. So in the rules for the S-matrix elements we replace the incoming and outgoing photon propagators with the factor

$$\pm \frac{ig(s)}{(2\pi)^4\lambda}\varepsilon^{\mu}(k,s) \tag{164}$$

where the incoming photon takes the sign $-$ and the outgoing photon takes the sign $+$.

6. Scattering cross-sections

In this section, we calculate the relationship of the scattering cross-section to the transition amplitudes calculated from the S-matrix expansions discussed in the previous section. In particular, we specialize the five-dimensional quantum theory to the case of scattering. The development here generally follows the presentation given in [3], with elements taken from [20] and [21].

The initial state will be represented by constructing wave packets of the form

$$|\Psi_p\rangle = \int \frac{d^4q}{(2\pi)^4}\tilde{\psi}_p(q)|q\rangle = \int \frac{d^4q}{(2\pi)^4}\tilde{\psi}(q-p)|q\rangle \tag{165}$$

in which we take $\tilde{\psi}_p(q) = \tilde{\psi}(q-p)$ to be a narrow distribution centered around the momentum p. These wave packets correspond to linear superpositions of solutions to the quantum mechanical eigenvalue equation

$$K\psi_q = \kappa_q\psi_q \tag{166}$$

where

$$K = \frac{p^2}{2M} + V \qquad \text{and} \qquad \kappa_q = \frac{q^2}{2M} + \delta\kappa_q. \tag{167}$$

The wavefunctions may be written in the form

$$\Psi_p(x) = \int \frac{d^4q}{(2\pi)^4}\tilde{\psi}(q-p)e^{iq\cdot x} = e^{ip\cdot x}\int \frac{d^4\rho}{(2\pi)^4}\tilde{\psi}(\rho)e^{i\rho\cdot x} = e^{ip\cdot x}G(x) \tag{168}$$

in which $G(x)$ is a slowly varying, localized function of x. The τ-dependent wavefunctions are

$$\Psi(x,\tau) = e^{-iK\tau}\Psi_p(x) = \int \frac{d^4q}{(2\pi)^4}\tilde{\psi}(q-p)e^{i(q\cdot x-\kappa_q\tau)}. \tag{169}$$

Since the momentum distribution is narrowly centered around p, the wavepacket in (169), may be expanded in $q = p + \rho$. To do this note that

$$\kappa_q = \kappa_p + \rho^{\mu}\partial_{p^{\mu}}\kappa_p + \frac{1}{2}\rho^{\mu}\rho^{\nu}\partial_{p^{\mu}}\partial_{p^{\nu}}\kappa_p + \cdots = \kappa_p + \rho\cdot u + o(\rho^2) \tag{170}$$

where $u^\sigma = \partial\kappa_p/\partial p_\sigma$ is the group 4-velocity. Then, to first order in ρ,

$$\begin{aligned}\Psi(x,\tau) &= \int \frac{d^4\rho}{(2\pi)^4} e^{i[(p+\rho)\cdot x-(\kappa_p+\rho\cdot u)\tau]}\tilde{\psi}(\rho) \\ &= e^{i[p\cdot x-\kappa_p\tau]}\int \frac{d^4\rho}{(2\pi)^4}\tilde{\psi}(\rho)e^{i\rho\cdot[x-u\tau]} \\ &= e^{i[p\cdot x-\kappa_p\tau]}G(x-u\tau) \end{aligned} \tag{171}$$

and the slowly varying function $G(x-u\tau)$ describes the Ehrenfest motion of the wavepacket. The event density corresponding to the wavefunction in (171) is given by the integral of the current

$$j^4(x,\tau) = \overline{\Psi}_p\Psi_p = |G(x-u\tau)|^2 . \tag{172}$$

In scattering, the initial state is a product of the target state and the beam state, and is represented by a wavepacket of the form

$$|i\rangle = |\Psi_{\text{target}}\rangle\,|\Psi_{\text{beam}}\rangle = \int \frac{d^4q_{\rm T}}{(2\pi)^4}\frac{d^4q_{\rm B}}{(2\pi)^4}\tilde{\psi}_{\rm T}(q_{\rm T}-p_{\rm T})\tilde{\psi}_{\rm B}(q_{\rm B}-p_{\rm B})|q_{\rm T}\; q_{\rm B}\rangle. \tag{173}$$

Writing the transition matrix T in terms of the scattering matrix S [18],

$$\begin{aligned} \text{S} &= 1+i\text{T} \\ \langle f|\text{S}|i\rangle &= \delta_{if}+i\langle f|\text{T}|i\rangle \\ &= \delta_{if}+i(2\pi)^5\delta(\kappa_f-\kappa_i)\delta^4(p_f-p_i)\langle f|\mathcal{T}|i\rangle \end{aligned} \tag{174}$$

we find for the initial state defined in (173),

$$\langle f|\mathcal{T}|i\rangle = \int \frac{d^4q_{\rm T}}{(2\pi)^4}\frac{d^4q_{\rm B}}{(2\pi)^4}\tilde{\psi}_{\rm T}(q_{\rm T}-p_{\rm T})\tilde{\psi}_{\rm B}(q_{\rm B}-p_{\rm B})\langle f|\mathcal{T}|q_{\rm T}q_{\rm B}\rangle. \tag{175}$$

The transition probability becomes

$$\begin{aligned} |\langle f|\text{T}|i\rangle|^2 = &\int \frac{d^4q_{\rm T}}{(2\pi)^4}\frac{d^4q'_{\rm T}}{(2\pi)^4}\frac{d^4q_{\rm B}}{(2\pi)^4}\frac{d^4q'_{\rm B}}{(2\pi)^4}\tilde{\psi}_{\rm T}(q_{\rm T}-p_{\rm T})\tilde{\psi}^*_{\rm T}(q'_{\rm T}-p_{\rm T})\tilde{\psi}_{\rm B}(q_{\rm B}-p_{\rm B})\tilde{\psi}^*_{\rm B}(q'_{\rm B}-p_{\rm B}) \\ &\langle f|\mathcal{T}|q_{\rm T}q_{\rm B}\rangle\langle f|\mathcal{T}|q'_{\rm T}q'_{\rm B}\rangle^*\delta(\kappa_f-\kappa_i)\delta^4(p_f-p_i)\delta(\kappa_f-\kappa'_i)\delta^4(p_f-p'_i). \end{aligned} \tag{176}$$

where

$$p_i = q_{\rm T}+q_{\rm B} = p_{\rm T}+p_{\rm B}+\rho_{\rm T}+\rho_{\rm B} \tag{177}$$
$$p'_i = q'_{\rm T}+q'_{\rm B} = p_{\rm T}+p_{\rm B}+\rho'_{\rm T}+\rho'_{\rm B} \tag{178}$$
$$\kappa_i = \kappa_{\rm T}+\kappa_{\rm B} = \frac{q^2_{\rm T}}{2M_{\rm T}}+\frac{q^2_{\rm B}}{2M_{\rm B}}. \tag{179}$$

We assume that the interaction occurs close to the central momenta, so that

$$\langle f|\mathcal{T}|q_{\rm T}q_{\rm B}\rangle\langle f|\mathcal{T}|q'_{\rm T}q'_{\rm B}\rangle^* \simeq |\langle f|\mathcal{T}|p_{\rm T}p_{\rm B}\rangle|^2 = |\mathcal{T}_{fi}|^2. \tag{180}$$

We may rewrite the δ-functions, replacing $q = p + \rho$ for each momentum:

$$\begin{aligned}
\delta^4(p_f - p_i)\delta^4(p_f - p_i') &= \delta^4(p_f - p_T - \rho_T - p_B - \rho_B)\delta^4(p_f - p_T - \rho_T' - p_B - \rho_B') \\
&= \delta^4(p_f - p_T - p_B - (\rho_T + \rho_B))\delta^4(p_f - p_T - p_B - (\rho_T' + \rho_B')) \\
&= \delta^4(p_f - p_T - p_B - (\rho_T + \rho_B))\delta^4(\rho_T + \rho_B - \rho_T' - \rho_B') \\
&\simeq \delta^4(p_f - p_T - p_B)\delta^4(\rho_T - \rho_T' + \rho_B - \rho_B'). \qquad (181)
\end{aligned}$$

Similarly,

$$\delta(\kappa_f - \kappa_i)\delta(\kappa_f - \kappa_i') \simeq \delta(\kappa_{p_f} - \kappa_{p_T} - \kappa_{p_B})\delta(\kappa_{\rho_T} + \kappa_{\rho_B} - \kappa_{\rho_T'} - \kappa_{\rho_B'}). \qquad (182)$$

so that, since $d^4q = d^4\rho$, (176) becomes

$$\begin{aligned}
|\langle f|T|i\rangle|^2 = (2\pi)^{10}\delta(\kappa_{p_f} - \kappa_{p_T} - \kappa_{p_B})\delta^4(p_f - p_T - p_B)|\mathcal{T}_{fi}|^2 \\
\int \frac{d^4\rho_T}{(2\pi)^4}\frac{d^4\rho_T'}{(2\pi)^4}\frac{d^4\rho_B}{(2\pi)^4}\frac{d^4\rho_B'}{(2\pi)^4}\tilde{\psi}_T(\rho_T)\tilde{\psi}_T^*(\rho_T')\tilde{\psi}_B(\rho_B)\tilde{\psi}_B^*(\rho_B')\delta^4(\rho_T - \rho_T' + \rho_B - \rho_B') \\
\delta(\kappa_{\rho_T} + \kappa_{\rho_B} - \kappa_{\rho_T'} - \kappa_{\rho_B'}). \qquad (183)
\end{aligned}$$

Notice that since $\kappa_\rho \simeq \rho \cdot u$,

$$\begin{aligned}
\kappa_{\rho_T} - \kappa_{\rho_T'} + \kappa_{\rho_B} - \kappa_{\rho_B'} &\simeq \rho_T \cdot u_T - \rho_T' \cdot u_T + \rho_B \cdot u_B - \rho_B' \cdot u_B \\
&= (\rho_T - \rho_T') \cdot u_T + (\rho_B - \rho_B') \cdot u_B \\
&= (u_B - u_T) \cdot (\rho_B - \rho_B') + u_T \cdot (\rho_B - \rho_B' + \rho_T - \rho_T') \\
&= (u_B - u_T) \cdot (\rho_B - \rho_B') \qquad (184)
\end{aligned}$$

where we have used (181) in the last line of (184). To continue, we write integral representations for the δ-functions. Thus,

$$\delta^4(\rho_T - \rho_T' + \rho_B - \rho_B') = \frac{1}{(2\pi)^4}\int d^4x e^{ix\cdot(\rho_T - \rho_T' + \rho_B - \rho_B')} \qquad (185)$$

and

$$\delta((u_B - u_T)\cdot(\rho_B - \rho_B')) = \int \frac{d\alpha}{2\pi} e^{i\alpha(u_B - u_T)\cdot(\rho_B - \rho_B')}. \qquad (186)$$

Inserting (184), (185) and (186) into (183), we have

$$\begin{aligned}
|\langle f|T|i\rangle|^2 = (2\pi)^{10}\delta(\kappa_{p_f} - \kappa_{p_T} - \kappa_{p_B})\delta^4(p_f - p_T - p_B)|\mathcal{T}_{fi}|^2 \\
\frac{1}{(2\pi)^5}\int d^4x d\alpha \frac{d^4\rho_T}{(2\pi)^4}\frac{d^4\rho_T'}{(2\pi)^4}\frac{d^4\rho_B}{(2\pi)^4}\frac{d^4\rho_B'}{(2\pi)^4}\tilde{\psi}_T(\rho_T)\tilde{\psi}_T^*(\rho_T')\tilde{\psi}_B(\rho_B)\tilde{\psi}_B^*(\rho_B') \\
e^{ix\cdot(\rho_T - \rho_T' + \rho_B - \rho_B')} e^{i\alpha(u_B - u_T)\cdot(\rho_B - \rho_B')} \\
= (2\pi)^5\delta(\kappa_{p_f} - \kappa_{p_T} - \kappa_{p_B})\delta^4(p_f - p_T - p_B)|\mathcal{T}_{fi}|^2 \\
\int d^4x d\alpha \, |G_T(x)|^2 \left|G_B\left(x + \alpha(u_B - u_T)\right)\right|^2. \qquad (187)
\end{aligned}$$

The transition probability per unit spacetime volume is then

$$\begin{aligned}
\frac{d}{dVdT}|\langle f|T|i\rangle|^2 &= (2\pi)^5\delta(\kappa_{p_f} - \kappa_{p_T} - \kappa_{p_B})\delta^4(p_f - p_T - p_B)|\mathcal{T}_{fi}|^2 \\
&\quad \int d\alpha \, |G_T(x)|^2 \left|G_B\left(x + \alpha(u_B - u_T)\right)\right|^2 \\
&= (2\pi)^5\delta(\kappa_{p_f} - \kappa_{p_T} - \kappa_{p_B})\delta^4(p_f - p_T - p_B)|\mathcal{T}_{fi}|^2 \, |G_T(x)|^2 \\
&\quad \int d\alpha \left|G_B\left(\alpha(u_B - u_T)\right)\right|^2. \qquad (188)
\end{aligned}$$

Journal of Physics: Conference Series **437** (2013) 012011 doi:10.1088/1742-6596/437/1/012011

We take the target-beam axis to be along the z-axis. Then, the relative group 4-velocity becomes

$$u_{\rm B} - u_{\rm T} = (u^0_{\rm B} - u^0_{\rm T}, 0, 0, u^3_{\rm B} - u^3_{\rm T}) \tag{189}$$

and so

$$\int d\alpha \left| G_{\rm B}\Big(\alpha(u_{\rm B} - u_{\rm T})\Big)\right|^2 = \int d\alpha \left| G_{\rm B}\Big(\alpha(u^0_{\rm B} - u^0_{\rm T}), 0, 0, \alpha(u^3_{\rm B} - u^3_{\rm T})\Big)\right|^2. \tag{190}$$

Making the change of variable $\xi = \alpha(u^3_{\rm B} - u^3_{\rm T})$, this becomes

$$\int d\alpha \left| G_{\rm B}\Big(\alpha(u_{\rm B} - u_{\rm T})\Big)\right|^2 = \frac{1}{(u^3_{\rm B} - u^3_{\rm T})} \int d\xi \left| G_{\rm B}\Big(\frac{\xi}{v}, 0, 0, \xi\Big)\right|^2 \tag{191}$$

where

$$v = \frac{u^3_{\rm B} - u^3_{\rm T}}{u^0_{\rm B} - u^0_{\rm T}} \tag{192}$$

corresponds to the relative speed ($\sim dx/dt$) of the target and beam events. By (172), and since $G(x - u\tau)$ satisfies

$$\partial_\tau G(x - u\tau) = -u^\mu \partial_{x^\mu} G(x - u\tau). \tag{193}$$

we may take j^μ to be

$$j^\mu = u^\mu G(x - u\tau) \tag{194}$$

In order to understand the integral over ξ in (191), we notice that for $u^3 = u^3_{\rm B} - u^3_{\rm T} > 0$, the relative motion will cross from $x^3 < 0$ to $x^3 > 0$ during the scattering. We define by $\mathrm{N}^+(\tau)$ the proportion of all events for which $x^3 > 0$, given by

$$\mathrm{N}^+(\tau) = \int_0^\infty dx^3 \int dt d^2x_\perp j^4(x,\tau) \tag{195}$$

where $x_\perp$ refers to the 1-2 plane. The rate at which events cross $x^3 = 0$ is then given by

$$\begin{aligned}
\frac{d}{d\tau}\mathrm{N}^+(\tau) &= \int_0^\infty dx^3 \int dt d^2x_\perp \partial_\tau j^4(x,\tau) \\
&= \int_0^\infty dx^3 \int_{-\infty}^\infty dt \int d^2x_\perp [-u^\mu \partial_\mu j^4(x,\tau)] \\
&= \int_0^\infty dx^3 \int dt d^2x_\perp [-u^3 \partial_3 j^4(x,\tau)]
\end{aligned} \tag{196}$$

where in the last line, we used $j^4(x,\tau) \to 0$ as $x^\mu \to \pm\infty$. Evaluating the integral on dx^3, we find

$$\frac{d}{d\tau}\mathrm{N}^+(\tau) = -u^3 \int dt d^2x_\perp j^4(t, x_\perp, x^3, \tau)\Big|_{x^3=0}^{x^3=\infty} = u^3 \int dt d^2x_\perp j^4(t, x_\perp, 0, \tau). \tag{197}$$

In terms of (196) and (172), the total number of events which cross $x^3 = 0$ for all τ is given by

$$\begin{aligned}
\mathrm{N}^+(\infty) - \mathrm{N}^+(-\infty) &= \int d\tau \frac{d}{d\tau}\mathrm{N}^+(\tau) \\
&= \int d\tau u^3 \int dt d^2x_\perp j^4(t, x_\perp, 0, \tau) \\
&= \int d\tau u^3 \int dt d^2x_\perp \left| G\Big((t, x_\perp, 0) - (u^0, 0, 0, u^3)\tau\Big)\right|^2 \\
&= \int d\tau u^3 \int dt d^2x_\perp \left| G\Big(t - u^0\tau, x_\perp, -u^3\tau\Big)\right|^2.
\end{aligned} \tag{198}$$

Journal of Physics: Conference Series **437** (2013) 012011 doi:10.1088/1742-6596/437/1/012011

Making the change of variables $\xi = u^3\tau$ puts (198) into the form

$$\begin{aligned} N^+(\infty) - N^+(-\infty) &= \int d\xi \int dt d^2x_\perp \left| G\left(t - \frac{u^0}{u^3}\xi, x_\perp, -\xi\right)\right|^2 \\ &= \int d\xi \int dt d^2x_\perp \left| G\left(t + \frac{1}{v}\xi, x_\perp, \xi\right)\right|^2 . \end{aligned} \tag{199}$$

in which v is given by (192). The total number of events which cross $x^3 = 0$, per unit area (in the 1-2 plane) per unit time (which defines the 3-flux $F^{(3)}$ of the beam) is just,

$$F^{(3)} = \int d\xi \left| G\left(\frac{1}{v}\xi, 0, 0, \xi\right)\right|^2 . \tag{200}$$

Comparison with (191) shows that

$$\int d\alpha \left| G_B\Big(\alpha(u_B - u_T)\Big)\right|^2 = \frac{1}{|\vec{u}_B - \vec{u}_T|} F^{(3)} \tag{201}$$

Therefore, we may rewrite (188) as

$$\begin{aligned} \frac{d}{dVdT} |\langle f|T|i\rangle|^2 &= (2\pi)^5 \delta(\kappa_{p_f} - \kappa_{p_T} - \kappa_{p_B}) \delta^4(p_f - p_T - p_B) |\mathcal{T}_{fi}|^2 \, |G_T(x)|^2 \\ & \quad \frac{1}{|\vec{u}_B - \vec{u}_T|} F^{(3)} . \end{aligned} \tag{202}$$

The scattering 3-cross-section is defined through

$$\frac{d}{dVdT} |\langle f|T|i\rangle|^2 = d\sigma^{(3)} \times \text{beam flux} \times \text{target density} \tag{203}$$

and so, we find that

$$d\sigma^{(3)} = (2\pi)^5 \delta(\kappa_{p_f} - \kappa_{p_T} - \kappa_{p_B}) \delta^4(p_f - p_T - p_B) \frac{|\langle f|\mathcal{T}|p_T p_B\rangle|^2}{|\vec{u}_B - \vec{u}_T|} . \tag{204}$$

The Process $B + T \to 1 + 2$

We will find it convenient to treat the case of two particle final states in relative coordinates, and we make the following definitions

$$P = \frac{1}{2}(p_T + p_B) \qquad p = p_T - p_B \tag{205}$$

$$P' = \frac{1}{2}(p_1 + p_2) \qquad p' = p_1 - p_2 \tag{206}$$

which have the inverse relations

$$p_T = P + \frac{1}{2}p \qquad p_B = P - \frac{1}{2}p \tag{207}$$

$$p_1 = P' + \frac{1}{2}p' \qquad p_2 = P' - \frac{1}{2}p' \tag{208}$$

In the center of mass system, we have that

$$\vec{p}_B + \vec{p}_T = \vec{p}_1 + \vec{p}_2 = 0 \tag{209}$$

so that

$$P = \frac{1}{2}\begin{bmatrix} E(\vec{p}_B) + E(\vec{p}_T) \\ \vec{p}_B + \vec{p}_T \end{bmatrix} = \begin{bmatrix} \frac{1}{2}[E(\vec{p}_B) + E(\vec{p}_T)] \\ \vec{0} \end{bmatrix} = \begin{bmatrix} \frac{1}{2}\sqrt{s} \\ \vec{0} \end{bmatrix} \tag{210}$$

where

$$s = -(p_B + p_T)^2 \tag{211}$$

is the usual Mandelstam parameter in this metric (and so $P^2 = -s/4$), and

$$E(\vec{p}) = \sqrt{(\vec{p})^2 + m^2}. \tag{212}$$

Similarly,

$$p' = p_1 - p_2 = \begin{bmatrix} E(p_1) - E(p_2) \\ 2\vec{p}_1 \end{bmatrix} \tag{213}$$

and

$$\begin{aligned} p_1^2 &= P^2 + \frac{1}{4}(p')^2 + P \cdot p' = -\frac{1}{4}s + \frac{1}{4}(p')^2 - \frac{1}{2}\sqrt{s}\,[E(p_1) - E(p_2)] \\ p_2^2 &= P^2 + \frac{1}{4}(p')^2 - P \cdot p' = -\frac{1}{4}s + \frac{1}{4}(p')^2 + \frac{1}{2}\sqrt{s}\,[E(p_1) - E(p_2)]. \end{aligned} \tag{214}$$

Since the relative momentum is spacelike, we may introduce the parameterization

$$p' = \rho \begin{bmatrix} \sinh\beta \\ \cosh\beta\hat{n} \end{bmatrix} \tag{215}$$

so that (214) becomes

$$p_1^2 = -\frac{1}{4}(s - \rho^2 + 2\sqrt{s}\rho\sinh\beta) \qquad p_2^2 = -\frac{1}{4}(s - \rho^2 - 2\sqrt{s}\rho\sinh\beta). \tag{216}$$

We may see the utility of this approach in the conventional description of relativistic scattering, where the cross-section has the O(3,1) invariant measure

$$dR_2 = \delta^4(p_1 + p_2 - p_T - p_B)\delta(p_1^2 + m_1^2)\delta(p_2^2 + m_2^2)\frac{d^4p_1}{(2\pi)^4}\frac{d^4p_2}{(2\pi)^4} \tag{217}$$

which in relative coordinates becomes

$$\begin{aligned} dR_2 &= \frac{1}{2(2\pi)^8}\delta^4(P - P')\delta(p_1^2 + m_1^2)\delta(p_2^2 + m_2^2)d^4P'd^4p' \\ &= \frac{1}{2(2\pi)^8}\delta(p_1^2 + m_1^2)\delta(p_2^2 + m_2^2)d^4p' \\ &= \frac{1}{2(2\pi)^8}\cdot 4\cdot 4\cdot \delta(s - \rho^2 + 2\sqrt{s}\rho\sinh\beta - 4m_1^2)\delta(s - \rho^2 - 2\sqrt{s}\rho\sinh\beta - 4m_2^2) \\ &\qquad \rho^3\cosh^2\beta d\rho d\beta d\Omega \\ &= \frac{4}{(2\pi)^8}\rho^3\cosh^2\beta d\rho d\beta d\Omega\delta(\rho^2 - s + (m_1^2 + m_2^2))\delta(s - \rho^2 - 2\sqrt{s}\rho\sinh\beta - 4m_2^2) \\ &= \frac{\rho}{(2\pi)^8\sqrt{s}}\cosh^2\beta d\rho d\beta d\Omega\delta\left[\rho - \sqrt{s - 2(m_1^2 + m_2^2)}\right]\delta\left[\sinh\beta - \frac{m_1^2 - m_2^2}{\sqrt{s}\rho}\right] \end{aligned} \tag{218}$$

Changing variables as

$$\rho \cosh^2 \beta d\beta = \rho \cosh \beta d(\sinh \beta) \tag{219}$$

and recognizing $\rho \cosh \beta$ as $|\vec{p}\,'| = 2|\vec{p}_f|$, this becomes,

$$dR_2 = \frac{2|\vec{p}_f|}{\sqrt{s}} d\Omega \tag{220}$$

with

$$\rho = \sqrt{s - 2(m_1^2 + m_2^2)} \tag{221}$$

so that

$$|\vec{p}_f| = \frac{\rho}{2}\sqrt{1 + \left[\frac{m_1^2 - m_2^2}{\sqrt{s}\rho}\right]^2} = \frac{1}{2\sqrt{s}}\sqrt{s^2 - 2s(m_1^2 + m_2^2) + (m_1^2 - m_2^2)^2} = \frac{\lambda^{\frac{1}{2}}(s, m_1^2, m_2^2)}{2\sqrt{s}} \tag{222}$$

in agreement with the usual derivation [3].

Returning to the off-shell theory, we first consider scattering of indistinguishable particles, which means that we may take the invariant mass parameter M to be the same for all events. The invariant measure for the matter field is given by

$$\begin{aligned} dR_2 &= \delta^4(p_1 + p_2 - p_T - p_B)\delta\left(\frac{(p_1)^2}{2M} + \frac{(p_2)^2}{2M} - \frac{(p_B)^2}{2M} - \frac{(p_T)^2}{2M}\right)\frac{d^4p_1}{(2\pi)^4}\frac{d^4p_2}{(2\pi)^4} \\ &= \frac{1}{2(2\pi)^8}\delta^4(P - P')d^4P'\ 2M\delta\left((p_1)^2 + (p_2)^2 - (p_B)^2 - (p_T)^2\right) d^4p' \\ &= \frac{2M}{(2\pi)^8}\delta(p^2 - p'^2)\, d^4p' \end{aligned} \tag{223}$$

where we have used

$$(p_1)^2 + (p_2)^2 = P'^2 + \frac{1}{4}p'^2 + P' \cdot p' + P'^2 + \frac{1}{4}p'^2 - P' \cdot p' = 2P'^2 + \frac{1}{2}p'^2 \tag{224}$$

$$(p_B)^2 + (p_T)^2 = 2P^2 + \frac{1}{2}p^2 \tag{225}$$

Now using the parameterization in (215), we find

$$dR_2 = \frac{2M}{(2\pi)^8}\delta(p^2 - \rho^2)\,\rho^3\cosh^2\beta d\rho d\beta d\Omega = \frac{M}{(2\pi)^8}\delta\left(\sqrt{p^2} - \rho\right)(2|\vec{p}_f|)^2 d\rho d\beta d\Omega \tag{226}$$

where the absolute value of

$$\vec{p}_f = \vec{p}_1 = -\vec{p}_2 = \frac{1}{2}\rho\cosh\beta\hat{n} \tag{227}$$

is undetermined in the off-shell theory, because the value of β is not fixed by conservation of mass-momentum. But notice that ρ is determined by

$$\rho^2 = p^2 = (p_B - p_T)^2 = (p_B)^2 + (p_T)^2 - 2p_B \cdot p_T = s - 2[(m_B)^2 + (m_T)^2] \tag{228}$$

which is identical to the usual on-shell value given in (221). Combining (226) with (204), we have for the scattering cross-section (see also [18])

$$d\sigma^{(3)} = \frac{(2\pi)^5}{(2\pi)^8}\frac{|\langle p_1 p_2|T|p_T p_B\rangle|^2}{|\vec{u}_B - \vec{u}_T|}\, 4M|\vec{p}_f|^2 d\beta d\Omega. \tag{229}$$

Since $p = Mu$ for the asymptotic free matter field, we may write this in the form

$$d\sigma^{(3)} = \frac{1}{(2\pi)^3} M^2 \frac{|\vec{p}_f|^2}{|\vec{p}_i|} |\langle p_1 p_2 | \mathcal{T} | p_\mathrm{T} p_\mathrm{B} \rangle|^2 d\beta d\Omega. \tag{230}$$

where $\vec{p}_i = \vec{p}_\mathrm{T} - \vec{p}_\mathrm{B}$, and we must keep in mind that $\vec{p}_f$ is dependent on β.

We denote by $\Delta m^2 = (m_1)^2 - (m_2)^2$, and notice that from (216),

$$\Delta m^2 = (p_2)^2 - (p_1)^2 = \sqrt{s}\,\rho \sinh \beta \tag{231}$$

so that

$$d(\Delta m^2) = \sqrt{s}\,\rho \cosh \beta d\beta = \sqrt{s} 2 |\vec{p}_f| d\beta. \tag{232}$$

Therefore, the cross-section can be written as

$$d\sigma^{(3)} = \frac{1}{(2\pi)^3} M^2 \frac{|\vec{p}_f|}{2\sqrt{s}\,|\vec{p}_i|} |\langle p_1 p_2 | \mathcal{T} | p_\mathrm{T} p_\mathrm{B} \rangle|^2 d\Omega d(\Delta m^2). \tag{233}$$

We may compare the cross-section with the comparable expression in the usual on-shell theory, which can be written in the center of mass as [3]

$$\frac{d\sigma^{(2)}}{d\Omega} = \frac{1}{64\pi^2} \frac{|\vec{p}_f|}{s\,|\vec{p}_i|} |\langle p_1 p_2 | \mathcal{T} | p_\mathrm{T} p_\mathrm{B} \rangle|^2. \tag{234}$$

Putting (233) into a form similar to (234), we have

$$\frac{d\sigma^{(3)}}{d\Omega} = \frac{1}{64\pi^2} \frac{|\vec{p}_f|}{s\,|\vec{p}_i|} |\langle p_1 p_2 | \mathcal{T} | p_\mathrm{T} p_\mathrm{B} \rangle|^2 d(\frac{4M^2\sqrt{s}}{\pi} \Delta m^2). \tag{235}$$

Examining the Feynman rules for the single photon interaction in Section 6, we see that the usual vertex factor of e is replaced by $e_0/2M = \lambda e/2M$ and photon propagator has an extra factor of $1/\lambda$, so that the squared transition amplitude will have a factor of $\lambda^2/16M^4$ in relation to the usual on-shell case. Combining this factor with the extra factor in (235), we have an overall extra factor of

$$\frac{\lambda^2}{16M^4} d(\frac{4M^2\sqrt{s}}{\pi} \Delta m^2) = d(\frac{\sqrt{s}\lambda^2}{4\pi M^2} \Delta m^2) \tag{236}$$

which we see has the expected units of length. We may understand this factor in the following way: During the scattering, the event will propagate a distance $\Delta x \sim (p/M)\Delta\tau$, so that the uncertainty relation tells us that

$$\Delta\tau \sim \frac{\Delta x}{(p/M)} = \frac{M}{p} \Delta x \sim \frac{1}{2\Delta p} \frac{M}{p} \tag{237}$$

So from $(\vec{p})^2 = E^2 - m^2$, we find that

$$\vec{p} \cdot d\vec{p} \sim m\, dm \sim \frac{1}{2} \Delta(m^2) \tag{238}$$

and combining with (237),

$$\Delta\tau \sim \frac{M}{\Delta(m^2)}. \tag{239}$$

Shnerb and Horwitz [10] have shown that for $\Delta\tau > \lambda$, the photon-current interaction becomes uncorrelated, so we may take $\lambda \sim \Delta\tau$. Thus, the factor in (236) becomes

$$\begin{aligned} d\left(\frac{\sqrt{s}\lambda^2}{4\pi M^2}\Delta(m^2)\right) &\sim d\Big(\frac{\sqrt{s}(\Delta\tau)^2}{4\pi M}\frac{\Delta(m^2)}{M}\Big) \sim d\Big(\frac{\sqrt{s}(\Delta\tau)^2}{4\pi M}\frac{1}{\Delta\tau}\Big) = d\Big(\frac{\sqrt{s}(\Delta\tau)}{4\pi M}\Big) \\ &\sim d\Big(\Delta\tau\frac{E}{4\pi M}\Big) \sim \frac{1}{4\pi}\frac{dt}{d\tau}d\tau \sim dt_{\text{shift}}\,, \end{aligned} \tag{240}$$

where we understand dt_{shift} as the change in the relative time coordinate of the two particles during the scattering. Thus, the usual on-shell scattering cross-section corresponds to a specific dt_{int}, and we may compare the off-shell cross-sections with the usual results through the expression.

$$\frac{d\sigma^{(3)}}{d\Omega\, dt_{\text{int}}} = \frac{1}{64\pi^2\, s}\frac{|\vec{p}_f|}{|\vec{p}_i|}|\langle p_1 p_2|\tilde{\mathcal{T}}|p_{\mathrm{T}} p_{\mathrm{B}}\rangle|^2\,, \tag{241}$$

where $\tilde{\mathcal{T}} = \left(\frac{\lambda}{4M^2}\right)^2\mathcal{T}$.

7. Møller scattering

We consider the scattering of two identical scalar particles to first order. The diagrams which contribute are

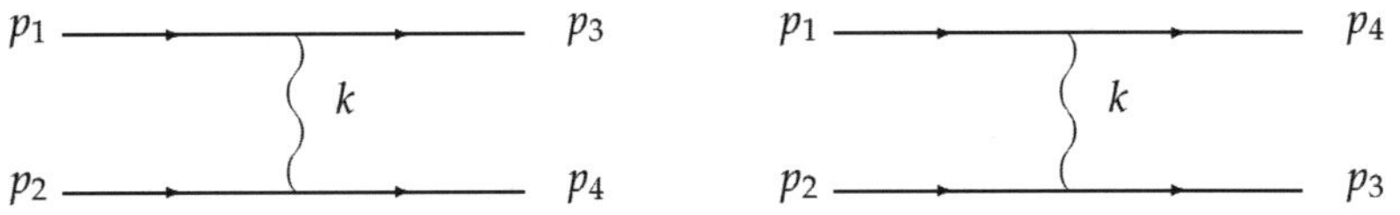

which contribute

$$\begin{aligned} &\langle 3\,4|iT|1\,2\rangle = \\ &= \int d^4q d\kappa_q (1)^4 \left[\frac{ie_0}{2M}(p_1+p_3)^\mu (2\pi)^5 \delta^4(p_1+q-p_3)\delta(\kappa_{p_1}-\kappa_{p_3}-\sigma\kappa_q)\right] \\ &\quad \left[\frac{ie_0}{2M}(p_2+p_4)^\nu (2\pi)^5 \delta^4(p_2-q-p_4)\delta(\kappa_{p_2}-\kappa_{p_4}+\sigma\kappa_q)\right] \\ &\quad \frac{1}{\lambda}\frac{-i}{q^2+\sigma\kappa_q^2-i\epsilon}\mathcal{P}_{\mu\nu} + (3,4)\leftrightarrow(4,3) \\ &= \left\{i(2\pi)^5\delta^4(p_1+p_2-p_3-p_4)\delta(\kappa_{p_1}+\kappa_{p_2}-\kappa_{p_3}-\kappa_{p_4})\right\}(2\pi)^5\frac{e_0 e}{(2M)^2} \\ &\quad \left\{(p_1+p_3)^\mu (p_2+p_4)^\nu \left[g_{\mu\nu} + \frac{(p_1-p_3)_\mu (p_2-p_4)_\nu}{(p_1-p_3)^2}\right]\right. \\ &\quad \left.\frac{1}{(p_1-p_3)^2+\sigma(\kappa_{p_1}-\kappa_{p_3})^2-i\epsilon} + (3,4)\leftrightarrow(4,3)\right\}, \end{aligned} \tag{242}$$

where $\kappa_p = \frac{p^2}{2M}$. From (174) which defines $\mathcal{T}$, we have

$$\begin{aligned} \langle 3\,4|\mathcal{T}|1\,2\rangle &= (2\pi)^5\frac{e_0 e}{(2M)^2}\left\{\left[(p_1+p_3)\cdot(p_2+p_4) + \frac{(p_1^2-p_3^2)(p_2^2-p_4^2)}{(p_1-p_3)^2}\right]\times\right. \\ &\quad \left.\times\frac{1}{(p_1-p_3)^2+\sigma(\kappa_{p_1}-\kappa_{p_3})^2-i\epsilon} + (3,4)\leftrightarrow(4,3)\right\}. \end{aligned} \tag{243}$$

At this stage it is convenient to introduce the Mandelstam parameters

$$t = -(p_1 - p_3)^2 = -\frac{1}{4}(p - p')^2 \qquad\qquad u = -(p_1 - p_4)^2 = -\frac{1}{4}(p + p')^2 \tag{244}$$

which complement the definition of s in (211), and where p, p' refer to the relative coordinates defined in (205) and (206). As in the on-shell case, we have

$$\begin{aligned} s + t + u &= -[p_1^2 + p_2^2 + p_1^2 + p_3^2 + p_1^2 + p_4^2 + 2(p_2 - p_3 - p_4) \cdot p_1] \\ &= -[p_1^2 + p_2^2 + p_3^2 + p_4^2] \\ &= m_1^2 + m_2^2 + m_3^2 + m_4^2. \end{aligned} \tag{245}$$

Similarly, since $p_1^2 + p_2^2 = p_3^2 + p_4^2$ is guaranteed by the 'fifth' δ-function, but the masses are not in general invariant, we have

$$\begin{aligned} (p_1 + p_2)^2 &= (p_3 + p_4)^2 \Rightarrow p_1 \cdot p_2 = p_3 \cdot p_4 = -\frac{1}{2}(s + p_1^2 + p_2^2) = -\frac{1}{2}(s + p_3^2 + p_4^2) \\ p_1 \cdot p_3 &= \frac{1}{2}(p_1^2 + p_3^2 + t) \neq p_2 \cdot p_4 = \frac{1}{2}(p_2^2 + p_4^2 + t) \\ p_1 \cdot p_4 &= \frac{1}{2}(p_1^2 + p_4^2 + u) \neq p_2 \cdot p_3 = \frac{1}{2}(p_2^2 + p_3^2 + u) \end{aligned} \tag{246}$$

Using these relations among the Mandelstam parameters, we may rewrite the terms of (243) in the form,

$$(p_1 + p_3) \cdot (p_2 + p_4) = -s + u \qquad\qquad (p_1 + p_4) \cdot (p_2 + p_3) = -s + t \tag{247}$$

so that

$$\langle 3\, 4 | \mathcal{T} | 1\, 2 \rangle = (2\pi)^5 \frac{e_0 e}{(2M)^2} \left\{ \frac{s - u - \frac{(p_1^2 - p_3^2)^2}{t}}{t - \sigma(\kappa_{p_1} - \kappa_{p_3})^2} + \frac{s - t - \frac{(p_1^2 - p_4^2)^2}{u}}{u - \sigma(\kappa_{p_1} - \kappa_{p_4})^2} \right\}. \tag{248}$$

We notice that for the case of on-shell scattering ($m_1^2 = m_3^2$) the amplitude becomes

$$\langle 3\, 4 | \mathcal{T} | 1\, 2 \rangle = (2\pi)^5 \frac{e_0 e}{(2M)^2} \left\{ \frac{s - u}{t} + \frac{s - t}{u} \right\} \tag{249}$$

which is the familiar result for identical Klein-Gordon particles in scalar QED [3].

We consider particles in their center of mass, so that $m_1 = m_2 = m$ and $E(p_1) = E(p_2)$. Then

$$p = p_1 - p_2 = (0, 2\vec{p}_1) \tag{250}$$

and $|p| = |p'| = \rho = 2|\vec{p}_1|$. In terms of the parameterization (215), the β of the incoming system is just zero. The relative momentum of the outgoing system is then

$$p' = p_3 - p_4 = 2|\vec{p}_1|(\sinh\beta, \cosh\beta\, \hat{n}) \tag{251}$$

and so the Mandelstam parameters become

$$t = -\frac{1}{4}(p - p')^2 = -\frac{1}{4}(2\rho^2 - 2\rho^2 \cosh\beta\, \cos\theta) = -2|\vec{p}_1|^2(1 - \cosh\beta\, \cos\theta) \tag{252}$$

and

$$u = -\frac{1}{4}(p + p')^2 = -2|\vec{p}_1|^2(1 + \cosh\beta\,\cos\theta). \tag{253}$$

These expressions agree with the usual on-shell expressions for t and u when $\beta = 0$. Since $p_1^2 = p_2^2$, we may write

$$p_1^2 + p_2^2 = 2p_1^2 = p_3^2 + p_4^2 \tag{254}$$

so that

$$p_1^2 - p_3^2 = \frac{1}{2}(p_3^2 + p_4^2) - p_3^2 = -\frac{1}{2}(p_3 - p_4)\cdot(p_3 + p_4) = -\frac{1}{2}p'\cdot P' = \frac{1}{4}\sqrt{s}\rho\sinh\beta \tag{255}$$

where we have used $P' = P$, (210) and (215). The amplitude in (248) contains the term

$$\frac{(p_1^2 - p_3^2)^2}{t} = -\frac{\frac{1}{16}s\rho^2\sinh^2\beta}{\frac{1}{2}\rho^2(1-\cosh\beta\cos\theta)} = -\frac{s\sinh^2\beta}{8(1-\cosh\beta\cos\theta)} \tag{256}$$

and the denominator

$$t - \sigma(\kappa_{p_1} - \kappa_{p_3})^2 = t - \frac{\sigma}{4M^2}(p_1^2 - p_3^2)^2 = -\frac{\rho}{2}\left[(1-\cosh\beta\cos\theta) + \frac{\sigma s}{32M^2}\sinh^2\beta\right]. \tag{257}$$

Therefore while on-shell scattering has a single forward direction pole at $\cos\theta = 1 \Rightarrow t = 0$, the off-shell scattering amplitude has two forward direction poles:

$$t = 0 \quad\Rightarrow\quad \cos\theta = \frac{1}{\cosh\beta} \tag{258}$$

$$t - \sigma(\kappa_{p_1} - \kappa_{p_3})^2 = 0 \quad\Rightarrow\quad \cos\theta = \frac{1}{\cosh\beta}\left[1 + \frac{\sigma s}{32M^2}\sinh^2\beta\right]. \tag{259}$$

The pole structure for a 1% mass shift is shown in Figure 1.

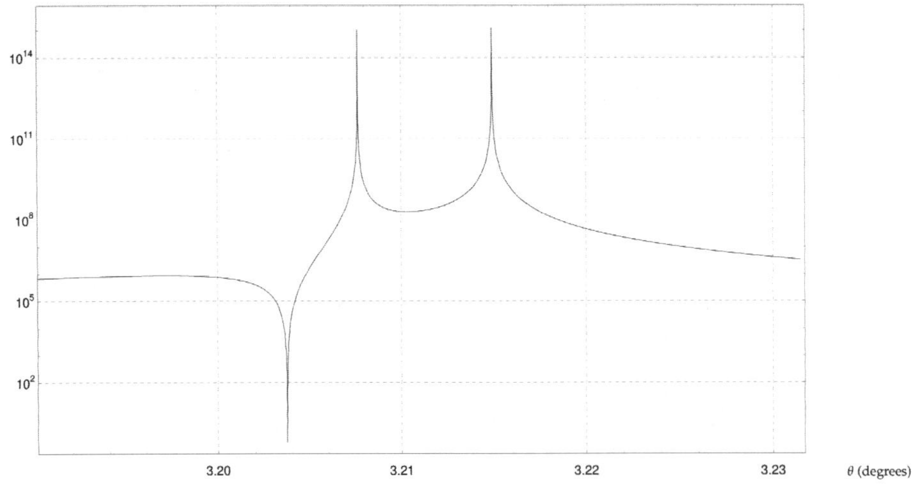

Figure 1 — Møller cross-section $\frac{d\sigma^{(3)}}{d\Omega\, dt_{\text{int}}}$ versus angle for 1% mass exchange

The appearance of two distinct poles in Møller scattering would be a consequence of $\Delta m^2 \neq 0$ and provide an experimental signature for off-shell phenomena.

8. Renormalization

Unlike conventional relativistic quantum field theories, the off-shell matter field corresponds to an underlying evolution mechanics, and the field undergoes retarded propagation from τ_1 to $\tau_2 > \tau_1$. Since a closed loop would require the field to propagate first from τ_1 to τ_2 and then from τ_2 to τ_1, there can be no matter field loops in off-shell QED. The absence of matter field loops leads us to expect that the charge e_0 will not be renormalized, and as we demonstrate below, this follows from the absence of wave function renormalization for the photon.

The Vector Ward Identity

The Ward identity in on-shell (Klein-Gordon) scalar quantum electrodynamics expresses the symmetry associated with the conservation of the four-current as a relationship between the vertex function of the 3-particle interaction and the single particle propagators. Since the Ward identity is preserved at all orders of perturbation theory, it leads to the universality of charge renormalization.

In off-shell quantum electrodynamics, the conservation of current is expressed as a vanishing *five-divergence* (12). Recalling equation (153)

$$\mathcal{L}_{\text{int}} = -\frac{ie_0}{2M} a_\mu (\psi^* \partial^\mu \psi - (\partial^\mu \psi^*)\psi) - \frac{e_0^2}{2M} a_\mu a^\mu |\psi|^2 \,, \tag{260}$$

we may express the interaction in terms of the current as

$$\mathcal{L}_{\text{int}} = e_0 a_\mu j^\mu - \frac{e_0^2}{2M} a_\mu a^\mu j^5 \tag{261}$$

where (j^μ, j^5) is the current for the free matter field. Since the entire interaction Lagrangian is in the form of *photon* $\times$ *current*, we shall see that the Ward identity is a relationship among the single particle propagators and the vertex functions of *both* the 3-particle interaction and the 4-particle interaction.

We begin with the three-point Green's function associated with the diagram

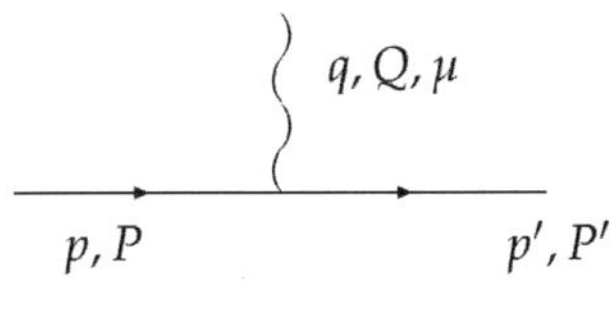

where $p' = q + p$ and $P' = P - \sigma Q$. To all orders in perturbation theory, the vertex function (the amputated Green's function) is given by $\Gamma_\mu^{(3)}(p, P; q, Q)$, where

$$G^{(3)\,\mu}(p, P; q, Q) = G^{(2)}(p, P)\; G^{(2)}(p', P')\; d^{\mu\nu}(q, Q)\; \Gamma_\mu^{(3)}(p, P; q, Q) \,. \tag{262}$$

To calculate the vertex function, we may write [22]

$$\begin{aligned} G_\mu^{(3)}(p, P; q, Q) &= \mathcal{F}\Big\{ \langle 0|\mathrm{T}\; a_\mu(x_1, \tau_1)\; \psi^*(x_2, \tau_2)\; \psi(x_3, \tau_3)|0\rangle \Big\} \\ &= \mathcal{F}\Big\{ \langle 0|\mathrm{T}\; a_\mu(x_1, \tau_1)\; \psi^*(x_2, \tau_2)\; \psi(x_3, \tau_3) e^{ie_0 \int d^4x_4 d\tau_4 a_\nu j^\nu} |0\rangle_{\text{free}} \Big\} \end{aligned}$$

$$= ie_0\mathcal{F}\left\{\int d^4x_4 d\tau_4 \langle 0|\mathrm{T}\, a_\mu(x_1,\tau_1)\, a_\nu(x_4,\tau_4)|0\rangle_{\text{free}} \right.$$
$$\left. \langle 0|\mathrm{T}\, \psi^*(x_2,\tau_2)\, \psi(x_3,\tau_3)\, j^\nu(x_4,\tau_4)|0\rangle_{\text{free}}\right\} \tag{263}$$

where $\mathcal{F}$ represents the Fourier transform. The vertex function becomes,

$$\Gamma^{(3)}_\mu(p,P;q,Q) = \frac{1}{G^{(2)}(p)G^{(2)}(p')}\int d^4x d\tau d^4x' d\tau' e^{-i[p\cdot x'-P\tau']}e^{-i[q\cdot x+\sigma Q\tau]}$$
$$\langle 0|\mathrm{T}\, ie_0 j_\mu(x,\tau)\, \psi(x',\tau')\, \psi^*(0)|0\rangle \tag{264}$$

where we have used translation invariance of the Green's functions to shift one of the field points to zero. Contracting with q^μ, we obtain

$$q^\mu\Gamma^{(3)}_\mu(p,P;q,Q) = e_0\frac{1}{G^{(2)}(p)G^{(2)}(p')}\int d^4x d\tau d^4x' d\tau' \left[(-\partial^\mu_x)\, e^{-i[p\cdot x'-P\tau']}e^{-i[q\cdot x+\sigma Q\tau]}\right]$$
$$\langle 0|\mathrm{T}\, j_\mu(x,\tau)\, \psi(x',\tau')\, \psi^*(0)|0\rangle$$
$$= e_0\frac{1}{G^{(2)}(p)G^{(2)}(p')}\int d^4x d\tau d^4x' d\tau' e^{-i[p\cdot x'-P\tau']}e^{-i[q\cdot x+\sigma Q\tau]}$$
$$\langle 0|\mathrm{T}\, \partial^\mu_x\, j_\mu(x,\tau)\, \psi(x',\tau')\, \psi^*(0)|0\rangle\ , \tag{265}$$

where we have performed one integration by parts. Using current conservation, we may make the replacement

$$\partial^\mu_x j_\mu(x,\tau) = -\partial_\tau j^5(x,\tau)\ . \tag{266}$$

Since the products are τ-ordered, we must carefully differentiate the implied θ-functions to find

$$\partial_\tau\langle 0|\mathrm{T}j^5(x,\tau)\, \psi(x',\tau')\, \psi^*(0)|0\rangle = \langle 0|\mathrm{T}\partial_\tau j^5(x,\tau)\, \psi(x',\tau')\, \psi^*(0)|0\rangle$$
$$+\delta(\tau-\tau')\langle 0|\mathrm{T}\, [j^5(x,\tau),\psi(x',\tau')]\, \psi^*(0)|0\rangle$$
$$+\delta(\tau)\langle 0|\mathrm{T}\, [j^5(x,\tau),\psi^*(0)]\, \psi(x',\tau')|0\rangle$$
$$= \langle 0|\mathrm{T}\partial_\tau j^5(x,\tau)\, \psi(x',\tau')\, \psi^*(0)|0\rangle$$
$$-\delta(\tau-\tau')\delta^4(x-x')\langle 0|\mathrm{T}\, \psi(x,\tau)\, \psi^*(0)|0\rangle$$
$$+\delta(\tau)\delta^4(x)\langle 0|\mathrm{T}\, \psi(x',\tau')\, \psi^*(0)|0\rangle \tag{267}$$

where we have used the commutation relations (52). Using (266) and (267) in (265), we find

$$q^\mu\Gamma^{(3)}_\mu(p,P;q,Q) = e_0\frac{1}{G^{(2)}(p)G^{(2)}(p')}\int d^4x d\tau d^4x' d\tau' \left[e^{-i[p\cdot x'-P\tau']}e^{-i[q\cdot x+\sigma Q\tau]}\right]$$
$$\left[-\partial_\tau\langle 0|\mathrm{T}j^5(x,\tau)\, \psi(x',\tau')\, \psi^*(0)|0\rangle\right.$$
$$-\delta(\tau-\tau')\delta^4(x-x')\langle 0|\mathrm{T}\, \psi(x,\tau)\, \psi^*(0)|0\rangle$$
$$\left.+\delta(\tau)\delta^4(x)\langle 0|\mathrm{T}\, \psi(x',\tau')\, \psi^*(0)|0\rangle\right]$$
$$= e_0\frac{1}{G^{(2)}(p)G^{(2)}(p')}\int d^4x d\tau d^4x' d\tau' \left[e^{-i[p\cdot x'-P\tau']}e^{-i[q\cdot x+\sigma Q\tau]}\right]$$
$$\left[-i\sigma Q\ \langle 0|\mathrm{T}j^5(x,\tau)\, \psi(x',\tau')\, \psi^*(0)|0\rangle\right.$$
$$\left.-\delta(\tau-\tau')\delta^4(x-x')G^{(2)}(x,\tau)+\delta(\tau)\delta^4(x)G^{(2)}(x',\tau')\right]$$
$$\tag{268}$$

where we have performed one integration by parts in the last line. Now, carrying out the integrations, we find that

$$\begin{aligned}\int d^4x d\tau d^4x' d\tau' & \left[e^{-i[p\cdot x' - P\tau']} e^{-i[q\cdot x + \sigma Q\tau]} \right] \delta(\tau - \tau')\delta^4(x - x') G^{(2)}(x,\tau) \\ &= \int d^4x d\tau \, e^{-i[(q+p)\cdot x - (P - \sigma Q)\tau]} G^{(2)}(x,\tau) \\ &= G^{(2)}(q + p, P - \sigma Q) \\ &= G^{(2)}(p', P') \end{aligned} \tag{269}$$

and

$$\int d^4x d\tau d^4x' d\tau' \left[e^{-i[p\cdot x' - P\tau']} e^{-i[q\cdot x + \sigma Q\tau]} \right] \delta(\tau)\delta^4(x) G^{(2)}(x',\tau') = G^{(2)}(p,P) . \tag{270}$$

We notice that the remaining Green's function is proportional to the vertex function for the 4-particle interaction,

$$\begin{aligned}\Gamma^{(4)}(p,P;q,Q) &= \frac{1}{G^{(2)}(p)G^{(2)}(p')} \int d^4x d\tau d^4x' d\tau' e^{-i[p\cdot x' - P\tau']} e^{-i[q\cdot x + \sigma Q\tau]} \\ & \langle 0|\mathrm{T}\Big(-i\frac{e_0^2}{2M} j^5(x,\tau)\Big) \psi(x',\tau') \, \psi^*(0)|0\rangle . \end{aligned} \tag{271}$$

Thus, we find

$$\begin{aligned} q^\mu \Gamma^{(3)}_\mu(p,P;q,Q) &= \frac{e_0}{G^{(2)}(p)G^{(2)}(p')} \int d^4x d\tau d^4x' d\tau' \left[e^{-i[p\cdot x' - P\tau']} e^{-i[q\cdot x + \sigma Q\tau]} \right] \\ & (-i\sigma Q) \, \langle 0|\mathrm{T} j^5(x,\tau) \, \psi(x',\tau') \, \psi^*(0)|0\rangle + \frac{e_0}{G^{(2)}(p',P')} - \frac{e_0}{G^{(2)}(p,P)} \\ &= e_0(-i\sigma Q)\left(-\frac{2M}{ie_0^2}\right)\Gamma^{(4)}(p,P;q,Q,k,k') + \frac{e_0}{G^{(2)}(p',P')} - \frac{e_0}{G^{(2)}(p,P)} \\ &= \frac{\sigma Q}{e_0/(2M)} \Gamma^{(4)}(p,P;q,Q,k,k') + \frac{e_0}{G^{(2)}(p',P')} - \frac{e_0}{G^{(2)}(p,P)} \end{aligned} \tag{272}$$

and the Ward identity takes the form,

$$e_0 \, q^\mu \Gamma^{(3)}_\mu(p,P;q,Q) - \sigma Q \, 2M \Gamma^{(4)}(p,P;q,Q,k,k') = e_0^2 \left[\frac{1}{G^{(2)}(p',P')} - \frac{1}{G^{(2)}(p,P)} \right] . \tag{273}$$

As expected, in the case that $Q = 0$, (273) reduces to the Ward identity for on-shell Klein-Gordon scalar QED [22]. At the tree level, where

$$\Gamma^{(4)}(p,P;q,Q,k,k') = \frac{-ie_0^2}{2M} \quad G^{(2)}(p,P) = \frac{-i}{\frac{p^2}{2M} - P} \quad \Gamma^{(3)}_\mu(p,P;q,Q) = i\frac{e_0}{2M}(p + p')_\mu \tag{274}$$

we verify that (273) is satisfied:

$$\begin{aligned} & e_0(p' - p)^\mu \cdot \frac{ie_0}{2M}(p' + p)_\mu - \sigma Q \, M \cdot \frac{-ie_0^2}{M} - e_0^2 \left[i\left(\frac{p'^2}{2M} - P'\right) - i\left(\frac{p^2}{2M} - P\right) \right] \\ &= i\frac{e_0^2}{2M} (p'^2 - p^2) + ie_0^2(P - P') - e_0^2 \left[i\left(\frac{p'^2}{2M} - P'\right) - i\left(\frac{p^2}{2M} - P\right) \right] = 0 \end{aligned} \tag{275}$$

Since the invariance of the Lagrangian is not changed by multiplying each invariant term by a constant, the most general gauge invariant form of the Lagrangian of (22), written in terms of renormalized quantities is given by

$$\begin{aligned}\mathcal{L} &= Z_2\,\psi^*\left(i\partial_\tau + (\frac{Z_1}{Z_2})\,e_0 a_5\right)\psi \\ &\quad - \frac{1}{2Z_4 M} Z_2\,\psi^*\left(-i\partial_\mu - (\frac{Z_1}{Z_2})\,e_0 a_\mu\right)\left(-i\partial^\mu - (\frac{Z_1}{Z_2})\,e_0 a^\mu\right)\psi \\ &\quad - \frac{\lambda}{4}\left[Z_3\, f_{\mu\nu} f^{\mu\nu} + 2Z_5\, f_{5\nu} f^{5\nu}\right] \end{aligned} \tag{276}$$

where we have written the bare field operators in terms of the renormalized field operators as

$$f_B^{\mu\nu} = Z_3 f^{\mu\nu} \qquad a_B^\mu = Z_3^{1/2} a^\mu \qquad f_B^{5\nu} = Z_5 f^{5\nu} \qquad a_B^5 = Z_5^{1/2} a^5 \tag{277}$$

$$\psi_B = Z_2^{1/2}\psi\,. \tag{278}$$

Any renormalization of the coupling λ may be absorbed into the wave function renormalizations Z_3 and Z_5. Consistency requires that

$$(\frac{Z_1}{Z_2})\,e_0 a_5 = (\frac{Z_1}{Z_2})\,e_0 Z_5^{1/2} a_5 \equiv e_0^B a_5^B \quad \text{and} \quad (\frac{Z_1}{Z_2})\,e_0 a_\mu = (\frac{Z_1}{Z_2})\,e_0 Z_3^{1/2} a_\mu \equiv e_0^B a_\mu^B \tag{279}$$

so we must have

$$Z_5 = Z_3 \qquad e_0 = \frac{Z_2 Z_3^{1/2}}{Z_1} e_0^B\,. \tag{280}$$

We may write the bare Green's functions in terms of the renormalized Green's functions as

$$\begin{aligned}\Gamma^{(3)}_{\mu\,B} &= \frac{1}{\langle 0|\mathrm{T}a_B a_B|0\rangle}\,\frac{1}{\langle 0|\mathrm{T}\psi_B\psi_B^*|0\rangle}\,\frac{1}{\langle 0|\mathrm{T}\psi_B\psi_B^*|0\rangle}\,\langle 0|\mathrm{T}a_B\psi_B\psi_B^*|0\rangle \\ &= \frac{1}{Z_3\langle 0|\mathrm{T}\,a\,a|0\rangle}\,\frac{1}{Z_2\langle 0|\mathrm{T}\,\psi\,\psi^*|0\rangle}\,\frac{1}{Z_2\langle 0|\mathrm{T}\,\psi\,\psi^*|0\rangle}\,Z_3^{1/2} Z_2\langle 0|\mathrm{T}\,a\psi\,\psi^*|0\rangle \\ &= \frac{1}{Z_3^{1/2} Z_2}\Gamma^{(3)}_\mu \end{aligned} \tag{281}$$

and similarly

$$\Gamma^{(4)}_B = \frac{1}{Z_3 Z_2}\Gamma^{(4)}\,. \tag{282}$$

Since the Ward identity must be valid for the renormalized quantities as well as the unrenormalized quantities, we may compare

$$\begin{aligned} e_0\, q^\mu \Gamma^{(3)}_\mu(p,P;q,Q) &- \sigma Q\, 2M\Gamma^{(4)}(p,P;q,Q,k,k') = \\ &\qquad e_0^2\left[\frac{1}{G^{(2)}(p',P')} - \frac{1}{G^{(2)}(p,P)}\right] \\ \frac{Z_1 Z_3^{1/2}}{Z_2} e_0^B\, q^\mu Z_3^{1/2} Z_2 \Gamma^{(3)}_{\mu\,B}(p,P;q,Q) &- \sigma Q\, 2Z_4 M_B Z_2 Z_3 \Gamma^{(4)}_B(p,P;q,Q,k,k') = \\ &\qquad (\frac{Z_1 Z_3^{1/2}}{Z_2} e_0^B)^2 Z_2\left[\frac{1}{G^{(2)}_B(p',P')} - \frac{1}{G^{(2)}(p,P)_B}\right] \end{aligned}$$

$$\begin{aligned} Z_1 Z_3 e_0^{\mathrm{B}}\, q^\mu \Gamma^{(3)}_{\mu\,\mathrm{B}}(p,P;q,Q) \quad - \quad & Z_2 Z_3 Z_4 \sigma Q\, 2M_{\mathrm{B}} \Gamma^{(4)}_{\mathrm{B}}(p,P;q,Q,k,k') = \\ & \frac{Z_1^2 Z_3}{Z_2} (e_0^{\mathrm{B}})^2 \left[\frac{1}{G^{(2)}_{\mathrm{B}}(p',P')} - \frac{1}{G^{(2)}(p,P)}_{\mathrm{B}} \right] \\ e_0^{\mathrm{B}}\, q^\mu \Gamma^{(3)}_{\mu\,\mathrm{B}}(p,P;q,Q) \quad - \quad & \frac{Z_2 Z_4}{Z_1} \sigma Q\, 2M_{\mathrm{B}} \Gamma^{(4)}_{\mathrm{B}}(p,P;q,Q,k,k') = \\ & \frac{Z_1}{Z_2} (e_0^{\mathrm{B}})^2 \left[\frac{1}{G^{(2)}_{\mathrm{B}}(p',P')} - \frac{1}{G^{(2)}(p,P)}_{\mathrm{B}} \right] \end{aligned} \tag{283}$$

and find that

$$Z_1 \equiv Z_2 \qquad\qquad Z_4 \equiv 1\,. \tag{284}$$

Notice that although the charge e_0 appears linearly and quadratically in the Ward identity, (283) makes no restriction on Z_3, which, from is seen to determine the charge renormalization by inserting $Z_1 = Z_2$ into (280). Thus, the appearance of $\Gamma^{(4)}$ in the Ward identity does not change the universality of charge renormalization found in on-shell QED since the conserved five-current is a consequence of gauge invariance. Nevertheless, since there are no matter field loops in off-shell QED, there are no possible contributions to photon renormalization, and we take $Z_3 \equiv 1$. Therefore, the charge e_0 is not renormalized.

Renormalizability

The remaining renormalization factor is Z_2, which derives from the renormalization of the matter field by photon loops. These photon loops (matter field self-energy diagrams) also contribute to the mass renormalization of the matter field, however, the mass term $(\psi^* i\partial_\tau \psi)$ absorbs these contributions. Nevertheless, by examining the primitive divergent self-energy diagrams of successively higher order, it may be seen that in order to make the theory counter-term renormalizable, a cut-off must be applied to the integrations over mass in loop diagrams. For example, at second order, the self-energy diagram with two overlapping photon loops is given by

$$\begin{aligned} G_2^{(2)}(p) = G_0^{(2)}(p) G_0^{(2)}(p) \int d^4q dQ d^4q' dQ' \left((2\pi)^5 \frac{ie_0}{2M} \right)^4 (2p-q)_\mu (2p-q-2q')_\nu d^{\mu\nu}(q) \\ (2p-2q-q')_\lambda (2p-q')_\sigma d^{\lambda\sigma}(q') G_0^{(2)}(p-q) G_0^{(2)}(p-q') G_0^{(2)}(p-q-q') \end{aligned} \tag{285}$$

which contains the following term proportional to p^4

$$\begin{aligned} & G_0^{(2)}(p)\, G_0^{(2)}(p) \left((2\pi)^5 \frac{ie_0}{2M} \right)^4 16 p^4 \int d^4q dQ d^4q' dQ' \\ & \quad \frac{1}{\lambda^2} \frac{1}{q^2 + \sigma Q^2} \frac{1}{q'^2 + \sigma Q'^2} \frac{2M}{(p-q)^2 - 2M(P+\sigma Q)} \\ & \quad \frac{2M}{(p-q')^2 - 2M(P+\sigma Q')} \frac{2M}{(p-q-q')^2 - 2M[P+\sigma(Q+Q')]} \end{aligned} \tag{286}$$

and diverges if the mass integrations on dQ and dQ' are taken to infinity. Since this term is proportional to p^4 it may not be renormalized by a counter term of a form which appears in the original Lagrangian. This divergence may be controlled by inserting a mass cut-off into the photon propagator as

$$\frac{1}{\lambda} \mathcal{P}^{\mu\nu} \frac{-i}{k^2 + \sigma\kappa^2 - i\epsilon} \quad \longrightarrow \quad \frac{1}{\lambda} \mathcal{P}^{\mu\nu} \frac{-i}{k^2 + \kappa^2 - i\epsilon} \frac{1}{1 + \lambda^2 \kappa^2} \tag{287}$$

which can be accomplished by adding the gauge-invariant scalar term

$$-\frac{\lambda^3}{4}\left[\partial_\tau f^{\alpha\beta}(x,\tau)\right]\left[\partial_\tau f_{\alpha\beta}(x,\tau)\right] \tag{288}$$

to kinetic term for the electromagnetic field. The action for the field is then

$$S_{em} = -\frac{\lambda}{4}\int d^4x\, d\tau\, ds\, f^{\alpha\beta}(x,\tau)\,\Phi(\tau - s) f_{\alpha\beta}(x,s) \tag{289}$$

where

$$\Phi(\tau) = \delta(\tau) - \lambda^2\delta''(\tau) = \frac{1}{2\pi}\int d\kappa\,\left[1 + (\lambda\kappa)^2\right] e^{-i\kappa\tau}\,. \tag{290}$$

This cut-off is also required at the classical level [23] in order to smooth the event current as

$$\partial_\beta f^{\alpha\beta}(x,\tau) = e j^\alpha_\varphi(x,\tau) = e\int ds\, \varphi(\tau - s)\, j^\alpha(x,s) \tag{291}$$

where

$$\int_{-\infty}^{\infty} ds\, \Phi(\tau - s)\varphi(s) = \delta(\tau) \rightarrow \varphi(\tau) = \int \frac{d\kappa}{2\pi}\frac{e^{-i\kappa\tau}}{1 + (\lambda\kappa)^2} = \frac{1}{2\lambda}e^{-|\tau|/\lambda}\,. \tag{292}$$

The set of divergent diagrams in the resulting theory is just the subset of the divergent diagrams in the on-shell theory which contain no matter field loops. Therefore, with the mass integrations made finite, the off-shell theory is seen to be renormalizable by the same arguments given by Rohrlich [24] for the on-shell theory. Unlike a momentum cut-off, the mass cut-off does not affect the invariances of the original theory.

References

[1] Stueckelberg E C G 1941 *Helv. Phys. Acta* **14** 322
Stueckelberg E C G 1941 *Helv. Phys. Acta* **14** 588
[2] Fock V A Phys 1937 *Z. Sowjetunion* **12** 404
[3] Itzykson C and Zuber J-B 1980 *Quantum Field Theory* (New York: McGraw-Hill)
[4] Saad D, Horwitz L P and Arshansky R I 1989 *Found. of Phys.* **19** 1126
[5] Land M C 1996 *Found. of Phys.* **27** 19
[6] Arshansky R L, Horwitz P and Lavie Y 1983 *Found. of Phys.* **13** 1167
[7] Land M C 2013 *J. Phys.: Conf. Ser.*
[8] R. Jackiw, *(Constrained) Quantization Without Tears* MIT CTP#2215 *Preprint* arXiv:hep-th/9306075
[9] Land M C and Horwitz L P 1998 Land M C **A239** 135
[10] Shnerb N and Horwitz L P 1993 *Phys. Rev.* **A 48** 4068
[11] Dirac P 1950 *Canad. J. Math.* **2** 129
Dirac P 1959 *Phys Rev* **114** 924
Dirac P 1964 *Lectures on Quantum Mechanics* (New York: Yeshiva University)
[12] Sakurai J J 1980 *Advanced Quantum Mechanics* (Reading: Addison-Wesley)
[13] Schwinger J 1951 *Phys. Rev.* **82** 664
[14] DeWitt B S 1965 *Dynamical Theory of Groups and and Fields* (New York: Gordon and Breach)
[15] Feynman R P 1950 *Phys. Rev.* **80** 440
Feynman R 1948 *Rev P. Mod. Phys.* **20** 367
[16] Nambu Y 1950 *Prog. Theor. Phys.* **5** 82
[17] Shnerb N and Horwitz L P 1994 *Jour. Phys. A: Math. and Gen.* **27** 3565
[18] Horwitz L P and Lavie Y 1982 *Phys. Rev.* D **26** 819
[19] Ryder L H 1986 *Quantum Field Theory* (Cambridge: Cambridge University Press)
[20] Goldberger M L and Watson K M 1964 *Collision Theory* (New York: Wiley)
[21] Novozhilov Y V 1975 *Introduction to Elementary Particle Theory* (Oxford: Pergamon Press)
[22] Cheng T-P and Li L-F 1989 *Gauge Theory of Elementary Particle Physics* (Oxford: Clarendon Press)
[23] Land M C 2003 *Found. of Phys.* **33** 1157
[24] Rohrlich F 1950 *Phys. Rev.* **80** 666

Electrostatics in Stueckelberg-Horwitz electrodynamics

Martin Land
Department of Computer Science, Hadassah College, 37 HaNeviim Street, Jerusalem
E-mail: martin@hadassah.ac.il

Abstract. In this paper, we study fundamental aspects of electrostatics as a special case in Stueckelberg-Horwitz electromagnetic theory. In this theory, spacetime events $x^\mu(\tau)$ evolve in an unconstrained 8-dimensional phase space, interacting through five τ-dependent gauge fields induced by the current densities associated with their evolutions. The chronological time τ was introduced as an independent evolution parameter in order to free the laboratory clock x^0 to evolve alternately 'forward' and 'backward' in time according to the sign of the energy, thus providing a classical implementation of the Feynman-Stueckelberg interpretation of pair creation/annihilation. The resulting theory differs in its underlying mechanics from conventional electromagnetism, but coincides with Maxwell theory in an equilibrium limit.

After a brief review of Stueckelberg-Horwitz electrodynamics, we obtain the field produced by an event in uniform motion and verify that it satisfies the field equations. We study this field in the rest frame of the event, where it depends explicitly on coordinate time x^0 and the parameter τ, as well as spatial distance R. Calculating with this generalized Coulomb field, we demonstrate how Gauss's theorem and Stoke's theorem apply in 4D spacetime, and obtain the fields associated with a charged line and a charged sheet. Finally, we use the field of the charged sheet to study a static event in the vicinity of a potential barrier. In all of these cases, we observe a small transfer of mass from the field to the particle. It is seen that for an event in the field of an oppositely charged sheet of sufficient density, the event can reverse time direction, providing a specific model for pair phenomena.

1. Introduction

As described by Born and Wolf, Maxwell's equations were not immediately accepted as a general theory of classical electrodynamics [1]. Maxwell's 1864 formulation summarized all prior research in electricity and magnetism, including Cavendish's 1771 - 1773 experiments in electrostatics, Coulomb's characterization of electric and magnetic forces in the 1780s and Faraday's investigations of time-varying fields in the 1830s. Nevertheless, this mathematically concise description encountered resistance until its prediction of electromagnetic waves traveling at the speed of light was verified by Hertz in 1888. Ultimately, the successful incorporation of optics into electrodynamics through Maxwell's equations inspired Einstein's 1906 study of the spacetime symmetries underlying the theory and Fock's 1929 association of potential theory with gauge symmetry in the quantum mechanics of charged particles [2]. Since that time, these symmetry considerations have taken the lead role in extending classical and quantum field theory. Thus, the Standard Model of strong and electroweak interactions, which was historically pieced together from often *ad hoc* models of particle behavior, is often presented in contemporary pedagogical treatments as a general implementation of *a priori* principles of relativistic and gauge invariance.

An important stage in the development of the Standard Model was the observation by Stueckelberg [3] that the classical Maxwell theory did not implement all Lorentz symmetries apparent in relativistic quantum mechanics, and in particular could not provide a classical account of pair creation/annihilation processes. Stueckelberg proposed a reformulation of the classical Lorentz force in which a particle worldline is traced out dynamically by an evolving spacetime event, and pair processes may be described by a single event evolving forward and backward though time. Since the coordinate time is not single-valued in this framework, Stueckelberg introduced a monotonically increasing Poincaré invariant evolution parameter τ, which plays a role similar to Newtonian time in nonrelativistic mechanics. In subsequent work on QED, Stueckelberg's approach was adopted by Feynman [4] and Schwinger [5], who employed an invariant time but de-emphasized its status. The parameterized canonical formalism was extended by Horwitz and Piron [6] to the relativistic classical and quantum mechanics of many particles with interactions. Consistency of this approach requires that the gauge symmetry include the parameter [7, 8], leading to a τ-dependent electrodynamics derived from five gauge fields. Although the resulting theory coincides with Maxwell theory in its equilibrium limit, it differs in its underlying mechanics from conventional electromagnetism, in particular that the total mass of particles and fields is conserved, but not the masses of individual interacting particles [9]. By overcoming the mass-shell constraint, timelike events may evolve continuously through the spacelike region on their way to reversed-time timelike evolution, and the theory thus provides a solution to a fundamental difficulty facing Stueckelberg's classical description of pair processes. Because the formalism contains a 4-vector potential and a fifth scalar potential, it provides a framework for relativistic generalizations of the classical central force problems [10, 11, 12, 13, 14]. Possible experimental signatures have been found in classical and quantum scattering [15, 16]. Nevertheless, relatively little work has been done in the areas of classical electrostatics and electrodynamics that first led to the Maxwell theory.

In this paper, we study aspects of laboratory electrostatic phenomenology as a limiting case of Stueckelberg-Horwitz electrodynamics. This study differs from Maxwell electrostatics in two essential ways: first, reversing the direction of 18th and 19th century research, we present a theory constructed as the implementation of symmetry principles and seek consequences that may lead to experimental observation. Second, although we borrow the notion of electrostatics as the study of events in their rest frames, only spatial coordinates remain fixed. The advance (or retreat) of time is an explicit form of motion for the events described by this theory, and these events may undergo dynamical forces in this dimension that affect their measurable characteristics but not their position. We begin by reviewing the essential features of Stueckelberg-Horwitz electrodynamics in section 2. In section 3 we obtain the field produced by an event in uniform motion and verify that it satisfies the field equations. In section 4 we specify the field of a uniformly moving event to the observations in the event's rest frame, which expresses the generalization of the Coulomb field. This field is explicitly dependent on coordinate time x^0 and the parameter τ, as well as spatial distance R. Applying this Coulomb field, we show how Gauss's theorem and Stoke's theorem apply in 4D spacetime, and also obtain the fields associated with a charged line and a charged sheet. In section 5 we treat the field of the charged sheet as an external force acting on an event in order to formulate the Lorentz force equations for an event in the vicinity of a potential barrier. In particular, we study the behavior of a 'static' event (held fixed in space) in this field and demonstrate that its time acceleration produces a small transfer of mass from the field to the particle. It is seen that for an event in the field of an oppositely charged sheet of sufficient density, the event can reverse time direction, providing a specific model for pair phenomena.

2. Stueckelberg off-shell electrodynamics

In seeking a classical description of pair creation/annihilation as a single worldline generated dynamically by the evolution of an event $x^\mu(\tau)$, Stueckelberg proposed [3] a generalized Lorentz force of the form

$$M\left(\ddot{x}^\mu + \Gamma^\mu_{\nu\rho}\dot{x}^\nu\dot{x}^\rho\right) = eF^{\mu\nu}g_{\nu\rho}\dot{x}^\rho + K^\mu \qquad \ddot{x}^\mu = \frac{d^2x^\mu}{d\tau^2} \qquad \dot{x}^\mu = \frac{dx^\mu}{d\tau} \tag{1}$$

with metric $g_{\mu\nu}(x)$ and connection $\Gamma^\mu_{\nu\rho}$. The electromagnetic field includes the familiar tensor part $F^{\mu\nu}(x)$, as well as a vector part $K^\mu(x)$ whose role is to overcome the mass-shell constraint

$$K^\mu = 0 \Rightarrow \begin{cases} g_{\nu\rho}\dot{x}^\nu\dot{x}^\rho = \text{constant} \\ \text{Timelike event remains timelike} \end{cases}$$

that prevents the event from smoothly entering the spacelike region and thus precludes pair processes. But finding no physical justification for introducing the vector field $K^\mu \neq 0$, Stueckelberg turned to τ-parameterized covariant canonical quantum mechanics in flat space with $g_{\mu\nu} \to \eta_{\mu\nu} = \text{diag}(-1,1,1,1)$

$$i\partial_\tau\psi(x,\tau) = \frac{1}{2m}\left[p^\mu - eA^\mu(x)\right]\left[p_\mu - eA_\mu(x)\right]\psi(x,\tau) \tag{2}$$

which permits an event to tunnel from the timelike to spacelike region. This quantum theory enjoys the standard U(1) gauge invariance under local transformations of the type

$$\psi(x,\tau) \longrightarrow \exp\left[ie\Lambda(x)\right]\psi(x,\tau) \tag{3}$$
$$A_\mu \longrightarrow A_\mu + \partial_\mu\Lambda(x). \tag{4}$$

The global gauge invariance associated with this gauge symmetry is the conserved current

$$\partial_\mu j^\mu + \partial_\tau\rho = 0 \tag{5}$$

where

$$\rho = \left|\psi(x,\tau)\right|^2 \qquad j^\mu = -\frac{i}{2M}\left\{\psi^*(\partial^\mu - ieA^\mu)\psi - \psi(\partial^\mu + ieA^\mu)\psi^*\right\}. \tag{6}$$

Stueckelberg [3] regarded (6) as a true current, leading to the interpretation of $\left|\psi(x,\tau)\right|^2$ as the probability density at τ of finding the event at the spacetime point x. However, under this interpretation, the non-zero divergence of the four-vector current $j^\mu(x,\tau)$ prevents its identification as the source of the field $A^\mu(x)$. As a remedy, Stueckelberg observed that assuming $\rho \to 0$ pointwise as $\tau \to \pm\infty$, integration of (6) over τ leads to

$$\partial_\mu J^\mu = 0 \qquad \text{where} \qquad J^\mu(x) = \int_{-\infty}^{\infty} d\tau\, j^\mu(x,\tau)\,. \tag{7}$$

However, in the resulting dynamical picture, the fields $A^\mu(x)$ that mediate particle interaction instantaneously at τ are induced by currents $J^\mu(x)$ whose support covers the particle worldlines, past and future. There is no *a priori* assurance that the particles moving in these Maxwell fields will trace out precisely the worldlines that induce the fields responsible for their motion.

In order to obtain a well-posed theory, Sa'ad, Horwitz, and Arshansky [7] generalized (2) by introducing a τ-dependent gauge field and a fifth gauge compensation field. Writing $x^5 = \tau$ and adopting the index convention

$$\alpha, \beta, \gamma = 0,1,2,3,5 \qquad \lambda, \mu, \nu = 0,1,2,3 \qquad i,j,k = 1,2,3 \tag{8}$$

the Stueckelberg-Schrodinger equation

$$\left[i\partial_\tau + e_0 a_5\left(x,\tau\right)\right]\psi\left(x,\tau\right) = \frac{1}{2M}\left[p^\mu - e_0 a^\mu\left(x,\tau\right)\right]\left[p_\mu - e_0 a_\mu\left(x,\tau\right)\right]\psi\left(x,\tau\right) \tag{9}$$

becomes locally gauge invariant under τ-dependent gauge transformations of the type

$$\psi \to e^{ie_0\Lambda(x,\tau)}\psi \qquad a_\mu \to a_\mu + \partial_\mu\Lambda\left(x,\tau\right) \qquad a_5 \to a_5 + \partial_\tau\Lambda\left(x,\tau\right). \tag{10}$$

Writing the classical Lagrangian [9] as

$$L = \dot{x}^\mu p_\mu - K = \tfrac{1}{2}M\dot{x}^\mu\dot{x}_\mu + e_0\dot{x}^\alpha a_\alpha \tag{11}$$

the Lorentz force

$$\frac{d}{d\tau}\frac{\partial L}{\partial \dot{x}_\mu} - \frac{\partial L}{\partial x_\mu} = 0 \quad \longrightarrow \quad \frac{d}{d\tau}\left[M\dot{x}^\mu + e_0 a^\mu(x,\tau)\right] = e_0\dot{x}^\alpha\partial^\mu a_\alpha(x,\tau) \tag{12}$$

takes the form

$$M\ddot{x}^\mu = e_0 f^\mu{}_\alpha(x,\tau)\dot{x}^\alpha = e_0\left[f^\mu{}_\nu(x,\tau)\dot{x}^\nu + f^\mu{}_5(x,\tau)\right] \tag{13}$$

where

$$f^\mu{}_\alpha = \partial^\mu a_\alpha - \partial_\alpha a^\mu \qquad \dot{x}^5 = \dot{\tau} = 1 \tag{14}$$

and Stueckelberg's vector field — now τ-dependent — may be identified as the term

$$K^\mu = f^\mu{}_5(x,\tau) = \partial^\mu a_5(x,\tau) - \partial_\tau a^\mu(x,\tau). \tag{15}$$

In this formalism, the mass shell constraint is overcome by the exchange of mass between particles and fields seen in

$$\frac{d}{d\tau}\left(-\frac{1}{2}M\dot{x}^2\right) = -M\dot{x}^\mu\ddot{x}_\mu = -e_0\,\dot{x}^\mu(f_{\mu5} + f_{\mu\nu}\dot{x}^\nu) = -e_0\,\dot{x}^\mu f_{\mu5} \tag{16}$$

but total mass-energy-momentum of particle and fields is conserved [9, 17]. This relaxation of the mass-shell constraint breaks general reparameterization invariance, but under the boundary conditions

$$\tau \to \pm\infty \qquad \Rightarrow \qquad a^5\left(x,\tau\right) \to 0 \qquad j^5\left(x,\tau\right) \to 0 \tag{17}$$

the remaining τ-translation symmetry is associated, via Noether's theorem, with dynamic conservation of mass. Since the gauge fields possess a non-zero mass spectrum, this theory has been called off-shell electrodynamics.

The electromagnetic field $f_{\alpha\beta}(x,\tau)$ is made a dynamical quantity by adding a kinetic term to the action [7]. In analogy to standard Maxwell theory, one may adopt the formal designation $f^{\mu 5} = \eta^{55} f^{\mu}{}_5 = -f^{\mu}{}_5$ and choose the form

$$-\frac{\lambda}{4} f^{\alpha\beta}(x,\tau)\, f_{\alpha\beta}(x,\tau) \tag{18}$$

leading to the classical action

$$S = \int d\tau\, \frac{1}{2} M \dot{x}^{\mu} \dot{x}_{\mu} + \int d^4 z d\tau \left\{ e_0 a_{\alpha}(z,\tau) j^{\alpha}(z,\tau) - \frac{\lambda}{4} f_{\alpha\beta}(z,\tau) f^{\alpha\beta}(z,\tau) \right\} \tag{19}$$

with locally conserved five-component event current

$$j^{\alpha}(z,\tau) = \dot{x}^{\alpha}(\tau) \delta^4 (z - x(\tau)) . \tag{20}$$

The homogeneous field equation

$$\epsilon^{\alpha\beta\gamma\delta\epsilon} \partial_{\alpha} f_{\beta\gamma} = 0 \tag{21}$$

follows automatically from the definition (14) of the fields $f^{\alpha\beta}$. The inhomogeneous field equations are obtained as

$$\partial_{\beta} f^{\alpha\beta}(x,\tau) = \frac{e_0}{\lambda} j^{\alpha}(x,\tau) = e j^{\alpha}(x,\tau) = e \dot{z}^{\alpha}(\tau)\, \delta^4 [x - z(\tau)] \tag{22}$$

by variation of the action. Under the boundary conditions (17) the standard Maxwell theory is extracted as the equilibrium limit of (21) and (22) by integration over the worldline

$$\left.\begin{array}{c} \partial_{\beta} f^{\alpha\beta}(x,\tau) = e j^{\alpha}(x,\tau) \\ \\ \partial_{[\alpha} f_{\beta\gamma]} = 0 \end{array}\right\} \quad \xrightarrow[\int d\tau]{} \quad \left\{\begin{array}{c} \partial_{\nu} F^{\mu\nu}(x) = e J^{\mu}(x) \\ \\ \partial_{[\mu} F_{\nu\rho]} = 0 \end{array}\right. \tag{23}$$

where

$$F^{\mu\nu}(x) = \int_{-\infty}^{\infty} d\tau\, f^{\mu\nu}(x,\tau) \qquad A^{\mu}(x) = \int_{-\infty}^{\infty} d\tau\, a^{\mu}(x,\tau) . \tag{24}$$

This integration has been called concatenation [18] and as in (7) links the event current $j^{\mu}(x,\tau)$ with the particle current $J^{\mu}(x)$ defined on the entire particle worldline, which by (20) recovers the standard covariant expression. It is seen from (9) and (24) that e_0 and λ must have dimensions of time, so that the dimensionless ratio $e = e_0/\lambda$ can be identified as the Maxwell charge. The microscopic τ-dependent fields have been called pre-Maxwell fields.

The wave equation derived from (22) is

$$\partial_{\alpha}\partial^{\alpha} a^{\beta}(x,\tau) = \left(\partial_{\mu}\partial^{\mu} - \partial_{\tau}^2\right) a^{\beta}(x,\tau) = -e j^{\beta}(x,\tau) \tag{25}$$

and the principal part Green's function [19] found from

$$\left(\partial_{\mu}\partial^{\mu} - \partial_{\tau}^2\right) G(x,\tau) = -\delta^5(x,\tau) \tag{26}$$

is

$$G(x,\tau) = -\frac{1}{2\pi}\delta(x^2)\delta(\tau) - \frac{1}{2\pi^2}\frac{\partial}{\partial x^2}\frac{\theta(x^2-\tau^2)}{\sqrt{x^2-\tau^2}} = D(x)\,\delta(\tau) - G_{correlation}(x,\tau) . \tag{27}$$

The first term has support on the lightcone at instantaneous τ, and recovers the standard Maxwell Green's function under concatenation. The second term has spacelike support ($x^2 > \tau^2 \geq 0$) and vanishes under concatenation, so it may contribute to correlations but not to Maxwell potentials. Terms of this type have been studied in [20]. The conserved current (5), electromagnetic action (18) and wave equation (25) suggest an underlying five-dimensional symmetry — O(3,2) or O(4,1) depending on the sign of η^{55} — however the equal-τ part of (27) breaks this symmetry to a vector-plus-scalar representation of O(3,1). Stueckelberg-Horwitz electrodynamics may be called a 5D theory of five gauge fields, as a shorthand for the 4+1 implementation of O(3,1) Lorentz symmetry. The equal-τ part is seen to satisfy

$$\left(\partial_\mu \partial^\mu - \partial_\tau^2\right) D\left(x\right) \delta(\tau) = \delta^4\left(x\right) \delta(\tau) - D\left(x\right) \delta''(\tau) \tag{28}$$

and may be treated as providing solutions the wave equation by neglecting terms associated with $\delta''(\tau)$. An integrated study of the relation between the two pieces of the Green's function is forthcoming.

Because a static event — one whose position is space remains unchanged — evolves at a timelike velocity along the t-axis in its rest frame, pre-Maxwell theory does not contain any precise equivalent of the motionless charge that produces the Coulomb force in Maxwell electrostatics. Low energy Coulomb scattering is studied by solving the Lorentz force (13) in the field of a spatially static event

$$x\left(\tau\right) = \left(t, \mathbf{x}\right) = \left(\tau, \mathbf{0}\right) \quad \longrightarrow \quad \begin{cases} j^0\left(x,\tau\right) = j^5\left(x,\tau\right) = \delta(t-\tau)\,\delta^3(x) \\ \mathbf{j}\left(x,\tau\right) = 0 \end{cases} \tag{29}$$

where boldface signifies 3D spatial quantities. Applying (27) to this current leads to

$$a^0\left(x,\tau\right) = a^5\left(x,\tau\right) = \frac{e}{4\pi} \frac{\delta\left(x^0 - |\mathbf{x}| - \tau\right)}{|\mathbf{x}|} \tag{30}$$

which recovers the correct Coulomb potential under concatenation

$$A^0(x) = \int d\tau\, a^0\left(x,\tau\right) = \frac{e}{4\pi R} \tag{31}$$

but incorrectly describes the microscopic dynamics

$$M\ddot{\mathbf{x}} = -e_0 \nabla \left[a^0\left(x,\tau\right) + a^5\left(x,\tau\right)\right] = -\frac{e_0}{4\pi} \nabla \frac{\delta(x^0 - |\mathbf{x}| - \tau)}{|\mathbf{x}|}. \tag{32}$$

Since the τ-translation invariance of the theory leaves the Coulomb interaction invariant under shift of the worldline origin $x\left(\tau\right) = \left(\tau, \mathbf{0}\right) \to \left(\tau + \delta\tau, \mathbf{0}\right)$, it was suggested in [15] to relax the synchronization of interacting events by taking the induced current to be

$$j^\alpha\left(x,\tau\right) \longrightarrow j^\alpha_\varphi\left(x,\tau\right) = \int ds\, \varphi\left(\tau - s\right) j^\alpha\left(x,s\right) \qquad \varphi(\tau) = \frac{1}{2\lambda} e^{-|\tau|/\lambda} \tag{33}$$

leaving the concatenated current unchanged

$$J^\mu\left(x\right) = \int d\tau\, j^\mu_\varphi\left(x,\tau\right) = \int ds\, d\tau\, \varphi\left(\tau - s\right) j^\mu\left(x,s\right) = \int ds\, j^\mu\left(x,s\right). \tag{34}$$

This modification leads to a Yukawa potential with reasonable low energy limit

$$M\ddot{\mathbf{x}} = -e_0\nabla\left[a^0(x,\tau) + a^5(x,\tau)\right] = -2e_0\nabla a^0(x,\tau) = e^2\nabla\left[\frac{e^{-|\mathbf{x}|/\lambda}}{4\pi|\mathbf{x}|}\right]. \tag{35}$$

The source $j^\alpha_\varphi(x,\tau)$ for the pre-Maxwell field is interpreted as a smoothed current density induced by an ensemble of events $x^\alpha(\tau+\delta\tau)$ along a particle worldline, where $\delta\tau$ is given by the normalized distribution $\varphi(\tau)$. The distribution $\varphi(\tau)$ provides a cutoff for the photon mass spectrum, which we take to be the conventional experimental uncertainty in photon mass ($\Delta m_\gamma \simeq 10^{-17}$ eV [21]), leading to a value of about 400 seconds for λ. The limit $\lambda \to 0$ restores $\varphi(\tau) \to \delta(\tau)$ and the limit $\lambda \to \infty$ restores standard Maxwell theory. Since the form of $\varphi(\tau)$ given in (33) represents the distribution of interarrival times of events in a Poisson-distributed stochastic process, this choice suggests an information-theoretic interpretation for the underlying the relationship between the current density and the ensemble of events from which it is induced.

The smoothed current can be introduced through the action [22], by adding a higher τ-derivative term to the electromagnetic part. The substitution

$$S_{em} \to \int d^4x d\tau \left[e_0 j^\alpha a_\alpha - \frac{\lambda}{4} f^{\alpha\beta}(x,\tau) f_{\alpha\beta}(x,\tau) - \frac{\lambda^3}{4}\left[\partial_\tau f^{\alpha\beta}(x,\tau)\right]\left[\partial_\tau f_{\alpha\beta}(x,\tau)\right]\right] \tag{36}$$

preserves Lorentz and gauge invariance, and leaves the action first order in spacetime derivatives. Defining a field interaction kernel

$$\Phi(\tau) = \delta(\tau) - \lambda^2\delta''(\tau) = \frac{1}{2\pi}\int d\kappa \left[1+(\lambda\kappa)^2\right] e^{-i\kappa\tau} \tag{37}$$

which is seen from

$$\int_{-\infty}^{\infty} ds\, \Phi(\tau-s)\varphi(s) = \delta(\tau) \to \varphi(\tau) = \int \frac{d\kappa}{2\pi}\frac{e^{-i\kappa\tau}}{1+(\lambda\kappa)^2} = \frac{1}{2\lambda}e^{-|\tau|/\lambda} \tag{38}$$

to be the inverse function to $\varphi(\tau)$, the action becomes

$$S_{em} = \int d^4x d\tau\, e_0 j^\alpha a_\alpha - \frac{\lambda}{4}\int d^4x\, d\tau\, ds\, f^{\alpha\beta}(x,\tau)\,\Phi(\tau-s) f_{\alpha\beta}(x,s). \tag{39}$$

The Euler-Lagrange equations

$$\partial_\beta f^{\alpha\beta}_\Phi(x,\tau) = \partial_\beta \int ds\, \Phi(\tau-s) f^{\alpha\beta}(x,s) = e j^\alpha(x,\tau) \tag{40}$$

can be inverted to recover

$$\partial_\beta f^{\alpha\beta}(x,\tau) = e j^\alpha_\varphi(x,\tau) = e\int ds\, \varphi(\tau-s)\, j^\alpha(x,s) \tag{41}$$

using (38). The action (39), in which the statistical synchronization performed by $\Phi(\tau-s)$ is made explicit, has the advantage of permitting the usual study of symmetries and being amenable to second quantization, where the factor $\left[1+(\lambda\kappa)^2\right]^{-1}$ provides a natural mass cutoff for the off-shell photon that renders off-shell quantum field theory super-renormalizable at two-loop order [22].

3. Induced fields

3.1. Liénard-Wiechert potential

The potential and field strength induced by the motion of an arbitrary event were obtained in [17] in studying the self-interaction problem in the context of Stueckelberg-Horwitz electrodynamics. Writing the smoothed current for an arbitrary event $r^\alpha(\tau)$ as

$$j^\alpha_\varphi(x,\tau) = c\int ds\, \varphi(\tau - s)\, \dot{r}^\alpha(s)\, \delta^4[x - r(s)] \tag{42}$$

the Liénard-Wiechert potential found from the Maxwell part of the Green's function (27) is

$$\begin{aligned} a^\alpha(x,\tau) &= \frac{e}{2\pi c}\int d^4x' d\tau' \delta\left((x-x')^2\right)\theta^{ret}\delta(\tau-\tau')\, j^\alpha_\varphi(x',\tau') \\ &= \frac{e}{2\pi}\int ds\, \varphi(\tau-s)\, \dot{r}^\alpha(s)\, \delta\left((x-r(s))^2\right)\theta^{ret} \end{aligned} \tag{43}$$

where θ^{ret} imposes retarded x^0 causality and we write the speed of light c explicitly. Using the identity

$$\int d\tau f(\tau)\, \delta[g(\tau)] = \frac{f(\tau_R)}{|g'(\tau_R)|} \quad , \quad \tau_R = g^{-1}(0) \tag{44}$$

we obtain

$$a^\beta(x,\tau) = \frac{e}{4\pi}\varphi(\tau-\tau_R)\frac{\dot{r}^\beta(\tau_R)}{(x^\mu - r^\mu(\tau_R))\,\dot{r}_\mu(\tau_R)} \tag{45}$$

where the retarded time τ_R satisfies $[x - r(\tau_R)]^2 = 0$ and $\theta^{ret} = \theta\left(x^0 - r^0(\tau_R)\right) = 1$.

It is convenient to express the field quantities as elements of a Clifford algebra with basis vectors

$$e_\alpha \cdot e_\beta = \eta_{\alpha\beta} = \text{diag}(-1,1,1,1,-1) \qquad e_\alpha \wedge e_\beta = e_\alpha \otimes e_\beta - e_\beta \otimes e_\alpha \tag{46}$$

and Clifford product

$$e_\alpha e_\beta = e_\alpha \cdot e_\beta + e_\alpha \wedge e_\beta \tag{47}$$

where we refer to the index convention (8). Separating spacetime and scalar quantities as

$$r(\tau) = r^\mu(\tau) e_\mu \qquad r^5 = c\tau \tag{48}$$

$$d = \partial_\mu e^\mu \qquad \partial_5 = \frac{1}{c}\partial_\tau \tag{49}$$

the field strength tensors

$$f = \frac{1}{2} f^{\mu\nu}\, e_\mu \wedge e_\nu \qquad f^5 = f^{5\mu}\, e_5 \wedge e_\mu = e_5 \wedge \epsilon. \tag{50}$$

are expressed as Clifford bivectors and (14) takes the form

$$f = d \wedge a \qquad \epsilon = -\partial_5 a - da^5. \tag{51}$$

The covariant equivalent of a spatially static charge is a uniformly evolving event

$$r(\tau) = u\tau = \left(u^0\tau, \mathbf{u}\tau\right) \tag{52}$$

with constant timelike velocity $\dot{r} = u$, which in its rest frame simply advances along the time axis as $t = (u^0/c)\tau$. The induced potential is found from (45) to be

$$a(x,\tau) = \frac{e}{4\pi}\frac{u\varphi(\tau - \tau_R)}{|u\cdot z|} \qquad a^5(x,\tau) = \frac{e}{4\pi}\frac{c\varphi(\tau - \tau_R)}{|u\cdot z|} \tag{53}$$

along the line of observation

$$z = x - r(\tau_R) = x - u\tau_R \qquad z^2 = 0. \tag{54}$$

Writing the timelike velocity u in terms of the unit vector $\hat{u}$

$$u^2 < 0 \qquad u = |u|\,\hat{u} \qquad \hat{u}^2 = -1 \qquad u^2 = -\,|u|^2 \tag{55}$$

the observation line z can be separated into components

$$z_{\|} = -\hat{u}\,(\hat{u}\cdot z) \qquad z_{\perp} = z + \hat{u}\,(\hat{u}\cdot z) \tag{56}$$

which satisfy

$$z_{\|}^2 = \hat{u}^2\,(\hat{u}\cdot z)^2 = -\,(\hat{u}\cdot z)^2 \tag{57}$$

$$z_{\perp}^2 = z^2 + 2\,(\hat{u}\cdot z)^2 - (\hat{u}\cdot z)^2 = (\hat{u}\cdot z)^2 = -z_{\|}^2 \tag{58}$$

$$(u\cdot z)^2 = |u|^2\,(\hat{u}\cdot z)^2 = -\,|u|^2\,z_{\|}^2. \tag{59}$$

The condition of retarded causality

$$z^2 = \tau_R^2 u^2 - 2\tau_R u\cdot x + x^2 = 0 \tag{60}$$

relates the field to the location of the event along the backward lightcone of the observation point. This implicit choice of τ_R and its gradient

$$0 = d(z^2) = 2\left(\tau_R d\tau_R u^2 - \tau_R u - d\tau_R u\cdot x + x\right) = 2\left[-\,(u\cdot z)\,d\tau_R + z\right] \tag{61}$$

define the essential kinematics of the system as embodied in the following expressions:

$$d\tau_R = \frac{z}{u\cdot z} \qquad (u\cdot d)\,\tau_R = \frac{u\cdot z}{u\cdot z} = 1 \qquad (z\cdot d)\,\tau_R = \frac{z^2}{u\cdot z} = 0 \tag{62}$$

$$d\,(u\cdot z) = d\left(u\cdot x - u^2\tau_R\right) = \frac{(u\cdot z)\,u - u^2 z}{u\cdot z} = |u|^2\,\frac{z_{\perp}}{u\cdot z} \tag{63}$$

$$d\frac{1}{(u\cdot z)^n} = \frac{-n\left[(u\cdot z)\,u - u^2 z\right]}{(u\cdot z)^{n+2}} = \frac{-n\,|u^2|\,z_{\perp}}{(u\cdot z)^{n+2}} \tag{64}$$

$$d\cdot z = d\cdot(x - u\tau_R) = d\cdot x - u\cdot d\tau_R = 3 \tag{65}$$

$$d\wedge z = d\wedge(x - u\tau_R) = d\tau_R\wedge u = \frac{u\wedge z}{u\cdot z} \tag{66}$$

$$d\wedge\hat{z} = d\wedge\frac{z}{|z|} = -\frac{1}{|z|}\frac{z\wedge u}{u\cdot z} - \hat{z}\wedge\frac{z}{|z|^2} = \frac{u\wedge\hat{z}}{u\cdot z} \tag{67}$$

We will refer to equations (57) – (59) and (62) – (67) collectively as the kinematic equations.

3.2. Field strengths

To find the field strengths we must calculate field derivatives of the type $\partial^\alpha a^\beta(x,\tau)$. As in the Maxwell case, the spacetime derivative is most conveniently found by returning to (43) with $\dot{r} = u$ to write

$$\partial^\mu a^\beta(x,\tau) = \frac{e}{2\pi}\int d\tau_R\; \varphi(\tau-\tau_R)\; u^\beta\, \partial^\mu \delta\left[(x-r(\tau_R))^2\right] \tag{68}$$

and combining the expressions

$$\partial^\mu \delta\left[(x-r(\tau_R))^2\right] = 2\delta'\left[(x-r(\tau_R))^2\right](x^\mu - r^\mu(\tau_R)) \tag{69}$$

$$\frac{d}{d\tau_R}\delta\left[(x-r(\tau_R))^2\right] = -2\delta'\left[(x-r(\tau_R))^2\right]\; u\cdot(x-r(\tau_R)) \tag{70}$$

to obtain

$$\partial^\mu a^\beta(x,\tau) = -\frac{e}{2\pi}\int d\tau_R\; \varphi(\tau-\tau_R) u^\beta \frac{x^\mu - r^\mu(\tau_R)}{u\cdot(x-r(\tau_R))}\frac{d}{d\tau_R}\delta\left[(x-r(\tau_R))^2\right]. \tag{71}$$

Integrating by parts

$$\partial^\mu a^\beta(x,\tau) = \frac{e}{2\pi}\int d\tau_R \left[\frac{d}{d\tau_R}\varphi(\tau-\tau_R) u^\beta \frac{x^\mu - r^\mu(\tau_R)}{u\cdot(x-r(\tau_R))}\right]\delta\left[(x-r(\tau_R))^2\right] \tag{72}$$

and again using identity (44) we find

$$f = d\wedge a(x,\tau) = \frac{e}{4\pi}\frac{1}{|u\cdot z|}\frac{d}{d\tau_R}\left[\varphi(\tau-\tau_R)\frac{z\wedge u}{u\cdot z}\right] \tag{73}$$

for the spacetime components and

$$da^5(x,\tau) = \frac{e}{4\pi}\frac{1}{|u\cdot z|}\frac{d}{d\tau_R}\left[\varphi(\tau-\tau_R)\frac{z u^5}{u\cdot z}\right] \tag{74}$$

for the fifth gauge field. Since (45) shows that $a(x,\tau)$ depends on τ only through $\varphi(\tau-\tau_R)$, we may write

$$\partial^5 a = -\frac{1}{c}\partial_\tau a(x,\tau) = -\frac{e}{4\pi}\varphi'(\tau-\tau_R)\frac{u}{|u\cdot z|} \qquad\qquad \varphi'(\tau) = \frac{d}{d\tau}\varphi(\tau) \tag{75}$$

for the τ-derivative term required in f^5. Using (75) and applying

$$\frac{d}{d\tau_R}u\wedge z = u\wedge\frac{d}{d\tau_R}(x-u\tau_R) = -u\wedge u = 0 \tag{76}$$

$$\frac{d}{d\tau_R}u\cdot z = -u\cdot(x-u\tau_R) = -u^2 \tag{77}$$

$$\frac{d}{d\tau_R}\frac{z}{u\cdot z} = \frac{-u(u\cdot z)+zu^2}{(u\cdot z)^2} = -u^2\frac{z_\perp}{(u\cdot z)^2} \tag{78}$$

to (73) and (74) we are finally led to

$$f = -\frac{e}{4\pi}\left[\varphi(\tau-\tau_R)\frac{(z\wedge u)\,u^2}{(u\cdot z)^3} - \varphi'(\tau-\tau_R)\frac{z\wedge u}{(u\cdot z)^2}\right] \tag{79}$$

and

$$f^5 = e_5 \wedge \frac{ec}{4\pi} \left[\varphi(\tau - \tau_R) \frac{u^2 z_\perp}{(u \cdot z)^3} - \varphi'(\tau - \tau_R) \frac{z - u (u \cdot z)/c^2}{(u \cdot z)^2} \right] = e_5 \wedge \epsilon\,. \tag{80}$$

Notice that integration of (79) over τ involves

$$\int d\tau\ \varphi(\tau) = 1 \qquad \int d\tau\ \varphi'(\tau) = \varphi(\infty) - \varphi(-\infty) = 0 \tag{81}$$

so that taking $u^2 = -c^2$ under concatenation recovers

$$F(x) = \frac{e}{4\pi} \frac{(z \wedge u)\, c^2}{(u \cdot z)^3} \tag{82}$$

which is the standard Maxwell field for a uniformly moving particle.

3.3. *Pre-Maxwell equations*

Separating the four-vector and fifth scalar components, the pre-Maxwell equations (21) and (22) can be written in 4D component form as

$$\begin{array}{ll} \partial_\nu\, f^{\mu\nu} - \partial_\tau\, f^{5\mu} = e j^\mu_\varphi & \partial_\mu\, f^{5\mu} = e\rho_\varphi \\ \partial_\mu f_{\nu\rho} + \partial_\nu f_{\rho\mu} + \partial_\rho f_{\mu\nu} = 0 & \partial_\nu f_{5\mu} - \partial_\mu f_{5\nu} + \partial_\tau f_{\mu\nu} = 0 \end{array} \tag{83}$$

which exposes the analogy with 3-vector Maxwell equations

$$\begin{array}{ll} \nabla \times \mathbf{B} - \partial_0 \mathbf{E} = e\mathbf{J} & \nabla \cdot \mathbf{E} = eJ^0 \\ \nabla \cdot \mathbf{B} = 0 & \nabla \times \mathbf{E} + \partial_0 \mathbf{B} = 0 \end{array} \tag{84}$$

by observing that the field $f^{5\mu} = \epsilon^\mu$ plays a role analogous to the Maxwell electric field $F^{0i} = E^i$ while $f^{\mu\nu}$ plays the role of the magnetic field $F^{ij} = \varepsilon^{ijk} B_k$. In the index-free notation, with c written explicitly, the pre-Maxwell equations take the form

$$\begin{array}{lcl} \partial_\mu\, f^{5\mu} = e\rho & \longrightarrow & d \cdot \epsilon = \frac{e}{c} j^5_\varphi = e\rho_\varphi \\ \partial_\nu\, f^{\mu\nu} - \partial_\tau\, f^{5\mu} = e j^\mu & \longrightarrow & -d \cdot f - \frac{1}{c} \partial_\tau \epsilon = \frac{e}{c} j_\varphi \\ \partial_\mu f_{\nu\rho} + \partial_\nu f_{\rho\mu} + \partial_\rho f_{\mu\nu} = 0 & \longrightarrow & d \wedge f = 0 \\ \partial_\nu f_{5\mu} - \partial_\mu f_{5\nu} + \partial_\tau f_{\mu\nu} = 0 & \longrightarrow & d \wedge \epsilon + \frac{1}{c} \partial_\tau f = 0 \end{array} \tag{85}$$

which must apply to the solution for the uniformly moving event.

To verify the Gauss law we calculate the 4-divergence

$$d \cdot \epsilon = d \cdot \frac{ec}{4\pi} \left[\varphi(\tau - \tau_R) \frac{u^2 z_\perp}{(u \cdot z)^3} - \varphi'(\tau - \tau_R) \frac{z - u (u \cdot z)/c^2}{(u \cdot z)^2} \right] \tag{86}$$

using

$$d\varphi\left(\tau-\tau_R\right)=-\varphi'\left(\tau-\tau_R\right)d\tau_R\,. \tag{87}$$

From the kinematic equations we find

$$\begin{aligned} d\cdot\left(\varphi\frac{u^2 z_\perp}{(u\cdot z)^3}\right) &= -\varphi'\frac{u^2 z_\perp^2}{(u\cdot z)^4}+\varphi\frac{u^2 d\cdot z_\perp}{(u\cdot z)^3}+\varphi\frac{-3u^2\left|u^2\right|z_\perp^2}{(u\cdot z)^5} \\ &= -\varphi'\frac{u^2 z_\perp^2}{(u\cdot z)^4}+3\varphi\left(\frac{u^2\left(u\cdot z\right)^2}{(u\cdot z)^5}-\frac{u^2\left|u^2\right|z_\perp^2}{(u\cdot z)^5}\right) \\ &= -\varphi'\frac{u^2 z_\perp^2}{(u\cdot z)^4} \end{aligned} \tag{88}$$

for the first term, and

$$d\cdot\left(\varphi'\frac{z-u\left(u\cdot z\right)/c^2}{(u\cdot z)^2}\right)=-\varphi'\frac{|u|^2 z_\parallel^2}{(u\cdot z)^4}-\frac{1}{c^2}\varphi''\frac{1}{u\cdot z} \tag{89}$$

for the second term. Since $u^2 z_\perp^2=|u^2|z_\parallel^2$ we are led to

$$d\cdot\epsilon=-\varphi''\frac{1}{c^2\left(u\cdot z\right)}\sim 0 \tag{90}$$

where the second derivative φ'' is associated with the term $\delta''(\tau)$ in (27), which we take to be equivalent to zero. Thus, Gauss's law is verified in the source-free region. We return to the integral form of Gauss's law in section 4 to establish equality including the enclosed source.

The time derivative of ϵ is easily found to be

$$\frac{1}{c}\frac{\partial}{\partial\tau}\epsilon=\frac{e}{4\pi}\left[\varphi'\left(\tau-\tau_R\right)\frac{u^2 z_\perp}{(u\cdot z)^3}-\varphi''\left(\tau-\tau_R\right)\frac{z-u\left(u\cdot z\right)/c^2}{(u\cdot z)^2}\right]. \tag{91}$$

The divergence of the field f requires

$$d\cdot f=-\frac{e}{4\pi}d\cdot\left[\varphi\frac{\left(z\wedge u\right)u^2}{(u\cdot z)^3}-\varphi'\frac{\left(z\wedge u\right)\left(u\cdot z\right)}{(u\cdot z)^3}\right] \tag{92}$$

from which the first term leads to

$$\begin{aligned} d\cdot\left[\varphi\frac{\left(z\wedge u\right)u^2}{(u\cdot z)^3}\right] &= d\varphi\cdot\frac{\left(z\wedge u\right)u^2}{(u\cdot z)^3}+\varphi\frac{d\cdot\left(z\wedge u\right)u^2}{(u\cdot z)^3}+\varphi u^2\left[d\frac{1}{(u\cdot z)^3}\right]\cdot\left(z\wedge u\right) \\ &= \varphi'\frac{u^2 z}{(u\cdot z)^3} \end{aligned} \tag{93}$$

and the second term is

$$\begin{aligned} d\cdot\left[\varphi'\frac{\left(z\wedge u\right)\left(u\cdot z\right)}{(u\cdot z)^3}\right] &= d\varphi'\cdot\frac{\left(z\wedge u\right)\left(u\cdot z\right)}{(u\cdot z)^3}+\varphi' d\cdot\frac{\left(u\cdot z\right)\left(z\wedge u\right)}{(u\cdot z)^3} \\ &= -\frac{e}{4\pi}\left[\varphi'\left(\tau-s\right)\frac{u^2 z_\perp}{(u\cdot z)^3}-\varphi''\left(\tau-s\right)\frac{\left(u\cdot z\right)z}{(u\cdot z)^3}\right]. \end{aligned} \tag{94}$$

Combining (93) and (94) we find

$$d \cdot f = -\frac{e}{4\pi}\left[\varphi'(\tau-\tau_r)\frac{u^2 z_\perp}{(u\cdot z)^3} - \varphi''(\tau-\tau_r)\frac{z-u(u\cdot z)/c^2}{(u\cdot z)^2}\right] \tag{95}$$

which when compared with (91) verifies Ampère's law in the source-free region.

The exterior derivative of f

$$d\wedge f = -\frac{e}{4\pi} d\wedge\left[\varphi(\tau-s)\frac{(z\wedge u)u^2}{(u\cdot z)^3} - \varphi'(\tau-s)\frac{z\wedge u}{(u\cdot z)^2}\right] \tag{96}$$

produces three types of term:

$$d\varphi\wedge(z\wedge u) = -\varphi'\frac{z}{u\cdot z}\wedge(z\wedge u) = 0 \tag{97}$$

$$d\wedge(z\wedge u) = (d\wedge z)\wedge u = 0 \tag{98}$$

$$\left[d\frac{1}{(u\cdot z)^n}\right]\wedge(z\wedge u) = \frac{-n\left|u^2\right| z_\perp}{(u\cdot z)^{n+2}}\wedge(z_\perp\wedge u) = 0 \tag{99}$$

and thus we recover the absence of electromagnetic monopoles in the form

$$d\wedge f = 0. \tag{100}$$

In the pre-Maxwell theory, the 4-divergence of the field $\varepsilon^\mu(x,\tau)$ locates the event density $\rho(x,\tau)$ as its source. The Maxwell field F^{ij} is induced by motion $J^i(x)$ of the charge density $J^0(x)$ and has no monopole source — in the pre-Maxwell theory the field $f^{\mu\nu}$ is induced by the motion j^μ of the event density $\rho(x,\tau)$ and has no monopole source.

The time derivative of the f is easily found to be

$$\frac{1}{c}\frac{\partial}{\partial\tau}f = -\frac{e}{4\pi}\left[\varphi'\frac{(z\wedge u)u^2}{(u\cdot z)^3} - \varphi''\frac{z\wedge u}{(u\cdot z)^2}\right]. \tag{101}$$

Writing the exterior derivative of ϵ in the form

$$d\wedge\epsilon = \frac{ec}{4\pi} d\wedge\left[\varphi\frac{|u|^2 z_\perp}{(u\cdot z)^3} - \varphi'\frac{z}{(u\cdot z)^2} + \varphi'\frac{u}{c^2(u\cdot z)}\right] \tag{102}$$

three terms contribute

$$d\wedge\varphi\frac{|u|^2 z_\perp}{(u\cdot z)^3} = d\varphi\wedge\frac{|u|^2 z_\perp}{(u\cdot z)^3} + \varphi\, d\wedge\frac{|u|^2 z_\perp}{(u\cdot z)^3} = -\varphi'\frac{z\wedge u}{(u\cdot z)^3} \tag{103}$$

$$d\wedge\varphi'\frac{z}{(u\cdot z)^2} = d\varphi'\wedge\frac{z}{(u\cdot z)^2} + \varphi' d\wedge\frac{z}{(u\cdot z)^2} = \varphi'\frac{z\wedge u}{(u\cdot z)^3} \tag{104}$$

$$d\wedge\varphi'\frac{u}{c^2(u\cdot z)} = d\varphi'\wedge\frac{u}{c^2(u\cdot z)} + \varphi' d\frac{1}{(u\cdot z)}\wedge\frac{u}{c^2} = \varphi'\frac{u^2 z\wedge u}{c^2(u\cdot z)^3} - \varphi''\frac{z\wedge u}{c^2(u\cdot z)^2} \tag{105}$$

leading to

$$d \wedge \epsilon = \frac{e}{4\pi c}\left[\varphi' \frac{(z \wedge u)\, u^2}{(u \cdot z)^3} - \varphi'' \frac{z \wedge u}{(u \cdot z)^2}\right]. \tag{106}$$

Comparing (102) and (106) we recover

$$d \wedge \epsilon + \frac{1}{c}\frac{\partial}{\partial \tau} f = 0 \tag{107}$$

which we recognize as Faraday's Law. This confirms that the fields (79) and (80) are an essentially kinematic solution of the pre-Maxwell equations.

4. Electrostatics

In order to compare the phenomenology of 3D electrostatics with the standard Maxwell theory we describe the source event in its rest frame, so that the event evolves uniformly along its time axis. We take the event velocity to be

$$u = (c, 0) = c e_0 \qquad u^2 = -c^2 \tag{108}$$

and write the point of observation as

$$x = (ct, \mathbf{x}) = (ct, R\hat{\mathbf{x}}) \tag{109}$$

so that the line of observation satisfies

$$z^2 = (x - c e_0 \tau_R)^2 = 0 \quad \rightarrow \quad \tau_R = t - \frac{R}{c} \quad \rightarrow \quad z(\tau_R) = R(e_0 + \hat{\mathbf{x}}) \tag{110}$$

$$z_\perp = R\hat{\mathbf{x}} \qquad u \cdot z = -cR \qquad u \wedge z = cR\,(e_0 \wedge \hat{\mathbf{x}})\,. \tag{111}$$

In this frame (53) becomes

$$a^0(x,\tau) = \frac{e}{4\pi}\frac{\varphi(\tau - t + R/c)}{R} \qquad \mathbf{a}(x,\tau) = 0 \qquad a^5(x,\tau) = a^0(x,\tau) \tag{112}$$

so that

$$f = d \wedge a = e_0 \wedge \left(-\nabla a^0\right) \qquad \epsilon = \partial^5 a - d a^5 = -\left(e_0 \partial_5 a^0 + d a^5\right) \tag{113}$$

and we see that in addition to the 4-gradient of the scalar potential a^5, the x^0 evolution contributes motion terms a^0 to both the magnetic-type f field strength and the electric-type field ϵ. This situation differs structurally from Maxwell electrostatics

$$\left.\begin{array}{l} A^0(x) = \dfrac{e}{4\pi R} \\[2ex] \mathbf{A}(x) = 0 \end{array}\right\} \longrightarrow \left\{\begin{array}{l} \mathbf{E}(x) = -\nabla A^0 \\[2ex] \mathbf{B}(x) = 0 \end{array}\right. \tag{114}$$

in which the field is pure electric and derives entirely from the gradient of the "scalar" potential A^0. Separating the gradient into space and time components $d = (\partial_0, \nabla)$ we find from (113)

$$\boldsymbol{\epsilon} = f^{5i} e_i = -\nabla a^5 \qquad \epsilon^0 = f^{50} = \partial^5 a^0 - \partial^0 a^5 = -\frac{2}{c}\,\partial_\tau a^5 \tag{115}$$

so that the space part of ϵ has zero curl.

4.1. Coulomb law

From the potentials (112) or directly from (79) and (80) we are led to

$$\epsilon^0 = -\frac{e}{2\pi}\frac{\varphi'(\tau - t + R/c)}{cR} \qquad \boldsymbol{\epsilon} = \frac{e}{4\pi}\left[\frac{\varphi(\tau - t + R/c)}{R^2} - \frac{\varphi'(\tau - t + R/c)}{cR}\right]\hat{\mathbf{x}} \tag{116}$$

and

$$f = \frac{e}{4\pi}\left[\frac{\varphi(\tau - t + R/c)}{R^2} - \frac{\varphi'(\tau - t + R/c)}{cR}\right] e_0 \wedge \hat{\mathbf{x}} \tag{117}$$

so that in terms of the spatial components $e_i = f^{0i}$ and $b_i = \epsilon_{ijk} f^{jk}$ we find $\mathbf{e} = \boldsymbol{\epsilon}$ and $\mathbf{b} = 0$. Using (81) we see that the concatentated electric field is

$$\mathbf{E} = \frac{e}{4\pi R^2}\hat{x} \tag{118}$$

recovering the standard Coulomb law. From (13) we may write the Lorentz force experienced by a test event at rest, that is $(\dot{x}^0, \dot{\mathbf{x}}) = (c, \mathbf{0})$, as

$$M\ddot{t} = \frac{e_0}{c^2}\,\mathbf{e}\cdot\dot{\mathbf{x}} + \frac{e_0}{c}\,\epsilon^0 = \varepsilon(\tau - t + R/c)\frac{e^2}{4\pi\lambda Rc}e^{-|\tau - t + R/c|/\lambda} \tag{119}$$

$$M\ddot{\mathbf{x}} = e_0\,\mathbf{e}\,\dot{t} + \frac{e_0}{c}\,\dot{\mathbf{x}}\times\mathbf{b} + e_0\,\boldsymbol{\epsilon} = \frac{e^2}{4\pi R^2}e^{-|\tau - t + R/c|/\lambda}\hat{\mathbf{x}} \tag{120}$$

where we used

$$\varphi'(\tau) = \frac{1}{2\lambda}\frac{d}{d\tau}e^{-|\tau|/\lambda} = -\varepsilon(\tau)\frac{1}{2\lambda^2}e^{-|\tau|/\lambda} \qquad \varepsilon(\tau) = \text{signum}(\tau). \tag{121}$$

Placing the test charge at $t = \tau + R/c + \alpha$ with $|\alpha| \ll \lambda$, that is, along the forward lightcone of the event producing the field, the force becomes

$$M\ddot{t} = \varepsilon(\alpha)\frac{e^2}{4\pi\lambda Rc} \qquad M\ddot{\mathbf{x}} = \frac{e^2}{4\pi R^2}\hat{\mathbf{x}} \tag{122}$$

and we see that the temporal acceleration for this configuration depends on the sign of α. In the absence of some mechanism to fix $\dot{x}^0$, a test event slightly behind the lightcone with $\alpha < 0$ will be further decelerated on the time axis until the temporal separation weakens the force. Similarly, a test event with $\alpha > 0$ will be further accelerated on the time axis. For a test event at rest in space, these possible changes in $\dot{x}^0$ — the energy of the test event — imply a change of mass, which will be further studied in section 5.

Applying the x^0-derivative and 3-gradient to (116) we find

$$\frac{1}{c}\frac{\partial}{\partial t}\epsilon^0 = \frac{e}{2\pi c^2 R}\varphi''(\tau - t + R/c) \tag{123}$$

$$\nabla\cdot\boldsymbol{\epsilon} = \frac{e}{4\pi}\left[4\pi\varphi(\tau - t + R/c)\,\delta^3(\mathbf{x}) - \frac{1}{c^2 R}\cdot\varphi''(\tau - t + R/c)\right] \tag{124}$$

which combine to confirm Gauss's law in differential form

$$\frac{1}{c}\frac{\partial}{\partial t}\epsilon^0 + \nabla\cdot\boldsymbol{\epsilon} = e\varphi(\tau - t + R/c)\,\delta^3(\mathbf{x}) + \frac{e}{4\pi c^2 R}\varphi''(\tau - t + R/c) = e\rho_\varphi(x) \tag{125}$$

Journal of Physics: Conference Series **437** (2013) 012012 doi:10.1088/1742-6596/437/1/012012

where as in (90) we take $\varphi'' \sim 0$. To obtain the $\delta^3(x)$ term, we used

$$\nabla \cdot \frac{\hat{\mathbf{x}}}{R^2} = 4\pi\delta^3(\mathbf{x}) \tag{126}$$

which is proven by applying Gauss's theorem to the 3D volume integral of the right hand side. In 4D the integral form of Gauss's law (125) can be found by performing spacetime integral

$$\int \partial_\mu \epsilon^\mu \, d^4x = e \int \rho_\varphi(x) \, d^4x = e \tag{127}$$

over the volume formed as the product of the time axis and a 3-sphere in space. To demonstrate the method, we consider the standard Coulomb field for a point charge and directly calculate

$$\int \nabla \cdot \mathbf{E} \, d^3x = \int \nabla \cdot \frac{\hat{\mathbf{x}}}{4\pi R^2} \, d^3x = e \int \delta^3(\mathbf{x}) \, d^3x = e \tag{128}$$

to obtain the total charge. Instead of the usual application of Gauss's theorem in spherical coordinates, we express the field in cylindrical coordinates

$$\mathbf{E}(\rho, \phi, z) = \frac{e\,(\rho\hat{\mathbf{n}} + z\hat{\mathbf{z}})}{4\pi\,(\rho^2 + z^2)^{3/2}} \qquad \hat{\mathbf{n}} = (\cos\phi, \sin\phi) \qquad \rho = \sqrt{x^2 + y^2} \tag{129}$$

and enclose the charge in a long thin cylinder. Neglecting the flux through the vanishingly small ends of the cylinder, we take the surface element to be

$$d\mathbf{S} = \hat{\mathbf{n}} dS = \hat{\mathbf{n}} \rho d\phi dz \tag{130}$$

so that the surface integral is easily evaluated as

$$\int \mathbf{E} \cdot d\mathbf{S} = \int_0^{2\pi} d\phi \int_{-\infty}^{\infty} dz \, \rho \, \frac{e\,(\rho\hat{\mathbf{n}} + z\hat{\mathbf{z}})}{4\pi\,(\rho^2 + z^2)^{3/2}} \cdot \hat{\mathbf{n}} = \frac{e\rho^2}{2} \left. \frac{z}{\rho^2\,(\rho^2 + z^2)^{1/2}} \right|_{-\infty}^{\infty} = e. \tag{131}$$

For the pre-Maxwell field, (127) becomes

$$e = \int \partial_\mu \epsilon^\mu \, d^4x = \int \nabla \cdot \boldsymbol{\epsilon} \, d^4x \tag{132}$$

because $\varphi'(\pm\infty) = 0$ assures

$$\int dx^0 \, \partial_0 \epsilon^0 = \frac{e}{2\pi} \frac{1}{cR} \int dt \, \varphi''\,(\tau - t + R/c) = 0. \tag{133}$$

Again neglecting the flux through the ends and taking the 3D boundary element to be

$$d\mathbf{S} = \hat{\mathbf{x}} R^2 \, d\Omega \, dt \tag{134}$$

so that using $\varphi(-x) = \varphi(x)$ and $\varphi(\pm\infty) = 0$, we obtain the surface integral

$$\int \boldsymbol{\epsilon} \cdot d\mathbf{S} = \int \frac{e}{4\pi} \left[\frac{\varphi\,(\tau - t + R/c)}{R^2} - \frac{\varphi'\,(\tau - t + R/c)}{cR} \right] \hat{\mathbf{x}} \cdot \hat{\mathbf{x}} R^2 \, d\Omega \, dt = e \tag{135}$$

verifying Gauss's law in integral form for the space-static solution. In the pre-Maxwell theory, the Gauss law expresses the equality of the total charged event density contained in a region of 4D spacetime with the total flux passing through a 3D boundary surrounding that region.

As in Maxwell theory, the field ϵ is characterized by its divergence and exterior derivative. In 3+1 components,

$$d \wedge \epsilon = -e_0 \wedge \left(\partial_0 \epsilon + \nabla \epsilon^0\right) + \nabla \wedge \epsilon = -e_0 \wedge \left(\partial_0 \epsilon + \nabla \epsilon^0\right) \tag{136}$$

so that from (116) we find

$$\partial_0 \epsilon = -\frac{e}{4\pi c}\left[\frac{\varphi'\left(\tau - t + R/c\right)}{R^2} - \frac{\varphi''\left(\tau - t + R/c\right)}{cR}\right]\hat{\mathbf{x}} \tag{137}$$

$$\nabla \epsilon^0 = \frac{e}{2\pi c}\left[\frac{\varphi'\left(\tau - t + R/c\right)}{R^2} - \frac{\varphi''\left(\tau - t + R/c\right)}{cR}\right]\hat{\mathbf{x}} \tag{138}$$

leading to

$$\partial_0 \epsilon + \nabla \epsilon^0 = \frac{e}{4\pi c}\left[\frac{\varphi'\left(\tau - t + R/c\right)}{R^2} - \frac{\varphi''\left(\tau - t + R/c\right)}{cR}\right]\hat{\mathbf{x}}. \tag{139}$$

Using (117) we calculate

$$\frac{1}{c}\frac{\partial}{\partial \tau} f = \frac{e}{4\pi c}\left[\frac{\varphi'\left(\tau - t + R/c\right)}{R^2} - \frac{\varphi''\left(\tau - t + R/c\right)}{cR}\right] e_0 \wedge \hat{\mathbf{x}} \tag{140}$$

to recover

$$d \wedge \epsilon + \frac{1}{c}\frac{\partial}{\partial \tau} f = 0 \tag{141}$$

as Faraday's law.

In Maxwell electrostatics, Stokes theorem along with (114) establishes the electric field as conservative through

$$\oint \mathbf{E} \cdot \mathbf{dl} = \int \left(\nabla \times \mathbf{E}\right) \cdot \mathbf{dS} = \int \left(\nabla \times \nabla A^0\right) \cdot \mathbf{dS} = 0. \tag{142}$$

This integral is seen to represent work by writing

$$\oint \mathbf{E} \cdot \mathbf{dl} = \oint \mathbf{E} \cdot \frac{\mathbf{dl}}{dt} dt = \oint \mathbf{E} \cdot \mathbf{v}\, dt \tag{143}$$

and noting that the integrand is the time-derivative of energy in the covariant Lorentz force for Maxwell's equations. Similarly writing

$$\oint \epsilon \cdot dx = \oint \epsilon \cdot \dot{x}\, d\tau \tag{144}$$

we recognize the integrand as the *fifth* Lorentz force equation (15) so that the integral represents the total mass change to the event. Considering the space part of the pre-Maxwell field, it follows from (115) that

$$\oint \epsilon \cdot \mathbf{dl} = \int \left(\nabla \wedge \epsilon\right) \cdot \mathbf{dS} = \int \left(\nabla \wedge \nabla a^5\right) \cdot \mathbf{dS} = 0 \tag{145}$$

for any closed path through space. But for a general closed path in spacetime, the Faraday law (141) leads to

$$\oint \epsilon \cdot dl = \int \left(d \wedge \epsilon\right) \cdot dS = -\frac{1}{c}\frac{\partial}{\partial \tau}\int f \cdot dS \tag{146}$$

which need not vanish. For example, representing a surface in spacetime as

$$dS = (e_0 \wedge \hat{\mathbf{x}})\ dt\, dR \tag{147}$$

and using the Clifford identity

$$(y \wedge x) \cdot (a \wedge b) = y \cdot [x \cdot (a \wedge b)] = (x \cdot a)\,(y \cdot b) - (x \cdot b)\,(y \cdot a) \tag{148}$$

to find

$$(e_0 \wedge \hat{\mathbf{x}}) \cdot (e_0 \wedge \hat{\mathbf{x}}) = 1 \tag{149}$$

we are led from (117) to

$$\oint \epsilon \cdot dl = -\frac{1}{c}\frac{\partial}{\partial \tau} \int \frac{e}{4\pi} \left[\frac{\varphi\,(\tau - t + R/c)}{R^2} - \frac{\varphi'\,(\tau - t + R/c)}{cR} \right] dt\, dR \tag{150}$$

which again using $\varphi'(\pm\infty) = 0$ becomes

$$\oint \epsilon \cdot dl = \frac{e}{4\pi c} \int \frac{\varphi'\,(\tau - t + R/c)}{R^2}\, dt\, dR. \tag{151}$$

For a sufficiently long time interval

$$\int \varphi'\,(\tau - t + R/c)\ dt \longrightarrow \varphi\,(-\infty) - \varphi\,(\infty) = 0 \tag{152}$$

and so the ϵ field is seen to be mass-conservative. But if we consider the short closed, timelike path illustrated in Figure 1,

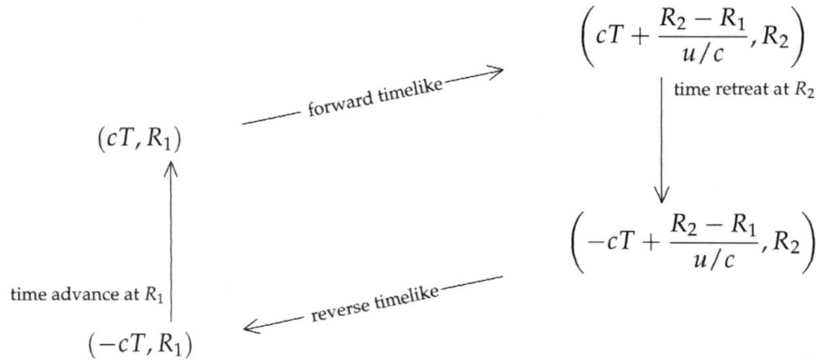

Figure 1

we find that for $R_2 - R_1 \ll cT$

$$\begin{aligned}\oint \epsilon \cdot dl \ &\simeq\ \frac{e}{4\pi c R_1} \left[\varphi\left(\tau - T + \frac{R_1}{c}\right) - \varphi\left(\tau + T + \frac{R_1}{c}\right) \right] \\ &\quad + \frac{e}{4\pi c R_2} \left[\varphi\left(\tau + T + \frac{R_2}{c}\right) - \varphi\left(\tau - T + \frac{R_2}{c}\right) \right]\end{aligned} \tag{153}$$

suggesting that net mass must be invested in moving an event around a closed timelike curve. Because this effect depends on the functional form of $\varphi(x)$ and the time parameter λ, it would provide an experimental signature for the theory.

4.2. Line charge

We now consider a long straight charged line oriented along the z-axis, with charge per unit length λ_e. In cylindrical coordinates

$$\mathbf{x} = (\boldsymbol{\rho}, z) \qquad \boldsymbol{\rho} = (x, y) = \rho\hat{\boldsymbol{\rho}} \qquad \rho = \sqrt{x^2 + y^2} \tag{154}$$

the fields ϵ and $\mathbf{e}$ are found by replacing $R = \sqrt{\rho^2 + z^2}$ in the fields (116) and (117) and integrating along the line to find

$$\mathbf{e} = \boldsymbol{\epsilon} = \frac{\lambda_e}{4\pi} \int dz \left(\frac{\varphi\left(\tau - t + \frac{\left(\rho^2+z^2\right)^{1/2}}{c}\right)}{\left(\rho^2 + z^2\right)^{3/2}} - \frac{\varphi'\left(\tau - t + \frac{\left(\rho^2+z^2\right)^{1/2}}{c}\right)}{c\left(\rho^2 + z^2\right)} \right) (\rho\hat{\boldsymbol{\rho}}, z) \tag{155}$$

$$\epsilon^0 = -\frac{\lambda_e}{4\pi} \int dz \frac{\varphi'\left(\tau - t + \frac{\left(\rho^2+z^2\right)^{1/2}}{c}\right)}{c\left(\rho^2 + z^2\right)^{1/2}}. \tag{156}$$

We may get a sense of these expressions by taking the sharp distribution $\varphi(x) = \delta(x)$ which permits us to easily carry out the z-integration to obtain

$$\mathbf{e} = \boldsymbol{\epsilon} = \frac{\lambda_e}{2\pi} \left(\frac{\theta\left(t - \rho/c - \tau\right)\rho}{c\left[\left(t - \tau\right)^2 - \rho^2/c^2\right]^{3/2}} - \frac{\delta\left(t - \rho/c - \tau\right)}{\sqrt{\left(t - \tau\right)^2 - \rho^2/c^2}} \right) \hat{\boldsymbol{\rho}}. \tag{157}$$

We observe the retarded causality in the vanishing of the field for $\tau > \tau_R = t - \rho/c$. Returning to (155) and using (81) to integrate over τ, the remaining z integration can be readily performed to obtain the concatenated electric field

$$\mathbf{E}(x) = \int d\tau\, \mathbf{e}(x, \tau) = \frac{\lambda_e}{4\pi} \int dz \frac{1}{\left(\rho^2 + z^2\right)^{3/2}} (\rho\hat{\boldsymbol{\rho}}, z) = \frac{\lambda_e}{2\pi\rho} (\hat{\boldsymbol{\rho}}, 0) \tag{158}$$

in agreement with the standard expression.

4.3. Charge Sheet

We finally consider a charged sheet in the $x - y$ plane with charge per unit area σ. Integrating the potential in (112) over x and y with $R = \sqrt{x^2 + y^2 + z^2}$ we find

$$a^0(x, \tau) = a^5(x, \tau) = \frac{\sigma c}{4\pi} \int dx' dy' \, \frac{\varphi\left(\tau - t + \frac{1}{c}\sqrt{\left(x - x'\right)^2 + \left(y - y'\right)^2 + z^2}\right)}{c\sqrt{\left(x - x'\right)^2 + \left(y - y'\right)^2 + z^2}}. \tag{159}$$

Changing to radial coordinates $(x, y) \to (\rho, \theta)$ we obtain

$$a^0(x, \tau) = a^5(x, \tau) = \frac{\sigma c}{4\pi} \int d\theta d\rho \, \frac{\varphi\left(\tau - t + \frac{1}{c}\sqrt{\rho^2 + z^2}\right)}{c\sqrt{\rho^2 + z^2}} \tag{160}$$

which by change of variable $\zeta = \frac{1}{c}\sqrt{\rho^2 + z^2}$ becomes

$$a^0(x,\tau) = a^5(x,\tau) = \frac{\sigma c}{2}\int_{|z|/c}^{\infty} \varphi(\tau - t + \zeta)\, d\zeta. \tag{161}$$

We calculate the fields from (115) and using $\varphi(\tau - t + \infty) = 0$ to find

$$\epsilon^0 = -\sigma \int_{|z|/c}^{\infty} \partial_\zeta \varphi(\tau - t + \zeta)\, d\zeta = \sigma\varphi\left(\tau - t + \frac{|z|}{c}\right) \tag{162}$$

and

$$\boldsymbol{\epsilon} = -\frac{\sigma c}{2}\nabla \int_{|z|/c}^{\infty} \varphi(\tau - t + \zeta)\, d\zeta = \frac{\sigma}{2}\varphi\left(\tau - t + \frac{|z|}{c}\right)\nabla|z| \tag{163}$$

so that

$$\boldsymbol{\epsilon}(x,\tau) = \mathbf{e}(x,\tau) = \frac{\sigma}{2}\varepsilon(z)\varphi\left(\tau - t + \frac{|z|}{c}\right)\hat{\mathbf{z}} \tag{164}$$

where $\varepsilon(z) = \text{signum}(z)$. By concatenation, we recover

$$\mathbf{E}(x) = \int d\tau\, \mathbf{e}(x,\tau) = \int d\tau\, \frac{\sigma}{2}\varepsilon(z)\varphi\left(\tau - t + \frac{|z|}{c}\right)\hat{\mathbf{z}} = \frac{\sigma}{2}\,\varepsilon(z)\,\hat{\mathbf{z}} \tag{165}$$

in agreement with the Maxwell field from a charged sheet. We notice that as expected, the space part of the electric fields change sign at the plane of the sheet, pointing out at each side. Consequently, an event passing through a charged sheet of equal sign will decelerate in space on its approach and then accelerate as it retreats. On the other hand, unlike the field of a point event, the temporal part ϵ^0 is an even function of spatial distance and so the event will accelerate along the time axis on both its approach to the charged sheet and its retreat.

5. Potential barrier

We consider the field produced by an infinite charge sheet in the $x - y$ plane at $z = 0$, for which $\sigma e > 0$ leads to a repulsive force on an event of charge e. Regarding this field as an external force provides a laboratory in which to study the behavior of an event approaching a potential barrier. Substituting (162) and (164) into (13) the Lorentz force on this event is found to be

$$M\ddot{t} = \frac{\lambda e}{c^2}\,\mathbf{e}\cdot\dot{\mathbf{x}} + \frac{\lambda e}{c}\,\epsilon^0 = \frac{\lambda e\sigma}{c}\left[\frac{\varepsilon(z)}{2c}\,\hat{\mathbf{z}}\cdot\dot{\mathbf{x}} + 1\right]\varphi\left(\tau - t + \frac{|z|}{c}\right) \tag{166}$$

$$M\ddot{\mathbf{x}} = \lambda e\,\mathbf{e}\,\dot{t} + \frac{\lambda e}{c}\,\dot{\mathbf{x}}\times\mathbf{b} + \lambda e\,\boldsymbol{\epsilon} = \frac{\lambda e\sigma}{2}\left[\varepsilon(z)\,\dot{t} + 1\right]\varphi\left(\tau - t + \frac{|z|}{c}\right) \tag{167}$$

which can be written

$$\frac{d}{d\tau}\begin{bmatrix} c\dot{t} \\ \dot{z} \end{bmatrix} = \begin{bmatrix} 2\Omega\left(\frac{1}{2}\varepsilon(z)\frac{\dot{z}}{c} + 1\right) \\ \Omega\,\varepsilon(z)\,(\dot{t} + 1) \end{bmatrix} = \frac{\varepsilon(z)}{c}\begin{bmatrix} 0 & \Omega \\ \Omega & 0 \end{bmatrix}\begin{bmatrix} c\dot{t} \\ \dot{z} \end{bmatrix} + \begin{bmatrix} 2\Omega \\ \Omega \end{bmatrix} \tag{168}$$

when the smoothing function $\varphi(\tau)$ is expressed in terms of the shape P_λ of the barrier potential

$$\varphi(\tau) = \frac{1}{2\lambda}P_\lambda(\tau) \tag{169}$$

and is given by

$$\Omega(x,\tau) = \frac{e\sigma}{4M} P_\lambda \left(\tau - t + \frac{|z|}{c}\right). \tag{170}$$

As an approximation to the smooth function given in (33) we begin with a rectangular barrier defined as

$$P_\lambda(\tau) = \theta(\tau+\lambda) - \theta(\tau-\lambda) = \begin{cases} 1, & -\lambda \leq \tau \leq \lambda \\ 0, & \text{otherwise} \end{cases} \tag{171}$$

as the shape function. The approaching event experiences force wherever

$$-\lambda \leq \tau - t + \frac{|z|}{c} \leq \lambda \tag{172}$$

leading to the conditions

$$t - \tau \geq \frac{|z|}{c} - \lambda \tag{173}$$

$$t - \tau \leq \frac{|z|}{c} + \lambda \tag{174}$$

which must be satisfied simultaneously for the barrier to affect the event. We consider an event approaching the barrier from the right with uniform velocity u so that

$$z(\tau) = Z_0 - u\tau \qquad Z_0, u > 0 \qquad Z_0 > \lambda c. \tag{175}$$

With these conditions, (173) becomes

$$t - \tau = \left(u^0 - 1\right)\tau \geq \frac{|z|}{c} - \lambda = \frac{Z_0 - u\tau}{c} - \lambda \tag{176}$$

so that the event reaches the potential barrier when

$$\tau = \tau_0 = \frac{Z_0 - \lambda c}{u + c\left(u^0 - 1\right)} \tag{177}$$

and condition (174) is automatically satisfied. We can solve (168) as

$$\begin{bmatrix} c\dot{t}(\tau) \\ \dot{z}(\tau) \end{bmatrix} = \begin{bmatrix} \cosh\frac{\Omega}{c}(\tau-\tau_0) & \sinh\frac{\Omega}{c}(\tau-\tau_0) \\ \sinh\frac{\Omega}{c}(\tau-\tau_0) & \cosh\frac{\Omega}{c}(\tau-\tau_0) \end{bmatrix} \begin{bmatrix} c\dot{t}(\tau_0) \\ -u \end{bmatrix} + \begin{bmatrix} 2\Omega \\ \Omega \end{bmatrix} (\tau - \tau_0) \tag{178}$$

from $\tau = \tau_0$ until the event reaches the charged sheet or reverses direction.

For low energy scattering with initial conditions

$$t(0) = \tau \qquad \dot{t}(0) = 1 \qquad -u = \dot{z}(0) = \frac{dz}{dt} \qquad |\dot{z}(0)| = |u| \ll c \tag{179}$$

we find from condition (173) that the approaching event experiences the potential barrier for times

$$\lambda \geq \frac{|z|}{c} = \frac{Z_0 - u\tau}{c} \longrightarrow \tau \geq \tau_0 = \frac{Z_0 - \lambda c}{u} \tag{180}$$

and notice that condition (174) is satisfied automatically because $t \simeq \tau$. The equations of motion (168) reduce to

$$\ddot{t} \simeq \frac{e\sigma}{2Mc} P_\lambda \left(\frac{|z|}{c} \right) \qquad \ddot{z} \simeq \frac{e\sigma}{2M} \varepsilon(z) \, P_\lambda \left(\frac{|z|}{c} \right) \tag{181}$$

so that the event will decelerate along the z-axis until it either reverses direction or passes through the charged sheet. For short times, such that $\frac{e\sigma}{2Mc}(\tau - \tau_0) < 1$, we may write the solutions

$$\dot{t}(\tau) \simeq 1 + \frac{e\sigma}{2Mc}(\tau - \tau_0) \qquad \dot{z}(\tau) \simeq -u + \frac{e\sigma}{2M}(\tau - \tau_0) \tag{182}$$

from which we notice that despite the spatial deceleration, the event accelerates in time. To compare this result with the energy of a particle evolving on its mass-shell we calculate the t-velocity

$$\dot{t}_{\text{on-shell}} = \frac{1}{\sqrt{1 - \frac{\dot{z}^2}{c^2}}} = \frac{1}{\sqrt{1 - \left[\frac{e\sigma}{2Mc}(\tau - \tau_0) - \frac{u}{c}\right]^2}} \simeq 1 + \frac{1}{2}\left[\frac{e\sigma}{2Mc}(\tau - \tau_0) - \frac{u}{c}\right]^2 \tag{183}$$

so that the t-acceleration

$$\ddot{t}_{\text{on-shell}} \simeq -\frac{e\sigma}{2Mc}\left[\frac{u}{c} - \frac{e\sigma}{2Mc}(\tau - \tau_0)\right] < 0 \tag{184}$$

is negative as expected. The t-acceleration of the off-shell event corresponds to a transfer of mass from the field to the event found from (16) to be

$$\frac{d}{d\tau}\left(-\frac{1}{2c^2} M\dot{x}^2\right) = -\frac{e_0}{c^2}\,\dot{x}^\mu f_{\mu 5} \simeq \frac{e_0}{c}\,\dot{t}\,\epsilon^0 \simeq \frac{e\sigma}{2c}\,P_\lambda. \tag{185}$$

If the charge density is low enough to permit the event to pass through the charge sheet, then neglecting the effect of deceleration on the transit time, the total mass shift of the event is on the order of $\lambda e\sigma/2u$.

As an interesting example of 5D electrostatics, we consider a test event at rest within the potential barrier at time $\tau = 0$

$$0 < z(0) = Z_0 < \lambda c \qquad \dot{z}(0) = 0 \qquad \dot{t}(0) = 1 \qquad t(0) = 0 \tag{186}$$

and bound in an insulator so that

$$\ddot{t} \simeq \frac{e\sigma}{2Mc} P_\lambda \left(\tau - t + \frac{Z_0}{c}\right) \qquad \ddot{z} = \frac{F_{\text{electric}} - F_{\text{binding}}}{M} = 0. \tag{187}$$

In this case, condition (173) is automatically satisfied so we may write

$$t = \begin{cases} \tau + \dfrac{1}{2}\dfrac{e\sigma}{2Mc}\tau^2, & 0 \le \tau \le \tau_c \\ t(\tau_c) + \left[1 + \dfrac{e\sigma}{2Mc}\right](\tau - \tau_c), & \tau > \tau_c \end{cases} \tag{188}$$

where, by condition (174),

$$t - \tau = \frac{Z_0}{c} + \lambda \qquad \Rightarrow \qquad \tau_c = \sqrt{\frac{4M(Z_0 + \lambda c)}{e\sigma}}. \tag{189}$$

Thus, the event experiences a short-lived acceleration which pushes it beyond the range of the potential barrier at τ_c. For $\tau > \tau_c$, the temporal velocity remains constant at

$$\dot{t}\left(\tau_c\right) = 1 + \frac{e\sigma}{2Mc}\tau_c = 1 + \sqrt{\frac{\left(Z_0 + \lambda c\right) e\sigma}{Mc^2}} \tag{190}$$

so that during the interaction with the potential barrier, the squared mass of the event grows from M^2 to

$$m^2\left(\tau_c\right) = M^2\left(\frac{\dot{x}}{c}\right)^2 = M^2\dot{t}^2 = M^2\left[1 + \sqrt{\frac{\left(Z_0 + \lambda c\right) e\sigma}{Mc^2}}\right]^2 . \tag{191}$$

However, if the event has opposite charge $e = -|e|$ then the event will decelerate in time as

$$\ddot{t} \simeq -\frac{|e|\sigma}{2Mc}P_\lambda\left(\tau - t + \frac{Z_0}{c}\right) \qquad \ddot{z} = \frac{F_{\text{electric}} - F_{\text{binding}}}{M} = 0 \tag{192}$$

and condition (173) leads to

$$\dot{t}\left(\tau\right) = 1 - \frac{|e|\sigma}{2Mc}\tau \tag{193}$$

suggesting that for a sufficient charge density the event could reverse direction in time. By condition (173) the time evolution is

$$t = \begin{cases} \tau - \dfrac{1}{2}\dfrac{|e|\sigma}{2Mc}\tau^2, & 0 \le \tau \le \tau_c \\ t(\tau_c) + \left[1 - \dfrac{e\sigma}{2Mc}\right]\left(\tau - \tau_c\right), & \tau > \tau_c \end{cases} \qquad \tau_c = \sqrt{\frac{4M\left(\lambda c - Z_0\right)}{e\sigma}} \tag{194}$$

so that for $\tau > \tau_c$ the time evolution is

$$\dot{t}\left(\tau_c\right) = 1 - \frac{e\sigma}{2Mc}\tau_c = 1 - \sqrt{\frac{\left(\lambda c - Z_0\right) e\sigma}{Mc^2}} \tag{195}$$

and

$$m^2\left(\tau \ge \tau_c\right) = M^2\left[1 - \sqrt{\frac{\left(\lambda c - Z_0\right) e\sigma}{Mc^2}}\right]^2 \tag{196}$$

is the squared mass. In the case that the event emerges from the interaction with the field as an anti-event with $\dot{t}\left(\tau_c\right) = -1$, it emerges with its initial mass.

For the case of an event held at rest in space, we are not restricted to the approximation of the rectangular barrier and can find a solution for the smooth potential given by

$$\varphi\left(\tau\right) = \frac{1}{2\lambda}e^{-|\tau|/\lambda} \quad \rightarrow \quad P_\lambda\left(\tau\right) = e^{-|T_0 - (t-\tau)|/\lambda} \qquad T_0 = \frac{Z_0}{c}. \tag{197}$$

The equations of motions are now nonlinear

$$\ddot{t} \simeq \frac{e\sigma}{2Mc}e^{-|T_0 - (t-\tau)|/\lambda} \qquad \ddot{z} = \frac{F_{\text{electric}} - F_{\text{binding}}}{M} = 0 \tag{198}$$

with the initial conditions $z(0) = Z_0$ and $t(0) = \tau = 0$. As the event accelerates in this field, the t will grow larger than τ and the field strength will increase to a maximum when $t - \tau = T_0$ and then decrease when $t - \tau > T_0$. Introducing the variables

$$s_1 = t - \tau < T_0 \qquad -s_2 = t - \tau > T_0 \tag{199}$$

the approach to the field maximum is described by

$$\ddot{s}_1 = \frac{e\sigma}{2Mc} e^{-T_0/\lambda}\, e^{s_1/\lambda} = \alpha\, e^{s_1/\lambda} \qquad \alpha = \frac{e\sigma}{2Mc} e^{-T_0/\lambda} \tag{200}$$

and the retreat from the field maximum is described by

$$\ddot{s}_2 = -\frac{e\sigma}{2Mc} e^{T_0/\lambda}\, e^{s_2/\lambda} = -\beta\, e^{s_2/\lambda} \qquad \beta = \frac{e\sigma}{2Mc} e^{T_0/\lambda}\,. \tag{201}$$

The generic equations (200) and (201) may be solved by writing

$$\ddot{s}_1 = \alpha\, e^{s_1/\lambda} \quad \rightarrow \quad \dot{s}_1\, \ddot{s}_1 = \frac{d}{d\tau}\left[\frac{1}{2}\left(\dot{s}_1\right)^2\right] = \dot{s}_1\, \alpha e^{s_1/\lambda} = \frac{d}{d\tau}\left[\lambda \alpha e^{s_1/\lambda}\right] \tag{202}$$

so that τ-integration leads to

$$\dot{s}_1(\tau) = \sqrt{2\alpha\lambda}\sqrt{e^{s_1/\lambda} - 1} \tag{203}$$

where we used $\dot{s}_1(0) = \dot{t} - 1 = 0$. Now integrating

$$\int \frac{1}{\sqrt{e^{s_1/\lambda} - 1}} \frac{ds_1}{d\tau} d\tau = 2\lambda \tan^{-1} \sqrt{e^{s_1/\lambda} - 1} = \sqrt{2\alpha\lambda}\tau + c_1 \tag{204}$$

and using $s_1(0) = t(0) - 0 = 0$ we find

$$s_1 = \lambda \log\left(\tan^2\left(\sqrt{\frac{\alpha}{2\lambda}}\tau\right) + 1\right) \tag{205}$$

which becomes

$$t = \lambda \log\left(\tan^2\left(\sqrt{\frac{\alpha}{2\lambda}}\tau\right) + 1\right) + \tau. \tag{206}$$

Designating τ_0 as the time when the field reaches its maximum, so that $s_1(\tau_0) = T_0$, we may invert (205) to find

$$\tau_0 = \sqrt{2\lambda/\alpha}\, \tan^{-1} \sqrt{e^{T_0/\lambda} - 1} \tag{207}$$

and

$$\dot{t}(\tau_0) = 1 + \sqrt{\frac{\lambda e\sigma}{Mc}\left(1 - e^{-T_0/\lambda}\right)}. \tag{208}$$

For $\tau > \tau_0$ we apply this generic solution to (201) using initial conditions

$$s_2(\tau_0) = T_0 \qquad \dot{s}_2(\tau_0) = 1 - \dot{t}(\tau_0) = -\sqrt{2\alpha\lambda\left(e^{T_0/\lambda} - 1\right)} \tag{209}$$

to find

$$\dot{s}_2 = -\sqrt{2\beta\lambda}\sqrt{C - e^{s_2/\lambda}} \tag{210}$$

with constant of integration

$$C = e^{-2T_0/\lambda}\left(2e^{T_0/\lambda} - 1\right). \tag{211}$$

Integrating again

$$\int \frac{1}{\sqrt{C - e^{s_2/\lambda}}} \frac{ds_2}{d\tau} d\tau = -\frac{2\lambda}{C} \tanh^{-1} \sqrt{\frac{C - e^{s_2/\lambda}}{C}} = -\sqrt{2\beta\lambda}\tau + C \tag{212}$$

we arrive at

$$s_2 = \lambda \log \left[C - C \tanh^2 \left(\sqrt{\frac{\beta C}{2\lambda}} \tau + \sqrt{2\beta\lambda} \tau_0 - \frac{2\lambda}{\sqrt{C}} \tanh^{-1} \sqrt{\frac{e^{T_0/\lambda} - 1}{2e^{T_0/\lambda} - 1}} \right) \right] \tag{213}$$

which becomes

$$t = \tau + \lambda \log \left[\frac{1}{C} \cosh^2 \left(\sqrt{\frac{\beta C}{2\lambda}} \tau + \sqrt{2\beta\lambda} \tau_0 - \frac{2\lambda}{\sqrt{C}} \tanh^{-1} \sqrt{\frac{e^{T_0/\lambda} - 1}{2e^{T_0/\lambda} - 1}} \right) \right]. \tag{214}$$

Inserting (213) into (209) we find

$$\dot{s}_2 = 1 - \dot{t} = -\sqrt{2\beta\lambda C} \tanh \left(\sqrt{\frac{\beta C}{2\lambda}} \tau + \sqrt{2\beta\lambda} \tau_0 - \frac{2\lambda}{\sqrt{C}} \tanh^{-1} \sqrt{\frac{e^{T_0/\lambda} - 1}{2e^{T_0/\lambda} - 1}} \right) \tag{215}$$

and

$$\dot{t} = 1 + \sqrt{2\beta\lambda C} \tanh \left(\sqrt{\frac{\beta C}{2\lambda}} \tau + \sqrt{2\beta\lambda} \tau_0 - \frac{2\lambda}{\sqrt{C}} \tanh^{-1} \sqrt{\frac{e^{T_0/\lambda} - 1}{2e^{T_0/\lambda} - 1}} \right). \tag{216}$$

To determine the asymptotic mass of the event we calculate

$$\dot{t} \xrightarrow[\tau \to \infty]{} \dot{t}_{\max} = 1 + \sqrt{\frac{\lambda e \sigma}{Mc}} \sqrt{2 - e^{-T_0/\lambda}} \tag{217}$$

so that

$$m^2 \longrightarrow M^2 \left(1 + \sqrt{\frac{\lambda e \sigma}{Mc}} \sqrt{2 - e^{-\frac{Z_0}{\lambda c}}} \right)^2 \simeq M^2 \left(1 + \sqrt{\frac{(Z_0 + \lambda c) e \sigma}{Mc^2}} \right)^2 \tag{218}$$

which is identical to (191) in the low energy approximation. As is previous examples, the field transfers mass to the event, despite the deceleration in space.

6. Discussion

In Stueckelberg-Horwitz electrodynamics, particle worldlines are traced out through the evolution of interacting spacetime events $x^\mu(\tau)$ whose associated current densities induce the τ-dependent fields that mediate their interactions. By introducing the chronological time τ as an independent evolution parameter, and freeing the laboratory clock x^0 to propagate alternately 'forward' and 'backward' in time according to the sign of its energy, this formalism provides a classical implementation of the Feynman-Stueckelberg interpretation of pair creation/annihilation. However, as Stueckelberg discovered, allowing $\dot{x}^0$ to evolve independently of $\dot{\mathbf{x}}$ is not sufficient to permit classical pair creation/annihilation because mass is conserved for interactions mediated by the antisymmetric electromagnetic tensor. To evolve outside the forward or reverse timelike region an event must undergo an exchange of mass with an external field.

It was shown in [8] that (11) is the most general classical Lagrangian consistent with the quantum commutation relations

$$[x^\mu, x^\nu] = 0 \qquad\qquad M[x^\mu, \dot{x}^\nu] = -i\hbar\eta^{\mu\nu} \tag{219}$$

among unconstrained phase space variables. Thus, Stueckelberg-Horwitz electrodynamics follows from two assumptions: a kinetic term of the type (18) for the fields modeled on Maxwell theory, and a phase space on which the usual mass-shell constraint $p^\mu p_\mu = -m^2$ is relaxed. By demoting mass conservation from an *a priori* constraint to a conservation law that applies for appropriate interactions, the formalism acquires its key features: integrability, local gauge symmetry with five τ-dependent gauge fields, classical implementation of negative energy evolution, and exchange of mass between particles and fields.

In a Lorentz-covariant generalization of the nonrelativistic picture, event evolution is associated with a current consisting of a scalar event density $j^5(x,\tau)$ that characterizes the event distribution in spacetime, and a vector current $j^\mu(x,\tau)$ that characterizes the motion of events through spacetime. In this picture, current conservation describes the 4-divergence of the spacetime current balancing the τ-variation of the event density. Even a 'static' event must move along its time axis, and so both the scalar current $j^5(x,\tau)$ and the temporal component $j^0(x,\tau)$ of the vector current contribute to gauge fields $a^5(x,\tau)$ and $a^0(x,\tau)$ that produce field strengths $e^i = f^{0i}$ and $\epsilon^\mu = f^{5\mu}$. As seen in the structure of the pre-Maxwell equations, the ϵ^μ field plays a role analogous to the Maxwell electric field — its divergence points to the scalar source inducing the field. Similarly, the field $f^{\mu\nu}$ plays a role analogous to the Maxwell magnetic field — it has no monopoles and is induced by the motion of the source through spacetime.

In this paper we studied simple solutions for timelike particles in uniform motion. Calculating the fields induced by these events, we confirmed that they satisfy the pre-Maxwell field equations. We found the τ-dependent potentials for the moving event and the field strengths $e^i = f^{0i}$ and $\epsilon^\mu = f^{5\mu}$ which take on a generalized Coulomb-like form. For a test event at rest with respect to the source event, a time acceleration was observed suggesting a transfer of mass between the field and test event. For this solution, the integral forms of Gauss's law and Faraday's law were found by extending the integrations to 4D volumes of spacetime. As in Maxwell theory, the evaluation of Stoke's theorem shows that the line integral of the electric field ϵ^μ around a closed path vanishes for paths in space or long paths in spacetime. However, it was seen that the smoothed structure of the current suggests that a net change in mass may result from the motion of an event around a short closed timelike path. These effects were the first of several indications of mass changing effects in 5D electrostatics. The fields produced by a line charge and a charged sheet were also calculated. The fields for the three source configurations were shown to recover the standard Maxwell fields under concatenation.

The field induced by the infinite charged sheet was used to study the evolution of a test charge approaching a potential barrier. It was seen that a low-energy event approaching the barrier would undergo a temporal acceleration (energy growth) while the event decelerated away from the barrier. We also studied the evolution of a test event held fixed in space in an insulator, and saw that from the onset of the configuration, the field would transfer mass to the test event as it gained energy (accelerates in t) with no corresponding change in spatial position or velocity. This effect was confirmed for both a rectangular barrier and the exponential shape associated with the smoothing function. Intriguingly, in these mass shift effects, the period of mass change is short-lived, because the field pushes the test event out of range, and so the total mass shift is small and finite. These mass shift effects suggest several questions related to the observed conservation of elementary particle mass. It would be interesting to find that just as a regular structure of particles remains at rest in an insulator, so the net effect of the ϵ^μ field in a regular structure produces a mass insulation effect. This speculation will be addressed in a forthcoming paper.

References

[1] Born and Wolf 1975 *Principles of Optics* fifth edition (Oxford: Pergamon Press) xxv-xxvi.
[2] For discussion of the conceptual history of gauge theory see J. D. Jackson 2001 Rev. Mod. Phys. **73** 663.
[3] Stueckelberg E C G 1941 *Helv. Phys. Acta* **14** 322
Stueckelberg E C G 1941 *Helv. Phys. Acta* **14** 588
[4] Feynman R P 1950 *Phys. Rev.* **80** 440
Feynman R 1948 *Rev P. Mod. Phys.* **20** 367
[5] Schwinger J 1951 *Phys. Rev.* **82** 664
[6] Horwitz L P and Piron C 1973 *Helv. Phys. Acta* **48** 316
[7] Saad D, Horwitz L P and Arshansky R I 1989 *Found. of Phys.* **19** 1126
[8] Land M C, Shnerb N and Horwitz L P 1995 *J. Math. Phys.* **36** 3263
[9] Land M C and Horwitz L P 1991 *Found. of Phys. Lett.* **4** 61
[10] Arshansky R I and Horwitz L P 1989 *J. Math. Phys.* **30** 66
[11] Arshansky R I and Horwitz L P 1989 *J. Math. Phys.* **30** 380
[12] Horwitz L P and Lavie Y 1982 *Phys. Rev.* D **26** 819
Arshansky R I and Horwitz L P 1989 *J. Math. Phys.* **30** 213
Arshansky R I and Horwitz L P 1988 *Phys. Lett* A **131** 222
[13] Land M C and Horwitz L P 1995 *J. Phys.* A **28** 3289
[14] Land M C 2001 *Found. of Phys.* **31** 967
[15] Land M C 1996 *Found. of Phys.* **27** 19
[16] Land M C and Horwitz L P 1998 Land M C **A239** 135
[17] Land M C 2011 *J. Phys.: Conf. Ser.* **330** 012015
[18] Arshansky R L, Horwitz P and Lavie Y 1983 *Found. of Phys.* **13** 1167
[19] Land M C and Horwitz L P 1991 *Found. of Phys.* **21** 299
[20] Aharonovich I and Horwitz L P 2012 *J. Math. Phys.* **53** 032902
[21] Yao W-M et. al. 2006 (Particle Data Group) *J. Phys.* G **33** 1
[22] Land M C 2003 *Found. of Phys.* **33** 1157

IARD2012 IOP Publishing
Journal of Physics: Conference Series **437** (2013) 012013 doi:10.1088/1742-6596/437/1/012013

On the possibility of a relativistic correction to the E and B fields around a current-carrying wire

Ron Folman
Department of Physics, Ben-Gurion University of the Negev, Be'er Sheva 84105, Israel

Abstract. It is well known that electric and magnetic fields may change when they are observed from different frames of reference. For example, the motion of a charged probe particle moving parallel to a current-carrying wire would be described by utilizing different electric or magnetic fields, depending on from which frame of reference the system is observed and described. To describe the situation in all frames by utilizing the theory of relativity, one has to first describe the situation in one particular frame, and this choice in the case of a current-carrying wire is the topic of this paper. Specifically, I consider the question of in which frame the current carrying wire is neutral. The importance of relaxation processes is emphasized. As an example, I examine a specific alternative to the standard choice, and consider its theoretical and experimental validity. An outcome of alternative approaches is that in the rest frame of a wire, running a current introduces also an electric field by giving rise to a minute charge. Present day experimental sensitivities, specifically those of cold ions, may be able to differentiate between the observable signatures predicted by the different approaches.

PACS: 37.10.Gh, 32.70.Cs, 05.40.-a, 67.85.-d

The theory of electromagnetism [1] and the special theory of relativity [2] have been confirmed in an abundance of experiments. In the following I describe an effect that is a result of these two theories and which may have been left unnoticed in the experiments performed due to its relatively weak nature.

The interplay between electric and magnetic fields when one moves from one frame of reference to another stands at the base of our understanding of how the theory of relativity affects these fields [3]. The special theory of relativity clearly explains how these fields are observed as changing as a function of the frame of observation, but in order to know what the fields are, the fields in one particular frame must first be determined by other theoretical or experimental means. It is the description of this particular frame, usually chosen to be the rest frame, that is at the core of this paper.

The dependence of an elementary charge (e.g. that of an electron or a proton) on its velocity would have been observed in particle accelerators, in the neutrality of atoms or even when heating a metal, and has been investigated by numerous methods ([3] and refs. therein). Specifically, experiments intended to measure a second order change of the charge with its velocity, seem to have observed no effect [4, 5]. It is therefore of common belief that at least in low energies, such a dependence has been ruled out. However, it seems that the dependence of charge density on the mean velocity (so-called drift velocity) in the rest frame of a wire has not been examined thoroughly and therefore cannot be ruled out. In this paper I explain why such a fundamental effect could be considered as possible and calculate the expected signal. (following the above, we shall assume the invariance of the elementary charge with respect to its velocity, meaning that if the total number of particles is conserved then the total charge is conserved.)

More specifically, the E and B fields are a result of the charge density ρ and the current density j. It is these physical observables that are analyzed henceforth. As is well known, the 4-vector $(\rho c, \mathbf{j})$, c being the speed of light, transforms according to the Lorentz transformations when the observer moves from one frame to the other. However, as noted, these transformations may only tell us what the charge and the current are in a specific frame if we have previously determined them in another frame. It is common practice to assume that in the rest frame of a wire (which we define here as the rest frame of the nuclei or the protons that form a wire) a current does not give rise to a charge. This comes about from the assumption that the electrons behave as free particles and therefore accelerating them to some finite velocity does not change the particle-particle distances. If indeed the effect we consider in this paper is real, the latter assumption is not correct.

The question of the charging of a current carrying wire in its rest frame, may be phrased in the following way: Albert Einstein, in his 1905 paper, wrote [6]: "At v=V, all moving objects - viewed from the "stationary" system - shrink into plane figures" (where V is the speed of light). It is therefore widely believed (although I am not sure if this has been verified) that in accelerators, when clouds of particles with strong enough restoring forces, e.g. an ensemble of nucleons in a nucleus, are accelerated to some finite speed, the cloud length is Lorentz contracted thus producing a pancake shaped cloud in the lab frame of the accelerator [7]. In the case of a nucleus, this comes about from the fact that the restoring forces keep its spherical shape in its rest frame. If such a pancake indeed exists, the question can then be asked as to whether such a contraction of an "electron cloud" may occur in a current carrying wire thereby producing a net charge in the wire rest frame.

Following this work [8], I became aware of numerous previous theoretical presentations of similar ideas and the criticism they received (see Refs. [9, 10] and references therein). In Appendix I we give an example of such a theory and the criticism it received. Contrary to some of the previous works (e.g. [11]), the analysis described here will adhere to standard electrodynamics. Also contrary to some previous works which state, to my mind without sufficient proof, that the wire in its rest frame is indeed charged (e.g. [17] and more recently [18]), it is claimed in this work that not enough is known to arrive at this conclusion. At the same time, it seems there is no theoretical or experimental proof in favor of the standard approach which holds that the wire is neutral in its rest frame.

This paper goes beyond the previous presentations as it emphasizes the importance of internal dynamics, giving rise to restoring (relaxation) processes (as in the pancake example above), and the fact that only a three dimensional many-body covariant analysis of the system can be expected to give a theoretical resolution regarding at which frame the equilibrium is at the $\rho_{net} = 0$ point. If this happens in several frames, a situation not compatible with relativity, the question most relevant may be in which frame the restoring dynamics are faster. In this context we analyze in detail Maxwell's equations and Ohm's law. This work also presents a new analysis of the experimental sensitivity required to observe the hypothesized effect and analyzes the feasibility of several novel experimental methods to make such an observation. The analysis indicates that cold ions may serve as a probe which is sensitive enough to differentiate between the different possibilities.

Let us begin directly with pointing out briefly one counter example to the standard approach, following which it is discussed in a detailed and consistent manner. If one assumes (as is commonly done) a zero net charge density for a current-carrying wire in its rest frame S ($\rho_+ = |\rho_-| = \rho_0$ where ρ_0 is the free electron density per unit length in the material when no current is applied), one must, due to Lorentz transformations, assume a charge density of $(\gamma - 1/\gamma)\rho_0$ in frame S' moving at the drift velocity of the electrons in frame S [where $\gamma = (1 - v^2/c^2)^{-\frac{1}{2}}$]. Namely, the wire, as observed in S', becomes charged. One intuitive explanation for this apparent charging is that somehow the net charge density is dependent on the velocity of the charges in S' so that $\rho'_+ \neq |\rho'_-|$. This cannot be the right explanation simply because the same velocity difference in S does not give rise to a similar difference of densities in frame S. As described below, while the common choice [$\rho_+ = |\rho_-|$ in S and $(\gamma - 1/\gamma)\rho_0$ in S'] is consistent with the special theory of relativity, it may not be the only plausible application of the theory. For example, one could have applied the theory and the arguments in a symmetric way in S and S'. As in S the electrons are moving and the positive charges are stationary, one could perhaps hypothesize to have in S a net charge of $\rho_{net} = \rho_0(1 - \gamma)$ while in S' one would expect $\rho'_{net} = \rho_0(\gamma - 1)$. As will be shown in the following, the latter choice is also compatible with the special theory of relativity. As for small velocities $\gamma = (1 + \frac{1}{2}v^2/c^2)$, the latter hypothesis gives rise to a second order effect in the electron velocity in the rest frame of the wire.

The difference between the two choices may perhaps be made more apparent by the following "exposition": we simulate a normal conductor with two infinite solid rods, one made of positive ions and one made of negative ions. As we are not interested in absolute charge but only in charge density (per unit length), the boundary conditions are of no importance. While the positive rod is at rest in S, a current in S is realized by moving the negative rod with a velocity equal to the electron drift velocity in a normal conductor. If we accept the standard choice for S, an observer in S would claim to see no charge. His friend is an observer in S^*, a frame in which both rods have the same velocity (in opposite directions). As in S there is a current while there is no charge, Lorentz transformations do not allow for any other frame to observe a charge of zero. This is well known to the observer in S and thus he communicates to his friend that he must be observing a non-zero charge, giving rise to a non-zero electric field. The friend, however, sees a completely symmetrical world. He observes two infinite rods, built exactly the same way with the exception of an opposite charge, going in opposite directions with equal velocity. As he believes the sign of the velocity vector should have no effect on how charge changes with velocity, he concludes that this symmetrical world must give rise to a charge of zero, in contradiction to what his friend communicated to him, and obviously in contradiction to the theory of relativity. Of course, if the non standard choice [i.e. $\rho_{net} = \rho_0(1-\gamma)$] was observed in S, Lorentz transformations would have predicted a zero charge density in the symmetrical world observed in S^*.

Let us emphasize that the following detailed analysis of the situation in which the S^* frame noted above is the neutral frame, is given here as a mere possible counter example to the standard approach. As in the real life situation of a current carrying wire the S^* frame is clearly not symmetric (masses and interactions are different), it may very well be that another frame eventually presents itself as the unique frame in which $\rho_{net} = 0$.

In his famous lectures on physics Feynman explains in detail an example which concerns the electric and magnetic fields produced by a current-carrying wire, as observed from different frames of reference [19]. Following Feynman's formalism, I examine the above alternative approach to the problem at hand. If one is not able to determine theoretically a single preferred option, then one will have to resort to experiment.

This paper is structured as follows: First I put forward in simple terms the difference between the two approaches regarding their underlying arguments. I then describe in more detail the standard approach followed by the alternative approach. I conclude with the theoretical and experimental consequences due to the difference between the two approaches.

Let me first describe the system we are addressing. As presented in Fig. , a current-carrying wire is placed in the lab frame (S). Due to the current, the electrons have some mean velocity v_e. A positively charged probe particle with charge $q = +1$ moves parallel to the wire axis

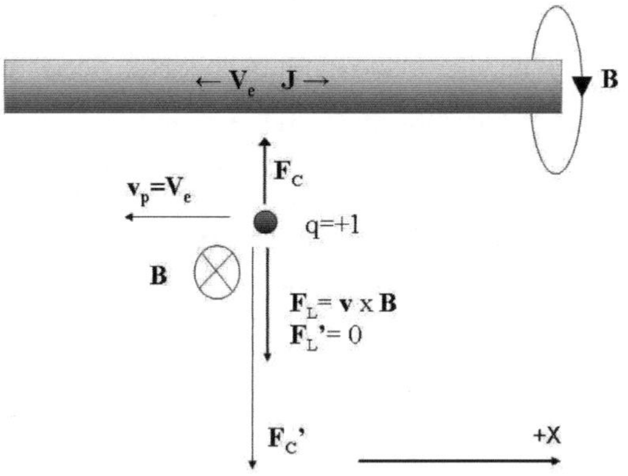

Figure 1. The system of the wire and probe particle. The wire is stationary in the lab frame S and the particle is stationary in its rest frame S'. The electric (F_C) and magnetic (F_L) forces in both frames are presented. Note that in S' there is no magnetic force as the particle does not move. Contrary to the standard approach, in the alternative approach presented in this work there is a Coulomb force also in S.

with velocity v_p and has the same velocity as the electrons so that $v_p = v_e$. It is well known that due to its charge and velocity in the magnetic field of the wire, the particle experiences a magnetic force, the Lorentz force (F_L), which is equal in this case to qv_pB and directed away from the wire. The question addressed by Feynman is what would be the forces experienced by the probe particle in its own frame of reference (S'), the so-called particle rest frame, namely, in the frame where the probe particle's velocity is zero. While it is clear that the magnitude of the above magnetic force is zero in S', the question is if any other forces are created? Specifically, concerning forces that are perpendicular to v_p, one would require, in order to be consistent with the theory of relativity, that the sum of forces in S' (F') would be larger than the sum of forces in S (F) by the relativistic Lorentz factor γ so that $F' = \gamma F$. This is so because the force is the derivative of the momentum in time, and as we do not expect the momentum in the direction transverse to the relative velocity of the frames (the latter being v_p) to change when moving from frame to frame, we expect $F'\Delta t' = F\Delta t$. As the theory of relativity dictates $\Delta t = \gamma\Delta t'$ we therefore require $F' = \gamma F$. Hence, although the magnetic force in S' is zero, the total force in S' cannot be zero.

In his usual brilliant manner, Feynman simply explains that as the wire becomes shorter in frame S', the density of the positive charges in the wire grows and this creates an electric field which gives rise to a Coulomb force (F'_C) which acts on the probe particle in S'. However, to give rise to an electric field Feynman has to differentiate between the relativistic effect due to the frame change as it acts on the positive and the negative charges. No quantified physical argument is put forward to justify the specific asymmetry of the $(\gamma - 1/\gamma)$ factor leading to a non-zero charge density in S'. In fact this factor is simply a Lorentz transformation of the assumption of neutrality in the rest frame of the positive charges. As will be explained in the following, the latter assumption is equivalent to the assumption that no restoring forces are at work on the electrons such that the electrons' inter-particle distance in some frame other than S is restored to their original distance in the wire frame before the current was turned on. Put in simple words this means that the electrons behave as free particles. This may be called "The Standard Assumption". As Feynman's treatment following this "assumption" is completely consistent with the theory of relativity, he indeed finds that as required $F' = \gamma F$.

In the alternative approach presented here we assume that both the positive and negative charges have the same restoring forces, so that they maintain their natural inter-particle distances in their rest frame. Following the above "pancake" example, in both frames the density is increased only for the moving particles due to Lorentz contraction. This imposed symmetry may be called "An Alternative Assumption". This approach also leads to $F' = \gamma F$ and hence it is also consistent with the theory of relativity. However, it leads to the charging of the current-carrying wire in its rest frame, a surprising result indeed. Thus, while both approaches are consistent with relativity, they start with very different assumptions and arrive at very different descriptions concerning the physical situation in frame S.

It seems likely to this author that both these assumptions present the extreme edges of a wide spectrum of possibilities. What is common to all models occupying this spectrum is that all, but the one in the standard edge, predict the charging of the wire in its rest frame, in varying magnitudes.

Let me now briefly describe the standard approach (in my own abbreviated way). A detailed account may be found in Ref. [19]. First, note that the magnetic force F_L experienced by the probe particle in S is

$$F_L = qv_pB = qv_p\frac{J}{2\pi rc^2\epsilon_0} = qv_p\frac{\rho_0 v_e}{2\pi rc^2\epsilon_0} = \frac{q\rho_0 v^2}{2\pi rc^2\epsilon_0}, \tag{1}$$

where r is the distance from the wire, ρ_0 is the free electron density (per unit length) in the material when no current is applied, and I have taken (following Feynman's example) $v = v_e = v_p$.

Next, Feynman calculates the Coulomb force F'_C in S' by integrating the net charge along the wire axis and its electric field at the position of the particle. Note that Feynman takes the positive charge density to be $\rho'_+ = \gamma\rho_0$ and the negative charge density to be $\rho'_- = -\rho_0/\gamma$ so that the net charge density is $(\gamma - 1/\gamma)\rho_0 = \gamma(1 - 1/\gamma^2)\rho_0 = \gamma\beta^2\rho_0$, where $\beta = v/c$. The Coulomb force comes out to be

$$F' = F'_C = \gamma\beta^2\rho_0 \int \frac{q}{4\pi\epsilon_0(r^2+x^2)}\frac{r}{\sqrt{(r^2+x^2)}}dx = \gamma\beta^2\rho_0\frac{q}{2\pi\epsilon_0 r} = \gamma F_L = \gamma F, \tag{2}$$

where, according to the standard approach, F and F' are the total forces in S and S', respectively.

Let me now describe in detail the alternative approach. As noted, the assumption regarding the charge neutrality of the wire in S is replaced with a symmetry assumption requiring that $\rho' = -\rho$, where ρ is the net charge density (noted previously as ρ_{net}). For simplicity, we shall not write the relevant parameters in units of ρ_0 as in the standard approach but rather in units of $\rho_0^{**} = \gamma\rho_0$, where $\gamma = (1 - v_e^2/c^2)^{-\frac{1}{2}}$.

To see exactly what the charge density should be in the two frames, let us remind ourselves of the four vector transformations. The vector $(\rho c, j_x)$ transforms under the Lorentz transformation matrix

$$\gamma \begin{pmatrix} 1 & -\beta \\ -\beta & 1 \end{pmatrix}.$$

We therefore see that $\rho' c = \gamma\rho c - \gamma\beta j$. As we are are discussing ρ and j in units of ρ_0^{**} (in which j is simply v_e), and as j is in the positive direction while the boost β from S to S' is in the negative direction, we find $\rho' = \gamma\rho + \gamma\beta^2$. Imposing symmetry, namely that $\rho' = -\rho$, we find that $\rho = (1/\gamma - 1)\rho_0^{**}$ and $\rho' = (1 - 1/\gamma)\rho_0^{**}$. Consequently, the net charge in S' is now $(\gamma - 1)\rho_0$ contrary to the standard $(\gamma - 1/\gamma)\rho_0$. That the positive charge density in S' is $\rho'_+ = \gamma\rho_0$ and the negative charge density is $\rho'_- = \rho_0$ is a natural outcome under the assumption of symmetry, as the positive charges are moving and the electrons are at rest. In the same way, in S we now have a net charge of $(1 - \gamma)\rho_0$ as the electrons are moving and the positive charges are at rest. Thus, our conclusion concerning the state of the wire in S must be different from that of standard approach that the wire in S is not charged. To make sure that also the alternative approach is consistent with the theory of relativity, let us now calculate the forces and see if we again find $F' = \gamma F$ as required.

In S we now expect to find both a Coulomb force and a magnetic force as depicted in Fig. . As the probe particle is positively charged and as the wire is negatively charged, and as the probe particle is moving in the same direction as the wire electrons, the electric and magnetic forces will be in directions opposite to each other. Note also that in order to be consistent, the ρ_0 appearing in the magnetic force in S must also be boosted by γ as it is due to the electrons. The total force in S will thus be

$$F = F_L + F_C = \gamma\rho_0\beta^2 \frac{q}{2\pi\epsilon_0 r} + (1-\gamma)\rho_0 \int \frac{q}{4\pi\epsilon_0(r^2+x^2)} \frac{r}{\sqrt{(r^2+x^2)}} dx = (\gamma\beta^2 + 1 - \gamma) \times \frac{q\rho_0}{2\pi\epsilon_0 r}. \tag{3}$$

In frame S' we find

$$F' = F'_L + F'_C = 0 + (\gamma - 1) \times \frac{q\rho_0}{2\pi\epsilon_0 r}, \tag{4}$$

and we therefore find

$$\frac{F'}{F} = \frac{\gamma - 1}{\gamma\beta^2 + 1 - \gamma} = \gamma \tag{5}$$

as required.

Let us now consider the theoretical consequences. One of the most important issues to clarify is how both approaches deal with charge conservation, which we assume to be correct. For example, as the alternative approach claims that in S a current induces a charge density, the conclusion must be that for an infinite wire, an infinite charge must be deposited. This is not very appealing indeed. However, this is also what happens in S' in the standard approach.

As it is not very clear what charge conservation means for an infinitely long wire, and as it is not clear what an experiment verifying any result concerning an infinitely long wire should look like, we shall consider a current in a conducting loop of finite dimensions.

Let us thus examine a loop in the shape of a square of unit length, so that in essence there are two wires of the form noted above in the $\hat{x}$ direction (see Fig.), and between them, at their ends, vertical connections [20]. It is clear that the total charge of the two horizontal wires A and

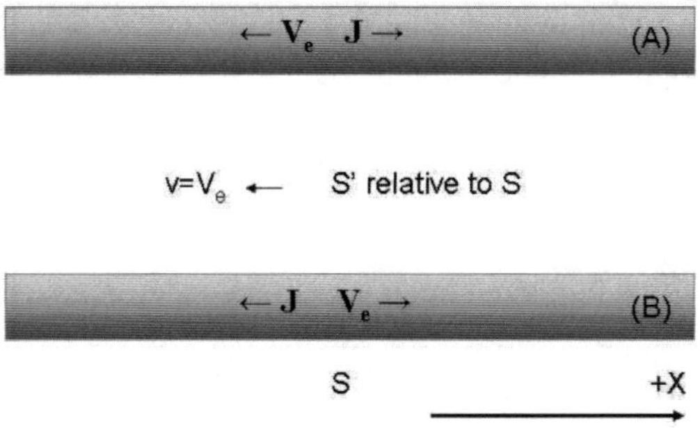

Figure 2. The current loop composed of two wires (presented) and vertical contacts at their ends (not presented). We also present the two frames of reference: S is the lab frame in which the wire is stationary and S' moves at the velocity of the electrons in wire A.

B is zero in S for the standard approach and negative, $2(1-\gamma)$, for the alternative approach. We of course expect that in S' both approaches would lead to the same charge the loop had in S, as is verified in appendix II. This is of course to be expected as in relativity the number of negative and positive charges is frame independent. But what about the situation in the alternative approach where the mere fact that we ran a current in S does not adhere to charge conservation? This is perhaps the most far reaching implication of the alternative approach.

In most experimental scenarios a current source (e.g. battery) must be embedded into the loop. An ad-hoc assumption may then be that the current source (acting as a charge reservoir) balances the hypothesized charging by being positively charged. Indeed, the previous analysis of charge conservation in the loop when changing between different frames, indicates, that the balance charge on the current source does not need to change from frame to frame. This is the same in both approaches.

However, in this work, as we aim at a feasible experimental configuration, we will not consider a loop with a current source, as this is known to give rise to a first order effect in the electron velocity [9, 21]. The latter charging is a much stronger effect and would mask the signal we are searching for [22].

When there is no current source (e.g. one introduces a current by a varying magnetic flux), and the loop charge must be preserved in S when inducing a current, two processes may take place. First, as by moving from an infinite wire to a finite loop without a charge reservoir we have changed the boundary conditions of the problem, it may be that the restoring forces are now such that the second order effect vanishes in the rest frame. Second, the current density may not be homogeneous throughout the cross-section of the wire such that some parts of the cross-section are negatively charged (where the current runs) and some positively charged, so that integrating over the cross section of the wire retains the charge neutrality. For example, if we examine the known fact that superconducting currents run along the wire edges, the expected increase in charge density where the current runs would mean that the center of the wire becomes positively charged while the edges become negatively charged. One should note that such a scenario of course creates electric fields which aim to counter-act the inhomogeneous charge distribution, but at least in the case of a superconductor, these edge currents are stable. Also in a regular conductor, the well known pinch effect, in which parallel currents attract each other thereby causing the electron density to be larger in the center of the wire, is eventually balanced by electric fields to create a stable non-homogeneous current density. In the experimental section of this paper, we will focus on superconductor loops due to the high electron velocity. High velocities may also be found in other conductors such as Graphene [23].

Let us now discuss the issue of internal dynamics. First, let us briefly remind ourselves of the importance of such internal dynamics in relativity. John Bell, perhaps best remembered for his brilliant work known as "Bell's inequalities", has clearly stated that if two particles (or space ships in his famous paradox [24]) are at rest in some system S with a distance L between them, and if in that system they are then accelerated with the same force and for the same time, the distance between them remains a constant. This must be so if relativity is to be consistent with the known laws of motion, namely, $x(t) = x(0) + \int v(t)dt$. So why are physicists, from the heavy ion collision community [7] or from other communities (e.g. [25]), confident that they should see Lorentz contraction in S? The answer is also given by Bell's paradox. If the above two particles retain their distance from one another as L, that distance in their new rest frame S' must be γL as relativity demands a factor γ between length measurements done in two different frames. So in fact, if these two particles would represent the edges of some system (such as a solid rod) this system would feel stretched in its rest frame, thus giving rise to internal dynamics such as relaxation or restoring forces. If these restoring forces shrink the system back to its equilibrium length L in the new rest frame S', then, the system must also shrink in S. A system restored to a sphere in S' (as in the heavy ion collision example) would thus look like a pancake in S.

Applying the above insight to the current carrying wire, it seems the solution to the debate would be found if one can decide if the electrons in the ion lattice are closer to a system of isolated (non-interacting) particles, in which case their distance from one another is increased in other frames (e.g. to γL in S'), or they are closer to an interacting system with strong restoring forces in some other frame, in which case we would expect their inter-particle distance in S to shrink (e.g. to L/γ if their distance in S' is restored to L) - once the current is turned on. How then can we try and describe the internal dynamics inside the wire, which we shall name "longitudinal relaxation"? One such description of the internal dynamics is given by Ohm's law, the latter being our next station.

So far, we have considered a symmetric physical system of positively and negatively charged particles (for final remarks regarding the symmetric system see [26]). This is perhaps the situation in a plasma of electrons and positrons (as well as a system of electrons and holes in a semiconductor), or perhaps, as noted in the introduction, in a wire made of two counter propagating solid rods, one made of negative ions and one of positive ions. However, for the case of a normal current carrying wire, one needs to take into account the possible asymmetry caused by the fact that the positive particles are much heavier and they are held together by strong bonds which form a stringent lattice [27]. One therefore needs to account for internal dynamics in the presence of the above asymmetry. Perhaps the most adequate formalism available for the movement of the electrons within a wire is Ohm's law. Let us therefore analyze Ohm's law in the context of the ideas presented in this paper. Could it be that the simple form of Ohm's law, $\vec{j} = \sigma\vec{E}$, is inconsistent with the $\rho \neq 0$ suggested by the alternative approach?

The covariant form of Ohm's law is [28]:

$$j^\alpha = \frac{\sigma}{c}F^{\alpha\beta}u_\beta + \frac{1}{c^2}(u_\beta j^\beta)u^\alpha \tag{6}$$

where j^α is the 4-current, $u^\alpha = \gamma(c, \vec{u})$ is the 4-velocity of the medium and $F^{\alpha\beta}$ is the field strength tensor.

This may also be written as (recalling that $j_0 = j^0$ and $j_i = -j^i$ for $i = 1-3$):

$$g_{\alpha\beta}j^\beta = \sigma F_{\alpha\beta}\beta^\beta + \beta_\alpha\beta_\beta j^\beta \tag{7}$$

where $g_{\alpha\beta}$ is the metric tensor and $\beta^\alpha = u^\alpha/c = \frac{\gamma}{c}(c, \vec{u})$.

In Appendix III we show that this reduces for $\vec{u} = 0$ to the known form of $\vec{j} = \sigma\vec{E}$. We also show in the appendix that if $u \ll c$, one finds to first order the current density $\vec{j} = \rho\vec{\beta} + \sigma(\vec{E} + \vec{\beta}\times\vec{B})$, which is the familiar form of the Hall effect.

However, most relevant for our discussion is the charge, namely, $\alpha = 0$. One then finds that Eq. 7 reduces to

$$\rho - \gamma(\gamma\rho - \gamma\vec{u}\cdot\vec{j}) = \sigma\gamma\vec{u}\cdot\vec{E} \tag{8}$$

In the rest frame of the wire, where $\vec{u} = 0$ and $\gamma = 1$, one finds the identity $0 = 0$ for any charge density ρ. This means that Ohm's law puts no constraints on what the charge density of the wire will be in the rest frame of the wire. Indeed, if we run a current through a charged capacitor plate (parallel to its plane) it will obey Ohm's law although the wire is charged. To conclude, it seems as if the alternative approach does not contradict Ohm's law in the simple way examined above, and more elaborate tests of the internal dynamics are required if one is to try and theoretically differentiate between the different approaches.

Let us now also analyze the situation in the direction transverse to the current, namely within the wire cross section, and attempt to analyze the processes dominating the "transverse relaxation". To begin with, one should note that together with the continuity equation Ohm's law (in the wire rest frame) gives

$$\begin{aligned}\nabla(\sigma\mathbf{E}) = (\nabla\sigma)\cdot\mathbf{E} + \sigma\nabla\cdot\mathbf{E} = \\ \nabla\cdot\mathbf{j} = -\dot{\rho},\end{aligned} \tag{9}$$

which together with Gauss's law $\nabla\cdot\mathbf{E} = \rho/\epsilon_0$ gives:

$$(\nabla\sigma)\cdot\mathbf{E} + \rho\sigma/\epsilon_0 = -\dot{\rho}, \tag{10}$$

and as long as σ is uniform one finds

$$\dot{\rho} = -\rho\sigma/\epsilon_0 \quad \rightarrow \quad \rho(t) = \rho(0)e^{-\sigma t/\epsilon_0}, \tag{11}$$

where this calculation does not hold for the surface where $\nabla\sigma \neq 0$. This seems to indicate that any charge in the bulk of the wire would decay to zero.

The above calculation should be met with some scepticism. First, the ϵ_0/σ time scale of roughly 10^{-20} seconds seems to indicate that the above model is not complete. Further doubt should arise when one notes there is no length scale in the equations. Such a time scale means for example that a wire with a radius smaller than the radius of a single atom would already require the electrons to move at a speed faster than that of light! More so, one would expect, for example, that there would be considerable difference in this time scale depending on whether the wire radius is much larger or much smaller than the electron mean free path of a few nano-meters. Furthermore, the idea that the spatial distribution of the extra charge would be infinitely thin (as is implied by the charging of the surface) would give rise to a very high electrostatic energy arising from the electron-electron repulsion. A more complete model would thus give a finite width distribution as a function of the charge. A similar result should come when one takes into account the finite temperature of the electrons from which one would assume that the thermal velocity would cause diffusion of the electrons against any tendency to concentrate them at a specific location.

A more complete description should include the magnetic force $ev \times B$ pushing the electrons towards the center of the wire (which may be termed a self induced Hall effect or a pinch as in plasma physics). One may combine between Eqs. 9-11 and the above, by utilizing $\mathbf{j} = \sigma(\mathbf{E} + \mathbf{v} \times \mathbf{B})$ and the continuity equation, assuming the conductivity is homogeneous, and noting that

$$\begin{aligned} -\dot{\rho} = \nabla \cdot \mathbf{j} = \\ \nabla \cdot \sigma(\mathbf{E} + \mathbf{v} \times \mathbf{B}) = \\ \rho\sigma/\epsilon_0 + \sigma\nabla \cdot \mathbf{v} \times \mathbf{B} = \\ \rho\sigma/\epsilon_0 - \sigma\mathbf{v} \cdot (\nabla \times \mathbf{B}). \end{aligned} \tag{12}$$

This gives rise in a steady state ($\dot{\rho} = 0$) to $\rho = \epsilon_0 \mathbf{v} \cdot \nabla \times \mathbf{B}$. The conclusion of Eq. 11 is thus not necessarily valid.

In appendix IV we analyze the role of the Lorentz force in different frames. It is shown that if we assume the bulk of the wire in its rest frame has no internal transverse electric field, the Lorentz force in other frames is typically not zero. Hence, if the Lorentz force is considered as the only force acting on the electrons and thus the only force responsible for the steady state, a zero transverse electric field in S does not lead to a transverse steady state in other frames. If in S there is a transverse electric field which leads to a steady state by equaling the above magnetic force, then some form of non zero transverse charge density distribution in S must result. Whether integrating over such a transverse distribution should give an overall neutrality - is another question.

Let us emphasize that Lorentz transformations should exist for every point in the wire cross section. Hence, as the transverse charge distribution changes with time in one frame, it should also change in time in other frames, and as the transverse distribution reaches a steady state in one frame its boosted distribution in another frame (boosting as usual parallel to the current or wire axis) should also present a steady state. Once this steady state is reached, there is no transverse current and $\mathbf{j} = \sigma(\mathbf{E} + \mathbf{v} \times \mathbf{B}) = \sigma\mathbf{E}_{\|}$. It remains to be seen if these transverse processes determine in some way what the allowed charging in the rest frame is or what frame is the neutral frame.

To conclude the theoretical analysis we have made, we may say the following: as a current is made to run in a wire, electrons are accelerated in the rest frame of the wire to their drift velocity. If they are to be considered as an ensemble of free particles, kinematics demands that their particle-particle distances would not change and no extra net charge density would appear. This requires the charging of the wire in other frames. If the electrons in a wire are subjected to significant longitudinal restoring forces, then the wire may be restored to neutrality faster in some other frame, in which case, the wire in its rest frame will have to be charged. Any solution would also need to take into account relaxation forces in the transverse direction. Thus a full three dimensional many-body covariant analysis is required.

Let us now examine the experimental implications with some detail. Let us start with a simple straight copper wire, being part of a square loop circuit, assuming the extra charge is provided by the current source. Copper has a density of 8.94g/cm, and an atomic number of 63.546g/mol, so there are $140685.5 \mathrm{mol/m^3}$. In 1 mole of copper there are 6.02×10^{23} atoms. Therefore in $1\mathrm{m}^3$ of copper there are about 8.5×10^{28} atoms. Copper has one free electron per atom, so the electron density is equal to $n = 8.5 \times 10^{28}$ electrons per m^3. Let us assume a typical laboratory situation in which we have a current of $I = 1$ Ampere in a wire of 1mm diameter (i.e. a cross section area of $A = 7.85 \times 10^{-7}\mathrm{m}^2$). The drift velocity can therefore be calculated to be

$$v_e = \frac{I}{nAq} = \frac{1}{(8.5 \times 10^{28}\ 7.85 \times 10^{-7}\ 1.6 \times 10^{-19})} = 9.367 \times 10^{-5} m/s \tag{13}$$

As $\beta = v/c = 9.367 \times 10^{-5}/2.99 \times 10^8 = 3.133 \times 10^{-13}$, we can now calculate the Lorentz factor γ to be equal to about

$$\gamma = \frac{1}{\sqrt{1-\beta^2}} \approx 1 + \frac{1}{2}\beta^2 \approx 1 + 4.5 \times 10^{-26} \tag{14}$$

In a 1m wire we expect to have $8.5 \times 10^{28} \times 7.85 \times 10^{-7} = 6.6725 \times 10^{22}$ electrons, so that the excess charge predicted by the alternative approach would be that of approximately 3×10^{-3} electrons per meter or $\rho_{net} = -5 \times 10^{-22}\mathrm{C/m}$. This is perhaps the strongest expected signal, namely that a positively charged probe particle should be attracted to a current-carrying wire, even when the particle is at rest relative to the wire.

It is very hard to answer the question of whether or not previous experiments should have observed such an attraction. Typically, charged particles in the vicinity of a current-carrying wire would be mostly affected by the magnetic Lorentz force. Following Eq. 3 and the approximation of γ in Eq. 14, one finds that the ratio R between the electric force and the magnetic force is

$$R = \frac{F_C}{F_L} = -\frac{v_e^2}{2c^2}/(\frac{v_e v_p}{c^2} + \frac{v_e^3 v_p}{2c^4}). \tag{15}$$

This ratio equals approximately half for small velocities and when $v_e = v_p$ and may therefore seem to point to a significant and observable effect. However, using the value for v_e calculated above and taking into account $\frac{1}{2}m_p v_p^2 = K_B T$, one finds that even for a heavy probe particle (e.g. an atomic mass of 100), the particle would have to be at a temperature of 0.1nK for the above ratio of half to be obtained. At a temperature of 1mK the effect visibility, namely R, is already at the 0.5% level. Hence very low velocity charged probe particles are needed and these have been made experimentally available only recently via the trapping of laser cooled ions [30]. However, these systems do not typically incorporate current-carrying wires close to the charged particles, and in any case, as we show in the following, the absolute effect for an ion of charge $|e|$ next to a normal conductor would be very small.

Such low velocities of probe particles, and indeed where $v_e = v_p$, may be found in a parallel current-carrying wire configuration (e.g. in the setup by which the Ampere standard is determined). However, the overall wire charge is again the very small quantity calculated above and the magnetic Lorentz force acting on all the moving charges in the probe wire would again dominate. Let us calculate the forces involved explicitly. The magnetic Lorentz force between two 1m wires is $\frac{\mu_0}{2\pi}\frac{1}{d}I_1 I_2$ which is for 1A and 1m distance 2×10^{-7}N. On the other hand, the expected electric field E(r) is (Eq. 3):

$$E(r) = (1-\gamma) \times \frac{\rho_0}{2\pi\epsilon_0 r} \approx \frac{\rho_{net}}{2\pi\epsilon_0 r} \tag{16}$$

where $\rho_{net} = -\frac{\beta^2}{2}\rho_0$. At 1m distance, the electric field is about $E = 9 \times 10^{-12}$V/m.

If the second parallel wire has the same current and therefore the same induced charge of ρ_{net}, the electric force between the wires will be on the order of 10^{-33}N, much smaller than the magnetic Lorentz force which the experiment is measuring. One may then wish to simply charge a disconnected piece of wire as a probe to the hypothesized induced charge in the current-carrying wire. Namely, to achieve via a macroscopic system $v_p = 0$.

Let us assume this probe charge to be $q_p = 1$C and see what forces we may expect. The expected electric force will be $F = Eq_p \approx 10^{-11}$N. On the other hand, the probe charge will induce a rather strong dipole in the current-carrying wire (as the internal field in the metallic wire, perpendicular to its longitudinal axis, is zero, we neglect the polarization of the atoms). An upper limit on the dipole d per unit length may be estimated as the diameter of the wire (1mm) multiplied by the charge of the free electrons per unit length which we have estimated above to be 6.6725×10^{22} $1.6 \times 10^{-19} \approx 10^4$C/m. As the force is simply the field gradient times the dipole, we find the force per unit length of the current-carrying wire to be (for a point like probe and without geometrical factors):

$$F = 10^4 \times 10^{-3} \times \frac{1}{4\pi\epsilon_0}\frac{d}{dr}\frac{q_p}{r^2} = \frac{10}{2\pi\epsilon_0}\frac{1}{r^3} \tag{17}$$

which is in the closest area to the probe (i.e. distance of 1m) of the order of $F = 1.8 \times 10^{11}$N, which is obviously huge compared to the induced electric force we would like to measure. In appendix V we show that a lower limit gives 7×10^9N.

However, if we choose $q_p = e$, the above lower limit value for the force reduces to about 10^{-28}N while the predicted electrical force is $q_p E = 1.6 \times 10^{-19} \times 9 \times 10^{-12} \approx 10^{-30}$N, not very much smaller. This may enable a relative measurement although the absolute values are extremely small. As the electric field of the wire scales as $\frac{1}{r}$ while the dipole force scales as $\frac{1}{r^3} - \frac{1}{r^2}$ for the upper and lower limits respectively, reducing the distance will not help.

One may also wish to try and measure the interaction between the electric field induced by the hypothesized charging of the current-carrying wire and the polarization it induces in atoms. In appendix VI, I explain why this is a difficult experiment to perform and consequently why previous experiments may have overlooked the effect. In appendix VII, I explicitly calculate the magnitude of this signal.

The above shows that the effect is quite elusive and it stands to reason that experiments so far have overlooked it. I present in the following an initial feasibility study of one idea that may enable the observation of the effect.

For the sake of this example, let us assume that the distribution of charges in a superconducting wire which is part of a persistent current loop, follows the current density distribution. Namely, that the excess in negative charge is taken from nearby parts of the wire through which current does not run, so that integrating over the cross section of the wire, charge neutrality is maintained. Such a charge distribution may be observable by cold ions. Inhomogeneous charge distributions in superconductors from other effects [31] may also be possible and should of course be analyzed.

As the mean velocity of electrons along the wire axis of a superconductor may be orders of magnitude higher than the typical electron drift velocity in normal conductors (simply because the cross section in which the super current runs is very small), one would expect that due to the alternative approach, there would be an excess of electrons in the edges of the wire. As noted, such an excess would create an excess of positive charges in the center of the wire. The exact current distribution in a superconductor and the calculation of the velocity is not a trivial matter [32] but let us make a simplified estimate. A superconductor may carry a current density of $10^6\mathrm{A/cm}^2$ (or $10^{-2}\mathrm{A}/\mu\mathrm{m}^2$), similar to that of a normal wire. Hence, a superconducting strip of 100μm width and 1μm thickness can carry 1A of current, similar to our example copper wire above. However, due to the concentration of current in the edges, one may roughly estimate that this 1A of current utilizes a wire cross section of about $1\mu\mathrm{m}^2$, six orders of magnitude less than in the above copper wire. This determines v_e and β to be six orders of magnitude larger than previously. Indeed, the literature discusses velocities as high as 1km/s [33]. Following Eq. 16, while using the same free electron density as in copper with the reduced cross section, the electric field also becomes six orders of magnitude larger. Hence, at 10μm distance from the wire, one would expect an electric field of about 1V/m. This field should reverse sign when moving from the edge of the wire to its center. The effect of such a field should be observable if the force of Eq it applies on the ion is similar or at least not much smaller in magnitude relative to the force applied on the ion by the trapping potential. The latter, in the harmonic approximation, is simply kx where $k = m\omega^2$ and x may be estimated as the ground state size $\sqrt{\hbar/m\omega}$. Indeed, for a cold ion trap of 10MHz frequency, both forces are on the order of 10^{-19}N. Recently, a force of about 10^{-22}N induced by an electric field of 1.8mV/m, was measured by cold ions [34]. One should note that such a measurement close to a surface may be hindered by so-called "patch potentials" [35]. Several methods may be used to bypass this problem; first, cooling to low temperatures has been found to reduce this hindering effect; second, as this effect seems to become weaker with the strong scaling of $1/d^4$, where d is the ion-surface distance, taking the ion-surface distance to $100 - 1000\,\mu$m would only reduce the signal electric field by one to two orders of magnitude, while reducing the masking fields by considerably more; third, one may oscillate the super current in direction or amplitude to form an oscillating force, and this could help purify the signal [36]. One should also take care that the ion should be in a state without a magnetic moment as the force due to the magnetic field induced by the 1A current, is significant. Finally, in the presence of an external magnetic field the Hall effect may induce a dipole field around the wire, and this should be taken into account (the Hall field is expected to be very different than the expected signal in its direction and distribution of magnitude). Noting the above considerations, it seems reasonable that adequate experimental parameter values may be found so as to ensure the visibility of the hypothesized effect.

To summarize, this paper did not attempt to give a final answer, but rather to lay down in a consistent manner (using known electrodynamics and charge conservation) the fundamental aspects of the problem, while introducing several insights beyond previous works concerning relaxation processes and experimental feasibility.

As an example of a possible alternative to the standard approach, I have presented a different approach as to how the physical situation concerning the electric and magnetic fields in the stationary frame S of a current-carrying wire, may be determined. While typically one determines the state of the system to be neutral in S (by assuming electrons are a free ensemble and therefore their particle-particle distances do not change upon acceleration), and makes use of the theory of relativity to calculate the state in other frames, one possible alternative approach begins with a physical symmetry based argument requiring that the restoring forces at work on both positive and negative charges are equal, and utilizes relativistic transformations to arrive at a different description of frame S.

In the symmetry based argument leading to neutrality in the "middle man" frame S^*, we modeled a current carrying wire as made of two counter propagating solid rods, one made of negative ions and one of positive ions, and where current comes from their relative speeds. It seems there would be a consensus that the system is completely symmetric and the charge density should be zero. Hence, if it is indeed found that the total charge density is zero in S or some frame other than S^* (in which the two rods have the same velocity), the theoretical analysis should be able to explain quantitatively what the difference is between the latter rod model and the real-life system of a current carrying wire in its rest frame and in the frame where it is neutral.

The main message of this work is three fold: First, it seems that both theoretical and experimental statements made so far, have not been able to sufficiently prove in favor of one resolution or another. Second, it seems that the rules of relativity alone may not be sufficient to solve the debate. A theoretical attempt to decide which approach is favorable, should take into account internal restoring (relaxation) processes, both transverse and longitudinal, in order to find out in which frame the equilibrium is at the $\rho_{net} = 0$ point. If this happens in several frames, a situation not compatible with the known transformations between frames, the question most relevant may be in which frame the restoring dynamics are faster. Only a full three dimensional many-body covariant analysis could perhaps decisively bring a theoretical resolution. The answer may very well be system and preparation dependent and will thus have to take into account specific boundary conditions. The third message is that an experimental feasibility study indicates that cold ions may serve as a probe which is sensitive enough to differentiate between different possibilities.

It may eventually be found that theory does not favor a specific approach [40], and choosing between them experimentally will perhaps be a formidable, yet as we have shown - possible, task.

Acknowledgments

I am sincerely thankful to Daniel Rohrlich, Yonathan Japha, Ferdinand Schmidt-Kaler, Michael Gedalin and Valery Dikovsky for their critical review of this work, and for helpful discussions. Special thanks to Ashok K. Singal, who was kind enough to repeat his arguments after nearly 20 years. Finally, I deeply thank my dear friend and colleague Carsten Henkel with whom I have had many enlightening discussions. This work was supported by the German-Israeli fund (GIF) as part of a project dedicated to cold ions near charge and current carrying structures.

Appendix I: A proposal from the 1990s

Following this work [8], I became aware of a previous presentation of a similar idea two decades ago [11] and the criticism it received [12]. In this paper we therefore re-visit the idea and introduce several novelties: first, while the previous work was criticized as being "unpalatable" [13] as it used theoretical and textual language beyond what is common practice, this work analyzes the problem within the established and accepted language and laws of physics (e.g. conservation laws). For example, the previous work states that standard electrodynamics is incorrect [14], while this work does not support such a claim. Furthermore, the significant difference between the previous formulation of the problem as well as the proposed solution, and the present one, results in different experimental predictions.

Let us note in some detail five differences between the present and the previous work:

- In [14] it is claimed that the standard electrodynamics theory is incorrect: "We have proved that in SET (standard electrodynamics) the law of invariance of charge is not well founded". The present paper makes no such claim. In this paper I claim SET is just as founded in both of the competing alternatives. Namely, I claim that there are several Lorentz invariant options to choose from.
- In [14] new and perhaps debatable theoretical constructions such as "Lorentz invariant charges" are defined, whereas in this paper only the standard elements of electrodynamics are used, and it is shown that the total charge of a current loop is conserved between frames in both theories. In the criticism made by [13] it is said: "We find Ivezic's modified definitions of the charge invariance and of the charge neutrality themselves to be unpalatable". Indeed while the previous proposal aims at charge neutrality for a wire in all frames, the present proposal makes no such claim.
- In [15] it is claimed that there is no electric force between two parallel current carrying wires in the rest frame. In this paper it is claimed that this force is not zero. Because Ref. [15] utilizes both Lorentz contraction of the wire and Lorentz enhancement of the electron density at the same time, the two effects cancel each other and the neutral wire cannot feel the electric field of the other wire. I believe that in [14, 15] there is confusion between the role of the two Lorentz effects. In the criticism made by Ref. [13] it is said: "But in order to measure the total charge in a section, he simultaneously employs two different stretches of the conductor length in any given reference frame, one for the ions and another for the electron subsystem". The same confusion is apparent in Eq. (3) of Ref. [15], where the magnetic force between two parallel wires is calculated.
- Due to the same confusion, when in Ref. [16] the situation in a stationary superconducting loop is addressed, it is said that: "This causes the moving-electron subsystem to shrink to a smaller length in the laboratory frame...the electric field caused by N moving electrons situated on a smaller, contracted, ring...". On the contrary, in this paper, no Lorentz contraction of the ring in which the electrons move is hypothesized; only the contraction of the mean distance between electrons. What is assumed is that as the current is concentrated at the edges of the superconductor, these edges will become negative while the center will become positively charged. As in the previous item, this is another experimental prediction that is different from that presented in [14, 15, 16].
- Contrary to the previous work, this paper does not require that the basic law of charge conservation be broken, as it does not claim that a bare loop becomes charged when current is made to flow.

It is therefore quite obvious that the formulation of the problem and the proposed solution presented in this paper are very different than those presented in the early 1990s.

In addition, this paper adds several new considerations. First, the previous work did not consider the asymmetry of the system due to the fact that the protons or positively charged centers that form the wire's solid lattice are much heavier than the conducting electrons. This work considers this through an analysis of Ohm's law. Most importantly, this work emphasizes the importance of longitudinal and transverse relaxation forces. Second, this work analyzes the feasibility of novel experimental techniques in observing the hypothesized effect.

Appendix II: Charge conservation

Let us describe what happens in S' in the standard approach. The vector $(\rho_{net}c, j_x)$ transforms under the Lorentz transformation matrix we have presented earlier so that in the standard approach, as $\rho_{net} = 0$ in S for any wire, one finds $\rho'_{net} = -\beta j_x \gamma / c$. As j_x for wires A and B in S is equal and opposite, it is clear that in any other frame the net charges on the two wires would be the same but with opposite sign, and so the total charge of zero is conserved in the standard approach. For completeness I also arrive at this conclusion by a direct calculation, presented below, where, as expected, the calculation reveals that the standard choice leads in S' to an equal but opposite charge density in the two wires, so that the total charge in the complete loop is maintained at zero.

Similarly, the alternative approach also leads in S' to the same charge observed in S. This is also presented below. The total charge density on both wires comes out to be $2(\gamma - \gamma^2)$ and taking into account Lorentz contraction, the total charge on the two wires is $2(1 - \gamma)$ which is equal to their charge in S. The vertical connections remain unchanged.

For a direct calculation, let us first remind ourselves why $\rho' = \rho\gamma'_0/\gamma_0$ where ρ and ρ' are the apparent charge densities, and where γ_0 and γ'_0 are the Lorentz factors due to the particles' velocity relative to the two frames of reference. Let us take the Lorentz transformation of the electric field stating $E'_\perp = \gamma^*(E_\perp + v \times B_\perp)$ [3] (in the case where β is perpendicular to E), where γ^* is due to the velocity $v = \beta c$ between the frames, and apply it to two oppositely charged parallel plates lying in the xz plane and moving with velocity $v_x = v_0 = \beta_0 c$. Let us also note that all Lorentz factors are due to movement in the x direction. The above transformation is now $E'_y = \gamma^*(E_y - vB_z)$ and as $B_z = \rho v_0/c^2\epsilon_0$, the transformation takes the form $E'_y = \rho'/\epsilon_0 = \gamma^*\rho/\epsilon_0(1 - vv_0/c^2)$. Hence $\rho'/\rho = \gamma^*(1 - vv_0/c^2)$. As $\gamma'_0 = \gamma_0\gamma^*(1 - \beta\beta_0)$ one finds that $\rho'/\rho = \gamma'_0/\gamma_0$ or $\rho' = \rho\gamma'_0/\gamma_0$.

In the standard approach, the charge density of the first wire (A) in S' is proportional to $(\gamma - 1/\gamma)$. The positive charges in the second wire (B) on the other side of the loop have exactly the same velocity relative to S' as those in wire A and so their charge density is proportional to γ just as it is in wire A. In S', the electrons in wire B have a velocity of $2v/(1 + \beta^2)$ relative to the electrons in wire A, and so one may describe the situation of the electrons in wire B as having $\gamma_0 = 1$ in their rest frame and $\gamma'_0 = (1 + \beta^2)/(1 - \beta^2)$ in S'. As $\rho = -1/\gamma$, their charge density in S' should be proportional to $-\gamma'_0/\gamma$. Hence the charge density in wire B on the other side of our loop is simply proportional to $\gamma - \gamma'_0/\gamma = 1/\gamma - \gamma$ which is just the negative of the charge density in wire A. Consequently, the total charge on the two wires in the standard approach is zero also in S'.

In the alternative approach the charge density of wire A in S' is $\gamma - 1$ (in units of ρ_0). Again, also the positive charges on wire B would experience a γ factor just as it is in wire A. In S', the electrons in wire B have again a velocity of $2v/(1+\beta^2)$ relative to the electrons in wire A, and so again $\gamma_0' = (1+\beta^2)/(1-\beta^2)$ and their charge density should thus be proportional to $-\gamma_0'$. Hence the charge density in wire B is simply $\gamma - \gamma_0'$. Adding the charge density of the two wires one finds that the total charge density is $2(\gamma - \gamma^2)$, and dividing by the Lorentz contraction one finds that the total charge is the same as it was in S.

Appendix III: Covariant Ohm's law

In this appendix we briefly examine the covariant form of Ohm's law.

Starting with,

$$g_{\alpha\beta} j^{\beta} = \sigma F_{\alpha\beta}\beta^{\beta} + \beta_{\alpha}\beta_{\beta} j^{\beta} \tag{18}$$

where $g_{\alpha\beta}$ is the metric tensor, j^{β} is the 4-current, $\beta^{\alpha} = u^{\alpha}/c = \frac{\gamma}{c}(c, \vec{u})$ is the 4-velocity of the medium and $F^{\alpha\beta}$ is the field strength tensor.

Taking $\alpha \neq 0$ one finds

$$\vec{j} - \gamma\vec{\beta}(\gamma\rho - \gamma\vec{\beta}\cdot\vec{j}) = \sigma\gamma(\vec{E} + \vec{\beta}\times\vec{B}) \tag{19}$$

If $\vec{u} = 0$ one finds $\vec{j} = \sigma\vec{E}$. If $|\vec{u}| \ll c$, and one takes the first order in $\gamma \approx 1 + \frac{1}{2}\beta^2$, one finds $\vec{j_{tot}} = \vec{j_e} + \vec{j_m} = \rho\vec{\beta} + \sigma(\vec{E} + \vec{\beta}\times\vec{B})$, where j_e and j_m are the dominant terms in the so-called Galilean electric ($E \gg cB$) and magnetic ($cB \gg E$) limits.

Appendix IV: The Lorentz force in different frames

Here, as an example, I calculate the Lorentz force on an electron inside a current carrying wire, in the direction perpendicular to the current, if there is no transverse electric field in the wire rest frame. The boost between frames is in the direction of the current. If we expect, in other frames, to reach a steady state in the transverse charge distribution, the transverse Lorentz force $F_{\perp}$, must be zero (if we assume it is the only force acting on the electrons).

Let us denote the wire axis as $\hat{x}$ so that the current density is $j\hat{x}$ (the total current is I). The wire has cylindrical symmetry and its radius is R. For simplicity (and hopefully without loss of generality) we look at an electron situated on the $\hat{y}$ axis having a distance $0 < r < R$ from the center of the wire, so that the probe point is $r\hat{y}$ ($z = 0$). This electron is moving with the current at a drift velocity. We will work in cgs units.

We first note that in the wire rest frame the electric fields at the probe point are (we denote x, y, z as $1, 2, 3$):

$$\begin{aligned} E_1 &= Const \\ E_2 &= 0 \\ E_3 &= 0 \end{aligned} \tag{20}$$

where E_1 drives the current and $E_{2,3} = 0$ because there is no charge inside the wire and due to Gauss's law.

Similarly, the magnetic fields are:

$$
\begin{aligned}
B_1 &= 0 \\
B_2 &= 0 \\
B_3 &= \frac{\mu_0}{2\pi} Ir/R^2
\end{aligned}
\tag{21}
$$

where from Ampere's law $\int \vec{B} \cdot \vec{dl} = \mu_0 I_c$ with $I_c/I = \pi r^2/\pi R^2$. As $\vec{I} = I\hat{x}$, the velocity of our probe electron is $\vec{v_e} = v_e(-\hat{x})$ and we find that the Lorentz force $ev_e \times B$ is *away* from the surface and towards the center of the wire i.e. in the direction of $-\hat{y}$. This is to be expected as what we have calculated is analogous to the attraction between two parallel currents. It is however an interesting result in the fact that the Lorentz force works against the charge tendency to concentrate at the edges.

Let us now find the electric and magnetic fields in other frames. We denote V (or $\beta = V/c$) as the boost velocity. Lorentz transformations of the fields are:

$$
\begin{aligned}
E'_1 &= E_1 = Const \\
E'_2 &= \gamma(E_2 - \beta B_3) = -\gamma\beta\frac{\mu_0}{2\pi} Ir/R^2 \\
E'_3 &= \gamma(E_3 + \beta B_2) = 0
\end{aligned}
\tag{22}
$$

and

$$
\begin{aligned}
B'_1 &= B_1 = 0 \\
B'_2 &= \gamma(B_2 + \beta E_3) = 0 \\
B'_3 &= \gamma(B_3 - \beta E_2) = \gamma\frac{\mu_0}{2\pi} Ir/R^2.
\end{aligned}
\tag{23}
$$

As the total Lorentz force is $F = q(E + \frac{1}{c} v \times B)$ we have

$$
F_\perp = e(E'_2 + \frac{1}{c} v'_e \times B'_3) = -\frac{e\mu_0}{2\pi}\frac{Ir}{R^2}\gamma(\frac{v'_e}{c} + \beta),
\tag{24}
$$

where v'_e is the electron velocity in the new frame.

According to relativity $v'_e = (v_e - V)/(1 - v_e V/c^2)$ and so for the Lorentz force to be zero we need to demand

$$
\begin{aligned}
-\frac{1}{c}\frac{v_e - V}{1 - v_e V/c^2} &= \frac{V}{c} \\
&OR \\
v_e &= 0.
\end{aligned}
\tag{25}
$$

This outcome means that if we have current and we want a steady state in all frames, we must have transverse electric fields in the wire rest frame.

What happens when our assumption $E_2 = 0$ is not made? From symmetry arguments one may conclude that there is a radial electric field all around the cross-section of the wire. Then, if we introduce a closed sphere of surface A into the bulk of the wire and recall that E_1 is homogeneous, we may use the integral form of Gauss's law $\epsilon_0 \int \mathbf{E} \cdot \mathbf{dA} = \rho$ to conclude that the region internal to the sphere is charged.

Assuming $E_\perp \neq 0$, the conclusion of a non uniform charge density can be made quantitative if a steady state exists in the rest frame, namely (up to factors of c), $E_\perp = -\mathbf{v} \times \mathbf{B}$. Since $\nabla \cdot \mathbf{E}_{||} = 0$ and $\mathbf{v}$ is constant, we may write:

$$\rho = \epsilon_0 \nabla \cdot \mathbf{E}_\perp = \epsilon_0 \mathbf{v} \cdot (\nabla \times \mathbf{B}). \tag{26}$$

where the last equality comes from the fact that $\rho = \epsilon_0 \nabla \cdot \mathbf{E}_\perp = -\epsilon_0 \nabla \cdot \mathbf{v} \times \mathbf{B} = -\epsilon_0 \mathbf{B} \cdot \nabla \times \mathbf{v} + \epsilon_0 \mathbf{v} \cdot \nabla \times \mathbf{B}$ [29]. $\nabla \times \mathbf{v} = 0$ is termed an "irrotational flow" (no turbulence) and may also come from the assumption that v is proportional to E ($j = \sigma E$) and that $\nabla \times \mathbf{E}$ is the time derivative of the magnetic field which is zero in steady state.

As Ampere's law implies $\nabla \times \mathbf{B} = \mu_0 \mathbf{J} = \mu_0 \rho_- \mathbf{v}$, where ρ_- is the charge density of the free electrons in the steady state, we have:

$$\rho = \rho_+ + \rho_- = \epsilon_0 \mu_0 \rho_- v^2 = \rho_- v^2/c^2, \tag{27}$$

namely $\rho_- = -\rho_+/(1 - v^2/c^2) = -\gamma^2 \rho_+$, which means that there is charging, or at least a non-uniform transverse charge density.

Appendix V: A lower limit on the induced dipole of a wire

We have made a rough estimate for an upper limit of the dipole of a metallic wire. This estimate is perhaps too large as it takes the maximal possible value independent of the size of the inducing charge. Let us now derive a lower limit by making use of the known induced dipole in a grounded metallic sphere. We simulate our 1mm diameter wire by a line of spheres with radius $a = 0.5$mm, with their centers along the x axis at locations x_i. We place the charge on the z axis at z_0. According to Eq. 2.6 in Jackson [37] the force between a probe charge q_p and a single sphere is expected to be:

$$|F| = \frac{q_p}{4\pi\epsilon_0} \frac{a q_p}{r_i} \frac{1}{[r_i^2 (1 - \frac{a^2}{r_i^2})^2]} \tag{28}$$

where $r_i = \sqrt{x_i^2 + z_0^2}$ is the distance between the probe charge and the center of the sphere, the second ratio is the image charge, and the last ratio is simply the Coulomb force factor of one over distance square, where the distance is calculated between the probe charge and the image charge in the sphere.

If we omit terms of a/r_i with a high power, as we assume $a \ll z_0$, and we also position spheres symmetrically along the x axis so that only the vertical force component counts, we find the total force to be:

$$F_{tot} \approx \Sigma_i \frac{q_p^2}{a^2 4\pi\epsilon_0} \frac{a^3}{r_i^3} \frac{z_0}{r_i} = \Sigma_i \frac{q_p^2}{4\pi\epsilon_0} \frac{a z_0}{r_i^4} \approx \int_{-\infty}^{\infty} \frac{dx}{2a} \frac{q_p^2}{4\pi\epsilon_0} \frac{a z_0}{r_i^4}. \tag{29}$$

We then find that

$$F_{tot} \approx \frac{q_p^2 z_0}{8\pi\epsilon_0} \int_{-\infty}^{\infty} \frac{dx}{(x^2 + z_0^2)^2} = \frac{q_p^2}{8\pi\epsilon_0 z_0^2} \int_{-\infty}^{\infty} \frac{du}{(1 + u^2)^2} = = \frac{q_p^2}{16\epsilon_0 z_0^2}. \quad (30)$$

Putting in a probe charge of 1C at a distance of 1m, we find a force of 7×10^9N, only about two orders of magnitude smaller than our upper limit value. Note that in this approximation the force is not dependent on the diameter of the wire.

Appendix VI: The experimental challenge of induced polarization in the electric field of a wire

Atom chips are devices in which isolated ultra cold atoms (in vacuum) are typically trapped in magnetic traps, microns away from the surface of the chip. Current-carrying wires on the surface of the chip provide the trapping magnetic fields. The magnetic fields interact with the magnetic moment of the neutral atoms (see our review [38]).

In a recent experiment we have carried out with an atom chip, we have also induced an electric field which interacts with the induced polarizability of the atoms, so that the neutral atoms interact with both fields simultaneously [39].

In general, aside from simple situations such as between capacitor plates, electric fields are quite hard to accurately engineer as they very much depend on where the ground is or in general, what are the potential surfaces in the vicinity of the charge. In our atom chip work where we combined electric and magnetic fields to create a lattice of traps, we have concentrated thousands of electrons on our surface electrodes in the vicinity of the neutral atoms. The charging and relatively high electric field were achieved by utilizing a capacitor configuration on the surface of the chip. The typical parameters of the setup we have used were $100 - 500$V on a capacitance of about 10^{-17}F coming from an electrode area of $2 \times 100\mu\text{m}^2$ and an electrode distance of a few tens of microns. The electric force is attractive and attempts to pull the atoms towards the surface while the magnetic force creates a barrier against such a crash. The magnetic force is simply $F_m = \mu_B \times dB/dr$, where $\mu_B = 1.4 \times 10^6\text{Hz/G}$ is the Bohr magneton and the typical magnetic gradient at an atom-surface distance of $50 - 100\mu$m is $1 - 10$kG/cm. This gives a magnetic force of about 10^{-22}N. As the electric force was seen to change the magnetic potentials but on the other hand was not able to crash the atoms to the chip surface, it is clear that its magnitude was only about 10^{-22}N or less.

As noted, the charging was made possible due to a capacitor configuration. The capacitor also created quite a strong field of about 10^6V/m, which gave rise to a potential modulation of about 100μK in the atom magnetic trapping potential. Having in a similar configuration (e.g. a current-carrying wire close to a ground potential) an excess of 10^{-3} electrons, as expected by the alternative approach, would give rise to a field which is smaller by $6 - 7$ orders of magnitude and consequently to a potential modulation which is smaller by $12 - 14$ orders. This would not be observable.

Appendix VII: Possible experiments involving induced polarization

Let us now explicitly calculate the electric force acting on a neutral atom next to a current-carrying wire (in SI units). We use the previous calculations of the hypothesized field induced by a wire carrying 1A of current. The field we found in Eq. 16 was $E(r) = 9 \times 10^{-12}\frac{1}{r}$V/m. As the electric potential is just $V(r) = -\frac{1}{2}\alpha E^2(r)$, where α is the atomic polarizability which is typically 5×10^{-39}Cm^2/V and αE is the induced dipole, we expect a force of:

$$F = -\frac{1}{2}\alpha\frac{dE^2(r)}{dr} = 81 \times 10^{-24}\ 2.5 \times 10^{-39}\frac{1}{r^3} \tag{31}$$

which, at an atom-distance of 1μm, should amount to about 2×10^{-43}N, an extraordinarily small number. The potential modulation (in degrees Kelvin), mentioned in appendix VI and equal to $\Delta T = -\frac{1}{2}\alpha E^2(r)/k_B$, where $k_B = 1.38 \times 10^{-23}$J/K is the Boltzman factor, would also be much too small to be detectable, even with ultra cold atoms.

Let us make the same calculation for the induced dipole of a neutral metallic wire of unit length. In Eq. 17 we have found that the induced dipole d of such a wire is bounded from above by d = 1Cm per unit length. For the purposes of this example, we use this value as our estimate for the dipole strength. Using the same E(r) noted above we find the force on a piece of wire of unit length to be:

$$F = d\frac{dE(r)}{dr} = 9 \times 10^{-12}\frac{1}{r^2} \tag{32}$$

which, at a distance of say 1m (the Ampere standard setup), amounts to about 10^{-11}N, four orders of magnitude smaller than the force the Ampere standard setup was meant to measure. However, moving the wires to a distance of 1mm (again, current is running in only one wire), the force would now have a magnitude of 10^{-5}N, and this may be measurable. However, while the upper limit used here scales as q, the lower limit (as calculated in appendix V) scales as q^2, thus giving rise for the expected 10^{-3} electrons to a difference of 10^{-22} in the magnitude of the force.

[1] James Clerk Maxwell, A dynamical theory of the electromagnetic field, Philosophical Transactions of the Royal Society of London **155**, 459512, (1865). (This article accompanied a December 8, 1864 presentation by Maxwell to the Royal Society.)
[2] Albert Einstein, Zur Elektrodynamik bewegter Körper, Annalen der Physik **17**, 891 (1905). English translation: On the Electrodynamics of Moving Bodies, by George Barker Jeffery and Wilfrid Perrett (1923); Another English translation On the Electrodynamics of Moving Bodies by Megh Nad Saha (1920).
[3] John David Jackson, Classical Electrodynamics (3rd edition), John Wiley and Sons Inc. (1999), Sections 11.9-11.10.
[4] D. F. Bartlett and B. F. L. Ward,Is an electron's charge independent of its velocity?, Phys. Rev. D **16**, 3453 (1977).
[5] C.S. Kenyon and W.F. Edwards, Test of current-dependent electric fields, Physics Letters A **156**, 391 (1991).
[6] See Ref. [2]. A nice describtion of the difference between the measured reality or contraction which is of interest to us here and seeing contraction may be found at: http://www.spacetimetravel.org/tompkins/node1.html.
[7] A nice illustration of the pancake shape of an accelerated nucleus may be found in: http://www.bnl.gov/rhic/physics.asp
[8] The idea that $\rho = 0$ is not a unique solution for the rest frame of the wire and that the Lorentz transformations could be, for example, applied symmetrically, came to my mind while teaching in 2009-2010 an undergraduate course in electromagnetism. I submitted the work for publication in November 2010.
[9] Andre Koch Torres Assis and Julio Akashi Hernandes, *The Electric Force of a Current*, Apeiron Montreal (2007).
[10] Dragan V. Redzic, Charges and fields in a current-carrying wire, Eur. J. Phys. **33**, 513 (2012); arXiv:1201.0918 (31 December 2011).
[11] Tomislav Ivezić, The "relativistic" electric fields arising from steady conduction currents, Phys. Lett. A **144**, 427 (1990); Tomislav Ivezić, The "relativistic" electric fields produced by steady currents: comparison

to experiments, Phys. Lett. A **156**, 27 (1991); Tomislav Ivezić, Electric fields from steady currents and unexplained electromagnetic experiments, Phys. Rev. A **44**, 2682 (1991); Tomislav Ivezić, Reply to 'Comment on "Electric fields from steady currents and unexplained electromagnetic experiments" ',Phys. Rev. E **48**, 4140 (1993); Tomislav Ivezić, The charge neutrality of a current-carrying consuctor, Phys. Lett. A **175**, 265 (1993).

[12] Ashok K. Singal, On the charge invariance and relativistic electric fields from a steady conduction current, Phys. Lett. A **162**, 91 (1992); Ashok K. Singal, The charge neutrality of a conductor carrying a steady current, Phys. Lett. A **175**, 261 (1993).

[13] Ashok K. Singal, Comment on "Electric fields from steady currents and unexplained electromagnetic experiments", Phys. Rev. E **48**, 4138 (1993).

[14] Tomislav Ivezić, The "relativistic" electric fields arising from steady conduction currents, Phys. Lett. A **144**, 427 (1990).

[15] Tomislav Ivezić, The "relativistic" electric fields produced by steady currents: comparison to experiments, Phys. Lett. A **156**, 27 (1991).

[16] Tomislav Ivezić, Electric fields from steady currents and unexplained electromagnetic experiments, Phys. Rev. A **44**, 2682 (1991).

[17] P. C. Peters, In what frame is a current-carrying conductor neutral?, Am. J. Phys. **53**, 1165 (1985).

[18] Michael H. Brill, David Dameron, and Thomas E. Phipps, Jr., A misapprehension concerning electric current neutrality, Physics Essays **24**, 325 (2011).

[19] Richard P. Feynman, Robert B. Leighton and Matthew Sands, The Feynman Lectures on Physics (1964), Volume II, Section 13.6.

[20] a nice introduction to the two wire scenario is given in: Edward M. Purcell, Electricity and Magnetism: Berkeley Physics Course Volume 2 (1965,85).

[21] W.G.V. Rosser, Classical electromagnetism via relativity, Butterworths, London (1968); Oleg Jefimenko, Demonstration of the Electric Fields of Current-Carrying Conductors, Am. J. Phys. **30**, 19 (1962); J. A. Hernandes and A. K. T. Assis, The potential, electric field and surface charges for a resistive long straight strip carrying a steady current, Am. J. Phys. **71**, 938 (2003), Manoj K. Harbola, Energy flow from a battery to other circuit elements: Role of surface charges, Am. J. Phys. **78**, 1203 (2010).

[22] If one does want to use a current source, there may be several way to differentiate between the first and second order signals. For example, charges due to the first order effect are expected to vary longitudinally and furthermore they are expected to change sign with current direction and hence average to zero in an AC current. On the contrary, the charges hypothesized by the second order effect should not vary longitudinally and should not change sign with current direction. In addition, if indeed first order surface charges are required to maintain a current or an energy flow, they are not expected to decay with time in any frame in which there is a current. On the contrary, the total charging per unit length hypothesized here should decay, at least in some frame, where the relaxation processes push the wire towards neutrality. These differences may enable to observe the much smaller relativistic effect.

[23] R.S. Deacon et al., Cyclotron Resonance study of the electron and hole velocity in graphene monolayers, arxiv 0704.0410 (2007).

[24] John Stewart Bell, Speakable and unspeakable in quantum mechanics, Cambridge: Cambridge University Press (1987). See also Bells spaceship paradox in wikipedia and references therein.

[25] Aleksandar Gjurchinovski, Reflection of light from a uniformly moving mirror, American Journal of Physics **72**, 1316 (2004).

[26] Imagine if you will the replacement of a normal metallic wire by two solid rods, one made of positive ions and the other of negative ions. A current carrying wire may then be simulated by simply moving the negative rod with the drift velocity v_e in the frame S of the positive rod. Now let us construct a frame S^* moving at velocity V relative to S, in which both rods move in opposite directions with the same velocity. Due to Lorentz transformations, this velocity will not exactly be $v_e/2$ but that is of no importance. (utilizing the rule for velocity transformations $u' = (u - V)/(1 - uV/c^2)$ the reader may easily convince himself that V is simply the solution of the quadratic equation $v_e/c^2 \times V^2 + 2 \times V + v_e = 0$). As the situation in S^* is completely symmetric, it does not matter if a velocity affects charge density or not; it is clear that what ever the physics is, the total charge density is zero. The only assumption required is that the sign of the velocity vector does not change the charge density in magnitude or sign. Utilizing the rule which has been presented in appendix II, namely $\rho' = \rho\gamma_0'/\gamma_0$, where ρ and ρ' are the apparent charge densities, and where γ_0 and γ_0' are the Lorentz factors due to the particles' velocity relative to the two frames of reference, one easily finds that transforming to S, the charge density there is proportional to $1 - \gamma$ as predicted by the alternative approach.

[27] Another type of asymmetry which should perhaps be considered lies in the fact that only for the solid lattice of the positive charges may it be claimed that there exists a frame in which all the particles in the closed

loop are at rest. No such frame exists for the electrons, when the current is non-zero. However, this is exactly the situation in a ring accelerator and nevertheless we expect to see Lorentz contraction of a nucleus everywhere along its circumference. In this sense, the loop wire may be viewed as a ring accelerator. It is therefore unclear why this fact should favor the standard approach over the alternative approach.

[28] John David Jackson, Classical Electrodynamics (3rd edition), John Wiley and Sons Inc. (1999), problem 11.16.

[29] W. G. V. Rosser, W. G. V., Magnitudes of surface charge distributions associated with electric current flow, Am. J. Phys., Vol. **38**, 265 (1970).

[30] Dietrich Leibfried, Rainer Blatt, Christopher Monroe, David Wineland, Quantum dynamics of single trapped ions, Review of Modern Physics **75**, 281 (2003).

[31] J. E. Hirsch, Explanation of the Tao Effect: Theory for the Spherical Aggregation of Superconducting Microparticles in an Electric Field, Phys. Rev. Lett. **94**, 187001 (2005); S. V. Yampolskii and B. J. Baelus and F. M. Peeters and J. Kolacek, Vortex charge in mesoscopic superconductors, Phys. Rev. B **64**, 144511 (2001).

[32] D. Saint-James, G. Sarma, and E.J. Tomas, Type II Superconductivity, Pergamon, Oxford (1969).

[33] Todor. M. Mishonov, Technique for measuring the Cooper-pair velocity, density and mass using Doppler splitting of the plasmon resonance in low-dimensional superconducting microstructures, Physical Review B **50**, 4004 (1993).

[34] Michael J. Biercuk, Hermann Uys, Joe W. Britton, Aaron P. VanDevender and John J. Bollinger, Ultrasensitive detection of force and displacement using trapped ions, Nature nanotechnology **5**, 646 (2010)

[35] N. Daniilidis, S. Narayanan, S. A. Möller, R. Clark, T. E. Lee, P. J. Leek, A. Wallraff, St. Schulz, F. Schmidt-Kaler and H. Häffner, Fabrication and heating rate study of microscopic surface electrode ion traps, New Journal of Physics **13**, 013032 (2011).

[36] S. Knünz, M. Herrmann, V. Batteiger, G. Saathoff, T. W. Hänsch, K. Vahala, and Th. Udem, Injection Locking of a Trapped-Ion Phonon Laser, Phys. Rev. Lett. **105**, 013004 (2010).

[37] John David Jackson, Classical Electrodynamics (3rd edition), John Wiley and Sons Inc. (1999).

[38] Ron Folman, Peter Krüger, Jörg Schmiedmayer, Johannes Denschlag and Carsten Henkel, Microscopic atom optics: from wires to an atom chip, Advances in Atomic, Molecular and Optical Physics **48**, 263 (2002).

[39] Peter Krüger, Xiao Luo, Matthias W. Klein, Karolina Brugger, Albrecht Haase, Stephan Wildermuth, Sönke Groth, Israel Bar-Joseph, Ron Folman and Jörg Schmiedmayer, Trapping and manipulating neutral atoms with electrostatic fields, Phys. Rev. Lett. **91**, 233201 (2003).

[40] The conclusion of Redzic's paper [10] that "it appears that the problem of in what frame is an infinite current-carrying wire neutral does not have a unique solution" emphasizes the importance of the experimental considerations detailed in this work.

The Lorentz transformations of the vectors E, B, P, M and the external electric fields from a stationary superconducting wire with a steady current and from a stationary permanent magnet

Tomislav Ivezić
Ruđer Bošković Institute, P.O.B. 180, 10002 Zagreb, Croatia
E-mail: ivezic@irb.hr

Abstract. In the first part of this paper we review the fundamental difference between the usual transformations of the three-dimensional (3D) vectors of the electric field **E**, the magnetic field **B**, the polarization **P**, the magnetization **M** and the Lorentz transformations of the 4D geometric quantities, vectors E, B, P, M, with many additional explanations and several new results. In the second part, we have discussed the existence of the electric field vector E outside a *stationary* superconducting wire with a *steady current* and also different experiments for the detection of such electric fields. Furthermore, a fundamental prediction of the existence of the external electric field vector E from a *stationary* permanent magnet is considered. These electric fields are used for the resolution of the "charge-magnet paradox" with 4D geometric quantities for a qualitative explanation of the Aharonov-Bohm effect in terms of fields and not, as usual, in terms of the vector potential and for a qualitative explanation that the particle interference is not a test of a Lorentz-violating model of electrodynamics according to which a magnetic solenoid generates not only a static magnetic field but also a static electric field.

1. Introduction

1.1 About this paper

Both, in the prerelativistic physics and in Einstein's formulation of special relativity (SR) [1] the electric and magnetic fields are represented by the 3-vectors $\mathbf{E}(\mathbf{r},t)$ and $\mathbf{B}(\mathbf{r},t)$. The notation is as in [2], i.e. **E** and **B** are called 3-vectors and they are designated in boldface type. In the whole physical literature after [1] the usual transformations of the 3-vectors **E** and **B**, the last equations in §6., II. Electrodynamical Part, [1] or, e.g., Eqs. (11.148) and (11.149) in [2], i.e., Eq. (10) here, are always considered to be the relativistically correct Lorentz transformations (LT) (boosts) of **E** and **B**. Here, in the whole paper, under the name LT we shall only consider boosts. They are first derived by Lorentz [3] and Poincaré [4] (see also two fundamental Poincaré's papers with notes by Logunov [5]) and independently by Einstein [1] and subsequently derived and quoted in almost every textbook and paper on relativistic electrodynamics. Henceforward, these usual transformations of the 3-vectors will be called the Lorentz-Poincaré-Einstein transformations (LPET), according to physicists who discovered them. The main feature of the LPET of **E** and **B** is that *the components of the transformed* $\mathbf{E}'$ *are expressed by the mixture of components of* **E** *and* **B**, *and similarly for* $\mathbf{B}'$, Eq. (11.148) in [2]. The electric field **E** in one frame is "seen" as slightly changed electric field $\mathbf{E}'$ and an *induced magnetic field* $\mathbf{B}'$ in a relatively moving inertial frame.

However, it has recently been proved both in the tensor formalism and in the geometric algebra formalism [6-11] that *these LPET ARE NOT the LT.* They drastically differ from the LT of the relativistically correct 4D geometric quantities, which represent the electric and magnetic fields. In contrast to the LPET, *the correct LT always transform the 4D algebraic object representing the electric field only to the electric field; there is no mixing with the magnetic field.* This fundamental difference between the LPET of the 3-vectors and the LT of the 4D geometric quantities is considered in the first part of this paper.

It is worth mentioning that some experimentally verifiable consequences of that fundamental difference between the LPET and the LT have been examined in several papers. Thus, it is shown in [7] that the conventional theory with the 3D **E** and **B** and their LPET (10) yields different values for the motional emf ε for relatively moving inertial observers, $\varepsilon = UBl$ and $\varepsilon = \gamma UBl$, whereas the approach with 4D geometric quantities and their LT (42), i.e., (43), always yields the same value for ε, which is defined as a Lorentz scalar, $\varepsilon = \gamma UBl$. This result is very strong evidence that the usual approach is not relativistically correct. If the experimentalists find a way to measure the emf ε for the considered problem of a conductor moving in a static magnetic field, not only for small velocities, $U \ll c$, they should see that in the laboratory frame $\varepsilon = \gamma UBl$ and not simply $\varepsilon = UBl$. That problem is of a considerable importance in practice. A similar discussion was presented for the Faraday disk in [8]. In [12, 13] the Trouton-Noble paradox is considered. It is shown that in the geometric approach with 4D quantities *the 4D torques* will not appear for the moving capacitor if they do not exist for the stationary capacitor, which means that with 4D geometric quantities the principle of relativity is naturally satisfied and there is not the Trouton-Noble paradox. The same conclusion holds in the low-velocity approximation $\beta \ll 1$, or $\gamma \simeq 1$. Very similar paradox to the Trouton-Noble paradox is Jackson's paradox. It is discussed in detail in [14]; the second paper is a simpler, more pedagogical, version of the first one. There, in [14], it is also shown that there is no paradox in the approach with 4D geometric quantities and their LT.

The most important experimentally verifiable consequence of the difference between the LPET and the LT refers to the existence of the electric field vector outside a *stationary* superconducting wire with a *steady current* and also outside a *stationary* permanent magnet. The second part of this paper, Secs. 7.1, 7.2, 8, is devoted to that problem. In Sec. 9.2, these electric fields are used for the resolution of the "charge-magnet paradox" in terms of 4D geometric quantities without introducing some "hidden" quantities and without changing the expression for the Lorentz force, but as a 4D geometric quantity, like the expression for the Lorentz force density k_L (67). Furthermore, in Sec. 10, these electric fields are used for a qualitative explanation of the Aharonov-Bohm effect in terms of fields and not, as usual, in terms of the vector potential and in Sec. 11, for a qualitative explanation that the particle interference is not a test of a Lorentz-violating model of electrodynamics according to which a magnetic solenoid generates not only a static magnetic field but also a static electric field.

The geometric approach to special relativity that is used in, e.g., [6-14], is called the invariant special relativity (ISR). In the ISR, it is considered that in the 4D spacetime the 4D geometric quantities are well-defined both theoretically and *experimentally;* they have an independent physical reality. The principle of relativity is automatically satisfied if the physical laws are expressed in terms of the 4D geometric quantities. It is not so in the SR [1] in which it is considered that the 3D quantities have an independent physical reality. There, the principle of relativity is postulated and it is supposed that it holds for physical laws expressed in terms of the 3D quantities, e.g., Maxwell's equations written in terms of the 3-vectors **E** and **B**. In the ISR, physical quantities are represented by the abstract, coordinate-free, 4D geometric quantities. In the papers [6-14] these quantities are treated as either tensors as geometric objects, e.g., in [6,

8, 10], or, multivectors in the geometric algebra formalism, e.g., in [7-9], [11-14]. If some basis has been introduced, these coordinate-free quantities are represented as 4D coordinate-based geometric quantities (CBGQs) comprising both components and *a basis*. Every 4D CBGQ is invariant under the passive LT; the components transform by the LT and the basis by the inverse LT leaving the whole CBGQ unchanged. This is the reason for the name ISR. The invariance of a 4D CBGQ under the passive LT reflects the fact that such mathematical, invariant, 4D geometric quantity represents *the same physical quantity* for relatively moving inertial observers, see, e.g., Eqs. (49), (73), (82) and (87) here. Hence, it can be stated that in the ISR only quantities that do not change upon the passive LT have an independent physical reality, both theoretically and *experimentally*. In contrast to it the SR [1] deals with the Lorentz contraction, the time dilation and the LPET of the 3D vectors **E** and **B**. However, e.g., the rest length and the Lorentz contracted length are not the same 4D quantity for relatively moving observers, since the transformed length $L_0(1-\beta^2)^{1/2}$ is different than the rest length L_0, see, e.g., Eq. (89) in Appendix. Rohrlich [15] named the Lorentz contraction and other transformations which do not refer to the same 4D quantity as the "apparent" transformations (AT), whereas the transformations which refer to the same 4D quantity as the "true" transformations, e.g., the LT. Hence, the other name for the ISR is the "True transformations relativity" ("TT relativity"), which is used, e.g., in [16, 17]. As proved in [6-11] and exposed here in Secs. 3 - 3.2 and Sec. 6, *the LPET are also the AT and not the LT.* In the 4D spacetime, as shown in detail in [16, 17], instead of the Lorentz contraction and the time dilation one has to consider the 4D geometric quantities, the distance vector l_{AB}, Eq. (85) here, and the spacetime length, Eq. (86) here, which properly transform under the LT, see Appendix here.

1.2 An outline of this paper

An outline of the present paper is as follows. In Sec. 2, a short review of the geometric algebra formalism is presented. For more detail see [18]. An important result from [19] is mentioned in that section. Namely, *what is essential for the number of components of a vector field is the dimension of its domain. Hence, the usual time-dependent* $\mathbf{E}(\mathbf{r},t)$, $\mathbf{B}(\mathbf{r},t)$ *cannot be the 3-vectors, since they are defined on the spacetime.* They are correctly defined geometric quantities, e.g., vectors (4-vectors in the usual notation) $E(x)$, $B(x)$, where x is the position vector.

Then, in Secs. 3 - 3.2, we discuss the traditional derivation of the LPET of the 3-vectors **E** and **B**, Eq. (10), and **P** and **M**, Eq. (16), or Eq. (17). As already stated, the main feature of the LPET is that, e.g., *the transformed* $\mathbf{E}'$ *is expressed by the mixture of the 3-vectors* **E** *and* **B**, *and similarly for* $\mathbf{B}'$. As shown in Sec. 3.1, for the derivation of (10) one first makes the identification of the six independent components of $F^{\alpha\beta}$ with six components of the 3-vectors **E** and **B**, Eq. (3). Then, it is simply argued that six independent components of $F'^{\alpha\beta}$ are the "Lorentz transformed" components E'_i and B'_i, Eq. (7), i.e., *the LPET of the components of* **E** *and* **B** *are derived assuming that they transform under the LT as the components of* $F^{\alpha\beta}$ *transform*, Eq. (11.148) in [2]. However, it is shown in that section that the identifications (3) and (7) depend on the chosen synchronization and that they are meaningless for some nonstandard synchronization, e.g., the "radio" synchronization, see Eqs. (4) and (5), which means that the LPET (10) (and (16), or (17)) are not the relativistically correct LT. In Sec. 3.2 the same discussion is presented for the derivation of the LPET of **P** and **M**, Eq. (16), or Eq. (17).

In Sec. 4, the definitions of vectors E, B, Eqs. (21) and (22), and P, M, Eqs. (25) and (26), in terms of F, v and $\mathcal{M}$, u, respectively, are examined; v is the velocity vector of the observers who measure E and B fields, while u is the velocity vector of a moving medium. It is visible from (22) that *in mathematically correct definitions the vectors* $E(x)$ *and* $B(x)$ *are derived from* F *AND* v, i.e., *they are defined with respect to the observer.* Similarly, it is visible from (26) that P and M depend not only on $\mathcal{M}$ but on u as well. Furthermore, the basic Lorentz invariant field equation for vacuum with F, Eq.(20), is written in terms of E and B, Eq. (27), i.e., Eqs. (28) and (29). The generalization of these field equations to the electromagnetic field equations for moving media is presented in [20] and also briefly considered in this section. The generalization of (20) to a moving medium is obtained simply replacing F by $F + \mathcal{M}/\varepsilon_0$, which yields Eqs. (30), the primary equations for the electromagnetism in moving media with bivectors $F(x)$ and $\mathcal{M}(x)$. Then, these equations are written with vectors $E(x)$, $B(x)$, $P(x)$ and $M(x)$, Eqs. (33) and (34). As stated in [20], Eq. (30), i.e., Eqs. (33) and (34), comprise and generalize all usual Maxwell's equations with 3-vectors for moving media. The equations (33) and (34) contain both the velocity vector u of a moving medium and the velocity vector v of the observers who measure E and B fields. They are first reported in [20] and do not appear in the previous literature.

In Sec. 5, the LT of vectors E and B, as 4D geometric quantities, are examined and compared with Minkowski's results. Note that Minkowski, Sec.11.6 in his famous paper [21], was the first who introduced vectors (in the usual notation 4-vectors) of the electric and magnetic fields and correctly defined their LT. It is shown that the LT of vectors E and B are obtained by a mathematically correct procedure in the 4D spacetime. As explained in [11], *Minkowski, in Sec. 11.6 in* [21], *showed that both factors of the vector E, as the product of one bivector and one vector, has to be transformed by the LT.* That fundamental Minkowski's result is reinvented and generalized in [6-11]. Thus, E from (22), $E = F \cdot v/c$, transforms under the active LT, e.g., Eqs. (39) and (40), in such a manner that both F and the velocity of the observer v are transformed by the LT, Eq. (41). These coordinate-free LT yield how vector E transforms under the active LT, Eq. (42). If these transformations are written in the standard basis then the transformations of the components are obtained, Eq. (43). The most important result is that *under the relativistically correct LT the electric field vector E transforms again to the electric field vector E'; there is no mixing with the magnetic field B.*

In Sec. 6, the LPET of the components of the 3-vectors **E** and **B** are retrieved using the geometric algebra formalism, i.e., the 4D geometric quantities. If in the transformation of $E = F \cdot v/c$ *only F is transformed by the LT, but not the velocity of the observer v*, then the LPET of the electric field vector E are obtained, Eqs. (44) and (45). These coordinate-free LPET are also written in the standard basis, Eq. (46), and it is visible that *the components of the transformed E'_F are expressed by the mixture of components of E and B*. As seen from Eq. (47), the same result is obtained for the magnetic field vector B. The comparison of the relation for the LPET of the components of E (46) with the LPET for the components $E_{x,y,z}$ of the 3-vector **E**, which are given, e.g. by Eq. (11.148) in [2], explicitly shows that *they are exactly the same transformations.* But, the LPET of the vector E, (44) and (45), are obtained by a mathematically incorrect procedure (*only F is transformed*), which means that *they are not the relativistically correct LT* and consequently, contrary to the general opinion, *the LPET of the 3-vector* **E** (and **B**, **P**, **M**) *ARE NOT THE LT but the AT.*

In Sec. 7.1, the second-order electric fields outside a stationary conductor with steady current are considered. In the usual approaches, e.g., [22-26], there is a magnetic field 3-vector outside a stationary (superconducting) wire with steady current, but, according to the LPET (10), there are both, the slightly changed magnetic field and an induced *second-order external electric field 3-vector for the same but moving wire with steady current.* Similarly, e.g. [22, 23], it is argued that a neutral stationary current loop has only a magnetic moment 3-vector. According to the LPET for the 3-vectors **p** and **m**, which are the same as (16), that current loop acquires an electric dipole moment (18) as well, if it is moving with uniform 3-velocity **U** ($\beta = \mathbf{U}/c$). However, *in the 4D spacetime, the electric and magnetic fields and the dipole moments are not the 3-vectors but the 4D vectors E, B, p, m, which transform under the LT* (42), *i.e.,* (43) *and not under the LPET of the 3-vectors* (10) *and* (16), *or* (17). The electric field vector E (the same for B, p, m) transforms by the LT again to the electric field vector without mixing with B and therefore if E exists for a moving wire with a steady current, or a moving current loop, it must exist for the same but *stationary* wire or current loop. The determination of the electric field vector for the stationary current-carrying conductor is investigated in detail in [27] and here it is briefly reviewed with some additional explanations. The expression for the current density vector in the rest frame of the wire is given by Eq. (54), whereas the expression for the external second-order electric field vector of the stationary wire with steady current is given by Eq. (55). Observe that in the second paper in [27], the incorrect quadrupole field of the stationary current loop from the published version is replaced by the dipole field. Therefore, henceforward, if referred to [27] I mean that the corrected version has to be taken into account.

In Sec. 7.2, the experiments for the detection of the second-order electric fields outside a stationary conductor with steady current are discussed. This is a new consideration that is not reported in my previous papers. In the measurements [28, 29] a direct contact with the superconducting coil is used and because of that they cannot either support or disprove the theory presented in [27]. In [30], a non-contact method of measuring is used, but in order to "see" the external second-order electric fields the coil used in their experimental setup would need to be a superconducting coil. Recently, [31], the most promising method is proposed and it deals with cold ions. The theory presented in [31] is essentially the same as in my paper [32]. However, both theories, [32] and [31], explicitly use the Lorentz contraction in the derivation of the expression (55) for the external second-order electric field and as such they are not the relativistically correct theories. In [27] the relativistically correct theory is presented, but it seems that Folman, [31], either was not aware of [27] or more believed in the usual approach with the Lorentz contraction and the LPET (10) than to the mathematically correct 4D geometric approach.

In Sec. 8, an essentially new prediction is presented that *a stationary permanent magnet possesses an intrinsic polarization, which induces the external electric field.* This prediction is supported both in the usual Ampèrian approach in which a permanent magnet is an assembly of small current loops and also using the recent fundamentally new results from [33], i.e., the relations (59) and (60), which show that any fundamental particle has not only the intrinsic magnetic dipole moment (MDM) m but also the intrinsic electric dipole moment (EDM) p. Then, in the same way as the MDMs determine the magnetization M of a stationary permanent magnet the EDMs determine its polarization P, which induces an electric field outside a permanent magnet (moving or *stationary*). We suggest that the experimental setup from [31] could be also used for the measurement of that electric field outside a stationary permanent magnet.

In Secs. 9 - 9.2 the "charge-magnet paradox" from [34] is discussed in detail together with the highlight of it from [35] and different resolutions from [36]. The paradox, that in a static electric field a MDM $\mathbf{m}$ is subject to a torque $\mathbf{N}$ in some frames and not in others is stated to be resolved in [34] replacing the conventional Lorentz force (density), Eq. (5) in [34], by Einstein-Laub law, Eq. (6) in [34], which predicts no torque in all frames. In [34], all quantities $\mathbf{E}$, $\mathbf{B}$, $\mathbf{P}$, $\mathbf{M}$, $\mathbf{F}$, $\mathbf{N}$, etc. are the 3-vectors and their transformations are given by the LPET (10) for $\mathbf{E}$ and $\mathbf{B}$ and (16), or (17) for $\mathbf{P}$ and $\mathbf{M}$, and the same for EDM $\mathbf{p}$ and MDM $\mathbf{m}$, including (18). All other approaches from [36] also deal with the 3D quantities and their AT and often introduce some "hidden" quantities. Moreover, the resolutions from [34-36] *depend on the chosen synchronization* and they are meaningless *if only the Einstein synchronization is replaced by the "radio" synchronization*, as can be concluded from Eqs. (4) and (5) in Sec. 3.1. This means that from the viewpoint of the ISR the resolutions from [34-36] are not relativistically correct resolutions. All these treatments from [34-36] are objected in Sec. 9.1.

The treatment of the interaction between a static electric field and a permanent magnet that is presented in Sec. 9.2 differs in two important respects relative to [34-36]. As explicitly shown in the very similar treatments of the Trouton-Noble paradox [12, 13] and Jackson's paradox [14] in the approach with 4D geometric quantities there is no paradox and the same is shown for the "charge-magnet paradox." Every 4D geometric quantity, e.g., the Lorentz force vector, the torque bivector is the same quantity for all relatively moving inertial observers and for all bases chosen in them. However, in contrast to the Trouton-Noble paradox [12, 13] and Jackson's paradox [14] in which the torque bivector is zero in all frames, in this case the torque bivector N is different from zero even in the common rest frame of the considered charge and magnet from [34]. The reason for that important difference is that according to the 4D geometric approach, i.e., the ISR, a stationary permanent magnet possesses not only the magnetization vector M but the polarization vector P as well. In the formulation with the torque bivector N the relations (71), (72) and (73) show that the principle of relativity is naturally satisfied and that there is no paradox. As can be seen from (78), (79) and (81), (82), the same holds for the formulation with the "space-space" torque N_s and the "time-space" torque N_t. It is visible from (78) and (79) that the torque N_s is always determined by the interaction of the EDM p of the permanent and the electric field E, whereas N_t is determined by the interaction of the MDM m and E.

In Sec. 10, it is suggested that the existence of the electric fields from a stationary permanent magnet yields the possibility for the explanation of the Aharonov-Bohm effect in terms of fields and not, as usual, in terms of the vector potential.

In Sec. 11, the results from [37] are discussed. In [37] it is argued that a static electric field generated by a magnetic solenoid is due to Lorentz-violating electrodynamics and that the particle interference is a test of that Lorentz-violating model of electrodynamics. However, according to Secs. 7.1 and 8 from this paper there is a static electric field generated by a magnetic solenoid, see Eq. (55), that is obtained in a completely Lorentz-covariant approach with the 4D geometric quantities, which means that the particle interference is not a test of a Lorentz-violating model of electrodynamics.

In Sec. 12, a discussion and conclusions are presented.

In Appendix, it is briefly described the essential difference between the 4D geometric approach, i.e., the ISR, with the spacetime length and Einstein's definition of the Lorentz contraction for a moving rod.

2. A brief summary of geometric algebra

The calculations in this paper are performed in the geometric algebra formalism. Physical quantities are represented by the abstract multivectors, or, if some basis has been introduced, these abstract quantities are represented as 4D CBGQs comprising both components and *a basis*. For simplicity and for easier understanding mainly the standard basis, described below, will be used.

Here, for the reader's convenience, we provide a brief summary of geometric algebra. Usually Clifford vectors are written in lower case (a) and general multivectors (Clifford aggregate) in upper case (A). The space of multivectors is graded and multivectors containing elements of a single grade, r, are termed homogeneous and usually written A_r. The geometric (Clifford) product is written by simply juxtaposing multivectors AB. A basic operation on multivectors is the degree projection $\langle A\rangle_r$ which selects from the multivector A its $r-$ vector part ($0=$ scalar, $1=$ vector, $2=$ bivector, ...). The geometric product of a grade-r multivector A_r with a grade-s multivector B_s decomposes into $A_rB_s=\langle AB\rangle_{r+s}+\langle AB\rangle_{r+s-2}\cdots+\langle AB\rangle_{|r-s|}$. The inner and outer (or exterior) products are the lowest-grade and the highest-grade terms respectively of the above series; $A_r\cdot B_s\equiv\langle AB\rangle_{|r-s|}$ and $A_r\wedge B_s\equiv\langle AB\rangle_{r+s}$. For vectors a and b we have: $ab=a\cdot b+a\wedge b$, where $a\cdot b\equiv(1/2)(ab+ba)$, $a\wedge b\equiv(1/2)(ab-ba)$.

In this paper the notation will not be the same as in the above mathematical presentation. Some vectors will be denoted in lower case, like u, v (the velocities), x (the position vector), whereas some others in upper case, like vectors of the electric and magnetic fields E and B respectively, the Lorentz force K_L. Bivectors will be denoted in upper case but without subscript that denotes the grade. Thus, for example, the electromagnetic field F is a bivector.

In, e.g., [18], one usually introduces the standard basis. The generators of the spacetime algebra (the Clifford algebra generated by Minkowski spacetime) are taken to be four basis vectors $\{\gamma_\mu\}, \mu = 0...3$, satisfying $\gamma_\mu \cdot \gamma_\nu = \eta_{\mu\nu} = diag(+---)$. This basis, the standard basis, is a right-handed orthonormal frame of vectors in the Minkowski spacetime M^4 with γ_0 in the forward light cone. The γ_k $(k = 1, 2, 3)$ are spacelike vectors. The γ_μ generate by multiplication a complete basis for the spacetime algebra: 1, γ_μ, $\gamma_\mu \wedge \gamma_\nu$, $\gamma_\mu\gamma_5$, γ_5 ($2^4 = 16$ independent elements). γ_5 is the right-handed unit pseudoscalar, $\gamma_5 = \gamma_0 \wedge \gamma_1 \wedge \gamma_2 \wedge \gamma_3$. Any multivector can be expressed as a linear combination of these 16 basis elements of the spacetime algebra. For all mathematical details regarding the spacetime algebra reader can consult [18]. It is worth noting that the standard basis $\{\gamma_\mu\}$ corresponds, in fact, to the specific system of coordinates, i.e., to Einstein's system of coordinates. In Einstein's system of coordinates the standard, i.e., Einstein's synchronization [1] of distant clocks and Cartesian space coordinates x^i are used in the chosen inertial frame. However different systems of coordinates are allowed in an inertial frame and they are all equivalent in the description of physical phenomena. For example, in [16] two very different, but physically completely equivalent, systems of coordinates, Einstein's system of coordinates and the system of coordinates with a nonstandard synchronization, the "everyday," i.e., the "radio" ("r"), synchronization, are exposed and exploited throughout the paper. The "r", synchronization is also used and explained in [38], [17], [27], [33], see also Sec. 3.1 here. For simplicity and for easier understanding we shall mainly deal with the standard basis, but remembering that the approach with 4D geometric quantities holds for any choice of basis in M^4. Observe that the usual covariant approach, e.g., from the well-known textbooks [2], [26] exclusively deals with *components* implicitly taken in a specific basis, the standard basis.

Here, it is worth mentioning an important result regarding the usual formulation of electromagnetism (as in [2], [26]), which is presented in [19]. This is also mentioned in [11]. It is explained in [19] that an individual vector has no dimension; the dimension is associated with the vector space and with the manifold where this vector is tangent. Hence, what is essential for the number of components of a vector field is the number of variables on which that vector field depends, i.e., the dimension of its domain. This means that the usual time-dependent $\mathbf{E}(\mathbf{r},t)$, $\mathbf{B}(\mathbf{r},t)$ cannot be the 3-vectors, since they are defined on the spacetime. That fact determines that such vector fields, when represented in some basis, have to have four components (some of them can be zero). Therefore, we use the term "vector" for the correctly defined geometric quantity, which is defined on the spacetime. However, an incorrect expression, the 3-vector or the 3D vector, will still remain for the usual $\mathbf{E}(\mathbf{r},t)$, $\mathbf{B}(\mathbf{r},t)$ from [2], see Eq. (10).

3. The LPET of the 3-vectors E and B, P and M

3.1 The 3-vectors **E** *and* **B** *and their LPET*

Firstly, we discuss and object the derivation of the LPET of the 3-vectors **E** and **B** by the identification of the components of **E** and **B** with the components (implicitly taken in the standard basis) of the electromagnetic field tensor $F^{\alpha\beta}$, as, e.g., in the usual covariant approach [2]. Einstein's derivation [1] of the LPET of the components of **E** and **B** is discussed and objected in Sec. 5.3 in [16]. (In [16] the LPET are called the "apparent" transformations.) In the usual covariant approach, e.g., [2], the field-strength tensor $F^{\alpha\beta}$ (only components in the standard basis and not the whole tensor as a 4D geometric quantity) is introduced and defined in terms of the vector potential A^μ, Eq. (11.136) in [2].

However, as already stated, the components are coordinate quantities and they do not contain the whole information about the physical quantity. They do not completely represent a physical quantity that is defined on the 4D spacetime, since a basis of the spacetime is not included. Furthermore, in such formulation the gauge dependent vector potential A^μ (not measurable quantity) is considered to be the primary quantity from which the field-strength tensor $F^{\alpha\beta}$ (a measurable quantity) is derived! In contrast to such usual approach, e.g., [2], it is shown in [12] that in the 4D spacetime the electromagnetic field, the bivector $F = F(x)$, can be taken as the primary quantity for the whole electromagnetism and there is no need for the electromagnetic potentials.

Then, in the usual approaches, the covariant form of the Maxwell equations is written with $F^{\alpha\beta}$ and its dual $^*F^{\alpha\beta}$

$$\partial_\alpha F^{a\beta} = j^\beta/\varepsilon_0 c, \quad \partial_\alpha\ ^*F^{\alpha\beta} = 0, \tag{1}$$

where $^*F^{\alpha\beta} = (1/2)\varepsilon^{\alpha\beta\gamma\delta}F_{\gamma\delta}$. In order to get the component form of the Maxwell equations with the 3D $\mathbf{E}$ and $\mathbf{B}$

$$\begin{aligned} \partial_k E_k - j^0/c\varepsilon_0 &= 0, \quad -\partial_0 E_i + c\varepsilon_{ikj}\partial_j B_k - j^i/c\varepsilon_0 = 0, \\ \partial_k B_k &= 0, \quad c\partial_0 B_i + \varepsilon_{ikj}\partial_j E_k = 0 \end{aligned} \tag{2}$$

from Eq. (1) one simply makes *the identification of* the six independent components of $F^{\alpha\beta}$ with six components of the 3-vectors $\mathbf{E}$ and $\mathbf{B}$. These identifications are

$$E_i = F^{i0}, \qquad B_i = (1/2)\varepsilon_{ijk}F_{kj} \tag{3}$$

(the indices $i,\ j,\ k,\ ... = 1, 2, 3$), Eq. (11.137) in [2]. The components of the 3D fields $\mathbf{E}$ and $\mathbf{B}$ are written with lowered (generic) subscripts, since they are not the spatial components of the 4D quantities. This refers to the third-rank antisymmetric ε tensor too. The super- and subscripts are used only on the components of the 4D quantities. Then the 3D $\mathbf{E}$ and $\mathbf{B}$, as *geometric quantities in the 3D space*, are constructed from these six independent components of $F^{\mu\nu}$ and *the unit 3D vectors* $\mathbf{i}$, $\mathbf{j}$, $\mathbf{k}$, e.g., $\mathbf{E} = F^{10}\mathbf{i} + F^{20}\mathbf{j} + F^{30}\mathbf{k}$.

It is worth noting that such an identification of the components of $\mathbf{E}$ and $\mathbf{B}$ with the components of $F^{\alpha\beta}$ is synchronization dependent as explicitly shown in [16]. This is also discussed in [33]. There, it is shown that the mentioned identifications are meaningless in the "r" synchronization, i.e., in the $\{r_\mu\}$ basis, in which only the Einstein synchronization is replaced by an asymmetric synchronization, the "radio" synchronization. At the end of this section, the "r" synchronization is explained in more detail. As explained in [16] (and [33]), in the "r" synchronization

$$F_r^{10} = E_1 + cB_3 - cB_2. \tag{4}$$

Hence, the identification $E_{1r} = F_r^{10}$, as in (3), shows that the component E_{1r} in the $\{r_\mu\}$ basis is expressed as the combination of E_i and B_i components from the $\{\gamma_\mu\}$ basis

$$E_{1r} = F_r^{10}, \quad E_{1r} = E_1 + cB_3 - cB_2. \tag{5}$$

This means that if the "r" synchronization is used, i.e., if the appropriate metric is used, then it is not possible to make the usual identifications (3). It follows that the usual identifications, Eq. (11.137) in [2], are meaningful *only* when the Minkowski metric, e.g., $diag(1, -1, -1, -1)$, is used. Thus, these identifications depend on the chosen synchronization, i.e., the metric. *But, different synchronizations are nothing else than different conventions and physics must not depend on conventions.*

In the usual covariant approach, e.g., [2], one transforms by the passive LT the covariant Maxwell equations (1) and finds

$$\partial'_\alpha F'^{a\beta} = j'^\beta/\varepsilon_0 c, \quad \partial'_\alpha\ ^*F'^{\alpha\beta} = 0. \tag{6}$$

Under the passive LT the set of components, e.g., j^μ from the S frame transforms to j'^μ in the relatively moving inertial frame of reference S', $j'^\mu = L^\mu{}_\nu j^\nu$, where, for the boost in the γ_1

direction, $L^0{}_0 = L^1{}_1 = \gamma$, $L^0{}_1 = L^1{}_0 = -\beta\gamma$, $L^2{}_2 = L^3{}_3 = 1$ and all other components are zero. Then, the same identification as in Eq. (3) is assumed to hold for the transformed components E'_i and B'_i

$$E'_i = F'^{i0}, \quad B'_i = (1/2c)\varepsilon_{ijk}F'_{kj}. \tag{7}$$

For example,

$$F'^{10} = F^{10},\ F'^{20} = \gamma(F^{20} - \beta F^{21}),\ F'^{30} = \gamma(F^{30} - \beta F^{31}), \tag{8}$$

which yields (by Eqs. (3) and (7)) that

$$E'_1 = E_1,\ E'_2 = \gamma(E_2 - \beta c B_3),\ E'_3 = \gamma(E_3 + \beta c B_2), \tag{9}$$

see Sec. 11.10 and Eq. (11.148) in [2]. The same remark about the (generic) subscripts holds also here. Thus in the usual covariant approach the components $F^{a\beta}$ are transformed by the passive LT into $F'^{a\beta}$ and then it is simply argued that six independent components of $F'^{a\beta}$ are the "Lorentz transformed" components E'_i and B'_i, i.e., *the LPET of the components of* **E** *and* **B** *are derived assuming that they transform under the LT as the components of* $F^{\alpha\beta}$ *transform*, Eq. (11.148) in [2]. Then, $\mathbf{E}'$ and $\mathbf{B}'$ are constructed in S' in the same way as in S, i.e. multiplying the components $E'_{x,y,z}$ and $B'_{x,y,z}$ by the unit 3-vectors $\mathbf{i}'$, $\mathbf{j}'$, $\mathbf{k}'$. This yields the LPET of the 3-vectors **E** and **B**, Eq. (11.149) in [2], i.e. Eq. (10) here

$$\begin{aligned} \mathbf{E}' &= \gamma(\mathbf{E} + \boldsymbol{\beta} \times c\mathbf{B}) - (\gamma^2/(1+\gamma))\boldsymbol{\beta}(\boldsymbol{\beta}\cdot\mathbf{E}), \\ \mathbf{B}' &= \gamma(\mathbf{B} - (1/c)\boldsymbol{\beta} \times \mathbf{E}) - (\gamma^2/(1+\gamma))\boldsymbol{\beta}(\boldsymbol{\beta}\cdot\mathbf{B}), \end{aligned} \tag{10}$$

where $\mathbf{E}'$, $\mathbf{E}$, $\boldsymbol{\beta}$ and $\mathbf{B}'$, $\mathbf{B}$ are all 3-vectors.

Observe that there are no LT, or any other transformations, that transform the unit 3-vectors **i**, **j**, **k** into the unit 3-vectors $\mathbf{i}'$, $\mathbf{j}'$, $\mathbf{k}'$. It is seen from Eqs. (11.148) and (11.149) in [2], i.e. from Eq. (10) here, that *the transformed* $\mathbf{E}'$ *is expressed by the mixture of the 3-vectors* **E** *and* **B**, *and similarly for* $\mathbf{B}'$. The electric field **E** in one frame is "seen" as slightly changed electric field $\mathbf{E}'$ and an *induced magnetic field* $\mathbf{B}'$ in a relatively moving inertial frame.

This type of the derivation of (10) was first presented in Sec. 3 in [21]. There, and in section 7.2 as well, Minkowski made the same identification of the components of $F^{\alpha\beta}$ with components of the 3-vectors **E** and **B** (his **M**), as in equation (11.137) in [2]. The equations (11.148) in [2] are nothing else but the equations (6) and (7) from Sec. 3 in Minkowski's paper [21]. Later, the same derivation is used in numerous textbooks and papers treating relativistic electrodynamics.

Minkowski's identifications, i.e., Eq. (11.137) in [2], refer, as already stated, only to the components implicitly taken in the standard basis, which means that they are not generally valid. Namely, in the 4D spacetime, physical quantities are represented by the abstract 4D geometric quantities that are basis independent. In some basis, as already mentioned, they are represented as CBGQs which contain *both* components *and the basis vectors* (4-vectors in the usual notation). The principle of relativity is naturally satisfied for physical laws written with such 4D geometric quantities, whereas in Einstein's formulation with the 3-vectors or with their components it has to be postulated outside the mathematical formulation of the theory. Components taken alone are only a part of a physical quantity; another, equally important, part are the basis vectors. The LPET (10) are derived using synchronization dependent identifications of *components* of $F^{\alpha\beta}$ with *components* of the 3-vectors **E** and **B** in both relatively moving inertial frames of reference. *This shows that both the 3-vectors* **E** *and* **B** *and their LPET* (10) *are determined in a relativistically incorrect way; the quantities entering into that derivation are not properly defined in the 4D spacetime.*

In almost all textbooks and papers on relativistic electrodynamics the LPET (10) (or (16), which are given below) are simply employed without any physical interpretation. It seems that the majority of physicists believe that it is physically justified to have, e.g., only magnetic field **B** in one frame, in S, which transforms into a slightly changed magnetic field $\mathbf{B}'$ and a *new electric field* $\mathbf{E}'$ in a relatively moving S' frame. In some textbooks the authors tried to give a physical "explanation" for the appearance of that *electric field* $\mathbf{E}'$ for the observers in S'. Thus, for example, Rosser [22] asked the reader (Problem 6.13) to interpret the origin of the electric field present in S'. Let us assume that the external magnetic field in S is due to a permanent magnet at rest in S. Then, as in Sec. 3.2 here, according to the LPET (16) ((17)) a moving magnet has an electric polarization $\mathbf{P}$ (19), which gives an electric field outside the moving magnet. However, as discussed below, in the relativistically correct 4D geometric approach the relation (19) for the polarization $\mathbf{P}$, which is induced by the movement of a permanent magnetization $\mathbf{M}'$, does not hold. It is not derived in a relativistically correct manner; the quantities entering into that derivation are not properly defined in the 4D spacetime.

Here, for the reader's convenience, we explain here the "r" synchronization. As explained, e.g., in [16], different systems of coordinates, including different synchronizations, are allowed in an inertial frame and they are all equivalent in the description of physical phenomena. Thus in [16], both, Einstein's synchronization [1] and the "r" synchronization are exposed in detail. The "r" synchronization is commonly used in everyday life and not Einstein's synchronization. *In the "r" synchronization there is an absolute simultaneity.* Hence, contrary to the common opinion, *the relativity of simultaneity is not an intrinsic relativistic effect.* As stated in [38]: "For if we turn on the radio and set our clock by the standard announcement "..."at the sound of the last tone, it will be 12 o'clock," then we have synchronized our clock with the studio clock according to the "r" synchronization. In order to treat different systems of coordinates on an equal footing it is presented, Eq. (4) in [16], the transformation matrix that connects Einstein's system of coordinates with another system of coordinates in the same reference frame. Furthermore, Eq. (2) in [27], Eq. (1) in [16], it is derived such form of the LT, which is independent of the chosen system of coordinates, including different synchronizations. The unit vectors in the standard basis $\{\gamma_\mu\}$ and the $\{r_\mu\}$ basis, i.e., with the "r" synchronization, [16], are connected as

$$r_0 = \gamma_0, \quad r_i = \gamma_0 + \gamma_i. \tag{11}$$

Hence, the components $g_{\mu\nu,r}$ of the metric tensor are $g_{ii,r} = 0$, and all other components are $= 1$. Remember that in the $\{\gamma_\mu\}$ basis $g_{\mu\nu} = diag(1, -1, -1, -1)$. (Note that in [16] and [17] the Minkowski metric is $g_{\mu\nu} = diag(-1, 1, 1, 1)$.) Then, according to (4) from [16], one can use $g_{\mu\nu,r}$ to find the transformation matrix $R^\mu_{\ \nu}$ that connects the components from the $\{\gamma_\mu\}$ basis with the components from the $\{r_\mu\}$ basis. The only components that are different from zero are

$$R^\mu_{\ \mu} = -R^0_{\ i} = 1. \tag{12}$$

The inverse matrix $(R^\mu_{\ \nu})^{-1}$ connects the "old" basis, $\{\gamma_\mu\}$, with the "new" one, $\{r_\mu\}$. Hence, the components of the position vector x are connected as

$$x^0_r = x^0 - x^1 - x^2 - x^3, \quad x^i_r = x^i. \tag{13}$$

Observe that vector x can be decomposed in both bases and it holds that

$$x = x^\mu \gamma_\mu = x^\mu_r r_\mu. \tag{14}$$

Obviously, the components of any vector transform in the same way as in (13), e.g., for the components of the electric field vector E it holds that

$$E^0_r = E^0 - E^1 - E^2 - E^3, \quad E^i_r = E^i. \tag{15}$$

In the same way as in (14), it can be written for the vector E that $E = E^{\mu}\gamma_{\mu} = E_r^{\mu}r_{\mu}$.

It is visible from Eqs. (11) and (13), i.e., from the fact that the metric tensor $g_{\mu\nu,r}$ is not diagonal, in the $\{r_{\mu}\}$ basis it is not possible to make the separation of the 4D spacetime into the time and the 3D space as it is possible in the standard basis $\{\gamma_{\mu}\}$. Thus *the space-time split is not possible in the* $\{r_{\mu}\}$ *basis.* In the first and the second paper (in that second paper the "r" synchronization is used as well) in [17] some of the well-known experiments: the "muon" experiment, the Michelson-Morley type experiments, the Kennedy-Thorndike type experiments and the Ives-Stilwell type experiments are analyzed using Einstein's formulations of SR, which deals with the Lorentz contraction and the time dilation, and the ISR, i.e., the approach with 4D geometric quantities, the position vector, the distance vector between two events and the spacetime length. It is shown that all experiments are in a complete agreement, independently of the chosen synchronization, with the 4D geometric approach, i.e., with the ISR, whereas it is not the case with the Einstein's approach with the Lorentz contraction and the time dilation if the "r" synchronization is used. In the third paper in [17] the same is shown considering in detail the Michelson-Morley experiment.

3.2 The 3-vectors **P** *and* **M** *and their LPET*

The LPET of the polarization and the magnetization 3-vectors **P** and **M** are also often derived from the covariant formulation using the mentioned, synchronization dependent, identifications of components (implicitly taken in the standard basis) of the magnetization-polarization tensor $\mathcal{M}^{\alpha\beta}$ with components of the 3-vectors **P** and **M** in both relatively moving inertial frames of reference. Thus, in S, these identifications are $P_i = \mathcal{M}^{i0}$, $M_i = (c/2)\varepsilon_{ijk}\mathcal{M}_{jk}$ and the same identifications hold in the relatively moving inertial frame of reference S', $P_i' = \mathcal{M}'^{i0}$, $M_i' = (c/2)\varepsilon_{ijk}\mathcal{M}'_{jk}$, see, e.g., Secs. 18-5 and 18-6 in [23]. The same remark about the (generic) subscripts holds also here. This procedure yields

$$\begin{aligned}\mathbf{P} &= \gamma(\mathbf{P}' + \boldsymbol{\beta}\times\mathbf{M}'/c) - (\gamma^2/(1+\gamma))\boldsymbol{\beta}(\boldsymbol{\beta}\cdot\mathbf{P}'),\\ \mathbf{M} &= \gamma(\mathbf{M}' - \boldsymbol{\beta}\times c\mathbf{P}') - (\gamma^2/(1+\gamma))\boldsymbol{\beta}(\boldsymbol{\beta}\cdot\mathbf{M}'),\end{aligned} \tag{16}$$

see e.g., Eqs. (18-68) - (18-71) in [23], or Eqs. (4.2) in [39], or Eqs. (6.78a) and (6.81a) in [22], etc. In the mentioned equations the transformations (16) are written in an equivalent form as

$$\begin{aligned}\mathbf{P}_{\parallel} &= \mathbf{P}'_{\parallel}, \quad \mathbf{P}_{\perp} = \gamma(\mathbf{P}' + \boldsymbol{\beta}\times\mathbf{M}'/c)_{\perp}\\ \mathbf{M}_{\parallel} &= \mathbf{M}'_{\parallel}, \quad \mathbf{M}_{\perp} = \gamma(\mathbf{M}' - c\boldsymbol{\beta}\times\mathbf{P}')_{\perp}.\end{aligned} \tag{17}$$

The inverse relations are obtained in the usual way by the exchange of the primed and unprimed quantities and by the replacement $\boldsymbol{\beta} \to -\boldsymbol{\beta}$. The main feature of the LPET of **P** and **M** (16), or (17), is the same as for the LPET of **E** and **B**, i.e., *the components of the transformed* **P**′ *are expressed by the mixture of components of* **P** *and* **M**, *and similarly for* **M**′.

Using completely the same procedure with the identifications of components one can derive the LPET of the 3-vectors of the electric **p** and magnetic **m** dipole moments from the tensor of the dipole moments $D^{\alpha\beta}$. Hence, these LPET for the 3-vectors **p** and **m** are the same as (16), or (17), but with **p** and **m** replacing **P** and **M**, respectively.

The interpretation and the derivation of the transformations (16) (or (17)) in terms of simplified classical models is presented in, e.g. [23] and [22]. It is stated in Sec. 18-6 in [23] that the relation $\mathbf{P}'_{\parallel} = \mathbf{P}_{\parallel}$ is expected, "since $\mathbf{P}_{\parallel}$ is the product of an (invariant) charge and a distance divided by a volume, *both contracted in the same ratio.*" (my emphasis) The calculation is given in section 6.7.1. in [22]. The S' frame is taken to be the rest frame of the material.

The classical model assumes that in S' the dielectric consists of n' stationary dipoles/m^3. If $\mathbf{P}'$ is parallel to $\boldsymbol{\beta}$, then $P' = n'p' = n'(ql_0)$, $P = np = (\gamma n')(\gamma^{-1}p') = P'$; $\gamma n'$ is *due to the contraction of the volume*, whereas $\gamma^{-1}p'$ is *due to the contraction of a distance*, $l = \gamma^{-1}l_0$. If the atomic electric dipoles are perpendicular to $\boldsymbol{\beta}$, then the first term $\gamma\mathbf{P}'_\perp$ simply follows from $p = ql = ql_0 = p'$ (*there is no Lorentz contraction if* $\mathbf{p}' \perp \boldsymbol{\beta}$), but $n = \gamma n'$, and thus the atomic electric dipoles give rise to a polarization $\gamma\mathbf{P}'_\perp$ in S.

It is argued, both in [23] and [22], that the extra term $(\boldsymbol{\beta} \times \mathbf{M}'/c)_\perp$ in (17) has no nonrelativistic counterpart. In the classical model, for purposes of calculating $\mathbf{M}'$, the magnetic dipoles are considered as little current loops. In section 18-4 in [23], see Fig. 18-4, and in Sec. 6.5 in [22], see Fig. 6.4 a,b, it is argued that a neutral stationary current loop, which has a magnetic moment $\mathbf{m}'$ in its rest frame S', acquires an electric dipole moment

$$\mathbf{p} = \boldsymbol{\beta} \times \mathbf{m}'/c \tag{18}$$

if it is moving with uniform 3-velocity $\mathbf{U}$ ($\beta = \mathbf{U}/c$) relative to the laboratory frame S. The result (18) also follows from the LPET for the 3-vectors $\mathbf{p}$ and $\mathbf{m}$, which are, as already stated, the same as (16) with $\mathbf{p}$ and $\mathbf{m}$ replacing $\mathbf{P}$ and $\mathbf{M}$, respectively. It is taken that *in the rest frame of the neutral current loop the electric moment* $\mathbf{p}'$ *is zero.*

In the Ampèrian approach a permanent magnet is essentially an assembly of current loops. Hence, if (18) holds for each atomic magnetic dipole in the moving magnet, then one has for the electric dipole moment per unit volume in S, $n\mathbf{p} = \gamma n'\mathbf{U} \times \mathbf{m}'/c^2 = \gamma(\mathbf{U} \times \mathbf{M}'/c^2)_\perp$. Thus, if a permanent magnetization $\mathbf{M}'$ is viewed from a moving frame it produces an electric polarization

$$\mathbf{P} = \gamma\mathbf{U} \times \mathbf{M}'/c^2. \tag{19}$$

In other words, *according to all usual approaches, if an observer moves with a 3-velocity* $\mathbf{U}$ *relative to a medium of magnetization* $\mathbf{M}'$ *that observer will observe an equivalent electric polarization given by* $\mathbf{P}$, (19). Adding this term to the term $\gamma\mathbf{P}'_\perp$ yields $\mathbf{P}_\perp$ from (17).

It can be seen from the mentioned textbooks, [23], [22], that the relations (18) and (19) are obtained using *the Lorentz contraction* and *the relativity of simultaneity.*

However, as shown in [16] and in the comparison with well-known experiments that test special relativity [17], *the relativity of simultaneity, the Lorentz contraction and the time dilation are not well-defined in the 4D spacetime. They are not intrinsic relativistic effects, because they depend on the chosen synchronization.* A clear presentation of the relativistic incorrectness of the Lorentz contraction is already given in Sec. 2.2 in [27]. Already in the year 1966, Rohrlich [15] clearly explained that *the Lorentz contraction* is not a true relativistic transformation, i.e., it *has nothing to do with the Lorentz transformation.* Similarly, in the next year, Gamba [40] stated for the Lorentz contraction: "Although it is a completely useless concept in physics, it will probably continue to remain in the books as an historical relic for the fascination of the layman." In the geometric approach, the ISR, in [16], [17], [27] it is *proved* that in the 4D spacetime two relatively moving observers cannot compare spatial lengths taken alone, which are synchronously determined for the observer. For the reader's convenience the relativistic incorrectness of the Lorentz contraction is explicitly shown in Appendix here. Also, it is proved in [16] and [17] that it is not correct to compare the temporal distances taken alone, since they are not well-defined quantities in the 4D spacetime. The properly defined quantities are the distance vector between two events A and B with the position vectors x_A and x_B and the spacetime length, which is a Lorentz scalar. Obviously, Gamba [40] was wrong with his statement about the Lorentz contraction. The papers [15] and [40] as well as my papers [16], [17], [27] remained almost completely overlooked and still the Lorentz contraction and the time dilation are treated in the leading physical journals and in the well-known textbooks, including [2] and [26], as intrinsic relativistic effects.

This consideration reveals in another way that the LPET of **P** and **M** (16), or (17), are not relativistically correct transformations, i.e., (16), *or* (17), *are not the LT but the AT.*

4. The definitions of vectors E, B and P, M in terms of F, v and $\mathcal{M}$, u, respectively. The Lorentz invariant field equations for vacuum and for a magnetized and polarized moving medium

Instead of dealing with quantities that are not well-defined in the 4D spacetime, like the 3-vectors **E** and **B** and their LPET (10), or with components implicitly taken in the standard basis as in the usual covariant approaches, [2], [26], we deal with 4D geometric quantities, which are properly defined in the 4D spacetime. Moreover, it is shown, particularly in [12], that the bivector $F = F(x)$, which represent the electromagnetic field, can be taken as the primary quantity for the whole electromagnetism and the field equation for F

$$\partial F = j/\varepsilon_0 c, \quad \partial \cdot F + \partial \wedge F = j/\varepsilon_0 c \tag{20}$$

is the basic equation. As shown in [12], the bivector field F yields the complete description of the electromagnetic field and, in fact, there is no need to introduce either the field vectors or the potentials. For the given sources the Clifford algebra formalism enables one to find in a simple way the electromagnetic field F, see Eqs. (7) and (8) in [12]. However, if one introduces the electric and magnetic fields, then they can be represented by different algebraic objects. *These fields are not determined by the usual identifications of the components, Eqs.* (3) *and* (7), *but they are derived in a mathematically correct way from* F, *as in Eqs.* (21) *and* (22) *here.*

In this geometric approach the electric and magnetic fields are represented by vectors $E(x)$ and $B(x)$. We deal with such representations of the electric and magnetic fields because they are simple and much closer to the classical representation of the electric and magnetic fields by the 3D vectors **E** and **B** than, e.g. the representations by bivectors, which are used in [18]. The decomposition of F in terms of vectors E, B and v is given as

$$F = (1/c)E \wedge v + (IB) \cdot v, \tag{21}$$

and E and B are determined as

$$E = (1/c)F \cdot v, \quad B = -(1/c^2)I(F \wedge v). \tag{22}$$

There is no rest frame for the field F, that is, for E and B, and therefore the vector v in the decomposition (21) is interpreted as the velocity vector of the observers who measure E and B fields. Then $E(x)$ and $B(x)$ are defined with respect to v, i.e., with respect to the observer. From (21) and (22) it also holds that $E \cdot v = B \cdot v = 0$; only three components of E and three components of B are independent since F is antisymmetric. The unit pseudoscalar I from (21) and (22) is defined algebraically without introducing any reference frame, as in Sec. 1.2. in the second reference in [18]. We choose I in such a way that when I is represented in the $\{\gamma_\mu\}$ basis it becomes $I = \gamma_0 \wedge \gamma_1 \wedge \gamma_2 \wedge \gamma_3 = \gamma_5$. With such choice for I, $\{\gamma_1, \gamma_2, \gamma_3\}$ form a right-handed orthonormal set, as usual for a 3D Cartesian frame. The LT do not change the orientation of the spacetime.

The equations that correspond to equations (22) (and (21)), but in the tensor formalism, with abstract indices a, b, c, .. , are $E^a = (1/c)F^{ab}v_b$, $B^a = (1/2c^2)\varepsilon^{abcd}F_{bc}v_d$ and $F^{ab} = (1/c)(E^a v^b - E^b v^a) + \varepsilon^{abcd}v_c B_d$, e.g., Eqs. (39) and (40) in [16]. They are based on the theorem that any second rank antisymmetric tensor can be decomposed into two vectors and a unit time-like vector (the velocity vector/c). These equations show that in the tensor formalism too both the electric and magnetic fields can be represented by vectors.

Let us introduce the frame of "fiducial" observers as the frame in which the observers who measure fields E and B are at rest. That frame with the standard basis $\{\gamma_\mu\}$ in it is called the γ_0-frame. In the γ_0-frame $v = c\gamma_0$ and therefore E from (22) becomes $E = F \cdot \gamma_0$ and $B = -(1/c)\gamma_5(F \wedge \gamma_0)$. Similarly, the decomposition (21) becomes $F = E \wedge \gamma_0 + c(\gamma_5 B) \cdot \gamma_0$. All these quantities can be written as CBGQs in the standard basis $\{\gamma_\mu\}$. This yields for E and B

$$\begin{aligned} E &= E^\mu \gamma_\mu = 0\gamma_0 + F^{i0}\gamma_i, \\ B &= B^\mu \gamma_\mu = 0\gamma_0 + (1/2c)\varepsilon^{0ijk}F_{kj}\gamma_i. \end{aligned} \tag{23}$$

Note that $\gamma_0 = (\gamma_0)^\mu \gamma_\mu$ with $(\gamma_0)^\mu = (1, 0, 0, 0)$. The components of F in the $\{\gamma_\mu\}$ basis give rise to the tensor (components) $F^{\mu\nu} = \gamma^\nu \cdot (\gamma^\mu \cdot F) = (\gamma^\nu \wedge \gamma^\mu) \cdot F$. It can be easily checked that in the γ_0-frame $E \cdot \gamma_0 = B \cdot \gamma_0 = 0$, which means that E and B are orthogonal to γ_0. Hence, the temporal components of E and B are zero $E^0 = B^0 = 0$ and only the spatial components remain

$$E^i = F^{i0}, \quad B^i = (1/2c)\varepsilon^{0ijk}F_{kj}. \tag{24}$$

Thus, E and B actually refer to the 3D subspace orthogonal to the specific timelike direction γ_0. In the γ_0-frame the remaining spatial components of E and B from (24) are the same as the components of the usual 3D **E** and **B**, Eq. (3), which are obtained by the usual identification of the components $F^{\mu\nu}$ (implicitly taken in the standard basis) with the components of the 3D vectors **E** and **B**. However, there is a very important difference between the identifications (3) and the above spatial components of E and B in the γ_0-frame, Eq. (24), which explicitly reveals that the usual procedure with the identifications (3) is not correct in the 4D spacetime. As explained above, the components of the 3D fields **E** and **B** in (3) are not the spatial components of the 4D quantities. They transform according to the LPET, Eq. (11.148) in [2]. Also, the antisymmetric ε tensor is a third-rank antisymmetric tensor. On the other hand, the components of E and B in (23), i.e., in (24), are the spatial components of the 4D geometric quantities that are taken in the standard basis. They transform according to the LT, which are given below, Eq. (43). Also, the antisymmetric ε tensor in (23) is a fourth-rank antisymmetric tensor. In the usual covariant approaches one forgets about the temporal components E^0 and B^0 and simply makes the identification of six independent components of $F^{\mu\nu}$ with three components E_i and three components B_i according to the relations (3).

In Eq. (12) in [20], in the same way as in (21), the generalized magnetization-polarization bivector $\mathcal{M}(x)$ is decomposed into two vectors, the polarization vector $P(x)$ and the magnetization vector $M(x)$ and the unit time-like vector u/c

$$\mathcal{M} = P \wedge u/c + (MI) \cdot u/c^2. \tag{25}$$

There is the rest frame for a medium, i.e., for $\mathcal{M}$, or P and M, and therefore the vector u in the decomposition (25) is identified with bulk velocity vector of the medium in spacetime. Then, $P(x)$ and $M(x)$ are defined with respect to u as

$$P = \mathcal{M} \cdot u/c, \quad M = cI(\mathcal{M} \wedge u/c) \tag{26}$$

and it holds that $P \cdot u = M \cdot u = 0$; only three components of P and three components of M are independent since $\mathcal{M}$ is antisymmetric. As in the case with F in (21) and (22), it is visible from (26) that P and M depend not only on $\mathcal{M}$ but on u as well.

Here, we briefly examine the Lorentz invariant field equations for vacuum and for a magnetized and polarized moving medium. Inserting (21) into the field equation for F (20) one finds the field equation in terms of E and B;

$$\partial(E \wedge (v/c) + (IB) \cdot v) = j/\varepsilon_0 c. \tag{27}$$

As explained in [8], Eq. (27) represents the Lorentz invariant generalization of the usual Maxwell equations. That form (27) is the most general form of the field equations with electric and magnetic fields as properly defined quantities in the 4D spacetime.

The equation (27) with the geometric product can be divided into the vector part (with sources)

$$\partial \cdot (E \wedge v/c + (IB) \cdot v) = j/\varepsilon_0 c \tag{28}$$

and the trivector part (without sources)

$$\partial \wedge (E \wedge v/c + (IB) \cdot v) = 0. \tag{29}$$

In that form it is clear that *it is not possible to separate the field equation with sources for the E field from that one for the B field.* Thus, *in the 4D spacetime, the generalizations with 4D geometric quantities of the usual Ampère-Maxwell law and Gauss's law are inseparably connected in one law* - Eq. (28). Similarly, *in Eq.* (29), *Faraday's law and the law that expresses the absence of magnetic charge are also inseparably connected in one law, which is expressed in terms of the 4D geometric quantities.* This is an essential difference relative to Maxwell's equations with the 3-vectors **E** and **B**.

The mathematical reason for such an inseparability is that, e.g., the gradient operator ∂ is a vector field defined on the 4D spacetime. If represented in some basis then its vector character remains unchanged only when *all its components together with associated basis vectors* are taken into account in the considered equation. The same holds for other vectors E, B, j, etc. and multivectors like F, $\mathcal{M}$, ... For example, in general, in the 4D spacetime, the current density vector j is a well-defined physical quantity, but it is not the case with the usual charge density ρ and the usual current density $\mathbf{j}$ as a 3-vector. Similarly, in general, the gradient operator ∂ cannot be divided into the usual time derivative and the spatial derivatives, e.g., in the $\{r_\mu\}$ basis with the "r" synchronization. In the 4D spacetime, an independent physical reality is attributed to the position vector x, the gradient operator ∂, the current density vector j, the vectors of the electric and magnetic fields E and B, respectively, etc., but not to the 3-vector $\mathbf{r}$ and the time t, to the 3-vectors $\mathbf{j}$, $\mathbf{E}$, $\mathbf{B}$, etc.

The generalization of (20) to a moving medium is presented in [20] and it is obtained simply replacing F by $F + \mathcal{M}/\varepsilon_0$, which yields the primary equations for the electromagnetism in moving media

$$\partial(\varepsilon_0 F + \mathcal{M}) = j^{(C)}/c; \quad \partial \cdot (\varepsilon_0 F + \mathcal{M}) = j^{(C)}/c, \ \partial \wedge F = 0, \tag{30}$$

where $j^{(C)}$ is the conduction current density of the *free* charges and $j^{(\mathcal{M})} = -c\partial \cdot \mathcal{M}$ is the magnetization-polarization current density of the *bound* charges. The total current density vector j is $j = j^{(C)} + j^{(\mathcal{M})}$. Inserting the decomposition (25) into $j^{(\mathcal{M})}$ it follows that

$$j^{(\mathcal{M})} = -(\partial \cdot P)u + (u \cdot \partial)P + (1/c)[u \wedge (\partial \wedge M)]I. \tag{31}$$

In the standard basis $\{\gamma_\mu\}$ and in the rest frame of the medium, $u = c\gamma_0$, $P^0 = M^0 = 0$,

$$j^{(\mathcal{M})\mu} = (-c\partial_k P^k, c\partial_0 P^i - \varepsilon^{0ijk}\partial_j M_k). \tag{32}$$

In the usual formulation with the 3-vectors these components correspond to, e.g., Eq. (18-61) in [23], i.e., $\rho = -\nabla\mathbf{P}$, $\mathbf{j} = \partial\mathbf{P}/\partial t + \nabla\times\mathbf{M}$. In (32), the components P^i, M^i with the upper indices correspond to the components of the 3-vectors. However, *in the 4D spacetime it is not correct to write the components of the properly defined vectors in terms of the 3-vectors and operations with them. There are no 3-vectors in the 4D spacetime.*

In most materials $\mathcal{M}$ is a function of the field F and this dependence is determined by the constitutive relations. In that case (30) are well-defined equations for F. Recently, the constitutive relations and the magnetoelectric effect for moving media are investigated in detail in [41].

Then, in [20], the general form of the field equation for a magnetized and polarized moving medium expressed in terms of $E(x)$, $B(x)$, $P(x)$ and $M(x)$ is obtained by the insertion of Eqs. (21) and (25) into the field equation (30). It is Eq. (15) in [20], which is the generalization of Eq. (27) (with the geometric product) to the moving media. The generalizations of Eq. (28) to the moving media is given by Eq. (16) in [20], the vector part (with sources), i.e., in the "source representation" by the equation

$$\partial \cdot \{\varepsilon_0[E \wedge v/c + (IB) \cdot v]\} = j^{(C)}/c - \partial \cdot [P \wedge u/c + (1/c^2)(MI) \cdot u], \tag{33}$$

according to which the sources of E and B fields are $j^{(C)}$ and P and M. Obviously, from (33), *it is not possible to separate the field equation with sources for the E field from that one for the B field.* This is an essential difference relative to Maxwell's equations with the 3-vectors **E**, **B**, **P** and **M**. The field equation without sources, the trivector part, remains unchanged relative to the corresponding equation for vacuum (29)

$$\partial \wedge [E \wedge v/c + (IB) \cdot v] = 0. \tag{34}$$

In Eq. (33), i.e., in Eq. (18) in [20], there are two different velocities u and v and such an equation is not previously reported in the physics literature.

As stated in [20], *Eq.* (30), i.e., *Eqs.* (33) *and* (34) *comprise and generalize all usual Maxwell's equations (with 3-vectors) for moving media.*

Again, as in the above discussion for the vacuum, *in the 4D spacetime*, in contrast to the usual formulation of electromagnetism with the 3-vectors **E**, **B**, **P**, **M**, **j**, ... , there are no two laws, the Ampèr-Maxwell law and Gauss's law, but *only one law*, that is expressed by Eq. (33) and the same for other two laws and Eq. (34).

In the same way as in [12] one can derive the expression for the Lorentz force density k_L,

$$k_L = F \cdot j/c = (1/c)F \cdot (j^{(C)} - c\partial \cdot \mathcal{M}). \tag{35}$$

Inserting the decompositions (21) and (25), i.e., (21) and (31), into k_L (35) one can find k_L expressed in terms of $E(x)$, $B(x)$, $P(x)$ and $M(x)$. This is discussed in Sec. 9 in connection with the "charge-magnet paradox."

5. The LT of vectors E and B; both F and the observer are transformed

As seen from (21) and (22) all quantities F, E, B and v are abstract 4D geometric quantities. If these geometric quantities from (21) and (22) are represented in some basis then they contain both components and basis vectors. In his fundamental work, Minkowski, in Sec. 11.6 in [21], wrote the relation (55) that corresponds to (21), but he considered that the quantities w, Φ and Ψ, which correspond to our v, E and B, are 1×4 matrices and that F is a 4×4 matrix. Their components are implicitly determined in the standard basis. In Sec. 11.6 in [21], the next paragraph below Eq. (44), Minkowski described how w and F separately transform under the LT A (the matrix of the LT is denoted as A in [21]) and then how the product wF transforms. Thus, he wrote

$$w' = wA \tag{36}$$

for the LT of the velocity vector w and

$$F' = A^{-1}FA \tag{37}$$

for the LT of the field-strength tensor. Then, the mathematically correct LT of $\Phi = wF$ are

$$\Phi = wF \longrightarrow \Phi' = (wA)(A^{-1}FA) = (wF)A = \Phi A, \tag{38}$$

which means that under the LT both quantities, the velocity w and F are transformed and their product transforms as any other vector (i.e., in [21], as an 1×4 matrix) transforms. The most important thing is that *the electric field vector Φ transforms by the LT again to the electric field vector Φ'; there is no mixing with the magnetic field Ψ.*

These correct LT of the electric and magnetic fields are reinvented and generalized in terms of 4D geometric quantities in [6-11]. In the geometric algebra, the LT (the active ones) are described by rotors R, $R\widetilde{R} = 1$, where the reverse $\widetilde{R}$ is defined by the operation of reversion according to which $\widetilde{AB} = \widetilde{B}\widetilde{A}$, for any multivectors A and B, $\widetilde{a} = a$, for any vector a, and it reverses the order of vectors in any given expression. For boosts in an arbitrary direction the rotor R is given by Eq. (8) in [7, 9], or Eq. (10) in [11], as

$$R = (1 + \gamma + \gamma\gamma_0\beta)/(2(1+\gamma))^{1/2}, \tag{39}$$

where $\gamma = (1-\beta^2)^{-1/2}$, the vector β is $\beta = \beta n$, β on the r.h.s. of that equation is the scalar velocity in units of c and n is not the basis vector but any unit space-like vector orthogonal to γ_0. Then, *any multivector N transforms by active LT in the same way,* i.e., as

$$N \to N' = RN\widetilde{R}. \tag{40}$$

Hence, vector E transforms by the LT R as $E \longrightarrow E' = RE\widetilde{R}$. In the γ_0-frame, $v = c\gamma_0$ is taken in (22). Then E becomes $E = F \cdot \gamma_0$ and it transforms under the LT in the same manner as in Minkowski's relation (38), i.e., that both F and v are transformed by the LT R as

$$E = F \cdot \gamma_0 \longrightarrow E' = R(F \cdot \gamma_0)\widetilde{R} = (RF\widetilde{R}) \cdot (R\gamma_0\widetilde{R}). \tag{41}$$

These correct LT give that

$$E' = E + \gamma(E \cdot \beta)\{\gamma_0 - (\gamma/(1+\gamma))\beta\}. \tag{42}$$

In the same way vector B transforms and vectors P, M as well, but for P and M the LT, like (42), *are the transformations from the rest frame of the medium* ($u = c\gamma_0$). For boosts in the direction γ_1 one has to take that $\beta = \beta\gamma_1$ (on the l.h.s. is vector β and on the r.h.s. β is a scalar) in the above expression for the rotor R (all in the standard basis). Hence, in the $\{\gamma_\mu\}$ basis and when $\beta = \beta\gamma_1$ Eq. (42) becomes

$$E'^{\nu}\gamma_\nu = -\beta\gamma E^1\gamma_0 + \gamma E^1\gamma_1 + E^2\gamma_2 + E^3\gamma_3. \tag{43}$$

As already mentioned in Sec. 1, the relations (42) and (43) are the fundamental results, which show that *under the relativistically correct LT the electric field vector E transforms again to the electric field vector E'; there is no mixing with the magnetic field B.* The same happens with vectors P and M. The same fundamental result can be obtained if electric and magnetic fields are represented, e.g., by bivectors as in [9]. In general, it can be stated that *the LT always transform the 4D algebraic object (vector, bivector) representing the electric field only to the electric field, and similarly for the magnetic field.*

It is important to note that E' *(and* B'*) from* (42) *and* (43) *are not orthogonal to* γ_0, i.e., *they have temporal components* $\neq 0$. They do not belong to the same 3D subspace as E and B, but they are in the 4D spacetime spanned by the whole standard basis $\{\gamma_\mu\}$.

The same components as in (43) would be obtained for $\Phi' = \Phi A$ in Minkowski's relation (38) if the components of w are $(0, 0, 0, ic)$ in his notation, which corresponds to $v = c\gamma_0$ in our formulation. It is worth noting that only in Sec. 11.6 in [21] Minkowski dealt with vectors (only components) w, Φ and Ψ, but in the rest of [21] he exclusively dealt with the usual 3-vectors $\mathbf{v}$, $\mathbf{E}$ and $\mathbf{B}$ (our notation) and not with correctly defined vectors w, Φ and Ψ.

In Sec. 11 under the title "Minkowski in 1908, and Ivezić Since 2003: Lorentz Covariance" in the third paper in [19] Oziewicz, from the mathematical point of view, nicely explains the results obtained in my papers [6-11]. (The references in the quoted part refer to the mentioned Oziewicz's paper.) He states:

"Ivezíć observed the logical and mathematical inconsistency of textbook treatments of the Lorentz-covariance since 2003. He noted that it is illogical to consider a closed differential biform F to be Lorentz-covariant, and at the same time, keep observer's time-like vector field, a '4-velocity', P≃(1,0,0,0), to be Lorentz-invariant-absolute. For example, compare how an absolute observer is hidden in calculations presented in (Misner, Thorne &Wheeler [31], Chapt. 3).

Minkowski [1], and then Ivezić [7-10], observed correctly that if a Lorentz transformation is an isomorphism of a vector space, then the entire algebra of tensor fields must be Lorentz-covariant. Every vector is Lorentz-covariant, and an observer-monad timelike vector field, also must be Lorentz-covariant. All tensor fields, F and P, must be Lorentz-covariant. An active Lorentz transformation must act on all tensor fields, including an observer's time-like vector field. Hence electromagnetic field F, potential A , and Paul P, must be Lorentz-covariant (Ivezić [7-10]).

Instead of Fock's and Jackson's transformations (10.2) - (10.3), (our Eqs. (44) and (45), my remark) Ivezić defined the Lorentz-covariance for the compound electric and magnetic fields, (7.2), (it corresponds to our Eq. (22), my remark) exactly as defined by Minkowski in [1], §11.6, just before formula (46). We stress that Minkowski in [1] does not in practice use his definition of Lorentz-covariance. Instead of (10.2) - (10.3), the Lorentz transformation of electric and magnetic concomitant vector fields according to the Minkowski and Ivezić definition of Lorentz covariance is: ..." given by Oziewicz's equations (11.4) - (11.9). His relation (11.5) is our Eq. (42).

Here, it is at place to give an interesting remark regarding Oziewicz's papers [19]. It has often been argued that it cannot be that the 3-vectors $\mathbf{E}$, $\mathbf{B}$, $\mathbf{P}$ and $\mathbf{M}$ and their transformations (10) and (16), or (17), have to be replaced by the 4D geometric quantities, e.g., vectors E, B, P and M and by their mathematically and relativistically correct LT (42) and (43). On the other hand Oziewicz, in difference to all others, correctly considers from the outset that *there are no 3-vectors in the 4D spacetime.* But, he incorrectly considers that, e.g., the transformations of the 3-vectors $\mathbf{E}$ and $\mathbf{B}$, Eq. (10), i.e., Jackson's Eq. (11.149) in [2], are his equations (10.2) - (10.3), i.e., our Eqs. (44) and (45). His equations (10.2) - (10.3) are the equations with 4D geometric quantities and they correspond to our Eqs. (44) and (45) given below, whereas Eq. (10) contains *only the 3-vectors* $\mathbf{E}$, $\mathbf{B}$, $\mathbf{E}'$, $\mathbf{B}'$ *and the velocity 3-vector.* Thus, his equations (10.2) - (10.3) are not "Fock's and Jackson's transformations."

There is a very important consequence of the LT (42) and (43), or the same for P and M. As mentioned above, *under the relativistically correct LT the polarization vector* P *transforms again to the polarization vector* P'*; there is no mixing with the magnetization vector* M. On the other hand, according to (10) if there is an external, static, magnetic field (3-vector) outside, e.g., a stationary current loop then there is the magnetic field *and a static electric field (3-vector)* outside the same current loop which moves with uniform 3-velocity $\mathbf{U}$.

According to the LT (42) and (43), if there is an electric field outside *moving* magnet *it would necessary need to exist outside the same but stationary magnet.* Thus, in the Ampèrian approach, there is a polarization vector P (remember that P is a 4D geometric quantity and not a 3-vector) for a stationary permanent magnet as an assembly of small current loops. As explained in Sec. 7.1 below, this happens because every current loop behaves like an electric dipole at points far from that loop. Then, that P induces an external electric field. This will be discussed in much more detail in Sec. 8 below.

6. The usual transformations of vectors E and B; only F is transformed but not the observer

Let us examine what will be obtained if in the transformation of $E = F \cdot \gamma_0$ only F is transformed by the LT R, Eq. (39), but not the velocity of the observer $v = c\gamma_0$. Of course, it will not be the LT of $E = F \cdot \gamma_0$, because they are given by Eq. (41). Thus

$$E = F \cdot \gamma_0 \longrightarrow E'_F = (RF\widetilde{R}) \cdot \gamma_0 = F' \cdot \gamma_0. \tag{44}$$

This procedure yields that

$$E'_F = \gamma\{E + (\beta \wedge \gamma_0 \wedge cB)I\} + (\gamma^2/(1+\gamma))\beta(\beta \cdot E). \tag{45}$$

If Eqs. (44) and (45) are written in the standard basis and if it is taken that $\beta = \beta\gamma_1$, then they become

$$\begin{aligned} E'_F &= F' \cdot \gamma_0 = 0\gamma_0 + F'^{i0}\gamma_i = E'^{\nu}_F\gamma_\nu \\ &= E^1\gamma_1 + \gamma(E^2 - c\beta B^3)\gamma_2 + \gamma(E^3 + c\beta B^2)\gamma_3. \end{aligned} \tag{46}$$

Similarly, we find for B'_F

$$\begin{aligned} B'_F &= -(1/c)\gamma_5(F' \wedge \gamma_0) = 0\gamma_0 + (1/2c)\varepsilon^{0ijk}F'_{kj}\gamma_i = B'^{\nu}_F\gamma_\nu \\ &= B^1\gamma_1 + (\gamma B^2 + \beta\gamma E^3/c)\gamma_2 + (\gamma B^3 - \beta\gamma E^2/c)\gamma_3. \end{aligned} \tag{47}$$

From the transformations (46) and (47) one simply finds the transformations of the spatial components E'^i_F and B'^i_F

$$E'^i_F = F'^{i0}, \quad B'^i_F = (1/2c)\varepsilon^{0ijk}F'_{kj}, \tag{48}$$

which is the relation (24) but with the primed quantities.

It is seen from (46) that *the components of the transformed E'_F are expressed by the mixture of components of E and B. The same conclusion follows for B'_F from (47).*

The transformation (45) can be compared with the LPET for the 3-vector **E** that are given by the first equation in (10), and Eq. (46) can be compared with Eq. (11.148) in [2], i.e., with Eq. (9) here. Remember that in Eq. (10) **E**′, **E**, **B**′, **B** and $\boldsymbol{\beta}$ are all the usual 3-vectors. The comparison of Eq. (46) with Eq. (11.148) in [2] shows that *the transformations of components (taken in the standard basis) of E'_F are exactly the same as the transformations of $E_{x,y,z}$ from Eq. (11.148) in [2].* The same conclusion holds for the comparison of Eq. (47) and $B_{x,y,z}$ from Eq. (11.148) in [2]. The result that the components in (46) are the same as the components of **E**′ from (10) is completely understandable. Namely, (45) and (46) are obtained by the application of the LT *only* to F. On the other hand, it is already stated in Sec. 3.1 that the LPET of the components of **E** and **B** are derived assuming that they transform under the LT as the components of $F^{\alpha\beta}$ transform, Eqs. (3) and (7).

In contrast to the LT of E (43) (and the same for B), it is visible from (46), (47) and (48) that *E'_F and B'_F are again in the 3D subspace of the γ_0 - observer,* as it holds for E and B

in the γ_0-frame, Eqs. (23) and (24). Thus for the transformed E'_F *and* B'_F again hold that $E'^0_F = B'^0_F = 0$, i.e., that $E'_F \cdot \gamma_0 = B'_F \cdot \gamma_0 = 0$ as for E and B in the γ_0-frame. This shows in another way that *the LPET* (46), (47) *and* (48) *are not the LT, since the LT cannot transform some quantity from the 3D subspace again only to the 3D subspace.*

The transformations (44) and (46)-(48) are first discussed in detail in [6-11] and compared with the LPET (11.148) and (11.149) from [2], whereas the general form of E'_F, Eq. (45), is first given in [19].

We now point out another difference between the LT and the LPET. If instead of the active LT we consider the passive LT then, e.g. the vector $E = E^\nu \gamma_\nu = E'^\nu \gamma'_\nu$ will remain unchanged, because the components E^ν transform by the LT and the basis vectors γ_ν by the inverse LT leaving the whole E invariant under the passive LT. Of course, the same holds for all bases including those with nonstandard synchronizations, as discussed, e.g., in [16] and [33]. For the $\{r_\mu\}$ basis, this can be easily proved using $R^\mu_{\ \nu}$, i.e., Eqs. (11) and (15);

$$E = E^\nu \gamma_\nu = E'^\nu \gamma'_\nu = E^\nu_r r_\nu = E'^\nu_r r'_\nu. \tag{49}$$

The primed quantities in both bases are the Lorentz transforms of the unprimed ones. For the general form of the LT, that is independent of the chosen system of coordinates, including different synchronizations, see, e.g., Eq. (1) in [16]. The LT in the $\{r_\mu\}$ basis are given in the same paper by Eq. (2) or Eq. (21) in [33].

This invariance of E means that the electric field E is the same physical quantity for all relatively moving observers and for all bases used by them. In the same way this requirement has to be fulfilled for any well-defined 4D quantity.

It is not so with the 3-vector **E** and its LPET, or, equivalently, with E'_F. Namely, $\mathbf{E} = E_x\mathbf{i} + E_y\mathbf{j} + E_z\mathbf{k}$ is completely different than $\mathbf{E}'$, $\mathbf{E}' = E'_x\mathbf{i}' + E'_y\mathbf{j}' + E'_z\mathbf{k}'$ from (10), $\mathbf{E} \neq \mathbf{E}'$, and the same holds for E'_F, i.e., $E^\nu \gamma_\nu \neq E'^\nu_F \gamma'_\nu$. This means that although **E** and **E**′ are measured by different observers *they are not the same quantity for such relatively moving observers.* The observers are not looking at the same physical object, here the electric field vector, but at two different objects. *Every observer makes measurement of its own 3-vector field,* **E** *and* **E**′, *and such measurements are not related by the LT.* As far as relativity is concerned the quantities, e.g., **E** and **E**′, i.e., $E^\nu \gamma_\nu$ and $E'^\nu_F \gamma'_\nu$, are not related to one another. Their identification is a typical case of *mistaken identity.*

In the 4D geometric approach, i.e., in the ISR, different relatively moving inertial 4D observers can compare only 4D quantities, here $E^\nu \gamma_\nu$ and $E'^\nu \gamma'_\nu$, because they are connected by the LT. The experimentalists have to measure *all components* of 4D quantities, here of E, in both frames S' and S. The observers in S' and S are able to compare only such complete set of data which corresponds to the *same* 4D geometric quantity. Hence, from the point of view of the ISR the transformations for E'^i_F and B'^i_F, Eq. (48), i.e., Eq. (11.148) in [2], are not the LT. Therefore, contrary to the general belief, *it is not true from the geometric approach viewpoint that,* e.g., Sec. 11.10 in [2]: "A purely electric or magnetic field in one coordinate system will appear as a mixture of electric and magnetic fields in another coordinate frame."

7. Clausius' hypothesis and the second-order electric field outside a stationary superconductor with steady current

7.1 The second-order electric field outside a stationary superconductor with steady current

As stated in Sec. 3.2, in the classical model, the magnetic dipoles are considered as little current loops. In Sec. 18-4 in [23] and Sec. 6.5. in [22], and in all other usual approaches, it is assumed that a stationary current loop *with steady current* is globally and *locally* charge neutral (Clausius' hypothesis, (1877)), i.e., *it is simply supposed* that in the ions' rest frame S the charge density of the moving electrons is $\rho_- = -\rho_0$, where ρ_0 is the positive charge density for the wire at rest but without a current. Clausius (see Ref. [4] in the first paper in [28]) stated that hypothesis in another but equivalent way, i.e., he stated, as an "experimental assumption" that a "closed current in a stationary conductor exerts no force on stationary electricity." But, it is also assumed that the same current loop does not remain locally neutral when observed from another inertial frame. Namely, in [23], [22], ..., it is argued that the legs parallel to the 3-velocity $\mathbf{U}$ will carry the charges equal in magnitude but opposite in sign, because they are Lorentz contracted, whereas the legs perpendicular to $\mathbf{U}$ remain uncharged, because there is no Lorentz contraction for them. Thus, an electric dipole moment given by the expression (18) is obtained. Hence, there is an external, static, magnetic field (3-vector) outside the stationary current loop, but there is the magnetic field *and a static electric field (3-vector)* outside the same current loop which moves with uniform 3-velocity $\mathbf{U}$. Note that the legs of the current loop are treated as that they are infinite wires with steady currents. It can be seen from the mentioned textbooks, [23], [22], that such result is obtained using *the Lorentz contraction* and *the relativity of simultaneity*. In all these derivations it is also used the conventional definition of charge in terms of 3D quantities,

$$Q = (1/c) \int_{V(t)} j^0(\mathbf{r}, t) dV. \tag{50}$$

In that definition the volume $V(t)$ is taken at a particular coordinate time t and it is stationary in some inertial frame of reference S. The values of the charge density $\rho(\mathbf{r}, t) = j^0(\mathbf{r}, t)/c$ are taken simultaneously for all $\mathbf{r}$ in $V(t)$. It is supposed in all usual treatments (see, particularly, the well-known Purcell's textbook [24]) that the volume elements dV' are Lorentz "contracted" in a relatively moving inertial frame of reference S' and all of them, i.e., the whole volume $V'(t')$, are taken simultaneously at some arbitrary t' in S'. *The coordinate time t' in S' is not connected in any way with t in S.* Also, it is assumed that j^0 from S is transformed (using the Lorentz "contraction") only to j'^0 in S' and all j'^0 are taken simultaneously at the same t' in S'. The new

$$Q' = (1/c) \int_{V'(t')} j'^0(\mathbf{r}', t') dV' \tag{51}$$

in S' is considered to be equal to the charge Q in S, $Q' = Q$, (the total charge is invariant). But we remark that the charge Q defined in such a manner cannot be invariant under the LT. The LT cannot transform one component j^0 from an inertial frame of reference S to the same component j'^0 in S'. Also, if all j^0 values are taken simultaneously at some t in S then the LT cannot transform them to the values j'^0 which are again all simultaneous, but now *at some arbitrary t'* in S'. This consideration shows that such usual definition of charge cannot be the relativistically correct definition.

In the 4D spacetime, two relatively moving observers cannot compare spatial lengths taken alone, which are synchronously determined for the observer, see Appendix here. Similarly, in the 4D spacetime, the temporal distances taken alone are not well-defined and two relatively moving observers cannot compare them. For the thorough discussion of these usual definitions of the Lorentz contraction and the time dilation see, e.g., Secs. 4. - 4.2 and Figs. 3 and 4 in [16]. Also, a clear presentation of the relativistic incorrectness of the Lorentz contraction is already given in Sec. 2.2 in [27]. In the ISR, the properly defined quantity is the distance vector and the spacetime length, which is a Lorentz scalar, see, e.g., Secs. 3 - 3.2 and Figs. 1 and 2 in [16] and also Sec. 2.1 in [27].

Furthermore, the above mentioned conventional definition of charge in terms of 3D quantities is objected in Sec. 3 in [27] and Sec. 5.3 in [16]. There, it is shown that such definition with

3D quantities has to be replaced by the definition in terms of 4D geometric quantities, i.e., the charge is defined as a Lorentz scalar. The total electric charge Q in a three-dimensional hypersurface H (with two-dimensional boundary δH), as a Lorentz scalar, is defined by the equation

$$Q_{\delta H} = (1/c)\int_H j \cdot n dH, \tag{52}$$

where j is the current density vector and the vector n is the unit normal to H.

In Secs. 3 and 3.1 in [27], the external electric field for an infinite wire with a steady current and the Clausius hypothesis are examined in detail.

In the prerelativistic physics and in Einstein's formulation of the relativistic physics the charge density is well-defined quantity both for charges at rest and for the moving charges. As discussed in Sec. 3 in [27], it is considered in the usual approaches, e.g., [24-26], that the charge density of the moving charges is properly defined; it is enhanced by $\gamma = (1 - \beta^2)^{-1/2}$ relative to the proper charge density due to *the Lorentz contraction of the moving volume.* Therefore, both in the prerelativistic physics and in Einstein's formulation of the relativistic physics the Clausius hypothesis is meaningful, i.e., it can be properly formulated. Hence, in the usual approaches, *there is no external electric field for a stationary current-carrying conductor.* Of course, the whole consideration refers to an ideal conductor or to a superconductor, because for a stationary *resistive* conductor carrying constant current there is always an external static magnetic field *and a time independent external electric field that is proportional to the current and which is caused by the distribution of surface charges on the conductor,* see, e.g., [42] and references therein.

On the other hand, as already mentioned several times, see, e.g., Sec. 3. in [27] and Sec. 4 here, in the 4D spacetime, only the current density vector j is a well-defined physical quantity, but not the usual charge density ρ and the usual current density $\mathbf{j}$ as a 3-vector. *In the 4D spacetime it is not possible to give a definite physical meaning to the charge density of moving charges.* As discussed in Appendix, the Lorentz contracted length is meaningless in the 4D spacetime. As shown in Secs. 3 - 3.3 in [27], for an infinite wire with a steady current (the wire is situated along the x^1 axis) and if the standard basis is introduced, one can take that in S', in which the drift velocity 3-vector of the electrons is zero, the current density vector of the electrons (in one spatial dimension) is

$$j'_{-} = (-c\rho_0)\gamma_0 + 0\gamma_1, \tag{53}$$

i.e., as that the proper charge density ρ'_{-} of the electrons ($j'^{i}_{-} = 0$) is equal to $-\rho_0$. This is completely different than the Clausius hypothesis. Then, by means of (53) and the LT one finds the current density vectors in S, the rest frame of the wire, i.e., the lab frame, as (only components)

$$j^{\mu}_{-} = (-c\gamma\rho_0, -c\gamma\beta\rho_0), \quad j^{\mu} = (c(1-\gamma)\rho_0, -c\gamma\beta\rho_0), \tag{54}$$

where $j = j^{\mu}\gamma_{\mu}$ is the total current density vector in S, i.e., (components in the standard basis) $j^{\mu} = j^{\mu}_{-} + j^{\mu}_{+}$ and $j^{\mu}_{+} = (c\rho_0, 0)$. Observe that it holds that $j = j^{\mu}\gamma_{\mu} = j'^{\mu}\gamma'_{\mu}$, where the primed quantities are the Lorentz transforms of the unprimed ones. The equations (53) and (54) are Eqs. (11) and (12) in [27], respectively.

The same equations, (53) (i.e., only components) and (54) were already obtained in [32]. There, in contrast to the usual approach, the Lorentz contraction is introduced not only for the mean spacing between moving ions in the S' frame, but also it is assumed that there is a Lorentz contraction of the mean spacing between moving electrons in the lab frame, i.e., in the stationary wire with steady current. This may seem surprising that the same equations exist in [32] in which the Lorentz contraction is used and in [27] and here, where the 4D geometric quantities are used. But, the results obtained in [32] are not actually based on the use of the Lorentz contraction, than on the *assumption* that in the electrons' rest frame S' the electrons' charge density ρ'_{-} is $= -\rho_0$. The Lorentz contraction of the mean distance between moving electrons was only taken as the interpretation for the assumption that $\rho'_{-} = -\rho_0$. In the 4D geometric approach from [27] and here Eq. (53) is neither *hypothesis* as in the traditional approach with the 3D quantities and the Lorentz contraction, nor the *assumption* as in [32], but it is a consequence of the covariant definition of an invariant charge (52) and of the invariance of the rest length (see Appendix), i.e., it is the result of the use of the correctly defined 4D geometric quantities.

The components in the standard basis of the electric and magnetic fields are determined from the known current density vector j^μ (54) in Secs. 3.2 - 3.3 in [27]. Taking that the rest frame of the wire is the γ_0-frame, i.e., that $E^0 = 0$, $E^i = F^{i0}$, then the external electric field is $E = E^\mu \gamma_\mu = 0\gamma_0 + E^i \gamma_i$, where the components are given by Eq. (22) in [27]; that field is in the plane orthogonal to the wire and in the radial direction in that plane (the wire is situated along the x^1 axis)

$$E^0 = E^1 = 0,\ E^2 = 2k(1-\gamma)\rho_0 y a^{-2},\ E^3 = 2k(1-\gamma)\rho_0 z a^{-2}, \tag{55}$$

where $k = 1/4\pi\varepsilon_0$, $a^2 = y^2 + z^2$, $(x^1, x^2, x^3) = (x, y, z)$. Then, it is concluded in Sec. 3.3. in [27]: "*The equation* (55) *shows that the observer who is at rest relative to a wire with steady current will see, i.e., measure, the second-order electric field outside such a current-carrying conductor.*" ("the second-order electric field" means that $E^i \propto U^2/c^2$, where U is the magnitude of the drift velocity 3-vector of the electrons.) Note that such fields, but as the 3-vectors, are first predicted on different grounds in [32], see the above discussion about the assumption from [32] that $\rho'_{-} = -\rho_0$. As already mentioned, the second-order electric field (55) exists in a resistive wire with a constant current as well, but there it is much smaller but the contribution to the external electric field that is caused by the quasistatic surface charges.

Recently, the same treatment with the Lorentz contraction and the same results as in [32] are presented in [31]. That work, [31], from the theoretical point of view is almost the same as the treatment in [32], i.e., it is not with 4D geometric quantities and thus it is not a mathematically and relativistically correct treatment. Several results from [32] are incorrectly understood and interpreted in [31]. This will not be discussed here since both papers deal with the Lorentz contraction. But [31] is an important progress in the investigation of the existence of the second-order electric field, because it presents, as asserted in [31]: "a new analysis of the experimental sensitivity required to observe the hypothesized effect and analyzes the feasibility of several novel experimental methods to make such an observation." This will be discussed below in Sec. 7.2 together with the discussion of the already performed experiments and some suggested experiments.

Under the passive LT it holds that $E = E^\mu \gamma_\mu = E'^\mu \gamma'_\mu$. This essential feature of the approach with 4D geometric quantities, i.e., of the ISR, shows that *if the electric field vector exists in one inertial frame of reference, say in the rest frame of the electrons, the S' frame, as in all usual approaches, then it must necessary exist in the rest frame of the ions, i.e., of the wire, the S frame.*

Similarly, it is obtained in [27] that the components in the standard basis of the magnetic field are $B^0 = 0$ and B^i, which are the same as for the usual expression for the magnetic field of an infinite straight wire with current (only the components j^i are γ times bigger).

$$B^0 = B^1 = 0,\ B^2 = \gamma\mu_0 I y a^{-2},\ B^3 = \gamma\mu_0 I z a^{-2}, \tag{56}$$

where $I = \rho_0 A U$, A is the cross-sectional area. The vectors of the electric and magnetic fields in some relatively moving frame, e.g., the rest frame of the electrons, can be obtained using (55), (56) and the LT (42) and (43). Again, as for E, it holds that under the passive LT B is unchanged, $B = B^\mu \gamma_\mu = B'^\mu \gamma'_\mu$, as can be easily checked.

In Sec. 4 in [27], the same consideration is presented for a current loop and it is shown that *the second-order external electric field exists not only for a moving current loop*, as in the usual approaches, but *for the stationary current loop as well.* There are opposite charges on opposite sides of a square loop with current, but the total charge of that loop is zero. All these charges are invariant charges, which means that they are the same for both, moving and stationary current loop. They are defined as the Lorentz scalars, i.e., as in Eq. (52) for $Q_{\delta H}$. *At points far from that current loop it behaves like an electric dipole, but as a 4D geometric quantity.* It is incorrectly asserted in the published paper in [27] that such a distribution of charges behaves like an electric quadrupole. This is corrected in the v2 of the second paper in [27]. However, it is worth noting that, in the same way as in the usual approaches, the legs of the current loop are treated as that they are infinite wires with steady currents.

Thus, in the 4D geometric approach, i.e., in the ISR, the relation (18), which is derived by the use of the Lorentz contraction and the time dilation, does not hold for the current loop and consequently the relation (19) for the polarization $\mathbf{P}$, which is induced by the movement of a permanent magnetization $\mathbf{M}'$, does not hold as well. The vector of the electric field outside a moving current loop is not caused by *the 3D polarization* $\mathbf{P}$, (19), but it could be determined by the LT, the same as (42) and (43), of *the vector of the electric field of the same but stationary current loop.*

7.2 The experiments for the detection of the second-order electric fields

The external second-order electric fields from a stationary, superconducting, current loop, i.e., coil, have not yet detected. However, in [28], an I^2 - dependent potential resulting from constant current in closed superconducting coils has been reported and the same happened in the first variation of experiments in [29], but not in the second one. It is worth mentioning that all these experiments are sensitive only to a monopole field and thus they cannot either support or disprove the theory presented in [27] and here, which predicts a dipole field; very small (second-order) external electric field. This happens because a direct contact with the superconducting coil is used in the measurements [28], [29] of the I^2 - dependent potential. In such measurements it can be only "seen" if some charge is created or destroyed. In order to directly measure the external electric fields it is necessary to use a non-contact method of measuring. Such a method is recently presented in [30]. The experiments [30] are "based essentially on the detection of a non-zero force between a circular steady current and a charge, both at rest in the Earth frame." For the experimental setup see Fig. 1. in both papers in [30]. Observe that a Helmholtz coil γ that is used in the experiments is a *normal metal with finite resistivity.* The authors of [30] considered that they obtained a positive evidence for a non-zero force and that such results "show that *local Lorentz invariance could in fact be broken even in electromagnetic experiments* .. ." However, their results have nothing to do with the breakdown of the local Lorentz invariance because, as mentioned above, the standard Maxwell theory predicts that there are always static magnetic fields and *a time independent electric field outside a normal conductor with a steady current.* Such an *external electric field* causes a non-zero force on a stationary charge q in the experiments in [30].

In order to "see" the existence of the external second-order electric fields the coil used in the experimental setup in [30] would need to be a superconducting coil. Hence, *we propose to experimentalists to make the similar measurements as in* [30], *but using a superconducting coil.* It would be an important test of the validity of the relations (53) and (54), i.e., of the relation (55), or the validity of the usual approaches which assume the Clausius hypothesis. Namely, it is often declared that the classical electromagnetic theory predicts a zero external electric field for a stationary superconducting coil. But, it is not true. Maxwell equations enable one to find fields in the case that the sources are known. Hence, Maxwell equations will give a zero external electric field only if one supposes that in a stationary superconductor with a steady current the local charge density (in the ions' rest frame and in the standard basis) is everywhere zero. Inside the classical electromagnetic theory this statement that the local charge density is everywhere zero is merely *a hypothesis*, the Clausius hypothesis. This means that the possible existence of the second-order external electric fields from steady currents in a stationary superconductor is not at all in contradiction with classical electromagnetism. However, as stated in [27] and mentioned above, in contrast to the usual approaches in which there is no either physical or mathematical justification for the Clausius hypothesis or for some other choice, e.g., $\rho'_{-} = -\rho_0$, in the 4D geometric approach Eqs. (53), (54) and (55) are the consequences of the use of correctly defined quantities in the 4D spacetime, i.e., the 4D geometric quantities. The fields E and B with components in the standard basis given by Eqs. (55) and (56), respectively, correctly transform under the LT (42) and (43), which is not the case with the 3-vectors **E** and **B** obtained in all previous approaches.

Recently, an interesting possibility to experimentally investigate these second-order external electric fields is proposed in [31]. As already mentioned above, the theory from [31] is not relativistically correct, but the proposed experiments with the cold ions could be the right way to detect such small electric fields. Moreover, the estimated size of that electric field could be even bigger than that one found in [31] if the experiments would be made using a superconducting coil that is wound using a large number of bifilar pairs.

Another possibility to study these second-order external electric fields by the use of a non-contact method of measuring could be similar to that one used to measure the Casimir force in [43], i.e., by the use of a torsion balance.

As stated in Sec. 3.3 in [27], the second-order electric fields could play an important role in many physical phenomena with steady currents, particularly in tokamaks and astrophysics, where high currents exist, and in superconductors, where the electric fields that are caused by surface charges are absent.

Regarding the role of the second-order electric fields in tokamaks, we have to mention that recently, [44] and references [1], [5] and [7] therein, the existence of the radial electric field $E_r(r)$ in quasi-neutral tokamak plasma is examined taking into account the Lorentz contraction of "an "electron ring" circumference in steady state tokamak plasma rotating in toroidal direction with current velocity $V_e(r)$ ... ." The similar consideration with the Lorentz contraction of an "electron ring" was already reported in [45]. But, as we mentioned several times, such theories with the Lorentz contraction are not the relativistically correct theories. It would be very important for physics that the experimentalists find the best way for the direct and precise measurements of the second-order electric fields that are predicted in [27] and here by the relativistically correct approach with 4D geometric quantities.

8. The electric field outside a stationary permanent magnet

An interesting consequence of the above consideration refers to the existence of *the electric field outside a stationary permanent magnet*. It was mentioned at the end of [20]. Namely, as already stated, in the Ampèrian approach a permanent magnet is essentially an assembly of current loops. However, as discussed above, *the second-order external electric field exists not only for every moving current loop, but also for the same stationary current loop*, which yields that *the electric field would need to exist not only outside a moving permanent magnet but outside a stationary permanent magnet as well.* This conclusion is also supported by the following argument. According to the LT (42) and (43), if there is an electric field outside the *moving* magnet *it would necessary need to exist outside the same but stationary magnet.* Thus, in the Ampèrian approach, there is a polarization vector P (remember that P is a 4D geometric quantity and not a 3-vector) for a stationary permanent magnet as an assembly of small current loops. As explained above, this happens because *every current loop behaves like an electric dipole at points far from that loop.* Then, that P induces an external electric field.

If we abandon the Ampèrian approach then, nevertheless, there is another explanation for the possibility that a stationary permanent magnet possesses an intrinsic polarization.

According to the well-known Uhlenbeck-Goudsmit hypothesis there is a connection between the 3-vectors of the magnetic moment $\mathbf{m}$ of an electron and its spin $\mathbf{S}$, $\mathbf{m} = \gamma_S \mathbf{S}$. However, in the 4D spacetime, i.e., in the approach with 4D geometric quantities, the 3-vectors $\mathbf{m}$ and $\mathbf{S}$ are not properly defined quantities. In the 4D spacetime, as explained in [33], and in [46-48], the primary quantity (with independent physical reality) for the dipole moments is the dipole moment bivector D (four-tensor D^{ab} in [33], [46-49]) of a fundamental particle. It is decomposed into the EDM vector p, the MDM vector m and the unit time-like vector u/c, where u is the velocity vector of the particle. Then, p and m are derived from D and the velocity vector of the particle u according to the equations

$$\begin{aligned} D &= (1/c)[p \wedge u + (mI) \cdot u/c], \\ p &= D \cdot u/c, \quad m = I(D \wedge u), \end{aligned} \tag{57}$$

Eq. (2) in [33] (but in the tensor notation). It holds that $p \cdot u = m \cdot u = 0$; only three components of p and three components of m are independent since $\mathcal{M}$ is antisymmetric. In the particle's rest frame (the K' frame) and the standard basis $\{\gamma'_\mu\}$, $u = c\gamma'_0$, and using (57), it follows that $p'^0 = m'^0 = 0$, $p'^i = D'^{i0}$, $m'^i = (c/2)\varepsilon^{0ijk} D'_{jk}$. Therefore p and m can be called the "time-space" part and the "space-space" part, respectively, of the dipole moment bivector D. But, these parts are written with quotation marks because in all other relatively moving frames and in all other bases the above identifications of the components of p and m with the components of D do not hold.

Similarly, it is shown in [33] (earlier, in [48] and Ref. [3] in [48]) that the primary quantity *with definite physical reality* for the *intrinsic* angular momenta is the spin bivector $\mathcal{S}$ (four-tensor S^{ab} in [33], [48]), which is decomposed into the usual "space-space" intrinsic angular momentum S, the "time-space" intrinsic angular momentum Z and the unit time-like vector u/c, where u is the velocity vector of the particle

$$\begin{aligned} \mathcal{S} &= (1/c)[Z \wedge u + (SI) \cdot u], \\ Z &= \mathcal{S} \cdot u/c, \quad S = I(\mathcal{S} \wedge u), \end{aligned} \tag{58}$$

Eq. (8) in [33]. It holds that $Z \cdot u = S \cdot u = 0$; only three components of Z and three components of S are independent since $\mathcal{S}$ is antisymmetric. S and Z depend not only on $\mathcal{S}$ but on u as well. Only in the particle's rest frame, the K' frame, and the $\{\gamma'_\mu\}$ basis, $u = c\gamma'_0$ and $S'^0 = Z'^0 = 0$, $S'^i = (1/2c)\varepsilon^{0ijk} \mathcal{S}'_{jk}$, $Z'^i = \mathcal{S}'^{i0}$. *According to Eq. (58), a new "time-space" spin Z is introduced and it is a physical quantity in the same measure as it is the usual "space-space" spin S.*

Then, in [33], it is suggested that instead of the connection between the 3-vectors $\mathbf{m}$ and $\mathbf{S}$ we need to have the connection between the dipole moment bivector D and the spin bivector $\mathcal{S}$, which is formulated in the form of the generalized Uhlenbeck-Goudsmit hypothesis

$$D = g_S \mathcal{S}, \tag{59}$$

Eq. (9) in [33]. Hence, inserting the decompositions of D (57) and $\mathcal{S}$ (58) into Eq. (59) we find the connections between the dipole moments m and p and the corresponding intrinsic angular momenta S and Z, respectively,

$$m = cg_S S, \; p = g_S Z, \tag{60}$$

Eq. (10) in [33]. In the particle's rest frame, the K' frame, and the $\{\gamma'_\mu\}$ basis, $u = c\gamma'_0$ and $p'^0 = m'^0 = 0$, $p'^i = g_S Z'^i$, $m'^i = cg_S S'^i$. Comparing this last relation with $\mathbf{m} = \gamma_S \mathbf{S}$, we see that $g_S = \gamma_S / c$. In contrast to all previous approaches with the 3-vectors in which both the MDM

$\mathbf{m}'$ and the EDM $\mathbf{p}'$ of an elementary particle are determined by the usual spin $\mathbf{S}'$, we find that the intrinsic MDM vector m of an elementary particle is determined by the "space-space" spin vector S, whereas *the intrinsic EDM vector p is determined by the "time-space" spin vector Z. The relations* (59) *and* (60) *show that any fundamental particle has not only the intrinsic MDM m, but also the intrinsic EDM p whose magnitude is* $(1/c)$ *of that for m.* The EDM p (see (60)) emerges from the connection with the intrinsic angular momentum Z, i.e. from (59) and (57), (58) in the same way as the MDM m emerges from the connection with the intrinsic angular momentum S. The EDM p is an intrinsic property of a fundamental particle in completely the same way as it is the intrinsic MDM m. As stated in [33]: "The EDM obtained in this way is of quite different physical nature than in the elementary particle theories, e.g., in the standard model and in SUSY. There, an EDM is obtained by a dynamic calculation and it stems from an asymmetry in the charge distribution inside a fundamental particle, which is thought of as a charged cloud."

These fundamental results for the generalized Uhlenbeck-Goudsmit hypothesis (59), i.e., for the new spin Z and the associated EDM p of a fundamental particle (60) are used in [33] and [49] for the discussion of the shortcomings in experimental searches for a permanent EDM of particles. In all these searches the AT of $\mathbf{E}$ and $\mathbf{B}$ (10) and the AT for the 3D EDM $\mathbf{p}$ and MDM $\mathbf{m}$ (18) are always considered to be relativistically correct, i.e., that they are the LT. Furthermore, it is supposed in the elementary particle theories, that not only the 3D MDM $\mathbf{m}$ is proportional to the usual 3D spin $\mathbf{S}$ ($\mathbf{m} = m(\mathbf{S}/S)$, the Uhlenbeck-Goudsmit hypothesis), but the 3D EDM $\mathbf{p}$ as well. The results from [33] and [49] explicitly show that from the ISR viewpoint this basic assumption is not true and that it has to be replaced by the relation (60) which deal with 4D geometric quantities that are correctly defined in the 4D spacetime.

If the spins and the dipole moments are quantized, i.e., if they become operators, then, in [48], the commutation relations for the components in the standard basis of the intrinsic angular momentums S and Z are given by Eq. (4) in [48]

$$\begin{aligned}
[S^{\mu}, S^{\nu}] &= (i\hbar/c)\varepsilon^{\mu\nu\alpha\beta}S_a u_{\beta},\ [Z^{\mu}, Z^{\nu}] = (-i\hbar/c)\varepsilon^{\mu\nu\alpha\beta}S_a u_{\beta}, \\
[S^{\mu}, Z^{\nu}] &= (i\hbar/c)\varepsilon^{\mu\nu\alpha\beta}Z_a u_{\beta}. \qquad (61)
\end{aligned}$$

Taking into account the relation (60) the commutation relations for the components of m and p, m^{μ} and p^{μ} respectively, are expressed in terms of those for S^{μ} and Z^{μ}, Eq. (61), and they are given by Eq. (5) in [48]

$$[m^{\mu}, m^{\nu}] = c^2 g_S^2[S^{\mu}, S^{\nu}],\ [p^{\mu}, p^{\nu}] = g_S^2[Z^{\mu}, Z^{\nu}],\ [m^{\mu}, p^{\nu}] = c g_S^2[S^{\mu}, Z^{\nu}]. \qquad (62)$$

As explained above, an electron possesses both intrinsic angular momentums, spins, S and Z and, according to (60), the associated dipole moments m and p, respectively. All these quantities, S and Z, m and p are vectors, properly defined geometric quantities in the 4D spacetime. They transform according to the LT, the same as (42) and (43). This means that *under the LT vector m transforms again to the magnetic dipole moment* and, contrary to the LPET of the 3-vectors $\mathbf{m}$ and $\mathbf{p}$ (the same as (16)) *there is no mixing with p*. The same holds for vectors S and Z.

At this point we present a simple discussion, only qualitative arguments, about the existence of an electric field outside a stationary permanent magnet and about the experimental detection of that field. It can be concluded that in the same way as the MDMs determine the magnetization M of a stationary permanent magnet the EDMs determine its polarization P, which induces an electric field outside a permanent magnet (moving or *stationary*). Note that according to the LT, the same as (42) and (43), if there is a polarization vector P in one inertial frame of reference it will exist in all other relatively moving inertial frames of reference. As $g_S = \gamma_S/c$, the EDM p is in magnitude $1/c$ of the MDM m. Consequently, the external electric field is much smaller than the magnetic field outside a stationary permanent magnet. It can be one of the reasons why such electric fields are not yet experimentally detected. Another, equally important reason, is that the experimentalists never looked for such fields because it was no theory that predicts them.

In order to directly measure the external electric fields from a stationary permanent magnet it is necessary to use a non-contact method of measuring. A possible experimental setup for the detection of that field could be again with the cold ions as in [31]. Another way of measuring such electric fields from a stationary permanent magnet would be some modification of torsion balance setup that is used in [43] for the detection of the Casimir force.

9. "Charge-magnet paradox" and its resolution with the 4D torques

9.1 Criticism of the "resolutions" [34-36] of Mansuripur's paradox

A nice example which clearly illustrates the essential difference between the 4D geometric approach, i.e., the ISR, with relativistically correct LT and Einstein's formulation of SR with the AT of the 3-vectors refers to the recently reported [34] "charge-magnet paradox," to the highlight of it [35] and different "resolutions" [36]. In [34], it is argued that in the presence of the magnetization **M** and the electric polarization **P** the usual expression for the Lorentz force with the 3-vectors fails to accord with the principle of relativity, because it leads to an apparent paradox involving a MDM **m** in the presence of an electric field **E**; in a static electric field a MDM **m** is subject to a torque **T** in some frames and not in others. (Henceforward we shall denote the usual 3D torque by **T**.) Mansuripur [34] argues that the conventional Lorentz force (density), Eq. (5) in [34], should be replaced by the Einstein-Laub law, Eq. (6) in [34], which predicts no torque **T** in all frames. The paradox is reminiscent of the Trouton-Noble paradox [50] and of the Jackson paradox [51]. For the resolution of these paradoxes with the 4D geometric quantities, i.e., in the ISR, see [12], [13] and [14], respectively. The paper [34] and the paper [35] that highlighted it and almost all papers in [36] that objected it deal with the 3-vectors and their LPET (10) for **E** and **B**, (16), or (17) for **P** and **M**, and the same for EDM **p** and MDM **m**, i.e., (18) or (19), as that these transformations are the relativistically correct LT. For example, in [34], it is stated: "Lorentz transformation to the xyz frame then yields a pair of point dipoles, ..," see also Eqs. (9 - 10b) and Eqs. (11 - 12b). Furthermore, Eq. (18), or (19), is taken as the main point in the majority of the expositions in [34-36]. But, these transformations are the AT and not the LT. This means that the requirement like (49) does not hold either for the conventional Lorentz force or the Einstein-Laub expression and also for the corresponding 3D torques. All these quantities are not well-defined quantities in the 4D spacetime. Similarly happens for all papers from [34-36]. Hence, as will be shown in this section, from the ISR viewpoint, the resolutions of the "charge-magnet paradox" that are proposed in [34-36] are not the relativistically correct resolutions.

In [34], in the discussion of another example, not only that LPET (10) and (16) are considered to be the LT, but the equation $\mathbf{F} = d\mathbf{p}/dt$ is considered to be "the relativistic version of Newton's law." There, it is stated ".. its relativistic momentum **p** increases with time, not because of a change of velocity but because of a change of mass." It is shown in the preceding sections, i.e., in [6-11], that the LPET ARE NOT the LT, and as shown in detail in, e.g., [14], [10], any 3D quantity cannot correctly transform under the LT; it is not the same quantity for relatively moving observers in the 4D spacetime. *The equation* $\mathbf{F} = d\mathbf{p}/dt$ *with* $\mathbf{p} = m\gamma_u\mathbf{u}$ *is not the relativistic equation of motion* since, contrary to the common assertions, e.g., [2], [22], [26], it does not retain the same form in two relatively moving inertial frames S and S', i.e., it is not covariant under the LT, see, e.g., the discussions in Sec. 3 in the first paper in [14] and particularly in [10]. In that equation the primed 3D quantities are not obtained by the LT from the unprimed ones, but they are obtained in terms of the AT for the 3D force **F**, e.g., Eqs.

(12.66) and (12.67) in [26], and the 3D momentum $\mathbf{p}$, i.e., the 3D velocity $\mathbf{u}$, Eq. (11.31) in [2]. The requirement like (49) does not hold for any of the mentioned 3D quantities. Instead of the equation with the 3D quantities one has to use the equation of motion with 4D geometric quantities, Eq. (10) in [14], $K = dp/d\tau$, $p = mu$, where p is the proper momentum vector and τ is the proper time, a Lorentz scalar. For the definition of the Lorentz force K_L

$$K_L = (q/c)F \cdot u = (q/c)\left[(1/c)E \wedge v + (IB) \cdot v\right] \cdot u, \tag{63}$$

see, e.g., the first paper in [14]. The quantities K (K_L), p, u *transform in the same way, like any other vector, i.e., according to the LT, the same as* (42), (43) and not according to the awkward, AT of the 3-force $\mathbf{F}$ and the 3-momentum $\mathbf{p}$, i.e., the 3-velocity $\mathbf{u}$. Moreover, only the rest mass is well defined quantity in the 4D spacetime and thus there is not "a change of mass."

In [35], it is tried to explain why the moving magnet appears to be electrically polarized. The explanation is completely the same as it is the derivation of the relations (18) and (19) in [23] and [22] and in Sec. 3.2 here, i.e., it explicitly uses *the Lorentz contraction*. But, as can be seen, e.g., from Appendix here, the Lorentz contraction is an AT, i.e., it is ill-defined in the 4D spacetime. Similar consideration with the use of *the Lorentz contraction* is presented in Unnikrishnan's paper in [36].

In some papers from [36], e.g., Griffiths and Hnizdo (GH), McDonald, Saldanha, the model of the "hidden" momentum and the "hidden" angular momentum is invoked in order to show that, contrary to the claims in [34], the usual Lorentz force law with the 3-vectors is consistent with SR. Their approach is also with the 3-vectors and the LPET, i.e., the AT, and, particularly, in all three papers it is taken that the moving magnetic dipole, i.e., an Ampèrian loop, acquires an electric dipole moment given by Eq. (18). Let us explain in some detail the approach from GH [36]. They find that if the magnetic dipole is considered to be an electric current loop then the net torque $\mathbf{T}$ (their $\mathbf{N}$) on the dipole, Eq. (4) in GH [36], is non-zero in the lab frame, i.e., the principle of relativity is violated. Observe that $\mathbf{T}$ from their Eq. (4) is obtained from the term $\mathbf{p} \times \mathbf{E}$, where $\mathbf{p}$ is the induced electric dipole moment, their Eq. (3), i.e., our Eq. (18), which means that *the violation of the principle of relativity is a direct consequence of the AT for the 3-vectors* $\mathbf{p}$ *and* $\mathbf{m}$. To resolve the problem they introduce two sorts of the angular momentums ".. $\mathbf{L}_o$ is the "overt" angular momentum (associated with actual rotation of the object), and $\mathbf{L}_h$ is the "hidden" angular momentum (so called because it is *not* associated with any overt rotation of the object)," $\mathbf{L} = \mathbf{L}_o + \mathbf{L}_h$, where $\mathbf{L}_h$ is given by their Eq. (10). Furthermore, they define the torque $\mathbf{T}$ by their Eq. (12), $\mathbf{T} = d\mathbf{L}/dt = d\mathbf{L}_o/dt + d\mathbf{L}_h/dt$ and state "*Physically* $d\mathbf{L}_h/dt$ *is not a torque*, (my emphasis) but (the rate of change of) a piece of the angular momentum .. ." In order to get the agreement with the principle of relativity they argue that it is not $\mathbf{T}$, which has to be zero in both frames, but it is the "effective" torque $\mathbf{T}_{\text{eff}}$ given by their Eq. (13). Thus, they assert: ".. it is the "overt" torque (given by Eq. (13), my remark) that must vanish to resolve the paradox, since the dipole is not rotating .. ."

Several objections to their treatment are at place already here. (i) First of all, as already mentioned, their approach deals with the 3-vectors and their AT and not with the 4D quantities and their LT. The requirement like (49) is not fulfilled for any 3D quantity. (ii) In such an approach with 3-vectors the introduction of the "hidden" quantities is very artificial and it is not justified either mathematically or physically. (iii) The derivatives in Eqs. (11-13) are over the *coordinate time* and not over the proper time, which means that neither of them can properly transform under the LT. (iv) Their assertion about $d\mathbf{L}_h/dt$ is meaningless. It cannot be that in an equation (their Eq. (13)) with physical quantities $\mathbf{T}$ is a torque and $d\mathbf{L}_o/dt$ is also a torque, but $d\mathbf{L}_h/dt$ *is not a torque*. Obviously, in the usual formulation with the 3-vectors there is no proper physical interpretation for $d\mathbf{L}_h/dt$, i.e., for different "hidden" quantities.

It can be seen from the resolution of Jackson's paradox [14], *the Trouton-Noble paradox* [12, 13] *and from Sec. 9.2 here that if the physical reality is attributed to the 4D geometric quantities which properly transform under the LT and not, as usual, to the 3D quantities which transform according to the AT, then all quantities are mathematically and physically correctly defined and there is no "hidden" quantity, i.e., the quantity for which there is not some proper physical interpretation in the 4D spacetime.* For example, the angular momentum is not the 3-vector $\mathbf{L} = \mathbf{r} \times \mathbf{p}$ and the torque is not $\mathbf{T} = \mathbf{r} \times \mathbf{F}$ with $\mathbf{T} = d\mathbf{L}/dt$, but they are the abstract 4D quantities, here the bivectors,

$$J = x \wedge p, \; N = x \wedge K; \quad N = dJ/d\tau, \tag{64}$$

where x is the position vector, p is the proper momentum vector, $p = mu$, u is the proper velocity vector $u = dx/d\tau$ of a charge q (it is defined to be the tangent to its world line), τ is the proper time and K is the force vector. If J and N are written as CBGQs in the $\{\gamma_\mu\}$ basis they are given by Eq. (12) in the first paper in [14],

$$\begin{aligned} J &= (1/2)J^{\mu\nu}\gamma_\mu \wedge \gamma_\nu, \; J^{\mu\nu} = m(x^\mu u^\nu - x^\nu u^\mu), \\ N &= (1/2)N^{\mu\nu}\gamma_\mu \wedge \gamma_\nu, \; N^{\mu\nu} = x^\mu K^\nu - x^\nu K^\mu. \end{aligned} \tag{65}$$

The components $J^{\mu\nu}$ from (65) are identical to the usual covariant angular momentum four-tensor and similarly for $N^{\mu\nu}$.

Cross [36], in the Comment on Mansuripur's paper, states: "The torque density is not a vector, but the antisymmetric tensor .. " and also ".. the angular momentum, which is the second rank tensor .. ." So, he correctly notices that the angular momentum and the torque density are not the 3D vectors. However, his quantities $J^{\alpha\beta}$ and $T^{\alpha\beta}$ also are not tensors, but components implicitly taken in the standard basis as in the usual covariant approaches, e.g., from [2], [22], [26]. They correspond to $J^{\mu\nu}$ and $N^{\mu\nu}$ from (65). Tensors are abstract, geometric quantities, e.g., in the abstact index notation M^{ab}, Eq. (4) in [33], which corresponds to the bivector J in (64) in the geometric algebra formalism. If tensors are represented in some basis then they become the CBGQs that contain components *and the basis vectors* as in Eq. (65). As already mentioned several times, if one does not use Einstein's synchronization but, e.g., the "r" synchronization then it is not possible to make the identification of the components of the 3-vectors $\mathbf{E}$ and $\mathbf{B}$ with the components $F^{\alpha\beta}$ of the electromagnetic field F, or the components of the 3-vectors $\mathbf{R}$ and $\mathbf{T}$ (in Cross's notation) with the components $T^{\alpha\beta}$ of the torque tensor, etc. This means that the physical laws have to be written with the abstract quantities and then to represent these quantities in some basis as the CBGQs. Furthermore, he states, "For short we write $M = (\mathbf{P}, -\mathbf{M})$ in terms of the defining 3-vectors." Similarly, he writes "$F = (-\mathbf{E}, -\mathbf{B})$ is the Faraday tensor .." and also "$m = (\mathbf{p}, -\mathbf{m})$ is the tensor of moments." Hence, in his covariant approach with "tensors," in the same way as in all traditional approaches, it is considered that the 3-vectors are the primary quantities with a definite physical reality, whereas $M^{\alpha\beta}$, $F^{\alpha\beta}$, $m^{\alpha\beta}$, etc. are only auxiliary mathematical quantities that are defined by the components of the 3-vectors.

The situation is completely different in the ISR in which the primary quantities are the abstract quantities, here the bivectors, thus not only components, $\mathcal{M}$, F, D (our notation), etc. As seen from Eqs. (25) and (26), Eqs. (21) and (22), Eq. (57), etc. P and M are derived from $\mathcal{M}$ and they depend not only on $\mathcal{M}$ but on u as well, and similarly for the bivector F and vectors E, B and v and also for D and p, m and u.

Note that Cross [36] also deals with the "hidden" quantities relating them to the time-space components, e.g., N^{0i} (T^{0i} in Cross's notation) of the torque tensor. He argues: ".. that the torque predicted in the moving frame is correct and necessary to balance the "hidden" angular momentum of the moving dipole rather than causing a precession of the spin." This means that under the torque he does not understand the whole torque tensor, but only the 3D torque $\mathbf{T}$ that is related with the rotation in the 3D space. Indeed, for the time-space components, i.e., for the components of his **R** it is stated: "These "torque" components are connected with the motion of the center of energy .. ."

In the 4D spacetime the center of energy is not well-defined quantity. As seen from [12-14] and as will be seen below such understanding is in a sharp contrast to ISR in which *only* the 4D geometric quantities are well-defined in the 4D spacetime, e.g., the torque bivector N and the torque vectors N_s and N_t, see Eq. (74).

Observe also that Cross [36], in contrast to all others from [34-36], showed that the offending torque $T'^x = \gamma vmE$ (in his notation) that is obtained in a non-covariant analysis, i.e., with the 3-vectors and their AT for EDM **p** and MDM **m**, Eq. (18), is determined as the Lorentz transformed time-space component that exists even in the rest frame. In his notation that nonvanishing component is $R^y \equiv T^{ty} = -mE$. Then, he states: "Under a Lorentz boost to the moving frame the space-space and time-space components mix." Hence, his 3D torque $T'^x = \Lambda_t^z T^{yt}$ is considered to be obtained by the LT. However, the comparison of his derivation of T'^x with the derivation of the AT (10), i.e., (9), in Sec. 3.1 reveals that both derivations are exactly the same. Indeed, Cross [36] simply assumes that the components of the 3-vectors **R** and **T** are identified with the components $T^{\alpha\beta}$ of the torque tensor *in both frames* (this corresponds to Eqs. (3) and (7)) and accordingly they transform like the components $N^{\alpha\beta}$ ($T^{\alpha\beta}$ in Cross's notation) transform. For comparison, the reader can look at the text in Sec. 3.1 between Eq. (9) and Eq. (10). Thus, contrary to the assertion in Cross [36] the transformations for T'^x are not the LT, but they are exactly the same as the LPET, i.e., the AT (10). The explicit AT of the components of the 3-vectors **T** and **R** are given, e.g., by Eq. (19) in [13], which is repeated here

$$\begin{aligned} T_1 &= T_1',\ T_2 = \gamma(T_2' - \beta R_3'),\ T_3 = \gamma(T_3' + \beta R_2'), \\ R_1 &= R_1',\ R_2 = \gamma(R_2' + \beta T_3'),\ R_3 = \gamma(R_3' - \beta T_2'). \end{aligned} \tag{66}$$

They are written for the motion along the x^1 axis and not along the x^3 axis as in Cross [36] and also the components $T_{t,i}$ from Eq. (19) in [13] are replaced by R_i. With these changes, the component in (66) that corresponds to T'^x in Cross [36] is $T_2 = -\gamma\beta R_3'$. It is visible from (66) that the transformations for R_i are the same as the AT for E_i in Eq. (9) and similarly is for T_i and B_i. The essential point is that, e.g., *the components T_i of the torque 3-vector* **T** *in the moving frame are expressed by the mixture of the components of the 3D vector* **T**′ *and of another 3D vector* **R**′ *from the rest frame.* This is the reason that the components of the usual 3D torque **T** will not vanish in the S frame even if they vanish in the S' frame, i.e., that there is the "charge-magnet paradox" in all usual approaches to special relativity that deal with the 3-vectors or with components implicitly taken in the standard basis. The same discussion as in Sec. 6 after Eq. (49) can be repeated for the 3-vectors **T** and **R**. Again it can be stated that *as far as relativity is concerned the quantities, e.g.,* **T** *and* **T**′ *are not related to one another* and that their identification is simply a *mistaken identity.*

However, as can be seen from Eq. (82) below, the situation is completely different for the 4D torque N_s, and the same for N_t. The physical 4D torque, the basis-free, abstract, N_s, simply has different representations, 4D CBGQs, in different bases.

Very similar procedure as in GH [36] is commonly used for the resolution of the Trouton-Noble paradox, see Refs. [3-7] in [13]. In the Trouton-Noble paradox, in the similar way as in [34], there is a 3D torque and so a time rate of change of 3D angular momentum in one inertial frame, but not in another relatively moving inertial frame. For the usual resolution of that paradox it is taken that *there is another 3D torque*, which is equal in magnitude but of opposite direction, giving that the total 3D torque is zero in order to have agreement with the principle of relativity. Different explanations have been offered for the existence of that additional 3D torque, e.g., the nonelectromagnetic forces with their additional torque, Refs. [3-6] in [13], or the field angular momentum and its rate of change, i.e., its additional torque, Ref. [7] in [13].

The resolution of Mansuripur's paradox that is presented by McDonald [36] is almost the same as in GH [36] with the difference that McDonald introduces the field angular momentum and states: "... the "paradoxical" nonzero torque is needed to change the "hidden" mechanical angular momentum of the system, such that this remains equal and opposite to the field angular momentum, .. ." The above objections (i) - (iii) hold in the same measure for McDonald's approach with an additional objection that the usual expressions for the electromagnetic momentum that are given by his Eq. (1) are all with the 3D quantities, which means that they also are not well-defined in the 4D spacetime; they do not properly transform under the LT.

Kholmetskii, Missevitch and Yarman [36] exclusively deal with 3D quantities and introduce the contribution of the hidden momentum as well.

Milton and Meille [36] also deal with the 3-vectors and the AT (10) for **E** and **B**. Particularly, for a point dipole they argue that: "... the torque is not zero, but is balanced by the rate of change of the angular momentum of the electromagnetic field, so there is no mechanical torque on the dipole." Their 3D quantities also do not properly transform under the LT.

Brachet and Tirapegui [36] combine the usual covariant approach with components implicitly taken in the standard basis with the formulation in terms of the 3-vectors writing, e.g., the energy density u and the Poynting 3-vector **S** in terms of the 3-vectors **E** and **B** and the integrations in $\mathcal{P}_{\text{field}}$ and $\mathcal{L}_{\text{field}}$ are over the 3D space. In the lab frame they have the torque, a 3D quantity, but they argue that "it is required to account for the motion at uniform speed of the missing momentum $\mathcal{P}_{\text{particle}}$." Thus, they also deal with quantities which do not correctly transform under the LT.

Boyer [36], as almost all others, defines the relativistic conservation laws, the Lorentz transformations of forces, the Lorentz transformations of energy and momentum, the angular momentum and torques, etc., all in terms of 3D quantities, e.g., the 3-vectors **E**, **B**, **r**, **p**, **F**, **L**, etc., considering that their AT are the LT. It is interesting that only Boyer in his Ref. 6 mentioned my Comment (arXiv: physics/0505013) to Jacson's paper [51] as that it was published. How can it be that *in the 4D spacetime* the 3D quantities are considered to be well-defined, whereas the 4D geometric quantities are considered to be ill-defined?

For the electromagnetic momentum that correctly transforms under the LT see Secs. 4, 5 - 5.3 in [52] (only components) and for the more general expressions with the 4D geometric quantities see Sec. 2.6 in [12]. There, in [12], an axiomatic geometric formulation of electromagnetism with the Faraday bivector field F as the primary quantity for the whole electromagnetism is presented and in Sec. 2.6 the observer independent expressions with the abstract quantities, Eqs. (37) - (43), for the stress-energy vector $T(n)$, the energy density U, the Poynting vector S and the momentum density g, the angular momentum density M and the Lorentz force K_L are directly derived from the field equation for F, Eq. (4) in [12] (the notation is that one from [12]). These quantities are also written as CBGQs in the standard basis, Eqs. (44) - (47). Furthermore, the local conservation laws are directly derived from that field equation for F and presented in Sec. 2.7 in [12], see Eqs. (48) - (51).

In [36], only in Vanzella's paper the CBGQs in the standard basis are used, but again the formulation with such quantities is combined with the formulation in terms of the 3-vectors. He "mimic the magnet by a neutral ring conducting an electric stationary current I." Although he does not explicitly use the AT of the 3D quantities, the assumption that $j^{\mu}u_{\mu} = 0$, "reflecting the neutrality of the ring according to $\mathcal{O}$," where $\mathcal{O}$ is an observer at rest in the frame of the ring, is equivalent to the Clausius hypothesis. According to the discussion in Sec. 7.1 that hypothesis is not well justified in the 4D spacetime. That assumption leads to the result that for a relatively moving observer the ring is polarized and there is an induced electric dipole moment 3-vector and a 3D torque $\tau = \mathbf{d} \times \mathbf{E}$ (his notation) on the ring. That torque is the same as that the AT (18) are explicitly used. Furthermore, he declares: "the torque exerted by the electric field is not used to rotate the ring but rather to move its asymmetric distribution induced by the very same the electric field. No paradox here." In principle, such a conclusion is very similar to that one as in GH [36] or Cross [36]. Griffiths and Hnizdo [36] and also Cross [36] consider that there is no paradox because the offending torque (3-vector) is not "associated with actual rotation of the object." In the 4D spacetime their resolution of the paradox is not correct, because it is based on the use of the 3D quantities and their AT. As shown below the relativistically correct torques are the bivector N and the vectors N_s and N_t.

Griffiths and Hnizdo [36] assert: "This "paradox" was resolved many years ago by Victor Namias [4]." But, it is not true. Namias, as almost all others, exclusively dealt with the 3D quantities and their AT. Similarly, Vanzella [36] declares: ".. there is a very similar and famous "paradox" (the Trouton-Noble "paradox" [2]) which was presented and resolved more than a hundred years ago [3] (see also Refs. [4, 5])." As can be seen from [12-14] that Vanzella's statement is not correct. The "resolutions" presented in Vanzella's references [3-5] dealt with different AT, the Lorentz contraction, the dilation of time and the LPET (10) and (16) considering them as that they are the relativistically correct LT.

In addition, it is worth mentioning that *all treatments from* [34-36] *are meaningless* if some other basis and not the standard basis is taken into account, e.g., *if only the Einstein synchronization is replaced by the "radio" synchronization.* This conclusion simply follows already from Eqs. (4) and (5).

9.2 The resolution of Mansuripur's paradox with the 4D torques

The geometric approach to the "charge-magnet paradox" that is presented here significantly differs from all formulations in [34-36]. First of all, in contrast to [34-36], this approach deals from the outset with 4D geometric quantities, their LT and equations with them. It is already shown in the similar treatments of the Trouton-Noble paradox [12, 13] and Jackson's paradox [14] that in the approach with 4D geometric quantities the principle of relativity is naturally satisfied and there is no paradox. Every 4D geometric quantity is the same quantity for all relatively moving inertial observers and for all bases chosen by them. The equation like (49) holds for any such quantity. Furthermore, the most important difference is that the torque bivector is different from zero even in the common rest frame of the considered charge and magnet from [34]. The reason for that important difference is that according to the 4D geometric approach, i.e., the ISR, a stationary permanent magnet possesses not only the magnetization vector M but the polarization vector P as well. Observe that, according to the preceding discussion, instead of to speak about a permanent magnet one can equivalently consider a current loop with a steady current.

Let us consider the system from [34], but, without loss of generality, the electric charge will be substituted by a uniform electric field. The common rest frame of the source of the electric field (a point charge q in [34]) and of the permanent magnet will be denoted as S', whereas the lab frame, in which the S' frame moves with uniform velocity $V = V\gamma_1$ along the common x_1, x_1' axes, will be denoted as S. (Here, such a choice is taken for easier comparison with [13] and [14].) The treatment of the interaction between a static electric field and a permanent magnet that is presented here is very similar to the formulations given in [12-14]. For this example, the Lorentz force density k_L is given by Eq. (35) in which it is taken that $j^{(C)} = 0$, i.e., $k_L = (1/c)F \cdot j^{(\mathcal{M})}$, where $j^{(\mathcal{M})}$ is given by Eq. (31) and $F = (1/c)E \wedge v$, what is Eq. (21) with $B = 0$. Remember that *the vector B is zero in all relatively moving inertial frames of reference* and therefore there is no reason for the appearance of the paradox. This is an essential difference between the approach with the 4D geometric quantities and their LT and the usual approach with the 3D quantities and their AT. Compare with Eqs. (11 - 12b) in [34]. Hence, k_L from (35) becomes

$$k_L = (1/c^2)(E \wedge v) \cdot [-(\partial \cdot P)u + (u \cdot \partial)P + (1/c)[u \wedge (\partial \wedge M)]I]. \tag{67}$$

Observe that the expression for k_L contains two velocity vectors, v - the velocity vector of the observers who measure E and B fields and u - the velocity vector of the permanent magnet, i.e., of the electric current loop. The Lorentz force density k_L (67) can be written as a CBGQ in the standard basis and its form can be easily inferred from the expression for the torque density n (69) that is given below. Here, it will be presented the expression for k_L in the S' frame, i. e., for the case that $u = v = c\gamma_0'$ and accordingly that $E'^0 = P'^0 = M'^0 = 0$. Hence, k_L as a CBGQ in the $\{\gamma_\mu'\}$ basis is

$$k_L = (-E'^k \partial_0' P_k' + (1/c)\varepsilon^{0jkl} E_j' \partial_k' M_l')\gamma_0' - E'^i(\partial_k' P'^k)\gamma_i'. \tag{68}$$

In the usual approaches with the 3-vectors and their AT, e.g., in [34] and in GH [36], the Lorentz 3-force density is zero in the S' frame; there is not the γ_0' term and there is not the polarization 3-vector $\mathbf{P}$. The components $k_L'^\mu$ in (68) correspond to the time and spatial components of f^α from [36] (Cross), i.e., to $f^0 = (1/c)\mathbf{E}\cdot(\partial\mathbf{P}/\partial t + \nabla \times \mathbf{M})$ and $f^i = -E^i(\nabla\mathbf{P})$. But, as already stated, it is not correct to write the components of the properly defined vector in the 4D spacetime in terms of the 3-vectors. The components $k_L'^\mu$ are multiplied by the unit basis vectors γ_μ', whereas the 3-vector, e.g., $\mathbf{E}$ is constructed from the components $E_{x,y,z}$ and *the unit 3-vectors* $\mathbf{i}$, $\mathbf{j}$, $\mathbf{k}$.

The torque density n is $n = x \wedge k_L$, where k_L is given by Eq. (67). If n is written as a CBGQ in the standard basis it becomes

$$\begin{aligned} n &= (1/c^2)x^\lambda\{(\partial_\mu P^\mu)[(E^\nu u_\nu)v^\rho - (v^\nu u_\nu)E^\rho] - (u^\mu \partial_\mu)P^\nu[E_\nu v^\rho - v_\nu E^\rho] \\ &\quad +(1/c)\varepsilon^{\mu\nu\alpha\beta}u_\mu(\partial_\alpha M_\beta)[E_\nu v^\rho - v_\nu E^\rho]\}(\gamma_\lambda \wedge \gamma_\rho). \end{aligned} \tag{69}$$

n from (69) will be determined in the S' frame in the same way as k_L is determined in (68), i.e., with $u = v = c\gamma_0'$, and $E'^0 = P'^0 = M'^0 = 0$,

$$\begin{aligned} n &= [-(\partial_k' P'^k)x'^0 E'^i + (\partial_0' P'^k)E_k' x'^i - (1/c)\varepsilon^{0jkl}E_j'(\partial_k' M_l')x'^i](\gamma_0' \wedge \gamma_i') - \\ &\quad (\partial_k' P'^k)x'^j E'^i(\gamma_j' \wedge \gamma_i') \end{aligned} \tag{70}$$

It is worth noting that in the approaches with the 3-vectors, as for k_L in (68), the torque density n is zero in the S' frame.

In S', the integrated torque N as a CBGQ is given as

$$N = -(1/c)E'^1 m'^2(\gamma_0' \wedge \gamma_3') - E'^1 p'^3(\gamma_1' \wedge \gamma_3'), \tag{71}$$

where m is the magnetic dipole moment vector and p is the electric dipole moment vector. All quantities in (71) are measured in the common rest frame S'. They are all properly defined in the 4D spacetime and they properly transform under the LT. Furthermore, in the considered case, the electric field vector $E = E'^1\gamma_1'$, the MDM $m = m'^2\gamma_2'$ and the EDM $p = p'^3\gamma_3'$. Using (71), the LT of the components $N'^{\mu\nu}$ and the inverse LT of the basis $\gamma_\mu' \wedge \gamma_\nu'$ from S' to S it can be shown that in S, the lab frame, the bivector N is given as

$$N = (-E^1 m^2/c + \beta E^1 p^3)(\gamma_0 \wedge \gamma_3) + (\beta E^1 m^2/c - E^1 p^3)(\gamma_1 \wedge \gamma_3), \tag{72}$$

and it holds that the whole 4D torque N is unchanged

$$N = (1/2)N'^{\mu\nu}\gamma_\mu' \wedge \gamma_\nu' = (1/2)N^{\mu\nu}\gamma_\mu \wedge \gamma_\nu, \tag{73}$$

where the LT of the electric field vector $E'^1 = \gamma(E^1 - \beta E^0)$, $E'^0 = \gamma(E^0 - \beta E^1) = 0$ are used to derive that $E'^1 = (1/\gamma)E^1$.

All quantities in (72) are measured in S in which the permanent magnet and the source of the electric field move with uniform velocity $V = V\gamma_1$ along the common x_1, x'_1 axes. The relations (71), (72) and (73) show that in the approach with the 4D torque N the principle of relativity is naturally satisfied and there is no paradox; *the 4D torque N is the same 4D quantity in all relatively moving inertial frames of reference.* Note that N will be the same 4D quantity, as in (73), for all bases, e.g., the $\{r_\mu\}$ basis, and not only for the standard basis.

In the same way as for F, Eqs. (21) and (22), or for $\mathcal{M}$, Eqs. (25) and (26), the bivector N can be decomposed into the "space-space" torque N_s and the "time-space" torque N_t, which *together* contain the same physical information as the bivector N, and the velocity vector v of a family of observers who measures N. In the geometric approach, i.e., in the ISR, *both N_s and N_t are equally physical 4D torques which taken together are equivalent to the 4D torque, the bivector N*,

$$\begin{aligned} N &= (v/c) \wedge N_t + (v/c) \cdot (N_s I), \\ N_t &= (v/c) \cdot N, \quad N_s = I(N \wedge v/c), \end{aligned} \tag{74}$$

with the condition

$$N_s \cdot v = N_t \cdot v = 0; \tag{75}$$

only three components of N_s and three components of N_t are independent since N is antisymmetric. *Both, N_s and N_t depend not only N but on v as well.* The primary physical quantity with definite physical reality for the torques is the 4D torque N, whereas N_s and N_t are derived from N *and* v. The equations. (74) and (75) are Eqs. (13) and (14) respectively in the first paper in [14], or Eq. (2) in [13]. If N_s and N_t are written as CBGQs in the $\{\gamma_\mu\}$ basis they become

$$N_s = (1/2c)\varepsilon^{\alpha\beta\mu\nu} N_{\alpha\beta} v_\mu \gamma_\nu, \; N_t = (1/c) N^{\mu\nu} v_\mu \gamma_\nu, \tag{76}$$

what is Eq. (3) in [13]. As seen from (76), in the frame of "fiducial" observers and in the $\{\gamma_\mu\}$ basis, $v^\mu = (c, 0, 0, 0)$, $N_s^0 = N_t^0 = 0$ and *only the spatial components* N_s^i and N_t^i remain

$$N_s^0 = 0, \; N_s^i = (1/2)\varepsilon^{0ijk} N_{jk}, \; N_t^0 = 0, \; N_t^i = N^{0i}, \tag{77}$$

which explains the names the "space-space" torque for N_s and the "time-space" torque for N_t. The quotation marks stand because such an identification of the components of the torques N_s and N_t with the components of the bivector N is not possible for some other bases. In the ISR, *both vectors N_s and N_t have to be treated on an equal footing.* It is worth noting that the whole discussion with the torque can be completely repeated for the angular momentum replacing N, N_s and N_t by J, J_s and J_t, see Eqs. (17) - (19) in the first paper in [14]. The Trouton-Noble paradox, Jackson's paradox and Mansuripur's paradox [34] , all of them stem from the fact that in the usual approaches an independent physical reality is attributed only to N_s and J_s, or better to say, to the 3D torque $\mathbf{T}$ and the 3D angular momentum $\mathbf{L}$, but not to N_t and J_t. Furthermore, in the usual approaches, the AT of $\mathbf{T}$ and $\mathbf{L}$ are considered to be the relativistically correct LT, see, e.g., Jackson's paper [51] and the discussion in Sec. 3 in the first paper in [14].

Let us determine N_s and N_t for our case as 4D CBGQs in the rest frame S'. In S', one finds from (71) and (76) that

$$N_s = N_s'^\mu \gamma'_\mu = (1/c) E'^1 p'^3 v'^0 \gamma'_2, \quad N_t = N_t'^\mu \gamma'_\mu = -(1/c^2) E'^1 m'^2 v'^0 \gamma'_3, \tag{78}$$

where, in S', $v'^{\mu} = (c, 0, 0, 0)$. The "time-space" torque N_t in (78), which comes from the first term in (71), corresponds to the expression $\mathbf{R} = \mathbf{m} \times \mathbf{E} = -mE\hat{\mathbf{y}}$ in Cross [36] that describes the interaction of the magnetic moment with the electric field in the rest frame S'. (Remember that in Cross [36] the rest frame is with unprimed quantities and the motion is along the x^3 axis.) The "space-space" torque N_s in (78), which comes from the second term in (71), does not appear in any previous paper since it emerges from the existence of the EDM p for a stationary permanent magnet, which is first predicted in Sec. 8 here. It describes the interaction of the EDM $p = p'^3\gamma'_3$ of the stationary permanent magnet with the electric field $E = E'^1\gamma'_1$ in the rest frame S'. In the usual formulation with the 3-vectors it would correspond to the usual 3D torque $\mathbf{T} = \mathbf{p} \times \mathbf{E}$, but, in contrast to all previous formulations, this torque is in the rest frame S'.

Next, we determine N_s and N_t in S, the lab frame. One way is to start with Eq. (78) and then to transform by the LT *all* quantities which determine N_s and N_t in (78), i.e., E'^{μ}, m'^{μ}, p'^{μ}, v'^{μ} and γ'_{μ}, from S' to S. This yields

$$N_s = N_s^{\mu}\gamma_{\mu} = (1/\gamma)E^1 p^3 \gamma_2, \quad N_t = -(1/c\gamma)E^1 m^2 \gamma_3. \tag{79}$$

Another way to determine N_s and N_t in S is to use the expression for N in S (72) and the relations (76). Note that "fiducial" observers are moving in S. Therefore the components v^{μ} of their velocity in S, which are obtained by the LT from $v'^{\mu} = (c, 0, 0, 0)$, are $v^{\mu} = (\gamma c, \gamma\beta c, 0, 0)$. Of course, for the whole CBGQ v it holds that $v = v'^{\mu}\gamma'_{\mu} = v^{\mu}\gamma_{\mu}$. Similarly, in this geometric approach, e.g., $N_s'^{\mu}\gamma'_{\mu}$ transforms under the LT as every vector transforms, i.e., as in LT (43), which means that *components* $N_s'^{\mu}$ *of the "space-space" torque* N_s *transform to the components* N_s^{μ} of the same torque N_s in the S frame; *there is no mixing with the components of the "time-space" torque* N_t

$$N_s^0 = \gamma(N_s'^0 + \beta N_s'^1), \; N_s^1 = \gamma(N_s'^1 + \beta N_s'^0), \; N_s^{2,3} = N_s'^{2,3} \tag{80}$$

and the same for N_t^{μ}. The LT (80) of the components of N_s and the same for N_t are obtained in the same way as the LT (43) are obtained, i.e., that both N and v from the definitions (74) are transformed by the LT R, Eq. (39), as in (40) and (41). This is in a sharp contrast to the AT (66) in which the transformed components T_i are expressed by the mixture of components T'_k of the 3D vector $\mathbf{T}'$ and of components R'_k of another 3D vector $\mathbf{R}'$ from the rest frame. The AT (66) can be obtained in the same way as the AT (46) and (47) in Sec. 6 are obtained, i.e., that only N from the definitions (74) is transformed by the LT R, Eq. (39), but not the velocity of the observer v. Furthermore, N_s and N_t are geometric quantities in the 4D spacetime since the components N_s^{μ} and N_t^{μ} are multiplied by the unit vectors γ_{μ}, whereas the 3D torque $\mathbf{T}$ is a geometric quantity in the 3D space. It is formed multiplying the components $N^{\mu\nu}$ (i.e., T_i determined by the same identification as in (3)) of a 4D geometric quantity, the bivector N, by the unit 3D vectors **i**, **j**, **k**.

It can be easily proved from (78) and (79) that the CBGQs $N_s'^{\mu}\gamma'_{\mu}$ and $N_s^{\mu}\gamma_{\mu}$ are *the same quantity* N_s in S' and S frames, and the same for N_t

$$N_s = N_s'^{\mu}\gamma'_{\mu} = N_s^{\mu}\gamma_{\mu}, \quad N_t = N_t'^{\mu}\gamma'_{\mu} = N_t^{\mu}\gamma_{\mu}; \tag{81}$$

remember that $E'^1 = (1/\gamma)E^1$. This again shows, as in [13] and [14], that *in the approach with the 4D torques* N_s *and* N_t *the principle of relativity is naturally satisfied and there is no paradox.* Observe that N_s is determined in all relatively moving inertial frames of reference by the interaction of the EDM p of the permanent magnet and the electric field E, whereas N_t is determined by the interaction of m and E.

It is worth mentioning that, in contrast to the 3D torque $\mathbf{T}$, the 4D torque N_s is the same quantity for observers in S' and S even if they use different bases, e.g., $\{\gamma_\mu\}$, $\{r_\mu\}$

$$N_s = N_s^\mu \gamma_\mu = N_s'^\mu \gamma_\mu' = N_{s,r}^\mu r_\mu = N_{s,r}'^\mu r_\mu', \tag{82}$$

where the primed quantities are the Lorentz transforms of the unprimed ones. The same holds for N_t. For the $\{r_\mu\}$ basis, this can be proved using the transformation matrix $R^\mu_{\ \nu}$ from Sec. 3.1 or, as discussed in connection with Eq. (49), using the LT in the $\{r_\mu\}$ basis.

Let us suppose for a moment that a permanent magnet possesses only a MDM m and not an EDM p. This is as in the usual approaches, but we deal with correctly defined vectors in the 4D spacetime and with their LT and not with the 3-vectors and their AT. In that case, as can be seen from (78), (79) and (81), (82) in the rest frame S' and in the lab frame S as well, i.e., *in all relatively moving inertial frames of reference and for all bases in them, the "space-space" torque $N_s = 0$ and only remains the "time-space" torque N_t,*

$$p = 0; \quad N_s = 0, \ N_t \neq 0. \tag{83}$$

As already stated, in S', the torque N_t corresponds to the expression $\mathbf{R} = \mathbf{m} \times \mathbf{E} = -mE\widehat{\mathbf{y}}$ in Cross [36]. In all other papers from [34-36] there is no 3D torque in the rest frame S'. However, as a result of the AT of the 3-vectors (18), or (10), i.e., the AT of components (66) as in Cross [36], there is the usual 3D torque $\mathbf{T}$ in S. In the considered case ($p = 0$), in the approach with N_s and N_t, according to (79) and (81) $N_s = 0$ and N_t is the same as in S', which means that there is no paradox. In the formulation with N, the relations (71), (72) and (73) again show that for $p = 0$ the torque N is the same 4D quantity in all relatively moving inertial frames of reference and for all bases chosen in them and again there is no paradox. Also, as in the case with $p \neq 0$, there is no need either for the "hidden" mechanical angular momentum or for the "hidden" torque.

10. On the Aharonov-Bohm effect in terms of fields. Is the AB effect purely quantum mechanical in nature?

If the existence of the electric fields from a stationary permanent magnet would be experimentally proved then it would enable a new interpretation of the particle interference experiments, particularly of the Aharonov - Bohm (AB) effect [53]. Such electric fields offer a new possibility for the explanation of the experimentally observed fringe shift for the magnetic AB effect even in Tonomura's experiments [54] *in terms of forces, which so far have been overlooked.* In this paper only a qualitative consideration will be presented.

Regarding the experiments with microscopic solenoids, e.g., [55], and also the recent experiment with macroscopic solenoid [56], they can be naturally explained by the fact that *always* there is an electric field outside stationary resistive conductors carrying constant currents, i.e., by *the existence of the electric force* acting on the particle. For the existence of such external electric fields from resistive conductors see, e.g., Sec. 4 in [42] and references therein. If the experiments would be made with superconducting solenoids then again there would be the external electric field, the second-order electric field (55). Thus, even in that case, it cannot be argued that there is no force acting on the particle and consequently that the observed phase shift is entirely due to nonzero vector potential.

Let us explain the above assertions in more detail. Aharonov and Bohm [53] theoretically predicted that there is a relative phase shift between two electron beams that pass on both sides of an infinitely long, stationary, coil with a steady current. The magnetic field does not exist outside that coil. But, using quantum mechanics, Aharonov and Bohm [53] showed that the interference fringes are displaced proportionally to the magnetic flux Φ_B flowing inside the coil even though neither electron beam touch the magnetic flux. They asserted: "We shall show that, contrary to the conclusions of classical mechanics, there exist effects of potentials on charged particles, even in the region where *all the fields* (my emphasis) (and therefore the forces on the particles) vanish." Thus, according to them, the electron wave packets are influenced although *it is supposed that they travel through regions entirely free from electromagnetic fields.* As the vector potential **A** exists in the considered case even in the magnetic field-free regions Aharonov and Bohm [53] proposed that "in quantum mechanics, the fundamental physical entities are the potentials." Their results became very important in the forefront of physics. Vector potentials are generalized to gauge fields and these fields are considered to be the fundamental physical quantities in the modern theories of gauge fields.

They, [53], have found that the relative phase shift, $\Delta\varphi$, is produced between the two wave packets due to the vector potential as

$$\Delta\varphi = (e/\hbar)\oint \mathbf{A}\cdot\mathbf{dl} = (e/\hbar)\int \mathbf{B}\cdot\mathbf{dS} = (e/\hbar)\Phi_B, \tag{84}$$

where **B** is the magnetic field of the solenoid.

Instead of an infinite coil with current the experiments, e.g., [55], dealt with microscopic solenoids. Möllenstedt and Bayh, [55], observed a fringe shift that is in agreement with the relation (84). The overlap between the incident electrons and the *magnetic* field strengths in their experiments was fairly small. Therefore, the experiments [55] are usually considered as a convincing demonstration of the existence of the AB effect, i.e., that the relative phase shift is due only to the vector potential and, according to (84), due to the quantum action of the magnetic flux enclosed between the two interfering electron trajectories. Thus, according to the presently accepted formulation the AB phase shift is caused by an enclosed magnetic flux and there is no need to examine any interaction between the passing charges and the sources of the magnetic flux.

However, several authors questioned the existence of the AB effect and tried to explain the observed fringe shift in a classical way, i.e., in terms of fields and the Lorentz force. One reason for such a possibility is the unphysical character of the infinitely long solenoid that is commonly used in the discussions of the AB effect. Hence, the experimental results with finite solenoids, e.g., in [55], and also with finite whiskers [57], could be attributed to the magnetic flux leaking outside such solenoids or whiskers.

In Tonomura's experiments [54] a tiny toroidal magnet is used instead of straight solenoids. The magnet was covered with a superconductor layer and further with a copper layer. The phase shift between two waves passing through the hole and outside of the toroid was measured by means of electron holography. The AB phase shift was detected even though the magnetic field was confined to the toroidal magnet; the Meissner effect prevented any flux from leaking out. The copper outer layer prevented any electrons from penetrating the magnet itself, i.e., there was no overlap of the incident electron wave with the magnetic fields inside the sample. Thus, it is generally accepted that Tonomura's experiments [54] give conclusive evidence for the AB effect, i.e., that there is no force acting on the particle and consequently that the observed phase shift is entirely due to nonzero vector potential. For a general review see, e.g., [58].

It is very interesting that both in the theoretical discussions and in the experiments with microscopic solenoids, and also in the recent experiment with macroscopic solenoid [56], it is never noticed that *always* there is an electric field outside stationary resistive conductors carrying constant currents. In such ohmic conductors there are quasistatic surface charges, which generate not only the electric field inside the wire driving the current, *but also a time independent electric field outside it,* see, e.g., Sec. 4 in [42] and references therein. There are no analytic solutions for these surface charges and the electric fields outside the wire for the case of finite solenoids. It is very difficult to determine the distribution of surface charges on the conductors of a circuit because it depends on the geometry of the circuit itself and of its surroundings. In order to have some qualitative orientation about that external electric field one can look at Eq. (16) in [42]. It is derived for a cylindrical wire of finite length. From that equation it is visible that *the electric field outside the wire is proportional to the current.* The radial component of that field falls as $1/r$. The proportionality of the external electric field with the current in the solenoid explains all important features of, e.g., the experiments from [55]. The magnetic flux Φ_B flowing inside the solenoid is also proportional to the current and it explains the generally accepted belief that the fringe shift in the experiments with microscopic solenoids is determined by the magnetic flux inside the solenoid. However, as seen from the above discussion, the relative phase shift *is not* due to the vector potential, i.e., according to (84), due to the quantum action of the magnetic flux, but *it is due to the existence of the electric field outside stationary solenoids with steady currents.*

From the viewpoint of the approach with 4D geometric quantities the *components* of the electric field 3-vector in Eq. (16) in [42] have to be understood as the *spatial components* in the standard basis of the electric field vector, a 4D geometric quantity; the rest frame of the considered wire is taken to be the γ_0-frame and therefore the temporal component $E^0 = 0$ (also $B^0 = 0$). Similarly, the surface charge density from Eq. (12) in [42] has to be understood as the *temporal component* in the standard basis of the current density vector. In the rest frame of the wire the *spatial components are zero*, $j^i = 0$. Also, in our approach, the vector of the Lorentz force K is given by the expression (63).

Observe that in [42] two other contributions to the external electric field are mentioned in Secs. 3 and 5, but they are of no concern for our consideration. In Sec. 3 in [42] the electric field from the induced charges is investigated, but it is the same whether or not there is current in the wire. In Sec. 5 in [42] the electric field proportional to the square of the current is derived using an action-at-a-distance Weber's electrodynamics, which is not in agreement with the field theory, i.e., with the special relativity. On the other hand, the second-order electric field (55) is proportional to the square of the current and it is derived in a relativistically correct way. In our approach it replaces that one from Sec. 5 in [42]. The electric field (55) exists in a stationary resistive conductor carrying constant current as well, but there it is negligible in comparison with the external electric field that is caused by the distribution of surface charges on the conductor, i.e., with the field which is proportional to the current.

From the consideration in Sec. 8 it follows that, both in the Ampèrian approach and in the approach in which the intrinsic EDM vector p is determined by the new "time-space" spin vector Z, there is a time independent electric field outside stationary permanent magnet. This yields the possibility to explain the fringe shifts in Tonomura's experiments [54] by the existence of the external electric fields from a stationary permanent magnet and not, as generally accepted, by the existence of a nonzero vector potential. The title of the paper [54] is "Evidence for Aharonov-Bohm Effect with Magnetic Field Completely Shielded from Electron Wave" and it is written in the Abstract: "A toroidal ferromagnet was covered with a superconductor layer to confine the field, and further with a copper layer for complete shielding from the electron wave." Thus, in [54], the experimental setup is designed in such a way that practically the overlap between the incident electrons and the *magnetic* field strengths is negligible.

However, both layers that are used in [54] do not prevent the overlap between the wave functions of the incident electrons and *the electric field from stationary, toroidal ferromagnet*, which is predicted in Sec. 8.

A strong theoretical argument that supports the interpretation of the particle interference experiments in terms of forces and not in terms of potentials comes from the fact that, as already stated, i.e., as shown in the axiomatic formulation of the electromagnetism [12], the bivector F can be taken as the primary quantity and the field equation for F (20) is the basic equation for the whole electromagnetism; the bivector field F yields the complete description of the electromagnetic field and there is no need to introduce either the field vectors or the potentials.

Furthermore, the qualitative theoretical explanations of the quantum phase shifts in terms of classical force vectors in the Aharonov-Casher and the Röntgen effects are already given in [46, 47]. In [48], the dipole moments are quantized according to (62) and it is shown that the expectation value for the quantum force vector is not zero in the case of the Aharonov-Casher and the Röntgen effects and in the neutron interferometry. This means that the phase shifts in these experiments are not due to force-free interaction of the dipole, i.e., they are not topological phase shifts.

11. Particle interference and Lorentz-violating electrodynamics

In the recent paper [37], under the title "Particle interference as a test of Lorentz-violating electrodynamics," it is argued that in a Lorentz-violating model of electrodynamics [59] a magnetic solenoid generates not only a static magnetic field but also a static electric field, which acts on interfering particles producing an extra path-dependent phase. That nontopological phase is considered to be the Lorentz-violating correction to the standard topological (path-independent) Aharonov-Bohm phase.

However, as shown in the consideration in Secs. 7.1, 7.2 and 8, there is a static electric field not only outside a magnetic solenoid but also outside a stationary permanent magnet and that field is obtained in a consistent *Lorentz-covariant approach* with 4D geometric quantities. Furthermore, it is worth mentioning that in [37], and [59] as well as in almost the whole physical literature, it is believed that Maxwell's equations with the 3-vectors are covariant under the LT, considering that the LPET (10) and (16) are the relativistically correct LT. But, the LPET (10) and (16) ARE NOT the LT and consequently, as explicitly proved in [8], Maxwell's equations with 3-vectors ARE NOT covariant under the relativistically correct LT (42) and (43). Hence, there is no sense to develop Lorentz-violating model of electrodynamics assuming that the LPET (10) and (16) are the LT and that Maxwell's equations with 3-vectors are covariant under the LT. It would be much more important that physics community stops to consider that in the 4D spacetime the Lorentz contraction and the dilation of time are the intrinsic relativistic effects and that the LPET (10) and (16) are the relativistically correct LT.

12. Discussion and Conclusions

The whole consideration explicitly shows that the 3D quantities $\mathbf{E}(\mathbf{r},t)$ and $\mathbf{B}(\mathbf{r},t)$, $\mathbf{P}(\mathbf{r},t)$ and $\mathbf{M}(\mathbf{r},t)$, $\mathbf{p}$ and $\mathbf{m}$, their LPET, i.e., the AT, (10), (16), or (17), (18), and the equations with them are not well-defined in the 4D spacetime. More generally, we can conclude that *the 3D quantities do not have an independent physical reality in the 4D spacetime.* Contrary to the general belief, we find that, in the 4D spacetime, *it is not true that observers in relative motion see different fields; the transformations, Eqs.* (10), (16), or (17), *or, equivalently,* (44) *and* (45), *i.e., in the standard basis Eqs.* (46) *and* (47), *are not the LT but the LPET, i.e., the AT. According to the LT; Eqs.* (40) - (43), *the electric field E transforms only to the electric field E' and the same holds for the magnetic field B, for the vectors of the polarization P and the magnetization M and for the EDM p and the MDM m.*

As already stated, the principle of relativity is automatically included in the approach with well-defined 4D geometric quantities, i.e., in the ISR, whereas in Einstein's formulation of the special relativity [1] the principle of relativity is postulated outside the framework of a correct mathematical formulation of the theory and it is supposed that it holds for the equations, the physical laws, which are expressed in terms of the 3D quantities.

Minkowski's great discovery of the correct LT, Sec. 11.6 in [21], their generalization and the explicit forms (42) and (43) that are found in [6-11] and also the mathematical argument from [19] that space and time dependent electric and magnetic fields cannot be the usual 3-vectors strongly suggest the need for further critical examination of the usual formulation of electromagnetism with 3-vectors $\mathbf{E}(\mathbf{r},t)$ and $\mathbf{B}(\mathbf{r},t)$, $\mathbf{P}(\mathbf{r},t)$ and $\mathbf{M}(\mathbf{r},t)$ and their LPET (10), (16), or (17). It also suggests the possibility for a complete and relativistically correct formulation of classical and quantum electromagnetism with multivector fields (as physically real fields), which are defined on the 4D spacetime and which transform according to the correct LT (40) - (43).

The advantages of such formulation with 4D geometric quantities, i.e., of the ISR, are already revealed in the cases of the interaction between the dipole moment tensor D^{ab} and the electromagnetic field F^{ab} in [48] and in much more detail in [33], in the discussion of quantum phase shifts in [46, 47], in the discussion of shortcomings in the current EDM searches in [49] and in the formulation of Majorana form of the Dirac-like equation for the free-photon [60].

Particularly important results of the 4D geometric approach, i.e., of the ISR, that are reported in this paper, Secs. 7.1, 7.2, 8, refer to the existence of the second-order electric field outside a superconducting loop with steady current and to the new prediction of the electric field outside a stationary permanent magnet, i.e. to the prediction that a stationary permanent magnet possesses an intrinsic polarization, which induces the external electric field. Also, it is suggested that the measurements of that external electric field from a stationary permanent magnet could be performed by the same method with cold ions as in [31] and possibly as in [43].

The investigation of the "charge-magnet paradox" from Secs. 9.1 and 9.2 again shows that the relativistically correct description of physical phenomena without any paradoxes can be achieved in the consistent way with 4D geometric quantities as physical quantities in the 4D spacetime. On the other hand, the use of 3D quantities and their AT necessarily leads to different ambiguities and inconsistencies.

A qualitative explanation of the AB effect from Sec. 10, together with possibly positive experimental results for the existence of the electric field vector from stationary permanent magnet, will surely be important for better understanding of the classical limit of the quantum physics. A more quantitative calculation of the AB effect in terms of fields will be done, e.g., using the method from [61] and the decomposition of F (21).

Acknowledgments

I am cordially thankful to Zbigniew Oziewicz for numerous and very useful discussions during years, which, among others, helped me to better understand that the relativistically correct mathematical formulation of the electromagnetism *requires* the representation of the electric and magnetic fields by the 4D geometric quantities and for the continuos support of my work. It is a pleasure to acknowledge to Larry Horwitz for inviting me to the IARD conferences, for the valuable discussions and for the continuos support of my work. I am also grateful to Alex Gersten for useful discussions and for the continuos support of my work.

Appendix

In this Appendix we shall briefly describe the essential differences between the 4D geometric approach, the ISR, and Einstein's definition of the Lorentz contraction, e.g., for a moving rod. This is explained in detail in Secs. 2 - 2.3 in [27] and Secs. 3.1, 4.1 and Figs. 1 and 3 in [16]. Here, the notation is slightly different than in [27] and [16]. In the geometric approach one deals with the abstract 4D geometric quantities, i.e., with the position vectors x_A, x_B, of the events A and B, respectively, with the distance vector $l_{AB} = x_B - x_A$ and with the spacetime length, $l = L_0$, see (86). The essential feature of the geometric approach is that *any abstract 4D geometric quantity*, e.g., the distance vector $l_{AB} = x_B - x_A$, *is only one quantity, the same quantity in the 4D spacetime* for all relatively moving frames of reference and for all systems of coordinates that are chosen in them. The abstract vector l_{AB} can be decomposed in different bases and then these representations, the CBGQs, of the same abstract 4D geometric quantity l_{AB} contain both the basis components and the basis vectors. Let us explain it taking a particular choice for l_{AB}, which in the usual "3+1" picture corresponds to a rod that is at rest in an inertial frame of reference (IFR) S (with the standard basis in it) and situated along the common x^1, x'^1 – axes. Its rest length is denoted as L_0. The situation is depicted in Fig. 1 in [16]. l_{AB} is decomposed, i.e., it is written as a CBGQ, in the standard basis and in S and S', where the rod is moving, as

$$l_{AB} = l^{\mu}_{AB}\gamma_{\mu} = 0\gamma_0 + L_0\gamma_1 = l'^{\mu}_{AB}\gamma'_{\mu} = -\beta\gamma L_0\gamma'_0 + \gamma L_0\gamma'_1, \qquad (85)$$

As already stated several times, the components l^{μ}_{AB} are transformed by the LT and the basis vectors γ_{μ} by the inverse LT leaving the whole CBGQ unchanged. In S, the position vectors $x_{A,B}$ are determined simultaneously, $x^0_B - x^0_A = l^0_{AB} = 0$, i.e., the temporal part of l^{μ}_{AB} is zero. In the standard basis, which is commonly used in the usual approaches, there is a dilation of the spatial part $l'^1_{AB} = \gamma L_0$ with respect to $l^1_{AB} = L_0$ and not the Lorentz contraction as predicted in Einstein's formulation of special relativity. Similarly, as explicitly shown in [27] and [16], in the $\{r_{\mu}\}$ basis, i.e., with the "r" synchronization, if only spatial parts of $l^{\mu}_{AB,r}$ and $l'^{\mu}_{AB,r}$ are compared then one finds the dilation $\infty \succ l'^1_{AB,r} \geq L_0$ for all β_r. However, the comparison of only spatial parts of the components of the distance vector l_{AB} in S and S' is physically meaningless in the geometric approach, since *some components of the tensor quantity, when they are taken alone, do not correspond to some definite 4D physical quantity.* Note that if $l^0_{AB} = 0$ then the LT yield that l'^{μ}_{AB} in any other IFR S' contains the time component as well, $l'^0_{AB} = x'^0_B - x'^0_A = -\beta\gamma L_0 \neq 0$. Hence, *the LT yield that the spatial ends of the rod are not determined simultaneously in S', i.e., the temporal part of l'^{μ}_{AB} is not zero.* For the spacetime length l it holds that

$$l^2 = \mid l^{\mu}_{AB} l_{AB,\mu} \mid = \mid l'^{\mu}_{AB} l'_{AB,\mu} \mid = \mid l^{\mu}_{AB,r} l_{AB,r,\mu} \mid = L_0^2. \qquad (86)$$

In S, the rest frame of the rod, where the temporal part of l^{μ}_{AB} is $l^0_{AB} = 0$, the spacetime length l is a measure of the spatial distance, i.e., of the rest spatial length of the rod, as in the prerelativistic physics. *The observers in all other IFRs will "look" at the same events A and B, the same distance vector l_{AB} and the same spacetime length l, but associating with them different coordinates; it is the essence of the geometric approach. They all obtain the same value l for the spacetime length, $l = L_0$.*

It is worth mentioning, once again, that the 4D geometric treatment with l_{AB} and l is a generalization and a mathematically better founded formulation of the ideas expressed by Rohrlich [15] and Gamba [40]. Indeed, Rohrlich [15] states: "A quantity is therefore physically meaningful (in the sense that it is of the same nature to all observers) if it has tensorial properties under Lorentz transformations." Similarly Gamba [40], when discussing the sameness of a physical quantity (for example, *a nonlocal quantity* $A_{\mu}(x_{\lambda}, X_{\lambda})$, which is a function of two points in the 4D spacetime x_{λ} and X_{λ}) for different inertial frames of reference S and S', declares: "The quantity $A_{\mu}(x_{\lambda}, X_{\lambda})$ for S is the same as the quantity $A'_{\mu}(x'_{\lambda}, X'_{\lambda})$ for S' when all the primed quantities are obtained from the corresponding unprimed quantities through Lorentz transformations (tensor calculus)." Rohrlich and Gamba worked with the usual covariant approach, i.e., with the components implicitly taken in the standard basis, which means that only Einstein's synchronization is considered to be physically admissible. The quantities $A_{\mu}(x_{\lambda}, X_{\lambda})$ and $A'_{\mu}(x'_{\lambda}, X'_{\lambda})$ refer to the same physical quantity, but they are not mathematically equal quantities since bases are not included. In the approach with the 4D geometric quantities, i.e., in the ISR, one deals with mathematically equal quantities, e.g., for a nonlocal quantity $l_{AB} = x_B - x_A$ it holds that

$$l_{AB} = l^{\mu}_{AB}\gamma_{\mu} = l'^{\mu}_{AB}\gamma'_{\mu} = l^{\mu}_{AB,r}r_{\mu} = l'^{\mu}_{AB,r}r'_{\mu} = .., \tag{87}$$

where the primed quantities are the Lorentz transforms of the unprimed ones. In order to treat different systems of coordinates on an equal footing we have derived a form of the LT that is independent of the chosen system of coordinates, including different synchronizations, see Eq. (2) in [27], or Eq. (1) in [16]. Also, Eq. (4) in [16], it is presented the transformation matrix that connects Einstein's system of coordinates with another system of coordinates in the same reference frame.

On the other hand, as shown in Sec. 2.2 in [27] and Sec. 4.1 and Fig. 3 in [16], in Einstein's formulation of SR, instead of to work with geometric quantities $x_{A,B}$, l_{AB} and l one deals only with the spatial, or temporal, *components* of their coordinate representations x^{μ}_{A}, x^{μ}_{B} and l^{μ}_{AB} *in the standard basis.* The geometric character of physical quantities, i.e., the basis vectors, and some asymmetric synchronization, e.g., the "r" synchronization, which is equally physical as the Einstein synchronization, are never taken into account. According to Einstein's definition [1] of the spatial length the spatial ends of the rod must be taken simultaneously for the observer, i.e., *he defines length as the spatial distance between two spatial points on the (moving) object measured by simultaneity in the rest frame of the observer.* In the 4D (here, for simplicity, as in [27] and [16], we deal only with 2D) spacetime and in the $\{\gamma_{\mu}\}$ basis the simultaneous events A and B (whose spatial parts correspond to the spatial ends of the rod) are the intersections of x^1 axis (that is along the spatial basis vector γ_1) and the world lines of the spatial ends of the rod that is at rest in S and situated along the x^1 axis. The components of the distance vector are $l^{\mu}_{AB} = x^{\mu}_{B} - x^{\mu}_{A} = (0, L_0)$; for simplicity, it is taken that $t_B = t_A = a = 0$. Then in S, the rest frame of the object, the spatial part $l^1_{AB} = L_0$ of l^{μ}_{AB} is considered to define the rest spatial length. Furthermore, one uses the inverse LT to express x^{μ}_{A}, x^{μ}_{B} and l^{μ}_{AB} in S in terms of the corresponding quantities in S', in which the rod is moving. This procedure yields

$$\begin{aligned} l^0_{AB} &= ct_B - ct_A = \gamma(l'^0_{AB} + \beta l'^1_{AB}), \\ l^1_{AB} &= x^1_B - x^1_A = \gamma(l'^1_{AB} + \beta l'^0_{AB}). \end{aligned} \tag{88}$$

Now, instead of to work with 4D tensor quantities and their LT, as in the 4D geometric approach, in the usual formulation one forgets about the transformation of the temporal part l^0_{AB}, the first equation in (88), and considers only the transformation of the spatial part l^1_{AB}, the second equation in (88). Furthermore, *in that relation for l^1_{AB} one assumes that $t'_B = t'_A = t' = b$*, i.e., *that x'^1_B and x'^1_A are simultaneously determined at some arbitrary $t' = b$ in S'.* However, *in 4D* (at us 2D) spacetime such an assumption means that *in S' one does not consider the same events A and B as in S* but some other two events C and D, which means that $t'_B = t'_A$ has to be replaced with $t'_D = t'_C = b$. The events C and D are the intersections of the line (the hypersurface $t' = b$ with arbitrary b) parallel to the spatial axis x'^1 (which is along the spatial base vector γ'_1) and of the above mentioned world lines of the spatial end points of the rod. Then, in the above transformation for l^1_{AB} (88) one has to write $x'^1_D - x'^1_C = l'^1_{CD}$ instead of $x'^1_B - x'^1_A = l'^1_{AB}$. The spatial parts l^1_{AB} and l'^1_{CD} are the *spatial distances* between the events A, B and C, D, respectively. *In Einstein's formulation, the spatial distance $l^1_{AB} = x^1_B - x^1_A = L_0$ defines the spatial length of the rod at rest in S, while $l'^1_{CD} = x'^1_D - x'^1_C$ is considered to define the spatial length of the moving rod in S'.* Hence, from the equation for l^1_{AB} (88) one finds the relation between $l'^1 = l'^1_{CD}$ and $l^1 = l^1_{AB} = L_0$ as the famous formula for the Lorentz contraction of the moving rod

$$l'^1 = x'^1_D - x'^1_C = L_0/\gamma = (x^1_B - x^1_A)/\gamma, \text{ with } t'_C = t'_D, \text{ and } t_B = t_A, \tag{89}$$

where $\gamma = (1 - \beta^2)^{-1/2}$, $\beta = U/c$ and $U = |\mathbf{U}|$; $\mathbf{U}$ is the 3-velocity of S' relative to S. As can be nicely seen from Fig. 3 in [16], the spatial lengths L_0 and l'^1_{CD} refer not to the same

4D tensor quantity, as in the 4D geometric approach, see Fig. 1 in [16], but to two different quantities in the 4D spacetime. These quantities are obtained by the same measurements in S and S'; the spatial ends of the rod are measured simultaneously at some $t = a$ in S and also at some $t' = b$ in S', and a in S and b in S' are not related by the LT or any other coordinate transformation. This means that the Lorentz contraction, as already shown by Rohrlich [15] and Gamba [40], is a typical example of an AT. *It has nothing in common with the LT of the 4D geometric quantities.* We see that in Einstein's approach [1] the spatial and temporal parts of events are treated separately, and moreover the time component is not transformed in the transformation that is called - the Lorentz contraction. Thus, contrary to the generally accepted opinion, *the Lorentz contraction is not a well-defined relativistic effect in the 4D spacetime.*

References

[1] A. Einstein, Annalen der Physik **17,** 891 (1905); Translated by W. Perrett and G. B. Jeffery in: *The Principle of Relativity* (Dover, New York, 1952).
[2] J. D. Jackson, *Classical Electrodynamics* 3rd ed. (Wiley, New York, 1998).
[3] H. A. Lorentz, Proceedings of the Royal Netherlands Academy of Arts and Sciences **6,** 809 (1904).
[4] H. Poincaré, Rend. del Circ. Mat. di Palermo **21,** 129 (1906).
[5] A. A. Logunov, Hadronic J. **19,** 109 (1996).
[6] T. Ivezić, Found. Phys. **33,** 1339 (2003).
[7] T. Ivezić, Found. Phys. Lett. **18,** 301 (2005).
[8] T. Ivezić, Found. Phys. **35,** 1585 (2005).
[9] T. Ivezić, Fizika A **17,** 1 (2008).
[10] T. Ivezić, arXiv: 0809.5277.
[11] T. Ivezić, Phys. Scr. **82,** 055007 (2010).
[12] T. Ivezić, Found. Phys. Lett. **18,** 401 (2005).
[13] T. Ivezić, Found. Phys. **37**, 747 (2007).
[14] T. Ivezić, Found. Phys. **36**, 1511 (2006); T. Ivezić, Fizika A **16,** 207 (2007).
[15] F. Rohrlich, Nuovo Cimento B **45**, 76 (1966).
[16] T. Ivezić, Found. Phys. **31,** 1139 (2001).
[17] T. Ivezić, Found. Phys. Lett. **15,** 27 (2002); arXiv: physics/0103026; arXiv: physics/0101091.
[18] D. Hestenes, *Space-Time Algebra (*Gordon & Breach, New York, 1966); D. Hestenes and G. Sobczyk, *Clifford Algebra to Geometric Calculus* (Reidel, Dordrecht, 1984); C. Doran and A. Lasenby, *Geometric algebra for physicists* (Cambridge University Press, Cambridge, 2003).
[19] Z. Oziewicz, J. Phys.: Conf. Ser. **330**, 012012 (2011); Z. Oziewicz, Rev. Bull. Calcutta Math. Soc. **16,** 49 (2008); Z. Oziewicz and C. K. Whitney, Proc. Nat. Phil. Alliance *(NPA)* **5,** 183 (2008) (also at http://www.worldnpa.org/php/).
[20] T. Ivezić, arXiv: 1101.3292.
[21] H. Minkowski, Nachr. Ges. Wiss. Göttingen, 53 (1908); Reprinted in: Math. Ann. **68,** 472 (1910); English translation in: M. N. Saha and S. N. Bose *The Principle of Relativity: Original Papers by A. Einstein and H. Minkowski* (Calcutta University Press, Calcutta, 1920).
[22] W. G. W. Rosser, *Classical Electromagnetism via Relativity*

(Plenum, New York, 1968).
[23] W. K. H. Panofsky and M. Phillips, *Classical electricity and magnetism* 2nd ed. (Addison-Wesley, Reading, 1962).
[24] E.M. Purcell, *Electricity and Magnetism* 2nd ed. (McGraw-Hill, New York, 1985).
[25] R. P. Feynman, R. B. Leighton and M. Sands, *The Feynman Lectures on Physics* Volume II (Addison-Wesley, Reading, 1964).
[26] D. J. Griffiths, *Introduction to Electrodynamics* 3rd ed. (Prentice-Hall, Upper Saddle River, 1999).
[27] T. Ivezić, Found. Phys. Lett. **12,** 507 (1999); arXiv: physics/0102014.
[28] W. F. Edwards, C. S. Kenyon and D. K. Lemon, Phys. Rev. D **14,** 922 (1976); D. K. Lemon, W. F. Edwards and C. S. Kenyon, Phys. Lett. A ***62,*** *105 (1992).*
[29] G. G. Shishkin et al., J. Phys. D: Appl. Phys. **35** 497 (2002).
[30] U. Bartocci, F. Cardone and R. Mignani, Found. Phys. Lett. **14,** 51 (2001); F. Cardone, R. Mignani and R. Scrimaglio, Found. Phys. **36,** 263 (2006).
[31] R. Folman, arXiv: 1109.2586.
[32] T. Ivezić, Phys. Lett. A **144**, 427 (1990).
[33] T. Ivezić, Phys. Scr. **81,** 025001 (2010).
[34] M. Mansuripur, Phys. Rev. Lett. **108**, 193901 (2012).
[35] A. Cho, Science **336**, 404 (2012).
[36] D. A. T. Vanzella, arXiv: 1205.1502; C. S. Unnikrishnan, arXiv: 1205.1080; D. J. Griffiths and V. Hnizdo, arXiv: 1205.4646; K. T. McDonald, www.physics.princeton.edu/mcdonald/examples/mansuripur.pdf; D. J. Cross, arXiv: 1205.5451; P. L. Saldanha, arXiv: 1205.6858; T. H. Boyer, arXiv: 1206.5322; M. Brachet and E. Tirapegul, arXiv: 1207.4613; Kimball A. Milton, Giulio Meille, arXiv: 1208.4826; A. L. Kholmetskii, O. V. Missevitch, T. Yarman, 1208.5296.
[37] A. Kobakhidze and B. H. J. McKellar, Phys. Rev. D **76**, 093004 (2007).
[38] C. Leubner, K. Aufinger and P. Krumm, Eur. J. Phys. **13,** 170 (1992).
[39] J. Van Bladel, *Relativity and Engineering* (Springer-Verlag, Berlin, 1984).
[40] A. Gamba, Am. J. Phys. **35**, 83 (1967).
[41] T. Ivezić, Int. J. Mod. Phys. B 26, 1250040 (2012).
[42] A. K. T. Assis, W. A. Rodrigues Jr. and A. J. Mania, Found. Phys. **29**, 729 (1999).
[43] W. J. Kim, A. O. Sushkov, D. A. R. Dalvit and S. K. Lamoreaux, Phys. Rev. Lett. **103**, 060401 (2009).
[44] A. Romannikov, Found. Phys. **41,** 1331 (2011).
[45] T. Ivezić, Phys. Lett. A **156**, 27 (1991).
[46] T. Ivezić, Phys. Rev. Lett. **98,** 108901 (2007).
[47] T. Ivezić, Phys. Rev. Lett. **98,** 158901 (2007).
[48] T. Ivezić, arXiv: hep-th/0705.0744.
[49] T. Ivezić, arXiv: 1005.3037; arXiv: 1006.4154.
[50] F. T. Trouton and H. R. Noble, Proc. Royal Soc. **74,** 132 (1903).
[51] J. D. Jackson, Am. J. Phys. **72**, 1484 (2004).
[52] T. Ivezić, Found. Phys. Lett. **12,** 105 (1999).
[53] Y. Aharonov and D. Bohm, Phys. Rev. **115**, 485 (1959).
[54] A. N. Tonomura, T. Osakabe, T. Matsuda, T. Kawasaki, J. Endo, S. Yano, and H. Yamada, Phys. Rev. Lett. **56**, 792 (1986).
[55] G. Möllenstedt and W. Bayh, Naturwissenschaften **49**, 81 (1962).

[56] A. Caprez, B. Barwick and H. Batelaan, Phys. Rev. Lett. **99**, 210401 (2007).
[57] R. G. Chambers, Phys. Rev. Lett. **5**, 3 (1960).
[58] M. Peshkin and A. Tonomura, *The Aharonov-Bohm Effect* (Springer, New York, 1989).
[59] V. A. Kostelecky and M. Mewes, Phys. Rev. D **66**, 056005 (2002).
[60] T. Ivezić, EJTP **10,** 131 (2006).
[61] J. Anandan, Int. J. Theor. Phys. **19**, 537 (1980).

On relativistic quantum mechanics in the rest-frame instant form of dynamics

Luca Lusanna
Sezione INFN di Firenze, Via Sansone 1, 50019 Sesto Fiorentino (FI), Italy

Abstract. A review is given of a new relativistic classical mechanics for N-particle systems compatible with relativistic bound states. It is formulated in global non-inertial frames in Minkowski space-time by means of parametrized Minkowski theories and then restricted to the intrinsic inertial rest frame of the isolated system. There is a complete control on the relativistic collective variables (Newton-Wigner center of mass, Fokker-Pryce center of inertia, M/oller center of energy) and on the realization of the Poincare' algebra (with the explicit form of the interaction-dependent Lorentz boosts). The particle world-lines are found to correspond to the ones of predictive mechanics and localization problems are clarified. The model can be consistently quantized avoiding the instantaneous spreading of the center-of-mass wave packets (Hegerfeldt theorem), because the non-local non-covariant center of mass is a non-measurable quantity. This implies that the Hilbert space must be the tensor product of the center-of-mass Hilbert space with the one of relative motions. This spatial non-separability (due to the Lorentz signature of space-time) makes relativistic entanglement much more involved than the non-relativistic one.

1. Introduction

There is no intrinsic notion of 3-space, simultaneity, 1-way velocity of light (two distant clocks are involved) in special relativity (SR): in the absolute Minkowski space-time, only the *conformal structure* (the light-cone) is intrinsically given as the locus of incoming and outgoing radiation. The light postulate says that the 2-way (only one clock is involved) velocity of light c is isotropic and constant. Its value replaces the rods (i.e. the standard of length) in modern metrology [1], where an atomic clock gives the standard of time and a conventional reference frame centered on a given observer is chosen as a standard of space-time (GPS is an example of such a standard). Instead in Galilei space-time there is no problem: time and Euclidean 3-space are absolute.

A well defined 3-space (i.e. a clock synchronization convention) is needed to be able to give Cauchy data for the relevant partial differential equations: only in this way one can use the existence and unicity theorem for their solutions to get predictability in classical and quantum physics.

The standard way out from the problem of 3-space is to choose the Euclidean 3-space of an inertial frame centered on an inertial observer and then use the kinematical Poincaré group to connect different inertial frames. This is done by means of *Einstein convention for the synchronization of clocks*: the inertial observer A sends a ray of light at x_i^o towards the (in general accelerated) observer B; the ray is reflected towards A at a point P of B world-line and then reabsorbed by A at x^o_f; by convention P is synchronous with the mid-point between emission and absorption on A's world-line, i.e. $x^o_P = x^o_i + \frac{1}{2}(x^o_f - x^o_i) = \frac{1}{2}(x^o_i + x^o_f)$. This convention selects the Euclidean instantaneous 3-spaces $x^o = ct = const.$ of the inertial frames centered on A. Only in this case the one-way velocity of light between A and B coincides with the two-way one, c. However, if the observer A is accelerated, the convention breaks down, because if *only* the world-line of the accelerated observer A (the *1+3 point of view*) is given, then the only way for defining instantaneous 3-spaces is to identify them with the Euclidean tangent planes orthogonal to the 4-velocity of the observer (the local rest frames). But these planes (they are tangent spaces not 3-spaces!) will intersect each other at a distance from A's world-line of the order of the acceleration lengths of A, so that all the either linearly or rotationally accelerated frames, centered on accelerated observers, based either on Fermi coordinates or on rotating ones, will develop *coordinate singularities*. Therefore their approximated notion of instantaneous 3-spaces cannot be used for a well-posed Cauchy problem for Maxwell equations.

In Refs. [2, 3] a general theory of non-inertial frames in Minkowski space-time was developed. As shown in Refs. [3, 4] this formulation allowed to develop a Lagrangian description (*parametrized Minkowski theories*) of isolated systems (particles, strings, fluids, fields) in which the transition from a non-inertial frame to another one is formalized as a gauge transformation. Therefore the inertial effects only modify the appearances of phenomena but not the physics.

Then the restriction to inertial frames allowed to define the Poincaré generators starting from the energy-momentum tensor. The intrinsic inertial rest frame of every isolated system allowed the definition of the *rest-frame instant form of dynamics* where it is possible to study the problem of the separation of the collective relativistic variable of an isolated system from the relative ones living in the so called Wigner 3-spaces in accord with the theory of relativistic bound states [5, 6, 7]. The ordinary world-lines of the particles can then be reconstructed [8].

The rest-frame instant form allows to give a formulation of *relativistic quantum mechanics* (RQM) [9] which takes into account of the problems of relativistic causality and relativistic localization implied by the Lorentz signature of Minkowski space-time. Also a definition of relativistic entanglement can be given: the Lorentz signature implies features of *non-locality and spatial non-separability* absent in the non-relativistic formulation of entanglement.

A complete review of the approach in SR and of its extension to non-inertial frames in general relativity (GR) in asymptotically Minkowskian space-times [10] is given in Ref.[11]. In GR there is no absolute notion since also space-time becomes dynamical (with the metric structure satisfying Einstein's equations): however in this class of space-times it is possible to make a Hamiltonian formulation and to separate the inertial degrees of freedom of the gravitational field from the tidal ones (the gravitational waves of the linearized theory).

2. Classical relativistic isolated systems in non-inertial frames in Minkowski space-time

A metrology-oriented description of non-inertial frames in SR can be done with the *3+1 point of view* and the use of observer-dependent Lorentz scalar *radar 4-coordinates*. Let us give the world-line $x^\mu(\tau)$ of an arbitrary time-like observer carrying a standard atomic clock: τ is an arbitrary monotonically increasing function of the proper time of this clock. Then we give an admissible 3+1 splitting of Minkowski space-time, namely a nice foliation with space-like instantaneous 3-spaces Σ_τ: it is the mathematical idealization of a protocol for clock synchronization. All the clocks in Σ_τ sign the same time of the atomic clock of the observer: it is the non-factual idealization required by the Cauchy problem generalizing the existing protocols for building a coordinate system inside the future light-cone of a time-like observer. On each 3-space Σ_τ we choose curvilinear 3-coordinates σ^r having the observer as origin. These are the *radar 4-coordinates* $\sigma^A = (\tau; \sigma^r)$.

If $x^\mu \mapsto \sigma^A(x)$ is the coordinate transformation from the Cartesian 4-coordinates x^μ of a reference inertial observer to radar coordinates, its inverse $\sigma^A \mapsto x^\mu = z^\mu(\tau, \sigma^r)$ defines the *embedding* functions $z^\mu(\tau, \sigma^r)$ describing the 3-spaces Σ_τ as embedded 3-manifold into Minkowski space-time. The induced 4-metric on Σ_τ is the following functional of the embedding ${}^4g_{AB}(\tau, \sigma^r) = [z^\mu_A\, \eta_{\mu\nu}\, z^\nu_B](\tau, \sigma^r)$, where $z^\mu_A = \partial\, z^\mu / \partial\, \sigma^A$ and ${}^4\eta_{\mu\nu} = \epsilon\,(+\,-\,-\,-)$ is the flat metric ($\epsilon = \pm 1$ according to either the particle physics $\epsilon = 1$ or the general relativity $\epsilon = -1$ convention). While the 4-vectors $z^\mu_r(\tau, \sigma^u)$ are tangent to Σ_τ, so that the unit normal $l^\mu(\tau, \sigma^u)$ is proportional to $\epsilon^\mu{}_{\alpha\beta\gamma}\, [z^\alpha_1\, z^\beta_2\, z^\gamma_3](\tau, \sigma^u)$, we have $z^\mu_\tau(\tau, \sigma^r) = [N\, l^\mu + N^r\, z^\mu_r](\tau, \sigma^r)$ with $N(\tau, \sigma^r) = \epsilon\, [z^\mu_\tau\, l_\mu](\tau, \sigma^r)$ and $N_r(\tau, \sigma^r) = -\epsilon\, g_{\tau r}(\tau, \sigma^r)$ being the lapse and shift functions.

The foliation is nice and admissible if it satisfies the conditions:

1) $N(\tau, \sigma^r) > 0$ in every point of Σ_τ (the 3-spaces never intersect, avoiding the coordinate singularity of Fermi coordinates);

2) $\epsilon\, {}^4g_{\tau\tau}(\tau, \sigma^r) > 0$, so to avoid the coordinate singularity of the rotating disk, and with the positive-definite 3-metric ${}^3g_{rs}(\tau, \sigma^u) = -\epsilon\, {}^4g_{rs}(\tau, \sigma^u)$ having three positive eigenvalues (these are the Møller conditions);

3) all the 3-spaces Σ_τ must tend to the same space-like hyper-plane at spatial infinity (so that there are always asymptotic inertial observers to be identified with the fixed stars).

These conditions imply that *global rigid rotations are forbidden in relativistic theories*. In Ref.[3] there is the expression of the admissible embedding corresponding to a 3+1 splitting of Minkowski space-time with parallel space-like hyper-planes (not equally spaced due to a linear acceleration) carrying differentially rotating 3-coordinates without the coordinate singularity of the rotating disk. It is the first consistent global non-inertial frame of this type.

Each admissible 3+1 splitting of space-time allows to define two associated congruences of time-like observers: a) the Eulerian observers with the unit normal to Σ_τ as 4-velocity; b) the non surface forming observers with 4-velocity proportional to $z^\mu_\tau(\tau, \sigma^r)$.

Therefore starting from the four independent embedding functions $z^\mu(\tau, \sigma^r)$ we obtain the ten components ${}^4g_{AB}(\tau, \sigma^u)$ of the 4-metric: they play the role of the *inertial potentials* generating the *relativistic apparent forces* in the non-inertial frame (the usual Newtonian inertial potentials can be recovered by doing the non-relativistic limit [3]). The extrinsic curvature tensor ${}^3K_{rs}(\tau, \sigma^u) = [\frac{1}{2N}\, (N_{r|s} + N_{s|r} - \partial_\tau\, {}^3g_{rs})](\tau, \sigma^u)$, describing the *shape* of the instantaneous 3-spaces as embedded 3-sub-manifolds of Minkowski space-time, is a secondary inertial potential functional of the inertial potentials ${}^4g_{AB}$.

In these global non-inertial frames of Minkowski space-time it is possible to describe isolated systems (particles, strings, fields, fluids) admitting a Lagrangian formulation by means of *parametrized Minkowski theories* [3, 4]. The matter variables are replaced with new ones knowing the clock synchronization convention defining the 3-spaces Σ_τ. For instance a Klein-Gordon field $\tilde{\phi}(x)$ will be replaced with $\phi(\tau, \sigma^r) = \tilde{\phi}(z(\tau, \sigma^r))$; the same for every other field. Instead for a relativistic particle with world-line $x^\mu(\tau)$ we must make a choice of its energy sign: then the positive- (or negative-) energy particle will be described by 3-coordinates $\eta^r(\tau)$ defined by the intersection of its world-line with Σ_τ: $x^\mu(\tau) = z^\mu(\tau, \eta^r(\tau))$. Differently from all the previous approaches to relativistic mechanics, the dynamical configuration variables are the 3-coordinates $\eta^r(\tau)$ and not the world-lines $x^\mu(\tau)$ (to rebuild them in an arbitrary frame we need the embedding defining that frame!).

Then the matter Lagrangian is coupled to an external gravitational field and the external 4-metric is replaced with the 4-metric $g_{AB}(\tau, \sigma^r)$ of an admissible 3+1 splitting of Minkowski space-time. With this procedure we get a Lagrangian depending on the given matter and on the embedding $z^\mu(\tau, \sigma^r)$, which is invariant under *frame-preserving diffeomorphisms* [12] and Dirac theory of constraints [13] is needed to study it. As a consequence, there are four first-class constraints (an analogue of the super-Hamiltonian and super-momentum constraints of canonical gravity) implying that the embeddings $z^\mu(\tau, \sigma^r)$ are *gauge variables*, so that *all the admissible non-inertial or inertial frames are gauge equivalent*, namely physics does *not* depend on the clock synchronization convention and on the choice of the 3-coordinates σ^r. Even if the gauge group is formed by the frame-preserving diffeomorphisms, the matter energy-momentum tensor allows the determination of the ten conserved Poincare' generators P^μ and $J^{\mu\nu}$ (assumed finite) of every configuration of the system (in non-inertial frames they are asymptotic generators at spatial infinity like the ADM ones in GR [14]).

If we restrict ourselves to inertial frames, we can define the *inertial rest-frame instant form of dynamics for isolated systems* by choosing the 3+1 splitting corresponding to the intrinsic inertial rest frame of the isolated system centered on an inertial observer: the instantaneous 3-spaces, named *Wigner 3-spaces* due to the fact that the 3-vectors inside them are Wigner spin-1 3-vectors, are orthogonal to the conserved 4-momentum P^μ of the configuration. The embedding corresponding to the inertial rest frame is $z^\mu(\tau, \vec{\sigma}) = Y^\mu(\tau) + \epsilon^\mu_r(\vec{h})\, \sigma^r$, where $Y^\mu(\tau)$ is the Fokker-Pryce center-of-inertia 4-vector, $\vec{h} = \vec{P}/\sqrt{\epsilon\, P^2}$ and $\epsilon^\mu{}_{A=\nu}(\vec{h})$ is the standard Wigner boost for time-like orbits sending $P^\mu = \sqrt{\epsilon\, P^2}\, (\sqrt{1 + \vec{h}^2}; \vec{h})$ to $(1; 0)$.

In Ref.[3] there is also the definition of the admissible *non-inertial rest frames*, where P^μ is orthogonal to the asymptotic space-like hyper-planes to which the instantaneous 3-spaces tend at spatial infinity. This non-inertial family of 3+1 splittings is the only one admitted by the asymptotically Minkowskian space-times without super-translations in GR [10]. Finally in Ref.[15] there is the definition of *parametrized Galilei theories*, non relativistic limit of the parametrized Minkowski theories. Also the inertial and non-inertial frames in Galilei space-time are gauge equivalent in this formulation.

3. Relativistic center of mass, particle world-lines and bound states

The framework of the inertial rest frame allowed the solution of the following old open problems:

A) The classification [16] of the relativistic collective variables (the canonical non-covariant Newton-Wigner center of mass (or center of spin), the non-canonical covariant Fokker-Pryce center of inertia and the non-canonical non-covariant Møller center of energy), replacing the Newtonian center of mass (and all tending to it in the non-relativistic limit), that can be built only in terms of the Poincare' generators: they are *non measurable* quantities due to the *non-local* character of such generators (they know the whole 3-space Σ_τ). There is a Møller world-tube around the Fokker-Pryce 4-vector containing all the possible pseudo-world-lines of the other two, whose Møller radius $\rho = |\vec{S}|/\sqrt{\epsilon\, P^2}$ ($\vec{S}$ is the rest angular momentum) is determined by the Poincaré Casimirs of the isolated system. This non-covariance world-tube is a non-local effect of Lorentz signature of the space-time absent in

Euclidean spaces. The world-lines $x_i^\mu(\tau)$ $(= Y^\mu(\tau) + \epsilon_r^\mu(\vec{h})\, \eta_i^r(\tau)$ in the rest-frame instant form) of the particles are derived (interaction-dependent) quantities and in general they do not satisfy vanishing Poisson brackets [8]: already at the classical level a *non-commutative structure* emerges due to the Lorentz signature!

B) The description of every isolated system as a decoupled (non-measurable) canonical non-covariant (Newton-Wigner) external center of mass (described by frozen Jacobi data) carrying a pole-dipole structure: the invariant mass and the rest spin of the system expressed in terms of suitable Wigner-covariant relative variables of the given isolated system inside the Wigner 3-spaces (after the elimination of the internal center of mass with well defined rest-frame conditions). The invariant mass is the effective Hamiltonian inside these 3-spaces. See Refs.[17] for all the interacting systems for which the explicit form of the interaction-dependent Lorentz boosts is known.

C) The formulation of classical relativistic atomic physics [5] (the electro-magnetic field in the radiation gauge plus charged scalar particles with Grassmann-valued electric charges to make a ultraviolet and infrared regularization of the self-energies and with mutual Coulomb potentials) and the identification of the Darwin potential at the classical level by means of a canonical transformation transforming the original system in N charged particles interacting with Coulomb plus Darwin potentials and a free radiation field (absence of Haag's theorem at least at the classical level). Therefore the Coulomb plus Darwin potential is the description as a Cauchy problem of the interaction described by the one-photon exchange Feynman diagram of QED (all the radiative corrections and photon brehmstrahlung are deleted by the Grassmann regularization).

D) A new formulation of relativistic kinetic theory and of the relativistic micro-canonical ensemble both in inertial and non-inertial frames [18].

E) The formulation of relativistic perfect fluids in the rest-frame instant form of dynamics [19].

4. Relativistic quantum mechanics, relativistic entanglement and localization problems

In Ref.[9] there is a new formulation of *relativistic quantum mechanics* in the inertial rest frame englobing all the known results about relativistic bound states due to the use of clock synchronization for the definition of the instantaneous 3-spaces, which implies the absence of relative times in their description.

In Galilei space-time non-relativistic quantum mechanics, where all the main results about entanglement are formulated, describes a composite system with two (or more) subsystems with a Hilbert space which is the tensor product of the Hilbert spaces of the subsystems: $H = H_1 \otimes H_2$. This type of spatial separability is named *the zeroth postulate* of quantum mechanics. However, when the two subsystems are mutually interacting, one makes a unitary transformation to the tensor product of the Hilbert space H_{com} describing the decoupled Newtonian center of mass of the two subsystems and of the Hilbert space H_{rel} of relative variables: $H = H_1 \otimes H_2 = H_{com} \otimes H_{rel}$. This allows to use the method of separation of variables to split the Schroedinger equation in two equations: one for the free motion of the center of mass and another, containing the interactions, for the relative variables (this equation describes both the bound and scattering states). A final unitary transformation of the Hamilton-Jacobi type allows to replace H_{com} with $H_{com,HJ}$, the Hilbert space in which the decoupled center of mass is frozen and described by non-evolving Jacobi data. Therefore we have $H = H_1 \otimes H_2 = H_{com} \otimes H_{rel} = H_{com,HJ} \otimes H_{rel}$.

While at the non-relativistic level these three descriptions are unitary equivalent, this is no more true in RQM. The non-local and non-covariant properties of the decoupled relativistic center of mass, described by the frozen Jacobi data $\vec{z}$ and $\vec{h} = \vec{P}/\sqrt{\epsilon P^2}$, imply that the only consistent relativistic quantization is based on the Hilbert space $H = H_{com,HJ} \otimes H_{rel}$ (in the non-relativistic limit it goes into the corresponding Galilean Hilbert space). We have $H \neq H_1 \otimes H_2$, because, already in the non-interacting case, in the tensor product of two quantum Klein-Gordon fields, $\phi_1(x_1)$ and $\phi_2(x_2)$, most of the states correspond to configurations in Minkowski space-time in which one particle may be present in the absolute future of the other particle. This is due to the fact that the two times x_1^o and x_2^o are totally uncorrelated, or in other words there is no notion of instantaneous 3-space (clock synchronization convention). Also the scalar products in the two formulations are completely different.

We have also $H \neq H_{com} \otimes H_{rel}$, because if instead of $\vec{z} = Mc\,\vec{x}_{NW}(0)$ we use the evolving (non-local and non-covariant) Newton-Wigner position operator $\vec{x}_{NW}(\tau)$, then we get a violation of relativistic causality because the center-of-mass wave packets spread instantaneously as shown by the Hegerfeldt theorem [20].

Therefore the only consistent Hilbert space is $H = H_{com,HJ} \otimes H_{rel}$. The main complication is the definition of H_{rel}, because we must take into account the three pairs of (interaction-dependent) second-class constraints eliminating the internal 3-center of mass inside the Wigner 3-spaces. When we are not able to make the elimination at the classical level and formulate the dynamics only in terms of Wigner-covariant relative variables, we have to quantize the particle Wigner-covariant 3-variables η_i^r, κ_{ir} and then to define the physical Hilbert space by adding the quantum version of the constraints a la Gupta-Bleuler.

Relativistic quantum mechanics in rotating non-inertial frames by using a multi-temporal quantization scheme is defined in Ref.[21]. In it *the inertial gauge variables are not quantized but remain c-numbers*; the known results in atomic and nuclear physics are reproduced.

Since we have that the Hilbert space $H = H_{com,HJ} \otimes H_{rel}$ is not unitarily equivalent to the one $H_1 \otimes H_2 \otimes ...$, where H_i are the Hilbert spaces of the individual particles, at the relativistic level the zeroth postulate of non-relativistic quantum mechanics does not hold (the necessity to use H_{rel} implies a type of *weak form of relationism* different from the formulations, like the one of Ref.[22], having their origin in the Mach principle). Since the Hilbert space of composite systems is not the tensor product of the Hilbert spaces of the sub-systems, we need a formulation of *relativistic entanglement* taking into account this spatial non-separability and the non-locality of the center of mass, both coming from the Poincare' group, i.e. from the Lorentz signature of space-time.

Since the center of mass is decoupled, its *non-covariance* is irrelevant. However its *non-locality* is a source of open problems: do we have to quantize it (in the preferred momentum basis implied by the quantum Poincaré algebra) and if yes is it meaningful to consider center-of-mass wave packets or we must add some super-selection rule? The wave function of the non-local center of mass is a kind of wave function of the 3-universe: who will observe it?

As a consequence in SR there are open problems on which type of *relativistic localization* is possible. There are strong indications that the Newton-Wigner position operator *cannot be self-adjoint but only symmetric* with the implication of a bad localization of relativistic particles [23]. The existing problems with relativistic position operators (like the Newton-Wigner one), deriving from the Lorentz signature of Minkowski space-time which forbids the existence of a unique collective variable with all the properties of Newton center of mass , point towards a *non-measurability of absolute positions* (but not of the relative variables needed to describe the spectra of bound states). This type of *un-sharpness* is induced also in non-relativistic quantum mechanics if one takes into account the crucial relativistic electro-magnetic effects of order $1/c$ in atomic physics. Moreover see Refs. [24] for another type of non-relativistic un-sharpness of positions. Experiments in atomic and molecular physics are beginning to explore these localization problems, in frameworks dominated by a *Newtonian classical intuition* and taking into account the experimental quantum limits for atom localization.

Therefore SR introduces a *kinematical non-locality* and a *kinematical spatial non-separability*, which reduce the relevance of *quantum non-locality* in the study of the foundational problems of quantum mechanics (see for instance Refs.[25]). Relativistic entanglement will have to be reformulated in terms of relative variables also at the non-relativistic level. Therefore the control of Poincare' kinematics will force to reformulate the experiments connected with Bell inequalities and teleportation in terms of the relative variables of isolated systems containing: a) the observers with their measuring apparatus (Alice and Bob as macroscopic quasi-classical objects); b) the particles of the protocol (but now the ray of light, the "photons" carrying the polarization, move along null geodesics); c) the environment (a macroscopic either quantum or quasi-classical object).

5. Open problems

In conclusion we now have a good control on both classical and quantum relativistic mechanics.

The main open problem in SR is the quantization of *fields* in non-inertial frames due to the no-go theorem of Ref.[26] showing the existence of obstructions to the unitary evolution of a massive quantum Klein-Gordon field between two space-like surfaces of Minkowski space-time. Its solution, i.e. the identification of all the 3+1 splittings allowing unitary evolution, will be a prerequisite to any attempt to quantize canonical gravity taking into account the equivalence principle (global inertial frames do not exist!). Moreover entanglement in non-inertial frames without Rindler observers is still to be formulated.

Finally in Ref.[27] there is a spatio-temporal approach to the emergence of classical notions (effective classical Newton trajectories for quantum massive particles) from non-relativistic and relativistic quantum mechanics. This justifies Bohr's point of view that it must be possible to describe every experiment in classical terms.

References

[1] L. Lusanna, *Relativistic Metrology: from Earth to Astrophysics*, In-Tech E-Book *Modern Metrology Concerns*, 2012 (ISBN 978-953-51-0584-8).
M. Soffel, S. A. Klioner, G. Petit, P. Wolf, S. M. Kopeikin, P. Bretagnon, V. A. Brumberg, N. Capitaine, T. Damour, T. Fukushima, B. Guinot, T. Huang, L. Lindegren, C. Ma, K. Nordtvedt, J. Ries, P. K. Seidelmann, D. Vokroulicky', C. Will and Ch. Xu, *The IAU 2000 Resolutions for Astrometry, Celestial Mechanics and Metrology in the Relativistic Framework: Explanatory Supplement* Astron. J., **126**, pp.2687-2706, (2003) (arXiv astro-ph/0303376).
IERS Conventions (2003), eds. D. D. McCarthy and G. Petit, IERS TN 32 (2004), Verlag des BKG. G. H. Kaplan, *The IAU Resolutions on Astronomical Reference Systems, Time Scales and Earth Rotation Models*, U. S. Naval Observatory circular No. 179 (2005) (arXiv astro-ph/0602086).
T. D. Moyer, *Formulation for Observed and Computed Values of Deep Space Network Data Types for Navigation* (John Wiley, New York, 2003).
[2] D. Alba and L. Lusanna, *Generalized Radar 4-Coordinates and Equal-Time Cauchy Surfaces for Arbitrary Accelerated Observers*, Int. J. Mod. Phys. **D16**, 1149 (2007) (arXiv gr-qc/0501090).

[3] D. Alba and L. Lusanna, *Charged Particles and the Electro-Magnetic Field in Non-Inertial Frames: I. Admissible 3+1 Splittings of Minkowski Spacetime and the Non-Inertial Rest Frames*, Int. J. Geom. Methods in Physics **7**, 33 (arXiv 0908.0213) and *II. Applications: Rotating Frames, Sagnac Effect, Faraday Rotation, Wrap-up Effect*, Int. J. Geom. Methods in Physics, **7**, 185 (2010) (arXiv 0908.0215).

[4] L. Lusanna, *The N- and 1-Time Classical Descriptions of N-Body Relativistic Kinematics and the Electromagnetic Interaction*, Int. J. Mod. Phys. **A12**, 645 (1997).
L. Lusanna, *The Chrono-Geometrical Structure of Special and General Relativity: A Re-Visitation of Canonical Geometrodynamics*, lectures at 42nd Karpacz Winter School of Theoretical Physics: Current Mathematical Topics in Gravitation and Cosmology, Ladek, Poland, 6-11 Feb 2006, Int. J. Geom. Methods in Mod. Phys. **4**, 79 (2007). (arXiv gr-qc/0604120).

[5] D. Alba, H. W. Crater and L. Lusanna, *Towards Relativistic Atom Physics. I. The Rest-Frame Instant Form of Dynamics and a Canonical Transformation for a system of Charged Particles plus the Electro-Magnetic Field*, Canad. J. Phys. **88**, 379 (2010) (arXiv 0806.2383) and *II. Collective and Relative Relativistic Variables for a System of Charged Particles plus the Electro-Magnetic Field*, Canad. J. Phys. **88**, 425 (2010) (arXiv 0811.0715).

[6] H. W. Crater and L. Lusanna, *The Rest-Frame Darwin Potential from the Lienard-Wiechert Solution in the Radiation Gauge*, Ann. Phys. (N. Y.) **289**, 87 (2001).

[7] D. Alba, H. W. Crater and L. Lusanna, *The Semiclassical Relativistic Darwin Potential for Spinning Particles in the Rest Frame Instant Form: Two-Body Bound States with Spin 1/2 Constituents*, Int. J. Mod. Phys. **A16**, 3365 (2001) (arXiv hep-th/0103109).

[8] D. Alba, H. W. Crater and L. Lusanna, *Hamiltonian Relativistic Two-Body Problem: Center of Mass and Orbit Reconstruction*, J. Phys. **A40**, 9585 (arXiv gr-qc/0610200) (2007).

[9] D. Alba, H. W. Crater and L. Lusanna, *Relativistic Quantum Mechanics and Relativistic Entanglement in the Rest-Frame Instant Form of Dynamics*, J. Math. Phys. **52**, 062301 (2011) (arXiv 0907.1816).

[10] D. Alba and L. Lusanna, Canad. J. Phys., 90 (2012) 1017-1076;
D. Alba and L. Lusanna, Canad. J. Phys., 90 (2012) 1077-1130;
D. Alba and L. Lusanna, Canad. J. Phys., 90 (2012) 1131-1178.

[11] L. Lusanna, *From Clock Synchronization to Dark Matter as a Relativistic Inertial Effect*, Lecture at the *Black Objects in Supergravity* School BOSS2011, Frascati, 9-13 May 2011 (arXiv 1205.2481).

[12] E. Schmutzer and J. Plebanski, *Quantum Mechanics in Noninertial Frames of Reference*, Fortschr. Phys. **25**, 37 (1978).

[13] P. A. M. Dirac, *Lectures on Quantum Mechanics*, Belfer Graduate School of Science, Monographs Series (Yeshiva University, New York, N. Y., 1964).
M. Henneaux and C. Teitelboim, *Quantization of Gauge Systems* (Princeton University Press, Princeton, 1992).
S. Shanmugadhasan, *Canonical Formalism for Degenerate Lagrangians*, J. Math. Phys. **14**, 677 (1973).
L. Lusanna, *The Shanmugadhasan Canonical Transformation, Function Groups and the Second Noether Theorem*, Int. J. Mod. Phys. **A8**, 4193 (1993); *The Relevance of Canonical Transformations in Gauge Theories and General Relativity*, Lecture Notes of "Seminario Interdisciplinare di Matematica" (Basilicata Univ.) **5**, 125 (2006).
. M. Chaichian, D. Louis Martinez and L. Lusanna, *Dirac's Constrained Systems: the Classification of Second-Class Constraints*, Ann. Phys. (N. Y.) **232**, 40 (1994).

[14] R. Arnowitt, S. Deser and C. W. Misner, *The Dynamics of General Relativity*, ch.7 of *Gravitation: an Introduction to Current Research*, ed. L. Witten (Wiley, New York, 1962)(arXiv gr-qc/0405109).
Lusanna L., *The Rest-Frame Instant Form of Metric Gravity*, Gen. Rel. Grav. **33**, 1579 (2001)(arXiv gr-qc/0101048).
Lusanna L. and Russo S., *A New Parametrization for Tetrad Gravity*, Gen. Rel. Grav. **34**, 189 (2002)(arXiv gr-qc/0102074).
D. Alba and L. Lusanna, *The York Map as a Shanmugadhasan Canonical Transformationn in Tetrad Gravity and the Role of Non-Inertial Frames in the Geometrical View of the Gravitational Field*, Gen. Rel. Grav. **39**, 2149 (2007) (arXiv gr-qc/0604086, v2).

[15] D. Alba, *Quantum Mechanics in Noninertial Frames with a Multitemporal Quantization Scheme: II. Nonrelativistic Particles*, Int. J. Mod. Phys. **A21**, 3917 (2006) (hep-th/0504060).

[16] D. Alba, L. Lusanna and M. Pauri, *New Directions in Non-Relativistic and Relativistic Rotational and Multipole Kinematics for N-Body and Continuous Systems* (2005), in *Atomic and Molecular Clusters: New Research*, ed. Y. L. Ping (Nova Science, New York, 2006) (arXiv hep-th/0505005).
D. Alba, L. Lusanna and M. Pauri, *Centers of Mass and Rotational Kinematics for the Relativistic N-Body Problem in the Rest-Frame Instant Form*, J. Math. Phys. **43**, 1677-1727 (2002) (arXiv hep-th/0102087).
D. Alba, L. Lusanna and M. Pauri, *Multipolar Expansions for Closed and Open Systems of Relativistic Particles* , J. Math. Phys. **46**, 062505, 1-36 (2004) (arXiv hep-th/0402181).

[17] D. Alba and L. Lusanna, *The Classical Relativistic Quark Model in the Rest-Frame Wigner-Covariant Coulomb Gauge*, Int. J. Mod. Phys. **A13**, 3275 (1998) (arXiv hep-th/9705156).
D. Alba, H. W. Crater and L. Lusanna, *Massless Particles plus Matter in the Rest-Frame Instant Form of Dynamics*, *J. Phys.* A **43** 405203 (arXiv 1005.5521).
D. Alba, H. W. Crater and L. Lusanna, *The Rest-Frame Instant Form and Dirac Observables for the Open Nambu String*, Eur. Phys. J. Plus **126**, 26 (2011)(arXiv 1005.3659).
D. Alba, H. W. Crater and L. Lusanna, *A Relativistic Version of the Two-Level Atom in the Rest-Frame Instant Form of Dynamics* (arXiv 1107.1669).
D. Bini and L. Lusanna, *Spin-Rotation Couplings: Spinning Test Particles and Dirac Fields*, Gen. Rel. Grav. **40**, 1145 (2008) (arXiv 0710.0791).

[18] D. Alba, H. W. Crater and L. Lusanna, *On the Relativistic Micro-Canonical Ensemble and Relativistic Kinetic Theory for N Relativistic Particles in Inertial and Non-Inertial Rest Frames* (arXiv 1202.4667).

[19] L. Lusanna and D. Nowak-Szczepaniak, *The Rest-Frame Instant Form of Relativistic Perfect Fluids with Equation of State $\rho = \rho(\eta, s)$ and of Nondissipative Elastic Materials.*, Int. J. Mod. Phys. **A15**, 4943 (2000).
D. Alba and L. Lusanna, *Generalized Eulerian Coordinates for Relativistic Fluids: Hamiltonian Rest-Frame Instant Form, Relative Variables, Rotational Kinematics* Int. J. Mod. Phys. **A19**, 3025 (2004) (arXiv hep-th/020903).
J. D. Brown, *Action Functionals for Relativistic Perfect Fluids*, Class. Quantum Grav. **10**, 1579 (1993).
D. Alba and L. Lusanna, *Dust in the York Canonical Basis of ADM Tetrad Gravity: the Problem of Vorticity* (arXiv 1106.0403).

[20] G. C. Hegerfeldt, *Remark on Causality and Particle Localization*, Phys. Rev. **D10**, 3320 (1974); *Violation of Causality in Relativistic Quantum Theory?*, Phys. Rev. Lett. **54**, 2395 (1985).

[21] D. Alba and L. Lusanna, *Quantum Mechanics in Noninertial Frames with a Multitemporal Quantization Scheme: I. Relativistic Particles*, Int. J. Mod. Phys. **A21**, 2781 (2006) (arXiv hep-th/0502194).

[22] C. Rovelli, *Relational Quantum Mechanics*, Int. J. Theor. Phys. **35**, 1637 (1996) (quant-ph/9609002).
C. Rovelli and M. Smerlak, *Relational EPR* Found. Phys. **37**, 427 (2007) (quant-ph/0604064).

[23] J. Butterfield and G. Fleming, *Strange Positions*, in *From Physics to Philosophy*, eds. J. Butterfield and C. Pagonis, pp.108-165 (Cambridge Univ. Press, Cambridge, 1999).
A. Kempf, *Unsharp Degrees of Freedom and the Generating of Symmetries*, Phys. Rev. **D63**, 024017 (2000) (arXiv hep-th/9907160).
R. T. W. Martin and A. Kempf, *Quantum Uncertainty and the Spectra of Symmetric Operators*, Acta Appl. Math. **106**, 349 (2009).

[24] P. Busch and L. Loveridge, *Position Measurements Obeying Momentum Conservation*, Phys. Rev. Lett. **106**, 110406 (2011) (arXiv 1005.0569).
P. Busch and L. Loveridge, *"Measurement of Quantum Mechanical Operators" Revisited*, to appear in Eur. Phys. J. D (arXiv 1012.4362).
P. Busch, P. Lathi and P. Mittelstaedt, *The Quantum Theory of Measurement* (Springer, Berlin, 1996).
P. Busch, M. Grabowski and P. Lahti, *Operational Quantum Physics*, Lecture Notes in Physics 31 (Springer, Berlin, 1995; 1997 2nd corrected printing).
P. Busch, *Unsharp Localization and Causality in Relativistic Quantum Theory*, J. Phys. **A32**, 6535 (1999).

[25] A. Einstein, B. Podolski and N. Rosen, *Can Quantum-Mechanical Description of Physical Reality be Considered Complete?*, Phys. Rev. **47**, 777 (1935).
H. R. Brown, *Physical Relativity. Space-Time Structure from a Dynamical Perspective* (Oxford Univ. Press, Oxford, 2005).

[26] C. G. Torre and M. Varadarajan, *Functional Evolution of Free Quantum Fields*, Class. Quantum Grav. **16**, 2651 (1999).

[27] L. Lusanna and M. Pauri, *On the Transition from the Quantum to the Classical Regime for Massive Scalar Particles: A Spatiotemporal Approach* (arXiv 1207.1248).

Inner symmetries of relativistic systems, the Laplace-Runge-Lenz symmetry, and the relativistic centre-of-mass

Uri Ben-Ya'acov
School of Engineering, Kinneret Academic College on the Sea of Galilee, D.N. Emek Ha'Yarden 15132, Israel
E-mail: uriby@kinneret.ac.il

Abstract. The talk concerns the inner symmetries of composite relativistic systems, their (generic) relation with the Laplace-Runge-Lenz (LRL) symmetry and the definition of the relativistic centre-of-mass. Global Lorentz-Poincaré symmetry implies the existence of LRL symmetry, in a naturally generalized sense, as part of the inner symmetry of these systems, and in a manner which is independent of the internal interaction. The corresponding LRL vectors form the internal moments associated with the Lorentz boost, which, in turn, determines the centre-of-mass.

1. Introduction

As the title suggests, the talk concerns the intrinsic relationship between the inner symmetries of composite relativistic systems, the Laplace-Runge-Lenz (LRL) symmetry, and the definition of the relativistic centre-of-mass (CM). The prime motive for the exploration that led to these results was the wish to find yet another way to solve and describe relativistic dynamics. The idea is that if Kepler-Coulomb systems, which are the non-relativistic limit of gravitational or electromagnetic systems, are endowed with the LRL symmetry (at least for 2-body systems) which completes the dynamical analysis of the system [1, 2, 3, 4], then perhaps the corresponding relativistic systems are endowed with a generalization of the LRL symmetry [5, 6, 7] which enables a better insight towards the full solution of the relativistic dynamics. For many years this issue puzzled me; recently many pieces of the puzzle came together and fell naturally and amazingly into their place.

The purpose of the talk is to show :

(i) That integration of the relativistic CM coordinate is not unique (as opposed to Newtonian systems), and the LRL vector appears naturally as part of the process. This is the fundamental observation upon which the talk is based.

(ii) The rôle of the LRL symmetry as inducing configuration-(or shape-)transformations which are energy preserving.

(iii) Since the definition of the relativistic CM involves the Lorentz boost, it follows that the LRL symmetry is naturally associated with Lorentz-Poincaré symmetry.

(iv) The LRL symmetry thus being an integral part of the internal symmetry of relativistic systems.

The following presentation is based on, and extends, recent publications by the present author [8, 9, 10]. The systems considered are Lorenz-Poincaré-symmetric, endowed with total linear momentum P^μ and total angular momentum $J^{\mu\nu}$ which are conserved and serve as generators of the corresponding global transformations. The metric convention is $g_{\mu\nu} = \mathrm{diag}(-c^2, 1, 1, 1)$. The relativistic invariant mass of the system is $M = \sqrt{-P^2}/c^2$.

2. The centre of mass

As a starting point, let us recall the way we arrive at the definition of the Newtonian CM in a general reference frame. The CM velocity, defined from the total linear momentum and total mass, is transformed into a derivative

$$\vec{V}_{CM} \equiv \frac{\vec{P}}{M_o} = \frac{\sum_a m_a \vec{v}_a}{M_o} = \frac{d}{dt}\left(\frac{\sum_a m_a \vec{x}_a}{M_o}\right), \tag{1}$$

($M_o \equiv \sum_a m_a$), leading to the identification and definition of the expression in the brackets as the CM coordinate,

$$\vec{X}_{\mathrm{N}} \equiv \frac{\sum_a m_a \vec{x}_a}{M_o} \tag{2}$$

Since any integration is determined up to an arbitrary additive constant, it might well be asked, why not add such a constant also to Eq.(2) :

$$\vec{X}'_{\mathrm{N}} = \frac{\sum_a m_a \vec{x}_a}{M_o} + \vec{C} \tag{3}$$

The argument is, of course, that if the particles are all located together at one point, say $\vec{x}_1 = \vec{x}_2 = ... = \vec{x}_o$, then also $\vec{X}_{\mathrm{N}}$ should coincide with the same point, requiring $\vec{C}$ to vanish. This requirement rules out any c-number $\vec{C}$, but what if $\vec{C}$ is not a c-number but a vector observable which is an integral-of-the-motion that vanishes when the particles' coordinate coincide ? Then a non-zero $\vec{C}$ cannot be ruled out. Newtonian dynamics lives very happily with definition (2), without the need for any addition. The case is essentially different in relativistic dynamics.

The relativistic 3-D CM velocity $\vec{V}_{CM} = \vec{P}/P^o$ may be expanded in a general inertial reference frame in powers of $1/c^2$ as

$$\vec{V}_{CM} = \frac{\sum_a m_a \vec{v}_a}{M_o} + O\left(c^{-2}\right) = \frac{d}{dt}\left[\frac{\sum_a m_a \vec{x}_a}{M_o} + O\left(c^{-2}\right)\right] \tag{4}$$

so that integration for the CM-coordinate itself yields

$$\vec{X}_{CM} = \frac{\sum_a m_a \vec{x}_a}{M_o} + O\left(c^{-2}\right) = \vec{X}_{\mathrm{N}} + O\left(c^{-2}\right) \tag{5}$$

The phenomenon that is revealed and discussed in the following, being the main theme of this presentation, an effect that is uniquely relativistic and disappears in the non-relativistic limit, is that the process of integration leading from Eq.(4) to Eq.(5) is not unique : While the integration of the Newtonian expression (1) involves only the kinematic relation $\vec{v}_a = d\vec{x}_a/dt$, the integration of the $O\left(c^{-2}\right)$-term in Eq.(4) requires, as becomes evident in the following, the dynamical equations of motion. As a consequence, the $O\left(c^{-2}\right)$-term in Eq.(5) is not an integral of the motion, and may be obtained in more than one form.

As an illustration, let us compute the CM coordinate of a 2-body electromagnetic system in the Post-Newtonian approximation. To keep things relatively simple let us consider the system in the CM reference frame ($\vec{P} = \vec{p}_1 + \vec{p}_2 = 0$). The total energy (keeping terms to order c^{-2}) is [11]

$$E = c^2 P^o \approx M_o c^2 + \frac{\vec{p}^{\,2}}{2\mu} - \left(\frac{1}{{m_1}^3} + \frac{1}{{m_2}^3}\right)\frac{p^4}{8c^2} + \frac{\kappa}{r} + \frac{\kappa}{2m_1 m_2 c^2 r}\left[\vec{p}^{\,2} + \frac{(\vec{p}\cdot\vec{r})^2}{r^2}\right] \tag{6}$$

with $M_o = m_1 + m_2$, $\mu = m_1 m_2/M_o$, $\vec{r} = \vec{x}_1 - \vec{x}_2$ and $\vec{p} = \vec{p}_1 = -\vec{p}_2$. The time derivative of the Newtonian CM coordinate (2) certainly does not vanish, so that, inserting the particles' velocities $\vec{v}_a = \partial H/\partial \vec{p}_a$, it becomes

$$\frac{d\vec{X}_{\rm N}}{dt} = \frac{m_1\vec{v}_1 + m_2\vec{v}_2}{M_o} \approx \frac{m_1 - m_2}{2m_1 m_2 M_o c^2}\left[\left(\frac{\vec{p}^{\,2}}{\mu} + \frac{\kappa}{r}\right)\vec{p} + \frac{\kappa\,(\vec{p}\cdot\vec{r})}{r^3}\vec{r}\right] \tag{7}$$

Since the rhs of Eq.(7) is already $O\left(c^{-2}\right)$, we may use the Newtonian equation of motion to express $d\vec{X}_{\rm N}/dt$ as a total derivative, and it turns out that this may be performed in two distinct ways :

$$\frac{d\vec{X}_{\rm N}}{dt} \approx \frac{m_1 - m_2}{2m_1 m_2 M_o c^2}\frac{d}{dt}\left[\left(\vec{p}^{\,2} + \frac{\mu\kappa}{r}\right)\vec{r}\right] \tag{8a}$$

$$\approx \frac{m_1 - m_2}{2m_1 m_2 M_o c^2}\frac{d}{dt}\left[(\vec{p}\cdot\vec{r})\,\vec{p}\right] \tag{8b}$$

with the immediate solutions

$$\vec{X}_{\rm N} \approx \frac{m_1 - m_2}{2m_1 m_2 M_o c^2}\left[\left(\vec{p}^{\,2} + \frac{\mu\kappa}{r}\right)\vec{r}\right] + \vec{C}_1 \tag{9a}$$

$$\approx \frac{m_1 - m_2}{2m_1 m_2 M_o c^2}\left[(\vec{p}\cdot\vec{r})\,\vec{p}\right] + \vec{C}_2 \tag{9b}$$

A-priori, both modes of integration have the same privileges, without any preference of one over the other. Evidently, the two constants of integration cannot vanish simultaneously, and, since the integration involved the equations of motion, their difference must be a dynamical constant of the motion. The surprising result is that the difference between these constants is

$$\vec{C}_2 - \vec{C}_1 = \frac{m_1 - m_2}{2m_1 m_2 M_o c^2}\left[\left(\vec{p}^{\,2} + \frac{\mu\kappa}{r}\right)\vec{r} - (\vec{p}\cdot\vec{r})\,\vec{p}\right] = \frac{m_1 - m_2}{2m_1 m_2 M_o c^2}\vec{K} \tag{10}$$

where

$$\vec{K} = \left(\vec{p}^{\,2} + \frac{\mu\kappa}{r}\right)\vec{r} - (\vec{p}\cdot\vec{r})\,\vec{p} = \vec{p}\times\vec{\ell} + \frac{\mu\kappa}{r}\vec{r} \tag{11}$$

is the classical LRL vector of the corresponding Kepler-Coulomb system !

The significance of this result is, of course, in demonstrating that the classical (Newtonian) LRL vector emerges naturally in a relativistic calculation involving the Lorentz boost. This state of affairs was originally observed by Dahl already years ago [12], but apparently passed almost unnoticed by the physics community. The purpose of the present talk is to discuss the meaning and applications of this observation to fully relativistic systems.

To get a first clue for the meaning of the constants $\vec{C}_1$ and $\vec{C}_2$, we recall that in the standard approach to the issue of the relativistic CM coordinate it is defined [13, 14] (in the CM reference frame) from the Lorentz boost $\vec{N} = J^{io}$ via

$$\vec{X}_{CM} = \frac{\vec{N}}{M} \tag{12}$$

where M is the invariant relativistic total mass of the system. With the PN Lorentz boost [11] (in the CM reference frame)

$$\vec{N} \approx M\vec{X}_{\rm N} + \frac{m_2 - m_1}{2m_1 m_2 c^2}\left(\vec{p}^2 + \frac{\mu\kappa}{r}\right)\vec{r} \tag{13}$$

and the post-Newtonian total mass

$$M = \frac{E}{c^2} \approx M_o + \frac{\vec{p}^2}{2\mu c^2} + \frac{\kappa}{rc^2}, \tag{14}$$

the standard approach yields, via Eq.(12),

$$\vec{X}_{CM} = \frac{\vec{N}}{M} \approx \vec{X}_{\rm N} + \frac{m_2 - m_1}{2m_1 m_2 M_o c^2}\left(\vec{p}^2 + \frac{\mu\kappa}{r}\right)\vec{r} = \vec{C}_1 \tag{15}$$

However, the second solution in (9*b*), which is a-priory valid just as the first one, suggests an alternative definition for the CM coordinate,

$$\vec{X}_{CM} = \frac{\vec{N}}{M} + \frac{m_1 - m_2}{2m_1 m_2 M_o c^2}\vec{K} \approx \vec{X}_{\rm N} + \frac{m_2 - m_1}{2m_1 m_2 M_o c^2}\left(\vec{p}\cdot\vec{r}\right)\vec{p} = \vec{C}_2 \tag{16}$$

The different constants of integration correspond, therefore, to different possible definitions of the CM coordinate.

The LRL vector is an internal vector, which manifests the internal configuration of the system. The standard approach as in Eq.(15) restricts the CM coordinate to dependence only upon the global configuration of the system (via the dependence upon $\vec{N}$ and M, or more generally upon P^μ and $J^{\mu\nu}$, alone). Eq.(16), on the other hand, opens the door for a broader definition, in which the CM coordinate depends also upon the internal configuration of the system.

3. Universality of the Laplace-Runge-Lenz symmetry

The above demonstration, of the appearance of the LRL vector in the computation of the post-Newtonian CM coordinate, calls for further inquiry regarding the appearance of the LRL symmetry in fully relativistic systems. Therefore, our intention is to explore the proposition that a generalization of this vector appears, in a similar fashion, in the computation of the CM coordinate for fully relativistic systems.

As a requisite, we need first to explain and define what is meant by LRL symmetry and its generalization to relativistic systems.

The LRL vector $\vec{K}$ (Eq.(11)) is a constant of the motion in classical Kepler-Coulomb systems with Hamiltonian

$$H = \frac{\vec{p}^2}{2\mu} + \frac{\kappa}{r} \tag{17}$$

On energy hyper-surfaces $H = E$ in phase-space, the internal angular momentum $\vec{\ell} = \vec{r}\times\vec{p}$ and $\vec{K}$ generate together, with the Poisson brackets (PB)

$$\begin{aligned} &\left\{\ell^i, \ell^j\right\} = \varepsilon^{ijk}\ell^k \quad , \quad \left\{K^i, \ell^j\right\} = \left\{\ell^i, K^j\right\} = \varepsilon^{ijk}K^k \\ &\left\{K^i, K^j\right\} = -2\mu E\varepsilon^{ijk}\ell^k \end{aligned} \tag{18}$$

a closed Lie-Poisson algebra which is $o(4)$ or $o(3,1)$ according to the sign of E [2, 3, 4]. Let $\vec{u}$ be an arbitrary constant unit vector and χ a dimensionless parameter, so that $\delta\mathcal{G} = \left(\vec{K}\cdot\vec{u}\right)\delta\chi$ is a generator of the infinitesimal transformations induced by $\vec{K}$. As is evident from Eq.(18), these transformations change $\vec{K}$ and $\vec{\ell}$, remaining on the same energy hyper-surface via the relation

$$\vec{K}^2 - 2\mu E\vec{\ell}^2 = \mu^2\kappa^2 \tag{19}$$

The shape of the orbits is characterized by $\vec{K}$ via the eccentricity $\varepsilon = \left|\vec{K}\right| / (\mu\,|\kappa|)$, while their spatial orientation is determined by the direction of $\vec{K}$. Elliptic closed orbits, for instance, with the same energy but differing by their value of $\left|\vec{K}\right|$ or $\left|\vec{\ell}\right|$ have the same major axis, and it is the minor axis that changes (Fig. 1) :

$$2a = \frac{|\kappa|}{|E|} \quad , \quad 2b = \sqrt{\frac{2\vec{\ell}^2}{\mu\,|E|}} \tag{20}$$

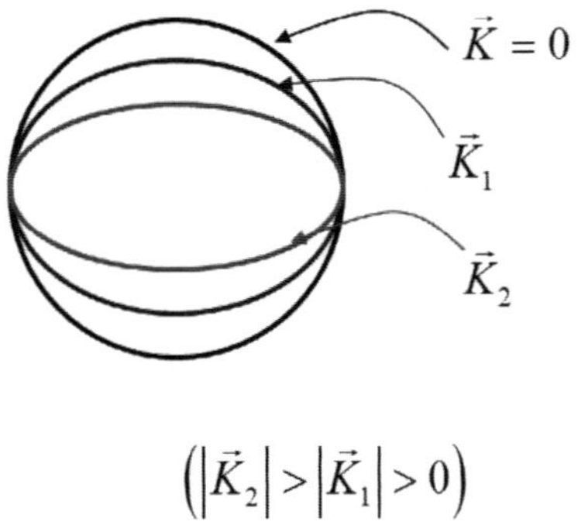

Figure 1. Elliptic configurations corresponding to different LRL vectors.

Hence, while $\vec{\ell}$ is responsible for internal rotations, $\vec{K}$-induced transformations are responsible for shape or configuration-changing, together composing the full internal symmetry of the system.

Although many still regard the LRL vector and the symmetry associated with it as 'accidental', corresponding only to Newtonian Kepler-Coulomb systems, it is known already since the mid-sixties that all rotationally invariant systems are endowed with a generalization of the LRL vector [5, 6, 7]. Consequently, generalized (constant) LRL vectors are definable for all rotationally symmetric systems, including systems with open orbits. The generalization of the LRL symmetry does not depend upon the type of interaction, implying its universality for all rotationally symmetric systems.

These properties of the LRL symmetry are well summarized by the following proposition [9]: The assumption of internal rotational symmetry implies the existence of an internal angular momentum $\vec{\ell}$, the generator of internal rotations (so that $\{\ell^i, V^j\} = \varepsilon^{ijk}V^k$ for all 3-vectors $\vec{V}$). Then, if $\vec{K}$ is a constant vector so that :

(i) *the scalar* $\vec{\ell}\cdot\vec{K}$ *is* $\vec{K}$*-invariant, in the sense that* $\left\{\vec{K}, \vec{\ell}\cdot\vec{K}\right\} = 0$,

(ii) $\vec{K}^2$, *being a constant scalar observable, is expressible as some function* $\vec{K}^2 = F\left(H, \vec{\ell}^2, \mathcal{A}\right)$ *only of* H, ℓ^2 *(which is always the case for 2-body systems), and possibly on some observables* $\mathcal{A}$ *that are also* $\vec{K}$*-invariant,* $\left\{\vec{K}, \mathcal{A}\right\} = 0$, *then the self PB of $\vec{K}$ are necessarily of the form*

$$\left\{K^i, K^j\right\} = -\frac{\partial\left(\vec{K}^2\right)}{\partial\left(\vec{\ell}^2\right)}\varepsilon^{ijk}\ell^k \tag{21}$$

The classical LRL vector (11) with $\vec{K}^2 = \mu^2\kappa^2 + 2\mu E\vec{\ell}^2$ and the self PB (18) is an immediate manifestation of Eq.(21). Any internal vector observable $\vec{K}$ that satisfies the conditions of the proposition may be regarded as a generalized Laplace-Runge-Lenz vector. Since these conditions are very broad, generalized LRL vectors are not unique; rather, for every system it is a whole family of vectors that satisfy the LRL conditions. Therefore, the LRL property does not refer to a specific vector, but it is the property of the whole system, stemming from its rotational symmetry.

The remarkable feature about Eq.(21) is its universality : It is based only upon the rotational symmetry of the system, and is independent of any particular form of the interaction. Also, it does not require any particular recipe for the computation of the PB (except for the general properties of Poisson brackets), and is therefore suitable also for systems which lack canonical or phase-space structure,

4. Relativistic Laplace-Runge-Lenz symmetry

Carrying-over the LRL symmetry to relativistic systems is now straight-forward. Even in the absence of clear and unique definition of canonical phase-space variables in classical relativistic systems, Poisson brackets $\{\cdot,\cdot\}$ may be postulated satisfying the standard rules of Lie-Poisson algebras, namely [15],

(i) Antisymmetry : $\{\mathcal{A},\mathcal{B}\} = -\{\mathcal{B},\mathcal{A}\}$
(ii) Jacoby identity : $\{\mathcal{A},\{\mathcal{B},\mathcal{C}\}\} + \{\mathcal{B},\{\mathcal{C},\mathcal{A}\}\} + \{\mathcal{C},\{\mathcal{A},\mathcal{B}\}\} = 0$
(iii) Product ("Leibnitz") rule : $\{\mathcal{A},\mathcal{B}\mathcal{C}\} = \{\mathcal{A},\mathcal{B}\}\,\mathcal{C} + \{\mathcal{A},\mathcal{C}\}\,\mathcal{B}$
(iv) Derivative rule: $\{\mathcal{A},f(\mathcal{B})\} = \{\mathcal{A},\mathcal{B}\}\,f'(\mathcal{B})$

If $\delta\mathcal{G}$ is the generator of an infinitesimal space-time transformation and $\delta\mathcal{A}$ is the variation of an observable $\mathcal{A}$ under that transformation, then the variation should define the PB via $\delta\mathcal{A} = \{\mathcal{A},\delta\mathcal{G}\}$. With well-defined and conserved total 4-linear momentum P^μ and 4-angular momentum $J^{\mu\nu}$, the fundamental PB are those of the Lorentz-Poincaré Lie-Poisson algebra,

$$\begin{aligned} \{P^\mu, P^\nu\} &= 0 \quad , \quad \left\{P^\mu, J^{\nu\lambda}\right\} = g^{\mu\lambda}P^\nu - g^{\mu\nu}P^\lambda \\ \left\{J^{\mu\nu}, J^{\lambda\rho}\right\} &= g^{\mu\rho}J^{\nu\lambda} - g^{\mu\lambda}J^{\nu\rho} + g^{\nu\lambda}J^{\mu\rho} - g^{\nu\rho}J^{\mu\lambda} \end{aligned} \tag{22}$$

Besides these relations, no specific canonical structure is assumed. The PB of any observable which is constructed from P^μ and $J^{\mu\nu}$ are easily computed from the fundamental PB (22), and the PB of other observables with P^μ and $J^{\mu\nu}$ are deduced from their transformation properties.

Since P^μ determines the orientation of the CM reference frame in space-time, we define internal observables by being :

(a) Invariant under uniform translations, $\{\mathcal{A}, P^\mu\} = 0$;
(b) If not scalars, all their components are confined to the 3D hyperplane perpendicular to P^μ.

The spatial internal angular momentum tensor $\ell^{\mu\nu}$, defined from P^μ and $J^{\mu\nu}$ via $\ell^{\mu\nu} \equiv \Delta^\mu_\lambda\Delta^\nu_\rho J^{\lambda\rho}$ with

$$\Delta^\mu_\nu \equiv \delta^\mu_\nu + \frac{P^\mu P_\nu}{M^2}\,, \tag{23}$$

is the generator of internal rotations in the CM reference frame. Dual to $\ell^{\mu\nu}$ is the vector (proportional to the Pauli-Lubanski vector) $\ell_\mu \equiv \varepsilon_{\mu\nu\lambda\rho}J^{\nu\lambda}P^\rho/(2M)$ with $\varepsilon_{1230} = +1$. The relativistic version of the LRL property is then
Let K^μ be an internal vector such that:

(i) $\{K^\mu, \ell \cdot K\} = 0$
(ii) $K^2 = F(M, \ell^2, \mathcal{A}) \qquad (\{K^\mu, \mathcal{A}\} = 0)$

Then the self-PB of K^μ satisfy

$$\{K^\mu, K^\nu\} = -\frac{\partial (K^2)}{\partial (\ell^2)}\ell^{\mu\nu} \tag{24}$$

where $\ell^2 = \ell_\mu \ell^\mu = \frac{1}{2}\ell_{\mu\nu}\ell^{\mu\nu}$.

5. Relativistic centre-of-mass integration

The preceding results brought together suggest the following picture for fully relativistic systems. With constant total linear momentum, the space-time trajectory of the CM coordinate is expected to be a straight line in the direction of P^μ, and it may always be written as the centroid

$$X^\mu = X_o^\mu + \frac{P^\mu}{M}\tau \tag{25}$$

where τ is the CM proper time and X_o^μ is a constant four-vector, identified as the spatial CM coordinate. Appropriately fixing the zero of τ, X_o^μ may be assumed orthogonal to P^μ without loss of generality, $X_o \cdot P = 0$.

Once the CM coordinate [Eq.(25)] is assumed to be known, the total angular momentum $J^{\mu\nu}$ may always be split into combination of orbital (CM) and internal parts,

$$J^{\mu\nu} = X^\mu P^\nu - X^\nu P^\mu + j^{\mu\nu} = X_o^\mu P^\nu - X_o^\nu P^\mu + j^{\mu\nu} \tag{26}$$

so that $j^{\mu\nu}$ is the (constant) internal angular momentum relative to the centre-of-mass. Out of the six components of $j^{\mu\nu}$ three are independent of the CM coordinate, fully determined by $J^{\mu\nu}$ and P^μ alone, constituting the spatial internal angular momentum tensor $\ell^{\mu\nu}$. The remaining three components of $j^{\mu\nu}$ determine X_o^μ. It is convenient to define the vector

$$Q^\mu \equiv \frac{j^{\mu\nu}P_\nu}{M^2} \tag{27}$$

which incorporates these components. If X_o^μ is required to be formed of P^μ and $J^{\mu\nu}$ alone in a frame-independent manner then the unique possibility for it is the so-called centre-of-inertia [13, 14],

$$X_{\rm I}^\mu \equiv -\frac{J^{\mu\nu}P_\nu}{M^2}, \tag{28}$$

which in the CM ref. frame is given by Eq.(12), and for which $Q^\mu = 0$. However, allowing a more general definition, X_o^μ is uniquely defined, inverting Eq.(26), in terms of P^μ, $j^{\mu\nu}$ and Q^μ,

$$X_o^\mu = -\frac{(J^{\mu\nu} - j^{\mu\nu})P_\nu}{M^2} = X_{\rm I}^\mu + Q^\mu \tag{29}$$

Q^μ, certainly an internal vector, is thus regarded as the *shift vector*, shifting X_o^μ from the centre-of-inertia (28) to the value given by Eq.(29).

In principle, the single particle trajectories may be parameterized each by a different timelike parameter, but for a common evolution picture a common parameter is required. Let σ be such a common evolution parameter. Then the single particle trajectories are $x^\mu = x_a^\mu(\sigma)$, with the particles' generalized velocities $\dot{x}_a^\mu \equiv dx_a^\mu/d\sigma$. Since the CM coordinate X^μ (25) is assumed to evolve in 4D space-time in the direction of P^μ, its projection $X_o^\mu = \Delta_\nu^\mu X^\nu$ onto the 3D CM-hyperplanes (perpendicular to P^μ, using the projection tensor (23)) should be constant, $\Delta_\nu^\mu dX^\nu/d\sigma = 0$. Introducing the Lorentz-covariant generalization of the Newtonian CM (2), the four-vector

$$X_{\mathrm{N}}^\mu \equiv \frac{\sum_a m_a x_a^\mu}{M_o}, \tag{30}$$

then its 3D-projected derivative,

$$\Delta_\nu^\mu \frac{dX_{\mathrm{N}}^\nu}{d\sigma} = \frac{\sum_a m_a \Delta_\nu^\mu \dot{x}_a^\nu}{M_o} = \frac{\sum_a m_a v_a^\mu}{M_o} \tag{31}$$

(with the spatial velocities $v_a^\mu \equiv \Delta_\nu^\mu \dot{x}_a^\nu$) vanishes in the nonrelativistic limit ($v_a/c \to 0$). Therefore, the time-varying part of $\Delta_\nu^\mu dX_{\mathrm{N}}^\nu/d\sigma$ is purely relativistic, and the difference $R^\mu = \Delta_\nu^\mu X_N^\nu - X_o^\mu$ is an internal vector which vanishes in the nonrelativistic limit. Hence, in order to determine X_o^μ, we look for the internal four-vector R^μ which satisfies

$$\frac{dR^\mu}{d\sigma} = \Delta_\nu^\mu \frac{dX_{\mathrm{N}}^\nu}{d\sigma} = \frac{\sum_a m_a v_a^\mu}{M_o} \tag{32}$$

and vanishes in the nonrelativistic limit. Eq.(32) is the exact relativistic analog of Eq.(1) and the full relativistic extension of Eq.(7). With its solution, the CM constant X_o^μ is given by

$$X_o^\mu = \Delta_\nu^\mu X_{\mathrm{N}}^\nu - R^\mu \tag{33}$$

In the standard approach the CM constant X_o^μ is given by the centre-of-inertia X_{I}^μ (28), suggesting an immediate solution of Eq.(32) in the form

$$R_1^\mu = \Delta_\nu^\mu X_{\mathrm{N}}^\nu - X_{\mathrm{I}}^\mu = \Delta_\nu^\mu X_{\mathrm{N}}^\nu + \frac{J^{\mu\nu}P_\nu}{M^2} \tag{34}$$

This is the relativistic extension of the post-Newtonian relation Eq.(15), and in the following is referred to as the trivial solution. But direct integration of Eq.(32) should yield also a different result, in analogy with Eq.(9*b*), which is denoted henceforth R_2^μ and referred to as the non-trivial solution, defining a different CM constant X_o^μ given by

$$X_{o,2}^\mu = \Delta_\nu^\mu X_{\mathrm{N}}^\nu - R_2^\mu \tag{35}$$

Eliminating $\Delta_\nu^\mu X_{\mathrm{N}}^\nu$ between the last two equations yields

$$X_{o,2}^\mu = R_1^\mu - R_2^\mu + X_{\mathrm{I}}^\mu \tag{36}$$

so comparison with Eq.(29) implies

$$Q^\mu = R_1^\mu - R_2^\mu \tag{37}$$

The process of solving Eq.(32) for the nontrivial solution R_2^μ, identifying the shift vector (37), and constructing consequently the spatial CM component X_o^μ via Eq.(36), is referred to as *integration of the relativistic centre-of-mass* [10].

In the post-Newtonian case discussed above (Eq.(16)) the shift vector is

$$\vec{Q} \approx \frac{m_1 - m_2}{2m_1 m_2 M_o c^2}\vec{K} \tag{38}$$

with $\vec{X}_{o,2} = \vec{C}_2$. In the general case, the existence of nontrivial solutions R_2^μ implies the existence of corresponding nonzero vectors Q^μ, which are in fact LRL vectors. A couple of simply solvable relativistic 2-body systems [8, 10] which serve here as a model, indicate the shift vector in the form

$$Q^\mu = \frac{2\,(m_1 - m_2)}{M_o\left[M^2 - (m_1 - m_2)^2\right]c^2} K^\mu \tag{39}$$

with K^μ the (fully relativistic) LRL vector with the self PB

$$\{K^\mu, K^\nu\} = -\frac{\left(M^2 - {M_o}^2\right)\left[M^2 - (m_1 - m_2)^2\right]c^2}{4M^2}\ell^{\mu\nu} \tag{40}$$

It is noted that Eqs.(39) and (40) are the same, for all the systems that were studied, regardless of the interaction. From Eqs.(39) and (40) then follows

$$\{Q^\mu, Q^\nu\} = -\frac{\left(M^2 - {M_o}^2\right)(m_1 - m_2)^2}{M^2 {M_o}^2\left[M^2 - (m_1 - m_2)^2\right]c^2}\ell^{\mu\nu} \tag{41}$$

The last relation bears consequence regarding the long-standing issue of the canonicity of the relativistic CM coordinate : Using the fundamental PB (22) it may be shown that the self-PB of X^μ satisfy

$$\Delta^\mu_\lambda \Delta^\nu_\rho \left\{X^\lambda, X^\rho\right\} = \{Q^\mu, Q^\nu\} + \frac{\ell^{\mu\nu}}{M^2c^2} \tag{42}$$

Combined with (41), it follows that $\{X^\mu, X^\nu\} \neq 0$, implying that X^μ cannot be canonical.

6. Applications

The scheme of integration of the relativistic CM has been already applied successfully to a number of models and systems, such as :

- Post-Newtonian extension of 2-body systems with arbitrary spherically-symmetric Newtonian potential [9]
- A fully relativistic two-body system without interactions) [8]
- A fully relativistic two-body system with special light-like, time-anti-symmetric, scalar-vector(EM-like) interaction (interaction retarded for one particle and advanced for the other) [10]

An interesting possible application stems from the fact that the CM integration scheme is not limited to 2-body systems, and, at least in principle, could be applied to systems with arbitrary number of particles or components. Therefore, it has the potential of defining a LRL-like vector which is an integral of motion for many-body systems. This has been done for post-Newtonian gravitational N-body systems with application to celestial mechanics [16], and for fully relativistic systems of many non-interacting bodies [10].

7. Extended Noether theorem

Finally, let us point out interesting associations that the foregoing results have with the extended Noether theorem regarding its application to the Lorentz boost. Consider a many-body system with Lagrangian $\mathcal{L}\left(\vec{x}_a, \vec{v}_a, t\right)$. Assume variation of the coordinates $\vec{x}_a \to \vec{x}_a + \delta\vec{x}_a$ with $\delta\vec{x}_a = \varepsilon\vec{\xi}_a\left(\vec{x}_b, \vec{v}_b, t\right)$, $\delta\vec{v}_a = \varepsilon\dot{\vec{\xi}}_a\left(\vec{x}_b, \vec{v}_b, t\right)$. Using Lagrange's equations of motion, the corresponding variation of the Lagrangian may always be put in the form $\delta\mathcal{L} = \varepsilon d\left(\sum_a \vec{p}_a \cdot \vec{\xi}_a\right)/dt$. Then, if $\delta\mathcal{L} = 0$, the classical Noether theorem states that $F = \sum_a \vec{p}_a \cdot \vec{\xi}_a$ is conserved. If, however, $\delta\mathcal{L} \neq 0$ but there exists a function $\phi(\vec{x}_a, \vec{v}_a, t)$ so that it is possible to write $\delta\mathcal{L}$ as $\delta\mathcal{L} = \varepsilon d\phi(\vec{x}_a, \vec{v}_a, t)/dt$ while $\sum_a \vec{p}_a \cdot \vec{\xi}_a \neq \phi(\vec{x}_a, \vec{v}_a, t)$ then the *extended Noether theorem* states that $F = \sum_a \vec{p}_a \cdot \vec{\xi}_a - \phi(\vec{x}_a, \vec{v}_a, t)$ *is a constant of the motion* [17, 18].

Particular examples of Newtonian symmetries that require the extended theorem are [17] the Galilei transformation and the LRL symmetry in Kepler-Coulomb systems. Similarly, the symmetry associated with the PN Lorentz transformation, and the corresponding conservation of the Lorentz boost, cannot be dealt with under the terms of the classical Noether theorem, and require the extended theorem. The phenomenon that is then revealed, much similar to the integration of the relativistic CM, is that the integration that leads to the function $\phi(\vec{x}_a, \vec{v}_a, t)$ involves the dynamical equations of motion and is not unique; and again, the difference between the integrations involves the classical LRL vector.

Thus we consider an infinitesimal Lorentz transformation from a simultaneity hyperplane in the CM reference frame to a simultaneity hyperplane in another inertial frame, moving with velocity $\delta\vec{V} = \varepsilon\vec{a}$ relative to the former. The corresponding variation vector is

$$\vec{\xi}_a(\vec{x}_a, \vec{v}_a, t) = \frac{1}{c^2}(\vec{x}_a \cdot \vec{a})\vec{v}_a - \vec{a}t \tag{43}$$

Then, while

$$\sum_a \vec{p}_a \cdot \vec{\xi}_a = \frac{\vec{p}^2}{\mu c^2}\left[\left(\vec{X}_{\rm N} \cdot \vec{a}\right) + \frac{m_2 - m_1}{M_o}(\vec{r} \cdot \vec{a})\right], \tag{44}$$

the corresponding variation of the 2-body PN Lagrangian [11] may be written as a derivative $\delta\mathcal{L} = \varepsilon d\phi/dt$ in two different modes, with either

$$\phi_1 = -\left(M_o - \frac{\vec{p}^2}{2\mu c^2} + \frac{\kappa}{rc^2}\right)\left(\vec{X}_{\rm N} \cdot \vec{a}\right) + \frac{m_2 - m_1}{2\mu M_o c^2}\left(\vec{p}^2 - \frac{\mu\kappa}{r}\right)(\vec{r} \cdot \vec{a}) \tag{45a}$$

or

$$\phi_2 = -\left(M_o - \frac{\vec{p}^2}{2\mu c^2} + \frac{\kappa}{rc^2}\right)\left(\vec{X}_{\rm N} \cdot \vec{a}\right) + \frac{m_2 - m_1}{2\mu M_o c^2}\left[(\vec{r} \cdot \vec{p})(\vec{p} \cdot \vec{a}) - \frac{2\mu\kappa}{r}(\vec{r} \cdot \vec{a})\right], \tag{45b}$$

thus defining, as $F_i = \sum_a \vec{p}_a \cdot \vec{\xi}_a - \phi_i$, two constants of motion,

$$F_1 = M\left(\vec{X}_{\rm N} \cdot \vec{a}\right) + \frac{m_2 - m_1}{2\mu M_o c^2}\left(p^2 + \frac{\mu\kappa}{r}\right)(\vec{r} \cdot \vec{a}) = \vec{N} \cdot \vec{a}, \tag{46a}$$

$$\begin{aligned} F_2 &= M\left(\vec{X}_{\rm N} \cdot \vec{a}\right) + \frac{m_2 - m_1}{2\mu M_o c^2}\left[-(\vec{r} \cdot \vec{p})(\vec{p} \cdot \vec{a}) + 2\left(\vec{p}^2 + \frac{\mu\kappa}{r}\right)(\vec{r} \cdot \vec{a})\right] = \\ &= \left(\vec{N} + \frac{m_2 - m_1}{2m_1 m_2 c^2}\vec{K}\right) \cdot \vec{a} \end{aligned} \tag{46b}$$

The constant F_1 is clearly related to the integration constant in Eq.(15), via $F_1 = M\vec{C}_1 \cdot \vec{a}$. The constant F_2 would be expected to be associated, in a similar way, to the integration constant in Eq,(16). Here, however, we encounter a bit of surprise, because while the expressions are similar, the coefficient of $\vec{K}$ in Eq,(46*b*) has the opposite sign than that in Eq.(16). The reason for this ambiguity is not completely clear, and the relationship between the integration of the relativistic CM and the extended Noether theorem will be further studied in a separate publication.

8. Conclusion

In this lecture we have demonstrated and discussed the relativistic origin and application of the LRL symmetry. The existence of LRL symmetry as an internal symmetry is implied, via the Lorentz boost, from the global Lorentz-Poincare' symmetry, in a way that is independent of internal interactions.

The LRL symmetry is therefore an integral part of the inner symmetry of relativistic systems. The internal angular momentum $j^{\mu\nu}$, defined in Eq.(26), may be expressed in terms of $\ell^{\mu\nu}$ and the shift vector Q^{μ} as

$$j^{\mu\nu} = \ell^{\mu\nu} - Q^{\mu}P^{\nu} + Q^{\nu}P^{\mu} \tag{47}$$

We have seen that Q^{μ} is proportional to the LRL vector, which in itself is responsible for changing and determining the internal configuration. Thus, while $\ell^{\mu\nu}$ is responsible for spatial internal rotations, the other part of $j^{\mu\nu}$, $Q^{\nu}P^{\mu} - Q^{\mu}P^{\nu}$, is responsible for determination of the internal configurations or state of motion. Since the shift vector Q^{μ} appears in the mixed space-time components of $j^{\mu\nu}$, it may be regarded as the internal moment corresponding to the Lorentz boost, while $\ell^{\mu\nu}$ contains the internal moments corresponding to spatial rotations (these are regarded as internal moments, being computed relative to the CM frame, as opposed to global moments, computed relative to arbitrary inertial reference frames). These relations are summed up in the following table :

	rotations	+	**change of configuration**
Global space-time symmetry :	global rotations	+	Lorentz transformations
	↕		↕
Internal symmetry :	internal rotations	+	LRL

[1] Goldstein H 1975 Prehistory of the "Runge-Lenz" vector *Am. J. Phys.* **43**, 737-8.
Goldstein H 1976 More on the prehistory of the Laplace or Runge-Lenz vector *Am. J. Phys.* **44**, 1123-4.
[2] Goldstein H 1980 *Classical Mechanics* (Reading, MA : Addison-Wesley)
[3] McIntosh H V 1971 Symmetry and degeneracy *Group Theory and its Applications* Vol. 2, ed E M Loebl (New York : Academic Press) pp 75–144
[4] Khachidze T T and Khelashvili A A 2008 *Dynamical Symmetry of the Kepler-Coulomb Problem in Classical and Quantum Mechanics* (New-York : Nova Science)
[5] Bacry H, Ruegg H and Souriau J 1966 Dynamical groups and spherical potentials in classical mechanics *Comm. Math Phys.* **3** 323-33
[6] Fradkin D M 1967 Existence of the dynamical symmetries O_4 and SU_3 for all classical central potential problems *Prog. Theor. Phys.* **37** 798-812
[7] Mukunda N 1967 Dynamical Symmetries and Classical Mechanics *Phys. Rev.* **155** 1383-6.
[8] Ben-Ya'acov U 2009 Localization of the relativistic centre-of-mass from internal dynamics *J. Phys. A: Math. Theor.* **42** (2009) 225401
[9] Ben-Ya'acov U 2010 Laplace-Runge-Lenz symmetry in general rotationally symmetric systems *J. Math. Phys.* **51** 122902
[10] Ben-Ya'acov U 2011 Center-of-mass and internal symmetries in classical relativistic systems *J. Math. Phys.* **52** 022903
[11] Landau L D and Lifshitz E M 1975 *The Classical Theory of Fields* (Oxford : Pergamon)
[12] Dahl J P 1968 Physical interpretation of the Runge-Lenz vector *Phys. Lett.* **27A** 62-3
Dahl J P 1997 Physical origin of the Runge-Lenz vector *J. Phys. A: Math. Gen.* **30** 6831-6
[13] Fleming G N 1965 Covariant position operators, spin and locality *Phys. Rev.* **137B** 188-97
[14] Alba D Lusanna L and Pauri M 2002 Centers of mass and rotational kinematics for the relativistic N-body problem in the rest-frame instant form *J. Math. Phys.* **43** 1677-727
[15] Marsden J E and Ratiu T S 1999 *Introduction to Mechanics and Symmetry* (New-York : Springer)

[16] Ben-Ya'acov U 2009 Laplace-Runge-Lenz-like new constant in many-body systems from post-Newtonian dynamics *J. Phys. A: Math. Theor.* **42** 375210

[17] Lévy-Leblond JM 1971 Conservation laws for gauge-variant Lagrangians in classical mechanics *Am. J. Phys.* **39** 502-6

[18] Sarlet W and Cantrijn F 1981 Generalization of Noether's theorem in classical mechanics *SIAM Review* **23** 467-94

Extending the Relativity of Time

Yaakov Friedman
Jerusalem College of Technology
Departments of Mathematics and Physics
P.O.B. 16031 Jerusalem 9116001, Israel

E-mail: friedman@jct.ac.il

Abstract. More than 100 years ago, Einstein's special relativity demonstrated that time is a relative notion. The observed rate of a moving clock differs from the rate of a stationary clock. In fact, the observed rate depends on the clock's velocity. All admissible velocities are bounded by the speed of light. These predictions of special relativity have been verified experimentally in several different ways. It is natural to ask whether acceleration also influences the observed rate of a moving clock in addition to the influence due to its velocity. Today, there are several existing experimental techniques to test whether acceleration influences the observed rate of a clock. We introduce here an extension of special relativity, which we call *extended relativity (ER)*, by assuming that acceleration effects the observed rate of a clock. We derive transformations between uniformly accelerated systems in ER. We show that ER predicts that there is a maximal acceleration. We obtain relativistic dynamics in ER. We show that Kundig's 1963 experiment indicates that acceleration does influence the rate of a clock, supporting the ER model and providing an estimate for the maximal acceleration. We will present an upcoming experiment which is designed to test whether acceleration influences the rate of a clock, and to determine the value of the maximal acceleration. A map for physics under ER will be presented. We will show how ER handles black-body radiation and some quantum properties of a Hydrogen-like atom.

1. Introduction

About 400 years ago, Galileo Galilei introduced the *Principle of Relativity*, which states that the laws of physics are the same in any inertial system. Let $(t, \mathbf{x})$ denote the space-time coordinates of an event A in an inertial frame K. It was clear in Galileo's day that the spatial coordinate $\mathbf{x}'$ of A in a second inertial frame K' will depend on t, $\mathbf{x}$ and the relative velocity $\mathbf{v}$ between the K and K'. But, what about the time component t' of the event A in K'? At the time of Galileo, there was no evidence to suggest that $t' \neq t$. Moreover, there was no mathematical model in Galileo's day to handle such an assumption. It was therefore assumed that $t' = t$. The ensuing space-time transformations are called *Galilean transformations* and are the basis for classical physics.

About 120 years ago, the validity of the Galilean transformations came into question. The results of the Michelson-Morley experiment could not be explained using the velocity addition based on Galilean transformations. It was also shown that the Maxwell equations are not covariant with respect to these transformations, while they *are* covariant under different transformations, called the *Lorentz transformations*. Albert Einstein, based on the principle of relativity and the constancy of speed of light, developed *special relativity* in which the transformation between inertial systems are Lorentz transformations. Under these transformations, the observed rate of a moving clock depends on its velocity. This is called the *time dilation* of a moving clock.

This raises the following question: "Does the observed rate of a moving clock also depend on its **acceleration**?" Currently, it is assumed that acceleration does not influence the observed rate of the clock. This is called the Clock Hypothesis or Clock Postulate. Till now this assumption was natural, as there was no physical evidence for this influence and no model to describe it. See a discussion of this question in article [19] "Does a clock's acceleration affect its timing rate?".

By the equivalence principle, gravitation can be interpreted as acceleration. In Einstein's General Relativity, gravitational fields are well described by the metric tensor on spacetime. Time dilation in general relativity is expressed by this metric. However, the time dilation of a clock due to gravitation cannot be observed, since both standard clocks and processes are affected in the same way. But, the difference in the time dilation between two different points could be observed and is called the gravitational redshift, see [33] pp.79-80. In general relativity the time dilation within an accelerated system with respect to an inertial lab system is assumed to be the same as in the instantaneous comoving inertial system. For uniformly accelerated systems, this is a part of the Hypothesis of Locality introduced by Mashhoon [25, 26] and of the Weak Hypothesis of Locality used in [18]. In these systems, the usual gravitational redshift occurs.

In this paper, we will show that there is a model for a relativity theory in which the observed time of a moving clock is also influenced by the acceleration of the clock. We will show that this theory implies the existence of a *maximal acceleration*. In special relativity, the time dilation can be measured by the transverse Doppler shift. Similarly, in this model, the additional time dilation predicted by this theory could be measured by the additional Doppler shift due to the acceleration of the clock. We will provide evidence supporting this extension of special relativity and also provide a numerical estimate of the value of the maximal acceleration. We call our extension of relativity, in which the observed time depends also on the acceleration, **Extended Relativity** (ER). We will describe feasible experiments to test ER and present a new map for physics under ER. Finally, we will show that some quantum effects can be understood within the framework of ER.

2. Kinematics of accelerated systems in ER

In this section, we will study accelerated systems within the the framework of Special Relativity. To study such accelerated systems, A. Einstein introduced the Clock Hypothesis, which states that the "rate of an accelerated clock is identical to that of the instantaneously comoving inertial clock." Not all physicists agree with this hypothesis. L. Brillouin ([6] p.66) wrote that "we do not know and should not guess what may happen to an accelerated clock." If we assume the validity of the Clock hypothesis, then the space-time transformation between accelerated systems are well known, see [28] and others.

In [14], we presented a systematic approach for transformations between accelerated systems *without* assuming the Clock Hypothesis. Our approach to describing transformations between two uniformly accelerated systems is based on the symmetry which follows from the general principle of relativity.

2.1. proper velocity - time description of events

The first step for describing transformations between two uniformly accelerated systems is to introduce a new *proper velocity - time description* of events. Proper velocity (*p-velocity* in short) is defined as

$$\mathbf{u} = \gamma(v)\mathbf{v} = \frac{d\mathbf{r}}{d\tau},$$

where $\mathbf{v} = d\mathbf{r}/dt$, $\gamma(v) = \frac{1}{\sqrt{1-v^2/c^2}}$ and τ is the proper time of the moving object. The proper velocity is also the canonically conjugate variable to the position in the relativistic phase-space.

The relativistic *acceleration* $\mathbf{g}$, which appears in the relativistic dynamic equation, is defined (see [32] p.71) to be the derivative of p-velocity with respect to time t:

$$\mathbf{g} = \frac{d\mathbf{u}}{dt}\,. \tag{1}$$

This acceleration coincides [18] with the acceleration in the comoving frame. Note that if an object moves with this constant acceleration, then its p-velocity satisfies the equation

$$\frac{d^2\mathbf{u}}{dt^2} = 0\,. \tag{2}$$

We will say that an object is *uniformly accelerated* if its acceleration is constant, or equivalently, satisfies (2). If the velocity of a uniformly accelerated object is parallel to the acceleration, then it moves with the well-known *hyperbolic motion* (see [28], [32] and [9]).

In the *p-velocity-time description*, an event is described by the time at which the event occurred and the p-velocity $\mathbf{u} \in R^3$ of the event. The evolution of an object in a system can be described by the p-velocity $\mathbf{u}(t)$ of the object at time t. The line $(t, \mathbf{u}(t))$ replaces the world-line of special relativity in this description. To obtain the position of the object at time t, we have to know the initial position of the object and then integrate its ordinary velocity (which is readily computed from the p-velocity) with respect to time. The following Table 1 shows the parallels between the space-time transformations for inertial systems and the p-velocity-time transformations between uniformly accelerated systems.

Table 1. Parallelism between the space-time transformations for inertial systems and the p-velocity-time transformations between uniformly accelerated systems.

Systems	Relative motion	Relative motion eqn.	Event descr.
inertial	uniform velocity	$\frac{d^2\mathbf{r}}{dt^2} = 0$	$(t, \mathbf{r})$
accelerated	constant acceleration	$\frac{d^2\mathbf{u}}{dt^2} = 0$	$(t, \mathbf{u})$

To obtain the Lorentz transformations in special relativity, it is essential that the relative position of the origins of the frames connected with the two inertial systems depends linearly on time. This linear map expresses the relative velocity between the systems. For two uniformly accelerated systems, if we assume that the systems are comoving, meaning having zero relative velocity at time $t = 0$, then the uniform acceleration between the systems, defined by (1), is a linear map from time to p-velocities.

Let T denote the transformation mapping the time and p-velocity (t, u) of an event in a uniformly accelerated system K_g to the time and p-velocity (t', u') of the same event measured in the uniformly accelerated system K_0. The situation is analogous to that of the space-time transformations between two inertial systems. In that case, the relative motion of one system with respect to the other is described by a uniform velocity, which is a linear map from time to space (or a line in the space-time continuum). For uniformly accelerated systems, the relative motion of one system with respect to the other is described by a uniform *acceleration*, which is a linear map from time to p-velocities (or a line in the p-velocity-time continuum). Since the space-time transformation between two inertial systems is linear, we will assume that the p-velocity-time transformation T between two uniformly accelerated systems is also *linear*. This assumption can be also justified by use of relativistic dynamics, as follows. Uniformly accelerated motion in an inertial system is described by a straight line in the p-velocity-time continuum and corresponds to a relativistic motion under a constant force. Any straight line in the p-velocity-time continuum in a uniformly accelerated system corresponds to relativistic motion under a constant force in this system and in any other uniformly accelerated system. Hence, the transformation T maps lines to lines and is, therefore, linear.

2.2. General proper velocity - time transformations between accelerated systems

To define the symmetry operator between two uniformly accelerated systems, we will use an extension of the principle of relativity, which we will call the *General Principle of Relativity.* This principle, as formulated by M. Born (see [4], p. 312), states that the "laws of physics involve only relative positions and motions of bodies. From this it follows that no system of reference may be favored *a priori* as the inertial systems were favored in special relativity." The principle of relativity from special relativity states that there is no preferred *inertial* system, and, therefore, the notion of rest (zero velocity) is a relative notion. From the general principle of relativity, it follows that there is no preference for inertial (zero acceleration) systems. Hence, when considering accelerated systems, we no longer give preference to free motion (zero force) over constant force motion. This makes all uniformly accelerated systems equivalent.

From the general principle of relativity, it is logical to assume that *the transformations between the descriptions of an event in two uniformly accelerated systems depend only on the relative motion between these systems.* Consider now two uniformly accelerated systems K_g and K_0, with a constant acceleration g between them. We choose reference frames in such a way that the description of relative motion of K_g with respect to K_0 coincides with the description of relative motion of K_0 with respect to K_g. The above principle implies that the transformation T mapping the description of an event in system K_g to the description of the same event in system K_0 will coincide with the transformation $\widetilde{T} = T^{-1}$ from system K_0 to K_g. This implies that T is a symmetry, or $T^2 = Id$.

The following derivation of the explicit form of the transformations between K_0 and K_g follows [10], where it is done for inertial systems. Similar derivations were obtained by several authors. The choice of the reference frames is as follows. We choose the origins O of K_g and O' of K_0 of the p-velocity axes to be the same at $t = 0$. We also synchronize the clocks positioned at the origins of the frames at time $t = 0$. We chose the direction of the first p-velocity axis in each system in such a way that the relative acceleration of the second system will be opposite to the direction of this axis. The corresponding axis will thus be reversed, as in Figure 1. The other two axes are chosen to be parallel.

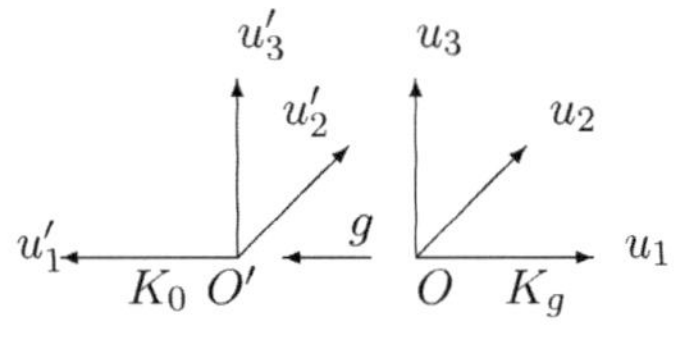

Figure 1. Symmetric uniformly accelerated systems

Note that with this choice of the axes, the acceleration g of O' in K_g is equal to the acceleration of O in K_0, and, thus, the p-velocity-time transformation problem is fully symmetric with respect to K_g and K_0. We will denote this transformation by S_g, since it is a symmetry and depends only on the acceleration **g** between the systems. By our choice of the p-velocity axes, the systems are symmetric with respect to the second and third coordinates. We may thus assume $u'_2 = u_2$

and $u_3' = u_3$. For the time and the first p-velocity coordinates, S_g has the form

$$\begin{pmatrix} t' \\ u_1' \end{pmatrix} = S_g \begin{pmatrix} t \\ u_1 \end{pmatrix} = \begin{pmatrix} S_{00} & S_{01} \\ S_{10} & S_{11} \end{pmatrix} \begin{pmatrix} t \\ u_1 \end{pmatrix}.$$

The world-line of the origin O in K' is $(S_{00}t, S_{10}t)$. From (1), the relative acceleration of system K with respect to K' is $g = S_{10}/S_{00}$. If we denote $\tilde{\gamma} = S_{00}$, we get

$$S_g = \tilde{\gamma} \begin{pmatrix} 1 & \kappa \\ g & \alpha \end{pmatrix}$$

for some constants α, κ. Now, from the symmetry $S_g^2 = I$, we have $\alpha = -1$ and $\tilde{\gamma} = \frac{1}{\sqrt{1+\kappa g}}$. Thus,

$$S_g = \frac{1}{\sqrt{1+\kappa g}} \begin{pmatrix} 1 & \kappa \\ g & -1 \end{pmatrix}.$$

Now we have two choices:

- If the observed time does not depend on the acceleration, then $\tilde{\gamma} = 1$ and $\kappa = 0$. This is true if the clock hypothesis is valid. In this case, the p-velocity-time transformations are Galilean.
- If the observed time depends on the acceleration, then $\tilde{\gamma} \neq 1$ and $\kappa \neq 0$. This is the assumption of extended relativity (ER).

To define the p-velocity-time transformations between accelerated systems under ER, it remains only to define the value of κ. To do this, we introduce a metric $\text{diag}(\mu^2, -1, -1, -1)$ on p-velocity-time $(t, \mathbf{u})$. We are looking for a μ which will make the symmetry S_g *self-adjoint.* In order for S_g to be self-adjoint, we must have

$$\begin{pmatrix} 0 \\ 1 \end{pmatrix} \cdot S_g \begin{pmatrix} 1 \\ 0 \end{pmatrix} = \begin{pmatrix} 1 \\ 0 \end{pmatrix} \cdot S_g \begin{pmatrix} 0 \\ 1 \end{pmatrix},$$

where $\cdot$ is the product corresponding to our metric. This identity yields $-g = \mu^2\kappa$. A satisfactory μ exists if $\kappa < 0$.

With the above choice of μ, the self-adjoint symmetry S_g becomes an *isometry.* This means that the interval $d\tilde{s}^2 = \mu^2 dt^2 - |du|^2$ is conserved under the transformation S_g. Consider a uniformly accelerated particle whose acceleration a has magnitude $|a| = \mu$ in system K. The interval $d\tilde{s}^2 = 0$ in system K, and thus the interval is also zero in system K'. This implies that there is a unique acceleration magnitude, μ which is conserved between K and K'. From the generalized principle of relativity, this unique acceleration magnitude, which we will denote $\mu(g)$, can depend only on the magnitude of the relative acceleration g between the accelerated systems.

The following argument shows that the unique acceleration magnitude $\mu(g)$ is a universal constant. Consider three uniformly accelerated systems K, K' and K''. Assume that the relative acceleration of K' with respect to K is $\mathbf{g}$ and that the relative acceleration of K'' with respect to K' is also $\mathbf{g}$. From the above argument, an acceleration with magnitude $\mu(g)$ in K will have the same magnitude in K'. Similarly, an acceleration with magnitude $\mu(g)$ in K' will have the same magnitude in K''. Hence, an acceleration with magnitude $\mu(g)$ in K will have the same magnitude in K''. But, the relative acceleration between K and K'' is $2\mathbf{g}$. Thus, $\mu(g) = \mu(2g)$. Repeating this argument, we see that this unique acceleration magnitude is independent of the relative acceleration between the systems. Thus, we have shown that in ER (if the observed time also depends on the acceleration), there is a unique universal acceleration magnitude such that any uniform acceleration with this magnitude in one uniformly accelerated system, will have the same magnitude in any other uniformly accelerated system. We will call it the **maximal acceleration** and denote its magnitude by a_m.

From the above arguments, $a_m = \mu$, which implies that $\kappa = -g/a_m^2$. Thus, the proper velocity-time transformation between two uniformly accelerated systems (for parallel axes, by replacing our previous u_1 with $-u_1$) in ER is

$$\begin{aligned} t' &= \tilde{\gamma}(t + gu_1/a_m^2) \\ u_1' &= \tilde{\gamma}(gt + u_1) \\ u_2' &= u_2 \\ u_2' &= u_2, \end{aligned} \tag{3}$$

with

$$\tilde{\gamma} = \frac{1}{\sqrt{1 - g^2/a_m^2}}. \tag{4}$$

This is a Lorentz-type transformation.

The existence of a maximal acceleration for massive objects has already been predicted by Caianiello, based on the time-energy uncertainty relation in quantum mechanics (see Caianiello [7], Papini and Wood [30] and Papini *et al.* [29] and references therein). Note that the acceleration used by Caianiello is the proper acceleration, defined as $\mathbf{g} = d\mathbf{u}/d\tau$ and not $\mathbf{g} = d\mathbf{u}/dt$, as defined in (1). In ER, the proper acceleration is unbounded, just as the proper velocity is unbounded. The estimate of the Caianiello maximal acceleration in Scarpetta [31] is $a_m = 5 \cdot 10^{50} g$, which is too large to have an effect in real physical processes. Boundedness of the proper acceleration excludes black-holes, since on the surface of a black-hole the proper acceleration has no bound. But in ER black-holes may exist, see end of next section for the meaning of the horizon in Schwarzschild universe.

The clock hypothesis has been tested and was found to be valid to great accuracy. See, for example, the experiment of Bailey J. *et al* [1] for measurements of the time dilation for muons. This paper claims that the experiment supports the validity of the clock hypothesis for accelerations $10^{18} g$. But the acceleration mentioned there is the proper acceleration and the magnitude of our acceleration in this experiment is significantly smaller then the claimed value, So, this experiment does not contradict the estimate that we will obtain later. For a long time, B. Mashhoon argued against the Clock Hypothesis and developed nonlocal transformations for accelerated observers (see the review article [24] and references therein). Our approach treats the problem differently.

3. Relativistic Dynamics in ER

Relativistic Dynamics in ER is a dynamics extending classical dynamics and preserving two limitations: the velocity $\mathbf{v} = d\mathbf{r}/dt$ is bounded by c, and the acceleration $\mathbf{g} = d\mathbf{u}/dt$ is bounded by a_m.

To derive the dynamics equations of ER, we will follow [13] and ideas which helped to change the first order dynamic system in classical mechanics to a corresponding system in special relativity. A first-order dynamic system in classical mechanics for the motion of an object of mass m under a force $\mathbf{F}$ can be written as

$$\left\{ \begin{array}{l} \frac{d\mathbf{x}}{dt} = \mathbf{v} \\ \frac{d\mathbf{v}}{dt} = \frac{\mathbf{F}}{m} \end{array} \right\}. \tag{5}$$

Einstein's relativistic dynamics equation in special relativity is $\frac{dp}{dt} = F$, or $m_0 \frac{d\gamma(v)v}{dt} = F$, where m_0 is the rest-mass of the object [8]. Thus, in special relativity the corresponding dynamic system is

$$\left\{ \begin{array}{l} \gamma\frac{d\mathbf{x}}{dt} = \mathbf{u} \\ \frac{d\mathbf{u}}{dt} = \frac{\mathbf{F}}{m_0} \end{array} \right\}, \tag{6}$$

where $\mathbf{u}$ is the p-velocity.

In special relativity, the time dilation (due to the velocity) factor γ enters in the first equation. Since $\gamma(\mathbf{v}(\mathbf{u})) = \sqrt{1+u^2/c^2}$, we can rewrite this system as

$$\left\{ \begin{array}{l} \frac{d\mathbf{x}}{dt} = \frac{\mathbf{u}}{\sqrt{1+u^2/c^2}} \\ \frac{d\mathbf{u}}{dt} = \frac{\mathbf{F}}{m_0} \end{array} \right\}.$$

Note that the first equation ensures that the magnitude of the velocity $\frac{d\mathbf{x}}{dt}$ during the evolution does not exceed c, as required in special relativity.

For the corresponding system in ER, in order to preserve the limitation of the acceleration $d\mathbf{u}/dt$ by a_m, we introduce the time dilation (due to the acceleration) factor $\tilde{\gamma}$, defined by (4), in the second equation. This leads to

$$\left\{ \begin{array}{l} \gamma\frac{d\mathbf{x}}{dt} = \mathbf{u} \\ \tilde{\gamma}\frac{d\mathbf{u}}{dt} = \frac{\mathbf{F}}{m_0} \end{array} \right\}.$$

As above, we can rewrite this system as

$$\left\{ \begin{array}{l} \frac{d\mathbf{x}}{dt} = \frac{\mathbf{u}}{\sqrt{1+|\mathbf{u}|^2/c^2}} \\ \frac{d\mathbf{u}}{dt} = \frac{\mathbf{F}}{\sqrt{m_0^2+|\mathbf{F}|^2/a_m^2}} \end{array} \right\}. \tag{7}$$

We will call this system the *relativistic dynamics equation in ER.*

The classical Hamiltonian for motion under a conserved force $F(x)$ with potential $V(x)$ is $H_{cl}(x,p) = \frac{p^2}{2m} + V(x)$, where x is the object's position, and $p = mv$ is the object's momentum. The Hamiltonian is the energy of an object, written as a function on the *phase space* (x,p), and remains constant throughout the motion. We will change the phase space to (x,v), where v is the velocity of the object and change the Hamiltonian to *energy per unit mass*:

$$H_{cl}(x,v) = \frac{v^2}{2} + V(x) = \int_0^v u du - \int_0^x a(y)dy\,, \tag{8}$$

where $V(x)$ is now the potential per unit mass, and $a(x) = F(x)/m$ is the acceleration at the point x. Note that the first integral, which represents the kinetic energy, depends on the velocity, while the second integral, which represents the potential energy, depends on a similar way on the acceleration. Under this modification, we can rewrite the system (5) as a Hamiltonian system

$$\left\{ \begin{array}{l} \frac{dx}{dt} = \frac{\partial H_{cl}}{\partial u} \\ \frac{du}{dt} = -\frac{\partial H_{cl}}{\partial x}. \end{array} \right\}. \tag{9}$$

Note that this Hamiltonian system is symmetric in x and u, as required by *Born Reciprocity* [5], which states that the "laws of nature are symmetrical with regard to space and momentum."

The symmetry becomes even more explicit in the case of the *classical harmonic oscillator.* For analysis of the harmonic oscillator, we use an inertial frame in which the attractive center is placed at rest at the origin. In this frame, the acceleration is $a(x) = -\frac{k}{m}x$. The motion of the oscillator is characterized by the *natural frequency* of the oscillator, defined as $\omega = \sqrt{\frac{k}{m}}$.

Substituting this into equation (8), we obtain the Hamiltonian for the classical harmonic oscillator:

$$H_{cl}(x,u) = \frac{u^2}{2} + \frac{(\omega x)^2}{2} = \int_0^u v dv + \int_0^{\omega x} y dy\,. \tag{10}$$

Note that both the kinetic energy and the potential energy are quadratic expressions in the variables u and ωx, respectively. Such expressions are natural in classical mechanics, which uses Euclidean geometry.

In special relativity, we have two candidates for the velocity component of the classical phase space: the usual velocity $v = dx/dt$ and the proper velocity $u = \gamma(v)v$. System (6) suggests using a phase space of (x, u), position and p-velocity. The Lagrangian formulation [20] of relativistic dynamics also implies that the canonical momentum in special relativity is $p = m_0 u$ and, for unit mass, should be u. Relativistic dynamics replaces the first term in (8) by the Einstein formula for the kinetic energy $E_k = mc^2 = m_0 c^2 \gamma$. Thus, using the connection between v and u, the Hamiltonian per unit rest-mass in special relativity is

$$H_{sr}(x,u) = c^2\gamma(v(u)) + V(x) = c^2\sqrt{1+\frac{u^2}{c^2}} + V(x). \tag{11}$$

This Hamiltonian reproduces the dynamics equation (6) and can be derived from the relativistic Lagrangian.

Note that this Hamiltonian breaks the symmetry in x and u. This can be seen explicitly for the harmonic oscillator. The Hamiltonian for the harmonic oscillator in special relativity is

$$H_{sr}(x,u) = c^2\sqrt{1+\frac{u^2}{c^2}} + \frac{\omega^2 x^2}{2}\,. \tag{12}$$

The kinetic energy E_k is now expressed by a hyperbola $\frac{E_k^2}{c^4} - \frac{u^2}{c^2} = 1$, which is natural for special relativity, which uses hyperbolic geometry. The asymptotes of this hyperbola are $E_k = cu$ for $u \to \infty$ and $E_k = -cu$ for $u \to -\infty$. Nevertheless, the second term (the potential energy) is still a parabola, as in the classical case.

The broken symmetry and Born reciprocity is restored by the relativistic dynamics in ER. The relativistic dynamics equation (7) in ER suggests the following Hamiltonian:

$$H_{er}(\mathbf{x},\mathbf{u}) = c^2\sqrt{1+|\mathbf{u}|^2/c^2} + U(\mathbf{x}), \tag{13}$$

with

$$\mathbf{a}(t) = \frac{d\mathbf{u}}{dt} = -\frac{\partial U(\mathbf{x})}{\partial x_j} = \frac{F_j(\mathbf{x})}{\sqrt{1+|\mathbf{F}(\mathbf{x})|^2/(ma_m)^2}}\,. \tag{14}$$

For the harmonic oscillator, the Hamiltonian becomes

$$H_{er}(x,u) = c^2\sqrt{1+\frac{u^2}{c^2}} + \frac{a_m^2}{\omega^2}\sqrt{1+\frac{\omega^4 x^2}{a_m^2}}\,. \tag{15}$$

This Hamiltonian has the same type of Born Reciprocity as in (10) for the classical harmonic oscillator. Note that the minimal energy of an oscillator $E_0 = m_0(c^2 + a_m^2/\omega^2)$ is chosen to be non-zero, in order to have a well-defined behavior at infinity for both kinetic and potential energies.

The relativistic dynamics equation (7) in ER can be rewritten as the usual Hamilton system:

$$\left\{ \begin{array}{l} \frac{d\mathbf{x}}{dt} = \frac{\partial H_{er}}{\partial \mathbf{u}} \\ \frac{d\mathbf{u}}{dt} = -\frac{\partial H_{er}}{\partial \mathbf{x}} \end{array} \right\}. \tag{16}$$

The solutions of equation (7) automatically satisfy the two limitations: the velocity $\mathbf{v} = d\mathbf{r}/dt$ is bounded by c, and the acceleration $\mathbf{g} = d\mathbf{u}/dt$ is bounded by a_m. The classical mechanics dynamics is obtained from this equation when $c \to \infty$ and $a_m \to \infty$, while the special relativity dynamics is obtained from this equation when $a_m \to \infty$.

If the mass of the moving object is positive, then its acceleration magnitude $|\mathbf{a}|$ is strictly less then a_m. On the other hand, any zero-mass particle, like a photon, accelerates with the maximal acceleration a_m. Note that in special relativity, a zero-mass particle has infinite acceleration, if this acceleration is defined properly [18]. This implies that the electromagnetic radiation which is generated by zero-mass particles is described in the p-velocity-time continuum by plane waves $f(\omega t - \tilde{\mathbf{k}}u)$, with $|\tilde{\mathbf{k}}|/\omega = 1/a_m$.

Consider now the freely falling object in a gravitational field outside a spherically symmetric mass M. If we use the classical description of the gravitation, the acceleration $\mathbf{g}$ due to the gravitation will be $\mathbf{g} = GM/r^2$, where r is the distance from the object to the center of the mass M. So, from (7) the dynamics equation for a freely falling object in ER will be

$$\frac{d\mathbf{u}}{dt} = \frac{\mathbf{g}}{\sqrt{1 + g^2/a_m^2}}.$$

In the Schwarzschild universe, the acceleration generated by the gravitation force is $\tilde{\mathbf{g}} = \mathbf{g}/\sqrt{1 - 2gr/c^2}$, see [32] p.230. So, from (7) the ER dynamics equation of a freely falling object in Schwarzschild universe is

$$\frac{d\mathbf{u}}{dt} = \frac{\tilde{\mathbf{g}}}{\sqrt{1 + \tilde{g}^2/a_m^2}} = \frac{g}{\sqrt{|1 - 2gr/c^2| + g^2/a_m^2}}.$$

This means that at the usual horizon $r = 2GM/c^2 = c^2/(2g)$ of this model, the magnitude of the acceleration is a_m, and $\tilde{\gamma}$, defined by (4), is infinite, implying that time stops at this horizon. As mentioned earlier, any massive object cannot reach the maximal acceleration in Minkowski space. But in the Schwarzschild universe each object at the horizon becomes light-like.

The largest accelerations observed in astronomy are the accelerations on neutron stars, where the acceleration is of order $10^{12} m/s^2$. Since the ER correction of the dynamics is of order $g^2/a_m^2 \leq 10^{-14}$ it will be hard to observe this correction in astronomy.

4. Testing Maximal Acceleration

4.1. Doppler type shift for an accelerated source

Extended relativity predicts an additional longitudinal Doppler shift of a source accelerating in the direction of the radiation. The observation of such a shift can be used to test extended relativity and to determine the value of the maximal acceleration.

Consider a source radiating at frequency ω with a "wave vector" $\tilde{\mathbf{k}}$ in the x-direction. We want to determine the frequency ω' of this radiation as observed in a system moving with acceleration $\mathbf{g}$ in the x-direction with respect to the source. By use of the p-velocity-time transformation (3) between these systems, we get

$$f(\omega t - \tilde{k}u) = f\left(\omega\tilde{\gamma}(t' + \frac{gu'}{a_m^2}) - \tilde{k}\tilde{\gamma}(gt' + u')\right) = f(\omega' t' - \tilde{k}' u').$$

Thus,

$$\omega' = \tilde{\gamma}(\omega - \tilde{k}g) = \omega\frac{1 - \frac{|\tilde{k}|}{\omega}g}{\sqrt{1 - \frac{g^2}{a_m^2}}} = \omega\frac{1 - \frac{g}{a_m}}{\sqrt{1 - \frac{g^2}{a_m^2}}}.$$

If $\frac{g}{a_m} \ll 1$, we get

$$\omega' = \omega\left(1 - \frac{g}{a_m}\right). \tag{17}$$

4.2. Kündig's experiment

After the discovery of the Mössbauer effect in 1958, quantitative measurements of relativistic time dilation were carried out in the 1960s based on this effect. The experiments reported full agreement with the time dilation predicted by Einstein's theory of relativity. In these experiments, the Mössbauer source was placed at the center of a fast rotating disk, and an absorber was placed at the rim of the disk. In the analyses of these experiments, it was assumed that the absorption line of the rotating absorber stays the same as at rest, and is only shifted by the time dilation factor. As it was shown in [15], this assumption is wrong.

Kündig's experiment (1963) [22] measured the transverse Doppler shift for a rotating disk by means of the Mössbauer effect, see Figure 2. In (only) this experiment, the absorption line of the

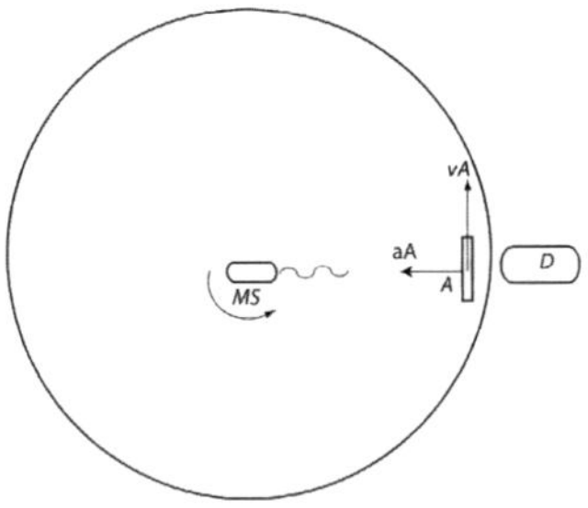

Figure 2. The rotating disk of the Kündig's experiment

rotating absorber was obtained by putting the source on a transducer. Thus, currently, Kündig's experiment is the only proper experiment to measure the transverse second order Doppler shift by Mössbauer spectroscopy using a rotating absorber.

For this experiment, Special relativity predicts a transverse Doppler effect of the value due to the time dilation of the absorber $\frac{\triangle E}{E} \approx -\frac{R^2\omega^2}{2c^2} = -b\frac{R^2\omega^2}{2c^2}$. The same value was obtained by Kündig analyzing the problem in the frame attached to the rotating disk and the accelerated absorber and treating the problem by the principle of equivalence and the general theory of relativity. The centrifugal force acting on the absorber is interpreted as a gravitational force with the potential $\Phi = -\frac{1}{2}R_A^2\omega^2$. The time of the absorber is slowed down by the gravitational potential in this analysis. Kündig in [22] claimed that the observed value of $b = 1.0065 \pm 0.011$, which is in full agreement with special relativity's prediction.

But Kholmetskii *et al* (2008) [21] found an error in the data processing of the results of Kündig's experiment. After the correction of the error, the correct value in this experiment is $b = 1.192 \pm 0.03$. Since the accuracy in the experiment was about 1%, this shows a significant deviation from special relativity's prediction. But, as we see from Figure 2, in this experiment there is also an acceleration between the source and the absorber. In [11], this deviation was explained by use of the additional longtitudal Doppler shift (17) due to the acceleration between the source and the absorber.

In Kündig's experiment, the acceleration $a = R\omega^2$ toward the source and $R = 9.3cm$. Hence, the additional shift with respect to that predicted by special relativity is $\frac{2c^2}{a_m R} \approx 0.192$ This gives an approximate value of $a_m = 10^{21}\frac{cm}{s^2}$ for the maximal acceleration. Notice that the calculated value of b is independent of the speed of rotation. This agrees approximately with the data in [21].

These observations lead us to the following conclusions:

- The observed rate of a clock also depends on the acceleration of the clock.

- The correcting time dilation, due to the acceleration, defines the value of the maximal acceleration to be $a_m \approx 10^{21}\frac{cm}{s^2}$.
- The value of the maximal acceleration is independent of the rotational velocity.

Note that Kholmetskii's correction of the result of the experiment can be interpreted as a change of the gravitational potential $\Phi = -\frac{1}{2}R_A^2\omega^2$ in the the frame attached to the disk to $\tilde{\Phi} = -\frac{1}{2}R_A^2\omega^2(1 + c^2\omega^2/a_m^2)$.

4.3. Future tests of maximal acceleration
An experiment is currently in process by the author and research teams from Hebrew University and Ben-Gurion University in Israel for testing ER and determining the value of the maximal acceleration. In the experiment, a Mössbauer source MS will be mounted, as usual, on a transducer. A semicircular absorber A will be placed on a disk of radius about 6 cm. The detector D will be diametrically opposed to the source, as shown in the Figure 3. Two collimators C will be placed between the source and the detector restricting the width of the measured radiation ray and insuring that the center of the disk is in this ray. The disk with the absorber will be rotated with a high-speed vibrationless spindle with several angular velocities ω, ranging up to 60,000 rpm. We will separate the counting of the detector for the times when the acceleration of the absorber is in the direction of the radiation, as in case (a) of Figure 3, and the times when the acceleration of the absorber is opposing the direction of the radiation, as in case (b) of Figure 3. Two absorption curves will be obtained for each case.

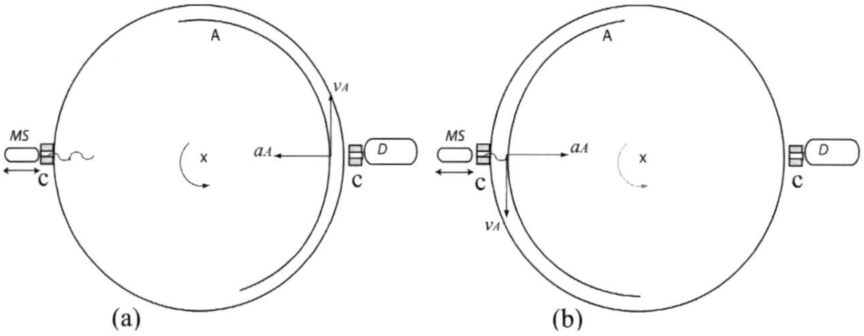

Figure 3. The proposed experiment, two positions of the absorber.

The radiation from the source MS undergoes two shifts, one due to the known transversal Doppler shift, and an expected second shift due to the acceleration of the absorber. The first shift should be the same for both cases, while the second shift changes sign from case (a) to (b). If the Clock Hypothesis is valid, no shift between the two absorption curves should be observed. On the other hand, if such a shift is observed and is significant, it will prove that the acceleration influences the observed rate of the clock and provide support for the validity of ER. Moreover, this experiment will enable us to calculate the value of the maximal acceleration.

Following is a list of other possible tests of ER and the maximal acceleration:

(i) Additional testing the Doppler shift due to the acceleration of a rotating disk by means of the Mössbauer effect

(ii) Muon-life experiment- improved (Bailey 77)- testing time dilation due to acceleration

(iii) Tests of the relativistic ER dynamics - Synchrotron

(iv) Tests of the relativistic ER dynamics by particle scattering

(v) Short-pulse laser acceleration-testing Doppler shift due to acceleration

(vi) Theoretical prediction of Quantum Mechanical effects based on ER

5. Extended relativity modification of physics

Extended relativity and the existence of a maximal acceleration change the entire map of physics. With the development of Einstein's theory of relativity and the development of quantum mechanics, the map of physics of the 20th century looks as follows. The areas of physics are determined by the magnitude of their velocities with respect to the maximal velocity (the speed of light c) and by their size. Classical mechanics is valid for large-sized objects and velocities significantly smaller than the speed of light. Relativistic dynamics is for large objects with speeds close to the speed of light. Quantum mechanics (non-relativistic) deals with small-sized objects with velocities significantly smaller than the speed of light, see Figure 4.

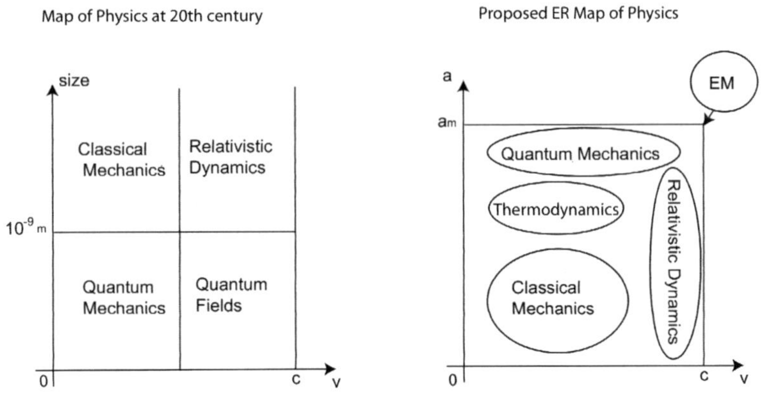

Figure 4. Two maps for physics

If the forthcoming experiments prove the validity of ER and the maximal acceleration turns out to be of the order of our current estimate, then the map of physics will need to be changed. First, we suggest to replace the size axis in the map by the acceleration. This will define a bounded domain of all ER admissible velocities and accelerations. This domain is the tangent domain to the relativistic phase-space. Classical mechanics is valid for both velocities and accelerations which are far from their maximal values. Relativistic dynamics is valid for velocities close to the speed of light and accelerations which are far from the maximal acceleration.

Quantum mechanics is not a model for small-sized objects, but for systems with accelerations close to the maximal one. For example, to describe the motion of electrons (small-sized objects) in a synchrotron, relativistic dynamics is used, and not quantum mechanics, since the velocities are close to the speed of light. On the other hand, the accelerations in the quantum region are extremely high. In [13] we asked "Can ER incorporate *quantum* phenomena?" Some partial results will be presented in Subsection 5.2.

Thermodynamics in this map is positioned between classical and quantum mechanics. Obviously, the velocity of thermal vibrations is far from the speed of light. Because of the high frequency of these oscillations, their acceleration can get very large. Indeed, for vibrations with small values of the natural frequency ω, the classical model describes well the radiation curves. But for large ω, only by assuming Planck's hypothesis, can one obtain the experimentally derived radiation curves. Note also that this hypothesis plays an important role for quantum mechanics. In the next subsections, we show that ER can explain the differences in thermal vibrations for small and large frequencies. Moreover, in this model, there is no "ultra-violet catastrophe" ([3] p.255).

Finally, electromagnetic fields (EM) are composed of photons which are zero-mass particles. As explained at the end of Section 3, in ER such particles move with maximal acceleration and reach velocities close to the speed of light in a very short time after emission. Thus, we positioned electromagnetism in the corner of our domain.

5.1. Quantum-like behavior of the harmonic oscillator in extended relativity

Following [12], we will describe now the motion of the harmonic oscillator in ER. The ER dynamic system for the harmonic oscillator is

$$\left\{ \begin{array}{l} \frac{d\mathbf{x}}{dt} = \frac{\mathbf{u}}{\sqrt{1+|\mathbf{u}|^2/c^2}} \\ \frac{d\mathbf{u}}{dt} = \frac{\omega^2 \mathbf{x}}{\sqrt{1+\frac{\omega^4 x^2}{a_m^2}}} \end{array} \right\}. \tag{18}$$

The Hamiltonian of this oscillator was given in (15). The effective potential energy $V_{eff} = \frac{a_m^2}{\omega^2}\sqrt{1+\frac{\omega^4 x^2}{a_m^2}}$ is now expressed by a hyperbola $\frac{V_{eff}^2}{\alpha^4} - \frac{(\omega x)^2}{\alpha^2} = 1$, with $\alpha = a_m/\omega$. The asymptotes of this hyperbola are $V = a_m x$ for $x \to \infty$ and $V = -a_m x$ for $x \to -\infty$ and are independent of the natural frequency ω of the harmonic oscillator. We normalized the potential to have the same limit at infinity and not by setting the value to zero when $x = 0$, as it is done for the classical harmonic oscillator. The value $V_{eff}(0) = a_m^2/\omega^2$ of the potential at the origin tends to zero as ω becomes extremely large, as shown in Figure 5. This will happen in

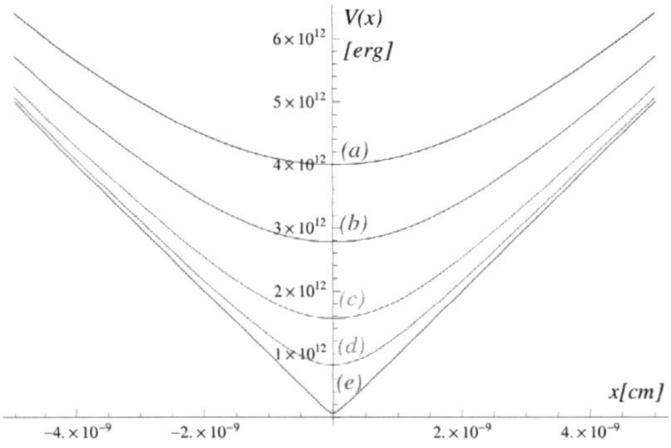

Figure 5. Effective potential $V_{eff}(x)$ for different ω :
(a)$5 \cdot 10^{14}s^{-1}$, (b)$6 \cdot 10^{14}s^{-1}$,(c)$8 \cdot 10^{14}s^{-1}$,(d)$11 \cdot 10^{14}s^{-1}$, (e)$40 \cdot 10^{14}s^{-1}$

the quantum region, where the ratio between the forces acting on the particles to their masses is extremely large.

From the Taylor expansion of V_{eff}, we see that if $\omega^4 x^2/a_m^2 \ll 1$, then $V_{eff} \approx \frac{\omega^2 x^2}{2} + const.$ This is the classical potential (per unit mass) for the harmonic oscillator. When $\omega^4 x^2/a_m^2 \gg 1$, on the other hand, which occurs for extremely large ω, the potential energy of an ER harmonic oscillator is approximately

$$V_q(x) \approx a_m |x| \,, \tag{19}$$

where the subscript q means "quantum."

We will solve first the evolution equation of the ER harmonic oscillator with this potential. Let A denote the *amplitude* of the vibrations. Assume that at time $t = 0$, the position of the oscillator was at $x(0) = -A$. At this time, the velocity was $u(0) = 0$. Denoting the total energy by E, the free energy is

$$E - E_0 = m_0 V_q(x(0)) = m_0 a_m A \,. \tag{20}$$

To solve the evolution of motion with potential $V_q(x)$, we start with the second equation of (18):

$$a(t) = \frac{du}{dt} = -\frac{\partial H}{\partial x} = -\frac{\partial V_q}{\partial x} = \begin{cases} a_m, & x < 0; \\ -a_m, & x > 0. \end{cases} \tag{21}$$

Thus, the acceleration $a(t)$ is a *square wave*, as shown in Figure 6. We interpret $a(t)$ as a *digitization* of the standard acceleration signal of the classical harmonic oscillator.

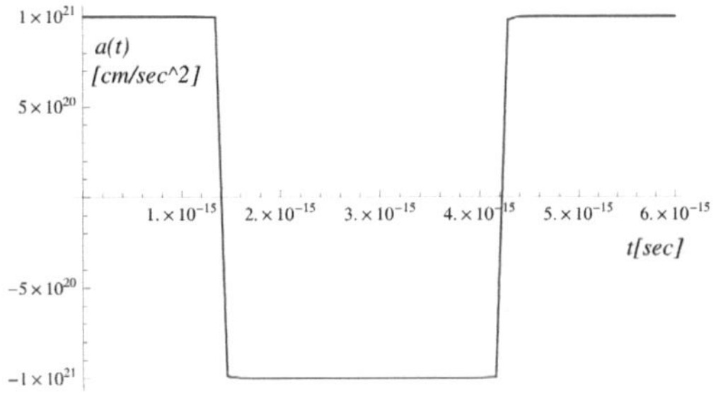

Figure 6. Square wave $a(t)$ for $A = 10^{-9} cm$

Note that this curve differs significantly from the acceleration curve of the classical harmonic oscillator. Since the radiation of the oscillating charge depends on its acceleration, the radiation in our model for extremely large ω will differ significantly from the radiation of a classical harmonic oscillator charge.

Equation (21) implies that $u(t) = a_m t + u_0$ for $x(t) < 0$, and $u(t) = -a_m t + u_0$ for $x(t) > 0$. After time $\frac{1}{4}T$, which is one fourth of the period T, the position of the oscillator will be $x(\frac{1}{4}T) = 0$. During the time interval $[0, \frac{1}{4}T]$, the proper velocity is $u(t) = a_m t$. Similarly, on the interval $[\frac{1}{4}T, \frac{3}{4}T]$, the proper velocity is $u(t) = a_m(\frac{1}{2}T - t)$. Figure 7 shows the function $u(t)$, which is called a *triangle wave*.

It is known (see for example: http://mathworld.wolfram.com /FourierSeriesTriangleWave) that the Fourier Series of a triangle wave $u(t)$ is

$$u(t) = \frac{2a_m T}{\pi^2} \sum_{k=0}^{\infty} \frac{(-1)^k}{(2k+1)^2} \sin\left(\frac{2\pi(2k+1)t}{T}\right) . \tag{22}$$

We will call the frequency of the leading term in this decomposition the *effective frequency* and denote it by ω_e. Thus

$$\omega_e = \frac{2\pi}{T}, \tag{23}$$

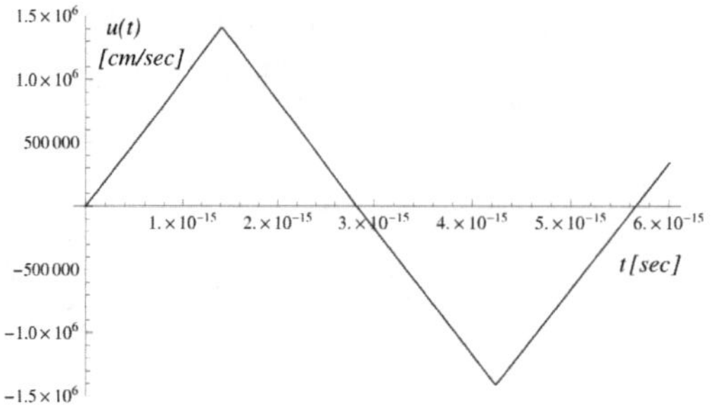

Figure 7. Triangle wave $u(t)$ for $A = 10^{-9}cm$

and we can rewrite (22) as

$$u(t) = \frac{4a_m}{\pi\omega_e}\sum_{k=0}^{\infty}\frac{(-1)^k}{(2k+1)^2}\sin\left((2k+1)\omega_e t\right) . \tag{24}$$

For the acceleration, one gets:

$$a(t) = \frac{du(t)}{dt} = \frac{4a_m}{\pi}\sum_{k=0}^{\infty}\frac{(-1)^k}{(2k+1)}\cos\left((2k+1)\omega_e t\right) . \tag{25}$$

This shows that the spectrum of these waves is supported at the points $\omega_e(2k+1)$, for $k = 0, 1, 2, \ldots$. This is similar to the spectrum of the quantum harmonic oscillator, which is known to be $\frac{\hbar\omega}{2}(2k+1)$, see [23] p.186. This indicates the connection of this new type of motion with the motion of the harmonic oscillator in the quantum region.

We use now the first Hamiltonian equation (9), which is also valid in ER, to calculate the displacement $x(t)$ during the time interval $[0, \frac{1}{4}T]$:

$$\frac{dx}{dt} = \frac{\partial H}{\partial u} = \frac{u(t)}{\sqrt{1+u(t)^2/c^2}} = \frac{a_m t}{\sqrt{1+(a_m t)^2/c^2}}. \tag{26}$$

Integrating this equation, we get $x(t) = \frac{c^2}{a_m}\left(\sqrt{1+\frac{(a_m t)^2}{c^2}} - 1\right) - A$.

Note that this graph is close to the displacement of the classical harmonic oscillator. This is due to the fact that the velocities are well below the speed of light and, therefore, $x(t)$ approximately equals to the integral of $u(t)$. From (24), we remark that the contributions of the non-leading terms in the Fourier decomposition will be relatively small.

Substituting $x(\frac{1}{4}T) = 0$, we get $A = \frac{c^2}{a_m}\left(\sqrt{1+\frac{(a_m T)^2}{16c^2}} - 1\right)$. From this, we can calculate the period T of oscillations:

$$T^2 = 16\frac{A^2}{c^2} + 32\frac{A}{a_m}. \tag{27}$$

The following Table 2 summarizes the behavior of the ER harmonic oscillator in the classical and quantum regions.

Note that the period T of the ER harmonic oscillator in the quantum region depends on the *amplitude* A of the motion and not on the natural frequency ω of the oscillator. This is in

Table 2. ER harmonic oscillator in two regions

ER oscillator.	classical region	quantum region
$a(t)$	$A\omega^2 \cos \omega t$	square wave (Figure 6)
$u(t)$	$A\omega \sin \omega t$	triangle wave (Figure 7)
$x(t)$	$-A \cos \omega t$	Figure 4
T	$2\pi/\omega$	$\sqrt{16\frac{A^2}{c^2} + 32\frac{A}{a_m}}$
$E - E_0$	$m_0 A^2 \omega^2/2$	$m_0 A a_m$
spectrum	$\{\omega\}$	$\frac{2\pi}{T}(2k+1) : k = 0, 1, 2, 3...$

contrast to the classical region, where T depends on ω and not on A. The period T can also be expressed in terms of the free energy by substituting $A = (E - E_0)/(m_0 a_m)$.

There is a difference in the energy spectrum of the quantum oscillator and the spectrum of the ER harmonic oscillator. The basic energy quanta in the quantum harmonic oscillator depend on the natural frequency of the corresponding classical harmonic oscillator. This makes it difficult to explain the fact that the blackbody spectrum is independent of the materials of which the walls are composed ([2] p. 20). On the other hand, the quanta of radiation of the ER harmonic oscillator depend only on the energy, which depends, in turn, on the temperature.

We have shown that the ER harmonic oscillator exhibits two significantly different behaviors. When ω is small, the oscillator behaves classically, while for large values of ω, the behavior is *quantum-like*. In the classical area, the spectrum of position, velocity, and acceleration oscillations consists of a single point ω, an inherent parameter of the oscillator. In the *quantum* area, on the other hand, the spectrum is similar to the spectrum of the energy of a *quantum harmonic oscillator*. In the *border area*, the classical signals become *digitized*. In Figure 8, we see the transition of the acceleration $a(t)$ from the classical to the quantum region.

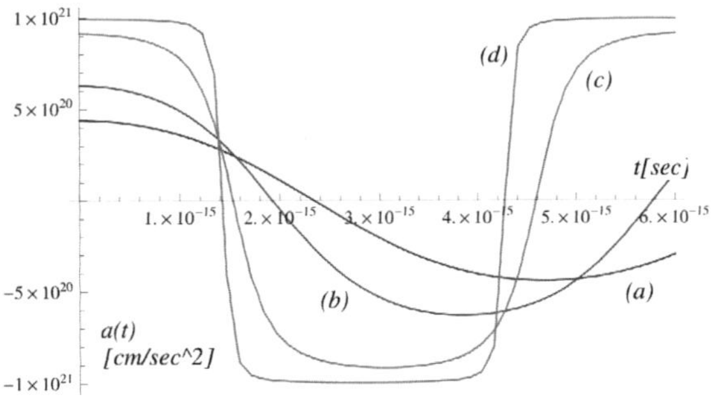

Figure 8. Evolution of $a(t)$ from the classical to the quantum region for ω : (a) $7 \cdot 10^{14} s^{-1}$,(b) $9 \cdot 10^{14} s^{-1}$, (c) $15 \cdot 10^{14} s^{-1}$, (d) $30 \cdot 10^{14} s^{-1}$

Figure 9 features the transition of the velocity $u(t)$ from the classical to the quantum region.

In Figure 10 we show the effective oscillation frequency ω_e as a function of the natural oscillator frequency $\omega = \sqrt{k/m}$ for oscillations with amplitude $A = 10^{-9} cm$. For the classical harmonic oscillator $\omega_e = \omega$, while for the ER harmonic oscillator this identity holds for small ω, while for large ω, ω_e approaches some limiting frequency, which could be calculated from (23) and (27).

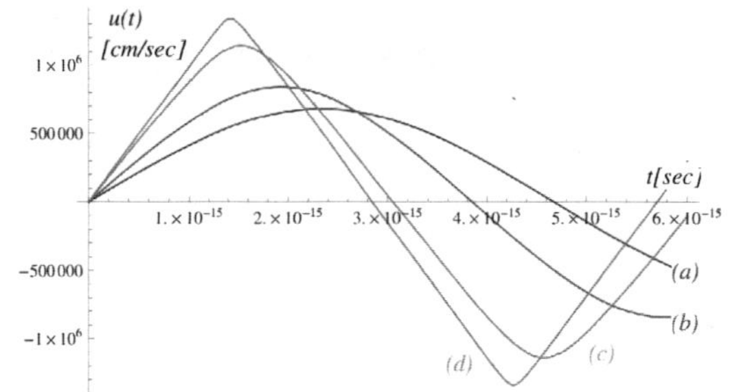

Figure 9. Evolution of $u(t)$ from the classical to the quantum region for ω : (a) $7 \cdot 10^{14} s^{-1}$,(b) $9 \cdot 10^{14} s^{-1}$, (c) $15 \cdot 10^{14} s^{-1}$, (d) $30 \cdot 10^{14} s^{-1}$

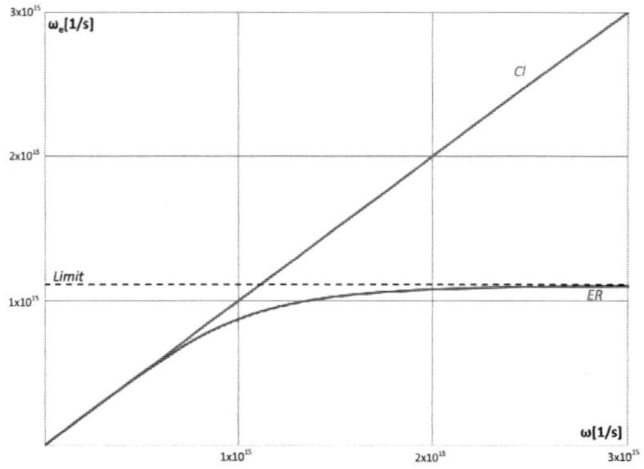

Figure 10. The effective oscillation frequency ω_e and the natural oscillator frequency ω for classical and ER harmonic oscillators

For an oscillator with a given ω, the signal will become digitized as the energy increases, which can be expressed by an increase of the amplitude A. In Figure 11, we show the transition of the acceleration $a(t)$ from the classical to the quantum region for an oscillator with $\omega = 10^{15} s^{-1}$, as the amplitude increases.

To understand the transition between the classical and quantum regions of the ER harmonic oscillator, we will compare it with the classical harmonic oscillator, as shown in Figure 12. We observe how gradually the signal gets digitized and the effective frequency ω_e of the ER harmonic oscillator becomes smaller than the corresponding frequency ω of the classical harmonic oscillator.

Since thermal vibrations can be represented as harmonic oscillators, the results of this paper may have applications to *blackbody radiation.* Blackbody radiation curves split into two regions: a classical region, corresponding to small values of ω, and a second region, in which ω is large. Note also that for large ω, the distances between the spectrum lines of the ER harmonic oscillator are an integer multiple of a constant. This is reminiscent of Planck's assumption. Since Relativistic Dynamics ($a_m = \infty$) does *not* predict different behaviors for different values of ω, our results provide more support, both to our conjecture that there *exists* a maximal acceleration a_m, and to our estimated *value* of a_m.

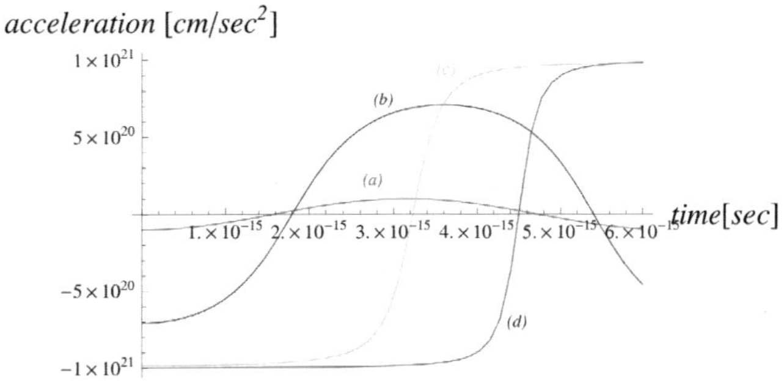

Figure 11. Evolution of $a(t)$ from the classical to the quantum region for $\omega = 10^{15}s^{-1}$ and amplitude A: (a) 10^{-10}cm,(b) 10^{-9}cm, (c) $5 \cdot 10^{-9}$cm, (d) 10^{-8}cm.

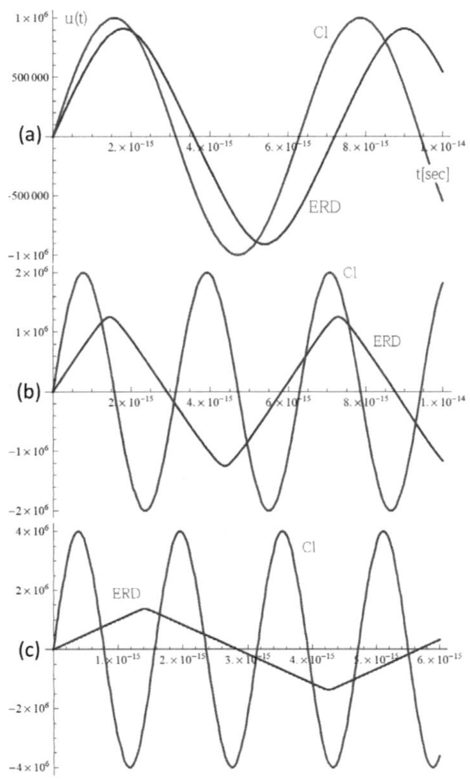

Figure 12. Velocity $u(t)$ for classical and ER harmonic oscillators for (a) $\omega = 10^{15}$, (b) $\omega = 2 \cdot 10^{15}$ and (c) $\omega = 4 \cdot 10^{15}$

Thus far, we have considered vibrations of the harmonic oscillator in ER without introducing *radiation*. In reality, when we consider thermal vibrations of atoms in a solid, the atom has to be considered positively charged moving in a negatively charged surrounding. In this case, the accelerated charged atom will radiate. The Abraham-Lorentz force caused by radiation is given by $\mathbf{F}_{rad} = \frac{2}{3}\frac{q^2}{c^3}\dot{\mathbf{a}}$. Thus, if the harmonic oscillator is in the quantum region, which is the usual state for high temperatures, most of the time the atom will not radiate, since the $|\mathbf{a}(t)| = a_m$. The radiation will be strong for the time when the acceleration changes the direction, which is the equilibrium position. So, the radiation will come in spikes. After each spike the oscillator will lose its energy, which implies that the following spike will be of a lower intensity and less sharp. This can be understood from Figure 11. In order to obtain blackbody radiation curves, we plan to continue the study of the electromagnetic radiation of the ER harmonic oscillator.

5.2. ER dynamics Hydrogen atom model

Here we present the results published in [16] Consider a system of two particles, a proton with mass $m_p = 1.7 \cdot 10^{-27} kg$ and an electron with mass $m_e = 9 \cdot 10^{-31} kg$. Denote the position and the proper velocity of the proton by $\mathbf{r}_p, \mathbf{u}_p$ and of the electron by $\mathbf{r}_e, \mathbf{u}_e$. At this point, we will restrict ourselves only to the Coulomb force, ignoring the interaction of the particles with the fields. The force of the proton acting on the electron is thus $\mathbf{F_1} = k(\mathbf{r}_p - \mathbf{r}_e)/|\mathbf{r}_p - \mathbf{r}_e|^3$, with $k = 2.3 \cdot 10^{-28} Nm^2$, while the electric force of the electron acting on the proton is $\mathbf{F_2} = k(\mathbf{r}_e - \mathbf{r}_p)/|\mathbf{r}_e - \mathbf{r}_p|^3 = -\mathbf{F_1}$.

Typical distances between the proton and the electron are of order $0.5A = 0.5 \cdot 10^{-10} m$. For both the proton and the electron, we have

$$\left|\frac{F_2}{a_m m_p}\right|^2 \approx 35 \gg 1, \quad \left|\frac{F_1}{a_m m_e}\right|^2 \approx 10^8 \gg 1. \tag{28}$$

Thus, the second equation of system (7) for both particles becomes

$$\frac{d\mathbf{u}_p}{dt} \approx a_m \frac{\mathbf{r}_e - \mathbf{r}_p}{|\mathbf{r}_e - \mathbf{r}_p|}, \quad \frac{d\mathbf{u}_e}{dt} \approx a_m \frac{\mathbf{r}_p - \mathbf{r}_e}{|\mathbf{r}_p - \mathbf{r}_e|}, \tag{29}$$

which implies that the magnitude of the acceleration of each particle will be close to the maximal one.

To estimate the velocities of the particles, we can use the connection of the acceleration and the velocity in circular motion. Since, in this case, $v^2 = aR$, in our system we will have $v \approx 10^{4.5} m/s \ll c$. Thus, in the first equation of system (7), we can ignore the denominator, and this equation becomes $\mathbf{u} \approx d\mathbf{r}/dt$. Substituting this into the second equation, we get an approximation of the ER dynamics equation for the particles in a hydrogen-like atom:

$$\frac{d^2\mathbf{r}_p}{dt^2} \approx a_m \frac{\mathbf{r}_e - \mathbf{r}_p}{|\mathbf{r}_e - \mathbf{r}_p|}, \quad \frac{d^2\mathbf{r}_e}{dt^2} \approx a_m \frac{\mathbf{r}_p - \mathbf{r}_e}{|\mathbf{r}_p - \mathbf{r}_e|}. \tag{30}$$

As usual for a two-body problem, in order to solve the system (30) of two particles, we decompose their motion into the motion of a "center of mass" and the motion of each particle with respect to the "center of mass." For our system, the average of these two equations gives

$$\frac{d^2(\mathbf{r}_p + \mathbf{r}_e)/2}{dt^2} \approx 0. \tag{31}$$

This implies that the point $\mathbf{R} := (\mathbf{r}_p + \mathbf{r}_e)/2$ moves freely, as a "center of mass." Thus, in this model, the definition of the "center of mass" differs from the definition in classical mechanics.

In a classical model for a hydrogen-like atom, the center of mass is positioned at or close to the proton, and the electron moves around the more or less stationary proton. Hence, classically, there should be a significant magnetic field for the atom. In our model, however, both the electron and the proton move around the the new "center" with similar trajectories. They thus produce the same magnetic field, but of opposite signs, due to their opposite charges. Thus, the total magnetic moment of our atom will be almost zero.

We introduce $\mathbf{r} = (\mathbf{r}_p - \mathbf{r}_e)/2$, so that $\mathbf{r}_p = \mathbf{R} + \mathbf{r}$ and $\mathbf{r}_e = \mathbf{R} - \mathbf{r}$. Subtracting the second equation of (30) from the first and dividing by 2, we obtain

$$\frac{d^2\mathbf{r}}{dt^2} \approx -a_m \frac{\mathbf{r}}{|\mathbf{r}|}, \tag{32}$$

which we call the "radial equation" describing the relative motion of the particles with respect to the center of mass. This is a typical equation for motion in a central field and can be solved by known methods. We will solve this equation by the method of [27], Chapter 14.

The solution of equation (32) is in a stable plane generated by the initial vectors $\mathbf{r}(0), \dot{\mathbf{r}}(0)$. Without loss of generality, we may assume that this plane is the x, y-plane. We complexify this plane by identifying the point (x, y) with the complex number $\zeta = x + iy = r(t)e^{i\varphi(t)}$. With this notation, $\frac{\mathbf{r}}{|\mathbf{r}|} = e^{i\varphi(t)}$, and equation (32) becomes

$$\ddot{r}e^{i\varphi} + 2i\dot{r}\dot{\varphi}e^{i\varphi} + i\ddot{\varphi}re^{i\varphi} - r\dot{\varphi}^2e^{i\varphi} = -a_m e^{i\varphi}.$$

Dividing by $e^{i\varphi}$, we have

$$\ddot{r} + 2i\dot{r}\dot{\varphi} + i\ddot{\varphi}r - r\dot{\varphi}^2 = -a_m. \tag{33}$$

The imaginary part of this equation is

$$2\dot{r}\dot{\varphi} + \ddot{\varphi}r = 0 \;\Rightarrow\; \frac{d}{dt}(r^2\dot{\varphi}) = 0,$$

implying that the angular momentum $r^2\dot{\varphi}$ in the center of mass system is conserved. Denote the angular momentum, defined from the initial conditions, by c_1. Then, the equation

$$\dot{\varphi} = c_1/r^2 \tag{34}$$

uniquely defines $\varphi(t)$ if we solve first the equation for $r(t)$ and use the initial conditions.

To find $r(t)$, substitute (34) into the real part of equation (33) to get $\ddot{r} - \frac{c_1^2}{r^3} + a_m = 0$. Multiplying this equation by $2\dot{r}$ and integrating by t, we get

$$\dot{r}^2 = c_2 - c_1^2 r^{-2} - 2a_m r,$$

with c_2 defined by the initial conditions. This equation shows that the radial part can be regarded as motion in one dimension in an effective field

$$U_{eff} = \frac{c_1^2 m}{2r^2} - m a_m r,$$

where $\frac{c_1^2 m}{2r^2}$ is called the centrifugal energy, and $U(r) = m a_m r$ is the potential energy.

Since

$$\dot{r} = \pm\sqrt{c_2 - c_1^2 r^{-2} - 2a_m r}, \tag{35}$$

in order that solutions will exist, the values of r must be restricted by

$$c_2 r^2 - c_1^2 + 2a_m r^3 \geq 0. \tag{36}$$

Journal of Physics: Conference Series **437** (2013) 012017 doi:10.1088/1742-6596/437/1/012017

The solutions of this inequality are $r_1 < r < r_2$, where r_1 and r_2 are the two positive roots of the cubic polynomial in (36), which always exist since this polynomial is negative at 0, negative towards ∞, and has at least one non-negative value at the initial state.

Thus, the solutions of the radial equation (32) of Extended Relativistic Dynamics for a hydrogen-like atom are obtained by solving the first-order differential equation (35) and then (34). It is known that only for central fields with potential energy proportional to r^2 or $1/r$ all finite motions take place in closed paths. The classical electromagnetic field is of this type, but under our dynamics, $U(r) = ma_m r$ is not. Hence, in general, our solution oscillates between the two radial values r_1 and r_2 and is not a closed path, see Figure 13. We can get a circular path

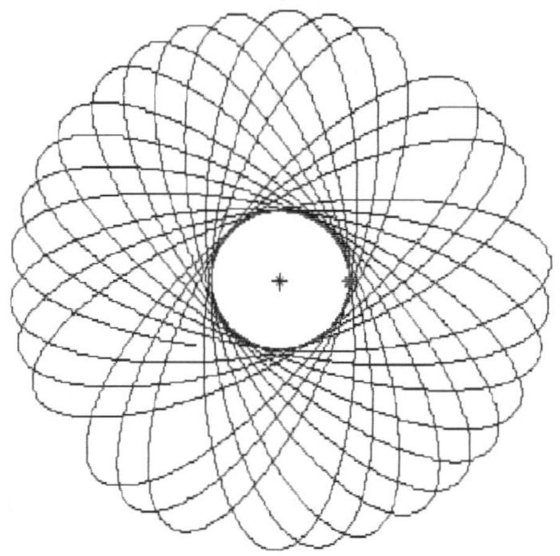

Figure 13. Electron and proton trajectories in ER dynamics Hydrogen like atom

at the minimum of the effective potential $r = \sqrt[3]{c_1^2/a_m}$, corresponding to the initial velocity $v = \sqrt{ra_m}$, when the velocity is perpendicular to $\mathbf{r}$. For certain discrete values, we can get closed path after n periods.

The frequency of these oscillations can be estimated from the fact that the magnitude of the acceleration is approximately a_m, and in approximately circular motion, we have $a = R\omega^2$. These considerations yield a frequency of $\nu \approx 10^{14} s^{-1}$. This implies that during one measurement time, the particle will cover a whole area in the annulus $r_1 < r < r_2$.

6. Summary and Discussion

Based on the symmetry following from the general principle of relativity, we have shown in Section 2 that if the observed time of an accelerated clock differs from the observed rate of a comoving inertial clock, then there is a universal maximal acceleration, which we denote by a_m. A relativity theory in which the observed time depends also directly on the acceleration we call **Extended Relativity** ER. We presented a systematic approach for transformations between two uniformly accelerated systems in ER. These transformations (3) are of Lorentz type. In Section 3 we presented an ER dynamics equation (7) which is an extension of the relativistic dynamics and in which all admissible solutions have a speed bounded by c, the speed of light, and an acceleration bounded by a_m, the maximal acceleration. We also obtained an ER Hamiltonian (13).

In Section 4, based on the results of the previous section, we have shown that ER predicts an additional Doppler shift due to the acceleration. We have shown that in Kündig's experiment (1963), which is the only proper experiment which measured time dilation by Mössbauer spectroscopy using a rotating absorber, this additional Doppler shift was observed. We described feasible experiments to test ER and to measure the value of the maximal acceleration.

An ER map for physics is proposed in Section 5. Extended relativity may provide a unifying framework for physics. It may also provide a new model for thermodynamics and quantum mechanics. We have shown that in ER, the harmonic oscillator has different behavior for different energies. For low energies its behavior is classical, while for high energies its behavior becomes quantum-like. Such energies do occur in thermal vibrations. In this model there is no "ultra-violet catastrophe." The model explains photon creation in thermal vibrations. For a hydrogen-like atom, we have shown that the classical electromagnetic force would generate accelerations above the maximal one. Thus, at the quantum level, ER dynamics differs significantly from relativistic dynamics. We obtained the first approximation of the solution for such systems, ignoring the interaction of the particles with the field. In a typical time that can be measured, the particle covers a whole area. This may provide an indication of the probabilistic description of particles in Quantum Mechanics. We have shown that in our model, the expression for the center of mass differs from the classical one. In our model, the total magnetic moment of a hydrogen atom is almost zero, which is not so in the classical (non-quantum) model. This observation also reveals the importance of the notion of symmetric velocity, which was introduced in Chapter 2 of [10]. This velocity is the relativistic half of the regular velocity. In our model, the velocity of both particles with respect to the new "center of mass" is the symmetric velocity of the velocity of the electron in the classical model. It is known that the transformations of the symmetric velocities are conformal [17]. Conformal transformations play an important role in the quantum region.

This is only the first step in analyzing the hydrogen-like atom by use of Extended Relativistic Dynamics. We plan to improve our model by: 1. Considering the next approximations of the model. 2. Incorporating the interaction of the charges with the fields. 3. Taking into the consideration the spin of the proton and the electron.

Acknowledgments

The author would like to thank the referee and F. W. Hehl for constructive remarks, T. Scarr for editorial proof and E. Yudkin for help with the final figures produced with the use of Mathematica. This research is supported in part by the German-Israel Foundation for Scientific Research and Development: GIF No. 1078-107.14/2009.

References

[1] Bailey J et al1977 *Nature*, **268**, 301-305
[2] Bohm D 1951 *Quantum Theory* (Dover Pub.)
[3] Born M 1962 *Atomic Physics* (Blackie & Son, London)
[4] Born M 1965 *Einstein's Theory of Relativity* (Dover Publications, New York)
[5] Born M 1938 *Proc. Roy. Soc.* A, **165**, 291
[6] Brillouin L 1970 *Relativity Reexamined* (Academic Press, New York)
[7] Caianiello E R 1981 *Lett. Nuovo Cimento* **32** 65
[8] Einstein A 1905 *Ann. Phys.* **17** 891
[9] Franklin J 2010 *Eur. J. Phys.* **31** 291-298
[10] Friedman Y 2004 *Homogeneous Balls and their Physical Applications* (Birkhäuser, Boston)
[11] Friedman Y 2009 The maximal acceleration, arXiv:0910.5629
[12] Friedman Y 2012 Quantum-like behavior of the harmonic oscillator in extended relativity, to appear
[13] Friedman Y 2011 *Ann. Phys. (Berlin)* **523** 408
[14] Friedman Y, Gofman Yu 2010 *Phys. Scr.* **82** 015004
[15] Friedman Y, Nowik I 2012 *Phys. Scr.* **85** 065702

[16] Friedman Y, Resin E 2012 *Phys. Scr.* **86** 015002
[17] Friedman Y, Semon M 2005 *Phys. Rev.* E **72**, 026603
[18] Friedman Y, Scarr T *Covariant Uniform Acceleration and Time Dilation*, arXiv:1105.0492 and article in current ussue.
[19] Don Koks 1998 *Does a clock's acceleration affect its timing rate?* http://math.ucr.edu/home/baez/physics/Relativity/SR/clock.html
[20] Goldstein H, Poole C and Safko J 2002 *Classical Machanics* (Pearson , New Jersey)
[21] Kholmetskii A L, Yarman T and Missevitch O V, 2008 *Phys. Scr.* **77** 035302.
[22] Kündig W 1963*Phys. Rev.* **126** 2371 .
[23] Liboff R L 1980 *Introductory Quantum Mechanics* (Holden-Day Inc. San Francisco)
[24] Mashhoon B 2008 *Ann. Phys.* **17** 705.
[25] Mashhoon B 1990 Limitations of spacetime measurements *Phys. Lett.* A **143** p. 176-82
[26] Mashhoon B 1990 The hypothesis of locality in relativistic physics *Phys. Lett.* A **145** p. 147-53
[27] Landau L D and Lifshitz E M 1978 *Mechanics* 3rd ed. (Pergamon Press)
[28] Moller C 1972 *The Theory of Relativity* (Clarendon Press, Oxford)
[29] Papini G, Scarpetta G, Bozza V, Feoli A, Lambiase G 2002 *Phys. Lett.* A **300** 603
[30] Papini G and Wood W R 1992 *Phys. Lett.* A **170** 409
[31] Scarpetta G 1984 *Lett. Nuovo Cimento* **41** 51
[32] Rindler W 2001 *Relativity: Special, General, and Cosmological* (Oxford University Press, New-York)
[33] Weinberg S 1972 *Gravitation and Cosmology* (John Wiley & Sons, New York)

The wave-function description of the electromagnetic field

Yaakov Friedman
Jerusalem College of Technology
Departments of Mathematics and Physics
P.O.B. 16031 Jerusalem 91160, Israel

E-mail: friedman@jct.ac.il

Abstract.
For an arbitrary electromagnetic field, we define a prepotential S, which is a complex-valued function of spacetime. The prepotential is a modification of the two scalar potential functions introduced by E. T. Whittaker. The prepotential is Lorentz covariant under a spin half representation. For a moving charge and any observer, we obtain a complex dimensionless scalar. The prepotential is a function of this dimensionless scalar. The prepotential S of an arbitrary electromagnetic field is described as an integral over the charges generating the field. The Faraday vector at each point may be derived from S by a convolution of the differential operator with the alpha matrices of Dirac. Some explicit examples will be calculated. We also present the Maxwell equations for the prepotential.

1. Introduction and motivation

In general, the electromagnetic field tensor F, expressed by a 4×4 antisymmetric matrix, is used to describe the electromagnetic field intensity. This description involves 6 parameters. An alternative way to describe an electromagnetic field is by use of the 4-potential. In a chosen gauge, the 4-potential transforms as a 4-vector. The electromagnetic field tensor is then recovered by differentiating the 4-potential. Choice of a gauge can reduce the degrees of freedom of this description to from 4 to 3.

In 1904 E. T. Whittaker introduced [24] two scalar potential functions. Thus, he was able to reduce the degrees of freedom of the electromagnetic field description to 2. He showed that the electromagnetic field can be expressed in terms of the second derivatives of these functions. However, he was not able to find the covariance of his scalar potentials. H. S. Ruse [21] improved the result of Whittaker. In 2009 Y. Friedman and S. Gwertzman showed [10] that it is possible to combine these two scalar potential functions into one complex-valued function $S(x)$ on Minkowski space, which we call the *prepotential of the electromagnetic field.* Moreover, they showed that this prepotential is invariant under a certain spin half representation of the Lorentz group. Thus, this prepotential provides a covariant description of an electromagnetic field with minimal degrees of freedom.

In 1953, influenced by M. Born, who was looking for the connection of the wave-function description of elementary particles in quantum mechanics and the electromagnetic field they generate, H. S. Green and E. Wolf introduced [15] a complex scalar potential for an electromagnetic field. They described the similarity of the expressions for energy and momentum densities between their potential and the wave function. They were unable, however, to find the connection between their potential and the Whittaker potentials.

An electromagnetic field is generated by a collection of moving charges. Thus, a description of an electromagnetic field can be obtained by integrating the fields of moving charges. For a moving charge and observer at point x, the position of the charge at the retarded time relative to the observer is a null-vector in spacetime. We show that for each such vector, there is a complex dimensionless scalar which is invariant under a certain representation of the Lorentz group. The prepotential $S(x)$ is defined to be the logarithm of this scalar.

The Aharonov-Bohm effect indicates that there is a multiple-valued prepotential of the 4-potential of an electromagnetic field. In their paper [2], a scalar function S such that $\nabla S = (e\hbar/c)A$, where A is the 4-potential of the electromagnetic field, was introduced. It has been shown that if ψ_0 is the solution of the Schrödinger equation in the absence of an electromagnetic field, then the function $\psi = \psi_0 e^{-iS/\hbar}$ is the solution of the equation in the presence of the field, at least in a simply connected region in which the electromagnetic field vanishes. In the magnetic Aharonov-Bohm experiment, however, the region outside the solenoid is not simply connected. This leads to a multi-valued ψ in this experiment. As it was shown in [11], our prepotential is of the same type, but is defined for any electromagnetic field. Our prepotential is also multi-valued.

In classical mechanics, the negative of the gradient of a scalar potential equals the force. This is true for forces which generate linear acceleration. Such forces are defined by a one-form (since their line integral gives the work). The derivative of a scalar function is also a one-form. Hence, the derivative of a potential can equal the negative of the force. But in classical mechanics we also have rotating forces, which are described by two-forms. Such forces cannot be expressed as derivatives of a scalar potential. For example, the electromagnetic field is not conservative in general. It generates both linear and rotational acceleration, and the electromagnetic tensor is a two-form. Hence, it is natural to assume that a kind of second derivative of a scalar potential will define the force tensor. Note that the usual differential of a gradient of a real-valued function is zero. Therefore, the prepotential must be complex-valued, and we will need to define a Lorentz invariant conjugation of the gradient of the prepotential in order to obtain the 4-potential of the field.

Another important property of a prepotential of an electromagnetic field is its *locality.* Note that the electromagnetic tensor $F_{\mu\nu}$ of a field of a moving charge depends on the position, velocity and acceleration, while the 4-potential A_μ depends only on the position and velocity of the source. Our prepotential S depends only on the position of the source.

In section 2, we obtain a Lorentz group representation based on the complex electromagnetic field tensor. We will show that the regular representation π of the Lorentz group can be decomposed as a product of two commuting representations $\tilde{\pi}$ and $\tilde{\pi}^*$. In section 3, we will introduce the prepotential of an electromagnetic field and show its geometric meaning. We will also show the covariance of this prepotential under the action of the representation $\tilde{\pi}$. In section 4, we will study the connection between the prepotential and the Faraday vector of a field. We will find the gauge of the prepotential. Explicit solutions for the prepotential and field of a charge at rest and a charged infinite rod will be found. Using computer algebra, we will derive the field of an arbitrary moving charge from its prepotential. The Maxwell equations for the prepotential will be derived in section 5 and we will demonstrate their use for deriving the prepotential and the field of current in straight wire. In section 6, we will show that the representation $\tilde{\pi}$ coincides with the spin half representation of the Lorentz group on the spinors. We also show that the matrices occurring in the description of the connection of the prepotential to the field are a representation of the Dirac α-matrices.

2. Lorentz group representation based on complex electromagnetic field tensor

In flat Minkowski space M, the spacetime coordinates of an event are denoted by x^μ ($\mu = 0,1,2,3$), with $x^0 = ct$. The Minkowski inner product is $x \cdot y = \eta_{\mu\nu}x^\mu y^\nu$, where the Minkowski metric is $\eta_{\mu\nu} = \mathrm{diag}(1,-1,-1,-1)$. The usual Lorentz group representation π can be associated with an electromagnetic field tensor $F_\alpha^\beta(\mathbf{E},\mathbf{B})$

$$F = F_\mu^\nu = \begin{pmatrix} 0 & E_1 & E_2 & E_3 \\ E_1 & 0 & cB_3 & -cB_2 \\ E_2 & -cB_3 & 0 & cB_1 \\ E_3 & cB_2 & -cB_1 & 0 \end{pmatrix}, \tag{1}$$

where $\mathbf{E}$ denote the electric field intensity and $\mathbf{B}$ the magnetic field intensity, as follows.

Since a magnetic field generates a rotation, a generator of a rotation about the direction $\mathbf{n} \in \mathbb{R}^3$ is defined by $F_\alpha^\beta(0,\mathbf{B})$, with $c\mathbf{B} = -\mathbf{n}$. Thus, the representation of the rotation about the direction $\mathbf{n}$ with angle ω is given by the operator $\exp(F_\alpha^\beta(0,-\mathbf{n})\omega)$. For example, a rotation $\mathfrak{R}^1$ about the x^1-axis (rotation in the x^2x^3 plane) is represented as

$$\pi(\mathfrak{R}^1)(\omega) = \Lambda_{23}(\omega) = \exp(\begin{pmatrix} 0 & 0 & 0 & 0 \\ 0 & 0 & 0 & 0 \\ 0 & 0 & 0 & -1 \\ 0 & 0 & 1 & 0 \end{pmatrix}\omega) = \begin{pmatrix} 1 & 0 & 0 & 0 \\ 0 & 1 & 0 & 0 \\ 0 & 0 & \cos\omega & -\sin\omega \\ 0 & 0 & \sin\omega & \cos\omega \end{pmatrix}. \tag{2}$$

Similarly, since an electric field generate boosts, a generator of a boost in the direction $\mathbf{n}$ can be identified with $F_\alpha^\beta(\mathbf{n},0)$. Thus, the representation of the boosts in the direction $\mathbf{n}$ with rapidity ω is given by the operator $\exp(F_\alpha^\beta(\mathbf{n},0)\omega)$. For example, boosts $\mathfrak{B}^1$ in the x-axis are represented as

$$\pi(\mathfrak{B}^1)(\omega) = \Lambda_{01}(\omega) = \exp(\begin{pmatrix} 0 & 1 & 0 & 0 \\ 1 & 0 & 0 & 0 \\ 0 & 0 & 0 & 0 \\ 0 & 0 & 0 & 0 \end{pmatrix}\omega) = \begin{pmatrix} \cosh\omega & \sinh\omega & 0 & 0 \\ \sinh\omega & \cosh\omega & 0 & 0 \\ 0 & 0 & 1 & 0 \\ 0 & 0 & 0 & 1 \end{pmatrix}. \tag{3}$$

Our complex prepotential $S(x)$ is a function $M \to \mathbb{C}$ on Minkowski space M. Its gradient at any spacetime point belongs to the complexified cotangent space at that point, which we identify with $\mathbb{C}^4$. Denote by M_c the complex space $\mathbb{C}^4$ endowed with the bilinear complex-valued form $x \cdot y = \eta_{\mu\nu}x^\mu y^\nu$, which can be considered as a complexification of Minkowski space. The bilinear form on M_c is an extension of the Minkowski inner product.

Obviously, the representation π of the Lorentz group on M can be extended linearly to a representation on M_c. We claim that this representation can be decomposed as follows.

Claim 1. *The representation π of the Lorentz group on M_c can be decomposed into a product*

$$\pi = \tilde{\pi}\tilde{\pi}^* \tag{4}$$

of two commuting representations $\tilde{\pi}$ and $\tilde{\pi}^*$ on M_c.

In order to prove this claim, we will decompose the generator F_β^α of the representation π into a sum

$$F_\beta^\alpha = \frac{1}{2}\mathcal{F}_\beta^\alpha + \frac{1}{2}\bar{\mathcal{F}}_\beta^\alpha, \tag{5}$$

where $\mathcal{F}_\beta^\alpha$ is the complex electromagnetic tensor and its complex adjoint $\bar{\mathcal{F}}_\beta^\alpha$, defined below.

An electromagnetic field can be defined by the *Faraday vector* $\mathbf{F} = \mathbf{E} + ic\mathbf{B}$. Note that since i is a pseudo-scalar and $\mathbf{B}$ is a pseudo-vector, the expression $i\mathbf{B}$ is a vector which is independent of the chosen orientation of the space. The Faraday vector is used to describe the Lorentz invariant field constants, see for example [18]. In [9], Friedman and Danziger introduced a *complex electromagnetic tensor* $\mathcal{F}^\alpha_\beta$ for the description of an electromagnetic field, similar to the one introduced by Silberstein in [22] and [23]. Complexified electromagnetic fields were also studied by A. Gersten [14] and played an important role in obtaining explicit solutions in [12] and [8] for motion of a charge in constant electromagnetic field.

Let $F^\alpha_\beta(\mathbf{E},\mathbf{B})$ be the usual electromagnetic tensor. The complex electromagnetic tensor $\mathcal{F}^\alpha_\beta(\mathbf{E},\mathbf{B})$ is defined by $\mathcal{F}^\alpha_\beta(\mathbf{E},\mathbf{B}) = F^\alpha_\beta(\mathbf{F},-i\mathbf{F})$ or

$$\mathcal{F}^\alpha_\beta = \begin{pmatrix} 0 & F_1 & F_2 & F_3 \\ F_1 & 0 & -iF_3 & iF_2 \\ F_2 & iF_3 & 0 & -iF_1 \\ F_3 & -iF_2 & iF_1 & 0 \end{pmatrix} = \sum_{j=1}^{3} (\rho^j)^\alpha_\beta F_j\,, \tag{6}$$

where $(\rho_j)^\alpha_\beta$ are the Majorana-Oppenheimer matrices (see [6])

$$(\rho^1)^\alpha_\beta = \begin{pmatrix} 0 & 1 & 0 & 0 \\ 1 & 0 & 0 & 0 \\ 0 & 0 & 0 & -i \\ 0 & 0 & i & 0 \end{pmatrix},\ (\rho^2)^\alpha_\beta = \begin{pmatrix} 0 & 0 & 1 & 0 \\ 0 & 0 & 0 & i \\ 1 & 0 & 0 & 0 \\ 0 & -i & 0 & 0 \end{pmatrix},$$

$$(\rho^3)^\alpha_\beta = \begin{pmatrix} 0 & 0 & 0 & 1 \\ 0 & 0 & -i & 0 \\ 0 & i & 0 & 0 \\ 1 & 0 & 0 & 0 \end{pmatrix}. \tag{7}$$

Note that the electromagnetic tensor F^α_β is the real part of the complex tensor $\mathcal{F}^\alpha_\beta$. Equation (5) holds now if we define the complex conjugate tensor to be $\bar{\mathcal{F}}^\alpha_\beta = \sum_{j=1}^3 (\bar{\rho}_j)^\alpha_\beta \bar{F}_j$.

The Majorana-Oppenheimer matrices can be derived from the connection of the 4-potential and the Faraday vector, as follows. It is known that if A_μ is the 4-potential of the electromagnetic field, then $E_j = F_{0j} = \partial_j A_0 - \partial_0 A_j = A_{0,j} - A_{j,0}$, where $\partial_\mu = \frac{\partial}{\partial x^\mu}$ and $B^j = (\nabla \times \mathbf{A})^j$, where $\nabla\times$ is the curl which is applied to the vector potential. Thus, the the j component of the Faraday vector is connected to the potential as

$$F^j = (\rho^j)^{\alpha\beta}\partial_\alpha A_\beta = (\rho^j)^{\alpha\beta} A_{\beta,\alpha}\,, \tag{8}$$

showing that the matrices ρ^j connect the 4-potential to the Faraday vector of the electromagnetic field. Using (6), we can define a differential operator which connects the 4-potential with the the complex electromagnetic tensor $\mathcal{F}^\alpha_\beta$:

$$\mathcal{F}^\alpha_\beta = \sum_{j=1}^{3} (\rho^j)^\alpha_\beta (\rho^j)^{\nu\mu}\partial_\nu A_\mu = \nabla \times A\,, \tag{9}$$

where the *curl* $\nabla\times$ on M_c is defined as

$$\nabla\times = \sum_{j=1}^{3} (\rho_j)^\alpha_\beta (\rho^j)^{\nu\mu}\partial_\nu\,. \tag{10}$$

We will now derive a Lorentz group representation on M_c based on the Majorana-Oppenheimer matrices. In Section 6 we will show that this representation can be identified with the known spin-half representation of the Lorentz group. To define this representation, it is enough to define the representation on the generators of boosts and rotations.

Definition 1. *Define a representation $\tilde{\pi}$ on M_c by defining generators $\tilde{B}^j$ of boosts $\mathfrak{B}^j$ in the direction of x^j to be $\frac{1}{2}\rho^j$ and generators $\tilde{R}^j$ of rotations $\mathfrak{R}^j$ about the direction x^j to be $\frac{i}{2}\rho^j$.*

Claim 2. *The representation $\tilde{\pi}$ is a Lorentz group representation on M_c.*

Direct calculation shows that the operators $\tilde{R}^j = \frac{i}{2}\rho^j$ obey the same commutator relations as the generators of the rotation group $SO(3)$:

$$[\tilde{R}^j, \tilde{R}^k] = -\varepsilon_l^{jk}\tilde{R}^l,$$

where ε^{ijk} is the Levi-Civita tensor. Moreover, the $\tilde{B}^j = \frac{1}{2}\rho^j$ matrices obey the same commutator relations as the generators of boosts in the Lorentz group:

$$[\tilde{B}^j, \tilde{B}^k] = \varepsilon_l^{jk}\tilde{R}^l, \quad [\tilde{R}^j, \tilde{B}^k] = \varepsilon_l^{jk}\tilde{B}^l$$

This proves Claim 2.

The complex adjoints $\bar{\sigma}^j = \frac{-i}{2}\bar{\rho}^j$ and $\frac{1}{2}\bar{\rho}^j$ satisfy similar relations. This leads to a second Lorentz group representation on M_c.

Definition 2. *Define a representation $\tilde{\pi}^*$ on M_c by defining generators $\tilde{B}^j$ of boosts $\mathfrak{B}^j$ in the direction of x^j to be $\frac{1}{2}\bar{\rho}^j$ and generators $\tilde{R}^j$ of rotations $\mathfrak{R}^j$ about the direction x^j to be $\frac{i}{2}\bar{\rho}^j$.*

Moreover, the two sets of operators $\{\rho^j\}$ and $\{\bar{\rho}^k\}$ commute:

$$[\rho^j, \bar{\rho}^k] = 0. \tag{11}$$

Thus, the representations $\tilde{\pi}$ and $\tilde{\pi}^*$ commute. Since

$$\exp F^\alpha_\beta = \exp(\frac{1}{2}\mathcal{F}^\alpha_\beta + \frac{1}{2}\bar{\mathcal{F}}^\alpha_\beta) = \exp(\frac{1}{2}\mathcal{F}^\alpha_\beta)\exp(\frac{1}{2}\bar{\mathcal{F}}^\alpha_\beta),$$

the representation π, associated with the tensor F^α_β, can be decomposed as a product (4) of representations $\tilde{\pi}$ and $\tilde{\pi}^*$. This proves Claim 1.

In addition to the above commutation relations, Majorana-Oppenheimer matrices also satisfy *anti-commutation relations*, which are very helpful for the calculation of the exponents of these matrices. The anti-commutation relations are

$$\{\rho^j, \rho^k\} = \frac{1}{2}(\rho^j\rho^k + \rho^k\rho^j) = \delta^{jk}I, \tag{12}$$

which implies that $(\rho^j)^2 = I$. Using these relation and the power series expansion of the exponent function, the exponents of these matrices are

$$\exp(\rho^j\frac{\omega}{2}) = \cosh\frac{\omega}{2}I + \rho^j\sinh\frac{\omega}{2}$$

$$\exp(\rho^j\frac{i\omega}{2}) = \cos\frac{\omega}{2}I + i\rho^j\sin\frac{\omega}{2}.$$

We can define now explicitly the representation under $\tilde{\pi}$ of the rotations and the boosts of the Lorentz group. For example, the rotation $\mathfrak{R}^1$ about the x^1-axis or rotation in the x^2x^3 plane by an angle ω is represented by

$$\tilde{\pi}(\mathfrak{R}^1)(\omega) = \Upsilon_{23}(\omega) = \exp(i\rho^1\frac{\omega}{2}) = \cos\frac{\omega}{2}I - i\rho^1\sin\frac{\omega}{2} \tag{13}$$

$$= \begin{pmatrix} \cos\omega/2 & -i\sin\omega/2 & 0 & 0 \\ -i\sin\omega/2 & \cos\omega/2 & 0 & 0 \\ 0 & 0 & \cos\omega/2 & -\sin\omega/2 \\ 0 & 0 & \sin\omega/2 & \cos\omega/2 \end{pmatrix}.$$

The representation $\tilde{\pi}^*$ of this rotation is given by $\bar{\Upsilon}_{23}(\omega)$, the complex conjugate of the above matrix. In the x^2x^3 plane, both $\tilde{\pi}(\mathfrak{R}^1)$ and $\tilde{\pi}^*(\mathfrak{R}^1)$ define a rotation by an angle $\omega/2$, and their product $\Upsilon_{23}(\omega)\bar{\Upsilon}_{23}(\omega) = \Lambda_{23}(\omega)$, defined by (2), is a rotation by an angle ω in this plane. However, in the plane x^0x^1, the rotation $\Lambda_{23}(\omega)$ under π is the identity, while both $\Upsilon_{23}(\omega)$ and $\bar{\Upsilon}_{23}(\omega)$ define a "complex rotation" in the plane $\{x^0 + ix^1 : \ x^0, x^1 \in \mathbb{R}\}$ by angles $-\omega/2$ and $\omega/2$, respectively, indicating the there is a rotation around the x^1 axis. This observation needs further understanding.

Similarly, the boost in the x^1-direction with parameter ω is

$$\tilde{\pi}(\mathfrak{B}^1)(\omega) = \Upsilon_{01}(\omega) = \exp(\rho^1\frac{\omega}{2}) = \cosh\frac{\omega}{2}I + \rho^1\sinh\frac{\omega}{2} \tag{14}$$

$$= \begin{pmatrix} \cosh\omega/2 & \sinh\omega/2 & 0 & 0 \\ \sinh\omega/2 & \cosh\omega/2 & 0 & 0 \\ 0 & 0 & \cosh\omega/2 & -i\sinh\omega/2 \\ 0 & 0 & i\sinh\omega/2 & \cosh\omega/2 \end{pmatrix}.$$

The representation $\tilde{\pi}^*$ of this boost is given by $\bar{\Upsilon}_{01}(\omega)$, the complex conjugate of the above matrix. In the x^0x^1 plane, both $\tilde{\pi}(\mathfrak{B}^1)$ and $\tilde{\pi}^*(\mathfrak{B}^1)$ define boosts in the x^1 direction with rapidity $\omega/2$, and their product $\Upsilon_{01}(\omega)\bar{\Upsilon}_{01}(\omega) = \Lambda_{01}(\omega)$, defined by (3), is a boost with rapidity ω in the x^1 direction. However, in the x^2x^3 plane, the boost $\Lambda_{01}(\omega)$ under π is the identity, while both $\Upsilon_{01}(\omega)$ and $\bar{\Upsilon}_{01}(\omega)$ define "complex boosts" in the plane $\{x^2 + ix^3 : \ x^2, x^3 \in \mathbb{R}\}$, with rapidity $-\omega/2$ and $\omega/2$, respectively, and the product of these boosts is the identity.

Note that the operators $\Upsilon_{\alpha\beta}$ are isometries on the space M_c. The representation $\tilde{\pi}$, defined by the operators $\Upsilon_{\alpha\beta}$, is a spin 1/2 representation of the Lorentz group on M_c. Similarly, the representation $\tilde{\pi}^*$ is defined by the operators $\bar{\Upsilon}_{\alpha\beta}$, which are obtained by taking the complex conjugate of $\Upsilon_{\alpha\beta}$. This is also a spin 1/2 representation of the Lorentz group on M_c, and it commutes with the representation $\tilde{\pi}$. In section 6, we will show that the representation $\tilde{\pi}$ can be identified with the usual representation of the Lorentz group on the *spinors*.

Let F_α^β be the electromagnetic tensor of some electromagnetic field. This tensor is covariant under the representation π of the Lorentz group. What is the covariance of the complex tensor $\mathcal{F}_\alpha^\beta$ associated with F_α^β? We will show the following.

Claim 3. *The covariance of the electromagnetic tensor F_α^β under the representation π is equivalent to the covariance of the corresponding tensor $\mathcal{F}_\alpha^\beta$ under the representation $\tilde{\pi}$.*

Let Λ be any element of the representation π of the Lorentz group. By (4), there is an element Υ of $\tilde{\pi}$ such that $\Lambda = \Upsilon\bar{\Upsilon}$. Using that $\mathcal{F}, \Upsilon \in \operatorname{span}\rho^j$, these operators commute with any $\bar{\rho}^k$ and thus with $\bar{\mathcal{F}}, \bar{\Upsilon}$. Thus,

$$F' = \Lambda F \Lambda^{-1} = \frac{1}{2}(\Upsilon\bar{\Upsilon}\mathcal{F}\bar{\Upsilon}^{-1}\Upsilon^{-1} + \Upsilon\bar{\Upsilon}\bar{\mathcal{F}}\bar{\Upsilon}^{-1}\Upsilon^{-1})$$

$$= \frac{1}{2}(\Upsilon\mathcal{F}\Upsilon^{-1} + \bar{\Upsilon}\bar{\mathcal{F}}\bar{\Upsilon}^{-1}) = \frac{1}{2}(\mathcal{F}' + \bar{\mathcal{F}}').$$

This proves Claim 3.

3. Definition of the prepotential

Denote by P a point in Minkowski space M at which we want to define the prepotential. We will call P the observer's point and denote his coordinates by $x = x^\mu$. Denote by $\breve{x}(\tau) = \breve{x}^\mu(\tau) : \mathbb{R} \to M$ the worldline of the charge q generating our electromagnetic field, as a function of its proper time. Let the point $Q = \breve{x}(\tau(x))$ be the unique point of intersection of the past light cone at P with the worldline $\breve{x}(\tau)$ of the charge. The time on the worldline of the charge corresponding to this intersection is uniquely determined by the point x. It is called the *retarded time* and will be denoted by $\tau(x)$. Note that radiation emitted at Q at the retarded time will reach P at time $t = x^0/c$ corresponding to this point. Our prepotential $S(x)$ will depend on the relative position

$$r(x) = x - \breve{x}(\tau(x)) \tag{15}$$

of the charge at retarded time $\tau(x)$, see Figure 1.

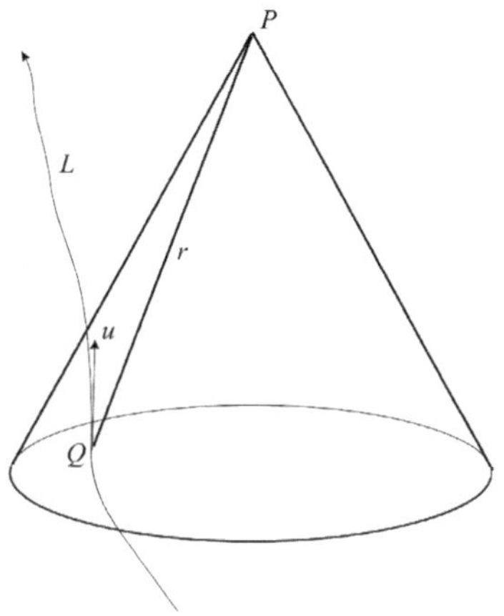

Figure 1. The four-vectors associated with an observer and a moving charge.

The relative position r is a null vector. Hence,

$$(r^0 + r^3)(r^0 - r^3) = (r^1 + ir^2)(r^1 - ir^2). \tag{16}$$

In this decomposition, we have split the four coordinates of r into two groups: the first group contains the time component r^0 and the third spatial component r^3. The second group contains the first two spatial components r^1 and r^2. We can now define a dimensionless complex constant ζ, as follows.

Definition 3. *For any null-vector r in M_c, we define a dimensionless complex scalar ζ by*

$$\zeta(r) = \frac{r^0 + r^3}{r^1 - ir^2} = \frac{r^1 + ir^2}{r^0 - r^3}. \tag{17}$$

For any $x \in M$ and any world-line $\breve{x}(\tau)$, we define $\zeta(x) = \zeta(r(x))$, where the relative position $r(x)$ is defined by (15).

The scalar $\zeta(r)$ can be identified as a simple function of the stereographic projection of the direction of the vector part $\mathbf{r}$ of r from the celestial sphere to the Argand plane, as follows. Since r is a null vector, it can be decomposed as $(|\mathbf{r}|, \mathbf{r})$ and is defined uniquely by its vector part $\mathbf{r}$. Define $\hat{\mathbf{r}} = \mathbf{r}/|\mathbf{r}|$, a unit vector in the direction vector of $\mathbf{r}$, and a null vector $\hat{r} = (|\hat{\mathbf{r}}|, \hat{\mathbf{r}})$. Since $\zeta(r)$ is the same for any multiple of r, we have $\zeta(r) = \zeta(\hat{r})$. Then, as shown in [20] p. 11, $\zeta(r) = \zeta(\hat{r}) = e^{i\varphi} cot\frac{\theta}{2}$, where φ, θ are the angles of standard spherical coordinates of $\hat{r}$. This shows that $\zeta(r)$ depends only on the direction from the observer to the charge at the retarded time.

There is no way to associate to every null vector r a non-trivial scalar which is invariant under the usual representation π. However, for the representation $\tilde{\pi}$ we have:

Claim 4. *For any null vector r, the complex constant $\zeta(r)$, defined by (17), is invariant under the representation $\tilde{\pi}$ on M_c.*

The boost Υ_{01} of $\tilde{\pi}$ in the x^1-direction is defined by (14). Applying this boost to the vector r yields

$$\Upsilon_{01}(r) = \cosh(\omega/2)(r^0, r^1, r^2, r^3) + \sinh(\omega/2)(r^1, r^0, -ir^3, ir^2).$$

Using (17), we have

$$\Upsilon_{01}(\zeta) = \zeta(\Upsilon_{01}(r)) = \frac{\cosh(\omega/2)(r^0 + r^3) + \sinh(\omega/2)(r^1 + ir^2)}{\cosh(\omega/2)(r^1 - ir^2) + \sinh(\omega/2)(r^0 - r^3)} =$$

$$\frac{r^0 + r^3}{r^1 - ir^2} \cdot \frac{1 + \tanh(\omega/2)(r^1 + ir^2)/(r^0 + r^3)}{1 + \tanh(\omega/2)(r^0 - r^3)/(r^1 - ir^2)} = \frac{r^0 + r^3}{r^1 - ir^2} \cdot \frac{1 + \tanh(\omega/2)/\bar{\zeta}}{1 + \tanh(\omega/2)/\bar{\zeta}} = \zeta,$$

showing that ζ is invariant under the boost Υ_{01}. Similarly, one may show the invariance of ζ under any boost and rotation of the representation $\tilde{\pi}$. This proves Claim 4.

Definition 4. *We define the prepotential $S(x)$ at point x of an electromagnetic field generated by a moving charge q as*

$$S(x) = \frac{q}{2} \ln \zeta(x) = \frac{q}{2} \ln \zeta(r(x)), \tag{18}$$

where $\zeta(r)$ is defined by (17) and the relative position of the charge $r(x)$ is defined by (15).

From Claim 4, it follows that the prepotential is invariant under the representation $\tilde{\pi}$ on M_c. Consider an electromagnetic field generated by any charge distribution $\sigma(x)$. The scalar prepotential for this field is defined by

$$S(x) = \frac{1}{2} \int_{K^-(x)} \zeta(r)\sigma(x + r) d^3 r, \tag{19}$$

where $K^-(x)$ denotes the backward light-cone at x, and the integration is by the 3D volume on this cone.

The geometric meaning of this prepotential can be understood if we introduce *relativistic bipolar coordinates* $(\varrho^0, \varrho^1, \theta, \varphi)$:

$$x^0 = \varrho^0 \cosh \theta, \;\; x^1 = \varrho^1 \cos \varphi, \;\; x^2 = \varrho^1 \sin \varphi, \;\; x^3 = \varrho^0 \sinh \theta.$$

The angle φ in these coordinates is the same angle φ occurring in polar and spherical coordinates. However, the angle θ differs from the angle θ of spherical coordinates. These coordinates fit better to our model as they give a simple form for the prepotential. The light-cone in these coordinates has the simple form $\varrho^0 = \varrho^1 = \varrho$, a hyperplane. For any null-vector r on the light-cone, the invariant constant $\zeta(r)$, defined by (17), in relativistic bipolar coordinates is

$$\zeta(r) = \frac{r^0 + r^3}{r^1 - ir^2} = \frac{\varrho e^\theta}{\varrho e^{-i\varphi}} = e^{\theta + i\varphi}.$$

The prepotential of a moving charge is

$$S(x) = \frac{q}{2}(\theta + i\varphi),$$

which is a multiple of the complex angle $\theta + i\varphi$ of the relative position $r(x)$ of the charge.

4. The prepotential and the Faraday vector of the field

4.1. Connection between the prepotential and the Faraday vector of a field

The 4-potential will be defined as a conjugate of the gradient of the prepotential $S(x)$. Since the prepotential $S(x)$ is invariant under the representation $\tilde{\pi}$, we want our conjugation to commute with this representation. By Definition 1 of the representation $\tilde{\pi}$, the conjugation needs to commute with any operator ρ^j. Using (11), we can choose any operator $\bar{\rho}^k$ to define the conjugation. Since, in (16), and in the definition of ζ, we choose the third spatial component to join the time component, also here we chose the $\tilde{\pi}$ *covariant conjugation* $\mathcal{C}$ on M_c to be given by multiplication by $\bar{\rho}^3$:

$$\mathcal{C} = (\bar{\rho}^3)^\alpha_\beta = \begin{pmatrix} 0 & 0 & 0 & 1 \\ 0 & 0 & i & 0 \\ 0 & -i & 0 & 0 \\ 1 & 0 & 0 & 0 \end{pmatrix}. \tag{20}$$

The following diagram shows the derivation of the complex tensor $\mathcal{F}$ of the field from its prepotential $S(x)$:

$$S \xrightarrow{\nabla} \nabla S \xrightarrow{\mathcal{C}} A = \mathcal{C}(\nabla S) \xrightarrow{\nabla\times} \mathcal{F}.$$

By differentiating the prepotential and taking its conjugate, we obtain the 4-potential A. Then we use (9) to get from A to the complex tensor $\mathcal{F}$ of the field. The components F_j of the Faraday vector can be found by (8) as

$$F_j = \partial_\nu (\rho^j)^{\mu\nu} (\bar{\rho}^3)^\lambda_\mu \partial_\lambda S\,. \tag{21}$$

Claim 5. *The Faraday vector F_j defined by (21) is invariant under the representation $\tilde{\pi}$*

This claim follows from the following facts. Claim 4 implies that $S(x)$ is $\tilde{\pi}$ invariant. By definition of the conjugation, $(\bar{\rho}^3)^\lambda_\mu$ is $\tilde{\pi}$ invariant. The remaining operators together form a $\tilde{\pi}$ invariant operator.

The last expression can be simplified, if for each $j = 1, 2, 3$, we introduce a new tensor $((\alpha^j)^{\mu\lambda} = (\rho^j)^{\mu\nu}(\bar{\rho}^3)^\lambda_\mu$. Then we can rewrite (21) as

$$F_j = (\alpha^j)^{\mu\lambda} \partial_\nu \partial_\lambda S\,. \tag{22}$$

The new tensors are

$$(\alpha^1)^{\mu\lambda} = \begin{pmatrix} 0 & 0 & -i & 0 \\ 0 & 0 & 0 & -1 \\ -i & 0 & 0 & 0 \\ 0 & -1 & 0 & 0 \end{pmatrix}, \quad (\alpha^2)^{\mu\lambda} = \begin{pmatrix} 0 & i & 0 & 0 \\ i & 0 & 0 & 0 \\ 0 & 0 & 0 & -1 \\ 0 & 0 & -1 & 0 \end{pmatrix}$$

$$(\alpha^3)^{\mu\lambda} = \begin{pmatrix} 1 & 0 & 0 & 0 \\ 0 & 1 & 0 & 0 \\ 0 & 0 & 1 & 0 \\ 0 & 0 & 0 & -1 \end{pmatrix}.$$

Thus, (22) defines the following explicit formulas for the connection between the prepotential $S(x)$ and the Faraday vector.

$$\begin{aligned} F_1 &= -2(S_{,13} + iS_{,02}) \\ F_2 &= -2(S_{,23} - iS_{,01}) \\ F_3 &= S_{,00} + S_{,11} + S_{,22} - S_{,33} \end{aligned} \tag{23}$$

Based on these formulas, we obtain the following gauges for the prepotential.

Claim 6. *The prepotential $S(x)$ is not determined uniquely by the electromagnetic field. Let $g(x)$ be a function on M_c from the following list, or a combination of such functions:*

(i) $g(x) = g(x^1, x^2)$ is harmonic, that is, $g_{,11} + g_{,22} = 0$,

(ii) $g(x) = g(x^0, x^3)$ satisfies the wave equation, that is, $g_{,00} - g_{,33} = 0$,

(iii) $g(x)$ is of order 1 in x^μ.

Then the transformation of the prepotential

$$S'(x) = S(x) + g(x) \tag{24}$$

does not affect the field. A function $g(x)$ of this type will be called the prepotential gauge.

To prove the claim, it is enough to check that the field of the gauge $g(x)$ defined by (23) is zero. In case (i), we have $g_{,0} = g_{,3} = 0$. Since each term of the components F_1, F_2 of the field in (23) involves differentiation by x^0 or by x^3, these components are zero. The third component F_3 is zero since $g_{,00} = g_{,33} = 0$ and $g_{,11} + g_{,22} = 0$. Similarly, for case (ii), we have $g_{,1} = g_{,2} = 0$. Since each term of the components F_1, F_2 of the field in (23) involve differentiation by x^1 or by x^2, these components are zero. The third component F_3 is zero since $g_{,11} = g_{,22} = 0$ and $g_{,00} - g_{,33} = 0$. Since each component of the field involves second derivatives, it will vanish for any function $g(x)$ of order 1 in x^μ. This completes the proof of the claim.

We demonstrate now the use of formulas (23) and Claim 6 in calculating the field of a rest charge at the origin from its prepotential. Consider a rest charge q at the origin. The worldline of this charge is $\breve{x}(\tau) = (\tau, 0, 0, 0)$. Thus, the relative position of the charge is $r = (|x|, x^1, x^2, x^3)$, where $|x| = \sqrt{(x^1)^2 + (x^2)^2 + (x^3)^2}$. From (17) and (18),

$$S'(x) = \frac{q}{2} \ln \frac{|x| + x^3}{x^1 - ix^2} = \frac{q}{2} \ln(|x| + x^3) - \frac{q}{2} \ln(x^1 - ix^2).$$

The function $\ln(x^1 - ix^2)$ is harmonic, and, by Claim 6(i), is a prepotential gauge.

Example 1. *For a rest charge q at the origin, we define the prepotential to be*

$$S(x) = \frac{q}{2} \ln(|x| + x^3). \tag{25}$$

Note that in this case the prepotential is well defined for all $x \neq 0$ and could be chosen to be real valued.

Claim 7. *The prepotential $S(x)$ defined by (25) defines the electromagnetic field of the rest charge q at the origin to be $\mathbf{E} = \frac{\mathbf{r}}{|\mathbf{r}|^3}$ and $\mathbf{B} = 0$. Moreover, $S(x)$ satisfies the wave equation $\Box S(x) = 0$.*

Since $S(x)$ is independent of x^0, we have $S_{,0} = 0$. Direct calculation shows that $\frac{\partial}{\partial x^j}|x| = \frac{x^j}{|x|}$. Thus, we get

$$S_{,1} = \frac{q}{2|x|}\frac{x^1}{|x| + x^3}, \quad S_{,2} = \frac{q}{|x|2}\frac{x^2}{|x| + x^3},$$

and

$$S_{,3} = \frac{q}{2}\frac{1}{|x| + x^3}\left(\frac{x^3}{|x|} + 1\right) = \frac{q}{2|x|}.$$

The 4-potential A in this case is

$$A = \mathcal{C}\nabla S = \frac{q}{2|x|}(1, -i\frac{x^2}{|x| + x^3}, i\frac{x^1}{|x| + x^3}, 0).$$

This potential behaves similarly to the usual 4-potential: when $|x|$ approaches infinity, it behaves like $\frac{1}{|x|}$.

By use of (23), we get

$$F_1 = -2(S_{,13} + iS_{,02}) = -2\frac{\partial}{\partial x^1}S_{,3} = -2\frac{\partial}{\partial x^1}\frac{q}{2|x|} = \frac{qx^1}{|x|^3},$$

$$F_2 = -2(S_{,23} - iS_{,01}) = -2\frac{\partial}{\partial x^2}S_{,3} = -2\frac{\partial}{\partial x^2}\frac{q}{2|x|} = \frac{qx^2}{|x|^3}.$$

To calculate F_3 using (23), we need first to calculate $S_{,\mu\mu}$.

$$S_{,11} = \frac{\partial}{\partial x^1}\frac{q}{2}\frac{x^1}{|x|(|x| + x^3)} = \frac{q}{2}\frac{|x|^2 + x^3|x| - 2(x^1)^2 - x^3(x^1)^2/|x|}{|x|^2(|x| + x^3)^2},$$

$$S_{,22} = \frac{\partial}{\partial x^2}\frac{q}{2}\frac{x^2}{|x|(|x| + x^3)} = \frac{q}{2}\frac{|x|^2 + x^3|x| - 2(x^2)^2 - x^3(x^2)^2/|x|}{|x|^2(|x| + x^3)^2}.$$

This implies that

$$S_{,11} + S_{,22} = \frac{qx^3}{2|x|^3}.$$

Since $S_{,00} = 0$ and $S_{,33} = \frac{-qx^3}{2|x|^3}$, equation (23) yields

$$F_3 = \frac{qx^3}{|x|^3}$$

and $\Box S(x) = S_{,00} - (S_{,11} + S_{,22}) - S_{,33} = 0$. This proves the claim.

As a second example, we will find now the prepotential of an infinitely long charged rod. We start with a prepotential of a finite rod on the x^3 axis, from $x^3 = -L$ to $x^3 = L$, with charge density σ. Denote by l the position of the charge on the rod, so that $-L < l < L$. The worldline of the charge is $\breve{x}(\tau) = (\tau, 0, 0, l)$, implying that $r = (\xi, x^1, x^2, x^3 - l)$, with

$$\xi(l) = \sqrt{(x^1)^2 + (x^2)^2 + (x^3 - l)^2} = \sqrt{\varrho^2 + (x^3 - l)^2}, \tag{26}$$

where $\varrho = \sqrt{(x^1)^2 + (x^2)^2}$. Thus, from (17), the zeta factor for the charge is

$$\zeta(r) = \frac{\xi(l) + x^3 - l}{x^1 - ix^2},$$

and the prepotential of the rod is

$$S(x)=\frac{\lambda}{2}\int_{-L}^{L}\ln\frac{\xi(l)+x^3-l}{x^1-ix^2}dl=\frac{\lambda}{2}\int_{-L}^{L}\ln(\xi(l)+x^3-l)dl-\frac{\lambda}{2}\int_{-L}^{L}\ln(x^1-ix^2)dl\,, \qquad (27)$$

with ξ defined by (26).

The second integral is equal to $2L\ln(x^1-ix^2)$, which is a harmonic function in x^1,x^2, and, by Claim 6(i), is a gauge and will be removed from the prepotential. For the first integral, we use a substitution $z=x^3-l$, implying $dl=-dz$. After the substitution, the first integral is

$$-\int_{x^3+L}^{x^3-L}\ln(\sqrt{\varrho^2+z^2}+z)dz=\sqrt{\varrho^2+z^2}-z\ln(\sqrt{\varrho^2+z^2}+z),$$

and the prepotential (27) is equivalent to

$$S(x)=\frac{\lambda}{2}\left((x^3+L)\ln(\xi(-L)+x^3+L)-(x^3-L)\ln(\xi(L)+x^3-L)+\xi(L)-\xi(-L)\right).$$

Using that $\xi(L)+x^3-L=\frac{\varrho^2}{\xi(L)-x^3+L}$ and $\xi(L)-\xi(-L)=\frac{-4x^3L}{\xi(L)+\xi(-L)}$, we get

$$S(x)=\frac{\lambda}{2}\left(-x^3\ln\varrho^2+L\ln\varrho^2-\frac{4x^3L}{\xi(L)+\xi(-L)}\right)$$
$$+\frac{\lambda}{2}\left((x^3+L)\ln(\xi(-L)+x^3+L)+(x^3-L)\ln(\xi(L)-x^3+L)\right).$$

The expression $L\ln\varrho^2=L\ln((x^1)^2+(x^2))$ is a harmonic function in x^1,x^2, and, by Claim 6(i), is a gauge and will be removed from the prepotential.

In order to be able to pass to the limit $L\to\infty$, we will subtract from the prepotential a gauge function with a behavior similar to $S(x)$ for large L. An example of such a function is $2x^3\ln 2L-2x^3$, which by Claim 6(iii) is a gauge. We obtain

$$S(x)=\frac{\lambda}{2}\left(-x^3\ln\varrho^2-\{\frac{4x^3L}{\xi(L)+\xi(-L)}-2x^3\}\right)$$
$$+\frac{\lambda}{2}\left((x^3+L)\ln(\xi(-L)+x^3+L)+(x^3-L)\ln(\xi(L)-x^3+L)-2x^3\ln 2L\right).$$

Since $\lim_{L\to\infty}\xi(\pm L)/L=1$, we have

$$\lim_{L\to\infty}\frac{4x^3L}{\xi(L)+\xi(-L)}-2x^3=\lim_{L\to\infty}\frac{4x^3}{\xi(L)/L+\xi(-L)/L}-2x^3=0$$

and

$$\lim_{L\to\infty}(x^3+L)\ln(\xi(-L)+x^3+L)+(x^3-L)\ln(\xi(L)-x^3+L)-2x^3\ln 2L=0$$

Thus, the prepotential of an infinite rod is

$$S(x)=-\frac{\lambda}{2}x^3\ln\varrho^2=-\lambda x^3\ln\sqrt{(x^1)^2+(x^2)^2}\,. \qquad (28)$$

By applying formulas (23) to this prepotential, we get

$$F_1=-2(S_{,13}+iS_{,02})=\lambda\tfrac{2x^1}{\varrho^2}$$
$$F_2=-2(S_{,23}-iS_{,01})=\lambda\tfrac{2x^2}{\varrho^2}$$
$$F_3=S_{,00}+S_{,11}+S_{,22}-S_{,33}=0\,,$$

which implies that the electric field is $\mathbf{E}(x)=\frac{2\lambda}{\varrho^2}(x^1,x^2,0)$ and that there is no magnetic field, as is known for the field of a charged rod.

4.2. The Faraday vector F from the prepotential of a moving charge

Consider now any charge q with worldline $\breve{x}(\tau)$. In order to be able to calculate the field of this charge from its prepotential $S(x)$ defined by (18), we will use formulas (23). To apply this formula, we have to calculate some second partial derivatives of $S(x)$. The difficulty is that S is a function of $\zeta(r(x))$, and $r(x)$ depends also on the worldline $\breve{x}(\tau)$. The idea is to present these derivatives in covariant form. This means that we want to express them by covariant vectors, namely, the relative position r of the observer with respect to the charge at the retarded time $\tau(x)$, the 4-velocity $u = \frac{d\breve{x}}{d\tau}(\tau(x))$ and the 4-acceleration $a = \frac{du}{d\tau}(\tau(x))$ at the retarded time.

From the definition (18) of the prepotential $S(x)$ its derivative by x^μ is

$$S(x)_{,\mu} = \frac{q}{2}\frac{\zeta_{,\mu}}{\zeta} = \left(\frac{r^0_{,\mu} + r^3_{,\mu}}{r^1 - ir^2} - \frac{(r^0 + r^3)(r^1_{,\mu} - ir^2_{,\mu})}{(r^1 - ir^2)^2}\right)\frac{r^1 - ir^2}{r^0 + r^3} = \frac{r^0_{,\mu} + r^3_{,\mu}}{r^0 + r^3} - \frac{r^1_{,\mu} - ir^2_{,\mu}}{r^1 - ir^2}. \quad (29)$$

Thus, to be able to calculate the derivatives of $S(x)$, we need a formula for the derivatives of the relative position $r(x)$ defined by (15). The partial derivative $r_{,\mu}$ of the position vector $r(x)$ by x^μ is

$$r_{,\mu} = x_{,\mu} - y_{,\mu}(\tau(x)) = x_{,\mu} - \partial y/\partial\tau\, \tau_{,\mu} = x_{,\mu} - u\tau_{,\mu}.$$

Since $r(x)$ is null, we have $r \cdot r_{,\mu} = 0$, where $a \cdot b = \eta_{\mu\nu}a^\mu b^\nu$ denote the Minkowski inner product. Thus,

$$0 = r \cdot r_{,\mu} = r \cdot x_{,\mu} - (r \cdot u)\tau_{,\mu},$$

which imply that the derivative of the retarded time is

$$\tau(x)_{,\mu} = \frac{r_\mu}{r \cdot u}. \quad (30)$$

This mean that varying the observers position by dx^μ will cause the parameter τ to change infinitesimally by $\frac{r_\mu}{r \cdot u}dx^\mu$. Since $(r)^2 = 0$ and $(u)^2 = 1$, from the Cauchy-Schwartz inequality it follows that $r \cdot u$ is always non-zero. So, equation (30) is always well defined.

Substituting this into the derivative of the position vector we get

$$r^\nu_{,\mu} = x^\nu_{,\mu} - \frac{r_\mu u^\nu}{r \cdot u} = \delta^\nu_\mu - \frac{r_\mu u^\nu}{r \cdot u}, \quad (31)$$

where δ^ν_μ denotes the Kronecker delta function. This formula could be understood as follows. An infinitesimal variation dx^μ of the observers position causes also a change of the charge position at the retarded time. The relative position r under such variation is effected by the two changes. The first term in (31) the effect of infinitesimal variation of the observers position, while the second one express change of the charge position at the retarded time.

Substituting derivative of the position vector (31) in (29) we obtain a covariant form for the derivative of the prepotential $S(x)$ as a function of the the relative position r and the 4-velocity of the charge u at the retarded time. To calculate the electromagnetic field of of a moving charge from its prepotential $S(x)$, according (23) we need to calculate some second derivatives of $S(x)$. This could be done by differentiating $S_{,\mu}$. The expression of $S_{,\mu}$ involve involve only non-constant vectors r and u. The derivatives of r are given by (31). The derivatives of u, by use of (30), are

$$u_{,\mu} = \frac{\partial u}{\partial\tau}\tau(x)_{,\mu} = \frac{ar_\mu}{r \cdot u}. \quad (32)$$

Since an infinitesimal change of the observers position by dx^μ causes the change of τ by amount $\frac{r_\mu}{r \cdot u}dx^\mu$, the variation of the 4-velocity u of the charge caused by this change is a multiple of the acceleration a of the charge by this amount.

To simplify the result of the differentiation by use of computer algebra we can introduce for any $j = 1, 2, 3$ an anti-symmetric scalar product on vectors of M_c by

$$[a, b]_j = (\rho^j)_{\alpha\beta} a^\alpha b^\beta = \rho^j a \cdot b, \quad a, b \in M_c \,. \tag{33}$$

By use of computer algebra (23) and using the above formulas for the derivatives we get

$$F^j = q \left(\frac{[r, u]_j}{(r \cdot u)^3} + \frac{[r, a]_j (r \cdot u) - [r, u]_j (r \cdot a)}{2(r \cdot u)^3} \right) \tag{34}$$

Using $1+3$ decomposition of 4-vectors it can be shown that these formulas coincide with the know field of a moving charge as derived from the Lienard-Wiecher potential, see [16] p.664. Thus, we have shown that the prepotential $S(x)$ of a moving charge defined by (18) produces by use of (23) the known electromagnetic field of such charge. Moreover, the superposition principle and that the connection of the prepotential to the field is linear, for a general electromagnetic field, which is produce by a collection of moving charges, by applying (23) to the prepotential $S(x)$ defined by formula (19) will recover properly this field.

5. The Maxwell equations of the prepotential

In this section we define the Maxwell equations for the prepotential $S(x)$ of the electromagnetic field. Consider an electromagnetic field generated by sources $J^\mu = (\rho, j^1, j^2, j^3)$. The pair of the scalar Maxwell equations

$$\nabla \cdot \mathbf{E} = \rho, \;\; \nabla \cdot \mathbf{B} = 0$$

for the Faraday vector $\mathbf{F} = \mathbf{E} + ic\mathbf{B}$ can be rewritten as one equation

$$\nabla \cdot \mathbf{F} = \rho \tag{35}$$

Substitute the values of $\mathbf{F}$ from (23) to get

$$F_{1,1} + F_{2,2} + F_{3,3} = -2S_{,113} - 2iS_{,102} - 2S_{,223} + 2iS_{,201} + S_{,300} + S_{,311} + S_{,322} - S_{,333}$$

$$= S_{,003} - S_{,113} - S_{,223} - S_{,333} = \rho \,.$$

So, the first Maxwell equation for $S(x)$ is:

$$\partial_3(\Box S) = \rho \tag{36}$$

The vector Maxwell equations are

$$\nabla \times \mathbf{E} + \dot{\mathbf{B}} = 0, \;\; \nabla \times \mathbf{B} - \dot{\mathbf{E}} = \mathbf{j}. \tag{37}$$

They can be rewritten by use of the Faraday vector as one equation

$$\mathbf{F}_{,0} + i\nabla \times \mathbf{F} = -\mathbf{j}. \tag{38}$$

Substitute the values of $\mathbf{F}$ from (23)) for the first component to get

$$F_{1,0} + iF_{3,2} - iF_{2,3} = -2S_{,013} - 2iS_{,002} + i(S_{,200} + S_{,211} + S_{,222} - S_{,233}) + 2i(S_{,323} - iS_{,301})$$

$$= i(-S_{,002} + S_{,211} + S_{,222} + S_{,233}) = -j^1 \,.$$

So, the second Maxwell equation for $S(x)$ is:

$$-i\partial_2(\Box S) = -j^1 \,. \tag{39}$$

For the second component we get

$$F_{2,0} + iF_{1,3} - iF_{3,1} = -2S_{,023} + 2iS_{,001} - 2i(S_{,313} + iS_{,302} - i(S_{,100} + S_{,111} + S_{,122} - S_{,133})$$
$$= i(S_{,100} - S_{,111} - S_{,122} - S_{,133}) = -j^2 .$$

So, the third Maxwell equation for $S(x)$ is:

$$i\partial_1(\Box S) = -j^2 . \quad (40)$$

For the third component we get

$$F_{3,0} + iF_{2,1} - iF_{1,2} = S_{,000} + S_{,011} + S_{,022} - S_{,033} - 2i(S_{,123} - iS_{,101}) + 2i(S_{,213} + iS_{,202})$$
$$= S_{,000} - S_{,011} - S_{,022} - S_{,033} = -j^3 .$$

So, the forth Maxwell equation for $S(x)$ is:

$$\partial_0(\Box S) = -j^3 . \quad (41)$$

Note since the left-hand side of these equations is a co-vector we lower the index of the sources of the field by $J_\mu = (\rho, -j^1, -j^2, -j^3)$. We will also apply the conjugation $\mathcal{C}$ defined by (20) to the gradient ∂ to get

$$\mathcal{C}_\mu^\nu \partial_\nu = (\partial_3, -i\partial_2, i\partial_1, \partial_0) .$$

With this notation we can rewrite our 4 Maxwell equations (36), (39),(40) and (41), as

$$\mathcal{C}_\mu^\nu \partial_\nu(\Box S(x)) = J_\mu(x) , \quad (42)$$

or in operator form

$$\mathcal{C}\nabla(\Box S(x)) = J(x) . \quad (43)$$

Our Maxwell equations (43)have some similarity to the ones obtained by use of Clifford algebras, see [4].

The Maxwell equation(s) for the prepotential we obtained is a third order system of equations. But this system can be decomposed into two systems of equations. The first system is a first order one $\mathcal{C}\nabla\Phi = J$. Using that $\mathcal{C}$ is a symmetry, meaning $\mathcal{C}^2 = I$, this system can be rewritten as $\nabla\Phi(x) = \mathcal{C}J(x)$. The meaning of such system is looking for a scalar potential $\Phi(x)$ for a vector-field $\mathcal{C}J(x)$. A similar formula is obtained in [3] p.98 for the spinor form of Maxwell equation describing the connection of the sources of the field to the Faraday vector. The second system is a known second order system $\Box S(x) = \Phi(x)$ for $\Phi(x)$, solved in the previous step. Thus, equation (43) is equivalent to

$$\nabla\Phi(x) = \mathcal{C}J(x), \;\; \Box S(x) = \Phi(x) . \quad (44)$$

Claim 7 show that $S(x)$ satisfy the wave equation $\Box S(x) = 0$ outside the rest charge. By using the covariance of such property, this will hold for the prepotential of any moving charge. Using the superposition principle the prepotential of any electromagnetic field will satisfy the wave equation

$$\Box S(x) = 0 \quad (45)$$

for x outside the sources of the field.

We will demonstrate now the use of the Maxwell equations (43) for derivation of the prepotential and the field of a current I in a thin long straight wire. We will assume that the wire is positioned on the x^3-axis. The field of such current is generated by negatively charged electrons with line density $\tilde{\lambda}$ moving with velocity $\mathbf{v}$ in the negative direction of the x^3-axis. These charges are compensated by positive charges of the same density resting in the wire. To express the fact that the wire is thin we will use 2D Dirac-delta $\delta(\varrho)$ for $\varrho = \sqrt{(x^1)^2 + (x^2)^2}$ with the property that the integral of this function over a domain D in x^1, x^2 plane is equal 1 if the point $(0,0)$ is in D and 0 otherwise.

The current density of the electron sources of the field are given by $j(e)^\mu = -\tilde{\lambda}\delta(\varrho)\gamma(1,0,0,-v/c)$, where $\gamma = \gamma(v) = (1 - v^2/c^2)^{-1/2}$. So, the effective density will be $\lambda = \tilde{\lambda}\gamma$. For the positive charges $j(p)^\mu = \lambda\delta(\varrho)(1,0,0,0)$. So, the total current of the sources of the field will be $j^\mu = \lambda\delta(\varrho)(0,0,0,\frac{v}{c})$ and using that $I = \lambda v$, $j^\mu = \delta(\varrho)(0,0,0,\frac{I}{c})$. The equation for $\Phi(x)$ from (44) become

$$\Phi_{,0} = -\frac{I}{c}\delta(\varrho), \quad \Phi_{,1} = \Phi_{,2} = \Phi_{,3} = 0\,.$$

Integrating the first equation and using that the average value of the Dirac delta is zero we get $\Phi(x) = -\frac{I}{c}\delta(\varrho)x^0$.

Now we need to solve the equation (44) $\Box S(x) = \Phi(x)$ for the prepotential $S(x)$. From [19] it follows that the electromagnetic field must be normal to the wire, implying $F_3 = 0$ and by use of (23) that $S_{,00} + S_{,11} + S_{,22} - S_{,33} = 0$. Subtracting from the last equation $\Box S(x) = \Phi(x)$ we get

$$S_{,11} + S_{,22} = -\frac{1}{2}\Phi(x) = \frac{I}{2c}\delta(\varrho)x^0 \tag{46}$$

and adding these equations we get

$$S_{,00} - S_{,33} = \frac{1}{2}\Phi(x) = -\frac{I}{2c}\delta(\varrho)x^0\,. \tag{47}$$

We can look for a solution for equations (46) and (47). We will decompose the prepotential $S(x)$ as a sum of two components $S(x) = S_1(x) + S_2(x)$, where each component will satisfy one of the equations and vanish by the differential operator of the second one. We will look for $S_1(x)$ of the form $S_1(x) = \frac{I}{2c}x^0 f(x^1,x^2)$ for some function $f(x^1,x^2)$. The equation (46) for f become $\nabla^2 f(x^1,x^2) = \delta(\varrho)$. It is known ([1] p.760) that the solution of such equation is $f(x^1,x^2) = \frac{-1}{2\pi}\ln\varrho$. Thus, $S_1(x) = \frac{I}{4\pi c}x^0\ln\varrho$ satisfy equation (46) and vanish by the differential operator of equation (47).

If we take $S_2(x)$ to be of the form $S_2(x) = \frac{I}{2c}\delta(\varrho)x^0 g(x^3)$, then such function will vanish differential operator of (46) and for (47) will give $g(x^3) = -(x^3)^2/2$. Thus the prepotential $S(x)$ for a strait wire is

$$S(x) = \frac{I}{4c}x^0\left(\frac{1}{\pi}\ln\varrho - (x^3)^2\delta(\varrho)\right)\,. \tag{48}$$

By used of (23) we can now calculate the Faraday vector of the field outside the wire. Since $\ln\varrho$ is harmonic outside the wire we have $F_3 = S_{,00} + S_{,11} + S_{,22} - S_{,33} = 0$. Direct calculation give $F_1 = -2(S_{,13} + iS_{,02}) = -i\frac{I}{2\pi c}\frac{x^2}{\varrho^2}$ and $F_2 = -2(S_{,23} - iS_{,01}) = i\frac{I}{2\pi c}\frac{x^1}{\varrho^2}$. Thus, the electric field is zero and the magnetic field is

$$B_1 = \frac{I}{2\pi c^2}\frac{-x^2}{\varrho^2}, \quad B_2 = \frac{I}{2\pi c^2}\frac{x^1}{\varrho^2}, \quad B_3 = 0.$$

This corresponds to the known formulas for such field (see [13] p.14-4).

6. Prepotential, spinors and the Dirac matrices

The prepotential $S(x)$ of a moving charge that we introduced was based on the decomposition (16) of any null vector in M_c. This decomposition involve four terms, which could be considered as coordinated of the null vector r in a new basis. This basis on M_c is called the *Bondi null tetrad* and is called also Newman-Penrose basis. It is defined as

$$\mathbf{n}_0 = \mathbf{n} = \frac{1}{\sqrt{2}}(\mathbf{e}_0 + \mathbf{e}_3),\ \mathbf{n}_1 = \bar{\mathbf{m}} = \frac{1}{\sqrt{2}}(\mathbf{e}_1 - i\mathbf{e}_2),$$

$$\mathbf{n}_2 = \mathbf{m} = \frac{1}{\sqrt{2}}(\mathbf{e}_1 + i\mathbf{e}_2),\ \mathbf{n}_3 = \mathbf{l} = \frac{1}{\sqrt{2}}(\mathbf{e}_0 - \mathbf{e}_3). \tag{49}$$

For the significance of the Bondi tetrad see [20], [17] and [7]. All vectors $\mathbf{n}$ are null vectors in M_c. Note that application of complex conjugation on M_c, which is equivalent to replacing i with $-i$, maps the Bondi tetrad into itself, but exchanges $\mathbf{n}_1$ with $\mathbf{n}_2$. This mean that also here i is a pseudo-scalar which with the change of orientation, that may be expressed by change of the order of the basis vectors, changes its sign.

We will denote by y^μ the coordinates of a vector in M with respect to the Bondi tetrad, meaning $x^\mu \mathbf{e}_\mu = y^\mu \mathbf{n}_\mu$. Then, the relation between the coordinates is

$$x^0 = \frac{1}{\sqrt{2}}(y^0 + y^3),\ x^1 = \frac{1}{\sqrt{2}}(y^1 + y^2),\ x^2 = i\frac{1}{\sqrt{2}}(y^2 - y^1),\ x^3 = \frac{1}{\sqrt{2}}(y^0 - y^3), \tag{50}$$

or inversely,

$$y^0 = \frac{1}{\sqrt{2}}(x^0 + x^3),\ y^1 = \frac{1}{\sqrt{2}}(x^1 + ix^2),\ y^2 = \frac{1}{\sqrt{2}}(x^1 - ix^2),\ y^3 = \frac{1}{\sqrt{2}}(x^0 - x^3). \tag{51}$$

The coordinate transformation could be expressed by the transfer matrix $L = L^k_j$ given by

$$L = \frac{1}{\sqrt{2}}\begin{pmatrix} 1 & 0 & 0 & 1 \\ 0 & 1 & i & 0 \\ 0 & 1 & -i & 0 \\ 1 & 0 & 0 & -1 \end{pmatrix},\quad L^{-1} = \frac{1}{\sqrt{2}}\begin{pmatrix} 1 & 0 & 0 & 1 \\ 0 & 1 & 1 & 0 \\ 0 & -i & i & 0 \\ 1 & 0 & 0 & -1 \end{pmatrix}. \tag{52}$$

Thus,

$$y^\nu = L^\nu_\mu x^\mu,\quad x^\nu = (L^{-1})^\nu_\mu y^\mu\,. \tag{53}$$

The metric tensor η in the Bondi tetrad is given by

$$\tilde{\eta} = \begin{pmatrix} 0 & 0 & 0 & 1 \\ 0 & 0 & -1 & 0 \\ 0 & -1 & 0 & 0 \\ 1 & 0 & 0 & 0 \end{pmatrix} = \tilde{\eta}^{-1}. \tag{54}$$

The bilinear symmetric scalar product of two 4-vectors $\mathbf{a} = a^\mu \mathbf{n}_\mu$ and $\mathbf{b} = b^\mu \mathbf{n}_\mu$ is given by

$$\mathbf{a}\cdot\mathbf{b} = \tilde{\eta}_{\mu\nu} a^\mu b^\nu.$$

This, for example, implies that

$$(\mathbf{a})^2 = \mathbf{a}\cdot\mathbf{a} = 0 \ \Leftrightarrow\ a^0 a^3 = a^1 a^2 \ \Leftrightarrow\ \frac{a^2}{a^0} = \frac{a^3}{a^1} \ \Leftrightarrow\ \frac{a^1}{a^0} = \frac{a^3}{a^2}\,. \tag{55}$$

In case $a^0 = 0$, the last identities need to be reversed $\frac{a^0}{a^2} = \frac{a^1}{a^3}$, $\frac{a^0}{a^1} = \frac{a^2}{a^3}$. In these coordinates, the lowering of indices is denoted by $a_\mu = \tilde{\eta}_{\mu\nu} a^\nu$. For example,

$$y_0 = y^3, \;\; y_1 = -y^2, \;\; y_2 = -y^1, \;\; y_3 = y^0. \tag{56}$$

To understand the connection with the spinors of the representation $\tilde{\pi}$, which was based on the complex electromagnetic tensor $\mathcal{F}$, defined by (6) we will present this tensor in Bondi tetrad. We will denote by $\widetilde{\mathcal{F}}$ the matrix of $\mathcal{F}$ in this representation. By the usual formula of basis transformation we get

$$\widetilde{\mathcal{F}}^\beta_\alpha = L\mathcal{F}^\beta_\alpha L^{-1} = \frac{1}{2}\begin{pmatrix} F_3 & F_1 - iF_2 & 0 & 0 \\ F_1 + iF_2 & -F_3 & 0 & 0 \\ 0 & 0 & F_3 & F_1 - iF_2 \\ 0 & 0 & F_1 + iF_2 & -F_3 \end{pmatrix} = \sum_{j=1}^{3} (\tilde{\rho}^j)^\alpha_\beta F_j. \tag{57}$$

This show that the complex electromagnetic tensor become decomposable in Bondi tetrad. This tensor has two invariant subspaces $M_0 = span_{\mathbb{C}}\{\mathbf{n}_0, \mathbf{n}_1\}$ and $M_1 = span_{\mathbb{C}}\{\mathbf{n}_2, \mathbf{n}_3\}$.

Claim 8. *The representation $\tilde{\pi}$ of the Lorentz group on M_c introduced in Definition 1 acts as a spin half representation on each of subspaces $M_0 = span_{\mathbb{C}}\{\mathbf{n}_0, \mathbf{n}_1\}$ and $M_1 = span_{\mathbb{C}}\{\mathbf{n}_2, \mathbf{n}_3\}$*

From the decomposition (57) of the complex electromagnetic tensor $\widetilde{\mathcal{F}}$, we see that in Bondi tetrad Majorana-Oppenheimer matrices $(\tilde{\rho}^j)^\alpha_\beta$ act on each M_k as the the usual Pauli matrices $(\tilde{\rho}^j)^\alpha_\beta M_k = \sigma^j M_k$, with

$$\sigma^1 = \begin{pmatrix} 0 & 1 \\ 1 & 0 \end{pmatrix}, \; \sigma^2 = \begin{pmatrix} 0 & -i \\ i & 0 \end{pmatrix}, \; \sigma^3 = \begin{pmatrix} 1 & 0 \\ 0 & -1 \end{pmatrix}. \tag{58}$$

Generators $\tilde{B}^j$ of boosts $\mathfrak{B}^j$ in direction of x^j in representation $\tilde{\pi}$ were defined (Definition 1) by $\frac{1}{2}\rho^j$. So, on the subspaces M_k they will act by $\tilde{\pi}(\tilde{B}^j) = \frac{1}{2}\sigma^k$. Similarly, generators $\tilde{R}^j$ of rotation $\mathfrak{R}^j$ about the direction x^j in representation $\tilde{\pi}$ were defined by $\frac{i}{2}\rho^j$. So, on the subspaces M_k they will act by $\tilde{\pi}(\tilde{R}^j) = \frac{i}{2}\sigma^k$. This identify the subspace M_k with spinors and the representation $\tilde{\pi}$ with the spin half representation of the Lorentz group on the spinors. This proves the Claim.

The following tensor decomposition can help us to understand the connection between the representations $\tilde{\pi}$ and $\tilde{\pi}^*$. We define a tensor decomposition of a 4×4 matrix as a tensor product of 2×2 matrices by use of the binary representation of numbers. Each of our indices $\mu = 0, 1, 2, 3$ can be considered as a pair of indices (μ_0, μ_1) with value in $\mu_k \in \{0, 1\}$ by

$$0 \longmapsto 00, \;\; 1 \longmapsto 01, \;\; 2 \longmapsto 10, \;\; 3 \longmapsto 11\,.$$

The *tensor decomposition* of a two tensor $D = D_{jk}$ is defined by

$$D = a \otimes b, \;\; D_{\mu\nu} = D_{(\mu_0,\mu_1)(\nu_0,\nu_1)} = a_{\mu_0\nu_0}\, b_{\mu_1\nu_1}\,. \tag{59}$$

For example, the tensor $\tilde{\eta}$ can be decomposed as $\tilde{\eta} = \begin{pmatrix} 0 & 1 \\ -1 & 0 \end{pmatrix} \otimes \begin{pmatrix} 0 & 1 \\ -1 & 0 \end{pmatrix} := \tilde{\eta}_2 \otimes \tilde{\eta}_2\,.$

The following properties of the tensor decomposition can be verified directly from the definition:

$$a \otimes (b + c) = (a \otimes b) + (a \otimes c), \;\; k(a \otimes b) = (ka) \otimes b = a \otimes kb$$

and

$$(a \otimes b)(c \otimes d) = ac \otimes bd, \tag{60}$$

where a, b, c, d are 2×2 matrices and k is a constant.

With this notation,the matrices ρ^j in Bondi tetrad therefore become

$$(\widetilde{\rho^j})^{\mu}_{\nu} = I_2 \otimes \frac{1}{2}\sigma^j. \tag{61}$$

where σ^j denote the Pauli matrices, and I_2 is the 2×2 identity matrix. In Bondi tetrad, the tensor $(\mathcal{F}^*)^{\alpha}_{\beta}$ becomes

$$(\widetilde{\mathcal{F}^*})^{\alpha}_{\beta} = L\mathcal{F}^*L^{-1} = \sum_j \bar{F}_j \sigma^*_j \otimes I_2\,.$$

and

$$(\widetilde{\rho^k})^{\mu}_{\nu} = \sigma^*_j \otimes I_2\,.$$

This and (60) explain why the representations $\tilde{\pi}$ and $\tilde{\pi}^*$ commute.

The connection (22) of the prepotential to the Faraday vector of the field is expressed by operators $\alpha^j := \rho^j \mathcal{C}$. Like the Majorana-Oppenheimer matrices ρ^j, also these matrices satisfy the canonical anti-commutation relations (12). In the Bondi tetrad, see [10]), the matrices α^j take the usual form of the Dirac's α-matrices $\begin{pmatrix} \sigma_j & 0 \\ 0 & -\sigma_j \end{pmatrix}$ where σ^j are the Pauli matrices (see, for example [5]).

Note that the matrices ρ^j, which define the representation $\tilde{\pi}$, also satisfy the canonical anti-commutation relations (12). However, the set ρ_1, ρ_2, ρ_3 cannot be completed by a forth anticommuting matrix, needed for the Dirac equation. Only after the conjugation they become Dirac's α-matrices.

Acknowledgments

Parts of this paper were done in final undergraduate projects at Applied Physics Department at Jerusalem College of Technology by students S. Gwertzman, D. C. Gootvilig and M. Eliyahu under the supervision of the author. The author would like to thank T. Scarr for editorial proof and E. Yudkin for help with the computer algebra use of Mathematica. This research was supported in part by German-Israel Foundation for Scientific Research and Development: GIF No. 1078-107.14/2009.

References

[1] Arfken G 1970 *Mathematical Methods for Physicists* (Academic pres)
[2] Aharonov Y and Bohm D 1959 *Phys. Rev.* **115**, No. 3 485-491
[3] Barut A O 1980 *Electrodynamicsand Classical Theory of Fields & Particles* (Dover Publ.)
[4] Baylis W E 1999 *Electrodynamics, A Modern Geometric Approach*, Progress in Physics **17**, (Birkhäuser, Boston)
[5] Berestetskii V B Lifshitz E M Pitaevskii L P 1982 *Quantum Electrodynamics. Vol. 4 (2nd ed.).* (Butterworth-Heinemann)
[6] Dvoeglazov V V 1998 *Apeiron* **5** 69-88
[7] Friedman Y 2009 arXiv:0901.1040
[8] Friedman Y 2004 *Physical Applications of Homogeneous Balls*, Progress in Mathematical Physics **40** (Birkhäuser,Boston)
[9] Friedman Y Danziger M 2008 PIERS Proceedings 529 - 533
[10] Friedman Y and Gwertzman S 2009 arXiv:0906.0930
[11] Friedman Y and Ostapenko V 2010 *Journal of Physics A:Mathematical and Theoretical* **43** 405305.
[12] Friedman Y and Semon M 2005 *Phys. Rev. E* **72**, No. 2, 026603.
[13] Feynman R P Leighton R B andSands M 1963 *The Feynman Lectures of Physics* v.2 (Addison-Wesley Pub., Reading)
[14] Gersten A 1999 *Found. Phys. Lett.* **12** 291-8, quant-ph/9911049.

[15] Green H S and Wolf E 1953 *Proc. Phys. Soc. A* **66** 1129
[16] Jackson J D 1999 *Classical Electrodynamics*, (Wiley & Sons, New York)
[17] O'Donnell P 2003 *Introduction to 2-Spinors in General Relativity*, (World Scientific Pub.)
[18] Landau L D and Lifschitz E M 1962 *The Classical Theory of Fields*, (Pergamon, Oxford)
[19] Landau L D and Lifschitz E M 1960 *Electrodynamics of Continuous Media* (Volume 8 of A Course of Theoretical Physics,) Pergamon Press)
[20] Penrose R and Rindler W 1986 *Spinors and space-time* v.1 (Cambridge Univercty Press)
[21] Ruse H S 1937 *The Quarterly Journal of Mathematics* **8** 148–160.
[22] Silberstein L 1907 *Ann. Phys. Lpz.* **24**, 783–784
[23] Silberstein L 1927 *The Theory of Relativity* (Macmillan and Co., London)
[24] Whittaker E T 1904 *Proc. London Math. Soc.* **2**, 367–372

Consistent quantization of massless fields of any spin

Alexander Gersten and Amnon Moalem
Department of Physics, Ben Gurion University of the Negev, Beer-Sheva, Israel

Abstract. Consistent quantization of massless fields of any spin is developed from first principles. Probability currents and Lagrangians are derived allowing a first quantized formalism. A simple procedure was derived of connecting the wavefunctions with potentials and gauge conditions. We emphasize the spin one and spin two cases, relating them to the quantum photon (via modified Maxwell's equations) and to the quantum graviton respectively.

Key words: massless fields, any spin, quantization.

1. Introduction

Quantum theory was first formulated via the first quantization of Schroedinger's, Klein-Gordon's and Dirac's equations. In the second quantization Maxwell's equations were treated as classical field equations without referring to the first quantization. Here we present a consistent first quantization of zero mass fields of any spin. The second quantization can be done by an existing standard formalism. We present a detailed analysis of the spin one case, with which we give a first quantized version of Maxwell's equations, and the spin two case relating it to the quantum graviton.

The present paper is a continuation of our previous work [1]. There we have used wavefunctions which formed a basis for the representations $D^{\left(s-\frac{1}{2},\frac{1}{2}\right)}$ of the Lorentz group. This representation describes the combination of two spins, spin s and spin $s-1$. It is $(2s+1)+(2s-1)=4s$ dimensional. By putting the spin $s-1$ components to zero we were able to get equations for spin s with proper subsidiary conditions , which led to only two projections of the helicities, in the forward and backward directions of the momenta, as is required for free massless particles. In the present paper we used non-zero spin $s-1$ components, thus giving a unified formalism for both spins. This has important advantages. It simplifies the formalism. Only after the final derivations the spin $s-1$ components are put to zero for free particles, which generates the proper subsidiary conditions and helicity projections. Moreover it allows to define potentials and their relations to the wavefunctions. The potentials are important, as in the case of Maxwell's equations, for defining interactions with other fields.

In the following we derive a consistent first quantization of massless fields of any spin. Detailed calculations are given for the spin 1 and spin 2 cases. For the spin 1 case generalized Maxwell's equations are derived and potentials defined, relating them to the wavefunction. The Lorenz condition on the potentials (named after Ludvig Lorenz, which is frequently called the Lorentz condition because of confusion with Hendrik Lorentz, after whom the Lorentz transformation and covariance are named) ensures that they describe a spin 1 field.

For the spin 2 case two-helicity wave equations are derived and potentials defined, relating them to the wavefunctions. Six gauge conditions are needed to ensure that they describe a spin 2 field.

2. Dirac's equations for massless particles of any spin

Dirac [2] has derived equations for massless particles with spin s, which in the ordinary vector notation are,

$$\frac{1}{s}\left[s\hat{p}_0 I^{(2s+1)} + S_x\hat{p}_x + S_y\hat{p}_y + S_z\hat{p}_z\right]\psi = \left[\frac{\hat{E}}{c}I^{(2s+1)} + \frac{\boldsymbol{S}}{s}\cdot\hat{\mathbf{p}}\right]\psi^{(2s+1)} = 0, \tag{1}$$

$$\left[s\hat{p}_x I^{(s)} + S_x\hat{p}_0 - iS_y\hat{p}_z + iS_z\hat{p}_y\right]\psi = 0, \tag{2}$$

$$\left[s\hat{p}_y I^{(s)} + S_y\hat{p}_0 - iS_z\hat{p}_x + iS_x\hat{p}_z\right]\psi = 0, \tag{3}$$

$$\left[s\hat{p}_z I^{(s)} + S_z\hat{p}_0 - iS_x\hat{p}_y + iS_y\hat{p}_x\right]\psi = 0, \tag{4}$$

where ψ is a $(2s+1)$ component wave function and S_n are the $(2s+1)\times(2s+1)$ spin matrices which satisfy,

$$[S_x, S_y] = iS_z, \quad [S_z, S_x] = iS_y, \quad [S_y, S_z] = iS_x, \quad S_x^2 + S_y^2 + S_z^2 = s(s+1)I^{(2s+1)}. \tag{5}$$

Above the $\hat{p}_n$ are the momenta, $\hat{p}_0 = \hat{E}/c$, $\hat{E}$ the energy, and $I^{(2s+1)}$ is a $(2s+1)\times(2s+1)$ unit matrix. Eqs. (1-4) were analyzed extensively by Bacry [3], who derived them using Wigner's condition [4] on the Pauli-Lubanski vector W^μ for massless fields,

$$W^\mu = s\hat{p}^\mu, \quad \mu = x_0, x, y, z. \tag{6}$$

Dirac has suggested to use Eq. (1) as the basic helicity equation and substitute from it the $\hat{p}_0\psi$ into the other 3 equations (2-4). Free massless particles may have only two helicity projections (forward and backward along the momrntum vector). Using the Dirac procedure one obtains the $2s+1$ component Eq. (1) and $2s-1$ independent subsidiary conditions which reduces the number of helicity projections to 2,

$$\left[\hat{E}I^{(2s+1)} + \frac{c}{s}\boldsymbol{S}\cdot\hat{\boldsymbol{p}}\right]\psi^{(2s+1)} = 0, \tag{7}$$

$$(\boldsymbol{\Pi}\cdot\hat{\mathbf{p}})\,\psi^{(2s+1)} = 0, \tag{8}$$

where $\boldsymbol{\Pi}$ is a $(2s-1)\times(2s+1)$ vector matrix of the subsidiary conditions. These two sets of equations can be combined into $(2s+1)+(2s-1) = 4s$ equations [1],

$$\left[\hat{E}I^{(4s)} + c\mathbf{\Gamma}^{(4s)} \cdot \hat{\boldsymbol{p}}\right] \Phi^{(4s)} = 0, \tag{9}$$

equivalent to the former equations, provided that the rows of the $\mathbf{\Pi}$ matrices are normalized so that the eigenvalues of $\mathbf{\Gamma}^{(4s)} \cdot \hat{\boldsymbol{p}}$ are $\pm p$ (two helicities) and,

$$\left(\hat{E}I^{(4s)} - c\mathbf{\Gamma}^{(4s)} \cdot \hat{\boldsymbol{p}}\right)\left(\hat{E}I^{(4s)} + c\mathbf{\Gamma}^{(4s)} \cdot \hat{\boldsymbol{p}}\right) = \left(\hat{E}^2 - c^2\hat{\boldsymbol{p}}^2\right) I^{(4s)}, \tag{10}$$

i.e. they factorize the d'Alembertian. The wave function $\Phi^{(4s)}$ is of the form [1] (with 2s-1 zeros),

$$\Phi^{(4s)} = \begin{pmatrix} 0 \\ \vdots \\ 0 \\ \psi_{2s+1}^{(2s+1)} \\ \vdots \\ \psi_{-2s-1}^{(2s+1)} \end{pmatrix}, \tag{11}$$

and the $\mathbf{\Gamma}^{(4s)}$ vector matrices have the form [1]:

$$\Gamma_0^{(4s)} = I^{(4s)} \quad \textit{(the unit matrix)}$$

$$\Gamma_k^{(4s)} = \frac{1}{s}\begin{pmatrix} \begin{pmatrix} -\tilde{S}_k^{(2s-1)} \end{pmatrix} & \begin{pmatrix} \Pi_k \end{pmatrix} \\ \begin{pmatrix} \Pi_k^H \end{pmatrix} & \begin{pmatrix} \tilde{S}_k^{(2s+1)} \end{pmatrix} \end{pmatrix}, \quad k = x, y, z \tag{12}$$

Lomont [5] has proved that Eq. (9) is Lorentz covariant and the zeros of the wave function are unchanged under the Lorentz transformation. We can interpret this result as follows, the wave functions $\Phi^{(4s)}$ of Eq. (11) belong to the basis of the $D^{\left(\frac{1}{2},s-\frac{1}{2}\right)}$ representation of the Lorentz group, which is $4s$ dimensional and its angular momentum decomposition is

$$4s = (2s-1) + (2s+1)\,, \tag{13}$$

i.e. a combination of spin $s-1$ and s. The absence of the spin $s-1$ components indicates that Eq. (9) is a pure spin s equation with two helicities. This should be correct for free particles, but interactions with other fields may introduce non-zero components, therefore the wavefunction $\Phi^{(4s)}$ of Eq. (11) should be complemented with its spin $s-1$ components,

$$\Phi^{(4s)} = \begin{pmatrix} \psi^{(2s-1)}_{2s-1} \\ \vdots \\ \psi^{(2s-1)}_{-2s+1} \\ \psi^{(2s+1)}_{2s+1} \\ \vdots \\ \psi^{(2s+1)}_{-2s-1} \end{pmatrix}. \tag{14}$$

For free particles equation Eq.(9), with $\Phi^{(4s)}$ of Eq. (14), will have a physical solution if,

$$\psi^{(2s-1)}_{\alpha} = 0, \quad \alpha = -2s+1, ..., 2s-1, \tag{15}$$

corresponding to a pure spin s state, moreover all the subsidiary conditions will be satisfied.

We should note that for spin $\frac{1}{2}$ Eq. (10) becomes,

$$[E\sigma_0 - c\mathbf{p}\cdot\sigma]\,[E\sigma_0 + c\mathbf{p}\cdot\sigma]\,\psi = \left(E^2 - c^2\mathbf{p}^2\right)\sigma_0\psi,$$

where the σ are the Pauli matrices

$$\sigma_0 = \begin{pmatrix} 1 & 0 \\ 0 & 1 \end{pmatrix}, \quad \sigma_x = \begin{pmatrix} 0 & 1 \\ 1 & 0 \end{pmatrix}, \quad \sigma_y = \begin{pmatrix} 0 & -i \\ i & 0 \end{pmatrix}, \quad \sigma_z = \begin{pmatrix} 1 & 0 \\ 0 & -1 \end{pmatrix}, \tag{16}$$

thus the Γ matrices form a representation of the Pauli matrices.

3. First quantization

As is in the case of the Klein-Gordon or Dirac equations, we substitute in Eq. (9)

$$\hat{E} \Longrightarrow i\hbar\frac{\partial}{\partial t}, \quad \hat{\mathbf{p}} \Longrightarrow -i\hbar\nabla, \tag{17}$$

and use the wavefunction of Eq. (14),

$$i\hbar\left(\Gamma^{(4s)}_0\frac{\partial}{\partial t} - c\mathbf{\Gamma}^{(4s)}\cdot\nabla\right)\Phi^{(4s)} = 0. \tag{18}$$

Similarly to the Schroedinger and Dirac equations we define the Hamiltonian $\mathcal{H}$ from Eq. (18),

$$i\hbar\frac{\partial}{\partial t}\Phi^{(4s)} = i\hbar c\mathbf{\Gamma}^{(4s)}\cdot\nabla\Phi^{(4s)} = \mathcal{H}\Phi^{(4s)}, \tag{19}$$

and find a conserved probability current (superscript H denotes the Hermitian conjugate),

$$\partial_t\left(\left(\Phi^{(4s)}\right)^H\Phi^{(4s)}\right)+\frac{c}{s}\nabla\cdot\left(\left(\Phi^{(4s)}\right)^H\tilde{\mathbf{S}}\Phi^{(4s)}\right)=0. \tag{20}$$

Thus the probability density (which should be normalized) is,

$$\rho=\left(\Phi^{(4s)}\right)^H\Phi^{(4s)}. \tag{21}$$

Having this result we can find a Lagrangian density,

$$\mathcal{L}=i\hbar\left(\Phi^{(4s)}\right)^H\left(\Gamma_0^{4s}\frac{\partial}{\partial t}-c\mathbf{\Gamma}^{(4s)}\cdot\nabla\right)\Phi^{(4s)}, \tag{22}$$

and using the definition of the energy momentum tensor $T^{\mu\nu}$, we find [1],

$$\int\int\int dxdydzT^{00}=\int\int\int dxdydz\left(\Phi^{(4s)}\right)^H\mathcal{H}\Phi^{(4s)}=<\mathcal{H}>, \tag{23}$$

$$\int\int\int dxdydzT^{0k}=\int\int\int dxdydz\left(\Phi^{(4s)}\right)^H(c\hat{p}_k)\,\Phi^{(4s)}=<c\hat{p}_k>,\quad k=1,2,3, \tag{24}$$

i.e. consistent with the expectation values of energy and momentum.

4. First quantized Maxwell's equations without sources

For spin=1, Eq. (18) becomes,

$$i\hbar\left(\Gamma_0^{(4)}\frac{\partial}{\partial t}-c\mathbf{\Gamma}^{(4)}\cdot\nabla\right)\Phi^{(4)}=0,\quad \Phi^{(4)}=\begin{pmatrix}\Psi_0\\ \Psi_x\\ \Psi_y\\ \Psi_z\end{pmatrix}, \tag{25}$$

$$\Gamma_0^{(4)}=\begin{pmatrix}1&0&0&0\\0&1&0&0\\0&0&1&0\\0&0&0&1\end{pmatrix},\quad \Gamma_1^{(4)}=\begin{pmatrix}0&-1\;0\;0\\-1&\\0&(S_x)\\0&\end{pmatrix}, \tag{26}$$

$$\Gamma_2^{(4)}=\begin{pmatrix}0&0\;-1\;0\\0&\\-1&(S_y)\\0&\end{pmatrix},\quad \Gamma_3^{(4)}=\begin{pmatrix}0&0\;0\;-1\\0&\\0&(S_z)\\-1&\end{pmatrix}$$

If we substitute,

$$\Phi^{(4)} = \begin{pmatrix} \Psi_0 \\ \Psi_x \\ \Psi_y \\ \Psi_z \end{pmatrix} = \mathcal{N} \begin{pmatrix} E_0 - iB_0 \\ E_x - iB_x \\ E_y - iB_y \\ E_z - iB_z \end{pmatrix}, \tag{27}$$

and assume $\Psi_0 = 0$, the free field Maxwell's equations are obtained for the electric $\boldsymbol{E}$ and magnetic field $\boldsymbol{B}$. Above $\mathcal{N}$ is a normalization factor. The subsidiary condition is equivalent to

$\Psi_0 = 0$, but Maxwell's Eqs. with sources may induce $\Psi_0 \neq 0$.

Four vectors belong to the $D^{\left(\frac{1}{2},\frac{1}{2}\right)}$ representation of the Lorentz group, its basis combines the spin 0 and spin 1 bases. Therefore $\Psi_0 = 0$ eliminates the spin 0 component.

The probability density becomes:

$$\rho = \left(\Phi^{(4)}\right)^H \Phi^{(4)} = \mathcal{N}^2 \left(E_0^2 + B_0^2 + \boldsymbol{E}^2 + \boldsymbol{B}^2\right), \tag{28}$$

and the energy density:

$$\left(\Phi^{(4)}\right)^H \mathcal{H}\Phi^{(4)} = \left(\Phi^{(4)}\right)^H \left(i\hbar c\boldsymbol{\Gamma}^{(4)} \cdot \nabla\right) \Phi^{(4)}. \tag{29}$$

Let us introduce the potentials A^μ,

$$A^{(4)} = \begin{pmatrix} A_0 \\ A_x \\ A_y \\ A_z \end{pmatrix},$$

Their relation to the wavefunction $\Phi^{(4)}$ can be obtained by requiring that in the Lorenz gauge they will satisfy the wave equation,

$$\partial_\mu \partial^\mu A^{(4)} = 0. \tag{30}$$

Using Eq.(10) we find that,

$$\Phi^{(4)} = \mathcal{N} \begin{pmatrix} E_0 - iB_0 \\ E_x - iB_x \\ E_y - iB_y \\ E_z - iB_z \end{pmatrix} = -\mathcal{N} \left(\Gamma_0^{(4)} \frac{\partial}{\partial t} + c\boldsymbol{\Gamma}^{(4)} \cdot \nabla\right) \begin{pmatrix} A_0 \\ A_x \\ A_y \\ A_z \end{pmatrix} \tag{31}$$

$$= -i\hbar\mathcal{N} \begin{pmatrix} \partial_0 A_0 + \partial_1 A_x + \partial_2 A_y + \partial_3 A_z \\ \partial_0 A_x + \partial_1 A_0 - i\left(\partial_2 A_z - \partial_3 A_y\right) \\ \partial_0 A_y + \partial_2 A_0 + i\left(\partial_1 A_z - \partial_3 A_x\right) \\ \partial_0 A_z + \partial_3 A_0 - i\left(\partial_1 A_y - \partial_2 A_x\right) \end{pmatrix}, \tag{32}$$

from which the relation between the fields and the potentials (in the Lorenz gauge) is,

$$E_0 = -\partial_0 A_0 - \nabla \cdot \mathbf{A}, \tag{33}$$

$$B_0 = 0, \tag{34}$$

$$\mathbf{E} = -\nabla A_0 - \partial_0 \mathbf{A}, \tag{35}$$

$$\mathbf{B} = \nabla \times \mathbf{A}. \tag{36}$$

The Lorenz condition

$$\partial_0 A_0 + \partial_1 A_x + \partial_2 A_y + \partial_3 A_z = 0, \tag{37}$$

eliminates the spin 0 component from the wave function.

5. First quantization of spin 2 (graviton?) equations

We proceed as in the section of first quantization of Maxwell's equations. The wavefunction is,

$$\Phi^{(8)} = \begin{pmatrix} \Psi_1^{(1)} \\ \Psi_0^{(1)} \\ \Psi_{-1}^{(1)} \\ \Psi_2^{(2)} \\ \Psi_1^{(2)} \\ \Psi_0^{(2)} \\ \Psi_{-1}^{(2)} \\ \Psi_{-2}^{(2)} \end{pmatrix}, \tag{38}$$

and the wave equation (see Eq.(18)) is,

$$i\hbar \left(\Gamma_0^{(8)} \frac{\partial}{\partial t} - c\mathbf{\Gamma}^{(8)} \cdot \nabla \right) \Phi^{(8)} = 0. \tag{39}$$

The $\mathbf{\Gamma}^{(8)}$ matrices are given in Eq.(12) with the subsidiary submatrices Π_k,

$$\Pi_x = \begin{pmatrix} \sqrt{3} & 0 & -\frac{1}{\sqrt{2}} & 0 & 0 \\ 0 & \sqrt{\frac{3}{2}} & 0 & -\sqrt{\frac{3}{2}} & 0 \\ 0 & 0 & \frac{1}{\sqrt{2}} & 0 & -\sqrt{3} \end{pmatrix}, \tag{40}$$

$$\Pi_y = i \begin{pmatrix} -\sqrt{3} & 0 & -\frac{1}{\sqrt{2}} & 0 & 0 \\ 0 & -\sqrt{\frac{3}{2}} & 0 & -\sqrt{\frac{3}{2}} & 0 \\ 0 & 0 & -\frac{1}{\sqrt{2}} & 0 & -\sqrt{3} \end{pmatrix}, \tag{41}$$

$$\Pi_z = \begin{pmatrix} 0 & -\sqrt{3} & 0 & 0 & 0 \\ 0 & 0 & -2 & 0 & 0 \\ 0 & 0 & 0 & -\sqrt{3} & 0 \end{pmatrix}. \tag{42}$$

If we impose,

$$\Psi_1^{(1)} = \Psi_0^{(1)} = \Psi_{-1}^{(1)} = 0, \tag{43}$$

the spin 1 contribution is eliminated and the free field equations for the the spin 2 field are obtained with the proper subsidiary conditions.

Let us introduce the potentials,

$$A^{(8)} = \begin{pmatrix} \mathcal{A}_1^{(1)} \\ \mathcal{A}_0^{(1)} \\ \mathcal{A}_{-1}^{(1)} \\ \mathcal{A}_2^{(2)} \\ \mathcal{A}_1^{(2)} \\ \mathcal{A}_0^{(2)} \\ \mathcal{A}_{-1}^{(2)} \\ \mathcal{A}_{-2}^{(2)} \end{pmatrix}. \tag{44}$$

Their relation to the wavefunction $\Phi^{(8)}$ can be obtained by requiring that in a certain gauge they will satisfy the wave equation,

$$\partial_\mu \partial^\mu A^{(8)} = 0. \tag{45}$$

This will happen if,

$$\Phi^{(8)} = \mathcal{N} \left(\Gamma_0^{(8)} \frac{\partial}{\partial t} + c\mathbf{\Gamma}^{(8)} \cdot \nabla \right) A^{(8)}, \tag{46}$$

where $\mathcal{N}$ is a normalization constant. The result is,

$$\frac{4\Phi^{(8)}}{\mathcal{N}} = \begin{pmatrix} 4\partial_0\mathcal{A}_1^{(1)} - \partial_z\mathcal{A}_1^{(1)} + \frac{1}{\sqrt{2}}\partial_x\mathcal{A}_0^{(2)} - \sqrt{3}\partial_x\mathcal{A}_2^{(2)} + \sqrt{3}\partial_z\mathcal{A}_1^{(2)} - \frac{1}{\sqrt{2}}\partial_x\mathcal{A}_0^{(1)} \\ 2\partial_z\mathcal{A}_0^{(2)} + 4\partial_0\mathcal{A}_0^{(1)} - \frac{1}{\sqrt{2}}\partial_x\mathcal{A}_{-1}^{(1)} - \frac{1}{\sqrt{2}}\partial_x\mathcal{A}_1^{(1)} - \sqrt{\frac{3}{2}}\partial_x\mathcal{A}_1^{(2)} + \sqrt{\frac{3}{2}}\partial_x\mathcal{A}_{-1}^{(2)} \\ 4\partial_0\mathcal{A}_{-1}^{(1)} + \partial_z\mathcal{A}_{-1}^{(1)} + \sqrt{3}\partial_x\mathcal{A}_{-2}^{(2)} + \sqrt{3}\partial_z\mathcal{A}_{-1}^{(2)} - \frac{1}{\sqrt{2}}\partial_x\mathcal{A}_0^{(2)} - \frac{1}{\sqrt{2}}\partial_x\mathcal{A}_0^{(1)} \\ 4\partial_0\mathcal{A}_2^{(2)} + \partial_x\mathcal{A}_1^{(2)} + 2\partial_z\mathcal{A}_2^{(2)} - \sqrt{3}\partial_x\mathcal{A}_1^{(1)} \\ 4\partial_0\mathcal{A}_1^{(2)} + \partial_x\mathcal{A}_2^{(2)} + \partial_z\mathcal{A}_1^{(2)} + \sqrt{3}\partial_z\mathcal{A}_1^{(1)} + \sqrt{\frac{3}{2}}\partial_x\mathcal{A}_0^{(2)} - \sqrt{\frac{3}{2}}\partial_x\mathcal{A}_0^{(1)} \\ 4\partial_0\mathcal{A}_0^{(2)} + 2\partial_z\mathcal{A}_0^{(1)} - \frac{1}{\sqrt{2}}\partial_x\mathcal{A}_{-1}^{(1)} + \frac{1}{\sqrt{2}}\partial_x\mathcal{A}_1^{(1)} + \sqrt{\frac{3}{2}}\partial_x\mathcal{A}_1^{(2)} + \sqrt{\frac{3}{2}}\partial_x\mathcal{A}_{-1}^{(2)} \\ 4\partial_0\mathcal{A}_{-1}^{(2)} + \partial_x\mathcal{A}_{-2}^{(2)} - \partial_z\mathcal{A}_{-1}^{(2)} + \sqrt{3}\partial_z\mathcal{A}_{-1}^{(1)} + \sqrt{\frac{3}{2}}\partial_x\mathcal{A}_0^{(2)} + \sqrt{\frac{3}{2}}\partial_x\mathcal{A}_0^{(1)} \\ 4\partial_0\mathcal{A}_{-2}^{(2)} + \partial_x\mathcal{A}_{-1}^{(2)} - 2\partial_z\mathcal{A}_{-2}^{(2)} + \sqrt{3}\partial_x\mathcal{A}_{-1}^{(1)} \end{pmatrix}$$

$$+i\partial_y \begin{pmatrix} \frac{1}{\sqrt{2}}\mathcal{A}_0^{(2)} - \frac{1}{\sqrt{2}}\mathcal{A}_0^{(1)} + \sqrt{3}\mathcal{A}_2^{(2)} \\ \sqrt{\frac{3}{2}}\mathcal{A}_{-1}^{(2)} + \sqrt{\frac{3}{2}}\mathcal{A}_1^{(2)} - \frac{1}{\sqrt{2}}\mathcal{A}_{-1}^{(1)} + \frac{1}{\sqrt{2}}\mathcal{A}_1^{(1)} \\ \frac{1}{\sqrt{2}}\mathcal{A}_0^{(1)} + \frac{1}{\sqrt{2}}{}_y\mathcal{A}_0^{(2)} + \sqrt{3}\mathcal{A}_{-2}^{(2)} \\ \mathcal{A}_1^{(2)} - \sqrt{3}\mathcal{A}_1^{(1)} \\ \sqrt{\frac{3}{2}}\mathcal{A}_0^{(2)} - \sqrt{\frac{3}{2}}\mathcal{A}_0^{(1)} - \mathcal{A}_2^{(2)} \\ \sqrt{\frac{3}{2}}\mathcal{A}_{-1}^{(2)} - \frac{1}{\sqrt{2}}\mathcal{A}_{-1}^{(1)} - \frac{1}{\sqrt{2}}\mathcal{A}_1^{(1)} - \sqrt{\frac{3}{2}}\mathcal{A}_1^{(2)} \\ \mathcal{A}_{-2}^{(2)} - \sqrt{\frac{3}{2}}\mathcal{A}_0^{(2)} - \sqrt{\frac{3}{2}}\mathcal{A}_0^{(1)} \\ -\mathcal{A}_{-1}^{(2)} + \sqrt{3}\mathcal{A}_{-1}^{(1)} \end{pmatrix}. \tag{47}$$

If the three upper components (for the real and imaginary parts) will be equal to zero, the spin 1 contribution will be eliminated including the correct subsidiary conditions. Thus we have six gauge conditions,

$$4\partial_0\mathcal{A}_1^{(1)} - \partial_z\mathcal{A}_1^{(1)} + \frac{1}{\sqrt{2}}\partial_x\mathcal{A}_0^{(2)} - \sqrt{3}\partial_x\mathcal{A}_2^{(2)} + \sqrt{3}\partial_z\mathcal{A}_1^{(2)} - \frac{1}{\sqrt{2}}\partial_x\mathcal{A}_0^{(1)} = 0, \tag{48}$$

$$2\partial_z\mathcal{A}_0^{(2)} + 4\partial_0\mathcal{A}_0^{(1)} - \frac{1}{\sqrt{2}}\partial_x\mathcal{A}_{-1}^{(1)} - \frac{1}{\sqrt{2}}\partial_x\mathcal{A}_1^{(1)} - \sqrt{\frac{3}{2}}\partial_x\mathcal{A}_1^{(2)} + \sqrt{\frac{3}{2}}\partial_x\mathcal{A}_{-1}^{(2)} = 0, \tag{49}$$

$$4\partial_0\mathcal{A}_{-1}^{(1)} + \partial_z\mathcal{A}_{-1}^{(1)} + \sqrt{3}\partial_x\mathcal{A}_{-2}^{(2)} + \sqrt{3}\partial_z\mathcal{A}_{-1}^{(2)} - \frac{1}{\sqrt{2}}\partial_x\mathcal{A}_0^{(2)} - \frac{1}{\sqrt{2}}\partial_x\mathcal{A}_0^{(1)} = 0, \tag{50}$$

$$\partial_y\left(\frac{1}{\sqrt{2}}\mathcal{A}_0^{(2)} - \frac{1}{\sqrt{2}}\mathcal{A}_0^{(1)} + \sqrt{3}\mathcal{A}_2^{(2)}\right) = 0, \tag{51}$$

$$\partial_y\left(\sqrt{\frac{3}{2}}\mathcal{A}_{-1}^{(2)} + \sqrt{\frac{3}{2}}\mathcal{A}_1^{(2)} - \frac{1}{\sqrt{2}}\mathcal{A}_{-1}^{(1)} + \frac{1}{\sqrt{2}}\mathcal{A}_1^{(1)}\right) = 0, \tag{52}$$

$$\partial_y\left(\frac{1}{\sqrt{2}}\mathcal{A}_0^{(1)} + \frac{1}{\sqrt{2}}\mathcal{A}_0^{(2)} + \sqrt{3}\mathcal{A}_{-2}^{(2)}\right) = 0, \tag{53}$$

which ensure the pure spin 2 properties.

6. Summary

Wavefunction equations were derived for massless fields of any spin. Consistent first quantization was worked out.

A simple procedure was derived of connecting the wavefunctions with potentials and gauge conditions.

Detailed calculations were worked out for the spin 1 and spin 2 cases. The results of the spin 1 case lead to a new quantization of Maxwell's equations. The Lorenz condition on the potentials ensures that they describe a spin 1 field.

We are continuing to work on the spin 2 case in order to connect it with quantum gravity. Six gauge conditions on the potentials were found to ensure that they describe a spin 2 field.

[1] A. Gersten and A. Moalem, Journal of Physics: Conference Series 330 (2011) 012010.
[2] P.A.M. Dirac, *Proc. Roy. Soc.* **A155**, 447-59 (1936).
[3] H. Bacry, *Nuovo Cimento*, **A32**, 448-60 (1976).
[4] E.P. Wigner, *Ann. Math.*, **40,** 149 (1939).
[5] J.S. Lomont, *Phys. Rev.* **111**, 1710-1716 (1958)

Functional methods underlying classical mechanics, relativity and quantum theory

A. Kryukov

Department of Mathematics, University of Wisconsin Colleges, 780 Regent Street, Madison, Wisconsin 53708, USA

Abstract.
The paper investigates the physical content of a recently proposed mathematical framework that unifies the standard formalisms of classical mechanics, relativity and quantum theory. In the framework states of a classical particle are identified with Dirac delta functions. The classical space is "made" of these functions and becomes a submanifold in a Hilbert space of states of the particle. The resulting embedding of the classical space into the space of states is highly non-trivial and accounts for numerous deep relations between classical and quantum physics and relativity. One of the most striking results is the proof that the normal probability distribution of position of a macroscopic particle (equivalently, position of the corresponding delta state within the classical space submanifold) yields the Born rule for transitions between arbitrary quantum states.

1. Introduction

Most of the modern theoretical physics is built on the following well accepted and experimentally verified theories: non-relativistic and relativistic classical mechanics, special relativity and Maxwell's electrodynamics, Einstein theory of gravity, quantum mechanics, quantum electrodynamics and gauge field theory. Despite the fact that each one of these theories embraces a wide range of phenomena, their interrelationship is far from being clear and leads to the well known difficulties and paradoxes. For instance, the special role played by the classical mechanical devices in quantum theory is a basis of the so-called measurement problem. The classical meaning of time in quantum mechanics makes it difficult to merge this theory with special relativity. The relationship of the quantum field theory and quantum mechanics is also far from being understood. General relativity so far has resisted all attempts of being quantized.

To thoroughly understand the relationship between these theories it is necessary to find a common mathematical language underlying them. Currently the classical mechanics, classical gauge theory and the theory of gravity employ the language of differential geometry of finite dimensional differential manifolds with a metric and possible additional structures, and fibre bundles over such manifolds with a connection. On the other hand, mathematics of the quantum theory is based on algebras of linear operators in Hilbert spaces and the theory of representations of groups. It is expected that, when properly understood, quantum theory will contain the classical theory in a certain limit. If so, then the problem of finding a common language for the theories is, loosely speaking, the problem of encoding the non-linear theory of finite dimensional manifolds into (an extension of) the linear theory of infinite-dimensional Hilbert spaces and linear operators on them.

A possible solution to the problem was recently proposed and investigated in Refs.[1]-[6]. The proposed formalism combines the strength of the functional analysis in Hilbert spaces, the theory of generalized functions, and differential geometry of Banach manifolds. It allows one to identify the classical space and space-time with submanifolds in a Hilbert space of quantum states. The classical mechanics and relativity theory are then reformulated in functional terms. The reformulation yields a new relativistic framework for quantum mechanics. In the process quantum observables are identified with vector fields on the space of states. Quantum commutators become the Lie brackets of the fields. Other important physically meaningful identifications follow.

The goal of this paper is to demonstrate that the classical particle mechanics, relativistic mechanics, dynamics of particles in gravitational field and quantum mechanics can be all formulated in a very elegant way using the new formalism and the principle of least action. The obtained unified treatment is then applied to shed new light into the nature of quantum uncertainty and the problem of the relationship between classical and quantum phenomena.

2. Classical mechanics in Hilbert space

The state of a spinless particle with a known position $\mathbf{a}$ in the Euclidean space $\mathbb{R}^3$ is described in quantum mechanics by the delta function $\delta^3_{\mathbf{a}}(\mathbf{x}) = \delta^3(\mathbf{x}-\mathbf{a})$. The map $\omega : \mathbf{a} \longrightarrow \delta^3_{\mathbf{a}}$ provides a one-to-one correspondence between points $\mathbf{a}$ and states $\delta^3_{\mathbf{a}}$. In the single-particle case the classical space $\mathbb{R}^3$ will be then identified with the set of all delta functions in the space of state functions of the particle. To begin with, we need a Hilbert space that contains delta functions.

Consider the usual Hilbert space $L_2(\mathbb{R}^3)$ of Lebesgue square-integrable functions on $\mathbb{R}^3$. Note that the inner product of functions $\varphi, \psi \in L_2(\mathbb{R}^3)$ can be formally written as follows:

$$(\varphi,\psi)_{L_2} = \int \delta^3(\mathbf{x}-\mathbf{y})\varphi(\mathbf{x})\overline{\psi}(\mathbf{y})d^3\mathbf{x}d^3\mathbf{y}. \tag{1}$$

In particular, the fact that delta functions are not in $L_2(\mathbb{R}^3)$ is related to the singularity of the kernel $\delta^3(\mathbf{x}-\mathbf{y})$ of the Hilbert metric. We can approximate this kernel by the Gaussian function $\left(\frac{L}{\sqrt{2\pi}}\right)^3 e^{-\frac{L^2}{2}(\mathbf{x}-\mathbf{y})^2}$ for some positive constant L. This yields the product

$$(\varphi,\psi)_{\mathbf{H}} = \left(\frac{L}{\sqrt{2\pi}}\right)^3 \int e^{-\frac{L^2}{2}(\mathbf{x}-\mathbf{y})^2}\varphi(\mathbf{x})\overline{\psi}(\mathbf{y})d^3\mathbf{x}d^3\mathbf{y}. \tag{2}$$

One can check that this is indeed an inner product on $L_2(\mathbb{R}^3)$. The Hilbert space $\mathbf{H}$ obtained by completing the space $L_2(\mathbb{R}^3)$ in this inner product contains delta functions $\delta^3(\mathbf{x}-\mathbf{a})$ and their derivatives. Moreover, by choosing L sufficiently large (or by choosing appropriate units), one can make the norm of any square-integrable function in this metric as close as desired to its $L_2(\mathbb{R}^3)$-norm. This relationship will be denoted by $\mathbf{H} \approx L_2(\mathbb{R}^3)$. Choosing for now $L=1$ and dropping the coefficient $(1/\sqrt{2\pi})^3$ we obtain the product

$$(\varphi,\psi)_{\mathbf{H}} = \int e^{-\frac{1}{2}(\mathbf{x}-\mathbf{y})^2}\varphi(\mathbf{x})\overline{\psi}(\mathbf{y})d^3\mathbf{x}d^3\mathbf{y}. \tag{3}$$

Formally,

$$\int e^{-\frac{1}{2}(\mathbf{x}-\mathbf{y})^2}\delta^3(\mathbf{x}-\mathbf{a})\delta^3(\mathbf{y}-\mathbf{a})d^3\mathbf{x}d^3\mathbf{y} = 1, \tag{4}$$

so that the norm of the delta function $\delta^3(\mathbf{x}-\mathbf{a})$ in the resulting space $\mathbf{H}$ is 1.

Consider the set M_3 of all delta functions $\delta^3_{\mathbf{a}}(\mathbf{x}) \equiv \delta^3(\mathbf{x}-\mathbf{a})$ in $\mathbf{H}$. To know position $\mathbf{a}$ of a classical particle in $\mathbb{R}^3$ is to know the corresponding point $\delta^3_{\mathbf{a}}$ in M_3. Consider a path $\mathbf{r} = \mathbf{a}(t)$ with values in $\mathbb{R}^3$ and the corresponding path $\varphi = \delta^3_{\mathbf{a}(t)}$ in M_3. With the use of the chain rule the velocity vector $d\varphi/dt$ can be written as

$$\frac{d\varphi}{dt} = -\frac{\partial}{\partial \mathbf{x}^i}\delta^3(\mathbf{x}-\mathbf{a})\frac{d\mathbf{a}^i}{dt}, \tag{5}$$

where the usual summation convention for repeating indices is accepted. It follows that the norm of the velocity in the space $\mathbf{H}$ is

$$\left\|\frac{d\varphi}{dt}\right\|^2_H = \int k(\mathbf{x},\mathbf{y})\frac{\partial}{\partial x^i}\delta^3(\mathbf{x}-\mathbf{a})\frac{d\mathbf{a}^i}{dt}\frac{\partial}{\partial y^k}\delta^3(\mathbf{y}-\mathbf{a})\frac{d\mathbf{a}^k}{dt}d^3\mathbf{x}d^3\mathbf{y}, \tag{6}$$

where $k(\mathbf{x},\mathbf{y}) = e^{-\frac{1}{2}(\mathbf{x}-\mathbf{y})^2}$. "Integration by parts" in the last expression gives

$$\left\|\frac{d\varphi}{dt}\right\|^2_H = \left.\frac{\partial^2 k(\mathbf{x},\mathbf{y})}{\partial x^i \partial y^k}\right|_{\mathbf{x}=\mathbf{y}=\mathbf{a}}\frac{d\mathbf{a}^i}{dt}\frac{d\mathbf{a}^k}{dt}. \tag{7}$$

Because

$$\left.\frac{\partial^2 k(\mathbf{x},\mathbf{y})}{\partial x^i \partial y^k}\right|_{\mathbf{x}=\mathbf{y}=\mathbf{a}} = \delta_{ik}, \tag{8}$$

where δ_{ik} is the Kronecker delta symbol, we obtain equality of the speeds

$$\left\|\frac{d\varphi}{dt}\right\|_H = \left\|\frac{d\mathbf{a}}{dt}\right\|_{\mathbb{R}^3}. \tag{9}$$

Accordingly, the set M_3 as a metric subspace of $\mathbf{H}$ is identical to the Euclidean space $\mathbb{R}^3$. Notice however that M_3 is not a vector subspace of $\mathbf{H}$. Rather, because the norm of delta functions in $\mathbf{H}$ is 1, the metric space M_3 is a submanifold of the unit sphere $S^{\mathbf{H}}$ in $\mathbf{H}$. Since delta functions $\delta^3_{\mathbf{a}_k}$ with different $\mathbf{a}_k$, $k = 1, ..., n$ are linearly independent, the manifold M_3 "spirals" through dimensions of the sphere, forming a complete subset of $\mathbf{H}$. This means that no function in $\mathbf{H}$ is orthogonal to the submanifold M_3.

Nevertheless, a vector structure on M_3 exists. Namely, the operations of addition $\oplus$ and multiplication by a scalar $\lambda\odot$ can be defined via $\omega(\mathbf{a}) \oplus \omega(\mathbf{b}) = \omega(\mathbf{a}+\mathbf{b})$ and $\lambda \odot \omega(\mathbf{a}) = \omega(\lambda\mathbf{a})$, where the map $\omega : \mathbb{R}^3 \longrightarrow \mathbf{H}$ is given by $\omega : \mathbf{a} \longrightarrow \delta^3_{\mathbf{a}}$. The resulting operations are continuous in the topology of $M_3 \subset \mathbf{H}$. That is, the metric space M_3 with this vector structure is isomorphic to the vector space $\mathbb{R}^3$ with the Euclidean metric.

Now that a functional realization of the classical space $\mathbb{R}^3$ and of position $\mathbf{a}$ and velocity $\frac{d\mathbf{a}}{dt}$ of a material point in the space are found, the dynamics of the point can be derived from the principle of least action. Namely, consider the action functional

$$S = \int e^{-\frac{1}{2}(\mathbf{x}-\mathbf{y})^2}\left[\frac{m}{2}\frac{d\varphi_t(\mathbf{x})}{dt}\frac{d\overline{\varphi}_t(\mathbf{y})}{dt} - V(x)\varphi_t(\mathbf{x})\overline{\varphi}_t(\mathbf{y})\right]d^3\mathbf{x}d^3\mathbf{y}dt, \tag{10}$$

where m is the mass of the particle, V is the potential, and φ_t is constrained to be on the submanifold $M_3 \subset \mathbf{H}$, i.e., $\varphi_t(\mathbf{x}) = \delta^3(\mathbf{x}-\mathbf{a}(t))$. Using (5) and integrating by parts as in (7), we immediately obtain

$$S = \int\left[\frac{m}{2}\left(\frac{d\mathbf{a}}{dt}\right)^2 - V(\mathbf{a})\right]dt, \tag{11}$$

i.e., the usual action functional for a material point in classical mechanics.

This demonstrates that the classical mechanics of a material point can be formulated in purely functional terms. Position and velocity of the point are now given by functions in a Hilbert space. The dynamics of the point is derived from the principle of least action by variation of paths in the Hilbert space. The condition that the position of the point at any moment of time is well defined is the constraint $\varphi_t(\mathbf{x}) = \delta^3(\mathbf{x} - \mathbf{a}(t))$. The relation between $d\varphi_t/dt$ and $d\mathbf{a}/dt$ is given by (5) and (9). That is, the velocity $d\varphi_t/dt$ of state in the particular case when the state travels along the submanifold M_3 gets identified with the usual velocity $d\mathbf{a}/dt$.

Note that the momentum of the delta state has infinite dispersion when the $L_2(\mathbb{R}^3)$ metric is used. The dispersion is finite in the $\mathbf{H}$ metric. Alternatively, the motion $\varphi_t(\mathbf{x}) = \delta^3(\mathbf{x} - \mathbf{v}t)$ in $\mathbf{H}$ can be identified with the motion of a Gaussian packet with group velocity $\mathbf{v}$ in the space $L_2(\mathbb{R}^3)$. The packet is not an eigenstate of the momentum operator, but its group velocity is well defined. The velocity $d\varphi_t/dt$ gets identified with the group velocity $\mathbf{v}$.

Note also that because state functions considered here are not the eigenstates of the momentum operator, a simultaneous description of position and velocity of the particles in the constructed realization does not contradict the uncertainty principle. Finally, as briefly discussed in section 12, systems of n interacting particles can be considered in a similar way.

3. Relativistic mechanics in Hilbert space

A functional realization of the relativistic mechanics requires a space of functions of four variables $\mathbf{x}, t$. Let's begin with the Hilbert space $L_2(\mathbb{R}^4)$ of square-integrable functions of four variables $x = (\mathbf{x}, t)$ and complete it in the metric given by the kernel $e^{-\frac{1}{2}(x-y)^2}$. As before, the resulting Hilbert space $\widetilde{H}$ contains delta functions δ^4_a and the set M_4 of all delta functions forms a submanifold of $\widetilde{H}$ with the induced metric of the Euclidean space $\mathbb{R}^4$.

To obtain the Minkowski space metric consider the Hermitian form $(f, g)_{H_\eta}$ given in the units where the speed of light c is 1 by

$$(f, g)_{H_\eta} = \int e^{-\frac{1}{2}(\mathbf{x}-\mathbf{y})^2 + \frac{1}{2}(t-s)^2} f(\mathbf{x}, t)\overline{g(\mathbf{y}, s)} d^3\mathbf{x} dt d^3\mathbf{y} ds. \tag{12}$$

Let $(f, f)_{H_\eta} \equiv \|f\|^2_{H_\eta}$ be the corresponding quadratic form, or the squared H_η-norm. Because of the positive term in the exponent of the kernel in (12), not every function in the Hilbert space $\widetilde{H}$ has a finite H_η-norm. In addition, the quadratic form $(f, f)_{H_\eta}$ is not positive-definite. More precisely, if $f \neq 0$ and $f(\mathbf{x}, t)$ is even in t, then $(f, f)_{H_\eta} > 0$. Likewise, if $f \neq 0$ and $f(\mathbf{x}, t)$ is odd in t, then $(f, f)_{H_\eta} < 0$. So, let H be the set of functions in $\widetilde{H}$ whose even and odd in t components have a finite H_η-norm. As shown in Ref.[2], H is exactly the set of all functions $f(\mathbf{x}, t) = e^{-t^2}\varphi(\mathbf{x}, t)$ with $\varphi \in \widetilde{H}$. Moreover, H furnished with the inner product $(f, g)_{H_+} = (\varphi, \psi)_{\widetilde{H}}$, where $f(\mathbf{x}, t) = e^{-t^2}\varphi(\mathbf{x}, t)$, $g(\mathbf{x}, t) = e^{-t^2}\psi(\mathbf{x}, t)$; is a Hilbert space. The Hermitian form (12) defines an indefinite, non-degenerate inner product on H. Finally, H contains the delta functions $\delta^4_a(x) = \delta^4(x - a)$ and their derivatives for all $a \in \mathbb{R}^4$.

The obtained Hilbert space H with the additional indefinite metric is an example of what is called a *Krein space.* Let M_4 be the set of all delta functions in H and let N be the Minkowski space-time. Then, similarly to the case of the Euclidean space $\mathbb{R}^3$, the set M_4 is a submanifold of the Hilbert space H that is diffeomorphic to N. Moreover, the indefinite metric on H yields the Minkowski metric on M_4. In fact, consider a path $x = a(\tau)$ with values in N and the corresponding path $\varphi = \delta^4_{a(\tau)}$ in M_4. As in (5), the velocity vector $d\varphi/d\tau$ can be written as

$$\frac{d\varphi}{d\tau} = -\frac{\partial}{\partial x^\mu}\delta^4(x-a)\frac{da^\mu}{d\tau}, \tag{13}$$

where summation goes over $\mu = 0, 1, 2, 3$. It follows that the norm of the velocity in the indefinite metric on the space H is

$$\left\|\frac{d\varphi}{d\tau}\right\|^2_{H_\eta} = \int k(x,y)\frac{\partial}{\partial x^\mu}\delta^4(x-a)\frac{da^\mu}{d\tau}\frac{\partial}{\partial y^\nu}\delta^4(y-a)\frac{da^\nu}{d\tau}d^4xd^4y, \tag{14}$$

where $k(x,y) = e^{-\frac{1}{2}(x-y)^2_\eta}$ and η stands for the Minkowski norm. "Integration by parts" in the last expression gives

$$\left\|\frac{d\varphi}{d\tau}\right\|^2_{H_\eta} = \left.\frac{\partial^2 k(x,y)}{\partial x^\mu \partial y^\nu}\right|_{x=y=a}\frac{da^\mu}{d\tau}\frac{da^\nu}{d\tau} \tag{15}$$

and since $\left.\frac{\partial^2 k(x,y)}{\partial x^\mu \partial y^\nu}\right|_{x=y=a}$ is equal to the Minkowski metric $\eta_{\mu\nu}$, we have

$$\left\|\frac{d\varphi}{d\tau}\right\|_{H_\eta} = \left\|\frac{da}{d\tau}\right\|_\eta, \tag{16}$$

i.e., the equality of the speed of evolution in the Hilbert space H to the usual 4-speed. In particular, if τ is the proper time parameter, $\left\|\frac{d\varphi}{d\tau}\right\|_{H_\eta}$ is equal to the speed of light (i.e., 1 in the chosen units).

A motion of a macroscopic particle in relativistic mechanics is now realized as a motion in $M_4 \subset H$. The dynamics of a free particle follow from the principle of the least action for the square-length action functional

$$S = \frac{m}{2}\int k(x,y)\frac{d\varphi_\tau(x)}{d\tau}\frac{d\overline{\varphi}_\tau(y)}{d\tau}d^4xd^4yd\tau, \tag{17}$$

$k(x,y) = e^{-\frac{1}{2}(x-y)^2_\eta}$, under the constraint $\varphi_\tau(x) = \delta^4(x-a(\tau))$. In fact, for S with this constraint we have

$$S = \frac{m}{2}\int \left\|\frac{da}{d\tau}\right\|^2_\eta d\tau, \tag{18}$$

which is the correct action functional for a free relativistic particle of mass m.

4. Gravity in Hilbert space

As demonstrated in Ref.[3], an arbitrary curved metric on space-time can be obtained in the described way. That is, for any Riemannian or pseudo-Riemannian metric g on space time N there exists a Krein space H that contains delta functions δ^4_a and such that the metric induced on the set M_4 of these delta functions coincides with g, at least locally.

It follows in particular that by an appropriate "curving" of the metric (12) in the action (17) it should be possible to derive the law of motion of particles in the field of gravity. In other words, geodesic motion in an arbitrary curved classical space-time can be identified with a geodesic motion on the submanifold M_4 with the induced metric.

To see how this can be done, consider the indefinite Hermitian metric with the kernel

$$k(x,y) = e^{-\frac{1}{2}(\mathbf{x}-\mathbf{y})^2+\frac{1}{2}(1+u(\mathbf{x})+u(\mathbf{y}))(t-s)^2}, \tag{19}$$

where u is an appropriate function (the symmetric sum $u(\mathbf{x})+u(\mathbf{y})$ is needed for Hermicity of the resulting 2-form). Note that when $u=0$ we get the already familiar metric. Under the constraint $\varphi_\tau(x) = \delta^4(x-a(\tau))$ the square-length action functional (17) with this metric yields

$$\begin{aligned} S &= \frac{m}{2}\int k(x,y)\frac{d\varphi_\tau(x)}{d\tau}\frac{d\overline{\varphi}_\tau(y)}{d\tau}d^4x d^4y d\tau = \frac{m}{2}\int \left.\frac{\partial^2 k(x,y)}{\partial x^\mu \partial y^\nu}\right|_{x=y=a}\frac{da^\mu}{d\tau}\frac{da^\nu}{d\tau}d\tau \\ &= \frac{m}{2}\int g_{\mu\nu}\frac{da^\mu}{d\tau}\frac{da^\nu}{d\tau}d\tau, \end{aligned} \tag{20}$$

where $g_{\mu\nu}$ is given by the length element $ds^2 = (1+2u(\mathbf{x}))dt^2 - d\mathbf{x}^2$. As is well known, Einstein equations for the metric ds^2 in the non-relativistic approximation yield the Newton's law of motion of the particle in gravitational potential u. In other words, variation of the functional (20) gives the equation

$$\frac{d\mathbf{v}}{dt} = -\nabla u. \tag{21}$$

5. Quantum mechanics in Hilbert space with indefinite metric

Recall that in Sec.3 the motion of a material point in relativistic mechanics was derived by variation of the square-length functional in a Hilbert space H with indefinite metric (Krein space), defined in Sec.3. An additional constraint was the condition that the particle has a well defined position at any time. In other words, the path of the particle in the space of states H has the form $\varphi_\tau(x) = \delta^4(x-a(\tau))$. Let's show that the non-relativistic quantum mechanics can be derived in a similar way, by dropping the condition of definiteness of the position, but preserving the condition of definiteness of time. Namely, we begin with the relativistic framework offered by the Krein space H, impose the constraint $\varphi_\tau(\mathbf{x},t) = \psi(\mathbf{x},t)\delta(t-\tau)$ on the path of a particle and derive the usual non-relativistic quantum mechanics of the particle.

To do this let's first relate the Hilbert space H of functions of four variables $\mathbf{x}, t$ with metric (12) to the usual Hilbert spaces of functions of three variables $\mathbf{x}$ with t as a parameter of evolution. For this consider the family of subspaces H_τ of H each consisting of all functionals $\varphi_\tau(\mathbf{x},t) = \psi(\mathbf{x},t)\delta(t-\tau)$ for some fixed $\tau \in \mathbb{R}$. The inner product of any two functionals in H_τ in either $\widetilde{H}$ or H_η metrics of Sec.3 is simply the inner product in the space $\mathbf{H}$ introduced in Sec.2. This is the case because the factor $\delta(t-\tau)$ eliminates integration in t and makes the term $(t-s)^2$ in the exponent of the kernel of the metric vanish. In the following, H_τ will be understood as a Hilbert space with this inner product. The map $I : H_\tau \longrightarrow \mathbf{H}$ defined by $I(\varphi_\tau)(\mathbf{x}) = \psi(\mathbf{x},\tau)$ is then an isomorphism of Hilbert spaces.

Because $\mathbf{H} \approx L_2(\mathbb{R}^3)$, the map I basically identifies each subspace H_τ with the usual space $L_2(\mathbb{R}^3)$ of state functions on $\mathbb{R}^3$ considered at time τ. The reason why these particular subspaces are physically meaningful becomes clear from the following result that relates the dynamics on the family of subspaces H_τ and the usual space $L_2(\mathbb{R}^3)$ of states of a spinless non-relativistic particle.

Theorem 1. *Let $\widehat{h} = -\Delta + V(\mathbf{x}, t)$ be the usual Hamiltonian of non-relativistic quantum mechanics of a single particle. Then the function $\psi(\mathbf{x}, t)$ satisfies the Schrödinger equation $\frac{\partial \psi(\mathbf{x},t)}{\partial t} = -i\widehat{h}\psi(\mathbf{x}, t)$ if and only if the path $\varphi_\tau(\mathbf{x}, t) = \psi(\mathbf{x}, t)\delta(t - \tau)$ in H satisfies the equation $\frac{d\varphi_\tau}{d\tau} = \left(-\frac{\partial}{\partial t} - i\widehat{h}\right)\varphi_\tau$.*

More generally, the result holds true when $\varphi_\tau(\mathbf{x}, t) = \psi(\mathbf{x}, t)\tilde{\delta}(t - \tau)$ for some function $\tilde{\delta}$ of $t - \tau$. In the following $\tilde{\delta}$ will be taken to be a Gaussian function that approximates delta function.

Note that an equation of the form $\frac{d\varphi_\tau}{d\tau} = \left(-\frac{\partial}{\partial t} - i\widehat{h}\right)\varphi_\tau$ is used in Floquet theory that deals with Hamiltonians periodic in time Ref.[12] and in problems with time-dependent Hamiltonians Ref.[13]. It is also a well known non-relativistic limit of the Stueckelberg-Schrödinger equation in the theory of Stueckelberg Ref.[7] and Horwitz & Piron Ref.[8]. This theory treats space and time symmetrically and predicts interference in time Refs.[9],[10]. The non-relativistic limit of Stueckelberg theory was investigated by Horwitz and Rotbart Ref.[11]. The approximate equality of the time variable t with the evolution parameter τ obtained in Ref.[11] is consistent with the definition of H_τ.

6. Arc-length action functional for quantum dynamics

Dynamics of a classical particle in a gravitational field follow from the variational principle for the arc length action functional. The resulting path is a geodesic. The following result demonstrates that the Schrödinger dynamics can be derived in a similar fashion.

Theorem 2. *Suppose that the evolution of a system satisfies the equation $\frac{d\varphi_\tau}{d\tau} = -i\widehat{A}\varphi_\tau$, where $\widehat{A}$ is an invertible self-adjoint operator in the space L_2 of Lebesgue square-integrable functions on a set. Then the operator $\widehat{K} = \widehat{A}^{-1}\left(\widehat{A}^{-1}\right)^* = \widehat{A}^{-2}$ defines an inner product on the image $\mathrm{R}(\widehat{A})$ of $\widehat{A}$. Let H be the Hilbert completion of $\mathrm{R}(\widehat{A})$. Then $\widehat{A}$ considered as a map into H is bounded and solutions of the evolution equation are geodesics on the sphere S^g of unit-normalized states in L_2 with Riemannian metric g defined by the inner product in H.*

For a finite dimensional example consider the space of spin states of a non-relativistic electron. Suppose the Hamiltonian is given by $\widehat{h}_M = M - \mu\widehat{\sigma} \cdot \mathbf{B}$, where $\mathbf{B}$ is a homogeneous magnetic field, μ is the electron's magnetic moment, $\widehat{\sigma}$ is given by Pauli matrices $\widehat{\sigma} = (\sigma_1, \sigma_2, \sigma_3)$ and $M = M \cdot I$ is a constant, such that $M \gg \mu B$.

The metric on the sphere of states S^3 in which solutions to the Schrödinger equation

$$\frac{d\psi_t(s)}{dt} = -i\widehat{h}_M\psi_t(s). \tag{22}$$

are geodesics is given the operator $\widehat{h}_M^{-2}$. Because $(\mu\widehat{\sigma} \cdot \mathbf{B})^2 = \mu^2 B^2$, where B is the norm of $\mathbf{B}$, we have:

$$\widehat{h}_M^2 = (M - \mu\widehat{\sigma} \cdot \mathbf{B})^2 = M^2 + \mu^2 B^2 - 2M\mu\widehat{\sigma} \cdot \mathbf{B}. \tag{23}$$

Suppose that the field $\mathbf{B}$ is directed along the Z-axis. Then $\widehat{\sigma} \cdot \mathbf{B} = \widehat{\sigma}_z B$ and

$$\widehat{h}_M^2 = \begin{bmatrix} (M - \mu B)^2 & 0 \\ 0 & (M + \mu B)^2 \end{bmatrix}. \tag{24}$$

Accordingly, the metric operator $\widehat{h}_M^{-2}$ has the form:

$$\widehat{h}_M^{-2} = \begin{bmatrix} \frac{1}{(M-\mu B)^2} & 0 \\ 0 & \frac{1}{(M+\mu B)^2} \end{bmatrix}. \tag{25}$$

Consider now the metric G induced on the unit sphere S^3 embedded into the space C^2 with metric given by (25). Since $M \gg \mu B$, this metric almost coincides with the standard metric on the sphere of radius $1/M$. More precisely, the sphere with the metric G is a 3-ellipsoid

$$\frac{x_1^2}{1/(M-\mu B)^2} + \frac{y_1^2}{1/(M-\mu B)^2} + \frac{x_2^2}{1/(M+\mu B)^2} + \frac{y_2^2}{1/(M+\mu B)^2} = 1 \tag{26}$$

in the Euclidean space $\mathbb{R}^4 = C^2$.

Solutions to the Schrödinger equation (22) are geodesics on the ellipsoid and are given by

$$\psi_t = \left[\ \psi_{0+}e^{-i(M+\mu B)t},\ \ \psi_{0-}e^{-i(M-\mu B)t}\ \right], \tag{27}$$

where $[\psi_{0+}, \psi_{0-}]$ is the initial spin state. The Schrödinger evolution is a slightly deformed phase motion. The curve (27) "spirals" along the sphere of states. In particular, after the time interval $[0, \frac{2\pi}{M}]$ the curve that starts at the point $[\psi_{0+}, \psi_{0-}]$ goes around the sphere and comes to a neighboring point

$$\left[\ \psi_{0+}e^{-i\frac{2\pi\mu B}{M}},\ \ \psi_{0-}e^{i\frac{2\pi\mu B}{M}}\ \right]. \tag{28}$$

Notice that the deformation of the metric on the sphere of states in the example is due to the interaction of the electron with the field.

It is also possible to find the length-type action functional on functions of four variables that yields the Schrödinger evolution. The idea is to apply Theorem 2 to the case of the self-adjoint operator $\widehat{A} = -i\frac{\partial}{\partial t} + \widehat{h}$, acting in the space $L_2(\mathbb{R}^4)$ of square-integrable functions of $\mathbf{x}, t$ (see Theorem 1). At first sight it seems that this will not work, because the kernel of $\widehat{A}$ is non-trivial. More precisely, the kernel consists of all solutions to the Schrödinger equation. However, it is easy to see that those solutions are not in the space $L_2(\mathbb{R}^4)$. In fact, since the Schrödinger evolution is unitary, the $L_2(\mathbb{R}^3)$-norm of solutions is preserved in time, so that the $L_2(\mathbb{R}^4)$-norm of any non-trivial solution is infinite. It follows that the operator $\widehat{A}$ is invertible and the Theorem 2 applies. The result is the following theorem

Theorem 3. *Let H be the Hilbert space in the Theorem 2 and suppose that the functions $\varphi_\tau(\mathbf{x}, t) = \psi(\mathbf{x}, t)\tilde{\delta}(t-\tau)$, $\psi(\mathbf{x}, t) \in L_2(\mathbb{R}^3)$ are in H. Then solutions to the Schrödinger equation are extrema of the square-length action functional $S = \int_a^b \left(\widehat{K}\frac{d\varphi_\tau}{d\tau}, \frac{d\varphi_\tau}{d\tau}\right) d\tau$, subject to the constraint $\varphi_\tau \in S^g$, $\varphi_\tau(\mathbf{x}, t) = \psi(\mathbf{x}, t)\tilde{\delta}(t-\tau)$, where S^g is the unit sphere in $L_2(\mathbb{R}^4)$ with the Riemannian metric given by the metric $\widehat{K} = \widehat{A}^{-2}$ on H.*

Note that the operator $\widehat{A}^{-1}$ considered on functions $\varphi_\tau(\mathbf{x}, t) = \psi(\mathbf{x}, t)\delta(t-\tau)$ acts as the Green function of the Schrödinger equation. In fact, if $x = (\mathbf{x}, t)$, $y = (\mathbf{y}, s)$ and $g(x; y)$ is the kernel of $\widehat{A}^{-1}$, then

$$\int g(x; y)\varphi_\tau(y)d^4y = \int g(\mathbf{x}, t; \mathbf{y}, \tau)\psi(\mathbf{y}, \tau)d^3\mathbf{y} \tag{29}$$

and

$$\left(-\frac{\partial}{\partial t} - i\widehat{h}\right) g(\mathbf{x}, t; \mathbf{y}, \tau) = \delta^3(\mathbf{x}-\mathbf{y})\delta(t-\tau). \tag{30}$$

So the Riemannian metric that makes Schrödinger evolution on the unit sphere in $L_2(\mathbb{R}^4)$ a geodesic motion is given by a propagator of the Schrödinger equation. Note that the free particle propagator

$$g(\mathbf{x}, t; \mathbf{y}, s) = \left(\frac{m}{2\pi i(t-s)}\right)^{\frac{3}{2}} e^{-\frac{m(\mathbf{x}-\mathbf{y})^2}{2i(t-s)}}, \tag{31}$$

has the exponential form similar to the metric (3) in the space $\mathbf{H}$. In particular, the resulting space contains delta functions and the metric induced on the three-dimensional manifold M_3 of all delta functions is Euclidean.

7. Vector representation in quantum mechanics

Quantum evolution is a path on the unit sphere of states in a Hilbert space L_2 (modulo the domain issues of involved operators). Given a self-adjoint operator $\widehat{A}$ on L_2, consider the vector field $A_\varphi = -i\widehat{A}\varphi$. This field is tangent to the sphere, because its integral curves are given by the family of unitary transformations applied to a point on the sphere: $\varphi_t = e^{-i\widehat{A}}\varphi_0$, so that the curves lie on the sphere. There is a simple relationship between the commutator of operators and the Lie bracket of the associated vector fields:

$$[A_\varphi, B_\varphi] = [\widehat{A}, \widehat{B}]\varphi. \tag{32}$$

Furthermore, a Hilbert metric on the space of states yields a Riemannian metric on the sphere. For this consider the realization $L_{2\mathbb{R}}$ of the Hilbert space L_2, i.e., the real vector space of pairs $X = (\mathrm{Re}\psi, \mathrm{Im}\psi)$ with ψ in L_2. If ξ, η are vector fields on S^{L_2}, define a Riemannian metric $G_\varphi : T_{\mathbb{R}\varphi}S^{L_2} \times T_{\mathbb{R}\varphi}S^{L_2} \longrightarrow \mathbb{R}$ on the sphere by

$$G_\varphi(X, Y) = \mathrm{Re}(\xi, \eta). \tag{33}$$

Here $X = (\mathrm{Re}\xi, \mathrm{Im}\xi)$, $Y = (\mathrm{Re}\eta, \mathrm{Im}\eta)$ and (ξ, η) denotes the L_2-inner product of ξ, η.

The value of the Lie bracket of vector fields associated with observables, that is, the commutator of observables, is related to the sectional curvature of the obtained Riemannian metric on S^{L_2}, Ref.[5]. The usual commutators are obtained when all sectional curvatures are equal to one in the corresponding Planck units. In such a way quantum non-commutativity gets encoded into the geometry of the sphere of states. Note that the minuscule size of the sphere $S^{\mathbf{H}}$ in spatial (i.e., tangent to M_3) directions does not contradict the fact that M_3 itself is the infinite Euclidean space. This is because M_3 gets large by "spiraling" through dimensions of $\mathbf{H}$, which has nothing to do with the size of the sphere.

The Riemannian metric on S^{L_2} yields a Riemannian (Fubini-Study) metric on the projective space CP^{L_2}, which is the base of the fibration $\pi : S^{L_2} \longrightarrow CP^{L_2}$. For this an arbitrary tangent vector $X \in T_{R\varphi}S^{L_2}$ is decomposed into two components: tangent and orthogonal to the fibre $\{\varphi\}$ through φ (i.e., to the plane C^1 containing the circle $S^1 = \{\varphi\}$). The differential $d\pi$ maps the tangential component to zero-vector. The orthogonal component of X can be then identified with $d\pi(X)$. If two vectors X, Y are orthogonal to the fibre $\{\varphi\}$, the inner product of $d\pi(X)$ and $d\pi(Y)$ in the Fubini-Study metric is equal to the inner product of X and Y in the metric G_φ:

$$(d\pi(X), d\pi(Y))_{FS} = G_\varphi(X, Y). \tag{34}$$

Using this, it is now easy to interpret geometrically the expected value and the uncertainty of an observable, Ref.[4]. Namely, the equality

$$\overline{A} \equiv (\varphi, \widehat{A}\varphi) = (-i\varphi, -i\widehat{A}\varphi), \tag{35}$$

(with the usual L_2 inner product) signifies that the expected value of an observable $\widehat{A}$ in the state φ is the projection of the vector $-i\widehat{A}\varphi \in T_\varphi S^{L_2}$ on the vector $-i\varphi = -iI\varphi \in T_\varphi S^{L_2}$, associated with the identity operator I and tangent to the phase circle through φ. Because

$$(\varphi, \widehat{A}^2\varphi) = (\widehat{A}\varphi, \widehat{A}\varphi) = (-i\widehat{A}\varphi, -i\widehat{A}\varphi), \tag{36}$$

the term $(\varphi, \widehat{A}^2\varphi)$ is just the norm of the vector $-i\widehat{A}\varphi$ squared. Note that the expected value $(\varphi, \widehat{A}_\perp\varphi)$ of the operator $\widehat{A}_\perp \equiv \widehat{A} - \overline{A}I$ in the state φ is zero. Therefore, the vector $-i\widehat{A}_\perp\varphi = -i\widehat{A}\varphi - (-i\overline{A}\varphi)$, which is the component of $-i\widehat{A}\varphi$ orthogonal to $-i\varphi$, is orthogonal to the entire fibre $\{\varphi\}$. Accordingly, the variance

$$\Delta A^2 = (\varphi, (\widehat{A} - \overline{A}I)^2\varphi) = (\varphi, \widehat{A}_\perp^2\varphi) = (-i\widehat{A}_\perp\varphi, -i\widehat{A}_\perp\varphi) \tag{37}$$

is the norm squared of the component $-i\widehat{A}_\perp\varphi$. Recall that the image of this vector under $d\pi$ can be identified with the vector itself. It follows that the norm of $-i\widehat{A}_\perp\varphi$ in the Fubini-Study metric coincides with its norm in the Riemannian metric on S^{L_2} and in the original L_2-metric.

Furthermore, the uncertainty relation

$$\Delta A \Delta B \geq \frac{1}{2}\left|\left(\varphi, [\widehat{A}, \widehat{B}]\varphi\right)\right| \tag{38}$$

follows now from the comparison of areas of rectangle $A_{|XY|}$ and parallelogram A_{XY} formed by vectors $X = -i\widehat{A}_\perp\varphi$ and $Y = -i\widehat{B}_\perp\varphi$:

$$A_{|XY|} \geq A_{XY}. \tag{39}$$

There is also an uncertainty identity, Ref.[4]:

$$\Delta A^2 \Delta B^2 = A_{XY}^2 + G_\varphi^2(X, Y). \tag{40}$$

Suppose that observables $\widehat{A}$, $\widehat{B}$ satisfy the canonical commutation relation $[\widehat{A}, \widehat{B}] = i$. A unitary transformation $\widehat{U}$ preserves the commutation relation, which together with (37) yields

$$\left\|-i(\widehat{U}^{-1}\widehat{A}\widehat{U})_\perp\varphi\right\| \left\|-i(\widehat{U}^{-1}\widehat{B}\widehat{U})_\perp\varphi\right\| \geq \frac{1}{2}. \tag{41}$$

Note that the uncertainty of the operators changes under a general unitary transformation $\widehat{U}$. In other words, the normal component of the associated vector field changes. At the same time the estimate for the product, i.e., the uncertainty relation itself, remains the same. The geometric nature of the uncertainty relation makes this invariance particularly transparent.

8. Geometry of the Schrödinger evolution

Any vector in the Hilbert space, attached to a point φ on the sphere, can be decomposed onto the radial component (parallel to the radius vector from the origin to the point φ, i.e., parallel to φ itself), and tangential component. Furthermore, the tangential component can be decomposed onto the one parallel to the fibre (i.e., tangent to the phase circle through φ) and the one orthogonal to the fibre. As already discussed, under the projection onto the projective space

CP^{L_2} the parallel tangential component gives zero, while the orthogonal tangential component can be identified with the vector tangent to CP^{L_2}.

Let's see what can be extracted from these simple geometric facts. Consider the Schrödinger equation

$$\frac{d\varphi}{dt} = -i\widehat{h}\varphi \tag{42}$$

for the state φ with the initial condition $\varphi|_{t=0} = \varphi_0$. Notice that

$$\mathrm{Re}(\varphi, -i\widehat{h}\varphi) = 0, \tag{43}$$

which means that the radial component of the velocity vector $\frac{d\varphi}{dt}$ is zero. This fact is clear already from its tangential nature. Let's decompose the tangent vector $-i\widehat{h}\varphi$ onto the components parallel $||$ and orthogonal $\perp$ to the fibre. The parallel component of $\frac{d\varphi}{dt}$ is numerically

$$\mathrm{Re}(-i\varphi, -i\widehat{h}\varphi) = \overline{E}, \tag{44}$$

i.e., the expected value of the energy. So the decomposition of the velocity vector $\frac{d\varphi}{dt}$ into the parallel and orthogonal components is then given by

$$\frac{d\varphi}{dt} = -i\overline{E}\varphi + \left(-i\widehat{h}\varphi - i\overline{E}\varphi\right) = -i\overline{E}\varphi - i\widehat{h}_{\perp}\varphi. \tag{45}$$

We see that the orthogonal component of the velocity $\frac{d\varphi}{dt}$ is equal to $-i\widehat{h}_{\perp}\varphi$, so we immediately conclude that: *The velocity of evolution of state in the projective space is equal to the uncertainty of energy.* Note that (45) also demonstrates that the physical state is driven by the operator $\widehat{h}_{\perp}$, which defines the uncertainty in energy rather than the energy itself.

Now let's decompose the acceleration vector $\frac{d^2\varphi}{dt^2} = \frac{d}{dt}\left(-i\widehat{h}\varphi\right) = -\widehat{h}^2\varphi$. Notice first of all that

$$\mathrm{Re}(-i\varphi, \widehat{h}^2\varphi) = 0, \tag{46}$$

so that the parallel tangential component of acceleration of Schrödinger evolution vanishes. This simply means that the phase component of the velocity (i.e., the expected value of energy, see above) does not change. In particular, the tangential component is purely orthogonal. The radial component is given by $-(\varphi, \widehat{h}^2\varphi)\varphi = -(-i\widehat{h}\varphi, -i\widehat{h}\varphi)\varphi$. Since $-i\widehat{h}\varphi$ is the velocity of evolution, we recognize in this term the centropidical acceleration ($-\frac{\mathbf{v}^2\mathbf{r}}{r^2}$ with $r = 1$).

The tangential component is therefore equal to

$$-\widehat{h}^2\varphi + (\varphi, \widehat{h}^2\varphi)\varphi = -\widehat{h}^2_{\perp}\varphi. \tag{47}$$

Therefore, the following result is obtained: *Acceleration of the Schrödinger evolution of state in the projective space is equal to the uncertainty of the square of energy.* Note also that the latter uncertainty can be interpreted as the fourth order central moment of the probability distribution associated with energy.

9. Geometry of transition from quantum to classical

Classical and quantum mechanics of a particle are now formulated within the same Hilbert space framework. Could it be that classical motion is a projection of some sort of the Schrödinger evolution on the submanifold M_3?

One standard way to describe the relation of the Schrödinger evolution with the classical Newtonian motion is via the Ehrenfest theorem (the expected value of the Heisenberg equation of motion):

$$\frac{d}{dt}(\varphi, \widehat{A}\varphi) = -i(\varphi, [\widehat{A}, \widehat{h}]\varphi). \tag{48}$$

Here $\widehat{A}$ does not depend on t. For example, for the momentum operator of a free particle we obtain

$$\frac{d\overline{\mathbf{p}}}{dt} = 0. \tag{49}$$

Recall that $\overline{\mathbf{p}}$ is the phase projection of the vector field p_φ. The equation (49) simply says that this projection is time-independent. Note that the orthogonal projection, i.e. the uncertainty $\Delta\mathbf{p}$ is also preserved in this case and this is not captured in (48).

Compare (48) to another equation that follows from the Schrödinger dynamics:

$$2\left(\frac{d\varphi}{dt}, -i\widehat{A}\varphi\right) = \left(\varphi, \{\widehat{A}, \widehat{h}\}\varphi\right) - \left(\varphi, [\widehat{A}, \widehat{h}]\varphi\right). \tag{50}$$

The Ehrenfest theorem (48) for a time-independent observable amounts to using the imaginary part of (50). The left hand side of (50) is the projection of the velocity of state onto the vector field associated with the observable $\widehat{A}$. It will be now argued that the real part of the projection may be even more relevant to the problem of transition from quantum to classical.

Recall that the classical Euclidean space $\mathbb{R}^3$ is realized by submanifold M_3 in the Hilbert space $\mathbf{H} \approx L_2(\mathbb{R}^3)$ defined by (3). The isomorphism $\widehat{\rho} : \mathbf{H} \longrightarrow L_2(\mathbb{R}^3)$ given by

$$\rho(\mathbf{x}, \mathbf{y}) = \left(\frac{1}{\pi\sigma^2}\right)^{3/4} e^{-\frac{(\mathbf{x}-\mathbf{y})^2}{2\sigma^2}} \tag{51}$$

identifies M_3 with the submanifold M_3^σ of Gaussian functions of width σ and norm 1 in $L_2(\mathbb{R}^3)$. In the following it will be assumed that the length in $\mathbb{R}^3$ is measured in the units of $\sqrt{2}\sigma$. Then, as before, the manifold M_3^σ is isometric to the Euclidean space $\mathbb{R}^3$.

Because all normalized Gaussian functions of a given width σ are obtained from a single one by translations in $\mathbf{x}$, the field $\mathbf{p}_\varphi = -i\widehat{\mathbf{p}}\varphi$ for $\varphi \in M_3^\sigma$ is tangent to M_3^σ. Consider a wave packet which is a solution of the uniform acceleration problem in potential V in one dimension

$$\varphi(x,t) = \frac{1}{(\pi)^{\frac{1}{4}}(\sigma^2 + \frac{t^2}{m^2\sigma^2})^{\frac{1}{4}}} e^{-\frac{(x-x_0-v_0t-\frac{1}{2}wt^2)^2}{2(\sigma^2+\frac{t^2}{m^2\sigma^2})}} e^{i\theta}, \tag{52}$$

where θ is a phase factor, v_0 is the initial group velocity of the packet and $w = -\nabla V/m$. The map $\pi : \varphi \longrightarrow r$, where r is the modulus of φ, projects the initial state $\varphi_0(x) = \varphi(x, 0)$ onto a point on M^σ, which is a one-dimensional version of M_3^σ. (If φ is represented by the pair (r, θ), then $\pi(r, \theta) = (r, 0)$. Physically, the phase factor plays no role in the probability of collapse to a position eigenstate. The spaces M_3^σ, M^σ must be therefore identified with the set of equivalence classes of states defined up to the phase θ. See below.) The derivative map $d\pi$ projects the vector $\frac{d\varphi}{dt}$ in the tangent space $T_\varphi S^{L_2}$ onto a vector in the tangent space $T_r S^{L_2}$. The component of this vector that is tangent to M^σ at $t = 0$ can be obtained using the Riemannian metric on the sphere and is equal to

$$\mathrm{Re}\left(\frac{d\varphi}{dt}e^{-i\theta}, -\frac{dr}{dx}\right) = \left(\frac{dr}{dt}, -\frac{dr}{dx}\right), \tag{53}$$

where parentheses denote the usual L_2 inner product. Note that the vector $-i\widehat{p}r = -\frac{dr}{dx}$ is unit in the chosen units. A direct calculation shows that at $t = 0$

$$\left(\frac{dr}{dt}, -\frac{dr}{dx}\right) = -v_0. \tag{54}$$

The left hand side of (54) is simply the M^σ-component of the velocity of the shadow $\pi(\varphi) = r$ of φ at $t = 0$. Note that the component of $d\varphi/dt = -i\widehat{h}\varphi$ that gives v_0 via (54) is orthogonal to the fibre through φ, i.e., it comes from the term $-i\widehat{h}_\perp\varphi$. So, *the velocity* v_0 *of the shadow is due to the uncertainty* Δh *in the energy.* Note also that the component in (53) is a part of the term

$$\mathrm{Re}\left(\frac{d\varphi}{dt}, -i\widehat{p}\varphi\right) = \left(\frac{dr}{dt}, -\frac{dr}{dx}\right) - \left(r\frac{d\theta}{dt}, r\frac{d\theta}{dx}\right), \tag{55}$$

which is given itself by the anticommutator term in (50).

Similarly, the acceleration of the shadow at $t = 0$ is given by

$$\left(\frac{d^2r}{dt^2}, -\frac{dr}{dx}\right) = -w. \tag{56}$$

These results were extended in Ref.[14].

The equations (54), (56) hint that the classical motion of a macroscopic particle may indeed be a projection of the Schrödinger evolution onto M_3^σ. To prove that this is the case one must first explain what happens at $t \neq 0$. The conjecture is that the state of the particle experiences a periodic "collapse" so that the "width" of the state following the collapse returns to the value σ. Suppose the state φ undergoes the Schrödinger evolution (52) between $t = 0$ and the collapse episode at $t = t_1$. To ensure the validity of (54), (56) at the moment $t = t_1$, suppose that the state $\psi(t_1)$ right after the collapse is the state $\varphi(t_1)$ given by (52), but with the width

$$\sigma^2 + \frac{t_1^2}{m^2\sigma^2} \tag{57}$$

in all terms (including phase) changed to the value σ^2. More precisely, the state $\psi(t_1)$ is simply the result of the Schrödinger evolution (52) of the initial state ψ_0, which is the original initial state φ_0 but with σ replaced with the smaller value $\tilde{\sigma}$ that satisfies the equation

$$\tilde{\sigma}^2 + \frac{t_1^2}{m^2\tilde{\sigma}^2} = \sigma^2. \tag{58}$$

The equations (54) (with $v(0) = v_0$ replaced by $v(t_1) = v_0 + wt_1$) and (56) are now satisfied for the state ψ at $t = t_1$ simply because they are satisfied for the state φ at $t = 0$.

To put this in a clear geometric context, consider the fibre bundle $p : E \longrightarrow M^\sigma$ over the manifold M^σ, whose fibre over $\tilde{\delta}_a \in M^\sigma$ consists of all normalized Gaussian functions with the mean a, multiplied by an arbitrary phase factor $e^{i\theta(x,t)}$. Let G be the group of transformations on E generated by scaling in x and multiplication by phase factors $e^{i\theta(x,t)}$. Define the (Ehresmann) connection on E by calling a tangent vector at a point φ horizontal if it is orthogonal in the L_2 metric to the fibre through φ. In particular, the tangent space $T_{\tilde{\delta}_a}M^\sigma$ to M^σ at $\tilde{\delta}_a$ is horizontal. The bundle projection p is a transformation in G. The connection is invariant under G, so that the vector $-dr/dxe^{i\theta}$ is horizontal along the fibre. Moreover, the projection

$$\mathrm{Re}\left(\frac{d\varphi}{dt}, -\frac{dr}{dx}e^{i\theta}\right) = \mathrm{Re}\left(\frac{d\varphi}{dt}e^{-i\theta}, -\frac{dr}{dx}\right) = \left(\frac{dr}{dt}, -\frac{dr}{dx}\right) \tag{59}$$

is constant along the fibre, so that the equation (54) with $-v(t)$ on the right remains true at all times.

When applied to microscopic particles, this result explains, in particular, why these particles move as if they follow a classical trajectory. Namely, the shadow $p(\varphi)$ of the state on the base M^{σ} moves classically. Note that this result is not about the change in the expected value of the position operator. It is about the change in the state itself and how this change is "observed" on M_3^{σ}.

What distinguishes macroscopic particles in this scenario is the mentioned process of collapse along the fibres of the bundle E. In this case the shadow of the state continues to satisfy the Newtonian equations of motion. Additionally, collapse ensures that the state remains "localized". Notice that the described collapse does not make the phase terms vanish. The classical space appears here as the set of equivalence classes of states of different width and phase. The question of what is a possible physical origin of collapse and how the above picture is consistent with the Born rule will be discussed in sections 10 and 11.

The Schrödinger equation in one dimensional case is equivalent to the following two equations:

$$\frac{dr}{dt} = -\frac{1}{m}\frac{dr}{dx}\frac{d\theta}{dx} - \frac{r}{2m}\frac{d^2\theta}{dx^2}, \tag{60}$$

$$\frac{d\theta}{dt} = -\frac{1}{2m}\left(\frac{d\theta}{dx}\right)^2 - V + \frac{1}{2mr}\frac{d^2r}{dx^2}. \tag{61}$$

The first equation yields the continuity equation (the quantum story) and the second yields Hamilton-Jacobi equation (the classical story). The obtained result suggests that the first equation may be even more relevant to transition from quantum to classical. Moreover, the equation for the state itself rather than for the associated probability density seems to be in the heart of the transition.

10. Quantum probability and the classical normal distribution

If a classical experiment is set with the goal of measuring the position of a macroscopic particle, the result is generically a normal probability distribution of the position variable. Recall now that the classical space $\mathbb{R}^3$ can be identified with the submanifold M_3^{σ} in the Hilbert space L_2 of states (equivalently, with the submanifold M_3 in the space $\mathbf{H}$). The experiment of measuring position of a macroscopic particle is then taking place within the sphere of states S^{L_2}. Because of this, the experiment has a huge impact on the possible outcomes of an experiment that measures the position of a microscopic particle.

In particular, suppose that results of a position measurement of a macroscopic particle are normally distributed. Because $\mathbb{R}^3$ is identified with M_3^{σ}, the normal distribution of position dictates the corresponding probability of transition between different states of the particle in M_3^{σ}. Suppose now that the probability for the particle in an arbitrary state φ to be found in a state ψ depends only on the distance $\rho(\pi(\varphi), \pi(\psi))$ between the states, in the Fubini-Study metric on the projective space CP^{L_2}. Then the Born rule for probability of transition between the states follows.

The proof of this claim follows immediately from the fact that the Born rule applied to the states in M_3^σ yields the normal probability distribution. In fact, for the state $\tilde{\delta}_{\mathbf{a}}$ in M_3^σ given by

$$\tilde{\delta}(\mathbf{x}-\mathbf{a}) = \left(\frac{1}{\pi\sigma^2}\right)^{3/4} e^{-\frac{(\mathbf{x}-\mathbf{a})^2}{2\sigma^2}}, \tag{62}$$

the probability density of finding the particle at $\mathbf{b}$ is equal to

$$\left(\frac{1}{\pi\sigma^2}\right)^{3/2} e^{-\frac{(\mathbf{a}-\mathbf{b})^2}{\sigma^2}}, \tag{63}$$

which is the normal distribution function. So, on the states $\tilde{\delta}_{\mathbf{a}}$ that form the classical space M_3^σ the quantum-mechanical Born rule makes the same prediction as the normal probability distribution. On the other hand, the corresponding probability of finding the particle in the state $\tilde{\delta}_{\mathbf{b}}$ is equal by the same Born rule to

$$P(\tilde{\delta}_{\mathbf{a}}, \tilde{\delta}_{\mathbf{b}}) = \cos^2 \rho(\tilde{\delta}_{\mathbf{a}}, \tilde{\delta}_{\mathbf{b}}), \tag{64}$$

where $\rho(\tilde{\delta}_{\mathbf{a}}, \tilde{\delta}_{\mathbf{b}})$ is the distance between the states $\tilde{\delta}_{\mathbf{a}}, \tilde{\delta}_{\mathbf{b}}$ in the projective space CP^{L_2}. Note that because M_3^σ can be identified with a submanifold in the base CP^{L_2} of the fibre bundle $\pi : L_2 \longrightarrow CP^{L_2}$, we set $\pi(\tilde{\delta}_{\mathbf{a}}) = \tilde{\delta}_{\mathbf{a}}$ in (64). Note also that the space of states here is the Hilbert space $L_2(\mathbb{R}^3)$. The projective space CP^{L_2} is only used in discussions related to the Fubini-Study distance between states.

Observe that an arbitrary state is a superposition of the states $\tilde{\delta}_{\mathbf{a}}$. Furthermore, the Fubini-Study distance between the states $\tilde{\delta}_{\mathbf{a}}$, $\tilde{\delta}_{\mathbf{b}}$ takes on all values from 0 to $\pi/2$, which is the largest possible distance between points in CP^{L_2}. In addition, the probability $P(\varphi, \psi)$ to find state φ in the state ψ was assumed to depend only on the Fubini-Study distance $\rho(\pi(\varphi), \pi(\psi))$ between the states. From this and (64) it follows that

$$P(\varphi, \psi) = \cos^2 \rho(\pi(\varphi), \pi(\psi)), \tag{65}$$

which is the Born rule for arbitrary states. In other words, the normal probability distribution for states in M_3^σ extends to the Born rule for superpositions of these states.

This beautiful result is based on a highly non-trivial way in which the classical space is embedded into the Hilbert space of states. Namely, because of the special properties of the embedding, the "classical law" (normal distribution of observation results) becomes a part of the quantum law, which simply extends the classical law to superpositions. The extension is unique if the assumption is made that the probability of transition must only depend on the distance between states in the Fubini-Study metric.

In more detail, assume the normal probability distribution of the position variable $\mathbf{a}$. Denote the distance between two points $\mathbf{a}, \mathbf{b}$ in $\mathbb{R}^3$ by $\|\mathbf{a}-\mathbf{b}\|_{\mathbb{R}^3}$. Under the embedding of the classical space into the space of states, the variable $\mathbf{a}$ is represented by the state $\tilde{\delta}_{\mathbf{a}}$. The set of states $\tilde{\delta}_{\mathbf{a}}$ form a submanifold M_3^σ in the Hilbert spaces of states $L_2(\mathbb{R}^3)$. The manifold M_3^σ is "twisted" in $L_2(\mathbb{R}^3)$, it belongs to the sphere S^{L_2} and spans all dimensions of $L_2(\mathbb{R}^3)$. Distance between the states $\tilde{\delta}_{\mathbf{a}}$, $\tilde{\delta}_{\mathbf{b}}$ in $L_2(\mathbb{R}^3)$ or in the projective space CP^{L_2} is not equal to $\|\mathbf{a}-\mathbf{b}\|_{\mathbb{R}^3}$. In fact, the former distance measures length of a geodesic between the states while the latter is obtained using the same metric on the space of states, but applied along a geodesic in the twisted manifold M_3^σ. In precise terms the relation between the two distances is given by

$$e^{-\frac{(\mathbf{a}-\mathbf{b})^2}{2\sigma^2}} = \cos^2 \rho(\tilde{\delta}_{\mathbf{a}}, \tilde{\delta}_{\mathbf{b}}). \tag{66}$$

This equation is what accounts for the relation between the normal probability distribution and the Born rule.

11. Collapse of the state function

By virtue of (66), the Born rule is an extension of the normal probability distribution from M_3^σ onto $L_2(\mathbb{R}^3)$. It is therefore natural to expect that quantum measurement and collapse of the state function can be explained via the corresponding extension of the classical measurement. An attempt to build a simple extension of this kind and to draw conclusions from it will now be presented. The results of this section, although relatively straightforward, require a rigorous mathematical foundation and will be presented in detail elsewhere.

Consider an experiment of measuring position of a macroscopic particle. Suppose that interaction in the experiment depends only on the distance between particles and that the resulting distribution of position $\mathbf{a}$ of the measured particle is normal. Using the embedding $\tilde{\omega} : \mathbf{a} \longrightarrow \tilde{\delta}_{\mathbf{a}}^3$, we can "transplant" the model of this experiment from $\mathbb{R}^3$ onto the submanifold M_3^σ in CP^{L_2}. Then we can extend the model to the projective space CP^{L_2} and to the space of states $L_2(\mathbb{R}^3)$ by replacing distance between "delta" states with the Fubini-Study distance between arbitrary states (which in particular gives correct distance between the "delta" states). By (66), the transition probability $P(\varphi, \psi)$ in the obtained model will satisfy the Born rule.

One simple method of obtaining a normal distribution of the position of a macroscopic particle is by the process of diffusion of the particle in an appropriate medium (Brownian motion). Suppose a macroscopic particle is placed in the medium and position of the particle a certain time interval τ later is recorded. The probability density function of the position variable is determined by the diffusion coefficient D. The process is isotropic: the coefficient D depends only on the frequency of various observed displacements of the particle in the medium, but not the direction of these displacements. Therefore, diffusion can be extended to a process on the projective space $CP^{L_2} = S^{L_2}/S^1$, with the diffused particle realized now by the state $\{\varphi\} \in CP^{L_2}$ and with displacements in $\mathbb{R}^3$ replaced by the like-displacements of the state in the Fubini-Study metric. As discussed, the probability of finding the state $\{\varphi\}$ of the "diffused" particle in a state $\{\psi\}$ (i.e., at the point $\{\psi\} \in CP^{L_2}$) is then given by the Born rule: $P(\{\varphi\}, \{\psi\}) = \cos^2 \rho(\{\varphi\}, \{\psi\})$.

Note that the obtained process, which we still call diffusion, or diffusion in the space of states, gives correct probabilities of outcomes of measurement of *any* observable. In fact, the eigenstates of observables are just points in the space of states. The diffusion can take the particle to an arbitrary eigenstate and the probability of the resulting transition is given by the Born rule. In this sense diffusion in the space of states serves as a model for an arbitrary quantum measurement. Suppose, for example, that we want to find the probability density of various positions of a particle. To measure position, we subject the particle (i.e., the state) to diffusion in the space of states. If in the process of diffusion the particle reaches the manifold M_3, then position $\mathbf{a} \in \mathbb{R}^3$ of the particle becomes meaningful and we say that collapse of the state has occurred. Note that the probability density function of finding the particle at $\mathbf{a}$ is the *conditional* probability density, given that the diffusive particle was found on M_3.

In this model the process of collapse is independent of a particular measurement performed on the particle. It is simply a random walk that leads to an eigenstate of the measured observable. In the case of position measurement the initial state simply happened to be "pushed" in the space of states in the direction of decreasing width σ, i.e., towards the submanifold M_3^σ.

Of course, collapse by diffusion does not preclude other forms of collapse. In fact, any process that involves a sufficient "squeezing" of the state in the position representation can be thought of as collapse to a point in $\mathbb{R}^3$. In particular, an appropriate potential well that narrows down in time will drive the state to the submanifold M_3 thus providing a model of controlled collapse to a position eigenstate. This latter example together with the Schrödinger equation itself prove that our usual instruments are capable of accessing points (i.e., states) outside the classical space M_3. In other words, the idea that a classical measuring device is responsible for a physical process on the space of states, and not just the classical space, seems to be realized in nature. Note also that because the sphere of normalized states can be arbitrarily small, the extension of a classical process to the space of states may be limited to an arbitrarily thin layer around the submanifold M_3. Furthermore, vector fields $dr/d\sigma$ and dr/dx are orthogonal, showing that collapse has nothing to do with motion in the classical space (see Sec.9). In particular, it does not make sense to impose a speed limit in the classical space on this process.

The introduced collapse by diffusion sheds new light into the relationship between classical and quantum measurements. If position of a macroscopic particle is measured, the diffusion process models a noise present in any measurement and the resulting normal probability distribution of measurement results. In this case the position of the particle at time τ can be observed no matter where the particle is located. On the other hand, in the experiment of measuring position of a microscopic particle, the position is not readily available. The state at time τ (i.e., the end-point of the random walk) does, of course, exist. But position in the usual sense is meaningful only when diffusion takes the state to the classical space M_3.

The situation can be elucidated by an example from the classical physics in which position measuring devices are on the X-axis in the Euclidean space $\mathbb{R}^3$. If this represents the only position measurement available to us, then a macroscopic particle has a definite position only if it is on the X-axis. The latter situation can be achieved, in particular, by subjecting the particle to diffusion in $\mathbb{R}^3$. When the drift motion in the diffusion can be neglected, the expected value of the measured position is simply the x-coordinate of the particle at the moment of measurement.

It was shown that diffusion can explain the outcomes of a measurement and the Born rule. Moreover, it is also capable of explaining why the state function of a macroscopic body does not spread. For this note first of all that a macroscopic body interacts continuously with the environment. This interaction is similar to the bombardment of a Brownian particle by molecules of the medium. However, because of the macroscopic character of the body, the diffusion process does not occur and position of the body remains unaffected by the interaction (e.g., a boat initially at rest in still water will not change its position). If now the interaction is extended to the space of states, then the conclusion is that the state of a macroscopic particle will not be the subject of a random walk. Since the increase in width is the increase in the distance from the state to the space M_3, we conclude that the width of the state function will not change.

Collapse as a physical process was discussed by several authors (see Ref.[15] for a review). In these works collapse is introduced by hand, either via a term in the stochastic Schrödinger equation or via random hits (consisting in multiplication by a narrow Gaussian function) that the state function of the particle undergoes. Even though no specific reference to the Fubini-Study metric is made in these models, it is implicitly present in the relationship between the norm of state-function and the probability of collapse. The dynamics of state under collapse in the existing models depends on the kind of measurement performed on the particle. This seems to be a negative feature of the models. As argued in Ref.[16], collapse to energy eigenstates via the stochastic equation in the projective space of states proposed in Ref.[17] may be sufficient to interpret an arbitrary quantum measurement. However, the proposed method may provide potentially a more appealing and physically motivated solution.

12. Conclusion and outlook

It was proved that the theories of classical and quantum mechanics can be considered within a single functional framework. The standard quantum mechanics in the framework is simply an extension of the classical mechanics to the space of states. The unusual properties of quantum mechanics are rooted in the strange way in which the classical space is embedded into the space of states. The classical mechanics is recovered from the condition that macroscopic bodies are constrained to the manifold M_3. The latter condition is supported by experiment and is likely to be due to interaction between macroscopic bodies and the environment. Radiation, other surrounding particles bombard a macroscopic body and subject it to diffusion in the space of states. However, as discussed in Sec.11, the macroscopic character of the body forestalls the diffusion and doesn't let the state leave the classical space M_3. As a result, position of the body is either defined precisely or normally distributed with a small standard deviation σ.

Contrary to this, free microscopic particles interact only occasionally with the environment and their state function can have any width. However, when a position measuring device is turned on, a microscopic object undergoes a bombardment similar to the one affecting a macroscopic body. Due to the microscopic nature of the object, its state does not remain static, but experiences a diffusion in the Hilbert space. Surrounding particles play billiard game with the object, and the rules of the game (i.e., the Born rule for collapse onto M_3) were shown to agree with the rules of the usual classical billiard.

The embedding of the classical space into the Hilbert space of states brings a totally new perspective into the observed phenomena and a new interpretation of the familiar experiments. Under the embedding the state gets "promoted" from a statistical notion to a physical concept that generalizes and replaces the notion of a material point. When the state is near the submanifold M_3 in the Hilbert space $\mathbf{H}$, the particle (i.e., the state!) behaves classically. When a particle goes "through" a screen with two slits, its state becomes, roughly speaking, a superposition of delta functions, which means that the particle is no longer a point of M_3. (Recall that the manifold M_3 is "made of" the delta functions, but not their nontrivial superpositions.) The entire process of passing "through" the slits is the process of refraction, when the path of the particle is "kicked out" of the classical space M_3 (while staying in the Hilbert space of states), and then returns to M_3 when position of the particle is measured.

It is rather straightforward to generalize the results to quantum systems of more than one particle. In the Hilbert space $\mathbf{H}^n = \otimes_{i=1}^n \mathbf{H}$ of states of an n-particle system, the states of non-interacting particles belong to the submanifold D of the product states. The state of a system of n classical particles is represented in D by the product of delta states. The set of all products of delta states forms a submanifold M_{3n} of D, which is diffeomorphic to the configuration space $\mathbb{R}^{3n}$. In the case of n non-interacting microscopic particles, the Schrödinger evolution of particles drives the state of the system out of the submanifold M_{3n}, although the path of state still takes values in D. Entanglement amounts to the state of the system leaving the manifold D and propagating in $\mathbf{H}^n$. If only two particles get entangled, the state of the system is the product of states of other particles, multiplied by the state of the entangled pair. This allows one to treat the state of the pair independently of other particles.

The obtained results are encouraging and appealing. However, the future work will show if these results lead in fact to a consistent theory of classical mechanics, relativity and quantum theory.

References

[1] A. Kryukov, *Int. J. Math. & Math. Sci.* **14**, 2241 (2005)
[2] —— *J. Math. Phys.* **49**, 102108 (2008)
[3] —— *J. Math. Phys.* **59**, 022110 (2010)
[4] —— *Phys. Lett. A* **370**, 419 (2007)
[5] —— *Found. Phys.* **37**, 3 (2007)
[6] —— *Found. Phys.* **41**, 129 (2011)
[7] E.C.G. Stueckelberg, *Helv. Phys. Acta* **14**, 372, 585 (1941); *Helv. Phys. Acta* **15**, 23 (1942)
[8] L.P. Horwitz and C. Piron, *Helv. Phys. Acta* **46**, 316 (1973)
[9] L.P. Horwitz and Y. Rabin, *Lett. Nuovo Cimento* **17**, 501 (1976)
[10] L.P. Horwitz, *Phys. Lett. A* **355**, 1 (2006)
[11] L.P. Horwitz and F. Rotbart, *Phys. Rev. D* **24**, 2127 (1981)
[12] H.L. Cycon, R.G. Froese, W. Kirsch, and B. Simon, *Schrödinger Operators with Application to Quantum Mechanics and Global Geometry*, Springer-Verlag, New York (1987)
[13] J.S. Howland, *Indiana Univ. Math. Jour.* **28**, 471 (1979)
[14] M. Forster and A. Kryukov, forthcoming
[15] A. Bassi, G. Ghirardi, *Phys. Rep.* **379**, 257 (2003)
[16] S. Adler and L.P. Horwitz, *J. Math. Phys.* **41**, 2485 (2000)
[17] L. P. Hughston, *Proc. Roy. Soc. Lond.* **A 452**, 953 (1996)

Neutrino oscillations in Stueckelberg semiclassical relativistic dynamics

L.P. Horwitz[a,b,c] **and I. Aharonovich**[b]

[a] School of Physics, Tel Aviv University, Ramat Aviv 69978, Israel
[b] Department of Physics, Bar Ilan University, Ramat Gan 52900, Israel
[c] Department of Physics, Ariel University Center of Samaria 40700, Israel

Abstract. The Stueckelberg formulation of a manifestly covariant relativistic classical and quantum mechanics is briefly reviewed and it is shown that in this framework a simple (semiclassical) model exists for the description of neutrino oscillations. The model is shown to be consistent with the field equations and the Lorentz force (developed here without and with spin by canonical methods) for Glashow-Salam-Weinberg type non-Abelian fields interacting with the leptons. We discuss a possible fundamental mechanism, in the context of a relativistic theory of spin for (first quantized) quantum mechanical systems, for CP violation. The model also predicts a possibly small "pull back,", *i.e.*, early arrival of a neutrino beam, for which the neutrino motion is almost everywhere within the light cone, a result which may emerge from future long baseline experiments designed to investigate neutrino transit times with significantly higher accuracy than presently available.

PACS 12.15.Ji,14.60.Pq,13.15.+g,12.90.+b,11.30.Er

1. Introduction

In 1941 Stueckelberg [1] considered the possibility that in classical mechanics, a timelike particle worldline can, due to interaction, curve sufficiently in spacetime to evolve continuously into a particle moving backward in time (see Figure 1). This configuration was interpreted as representing particle-antiparticle annihilation in classical mechanics. He parametrized this curve with an invariant parameter τ, which may be thought of as a world time[1] associated with the evolution. The time t of the *event* moving on a worldline of the type of Figure 1, corresponding to its identification by Einstein as a coordinate of the Minkowski spacetime manifold, is understood to be an observable in the same sense as the space coordinates $\mathbf{x}$,*i.e.*, as the outcome of a *measurement*, and the motion of an event in the eight dimensional phase space $\{x^\mu, p^\mu\}$ is assumed to be governed by a Hamiltonian dynamics. The Hamiltonian for a freely moving event was taken by Stueckelberg to be

$$K = \frac{p^\mu p_\mu}{2M}, \tag{1}$$

[1] In 1973, Piron and Horwitz [2] generalized Stueckelberg's theory to be applicable to many body systems by assuming τ to be a universal world time essentially identical to the time postulated by Newton. The classical Kepler problem was solved in [2] by replacing the two body nonrelativistic potential $V(r)$ by $V(\rho)$, where ρ is the invariant spacelike separation between the two particles, and Arshansky and Horwitz [3] solved the quantum bound state and scattering problems in this context; their method provides a solution, in principle, for any two body problem described nonrelativistically by a symmetric potential function $V(r)$.

where M is an intrinsic property of the event with dimension of mass; it may be thought of as the Galilean limiting mass of the particle [4]. The quantity $m^2 = -p^\mu p_\mu = E^2 - \mathbf{p}^2$ corresponds

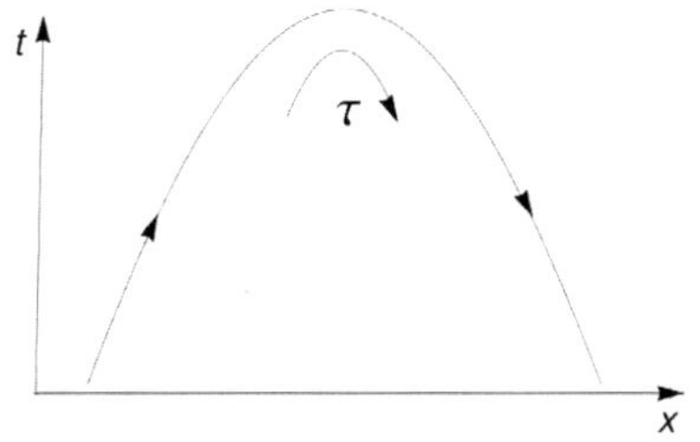

Figure 1. Basic Stueckelberg graph for classical pair annihilation. τ is the world time parameter along the curve.

to the actual measured mass of the associated particle. Since $t, \mathbf{x}$ are dynamical variables, $E, \mathbf{p}$ are also, and the theory is therefore intrinsically "off-shell".

The Hamilton equations are, quite generally,

$$\dot{x}^\mu = \frac{\partial K}{\partial p_\mu} \tag{2}$$

and

$$\dot{p}_\mu = -\frac{\partial K}{\partial x^\mu}, \tag{3}$$

where the dot corresponds to differentiation with respect to τ (thus, for the free particle, $d\mathbf{x}/dt = \mathbf{p}/E$, the standard relativistic definition of velocity).

Stueckelberg [1] wrote a Schrödinger type equation to describe the evolution of the corresponding quantum state for a general Hamiltonian function of the quantum variables x^μ, p^μ (which satisfy the commutation relations $[x^\mu, p^\nu] = ig^{\mu\nu}$, with metric signature $(-,+,+,+)$)

$$i\frac{\partial \psi_\tau(x)}{\partial \tau} = K\psi_\tau(x), \tag{4}$$

where we denote $x \equiv (t, \mathbf{x})$. Since the wavefunction is coherent in both space and time, the theory predicts interference in time [5] in the same way that the nonrelativistic quantum theory predicts interference in space. An experiment of this type has been done by Lindner *et al* [6] (this experiment was discussed in the context of this theory by Horwitz [7]). The extension of the theory to describe particles with spin [8,9] introduces a foliation of the quantum mechanical Hilbert space as well as the corresponding quantum field theory as a consequence of the spin-statistics relation [10] (Palacios *et al* [11] have proposed an experiment which could be able to detect the persistence of the entanglement of two particles with spin at different times, as is consistent with this theory).

This foliation, which can be put into correspondence with the orientation of the spacelike surfaces [10] forming the support of the complete set of local observables providing the representations of quantum fields [12], is involved in a fundamental way in CP (or T) conjugation,as we shall discuss below, and therefore intrinsically associated with the phenomenon of CP violation (see, for example, refs. [13] for a discussion of neutrino flavor oscillations and a comprehensive treatment of the CP violation problem). Similar considerations can be applied to CP and T violations in the K, B and D systems [14]. There have been many discussions in the framework of quantum field theory [15], one of the earliest that of Weinberg [16], providing a mechanism through coupling with a scalar field. It will be of interest to investigate the relation between our observations in the framework of the first quantized theory with these studies.

Under a gauge transformation, for which ψ is replaced by $e^{i\Lambda}\psi$ (see,*e.g.*, Yang and Mills [17]), this equation remains of the same form if one replaces p^μ by $p^\mu - ea^\mu$, where e is here the coupling to the 5D fields, and $i\frac{\partial}{\partial\tau}$ by $i\frac{\partial}{\partial\tau} + ea^5$, with a^5 a new scalar field, and, under gauge transformation, the gauge compensation field a^α is replaced by $a^\alpha - \partial^\alpha\Lambda$, for $\alpha = 0, 1, 2, 3, 5$ [18]. In the following, we describe a model non-Abelian gauge theory, in which the fifth gauge field plays a fundamental role, for neutrino oscillations. We study the structure of a Lorentz invariant and gauge covariant Hamiltonian and Lagrangian for a particle in interaction with a non-Abelian gauge field, and the resulting field equations and Lorentz force, both without and with spin. We point out mechanisms in this structure that can lead to CP symmetry violation on a fundamental level.

2. The model

In the flavor oscillations of the neutrino system, interactions with the vector bosons of the Glashow-Salam-Weinberg (GSW) theory [19] which induce the transition can produce pair annihilation-creation events. In the framework of Stueckelberg theory, pair annihilation and creation events can be correlated, as shown in Figure 2, by following the world line.[2] The

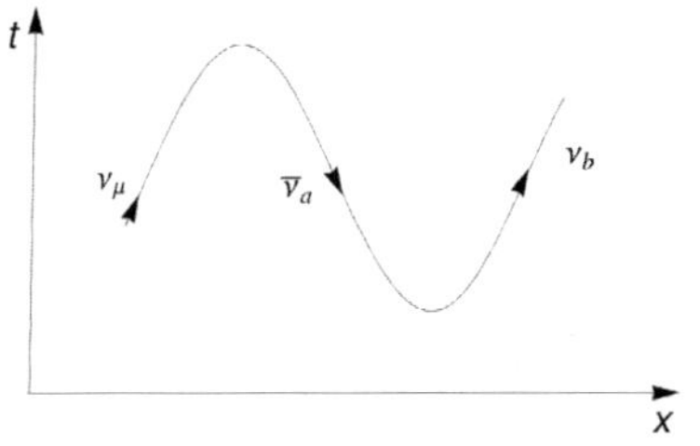

Figure 2. Annihilation-creation model for oscillation. ν_μ is the muon neutrino and ν_a, ν_b could be muon or electron neutrinos; the vector meson fields acting at the vertices can induce transitions.

methods of Feynman's original paper, based on a spacetime picture [20], closely related to Stueckelberg's earlier formulation, would admit such a construction as well. An "on-shell" version of our Figure 2 appears, with sharp vertices, in the first of ref. [20].

It may be noted from this figure that there is a net decrease in the time interval, possibly very small, observed for the particle to travel a certain distance. This mechanism is quite different from that discussed by Floyd [21], Matone [22], and Faraggi [23], discussing the phenomenon of early arrival as a purely quantum mechanical effect. The most recent experiments [24] have shown that the arrival times are consistent with light speed,*e.g.*, in the OPERA experient [24], over the 732 km distance form CERN to Gran Sasso, an arrival time of $6.5 \pm 7.4\{^{+8.3}_{-8.0}\}ns$ less than light speed arrival is reported, certainly consistent with light speed. [3]

[2] The curve shown in Figure 2 should be thought of as corresponding to the expectation values computed with the local density matrix associated with the gauge structured wave function of the neutrino beam.

[3] It has been observed in the Supernova 1987a that the neutrinos arrive about 3 hours before the light signal [25]. It has been argued that the light is delayed, for example, by gravitational effects on the virtual electron-positron pairs [26], but arguments of Lorentz invariance [27] and the apparently universal applicability of the geodesic behavior of light provide some difficulties for this view. On the other hand, an advanced arrival of the order

Suppose, for example, that such oscillations can occur twice during this transit [28] as in Figure 3. The particles (and antiparticles) have almost everywhere propagation speed less than

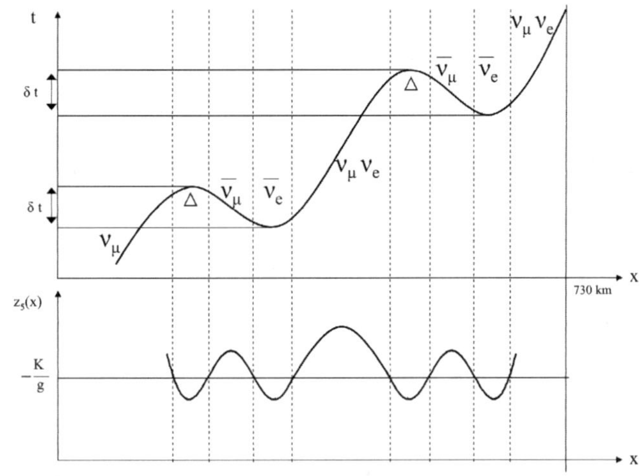

Figure 3. Two oscillation processes with sketch of required form of z^5 field, which must pass through the value K/g to admit passage through the light cone. δt indicates the "pull-back" time at each annihilation-creation event.

light velocity (except for the vertices, which we estimate, based on the Z, W lifetimes, to occur in about 10^{-22} seconds); it is clear from Figure 3 that an early arrival would *not* imply, in this model, that the neutrinos travel faster than light speed. The effect noted by Glashow and Cohen [29], indicating that Ĉerenkov radiation would be seen from faster than light neutrinos, would likely not be observed from the very short lived vertices, involving interaction with the W and Z fields, without sensitive detectors placed appropriately on the track [4]

A quantum mechanical counterpart of this model, in terms of Ehrenfest wave packets, is consistent with this construction. The derivation of the Landau-Peierls relation [31] $\Delta p \Delta t \geq \frac{\hbar}{2c}$ in the framework of the Stueckelberg theory [32] involves the assumption that the energy-momentum content of the propagating wave function contains predominantly components for

of 6.5 ns in each 730 km (consistent with this data) would result in approximately 3×10^3 hours early arrival. However, as we shall see below, the mechanism for the oscillations associated with such a "pull-back" involves the participation of the fifth field in an essential way, expected to fall off far from sources. One may estimate on the basis of a 3 hour early arrival the range of effectiveness of the fifth field, assuming an advance of 6.5 ns in each 730 km where effective. A simple estimate yields about 30 parsec(pc), as an effective size of the supernova. The Sun is about 10^4 pc from the center of the galaxy, so an effective range of about 30 pc is not unreasonable. This argument is certainly not a proof of a "pull back" ; it is meant to show that a small effect of this type could be consistent with the supernova 1987a data (see, moreover, further discussion in [25]).

[4] The neutrino arrivals detected at Gran Sasso appear to be almost certainly normal particles. The ICARUS detector (Antonello *et al* [30]) records no γ's or e^+e^- pairs which would be expected from Ĉerenkov radiation from faster than light speed neutrinos. Our model is consistent with the presence of neutrinos for which the total travel time is closely bounded by light velocity, since the "'pull-back" effect can be very small, limited (as in Figure 3) by the interval involved in the oscillation transitions.

which $\frac{p}{E} < 1$. Interactions, *e.g.*, at the vertices of the curve in Figure 3, can affect this distribution in such a way that, for some (small) interval of evolution, the wave packet can contain significant contributions to the expectation value of p/E much larger than unity, and thus the dispersion Δt in the Landau-Peierls relation can become very small without violating the uncertainty bound established by $< E/p >$. The interaction vertex may then be very sharp in t, admitting a precise manifestation of the deficit time intervals.

The upper part of Figure 3 shows schematically the orbit of a neutrino in spacetime during its transit, according to this theory, in which the first(annihilation) event results in the transition from a ν_μ to either a ν_μ or ν_e through interaction with a GSW boson (for this simple illustration we consider only the μ and e neutrinos, although there is no reason to exclude the τ neutrino) and the second (creation) event involves a transition from either of these states back to a ν_μ, ν_e state.

3. Formulation of the non-Abelian model

The gauge covariant form of the Stueckelberg Hamiltonian, valid for the non-Abelian case as well as for the Abelian, with coupling g to the 5D fields, is

$$K = \frac{(p^\mu - gz^\mu)(p_\mu - gz_\mu)}{2M} - gz^5(x), \tag{5}$$

where the z^μ fields are non-Abelian in the $SU(2)$ sector of the electroweak theory. Since, as we shall below, $\dot{x}^\mu$ is proportional to $p^\mu - gz^\mu$, the local expectation of the square of the "proper time" is proportional to that of the first term in the Hamiltonian. Therefore, see we see that the local expectation of z^5 must pass through that of the conserved value of $-K/g$ to admit passage of the orbit through the light cone. In the lower part of Figure 3, we have sketched a form for a smooth z^5 wave (in expectation value) that would satisfy this condition. Such a wave can be easily constructed as the superposition of a few harmonic waves with different wavelengths (originating in the spectral density of the neutrino wave functions [see Eq. (17)]).

The occurrence of such a superposition can be understood from the point of view of the structure of the $5D$ GSW fields. Working in the context of the first quantized theory, where the functions ψ belong to a Hilbert space $L^2(x, d^4x) \otimes d$, with d the dimensionality of the gauge fields ($d = 2$ corresponds to the Yang-Mills case [17] and the $SU(2)$ sector of the electroweak theory which we shall deal with here; our procedure for extracting the field equations and Lorentz force applies for any d), the field equations can be derived from the Lagrangian density (we consider the case of particles with spin in Sec. 5)

$$\begin{aligned} \mathcal{L} = &\frac{1}{2}\mathrm{Tr}(i\frac{\partial \psi}{\partial \tau}\psi^\dagger - i\psi\frac{\partial \psi^\dagger}{\partial \tau}) \\ &- \frac{1}{2M}\mathrm{Tr}\big[(p^\mu - gz^\mu)\psi((p_\mu - gz_\mu)\psi)^\dagger\big] \\ &+ g\mathrm{Tr}(z^5\psi\psi^\dagger) - \frac{\lambda}{4}\mathrm{Tr}f^{\alpha\beta}f_{\alpha\beta}, \end{aligned} \tag{6}$$

where ψ is a vector valued function representing the algebraic action of the gauge field, and $\psi^\dagger$ is a 2-component (row) conjugate vector valued function; $\mathcal{L}$ is a local scalar function. The operation Tr corresponds to a trace over the algebraic indices of the fields (the dimensional parameter λ arises from the relation of these fields to the zero mode fields of the usual $4D$ theory [18]). For the variation of the field strengths we take δz^α to be general infinitesimal Hermitian algebra-valued functions. Extracting the coefficients of these variations, with the definition of the non-Abelian gauge invariant field strength tensor [17]

$$f^{\alpha\beta} = \partial^\alpha z^\beta - \partial^\beta z^\alpha - ig[z^\alpha, z^\beta], \tag{7}$$

one obtains the field equations

$$\lambda\big[\partial^\alpha f_{\beta\alpha} - ig[z^\alpha, f_{\beta\alpha}]\big] = j_\beta \tag{8}$$

where

$$j_\mu = \frac{ig}{2M}\{(\partial_\mu - igz_\mu)\psi\psi^\dagger - \psi((\partial_\mu - igz_\mu)\psi)^\dagger\}, \tag{9}$$

and

$$j_5 = g\psi\psi^\dagger \equiv \rho_5. \tag{10}$$

Let us now impose, as done by Yang and Mills [17], the subsidiary condition

$$\partial^\alpha z_\alpha = 0. \tag{11}$$

We then obtain from (8)

$$(-\partial_\tau^2 + \partial_t^2 - \nabla^2)z_\beta = j_\beta/\lambda + ig[z^\alpha, f_{\beta\alpha}], \tag{12}$$

where we have taken the $O(4,1)$ signature for the fifth variable τ. Representing $z_\beta(x,\tau)$ in terms of its Fourier transfrom $z_\beta(x,s)$, with

$$z_\beta(x,\tau) = \int ds e^{-is\tau} z_\beta(x,s), \tag{13}$$

one obtains

$$(s^2 + \partial_t^2 - \nabla^2)z_\beta(x,s) = j_\beta(x,s)/\lambda + ig\int d\tau e^{is\tau}[z^\alpha(x,\tau), f_{\beta\alpha}(x,\tau)], \tag{14}$$

providing a relation between the off-shell mass spectrum of the z_β field and the sources including the quantum mechanical current as well as the non-linear self-coupling of the fields.

Since the behavior of the z_5 field plays an essential role in the immediately applicable predictions of our model, consider the equation (14) for $\beta = 5$,

$$(s^2 + \partial_t^2 - \nabla^2)z_5(x,s) = j_5(x,s)/\lambda + ig\int d\tau e^{is\tau}[z^\nu(x,\tau), f_{5\nu}(x,\tau)]. \tag{15}$$

In a zeroth approximation, neglecting the nonlinear coupling term, we can study the equation

$$(s^2 + \partial_t{}^2 - \nabla^2)z_5(x,s) \cong j_5(x,s)/\lambda. \tag{16}$$

The source term is a convolution of the lepton wave functions in the Fourier space, so that

$$(s^2 + \partial_t{}^2 - \nabla^2)z_5(x,s) \cong \frac{g}{2\pi\lambda}\int ds'\psi(x,s')\psi^\dagger(x,s'-s). \tag{17}$$

The Fourier representation over s of the wave function corresponds to the set of probability amplitudes for finding the particle in the corresponding mass states; we expect these functions to peak in absolute value, in free motion, at the measured neutrino masses. There is therefore the possibility of several mass values contributing to the frequency of the spectrum of the z_5 field (the diagonal contributions contribute only to its zero mode, a massless radiative field of essentially zero measure). It is of interest to note that in order for the sources to give rise to a form for the z_5 field of the type illustrated in Figure 3, there must be at least three peaks in the mass distribution of the wave functions, corresponding to three families of neutrinos. This condition has been noted in a somewhat different context in the last of [13] (p. 37) and in other studies (for example refs.[33], discussing the three family structure).

4. The non-Abelian Lorentz force

We now turn to study the quantum trajectories of the particles with non-Abelian gauge interactions to further check the consistency of our model. The Heisenberg equations of motion are associated with expectation values for which the classical motion is a good approximation if the wave packets are fairly well localized.

From the Hamiltonian (5) one obtains

$$\begin{aligned}\dot{x}^\lambda &= i[K, x^\lambda] \\ &= \frac{1}{M}(p^\lambda - g z^\lambda),\end{aligned} \tag{18}$$

of the same form as the classical result.

The second derivative is defined by

$$\ddot{x}^\lambda = i[K, \dot{x}^\lambda] + \frac{\partial \dot{x}^\lambda}{\partial \tau}, \tag{19}$$

where the last term is necessary because $\dot{x}^\lambda$ contains, according to (18), an explicit τ dependence which occurs in the fields z^λ. One then obtains (the Lorentz force for the non-Abelian case was also obtained, using an algebraic approach, in [34])

$$\ddot{x}^\lambda = -\frac{g}{2M}\{\dot{x}^\mu, f^\lambda{}_\mu\} - \frac{g}{M} f^{5\lambda}. \tag{20}$$

Let us make here the crude approximation that was used in obtaining (16), *i.e.*, neglecting the nonlinear coupling to the spacetime components of the field. Then, (20) becomes, for the time component,

$$\ddot{t} \cong -\frac{g}{M}\frac{\partial z^5}{\partial t}. \tag{21}$$

The rising z^5 field (Figure 3), before the first passage through the light cone, would imply a negative curvature, as required. This consistency persists through the whole process.

We further note that

$$-\frac{ds^2}{d\tau^2} = \frac{2}{M}(K + g z^5), \tag{22}$$

so that

$$\frac{d}{d\tau}\frac{ds^2}{d\tau^2} = -\frac{2g}{M}\frac{dz^5}{d\tau}, \tag{23}$$

consistent as well with the form of Figure 3.

5. The Hamiltonian for the spin $\frac{1}{2}$ neutrinos

The original construction of Wigner [35] for the description of relativistic spin, for application to a relativistic quantum theory, as explained in refs. [8] and [9], has a fundamental difficulty in that the resulting Wigner rotation [35] depends on momentum. The action of the resulting little group would depend on this momentum vector and the expectation value of the operator x^μ, $< x^\mu >$ would not be covariant; x^μ acts as a derivative of the momentum and this would destroy the unitarity of the little group action. Inducing the representation on a timelike four vector n^μ [8][9] preserves the covariance of $< x^\mu >$, and also admits the possibility of linear superpositions over momenta preserving the definition of the spin, for example, the construction of wave packets in spacetime of definite spin. It also implies a foliation of the Fock space for identical particles with strong implications for many body systems as well as quantum field theory [10].

Journal of Physics: Conference Series **437** (2013) 012021 doi:10.1088/1742-6596/437/1/012021

The wave functions, in coordinate or momentum space, now related in the usual way by Fourier transform, then transform , under a Lorentz transformation Λ, as

$$\hat{\psi}'_{\tau,\sigma}(x,n) = D(\Lambda,n)_{\sigma\sigma'}\hat{\psi}_{\tau,\sigma'}(\Lambda^{-1}x,\Lambda^{-1}n), \tag{24}$$

where $D(\Lambda,n)$ is the associated Wigner rotation represented here in a fundamental representation of $SL(2,C)$. There are two inequivalent representations of $SL(2,C)$; since the operator $\sigma^\mu p_\mu$ connects these two representations, and such an operator would certainly occur in almost any dynamical theory, the wave function must contain both fundamental representations. Within a simple transformation (see *e.g.*, [9],[10]), a vector-valued function combining the two forms of $L(n)\hat{\psi}(x)$ (where the $SL(2,C)$ matrix $L(n)$ acts to bring $(-1,0,0,0)$ to n^μ) of each type transforms like the Dirac wave function ψ, *i.e.*,

$$\psi'_\tau(x,n) = S(\Lambda)\psi_\tau(\Lambda^{-1}x,\Lambda^{-1}n), \tag{25}$$

where $S(\Lambda)$ is generated in the usual way by

$$\Sigma^{\mu\nu} = \frac{i}{4}[\gamma^\mu,\gamma^\nu]. \tag{26}$$

The norm, constructed of the sum of the norms of the two two-component representations, is then given by[5]

$$\|\psi\|_n{}^2 = \int d^4x\bar{\psi}_\tau(x,n)\gamma\cdot n\psi_\tau(x,n) \tag{27}$$

for each n in the negative light cone. The complete positive definite norm is given by the integral $\int\frac{d^3\mathbf{n}}{n^0}$ over the full foliation (which is required to be convergent on the full Hilbert space).

Following the method of ref. [9] for the non-Abelian case, we find a Hamiltonian of the form

$$K = \frac{1}{2M}(p-gz)_\mu(p-gz)^\mu - \frac{g}{2M}f_{\mu\nu}\Sigma_n{}^{\mu\nu} - gz^5, \tag{28}$$

where

$$\Sigma_n{}^{\mu\nu} = \Sigma^{\mu\nu} + K^\mu n^\nu - K^\nu n^\mu \equiv \frac{i}{4}[\gamma_n^\mu,\gamma_n^\nu], \tag{29}$$

with

$$\gamma_n^\mu = \gamma_\lambda\pi_n^{\lambda\mu}; \tag{30}$$

the projection

$$\pi_n^{\lambda\mu} = g^{\lambda\mu} + n^\lambda n^\mu \tag{31}$$

plays an important role in the description of the dynamics in the induced representation. In (28), the existence of projections on each index in the spin coupling term implies that $f^{\mu\nu}$ can be replaced by $f_n{}^{\mu\nu}$ in this term, a tensor projected into the foliation subspace (see ref. [10] for discussion of the properties of this foliation).

The $\Sigma_n{}^{\mu\nu}$ generate a Lorentz covariant form of the usual Pauli algebra (the compact $SU(2)$ part of the Lorentz algebra), and the K^μ generate the non-compact part of the Lorentz algebra [9] (since $n^\mu\Sigma_n{}^{\mu\nu} = K^\mu n_\mu = 0$, there are just three independent K^μ and three $\Sigma_n{}^{\mu\nu}$).

[5] Note that in the Stueckelberg theory, the wave functions provide *local* probability amplitudes associated with particles, unlike the non-local properties pointed out by Newton and Wigner [36] of the Klein-Gordon or Dirac (on shell) functions [2].

The quantities K^μ and $\Sigma_n^{\mu\nu}$ satisfy the commutation relations [9]

$$\begin{aligned}
[K^\mu, K^\nu] &= -i\Sigma_n^{\mu\nu} \\
[\Sigma_n^{\mu\nu}, K^\lambda] &= -i[(g^{\nu\lambda} + n^\nu n^\lambda)K^\mu - (g^{\mu\lambda} + n^\mu n^\lambda)K^\nu], \\
[\Sigma_n^{\mu\nu}, \Sigma_n^{\lambda\sigma}] &= -i[(g^{\nu\lambda} + n^\nu n^\lambda)\Sigma_n^{\mu\sigma} - (g^{\sigma\mu} + n^\sigma n^\mu)\Sigma_n^{\lambda\nu} \\
&\quad - (g^{\mu\lambda} + n^\mu n^\lambda)\Sigma_n^{\nu\sigma} + (g^{\sigma\nu} + n^\sigma n^\nu)\Sigma_n^{\lambda\nu}].
\end{aligned} \tag{32}$$

The last of (32) is the Lie algebra of $SU(2)$ in the spacelike surface orthogonal to n^μ. The three independent K^μ correspond to the non-compact part of the algebra which, along with the $\Sigma_n^{\mu\nu}$ provide a representation of the Lie algebra of the full Lorentz group. The covariance of this representation follows from

$$S^{-1}(\Lambda)\Sigma_{\Lambda n}^{\mu\nu} S(\Lambda)\Lambda_\mu^\lambda \Lambda_\nu^\sigma = \Sigma_n^{\lambda\sigma}. \tag{33}$$

In the special frame for which $n^\mu = (-1, 0, 0, 0))$, Σ_n^{ij} become the Pauli matrices $\frac{1}{2}\sigma^k$ with (i, j, k) cyclic, and $\Sigma_n^{0j} = 0$. In this frame there is no direct electric type interaction with the spin in the minimal coupling model (28) (the theory admits a covariant form of electric coupling of electric dipole type [9]; we will not consider this structure here). We remark that γ^5 commutes with this Hamiltonian, and therefore there is a chiral decomposition (independently of the mass of the neutrinos) that would admit the usual construction of the $SU(2) \times U(1)$ electroweak gauge theory. The $SU(2)$ sector that we discuss below would then apply to the left handed leptons. The asymptotic (free) solutions also admit a (foliated) helicity decomposition [9].

We record here the properties of the wave functions of a particle with spin one half under the discrete symmetries C, P and T, obtained from the Stueckelberg-Schrödinger Eq. (4), with the evolution generator K given by (5)(computed for simplicity for real Abelian gauge fields) [9]:

$$\begin{aligned}
\psi_{\tau n}^C &= C\gamma^0 \psi_{-\tau n}^*(x) \\
\psi_{\tau n}^P(x) &= \gamma^0 \psi_{\tau, -\mathbf{n}, n^0}(-\mathbf{x}, t), \\
\psi_{\tau n}^T &= i\gamma^1\gamma^3 \psi_{-\tau, \mathbf{n}, -n^0}^*(\mathbf{x}, -t), \\
\psi_{\tau n}^{CP}(x) &= C\psi_{-\tau, -\mathbf{n}, n^0}(-\mathbf{x}, t) \\
\psi_{\tau n}^{CPT}(x) &= i\gamma^5 \psi_{\tau, -n}(-\mathbf{x}, -t),
\end{aligned} \tag{34}$$

where $C = i\gamma^2\gamma^0$. The CPT conjugate wavefunction, according to its evolution in τ, moves backwards in spacetime relative to the motion of $\psi_{\tau n}$. For a wave packet with $E < 0$ components, which moves backwards in t as τ goes forward, it is the CPT conjugate wavefunction which moves forward with opposite charge, *i.e.*, the observed antiparticle.

We shall discuss the possibilities of CP violation provided by this structure below.

6. Lorentz force for spin 1/2 particle with non-Abelian gauge interactions

As in (18), one obtains the particle velocity

$$\dot{x}^\lambda = i[K, x^\lambda] = \frac{1}{M}(p^\lambda - gz^\lambda). \tag{35}$$

For the second derivative, from (19) and (32), we obtain

$$\begin{aligned}
\ddot{x}^\lambda &= -\frac{g}{2M}\{f^{\lambda\mu}, \dot{x}_\mu\} - \frac{g}{M} f^{5\lambda} \\
&\quad + \frac{g}{2M^2}\partial^\lambda f^n{}_{\mu\nu}\Sigma_n{}^{\mu\nu} + \frac{ig^2}{2M^2}[f^n{}_{\mu\nu}, z^\lambda]\Sigma_n{}^{\mu\nu}.
\end{aligned} \tag{36}$$

The third term of (36) corresponds to a Stern-Gerlach type force. Note that we have included the subscript or superscript n to the quantities that are transverse in the foliation.

Under the assumption that the fields are not too rapidly varying, and again neglecting coupling to the spacetime components of the field z^α, we see that the acceleration of the time variable along the orbit may again be approximated by (21).

7. Non-Abelian field equations with spin

We are now in a position to write the Lagrangian for the full theory with spin. We take for the Lagrangian the form (6) with an additional term for the spin interaction and factors of $\gamma^0(\gamma \cdot n)$ to assure covariance, yielding under variation of $\psi^\dagger$ the Stueckelberg equation for ψ with Hamiltonian (28):

$$\begin{aligned}\mathcal{L}_n &= \frac{1}{2}\mathrm{Tr}\big(i\frac{\partial\psi}{\partial\tau}\bar{\psi} - i\psi\frac{\partial\bar{\psi}}{\partial\tau}\big)(\gamma\cdot n)\\ &- \frac{1}{2M}\mathrm{Tr}\left[(p^\mu - gz^\mu)\psi\overline{(p_\mu - gz_\mu)\psi})(\gamma\cdot n)\right]\\ &+ g\mathrm{Tr}(z^5\psi\bar{\psi}(\gamma\cdot n)) - \frac{\lambda}{4}\mathrm{Tr}f^{\alpha\beta}f_{\alpha\beta}\\ &+ \frac{g}{2M}\mathrm{Tr}(f_{\mu\nu}\Sigma_n{}^{\mu\nu}\psi\bar{\psi}(\gamma\cdot n)).\end{aligned} \tag{37}$$

Defining j_α as in (9), (10), but with the factor $\gamma^0\gamma\cdot n$, required for covariance, i.e.,

$$j_{n\mu} = \frac{ig}{2M}\{(\partial_\mu - igz_\mu)\psi\bar{\psi} - \psi\overline{(\partial_\mu - igz_\mu)\psi)}\}(\gamma\cdot n), \tag{38}$$

and

$$j_{n5} = g\psi\bar{\psi}(\gamma\cdot n) \equiv \rho_n, \tag{39}$$

the variation of the Lagrangian with respect to the z-fields , where we have used the cyclic properties of matrices under the trace, yields, setting the coefficients of $\delta z^\nu, \delta z^5$ equal to zero, the field equations

$$\lambda(\partial_\beta f^{5\beta} - ig[z_\beta, f^{5\beta}]) = \rho_n \tag{40}$$

and

$$\begin{aligned}&\lambda(\partial_\beta f^{\nu\beta} - ig[z_\beta, f^{\nu\beta}])\\ &\quad = j_n{}^\nu + \frac{g}{M}\Sigma_n{}^{\mu\nu}\{\partial_\mu\rho_n - ig[z_\mu, \rho_n]\}.\end{aligned} \tag{41}$$

Eq. (41) corresponds to a Gordon type decomposition of the current, here projected into the foliation space (spacelike) orthogonal to n^μ. Note that the covariant derivative of ρ_n in the last term is also projected into the foliation space.

With the subsidiary condition $\partial^\beta z_\beta = 0$, as before, we may write the field equations as

$$\lambda(-\partial^\beta\partial_\beta z^5 - ig[z_\mu, f^{5\mu}]) = \rho_n \tag{42}$$

and

$$\lambda(-\partial^\beta\partial_\beta z^\nu - ig[z_\beta, f^{\nu\beta}]) = j_n{}^\nu + \frac{g}{M}\Sigma_n{}^{\mu\nu}\{\partial_\mu\rho_n - ig[z_\mu, \rho_n]\}. \tag{43}$$

Note that the spin coupling is not explicit in (42). Neglecting, as before, coupling to the spacetime components, one reaches the same conclusions for the approximate behavior of the z^5 field, *i.e.*, as determined by Eq.(17) with $\psi^\dagger$ replaced by $\bar{\psi}\gamma\cdot n$. The latter reduces to the same expression for $n^\mu \to (-1, 0, 0, 0)$.

Journal of Physics: Conference Series **437** (2013) 012021 doi:10.1088/1742-6596/437/1/012021

8. CP and T conjugation

The association of this timelike vector with the spacelike surfaces used by Schwinger and Tomonaga [12] for the quantization of field theories has been recently discussed [10]. These spacelike surfaces form the support of a complete set of commuting local observables on which the Hilbert space of states in constructed. It follows from the above properties of the wave functions for a particle with spin, that the CPT conjugate theory would be associated with the same spacelike surface, corresponding to $\pm n^\mu$. However, the CP conjugate, taking $\mathbf{n} \to -\mathbf{n}$ and $n^0 \to n^0$ refers to an entirely different spacelike surface (the time reversed states, for which $\mathbf{n} \to \mathbf{n}$ and $n^0 \to -n^0$ are associated with this spacelike surface as well, with reflected unit timelike vector). The equivalence of the physical processes described in these two frameworks would depend on the existence of an isometry (including both unitary and antiunitary transformations) changing the basis of the space from the set of local observables on the first spacelike surface to those defined on the conjugated surface as well as the equivalence of the physics evolving from it after the CP (or T) conjugation.

The spin coupling term in (28) contains the possibility of CP violation in generating a physics that is inequivalent on the new spacelike surface. The nonrelativistic quantum theory with Zeeman type $\sigma \cdot \mathbf{H}$ coupling is, of course, not invariant under T conjugation. Precisely the same situation is true in the corresponding relativistic equation (28); as we have pointed out, in the special frame in which $n^\mu = (-1, 0, 0, 0)$, the matrices $\Sigma_n^{\mu\nu}$ reduce to Pauli matrices. Under Lorentz transformation they still generate the algebra of $SU(2)$ in a fundamental representation, and therefore still contain the imaginary unit. Therefore, the physical evolution on the CP conjugate spacelike surface is not, in general, equivalent to the original evolution. For this phenomenon to occur, it is necessary that there be present an $f_{\mu\nu}$ field. In addition to self-interaction effects, for which the intrinsic CP violation can be expected to cancel, the Stueckelberg oscillation diagram of Figure 2 suggests the existence of fields present in the equations of motion of the second branch due to the proximity of the accelerated motion in the first branch, thus providing a fundamental mechanism for CP violation. A consequence of this structure is that the physics in the corresponding CP conjugated system of the quantum fields, evolving from the CP conjugate spacelike surface, could be inequivalent.

9. Conclusions

We have argued that, according to the derivation of the Landau-Peierls relation given in ref. [32], the vertices of the neutrino-antineutrino transitions may be very sharp, and provide for a rather precise "pull back" of the time interval. Significantly higher precision than available in the present experiments would be necessary to see such an effect.

We have worked out the equations describing the Lorentz forces and the field equations of the corresponding ($5D$) non-Abelian gauge theory, with the help of Stueckelberg type Hamiltonians both for the spinless case and for the case of relativistic particles with spin in interaction with such a nonabelian gauge field, and have shown that the conclusions reached are, in lowest approximation, consistent with our simple model. We emphasize that, in the framework of the Stueckelberg model, the dynamics of the fifth gauge field, modulated by the particle mass spectrum contained in the wave function (as in Eq.(17)), plays an essential role for the oscillation process.

The presence of spin, described in the relativistic framework of Wigner [35], as in refs. [9][10], introduces a foliation in the Hilbert space and in the structure of the fields, both classical and quantum. Since, in Tomonaga-Schwinger quantization of the fields, the spacelike surface constructed to define a complete set of local observables is characterized by being orthogonal to the timelike vector n of the foliation [10], the actions of the discrete CP or T transformations change the basis for the construction of the Hilbert space to essentially different spacelike surfaces. Along with the form of the spin coupling term in (28), this suggests a model for

CP or T violation on the first quantized level. Further consequences of this foliation will be explored elsewhere.

We furthermore remark that our model would be applicable to the K, B and D systems [14] as well, manifested by the quark gluon interactions in their substructure. This possibility is under study.

Acknowledgments
We wish to thank Asher Yahalom, Jacob Levitan, Avi Gershon, Andrew Bennett and David Owen for discussions. We also thank E.R. Floyd for bringing his work to our attention, and Steve Weinberg for a communication on the 1987a supernova. We are also grateful to the Physics Department at the Ben Gurion University of the Negev, and the Physics Department at Bar Ilan University for opportunities to give seminars on this work in its early stages. We would also like to thank Itzhak Rabin for his encouragment and Faina Shikerman for designing the figures and for discussions.

References

[1] E. C. G. Stueckelberg, *Helv. Phys. Acta* 14, 322-323,588-594 (1941); 15, 23-27 (1942).
[2] L. P. Horwitz and C. Piron, *Helv. Phys. Acta* 46, 316-326 (1973).
[3] R. I. Arshansky and L. P. Horwitz, *Jour. Math. Phys.* 30 66,380 (1989).
[4] L. P. Horwitz, W. C. Schieve and C. Piron, *Ann. Phys.*137, 306-340 (1981).
[5] L. P. Horwitz and Y. Rabin, *Lett. al Nuovo Cim* 17 501 (1976).
[6] F. Lindner, *et al Phys. Rev. Lett.* 95, 040401 (2005).
[7] L. Horwitz, *Phy. Lett. A* 355, 1-6 (2006).
[8] L. Horwitz, C. Piron and F. Reuse, *Helv. Phys. Acta* 48, 546 (1975)
[9] R. Arshansky and L. P. Horwitz, *J. Phys. A: Math. Gen.* 15, L659-662 (1982).
[10] L. P. Horwitz, *Jour. of Phys. A: Math. Theor.* 46 035305 (2013).
[11] A. Palacios, T. N. Rescigno, and C. W. McCurdy, *Phys. Rev. Lett.* 103, 253001-1-253001-4 (2009).
[12] S. Tomonaga, *Prog. Theor. Phys.*1, 27 (1946), J. Schwinger, Phys. Rev. 74 1439 (1948).
P. Weiss , Proc. Roy. Soc. A169
102 (1938), J. M. Jauch and F. Rohrlich, *The Theory of Photons and Electrons*, Springer-Verlag (1976).
[13] S. M. Bilenky and B. Pontecorvo, *Phys. Rep.* (Phys. Lett. C) 41, 225 (1978);
H. Nunokawa, S. Parke and W. F. J. Valle, *Prog. Part. Nucl. Phys.* 60, 338 (2008), arXiv:0710.0554 (2007).
[14] B. Kayser, *The frequency of neutral meson and neutrino oscillation*, SLAC-PUB-7123 (2005)
Neutrino Physics SLAC Summer Institute on Particle Physics, Aug. 2-13 (2004)
T. Aaltonen *et al*, *Phys. Rev.* *D*85, 012009 (2012)
See also, J. P. Lees *et al*, *Phys. Rev. Lett.* 109, 211801 (2012), arXiv:1207.5832
B. Bhattacharya, M. Gronau, and J. L. Rosner, *Phys. Rev. D* 85 054014 (2012), arXiv:1207.0761.
[15] For example, E. Sassaroli,*Neutrino flavor mixing and oscillation in Field theory*, arXiv:hep-ph/9805480 (1998)
Fayyazuddin and Riazuddin, *Modern Introduction to Particle Physics*, World Scientific, Singapore (2012)
R. N. Mohapatra, *Unification and Supersymmetry*, Springer, New York (1986).
[16] S. Weinberg, *Phys. Rev. Lett.* 37 657(1976)
R. D. Peccei and H. R. Quinn, *Phys. Rev. D* 16, 1791 (1977).
[17] C. N. Yang and R. Mills , *Phys. Rev.* 96, 191-195 (1954).
[18] D. Saad, L. P. Horwitz and R. I. Arshansky, *Found. Phys.* 19, 1125-1149 (1989)
I. Aharonovich and L. P. Horwitz, *Jour. Math. Phys.* 47, 122902-1-12290226 (2006), 51, 052903-1-052903-27 (2010), 52, 082901-1-08290111 (2011), 53, 032902-1-032902-29 (2012)
Eur. Phys. Lett. 97, 60001-p1-60001-p3 (2012).
[19] S. Glashow, J. Iliopoulis, and L. Maini, *Phys. Rev.* D2, 1285-1292 (1970)
A. Salam, *Elementary Particle Theory: Relativistic Groups and Analyticity*, 8th Nobel Symposium, 367-377, ed. Svartholm N, Almquist and Wiksell, Stockholm (1968)
S. Weinberg, *Phys. Rev. Lett.*19, 1264-1266 (1967).
[20] R. P. Feynman, Phys. Rev. 76, 749-784, 769-789 (1949)
J. Schwinger, *Phys. Rev.* 82, 664-679 (1951).
[21] E. R. Floyd, *Opera superluminal neutrinos per quantum trajectories* arXiv: 1112.4779 (2012).
[22] M. Matone, *Superluminality and a curious phenomenon in the relativistic Hamilton-Jacobi equation* arXiv: 1109.6631 (2011).

[23] A. Faraggi, *Opera data and the equivalence postulate of quantum mechanics*, arXiv:1110.1857 (2011).
[24] T. Adam *et al*,Jour. of High Energy Physics 10 093 (2012), arXiv:1109.4897
Agafonova N Yu *et al*, *Phys. Rev. Lett.* 109 070801 (2012) arXiv:1208.1392.
[25] W. D. Arnett, J. N. Bahcall, R. P. Kirschner, and S. E. Woosley, *Ann. Rev. Astron. Astrophys* 27, 629 (1989).
[26] J. D. Franson, *Reduced photon velocities in the OPERA neutrino experiment and Supernova 1978a*, arXiv:1111.6986 (2012).
[27] Personal communication, S. Weinberg.
[28] See, for example, J. W. F. Valle, *Journal of Physics: Conference Series* 56, 473-505 (2006).
[29] A. G. Cohen and S. L. Glashow, *Phys. Rev. Lett.* 107, 181803 (2011); arXiv:1109.6562 (2011).
[30] M. Antonello, *et al*, arXiv: 1110.3763 (2011).
[31] L. Landau and R. Peierls, *Z. Phys.* 69, 56-69 (1931).
[32] R. Arshansky and L. P. Horwitz, *Found. of Phys.* 15, 701-715 (1984).
[33] For example, G. L. Fogli, E. Lisi and G. Scioscia, *Accelerator reactor neutrinos oscillation experiments in a simple three-generation framework*, IASSNS-AST 95/28, BARI-TH/219-95, arXiv hep-ph/9506350 (1995)
A. Bandyopadahyay, S. Choubey, S. Goswami and K. Kamales, *Phys. Rev. D* 65 073031 (2002) and arXiv:hep-ph/0110307 (2002). See also ref [14].
[34] M. C. Land, N. Shnerb and L. P. Horwitz, *Jour. Math. Phys.* 36, 3263-3288 (1995).
[35] E. Wigner, *Ann. of Math.* 40, 149-204 (1939).
[36] T. D. Newton and E. Wigner, *Rev. Mod. Phys.* 21, 400-406(1949).

The geometric phase of a relativistically covariant four dimensional harmonic oscillator

Y Bachar[1], R I Arshansky[2], L P Horwitz[1,3,4] and I Aharonovich[1]

[1] Dept. of Physics, Bar Ilan Univeristy
[2] Givat Zorfatit, Etzel 12/14, Jerusalem 9754
[3] School of Physics and Astronomy, Tel-Aviv University
[4] Dept of Physics, Ariel University

E-mail: yossi_dk@yahoo.com

Abstract. We show that there is nontrivial Berry relativistically covariant phase generated by a perturbed relativistic oscillator. This phase is associated with a fractional perturbation of the azimuthal symmetry of the oscillator.

A manifestly covariant quantum mechanics was formulated by E. C. G. Stueckelberg [1] in 1941. He studied this theory for the case of a single particle in an external field. He considered the phenomenon of pair annihilation and creation as a manifestation of the development, in each case, of a single world line that curves in such a way that one part runs backwards in time, and above the turning point the line does not pass at all. This configuration was considered by Stueckelberg to represent pair annihilation. To describe such a curve, parametrization by the variable t is ineffective, since the trajectory is not single valued. He therefore introduced a parametric description, with parameter τ, along the world line. Hence one branch of the curve is generated by motion in the positive sense of t as a function of increasing τ, and the other branch by motion in the negative sense of t.

The motion, in space-time, of the point generating the world line, which we shall call an event (and has properties of space-time position and energy momentum), is governed in the classical case by the Hamiltonian equations in space-time

$$\frac{\mathrm{d}x^\mu}{\mathrm{d}\tau} = \frac{\partial K}{\partial p_\mu}, \qquad \frac{\mathrm{d}p^\mu}{\mathrm{d}\tau} = -\frac{\partial K}{\partial x_\mu} \tag{1}$$

where $x^\mu = (t, \vec{x})$, $p^\mu = (E, \vec{p})$ [we take $c = 1$ and $g_{\mu\nu} = (-1, 1, 1, 1)$] and the evolution generator K is a function of the canonical variables x_μ, p_μ. For the special case of free motion,

$$K_0 = \frac{p^\mu p_\mu}{2M} \tag{2}$$

where M is an intrinsic parameter assigned to the generic event, and hence

$$\frac{\mathrm{d}x^\mu}{\mathrm{d}\tau} = \frac{p^\mu}{M} \tag{3}$$

It then follows that

$$\frac{d\vec{x}}{dt} = \frac{\vec{p}}{E} \tag{4}$$

consistent with standard relativistic kinematics. We note, however, that the mass squared $m^2 = -p^\mu p_\mu$ is a dynamical variable since $\vec{p}$ and E are considered to be kinematically independent, and therefore it is not taken to be equal to a given constant. The set of values taken by m^2 in a particular dynamical context is determined by initial conditions and the dynamical equations.

In the quantum theory, $\vec{x}, t$ (and $\vec{p}, E$) denote operators satisfying the commutation relations (we take $\hbar = 1$)

$$[x^\mu, p^\mu] = ig^{\mu\nu} \tag{5}$$

The state of one-event system is described by a wave function $\psi_\tau(x) \in L^2(R^4)$, a complex Hilbert space with measure $d^4x = d^3x\ dt$ satisfying the equation

$$i\frac{\partial \psi_\tau(x)}{\partial \tau} = K\psi_\tau(x) \tag{6}$$

This equation, designed to provide a manifestly covariant description of relativistic phenomena, is similar in form to the non-relativistic Schrodinger equation. Although free motion is determined by the operator form of K_0 of Eq. (3), i.e., the d'Alembertian, which is hyperbolic ($p_\mu p^\mu \equiv -\partial_\mu \partial^\mu$ instead of the elliptic operator $\vec{p}^2 \equiv -\nabla^2$), the same methods may be used for studying Eq. (6) as for the non-relativistic Schrodinger equation.

In 1973 Horwitz and Piron [2] generalized the Stueckelberg theory by assuming that for the treatment of systems of more than one event (generating world lines of more than one particle), one assumes the unperturbed evolution generator to be of the form ($p_i^2 = p_{i\mu} p_i^\mu$)

$$K_0 = \sum_{i=1}^{N} \frac{p_i^2}{2M_i} \tag{7}$$

They further assumed that there is a single universal τ which parametrizes the motion of all of the particles of the many body system (we denote this generalaized theory by SHP.) There is a class of model systems, for which solutions can be achieved using straightforward methods, which involve only effective action-at-a-distance (direct action) potentials, where the evolution generator is of the form

$$K = \sum_{i=1}^{N} \frac{p_i^2}{2M_i} + V(x_1, x_2, ..., x_N) \tag{8}$$

Note that in this case the potential function enters into the dynamical evolution equation as a term added to the generator of the free motion, and therefore corresponds to a space-time coordinate-dependent interaction mass.
Equations (2) become

$$\frac{dx_i^\mu}{d\tau} = \frac{\partial K}{\partial p_{i\mu}}, \qquad \frac{dp_i^\mu}{d\tau} = -\frac{\partial K}{\partial x_{i\mu}} \tag{9}$$

In 1989 Horwitz and Arshansky [3] demonstrated the existence of bound state solutions for the quantum case by solving the dynamical equation (6) associated with the dynamical evolution

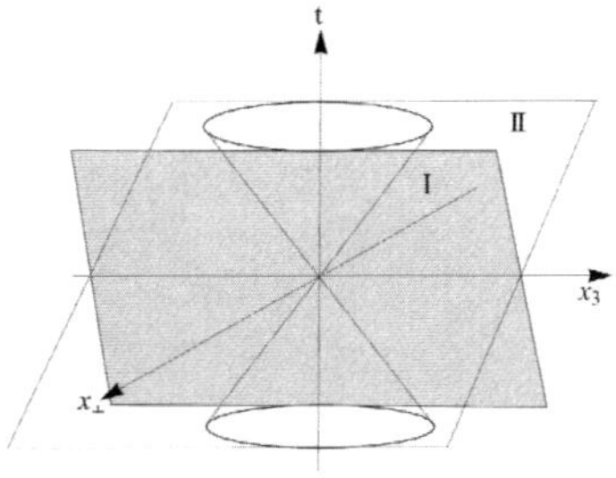

Figure 1. RMS

operator (8). The two-body potential function, which they chose for Poincaré invariance was of the form $V(\rho)$, where $x_1^\mu - x_2^\mu$ is spacelike,

$$\rho = \sqrt{(x_1^\mu - x_2^\mu)(x_{1\mu} - x_{2\mu})} \equiv \sqrt{(x_1 - x_2)^2} \tag{10}$$

Instead of the more widely used decomposition of Minkowski space into the timelike and full spacelike regions, they used the decomposition of the spacelike region into two subregions [invariant under an O(2,1) subgroup of O(3,1)].(see Fig.1) One of these (called the RMS - Reduced Minkowski Space) sectors (I) consists of the space-time points external (in spacelike directions) to two hyperplanes tanget to the light cone that are oriented along the z axis (the direction must be chosen to define this space). The second (II) consists of the space-time points in the sector interior (timelike direction) to these hyperplanes, but excluding the light cone. In the figure 1, this composition is shown schematically by folding the two space axes x,y together (defining the coordinate $x_\perp$); in the resulting three-dimensional space, the two hyperplanes become planes and intersect along the z axis.

where

$$x^0 = \rho \sin\theta \sinh\beta \quad x^1 = \rho \sin\theta \cos\phi \cosh\beta \quad x^2 = \rho \sin\theta \sin\phi \sinh\beta \quad x^3 = \rho\cos\theta \tag{11}$$

This procedure resulted in the correct (Schrodinger) spectrum for the bound state. When they applied the method of treating the relativistic quantum two body problem to the case of the four dimensional Harmonic Oscillator so that the reduced Hamiltonian is

$$\begin{aligned} K = \frac{p_\mu p^\mu}{2m} + \frac{1}{2}kx_\mu x^\mu &= -\frac{1}{2m}\partial_\mu \partial^\mu + \frac{1}{2}mw^2\rho^2 \\ &= \frac{1}{2m}[-\frac{\partial^2}{\partial\rho^2} - \frac{3}{\rho}\frac{\partial}{\partial\rho} + \frac{\Lambda}{\rho^2}] + \frac{1}{2}mw^2\rho^2 \end{aligned} \tag{12}$$

They obtained the complete set of eigenvectors which span the RMS, represented by the τ independent wave functions[1]

[1] Cook [4] solved this problem with full spacelike support for the wavefunction and obtained an incorrect spectrum. Note than Zmuidzinas [5] first showed that there is no complete set of orthogonal functions in the full spacelike region,and constructed the RMS

$$\psi_{mnln_a}(\phi,\beta,\theta,\rho) = \frac{1}{2\pi}e^{i(m+1/2)\phi}\cdot\sqrt{n}\sqrt{\Gamma(1+m+n)/\Gamma(1+m-n)}\cdot \\ \cdot(1-\tanh^2\beta)^{\frac{1}{4}}P_m^{-n}(\tanh\beta)\cdot(1-\cos^2\theta)^{-\frac{1}{4}}\cdot \\ \cdot P_l^n(\cos\theta)\frac{1}{\sqrt{\rho}}(\frac{mw\rho^2}{\hbar})^{l/2}e^{-\frac{mw\rho^2}{2\hbar}}L_{n_a}^{l+1/2}(\frac{mw\rho^2}{\hbar}) \tag{13}$$

The idea of the geometric phase proposed by Berry [6] in 1984 asserts that under adiabatic processes the wave function of a system picks up a phase factor that can be found in the nonrelativistic case by the integral

$$\gamma_n(t) \equiv i\int_0^t <\psi_n(t')\mid\frac{\partial\psi_n(t')}{\partial t}> dt' \tag{14}$$

where ψ_n is the nth eigenvector of the evolving Hamiltonian and t is the time between final and initial state of the time dependent parameters of the Hamiltonian (i.e. the evolution time). To construct a manifestly covariant form of (14), we have to replace the variable t by the variable τ which is the evolution time according to SHP quantum mechanics. Suppose there are N τ dependent parameters: $R_1(\tau), R_2(\tau), ...R_N(\tau)$ in the Hamiltonian of a given problem then

$$\frac{\partial\psi_n}{\partial\tau} = \frac{\partial\psi_n}{\partial R_1}\frac{dR_1}{d\tau} + \frac{\partial\psi_n}{\partial R_2}\frac{dR_2}{d\tau} + ... + \frac{\partial\psi_n}{\partial R_N}\frac{dR_N}{d\tau} = (\nabla_{\vec{R}}\psi_n)\cdot\frac{d\vec{R}}{d\tau} \tag{15}$$

where

$$\vec{R} = (R_1, R_2, ..., R_N) \tag{16}$$

is the gradient with respect to these parameters. We then have

$$\gamma_n(\tau) = i\int_{\vec{R}_i}^{\vec{R}_f} <\psi_n\mid\nabla_{\vec{R}}\psi_n> \cdot d\vec{R} \tag{17}$$

Now, if the Hamiltonian returns to its original form after a time $\tau = T$, then the geometric phase is

$$\gamma_n(T) = i\oint <\psi_n\mid\nabla_{\vec{R}}\psi_n> \cdot d\vec{R} \tag{18}$$

We wish to demonstrate the realization of the manifestly covariant Berry phase in the example of the four dimensional oscillator. For that propose we add a perturbation to the Hamiltonian (12) which breaks the hyperangular symmetry of this Hamiltonian. Since the complex valued matrix elements necessary to develop a dynamical phase arise in the example from the ϕ dependance, one must perturb the azimuthal symmetry with a fractional coefficient, as we shall see below.

$$K = \frac{1}{2m}\Big[-\frac{\partial^2}{\partial\rho^2} - \frac{3}{\rho}\frac{\partial}{\partial\rho} + \frac{\Lambda}{\rho^2}\Big] + \frac{1}{2}mw^2\rho^2 + \\ + 2\Big(\varepsilon_1\rho^2\sin^2\theta\cos^2((2/3)\phi)\cosh^2\beta + \\ + \varepsilon_2\rho^2\sin^2\theta\sin^2((2/3)\phi)\cosh^2\beta + \\ + \varepsilon_3\rho^2\cos^2\theta - \varepsilon_0\rho^2\sin^2\theta\sinh^2\beta\Big) \tag{19}$$

where $\varepsilon_0, \varepsilon_1, \varepsilon_2, \varepsilon_3$ are small parameters of the perturbation.
Now using degenerate time independent perturbation theory, we calculate the first order correction to the wave function (13).

The new wave function produced by the Hamiltonian (19) is equal to a linear combination of the wave function (13) and the first order correction to this wave function which is given by

$$\psi^{(1)}_{mnln_a} = \sum_{m'n'l'n'_a \neq mnln_a} \frac{< \Phi_{m'n'l'n'_a} \mid V \mid \Phi_{mnln_a} >}{K_a - K'_a} \Phi_{m'n'l'n'_a} \tag{20}$$

where V is the perturbation given in (19), K_a is the eigenvalue of the unperturbed Hamiltonian (12), and $\Phi_{m'n'l'n'_a}$ is a linear combination of the $\psi_{m'n'l'n'_a}$ due to the degeneracy of the energy eigenvalues

$$K_a = \hbar\omega(l + 2n_a + \frac{3}{2}) \tag{21}$$

Now suppose ε_0 and ε_3 are equal to zero so Eq.(20) will be

$$\begin{aligned} \psi^{(1)}_{mnln_a} = \varepsilon_1 & \sum_{m'n'l'n'_a \neq mnln_a} \frac{< \Phi_{m'n'l'n'_a} \mid \rho^2 \sin^2\theta \cos^2((2/3)\phi) \cosh^2\beta \mid \Phi_{mnln_a} >}{l - l' + 2(n_a - n'_a)} \Phi_{m'n'l'n'_a} \\ + \varepsilon_2 & \sum_{m'n'l'n'_a \neq mnln_a} \frac{< \Phi_{m'n'l'n'_a} \mid \rho^2 \sin^2\theta \sin^2((2/3)\phi) \cosh^2\beta \mid \Phi_{mnln_a} >}{l - l' + 2(n_a - n'_a)} \Phi_{m'n'l'n'_a} \end{aligned} \tag{22}$$

or

$$\psi^{(1)}_{mnln_a} = \varepsilon_1 \psi' + \varepsilon_2 \psi'' \tag{23}$$

The calculation of ψ' and ψ'' will done numerically by letting each of the indices $(m, n, l, n_a, m', n', l', n'_a)$ run from 2 to 5 and competing the ratios in (22) for each permutation of these indices. In this way the matrix of the perturbation will have an order 256x256 and each of the eigenvectors, $\psi_{m'n'l'n'_a}$, and the first order correction will be 256x1 column vector. The degeneracy is 16 fold beside the fact that we have additional degeneracy due to the linearity of the energy values in n and l (so the difference in eigenvalues can be zero even if $l \neq l'$ and $n_a \neq n'_a$ which requires a more complete treatment, now in preparation [7]) and the corrected wave function will be

$$\psi = \psi_{mnln_a} + \psi^{(1)}_{mnln_a} = \psi_{mnln_a} + \varepsilon_1 \psi' + \varepsilon_2 \psi'' \tag{24}$$

Eq. (24) can be written as

$$\psi = \psi_{mnln_a} + \varepsilon_1 \sum_{m'n'l'n'_a} c_{mm'} A_{mm'nn'll'n_an'_a} \Phi_{m'n'l'n'_a} + \varepsilon_2 \sum_{m'n'l'n'_a} d_{mm'} B_{mm'nn'll'n_an'_a} \Phi_{m'n'l'n'_a} \tag{25}$$

where $A_{mm'nn'll'n_an'_a}$ and $B_{mm'nn'll'n_an'_a}$ are the factors of the perturbation matrix elements which results from integration on β, ρ and θ divided by the energy differences (and hence are real) and the factors $c_{mm'}$ and $d_{mm'}$ are given by

$$c_{mm'} = \int_0^{2\pi} e^{i(m'-m)\phi} \cos^2((2/3)\phi) d\phi = \frac{\frac{3}{2} i(m - m') - \frac{1}{\sqrt{3}}}{\frac{4}{9} - (m' - m)^2} \tag{26}$$

and

$$d_{mm'} = \int_0^{2\pi} e^{i(m'-m)\phi} \sin^2((2/3)\phi) d\phi = \frac{\frac{\sqrt{3}}{2} i(m - m' + \frac{2}{3\sqrt{3}}) + \frac{2}{3}}{\frac{4}{9} - (m' - m)^2} \tag{27}$$

substitute (25) in Eq. (18) where $\vec{R} = (\varepsilon_1, \varepsilon_2)$ and expressing ε_1 and ε_2 in polar coordinates, then taking the line integral over the unit circle, we get

$$\begin{aligned} \gamma_{mnln_a}(T) &= i\pi(< \psi' \mid \psi'' > - < \psi'' \mid \psi' >) = i\pi \cdot 2iIm[< \psi' \mid \psi'' >] = \\ &= -2\pi Im[\sum_{m'n'l'n'_a} c_{mm'} A_{mm'nn'll'n_a n'_a} d^*_{mm'} B_{mm'nn'll'n_a n'_a}] \end{aligned} \tag{28}$$

We now substitute (26) and (27) into (28) and obtain

$$\gamma_{mnln_a}(T) = 3\pi \sum_{m'n'l'n'_a} A'_{mm'nn'll'n_a n'_a} B'_{mm'nn'll'n_a n'_a} (m' - m - \frac{2}{9\sqrt{3}}) \tag{29}$$

Which is the geometric phase due to the perturbation on the eigenvector (13) where

$$A'_{mm'nn'll'n_a n'_a} = \frac{A_{mm'nn'll'n_a n'_a}}{\frac{4}{9} - (m' - m)^2}; \qquad B'_{mm'nn'll'n_a n'_a} = \frac{B_{mm'nn'll'n_a n'_a}}{\frac{4}{9} - (m' - m)^2} \tag{30}$$

We have shown in this note that a nontrivial Berry phase can be associated with a perturbed relativistically covariant oscillator. The fractional breaking of the azimuthal symmetry was necessary to achieve this, and makes explicit, in this example, the association of the Berry phase with topological properties of the system (as in the Aharonov-Bohm effect [8], [9]). We remark that the Berry phase computed on this way is covariant, since the RMS coordinates in every Lorentz frame are isomorphic [10].

References

[1] E. C. G. Stueckelberg, *Helv. Phys. Acta*, 14:372, 588, 1941.
[2] L. P. Horwitz and C. Piron, *Helv. Phys. Acta.*, 46:316, 1973.
[3] R. Arshansky and L. P. Horwitz, *J. Math. Phys.*, 30(1):6680, 1989.
[4] J. L. Cook, *Aust. J. Phys*, 25:141, 1972.
[5] J. S. Zmuidzinas, *J. Math. Phys.*, 7:764, 1966.
[6] David J. Griffiths, *Introduction to Quantum Mechanics.* Benjamin Cummings, 2004.
[7] Y. Bachar, R. I. Arshansky, L. P. Horwitz, and I. Aharonovich, *Covariant Berry phase for a perturbed relativistic four dimensional harmonic oscillator*, in preparation, 2013.
[8] Y. Aharonov and D. Bohm, *Phys. Rev.*, 115:485491, 1959.
[9] M. V. Berry, *Proc. R. Lond.*, A392:4557, 1984.
[10] R. Arshansky and L. P. Horwitz, *J. Math. Phys*, 30(2):380392, 1989.

A classical path to unification

K Hasselmann
Max Planck Institute for Meteorololgy, Hamburg
E-mail: klaus.hasselmann@zmaw.de

Abstract. An overview is given of a classical unified theory of gravity, elementary particles and quantum phenomena based on soliton solutions of Einstein's vacuum equations in twelve dimensional space. Bell's theorem on the Einstein-Podolsky-Rosen experiment, which is widely interpreted as ruling out classical explanations of quantum phenomena, is shown to be non-applicable as violating time-reversal symmetry. Entanglement is a relativistic consequence of Newon's third law and a property of all time-symmetrical theories, whether classical or quantal. The *metr*ic solit*ons* (*metrons*) are composed of strongly nonlinear periodic core components, far fields corresponding to the classical gravitational and electromagnetic far fields of point-like particles, and further fields representing the weak and strong interactions. The core fields represent nonlinear eigenmodes trapped in a self-generated wave guide. Computations are presented for the first family of elementary particles corresponding to the lowest nonlinear eigenmodes; the second and third families are assumed to correspond to higher eigenmodes. It is shown that the periodicities of the soliton core modes produce the wave-particle duality paradoxes of quantum phenomena, as exemplified by single- and double-slit particle diffraction and the discrete structure of atomic spectra.

1. Introduction

The paradoxes faced by physicists in the beginning of the last century gave rise to two conceptual revolutions which resulted in two very different pictures of physical reality. The alternative views have still not been reconciled within a unified theory to this day.

The paradox of the Galilean invariance of the speed of light was resolved by Einstein (and later extended to arbitrary coordinate transformations) by the recognition that the concepts of space and time could not be accepted as given *a priori*, but must be defined explicitly, together with the properties of the other physical objects with which they were to be combined in a physical theory. In contrast, the paradox of the observation of both wave-like and corpusular properties in the same microphysical phenomena was resolved by Born, Heisenberg, Jordan, Bohr, Pauli and other proponents of the new quantum theory by the opposite approach of rejecting the concept of real existing microphysical objects as meaningless. Meaningful was only the prediction of the outcome of experiments. The experiments themselves must nonetheless still be described in terms of the microphysical objects which have been declared before to defy definition.

Despite the unresolved conceptual problems of the underlying tenets of quantum theory, the theory has been extremely successful in describing a wide spectrum of quantum phenomena, many with impressive accuracy. And it has withstood for more than eight decades all attempts to demonstrate logical inconsistencies in its axiomatic structure. Indeed, the critical analysis of

Einstein, Poldolsky and Rosen [16], suggesting that quantum theory was incomplete, is widely interpreted today, in conjunction with the experimental validation of Bell's [5] inequality theorem on the EPR experiment, as demonstration that classical theories are fundamentally unable to explain the quantum theoretical phenomenon of entanglement, as exemplified by the EPR experiment.

It is therefore understandable that most unification approaches, such as string theory [22],[39] or quantum loop gravity [42],[7], seek to incorporate gravity in a broader theory that accepts the basic tenets of quantum theory, and its relativistic generalization, quantum field theory. We explore nevertheless the alternative route of generalizing Einstein's theory of gravity in order to incorporate quantum phenomena. The motivation for our approach is threefold.

First, we show in the next section that the popular "no go" misnomer [49] of Bell's theorem ignores the essential locality condition that Bell himself stressed. Entanglement is found to be a fundamental property not only of quantum theory, but of all theories, whether quantal or classical, that satisfy microphysical time-reversal symmetry. Entanglement is a necessary relativistic consequence of Newton's third law *actio = reactio.*

Secondly, we present in Section 3 a classical theory of elementary particles that resolves the wave-particle duality paradox that was the original motivation for the creation of quantum theory. Fields and particles are derived as integral components of a single object, the *metr*ic solit*on* (*metron*) solutions of Einstein's vaccuum equations

$$E_{LM} = 0 \tag{1}$$

in twelve-dimensional space, where

$$E_{LM} = R_{LM} - \frac{1}{2} g_{LM} R_N^N \tag{2}$$

is the Einstein tensor and R_{LM} the Ricci curvature tensor (cf. [37]; details will be given later).

The soliton solutions of eq.(1) consist of a strongly nonlinear periodic eigenmode concentrated in the soliton core, far fields corresponding to the classical electromagnetic and gravitational fields of point-like particles, and further fields representing the weak and strong forces. The solutions reproduce the basic symmetries and other properties of elementary particles.

Thirdly, we show in the Section 4 that the metron model yields a straightforward explanation of the wave-like interference properties of particles observed in microphysical phenomena. The interferences arise through the interactions of the periodic core mode of the soliton solution with its own field, after the field has been scattered off other matter. As examples, we compute the diffraction of a stream of particles for the standard single and double-slit experiments and summarize the metron interpretation of atomic spectra as an amalgam of quantum electrodynamics and the original Bohr model of orbiting electrons.

Concepts related to the metron model have been proposed previously. The interpretation of particles as solitons was popular at the turn of the 19th century (see, for example, [1], or Lorentz's [35], [36] well-known model of the electron). The advent of quantum theory, however, in response to the mounting paradoxes of microphysics, deflated the belief that particles could be represented by something as simple as solitons, and interest in this approach faded. The attribution of the wave-like interference properties of particles to a guiding periodic field was proposed by de Broglie [12],[13] and developed further by Bohm [6] (see also Holland [29]). However, in contrast to the de Brogle-Bohm "pilot" wave, which was introduced *ad hoc* as an additional external field independent of the particles (which were also prescribed rather than derived), the relevant periodic core mode of the metron model represents an integral component of the soliton system itself.

The metron model was first developed in a four-part paper using inverse modelling methods ([23] - [26], referred to in the following as H; see also [27], [28] and [34]). The solitons were treated

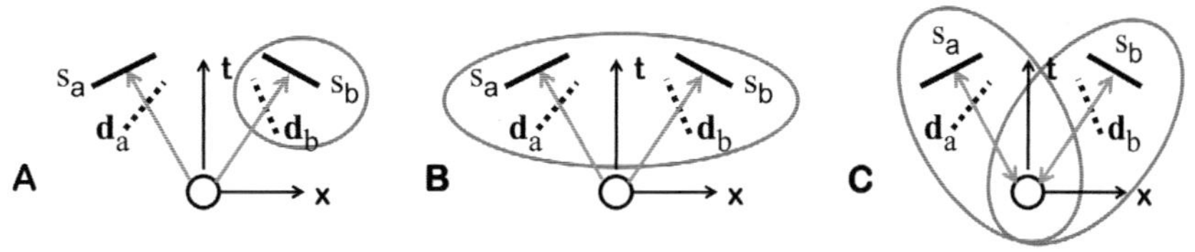

Figure 1. *Alternative interpretations of the EPR experiment. A: a hidden-variable local model (disentangled, in disageement with experiment); B: the quantum theoretical picture of space-like entanglement; C: a time-symmetrical classical model with past-future entanglement.*

as perturbations superimposed on a constant background metric. However, explicit numerical solutions, illustrating the nonlinear wave-guide/trapped-mode interactions that lead to solitons, were presented only for a simpler scalar analogue of the Einstein equation. The number of dimensions and metric signature of extra-space were also left open.

In the present paper, we present now numerical solutions for the full system (1). The solutions confirm the basic concepts of H, but yield also a number of new insights and important modifications of the original metron model. In particular, a background wave field has been introduced that defines the basic physical constants of the theory (the vacuum eqs. (1) contain no physical constants apart from the velocity of light), as well as resolving a number of other open questions listed in H.

Our overview in Sections 3 and 4 of the application of the metron model both for the computation of elementary particles and the interpretation of quantum phenomena at lower energy levels necessarily omits many technicalities. More detailed expositions are in preparation. The last Section 5 concludes with a summary of the basic metron concepts in relation to quantum field theory.

2. Entanglement

Entanglement – the apparent conflict with causality of the measured correlation between space-like separated particles, or other objects, emitted from a common state – is widely regarded as a defining property of quantum theory [41], [38], [20]. Particles originating from a common state will necessarily be correlated; they are termed entangled if the correlation violates causality in the sense assumed by Bell in the derivation of his inequality relation. Bell's theorem was derived for classical particles under the locality condition that measurements performed on space-like separated particles are independent. In the case of the EPR experiment, the quantum theoretical prediction $C_{ab} = -\cos(\mathbf{d}_a \mathbf{d}_b)$ of the correlation between the spins s_a, s_b of two particles measured with Stern-Gerlach magnets oriented in the directions $\mathbf{d}_a, \mathbf{d}_b$, respectively, violates Bell's inequality relation. Experiments [18], [2] have clearly confirmed the quantum theoretical prediction. Thus the EPR system is entangled, and cannot be explained by a classical model satisfying the locality condition.

Although locality appears to be a self-evident requirement of causality, it corresponds, in fact, to the traditional concept of causality only for macrophysical systems characterized by an arrow of time. On microphysical scales, locality is in conflict with the more fundamental condition of time-reversal symmetry. Entanglement is found to be a universal feature of time-symmetrical microphysics, whether quantal or classical. However, the implications of time-reversal symmetry for microphysical systems are rather counterintuitive and require a more careful inspection of the concept of causality.

The alternative interpretations of the EPR experiment are illustrated in Figure 1. The left panel A corresponds to the classical hidden-variable model discussed by Bell, assuming independence of the two spin measurements. The diagram is based on directed arrows-of-time connecting the emission and measurement processes.

The middle panel B corresponds to the standard quantum theoretical interpretation, in which the spin measurement s_a of particle a leads to a simultaneous collapse of the state vectors for both particles a and b. Entanglement appears in this representation as a causality-violating space-like phenomenon directly connecting the two measurement processes.

The right panel C represents the interpretation of entanglement for a classical model exhibiting time-reversal symmetry. A change in the orientation $\mathbf{d}_b$ of the Stern-Gerlach magnet for the second particle b affects not only the local measurement s_b, but the entire trajectory of particle b, from the time of emission to the time of detection. This modifies the emission process itself, and thereby also the emission of the first particle a and the resultant spin measurement s_a. Thus, the entire experiment, from the common emission process to the final particle measurements, is seen as a single, coupled, time-symmetrical system.

Time-reversal symmetry has been evoked by Costa de Beauregard [8], Davidson [10], Cramer [9] and Basini et al [4] to *interpret* quantum theoretical entanglement. However, it has apparently not been generally recognized that time-reversal symmetry implies entanglement for essentially all systems, whether classical or quantal. The debate on the foundations of quantum theory is thereby *circumvented*, opening the door to the development of a general classical theory unifying all forces of nature

The implications of time-reversal symmetry for classical microphysical systems are best illustrated by the photon version of the EPR experiment, in which two photons rather than particles are emitted from a common initial state. We assume that in the classical theory, the electromagnetic far fields of particles correspond (as in the case of the metron model) to the standard electromagnetic fields of classical point-like particles.

The requirement that total four momentum is conserved for a finite set of interacting point-like particles that are initially far separated, and diverge again after interacting via their far fields, implies that the interactions between the particles must be represented by time-symmetrical potential functions composed of equal advanced and retarded components. This was pointed out already by Einstein [14] in an exchange with Ritz [40], who had advocated augmenting Maxwell's equations by an outgoing radiation condition. Ritz's view is consistent with the observed radiation of accelerated charged particles to space, but violates the conservation of four-momentum for a closed system of interacting particles. Einstein's view was supported by Tetrode [44], Frenkel [19] and Fokker [17]. It was developed in detail later by Wheeler and Feynman [46], [47], who explained the observed radiative damping by interactions with a distant ensemble of absorbing particles. A generalization to gravitational interactions has been given by Hoyle and Narlikar [30]. Einstein's view, based on time-reversal symmetry, is now generally accepted.

Figure 2 illustrates the time-symmetrical electromagnetic coupling between two particles a and b travelling on trajectories $x_a^\lambda = \xi_a^\lambda(\sigma_a), x_b^\lambda = \xi_b^\lambda(\sigma_b)$. The coupling occurs along both branches of the light cone $\xi_{ab}^2 = \xi_{ab}^\lambda \xi_{ab}^\mu \eta_{\lambda\mu} = 0$, where $\xi_{ab}^\lambda = x_a^\lambda - x_b^\lambda$ and $\eta_{\lambda\mu} =$ diag $(+1, -1, -1, -1)$ (the summation convention is applied only to tensor indices, in this case λ, μ).

The coupled equations for the rate of change of the particle momenta p_a^λ, p_b^λ (ignoring the singular field-particle self-interactions at the locations of the particles, which in a complete theory, including the finite particle core, are treated as a regular part of the internal particle structure, cf Section 3) are given by [47]

$$\frac{dp_a^\lambda(x_a)}{d\sigma_a} = \int K_{ab} \frac{\partial}{\partial \xi_{a\lambda}} [\delta(\xi_{ab}^2) p_b^\mu p_{a\mu}] d\sigma_b \tag{3}$$

with a second equation in which a and b are interchanged. $K_{ab} = K_{ba}$ denotes a symmetrical coupling expression dependent on the individual particles' mass m and charge q (in cgs-

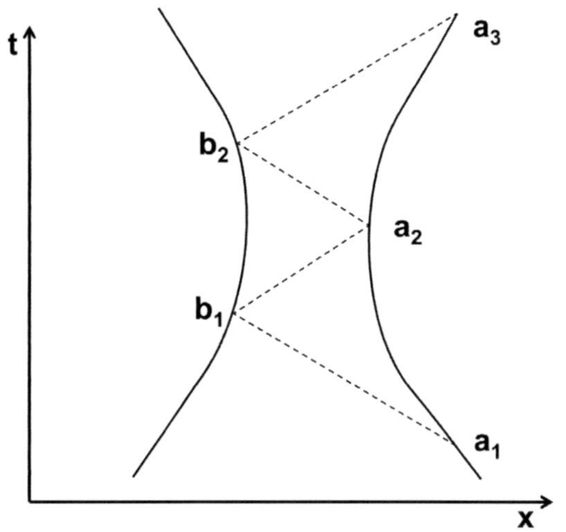

Figure 2. *Past-future entanglement in relativistic time-symmetrical systems. The acceleration of particle a at position a_2 depends on the states of particle b at both positions b_1 and b_2 on the backward and forward light cones (dashed lines), respectively. The interactions $a_1 \leftrightarrow b_1, b_1 \leftrightarrow a_2, a_2 \leftrightarrow b_2, b_2 \leftrightarrow a_3$ represent a local time-symmetrical transfer of four-momentum in accordance with the relativistic generalization of Newton's third law actio = reactio.*

ESU units, $K_{ab} = q_a q_b (m_a m_b)^{-1}$). The path variables are normalized by the condition $(d\xi^\lambda/d\sigma)(d\xi_\lambda/d\sigma) = 1$. For simplicity, the particle spin (included in the full metron model, Section 3) has been ignored.

Since

$$\frac{\partial}{\partial \xi_{a\lambda}}[\delta(\xi_{ab}^2) p_b^\mu p_{a\mu}] = -\frac{\partial}{\partial \xi_{b\lambda}}[\delta(\xi_{ab}^2) p_b^\mu p_{a\mu}], \tag{4}$$

it follows that

$$\Delta p_a^\lambda = \int \frac{dp_a^\lambda}{d\sigma_a} d\sigma_a = -\Delta p_b^\lambda = -\int \frac{dp_b^\lambda}{d\sigma_b} d\sigma_b \tag{5}$$

Thus $p_a^\lambda + p_b^\lambda$ is conserved – in accordance with the principle *actio = reactio*.

We note that the relativistic generalization of Newton's third law applies already at the local level: the conservation of momentum holds for the interactions at all particle-pair positions connected by positive or negative light cone segments (Figure 2):

$$\delta(\frac{dp_a^\lambda}{d\sigma_a}) d\sigma_a = -\delta(\frac{dp_b^\lambda}{d\sigma_b}) d\sigma_b \tag{6}$$

Thus, although it is customary to speak of a photon travelling from particle a to particle b (or from b to a), the interaction is, in fact, completely time symmetrical, the photon representing rather an entanglement between each pair of light-cone-connected particle states[1].

[1] Not addressed here is the basic quantum theoretical result that for a periodic electromagnetic field, the momentum transfer is quantised. Following Einstein [15], this can be attributed within the framework of a quasi-classical model (a fully classical interpretation is outlined in Section 4.2) to the discrete structure of the atomic states from which the electromagnetic radiation emanates. For the present discussion, however, the extension of the present two-particle interaction picture to include these interactions - which again underly time-reversal symmetry – is irrelevant.

The implications of time-reversal symmetry and past-future entanglement for the concept of causality have been discussed in detail by Wheeler and Feynman [47]. Although counter-intuitive, it leads to no logical contradictions, provided the application of the familiar concept of an arrow of time is systematically replaced by a time-symmetrical picture.

The application of these concepts to the photon version of the EPR experiment is straightforward. The "particles" a and b in Figure 1 represent photons coupling the intial and final electron states in the emission and detection systems, respectively, the transmitted photons being filtered by the polarization filters $\mathbf{d}_a$, $\mathbf{d}_b$. The initial and final states are thus directly "entangled", in accordance with Figure 1C.

The same picture applies also to the particle version of the EPR experiment, although in this case the time-symmetrical coupling of the initial and final states is less transparent. However, it is sufficient to recognize that also in this case the central assumption of Bell's analysis, namely the independence of the initial emission process on the later measurement process, implies an arrow of time, and is therefore inapplicable for time-symmetrical systems.

3. The metron model

3.1. The Einstein vacuum equations

Having established that Bell's theorem is no hindrance to the development of a classical theory of quantum phenomena, we turn now to the metron model. We seek a representation of elementary particles as soliton solutions of the twelve-dimensional Einstein vacuum equations (1), without a cosmological term. The equations may be written alternatively in the Ricci-flat form

$$R_{LM} = 0, \tag{7}$$

where

$$R_{LM} = \partial_M \Gamma^N_{LN} - \partial_N \Gamma^N_{LM} + U_{LM} \tag{8}$$

is the Ricci curvature tensor, with

$$U_{LM} = \Gamma^N_{LO}\Gamma^O_{MN} - \Gamma^N_{LM}\Gamma^O_{NO}, \tag{9}$$

$$\begin{aligned} \Gamma^L_{MN} &= \frac{1}{2} g^{LO}\left[\partial_M g_{ON} + \partial_N g_{OM} - \partial_O g_{MN}\right] \\ &= g^{LO}\Gamma_{OMN}, \end{aligned} \tag{10}$$

and where

$$\Gamma_{OMN} = \frac{1}{2}\left[\partial_M g_{ON} + \partial_N g_{OM} - \partial_O g_{MN}\right] \tag{11}$$

is the affine connection of the metric tensor.

Essential for our approach is the exclusion of the matter-dependent source terms. This enables particles to be *derived* as the soliton solutions of eq.(1), rather than being *specified a priori* through the source terms.[2]

We make a second important assumption, following Klein [33]: all fields are assumed to be periodic (or constant) with respect to extra-space and time (in the soliton restframe). The extra-space and temporal periodicities are prescribed; they will be identified with the particle masses and generalized charges. The periodicity assumption reduces the problem of solving the twelve-dimensional Einstein equations to a three-dimensional problem. In the particle models

[2] This is also in the spirit of Einstein, who is said to have the regarded the matter-dependent energy-momentum source term tensor of his equation as "plywood" compared to the "marble" of the geometrical curvature term.

presented below, the dimensionality is reduced further to one by assuming spherical symmetry with respect to physical space.

Formally, the periodicity assumption with respect to extra-space implies that the twelve-dimensional manifold can be represented as a bundle [43],[21],[31] composed of an eight-dimensional fibre (extra-space), dependent on a small number of discrete wavenumber parameters, defined over a four-dimensional base space (spacetime). The periodicity ansatz avoids also the need to compactify extra-space onto a very small Planck sphere in order to explain why we experience physics in only four dimensions. In this respect our approach differs from string theory and other higher-dimensional models.

We derive the metric soliton solutions as perturbations

$$g'_{LM} = g_{LM} - \eta^{(b)}_{LM} \tag{12}$$

of the metric g_{LM} relative to a homogeneous background metric $\eta^{(b)}_{LM}$. The latter is assumed to consist of a superposition

$$\eta^{(b)}_{LM} = \eta_{LM} + \tilde{\eta}_{LM} \tag{13}$$

of a constant diagonal metric

$$\eta_{LM} = \mathrm{diag}(+1, -1, -1, -1, -1, ..., -1, +1) \tag{14}$$

and a background free-wave field $\tilde{\eta}_{LM}$.

The first four components $\eta_{00}, \ldots, \eta_{33}$ of the mean background metric (14) refer to physical spacetime. We use natural units: $c = 1$, and (in wavenumber-dependent expressions below) $\hbar = 1$. The spacetime signature pattern is repeated in the metric components $\eta_{44}, \ldots, \eta_{11,11}$ of the fibre, the first seven of which have negative sign, while the sign of the last component is again positive.[3]

The spacetime components of the metric perturbation g'_{LM} represent the normal gravitational field, while metric-tensor components with fibre-only or mixed spacetime-fibre indices will be identified with fermion and boson fields, respectively.

The introduction of a background free-wave field $\tilde{\eta}_{LM}$ represents an important extension of the original metron model (H). The wave field consists of a continuum[4]

$$\tilde{\eta}_{LM} = \sum_{w} \eta^{(w)}_{LM} = \sum_{w} B^{(w)}_{LM} \exp i(k^{(w)}_N x^N) \tag{15}$$

of statistically independent wave components $\eta^{(w)}_{LM}$ of amplitude $B^{(w)}_{LM}$ whose wavenumbers satisfy the dispersion relation

$$k^{(w)}_L k^{(w)}_M \eta^{LM} = 0. \tag{16}$$

of linear waves on the twelve-dimensional bundel. The fibre wavenumber components k_L, $L = 4, \ldots, 11$ consist of a prescribed discrete set identical to the fibre wavenumber components of the soliton solutions. They represent the generalized particle charges. The spacetime wavenumbers k_L, $L = 0, \ldots, 3$ consist of a continuum lying on three-dimensional subspaces corresponding to the particle masses.

[3] In the inverse analysis of H, both a Euclidean and a Minkowski-type extra-space metric were considered. However, it was found that the Euclidean metric fails to yield soliton solutions of the full tensor equations (1). The possible occurrence of tachyons in non-Euclidean higher-dimensional gravity theories is irrelevant for the metron model, as the periodic higher dimensional space serves only as a template for deriving field equations in four-dimensional spacetime.

[4] Following common practice, we write the statistical continuum of wave components as a sum rather than as a formally more correct (but more cumbersome) Fourier-Stieltjes integral.

The inclusion of a background wave field resolves a number of open questions, listed partially already in H: 1) by assuming that the background wave field exhibits the same periodicities as the metron solutions, it enables the introduction of the basic physical constants lacking in the vacuum equations (1) themselves; 2) it ensures the uniqueness and stability of the metron solutions; 3) it removes, through selective attenuation, a spurious super strong far-field component that had concerned already Kaluza [32] in his five-dimensional generalization of Einstein's equation, and 4) it explains the left-handed polarity of the weak interactions and the finite masses of the weak-interaction gauge bosons in terms of interactions with a sub-set of background field components exhibiting these properties.

We consider the role of interactions with the background wave field later in Section 3.5, after we have presented the basic properties of the metron soliton solution for a constant background metric.

To obtain unique solutions, the Einstein equations, with appropriate boundary conditions, must be augmented by additional gauge conditions [37]. We choose the usual Lorentz gauge, the divergence condition

$$\partial_M h_L^M = 0, \tag{17}$$

where

$$h_{LM} = g'_{LM} - \frac{1}{2}\eta_{LM}(\eta^{NO} g'_{NO}). \tag{18}$$

represents the trace-modified metric perturbation.

Expressed in terms of h_{LM}, and applying (17), the Einstein vacuum equations (1) then take the general form

$$\Box_n h_{LM} = q_{LM}, \tag{19}$$

where $\Box_n = \partial_N \partial_O \eta^{NO}$ denotes the higher-dimensional box operator and q_{LM} are nonlinear source terms.

Formally, q_{LM} represent the gauge-field interactions that ensure the invariance of the Einstein equations with respect to coordinate transformations. However, this property, although important in comparing the formal structure of the higher-dimensional Einstein-Hilbert Lagrangian underlying the metron model with the Standard Model Lagrangian of quantum field theory (cf. H), is irrelevant for the present discussion. For our purposes, it is sufficient to note that the perturbation expansions of the source terms q_{LM} begin with quadratic terms and are homogeneous of second order with respect to the derivatives. The algebra of the perturbation expansion can then be summarized in terms of interaction diagrams, with associated simple rules for the computation of the various vertex contributions (cf. Section 3.3).

The traces of the fields g'_{LM} and h_{LM} are found to be very small, so that $g'_{LM} \approx h_{LM}$. Results will be presented for h_{LM}, but can equally be interpreted as representing g'_{LM}.

3.2. The soliton dynamics

The system (19) supports soliton solutions consisting of periodic nonlinear eigenmodes that are trapped in a wave guide. The soliton structure differs significantly from the familiar Korteweg-de Vries balance between linear dispersion and nonlinear amplification. The trapping wave guide h^g represents a distortion of the metric in which the eigenmodes h^c propagate (Figure 3). It is generated by the radiation stresses of the trapped eigenmodes. The eigenmodes (*core* modes) are concentrated in the strongly nonlinear metron core, decreasing exponentially for larger distances r from the core, while the non-periodic wave-guide far fields fall off as $1/r$.

For periodic, spherically symmetric soliton solutions, the system (19) reduces to a set of coupled equations of the general form

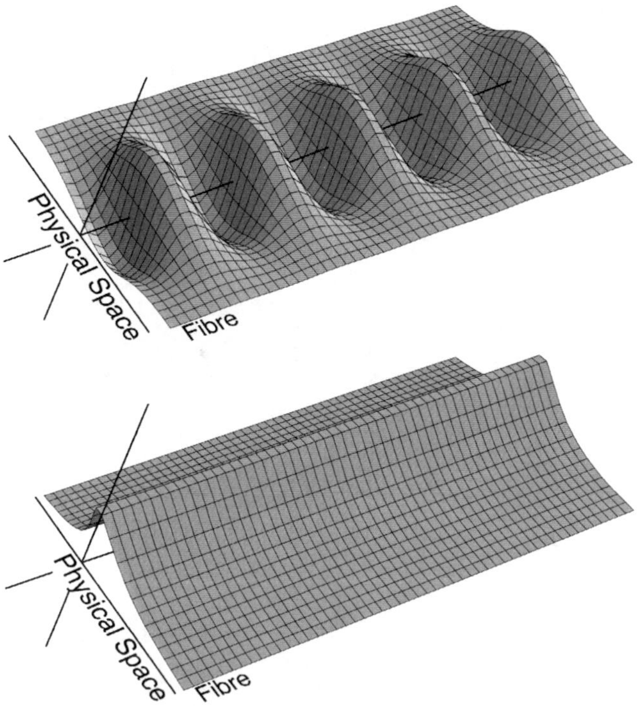

Figure 3. *Schematic diagram of the periodic trapped-mode (upper panel) and wave-guide (lower panel) components of a metric soliton in the product space physical-space × fibre (from H). The trapped core modes are periodic also with respect to time.*

$$\left(\frac{d^2}{dr^2} + \frac{2}{r}\frac{d}{dr} - (\kappa^{(c)})^2 + u^{(c)}(h^{(g)})\right) h^{(c)}(r) \quad = \quad 0 \qquad \text{(core modes)} \tag{20}$$

$$\left(\frac{d^2}{dr^2} + \frac{2}{r}\frac{d}{dr}\right) h^{(g)}(r) \quad = \quad q^{(g)}(h^{(c)}) \ \text{(wave guides)} \tag{21}$$

where $h^{(c)}$ denote the amplitudes of the fundamental periodic core modes of the soliton, $h^{(g)}$ are non-periodic wave-guide fields generated by quadratic difference interactions, $u^{(c)}(h^{(g)}), q^{(g)}(h^{(c)})$ represent nonlinear interaction terms, and

$$(\kappa^{(c)})^2 = -k_L^{(c)} k^{(p)L} = -(k_0^{(c)})^2 + \sum_{L=4}^{10} (k_L^{(c)})^2 - (k_{11}^{(c)})^2. \tag{22}$$

The parameter $\kappa^{(c)}$ determines the asymptotic exponential fall-off rate of the core mode and thus the effective spatial scale of the soliton solution.

We have suppressed the tensor structure of the fields, which slightly modifies the spherically symmetric form of the Laplace operator for the wave-guide components, and have similarly ignored small modifications of the spherically symmetric equations for solitons composed of several spinor core-mode components (cf. Section 3.7.1). Not included in eq (20) are also interactions with the second-harmonic fields and the direct cubic interaction term, both of which are formally of the same perturbation order as the interaction with the wave-guide field. These

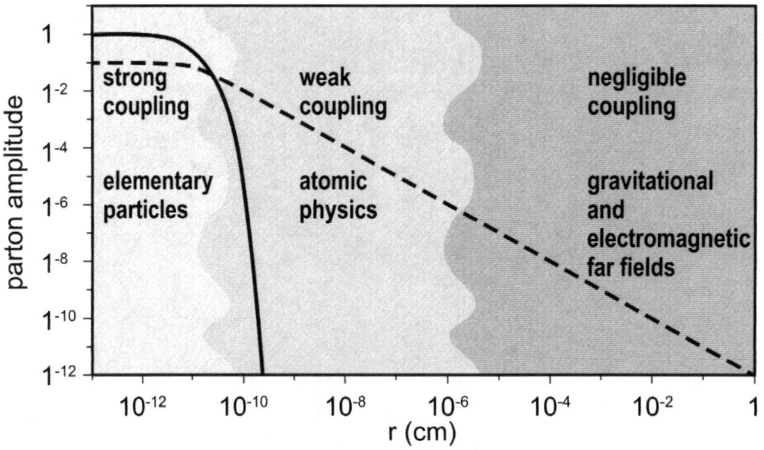

Figure 4. *Near- and far-field components of metron solutions. Full line: core modes (fermions), decreasing exponentially for large r; dashed line: wave-guide fields (massless bosons), decreasing asymptotically as* $1/r$.

interactions are neglible under the second of the following two inequality conditions, which are found to be essential for the existence of soliton solutions:

$$0 < (\kappa^{(c)})^2 \ll (k_L^{(c)})^2, \; L = 0, 4, 5, \dots 11, \tag{23}$$

The left inequality is necessary in order that the core modes are localized in the soliton core, falling off exponentially for large r. The right inequality implies that $|dh^{(c)}/dr| \ll |k_L^{(c)} h^{(c)}|$, $L = 0, 4, 5, \dots 11$. This turns out to be a key condition both for the convergence of the iterative construction of numerical solutions of the system (20), (21) and for the identification of the wave-guide far fields with the classical electromagnetic and gravitational far fields of point-like particles.

The basic core-mode/wave-guide interaction mechanism is independent of the details of the tensor and periodicity structure of the soliton solutions. However, these properties are important for the mapping of the different soliton components into the associated classes of elementary particles. We distinguish in the following between tensor indices $L, M, \dots$ of the full twelve-dimensional bundle, fibre tensor indices $l, m, \dots$, spacetime indices $\lambda, \mu, \dots$ and physical-space indices $\bar{i}, \bar{j}, \dots$.

Core-mode components are identified with fermions, wave-guide components with bosons (Figure 4). The mapping relations of the tensor components are listed in Table 1, while the associated wavenumber components are shown in Figure 5. Fermion fields h_{lm} lie in the fibre domain $8 \leq l, m \leq 10$; the associated wavenumbers k_L lie in the complementary index domain $L < 8$ and $L = 11$. This ensures that the fermion core-mode fields satisfy the gauge condition (17). Boson fields h_{0l} lie in the mixed-index domain $4 \leq l \leq 7$; the field $h_{0,0}$ corresponds to the standard gravitational field (in the particle restframe) of general relativity. Anti-particles are represented by an inversion of the fibre, i.e. by a change in sign of the fibre wavenumber components.

The four quantum-theoretical spinor components, represented by left and right handed Weyl two-spinors φ_1^L, φ_2^L and φ_1^R, φ_2^R, respectively, are mapped into metric tensor components in the fermion sector in accordance with the scheme (representing one particular realization of the

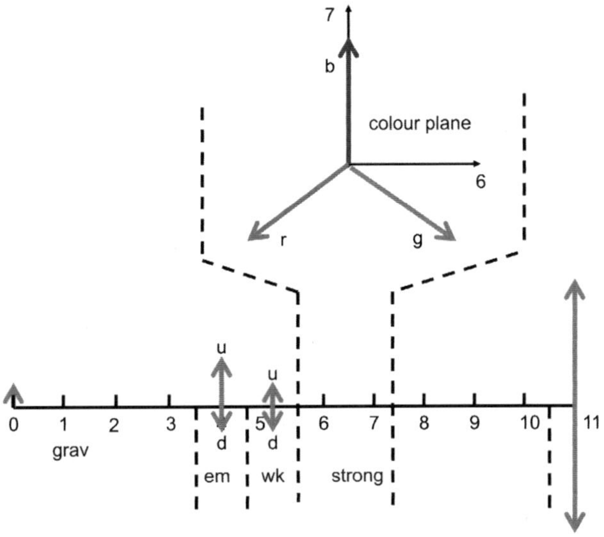

Figure 5. *Wavenumber components ($k_0, k_4 - k_7, k_{11}$) of fermions (leptons and up and down quarks). Each fermion is composed of mirrored core-mode pairs with equal wavenumber components k_L for $L<11$ and opposite wavenumber components $\pm k_{11}$.*

Indices	0-3	4	5	6,7	8,9,10	11
0 - 3	grav	el-magn	weak	strong	-	-
4	el-magn	KSS	KSS	KSS	-	KSS
5	weak	KSS	KSS	KSS	-	KSS
6,7	strong	KSS	KSS	KSS	-	KSS
8,9,10	-	-	-	-	fermion	-
11	-	KSS	KSS	KSS	-	KSS

Table 1. Mapping of metron components into gravitational fields, fermions, electromagnetic, weak and chromodynamic (strong) bosons, and KSS fields.

mapping of the five independent components of a traceless metric h_{lm}, $l, m = 8, 9, 10$ into four spinor components):

$$\begin{pmatrix} h_{88} & h_{89} & h_{8,10} \\ h_{98} & h_{99} & h_{9,10} \\ h_{10,8} & h_{10,9} & h_{10,10} \end{pmatrix} = \begin{pmatrix} 0 & \varphi_1^L & \varphi_2^L \\ \varphi_1^L & \varphi_2^R & \varphi_1^R \\ \varphi_2^L & \varphi_1^R & -\varphi_2^R \end{pmatrix} \tag{24}$$

The metron picture yields a simple explanation of the Fermi-Dirac and Bose-Einstein statistics of fermions and bosons, respectively. Since fermions represent strongly nonlinear fields localised in the particle core, they cannot be superimposed with the core fields of other particles within the same particle core. Bosons, in contrast, represent far fields, which can be readily superimposed beyond the particle core regions.

In addition to the standard gravitational and quantum theoretical fields, Table 1 includes the afore-mentioned spurious Kaluza Super Strong (KSS) fields in the non-mixed fibre-index sector

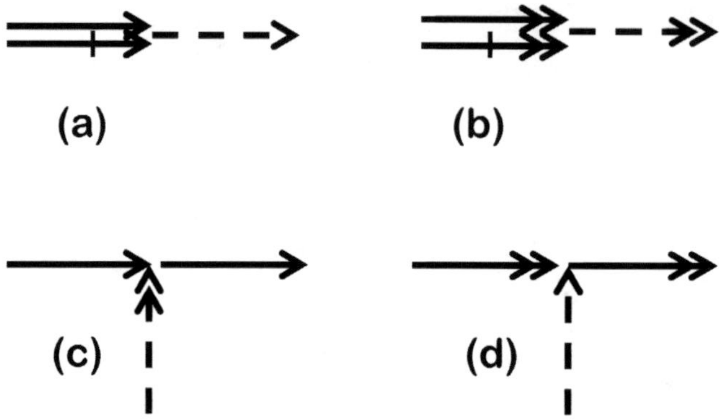

Figure 6. *Panels (a), (b): generation of wave-guide fields (dashed arrows) by quadratic interactions of core modes (full arrows) and mirrored core modes (double-headed arrows). Cross-bars denote complex-conjugate fields (not anti-particles). Panels (c), (d): feedback of wave-guide fields on the eigenvalues of core modes. The dominant feedbacks are cross-interactions between core modes and wave-guide fields generated by mirrored core modes.*

complementary to the fermion sector. It will be shown in Section 3.5 that these are effectively removed through interactions with the background wave field.

The wavenumbers k_L represent generalized particle charges. The wavenumbers k_0, k_4, correspond to the particle mass and electric charge, respectively, the associated wave-guide far fields h_{00}, h_{04} representing the classical gravitational and electromagnetic far fields of a point-like particle (cf. Section 3.4). The weak iso-charge and strong coupling coefficients are given, respectively, by the wavenumbers k_5 and wavenumber vectors $\mathbf{k}_r, \mathbf{k}_b, \mathbf{k}_g$ in the colour plane $k_6.k_7$, where $\mathbf{k}_r + \mathbf{k}_b + \mathbf{k}_g = 0$ (cf. Figure 5). In analogy with Einstein's four-dimensional general relativity, k_0 plays a dual role of representing both the gravitational and inert mass.

It is shown in the next section that the wave-guide far fields h_{0L}, $L = 0, 4, 5, \ldots, 10$ are proportional to the wavenumber products $k_0 k_L$, with a common proportionality factor. The empirical result that the electromagnetic forces greatly exceed the gravitational forces implies then $k_4 \gg k_0$ (for the electron, for example, $k_4/k_0 \approx 1.4\ 10^{21}$). The right-hand inequality (23) can be satisfied in this case only if the core-mode wavenumber vector includes a component, k_{11}, say, of approximately the same magnitude as k_4, but of opposite metric signature. This is the motivation for the introduction of a positive time-like sign for the last metric component $\eta_{11,11}$. Since a net force corresponding to $h_{0,11}$ is not observed, however, the soliton must contain two "mirrored" core-mode constituents of the same amplitude, whose last wavenumber components k_{11} have opposite signs, but whose wavenumbers are otherwise identical[5]. This results in a cancellation of the net boson far field $h_{0,11}$, while the gravitational, electromagnetic and other far fields are not affected.

3.3. Interaction diagrams

The interactions between core-mode and wave-guide fields can be conveniently summarized in terms of interaction diagrams (Figure 6). Core-mode fields are represented by full arrows, wave-guide fields by dashed arrows. A double-headed arrow represents the mirrored core

[5] Instead of introducing two mirrored components $h_{L,11} \sim \exp(\pm i k_{11} x^{11})$, one could, of course, equivalently consider a single component $h_{L,11} \sim \cos(k_{11} x^{11})$. For interaction computations, however, it is more convenient to work with two complex exponentials.

mode of a core-mode pair, or the wave-guide field generated by a mirrored core mode. Cross-bars denote complex-conjugate fields of the same particle (as opposed to the usual Feynman-diagram notation for anti-particles). Arrow directions indicate the sequence in the perturbation expansion; the interactions themselves are time-symmetrical.

The coupling coefficients associated with a vertex consist of contractions of tensor expressions composed of quadratic products of the wavenumber components of any of the incident fields multiplied by the incident field amplitudes. This yields a forcing of the gravitational, bosonic and KSS wave-guide fields $h_{00}^{(g)}$, $h_{0l}^{(g)}$ and $h_{lm}^{(g)}$ proportional to the square of the core-mode amplitude times k_0^2, $k_0 k_l$ and $k_l k_m$, respectively, with a common proportionality factor.

The very large ratios k_l/k_0, mentioned above as motivation for the introduction of a time-like fibre metric signature to satisfy the inequality (23), presents now a further problem: it implies very large KSS fields in the pure fibre sector compared with the bosonic fields of the mixed spacetime-fibre sector (cf. Table 1). The KSS fields are not observed; they will be removed in Section 3.5 through interactions with a background wave field. Before addressing this problem, however, we investigate first the implications of the wave-guide far fields for the interactions between well separated particles.

3.4. Far-field interactions

So far, we have considered only an isolated particle at rest at the location $\mathbf{x} = 0$. We consider now the forces produced by the interaction of the particle with the wave-guide far fields g_{LM}^F of another distant particle (or of several such particles).

According to higher-dimensional general relativity, the force experienced by a particle propagating with the velocity v^L in a perturbed higher-dimensional metric

$$g^{LM} = \eta^{LM} - g_{NO}^F(x)\eta^{LN}\eta^{MO} + \cdots \tag{25}$$

can be represented by the geodetic acceleration

$$\frac{dv^L}{ds} = -\Gamma_{MN}^L v^M v^N. \tag{26}$$

Applied to the spacetime velocity components (the rate of change of the fibre velocity components is irrelevant due to the assumed homogeneity on the fibre), this yields

$$\frac{dv^\lambda}{ds} = -\Gamma_{\mu\nu}^\lambda v^\mu v^\nu + F_{\mu l}^\lambda v^\mu v^l + K^\lambda, \tag{27}$$

where the first term on the right hand side represents the standard Einstein gravitational acceleration in physical spacetime, the second term the acceleration due to the mixed-index (i.e. bosonic, cf. Table 1) far fields

$$F_{\mu l}^\lambda = \eta^{\nu\lambda}\left(\partial_\nu g_{\mu l} - \partial_\mu g_{\nu l}\right), \tag{28}$$

or

$$F_{\mu l}^\lambda = \partial^\lambda A_{\mu l} - \partial_\mu A_l^\lambda \quad \text{with} \tag{29}$$

$$A_{\lambda l} = g_{\lambda l}, \tag{30}$$

and the third term represents the Kaluza super strong (KSS) acceleration

$$K^\lambda = \frac{1}{2}\eta^{\nu\lambda}\partial_\nu g_{lm} v^l v^m \tag{31}$$

produced by the far fields g^F_{lm} in the fibre sector.

The bosonic acceleration expressions (28) - (30) have a structure analagous to the Lorentz force

$$m\,dv^\lambda/ds = e\hat{F}^\lambda_\mu v^\mu (v_\nu v^\nu)^{1/2} \tag{32}$$

where m and e denote the particle's mass and charge, respectively, and

$$\hat{F}^\lambda_\mu = \partial^\lambda \hat{A}_\mu - \partial_\mu \hat{A}^\lambda \tag{33}$$

is the electromagnetic field, expressed in terms of the electromagnetic potential $\hat{A}_\lambda$.

Before proceeding, we must first define what is meant by the velocity v^L in twelve-dimensional space of a "particle" that corresponds to the classical concept of a localized point-like particle in physical spacetime, but represents a homogeneous periodic wave field with respect to the fibre. A consistent definition can be obtained by formally relaxing the homogeneity of the periodic field with respect to the fibre and considering instead a finite-extent wave group of scale R that satisfies the WKB conditions $|k_l|R \ll 1$. The metron particle can then be treated as a 12-dimensional wave group propagating with the group velocity

$$v^L = \partial\Omega/\partial k_L, \tag{34}$$

where $\Omega(x,k) = 0$ represents the dispersion relation of waves in the perturbed twelve-dimensional space:

$$\Omega(x,k) = g^{LM}(x) k_L k_M = 0, \tag{35}$$

with $g^{LM}(x)$ given by eq. (25).

The group velocity is accordingly

$$v^L = 2\eta^{LM} k_M \tag{36}$$

The metron relations (28) - (30) can then be recognized as the generalization of the Lorentz electromagnetic force to all bosonic forces, where the wavenumbers represent, as mentioned above, the generalized charges of the various fields. In particular, the metron expressions for the gravitational and electromagnetic accelerations map into the corresponding classical expressions by applying scaling factors ρ_1, ρ_2, ρ_3:

$$A_{\lambda 4} = \rho_1 \hat{A}_\lambda \tag{37}$$

$$F^\lambda_{\mu 4} = \rho_1 \hat{F}^\lambda_\mu \tag{38}$$

$$k_4 = \rho_2\, e \tag{39}$$

$$k_0 = \rho_3\, m, \tag{40}$$

where the scaling factors satisfy the condition

$$\rho_3 = \rho_1 \rho_2. \tag{41}$$

The scaling factors depend on the choice of units in the metron and standard classical theories. We identify the rest-frame frequency k_0 with the mass m via the Planck relation $k_0 = mc/\hbar$, or, in natural units, $k_0 = m$. Thus,

$$\rho_3 = c/\hbar = 1 \quad \text{in natural units.} \tag{42}$$

so that eq. (41) reduces to

$$\rho_1 \rho_2 = c/\hbar = 1 \quad \text{in natural units.} \tag{43}$$

Similar scaling factors apply for the relevant relations for the weak interactions (coupling constant k_5) and strong forces (coupling wavenumbers $\mathbf{k}_r, \mathbf{k}_b, \mathbf{k}_g$) that will be considered after we have presented numerical metron solutions in Section 3.6 (the remaining wavenumber k_{11} yields no net force, as pointed out above, since the mixed-index fields $g_{\mu,11}$ occur in cancelling mirrored pairs). It is found that the free parameters of the metron model (basically, the scale $\kappa^{(c)}$ of eq. (22) and the energy level of the background wave field) can be chosen such that the scaling parameters $\rho_1, \rho_2, \ldots$ for all bosonic forces are in agreement with experiment.

There remains the KSS force. A comparison of eqs.(31) and (29) reveals, as pointed out already by Kaluza, that, considering alone the contributions to K^λ from the metric components g_{44}, the KSS force exceeds the electromagnetic force by a factor of the order $g_{44}k_4/g_{04}k_0$. Since the ratio g_{44}/g_{04} for the fields alone is already of order k_4/k_0, as pointed out in the previous section, we obtain for the force ratio $K^\lambda/(F_\mu^{\lambda(em)} v^\mu) = O\{(k_4/k_0)^2\} \approx 2\,10^{42}$ for the electron. For a viable theory, the KSS fields must therefore clearly be removed by some mechanism. This is achieved through interactions with the background wave field.

3.5. The background wave field

A background wave field is required not only to attenuate the KSS fields, but also for the other important reasons mentioned earlier. We summarize in the following only the basic concepts.

Statistical homogeneity implies for the amplitudes of the wave-field representation (15)

$$< B_{LM}^{(w)} >= 0 \tag{44}$$

and

$$< B_{LM}^{(w)} B_{L'M'}^{(w')*} >= P_{LML'M'}(k^{(w)})\, \delta_{ww'}, \tag{45}$$

where the cornered parentheses denote statistical ensemble averages and $P_{LML'M'}$ represents the wave spectrum.

In contrast to the fibre wavenumbers k_l, $l = 4, 5, \ldots, 11$, which consist of a discrete set corresponding to the generalized particle charges, the spacetime wavenumber components $k_\lambda^{(w)}, \lambda = 0, \ldots, 3$ represent a three-dimensional continuum (since we can no longer remain in the restframe of a particular particle) satisfying the de Broglie dispersion relation

$$k_\lambda^{(w)} k_\mu^{(w)} \eta^{\lambda\mu} - m_w^2 = 0, \tag{46}$$

where the constants m_w represent (to within a small deviation of order $(\kappa^{(c)}/k_0^{(c)})^2$, cf. eq. (23), that is irrelevant here) the discrete masses of the particles. Applying eqs.(16) and (46), the fibre wavenumbers are seen to be inter-related via the particle masses:

$$k_l^{(w)} k_m^{(w)} \eta^{lm} + m_w^2 = 0. \tag{47}$$

We shall be concerned primarily with interactions between the background wave continuum and the metron soliton in the metron restframe, in which the metron spacetime wavenumber vector reduces to the metron frequency k_0 (i.e. the metron mass), with $\mathbf{k} = (k_1, k_2, k_3) = 0$. Important for the question of the particle stability is then the background wave spectral density $P_{LML'M'}(\mathbf{k}^{(w)})$ at $\mathbf{k}^{(\mathbf{w})} \approx 0$, where the frequencies $k_0^{(w)}$ of the background wave field and metron core mode are sufficiently close that the two fields can exchange energy through resonant interactions. However, other background wave components with $\mathbf{k} \neq 0$ can engage in a resonant energy transfer at higher interaction order. The full background spectrum enters also in the computation of the damping factors for the KSS far fields, which is not dependent on resonant interactions.

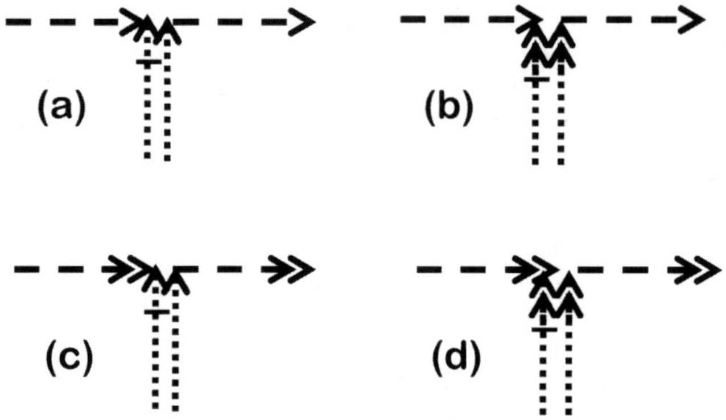

Figure 7. *Selective damping of KSS fields (dashed arrows) through interactions with background wave fields (dotted arrows). Panels (a), (b): damping of KSS fields by cubic interactions with the background wave fields and their mirrored partners. Panels (c),(d): analagous diagrams for the damping of the mirrored KSS fields.*

Important for the interactions with the background wave field are not only their wavenumbers, but also their polarization properties. We assume that these correspond to fermion fields, i.e. that the tensor indices of the background waves $B^{(w)}_{LM}$ lie in the domain $8 \leq L, M \leq 10$.

The general structure of the various interactions with the background wave field are summarized in the following in terms of interaction diagrams.

3.5.1. The KSS damping mechanism

Figure 7 depicts the interactions that lead to a selective attenuation of the KSS fields. All wave-guide fields, which are themselves generated by quadratic difference interactions of the basic core-modes (Figure 6), experience an additional quadratic difference interaction with the background wave field. This yields a modified field equation for the KSS wave-guide fields of the form

$$\left(\frac{d^2}{dr^2} + \frac{2}{r}\frac{d}{dr} \right) h_{lm} - D k_l k_m h^n_n = q_{lm}, \tag{48}$$

where D is a positive damping coefficient given by an integral over the background wave spectrum (45).

Essential for the selective damping of the KSS fields is the proportionality of the damping term to the fibre wavenumber product $k_l l_m$. This implies that the damping of the KSS fields, which is proportional to quadratic fibre wavenumber products, is very much larger than the analagous damping expressions for the boson fields, which contain at most one fibre wavenumber component. Thus, for the KSS field h_{44}, for example, the damping is a factor k_4/k_0 larger than the damping of the electromagnetic far field h_{04}, and of order k_4^2/k_0^2 larger than the damping of the gravitational field (cf. eq. 27). For a suitable choice of the the damping factor (i.e. of the energy density of the background wave spectrum) the KSS fields can thus be effectively attenuated within a scale of the order of the soliton-core scale $(\kappa^{(c)})^{-1}$, while the electromagnetic and gravitational far fields remain unaffected.

3.5.2. Uniqueness and stability of the metron solutions

The coupled core-mode/wave-guide system (20), (21) yields a continuum of possible soliton solutions, given an arbitrary set of fibre wavenumbers, provided only that the wavenumbers

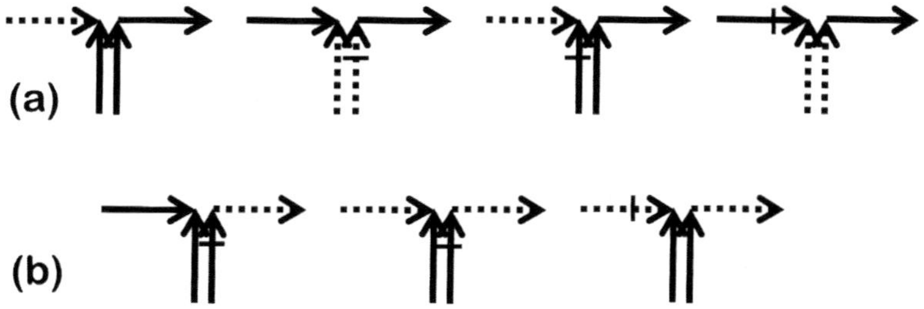

Figure 8. *Interactions ensuring stability and uniqueness of the soliton solutions.*

satisfy the inequality relations (23). In the absence of a background wave-field, however, the solutions are unstable and decay into free waves. But if there exists a background wave field, the sub-set of soliton solutions whose fibre wavenumbers are the same as those of the background wave field are stabilized through resonant interactions with the background field. The radiative losses are balanced in this case by the radiation pressure of the background waves.

The relevant interactions are summarized in Figure 8. Diagrams a show the interactions producing a net transfer of energy from the background wave field to the core mode, while diagrams b show the balancing interactions yielding a net transfer from the core mode to the background wave field (the number of interactions differ in the two cases, as it is assumed that the background wave field enters only to second order, while the core-mode interactions are retained to third order). To limit the number of diagrams, we have not differentiated in this case between mirrored partners. Starting from an arbitrary initial soliton solution with fibre wavenumber components corresponding to the background wave field, the soliton solution adjusts to a unique solution determined by the energy density of the background wave field.

A further important process is the interaction with an exceptional background wave field characterized by left-handed polarization and a relatively large frequency (mass) $m = (k_\lambda k^\lambda)^{1/2}$. This is the origin of the chirality of the weak interactions and the mass of the weak interaction gauge bosons. We return to the weak interactions in Section 3.7.4, after we have reviewed some further properties of the metron model.

3.6. A numerical metron solution

Numerical solutions of the system (20), (21) can be constructed iteratively. Starting from a first-guess core mode, one computes the wave-guide field that it generates. This enables the computation of a second-guess core mode, including the feedback from the wave-guide field, and so on. For a given background wave field and prescribed wavenumbers satisfying the inequalities (23), the solution typically converges within twenty to forty iterations.

Figures 9, 10 show the computed core-mode and wave-guide fields, respectively, for a scalar model containing only a single core mode, with wavenumbers $k_0 = 0.1, k_4 = 1000$ (corresponding to a gravitational and electromagnetic far field) and $k_{11} = \pm k_4(1 - 0.505\,10^{-8})$, yielding $\kappa = 0.01$. The model may be regarded as a rudimentary representation of the electron. It will be generalized in the next section to include the particle spin and weak interactions. The ratio $k_4/k_0 = 10^4$ (imposed by computer limitations) is considerably smaller than the real ratio $\approx 1.4\,10^{21}$ for the electron, but the computations can be readily scaled to realistic values using the scaling relations

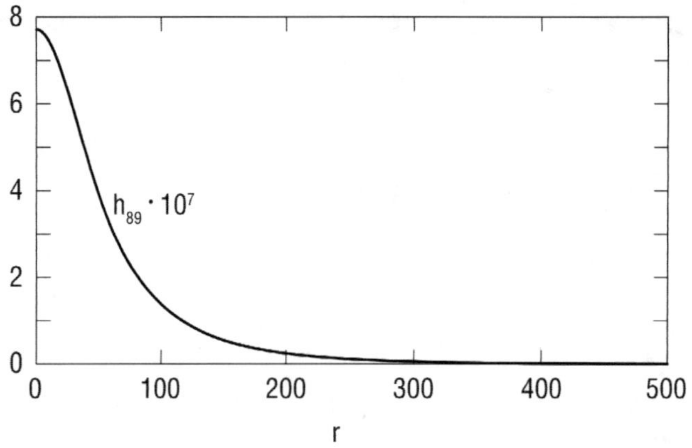

Figure 9. *Core-mode field of the scalar metron solution.*

$$h_{89}^{(c)} = \left(\frac{\kappa^2}{k_0\, k_4}\right) f_c(r) \quad \text{for the core-mode field} \tag{49}$$

$$h_{0l}^{(g)} = \left(\frac{\kappa^2 k_l}{k_0 k_4^2}\right) f_g(r) \quad \text{for gravitational and electromagnetic far fields} \tag{50}$$

$$h_{lm}^{(g)} = \frac{\kappa^2 k_l k_m}{k_0 k_4^3} \hat{f}_g(r) \quad \text{for KSS far fields.} \tag{51}$$

where

$$f_c(r) = \left\{ \begin{array}{ll} \alpha_c & \text{for} \quad r = 0 \\ \beta_c e^{-\kappa r} (\kappa r)^{-1} & \text{for} \quad \kappa r \gg 1 \end{array} \right\} \tag{52}$$

$$f_g(r) = \left\{ \begin{array}{ll} \alpha_g & \text{for} \quad r = 0 \\ \beta_g (\kappa r)^{-1} & \text{for} \quad \kappa r \gg 1 \end{array} \right\} \tag{53}$$

$$\hat{f}_g(r) = \left\{ \begin{array}{ll} \hat{\alpha}_g & \text{for} \quad r = 0 \\ \hat{\beta}_g e^{-\hat{\kappa} r} (\hat{\kappa} r)^{-1} & \text{for} \quad \hat{\kappa} r \gg 1 \end{array} \right\}, \tag{54}$$

$\alpha_c, \beta_c, \alpha_g, \beta_g, \hat{\alpha}_g, \hat{\beta}_g$ are constants of order unity, and the exponential fall-off rate $\hat{\kappa}$ of the KSS fields h_{lm}^g is given by $\hat{\kappa} = (D k_l k_m)^{1/2}$, cf. eq.(48).

Although the exponential fall-off rate of the KSS fields yields neglible fields beyond the particle core, the KSS fields nevertheless still exceed the gravitational and electromagnetic fields within the core region itself. This is a necessary condition for the trapping of the core mode, since the electromagnetic wave-guide component, the largest non-KSS wave-guide field, has the wrong sign for trapping.

The energy level of the background wave field and the particle scale κ^{-1} (the two free parameters of the model) can be chosen such that the scalar metron solution reproduces the charge and mass of the electron, while satisfying the trapping condition and the condition that the electron scale κ^{-1} is large compared with the Compton radius $\hbar/(mc) = 3.86\ 10^{-11}$ cm (eq. 23)). The parameters can furthermore be chosen such that κ^{-1} is of the same order as the Bohr radius $\hbar^2/(e^2\, m) = 5.3\ 10^{-9}$ cm, which will be found below (Section 4.2) to be a necessary condition for the metron interpretation of atomic spectra.

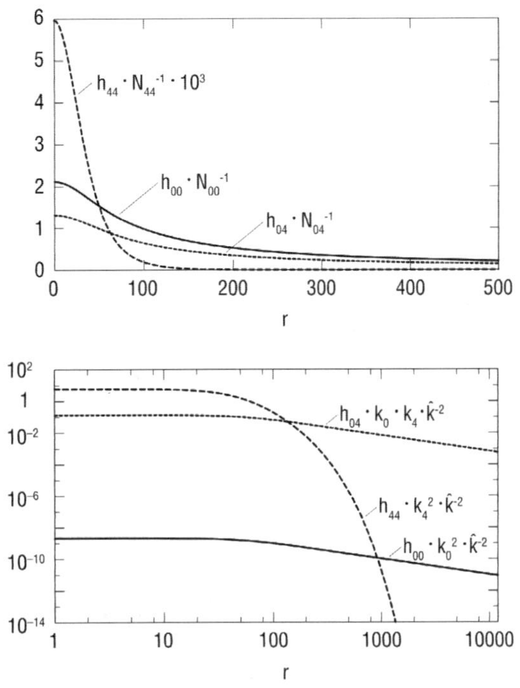

Figure 10. *Wave-guide fields of the scalar metron solution. Top panel: linear fields; bottom panel: log-log plot ($\hat{k} \equiv \kappa$; N_{44}, N_{00} are normalization factors).*

To obtain a realistic base model for the representation of the elementary particle spectrum, our model needs to be extended through the inclusion of the particle spin and the fibre wavenumber components k_5 and $\mathbf{k} = (k_6, k_7)$ representing the weak and strong interaction coupling constants, respectively. We introduce these extensions in the following section describing the metron representations of the electron, neutrino and proton/neutron system.

3.7. The first family of elementary particles

3.7.1. The electron

Spin can be introduced into the metron model by replacing the scalar core-mode component by a four-spinor field composed of left and right two-spinor fields $\psi^L = (\varphi_1^L, \varphi_2^L)$, $\psi^R = (\varphi_1^R, \varphi_2^R)$ representing metric components as defined in Table 1.

In the linear approximation, the two-spinors are coupled via the Weyl form

$$(\sigma^{\bar{i}}\partial_{\bar{i}} - \partial_t)\psi^L = i\hat{\omega}\psi^R \qquad (55)$$

$$(\sigma^{\bar{i}}\partial_{\bar{i}} + \partial_t)\psi^R = -i\hat{\omega}\psi^L. \qquad (56)$$

of the Dirac equation, where $\sigma^{\bar{i}}$ $(\bar{i} = 1, 2, 3)$ denote the Pauli matrices and

$$\hat{\omega}^2 = -k_l^{(c)} k_{(c)}^l = (\kappa^{(c)})^2 + (k_0^{(c)})^2, \qquad (57)$$

We seek as before a spherically symmetric solution. In view of the coupling relations (55), (56), we can no longer simply assume $\psi^{L,R} = \psi^{L,R}(r)$. Instead, the isotropy of a four-spinor field that is periodic in time t, $\psi^{L,R} \sim \exp(i\omega t)$ is represented by the general form

$$\psi^L = \{-(\hat{\omega}+\omega) + i\sigma^{\bar{i}}\partial_{\bar{i}}\} f(r)\, e^{i\omega t} A \tag{58}$$
$$\psi^R = \{(\hat{\omega}+\omega) + i\sigma^{\bar{i}}\partial_{\bar{i}}\} f(r)\, e^{i\omega t} A \tag{59}$$

where $A = (A_1, A_2)$ is an arbitrary constant 2-spinor factor and the function $f(r)$ satisfies the scalar spherically-symmetric Klein-Gordon equation

$$\left\{ \frac{d^2}{dr^2} + \frac{2}{r}\frac{d}{dr} + \omega^2 - \hat{\omega}^2 \right\} f(r) = 0 \tag{60}$$

The relations (58)-(60) represent the linearized four-spinor generalization of the scalar eigenvalue equation (20).

The nonlinear eigenvalue equations, including the coupling to the wave-guide fields, are obtained as in the scalar case by substituting the linear form (58)-(59) into the nonlinear Einstein equations and expanding in a perturbation series.

The analysis carries through effectively unchanged as before. The net wave-guide field generated by the quadratic interaction of the fermion fields is obtained simply as the sum of the separate wave-guide fields generated by the individual spinor components: there is no contribution to the wave-guide fields from the cross-coupling of spinor fields with different metric indices. The resultant wave-guide fields, including, in particular, the gravitational and electromagnetic far fields, are furthermore real and exhibit the same isotropic tensor forms as in the scalar case.

Finally, the nonlinear eigenvalue equation for the common scalar function $f(r)$ of the spinor fields is identical to the previous nonlinear eigenvalue equation (20) in the scalar case, with $(\kappa^{(c)})^2 = \hat{\omega}^2 - \omega^2$, and with a nonlinear eigenvalue correction consisting as before of a bi-linear product of the function $f(r)$ and the net wave-guide field.

In summary, the computations for the scalar metron solution carry over into the four-spinor case without change for the net wave-guide far fields, while the amplitudes of the four core-mode components (each of which contributes quadratically to the unchanged net wave-guide field) are halved relative to the values for the scalar model.

In the presence of an external magnetic field, the particle spin becomes aligned parallel or anti-parallel to the external field, as in quantum theory, with a resultant net force if the external field is non-homogeneous. However, these considerations, which become relevant mainly in the context of atomic spectra, lie beyond the scope of the present overview.

Missing, finally, in our model of the electron are the weak interactions. The weak charge is represented by the wavenumber component k_5 (Table 1). However, in view of the left-handed chirality of the weak interactions and the finite masses of the associated gauge bosons, the weak forces can clearly not be treated in analogy with the electromagnetic forces (or the strong forces discussed in Section 3.7.3) by simply replacing the electric change k_4 by k_5. Instead, we shall associate the weak interactions later in Section 3.7.4 with interactions with a background wave field that exhibits the exotic properties mentioned.

3.7.2. The neutrino

The metron model of the neutrino depends in detail on the just mentioned representation of the weak interactions later in Section 3.7.4. However, two alternative models can be indicated already here.

In contrast to quantum field theory, in which the breaking of parity symmetry is described by the weak interaction sector of the Standard Model Lagrangian, the metron model attributes these features to an external background wave field. Thus, there is no need to assume (although it can also not be excluded) a breaking of parity symmetry in the metron soliton itself.

Two alternative models of the neutrino can thus be constructed through minor modifications of the electron model. In the first model, we simply replace the electric-charge wavenumber k_4 of the electron by the weak-charge wavenumber k_5, while setting the mass k_0 to zero or a very small number. In the second model, we set also the right-handed core-mode components to zero, in accordance with the quantum theoretical picture that the neutrino itself breaks parity symmetry.

In both cases, we assume that the fibre wavenumber components consist only of the weak-interation component $k_5^{(\nu)}$ and $k_{11}^{(\nu)} \approx \pm k_5^{(\nu)}$.

3.7.3. The proton-neutron system

The metron picture of the proton-neutron system closely parallels the QFT representation: protons and neutrons are composed of three quarks, which come in two forms: an "up" quark with charge 2/3e and a "down" quark with charge -1/3e. Each quark is characterized by one of three different "colours" (blue, b, green, g and red, r)[6]. The net colour of each three-quark particle is colourless, i.e. each of the three quarks has a different colour, which cancel in the sum. Proton-neutron transformations are furthermore mediated by gauge bosons, that convert an up quark into a down quark, and vice versa.

In the following, we translate these abstract field-operator concepts of QFT into the corresponding metron picture of real existing particles and fields.

The metron model of the proton or neutron consists of three basic core-mode (quark) components, with associated zero-wavenumber wave-guide components. Each quark core-mode component consists of a mirrored pair of four-spinors. The individual metric tensor components of the core modes are assigned to spinor components in accordance with the mapping relations of eq.(24).

The associated wavenumbers of the quarks consist of the components $k_0^{(q)} = m^{(q)} =$ mass, $k_4^{(q)} =$ electric charge, $k_5^{(q)} =$ weak charge, $(k_6^{(q)}, k_7^{(q)}) =$ quark colour-charge vector, and an associated pair of wavenumber components of opposite sign, $\pm k_{11}^{(q)}$, assigned to the two components of a mirrored core-mode pair. The magnitudes of the fibre wavenumber components $\pm k_{11}^{(q)}$ are again chosen such that $(\kappa^{(q)})^2 = -k_L^{(q)} k^L_{(q)} \ll (k_M^{(q)})^2$ for all M. This ensures that $\partial_{\bar{i}} \ll k_M$, as required for the existence of soliton solutions.

The three colour wavenumber vectors, representing the three colour charges blue (b), green (g) and red (r), lie in the k_6, k_7 colour plane (Figure 5). The QFT property of colourless net particles translates into the condition that the vectors form a symmetrical star:

$$\sum_{q=b,g,r} k_6^{(q)} = \sum_{q=b,g,r} k_7^{(q)} = 0, \tag{61}$$

with

$$(k_6^{(b)})^2 + (k_7^{(b)})^2 = (k_6^{(g)})^2 + (k_7^{(g)})^2 = (k_6^{(r)})^2 + (k_7^{(r)})^2 = (k^{(c)})^2, \tag{62}$$

where $k^{(c)}$ represents the magnitude of the strong-force coupling constant.

The star symmetry implies, in particular, that the sum of the strong-interaction wave-guide boson fields generated by the three quarks vanishes: the individual wave-guide fields $h_{06}^{(q)}, h_{07}^{(q)}$ representing the strong force are proportional to the wavenumber product $k_0^{(q)} k_6^{(q)}$, $k_0^{(q)} k_7^{(q)}$, respectively (cf. eq.(50)), so that $\sum_{q=b,g,r} h_{06}^{(q)} = \sum_{q=b,g,r} h_{07}^{(q)} = 0$. Thus, there exist no far fields of the colour force, in agreement with experiment.

[6] The index g refers in this section to the colour green, as opposed to the previous reference to wave-guide fields.

For a rigorously spherically symmetric solution, in which all three quarks are located in the same position, the colour forces cancel not only in the far field, but identically[7]. There should therefore exist no inter-nuclear forces, which are normally attributed to the strong forces acting between closely neighbouring nuclei. However, due to the different electromagnetic charges of the quarks, two neighbouring nucleons become polarized, thereby breaking the spherical symmetry and producing a dipole-type near field that is responsible for the inter-nuclear forces.

The representation of the proton-neutron system in terms of real existing quark components explains also the phenomenon of confinement in particle collision processes. At low collision energies, the penetration of the particles into each other's core region is small, and the three quarks of a particle appear as a single particle. However, at higher collision energies with pronounced core penetration, the binding forces of the quarks within a particle become small compared with the forces between the individual quarks of different particles, and the relevant model is that of direct quark-quark interactions.

The nonlinear eigenmode equations for a set of three four-spinor core-mode pairs, with associated field equations for the forced wave-guide fields, follow by straightforward generalization of the scalar model of a single core-mode pair. Thus, eqs. (58)-(60) for a single four-spinor core mode apply in the three-quark case for each of the three four-spinor core modes, and the net wave-guide field is again given simply by the superposition of the wave-guide fields generated by each of the individual core modes.

An additional feature not present in the electron model of a single core-mode pair, however, is the coupling between the three core-mode pairs via their magnetic fields. Thus one needs to distinguish between the net spin 1/2 and (unstable) spin 3/2 configurations.

The magnitudes of the wavenumbers characterizing the various forces in the nuclear system satisfy the inequalities

$$k_0^{(q)} \ll |k_4^{(q)}| \ll k^{(c)} \approx |k_{11}^{(q)}| \tag{63}$$

As in the electron model, the relevant metron parameters can again be chosen to be consistent with the empirical data.

3.7.4. Weak interactions

In contrast to the metron representation of the strong and electromagnetic forces, which in many respects parallels that of quantum field theory, the metron and quantum-field representations of weak interactions differ significantly. In the Standard Model, the Higgs mechanism plays a central role in explaining the origin of mass, while in the metron model, mass appears naturally as an intrinsic property of the soliton solutions. Furthermore, in the metron model, the left-handed chirality of the weak interactions follows from interactions with a finite-mass, left-hand-polarized background wave field, rather than being defined as a property of the basic Lagrangian. The polarized background wave field, which interacts only with the the left-handed spinor components of the soliton solutions, replaces the finite-mass gauge bosons of the Standard Model.

Figure 11 shows the metron interaction diagram for the example of neutron-proton decay. The background wave field L^- (dotted double-pointed arrow) interacts simultaneously with both the leptons and quarks. Apart from this symmetry, and the fact that L^- represents an external background wave field component rather than a weak-interaction gauge boson W^-, the diagram has the same structure as the corresponding Feynman diagram of the Standard Model.

Whether or not the proposed metron picture is able to reproduce the extensive experimental data on weak interactions remains at this time speculative. The computation of interaction cross

[7] The conclusion that the net colour force must vanish also within the particle core has no impact for the trapping of the quark core modes, since the trapping is produced, as in the electron model, by the KSS fields, which, although neglible outside the core region, are dominant and do not cancel within the core.

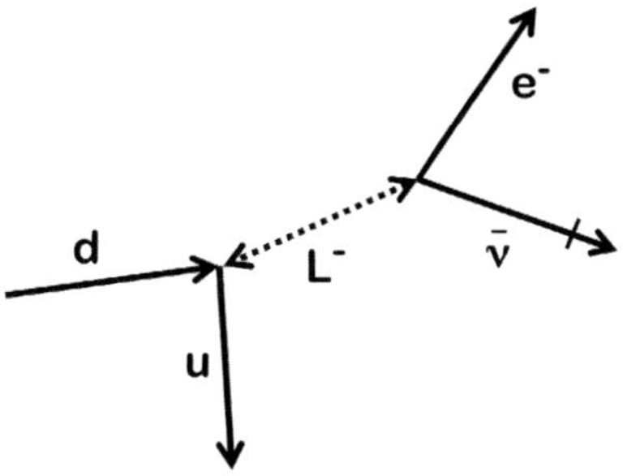

Figure 11. *Metron picture of weak interactions resulting from the coupling of a left-spinor background wave field L^- with left-spinor core-mode components, illustrated for neutron decay. Cross-bars denote complex conjugate fields (in the limit of zero mass, this is identical to the anti-particle).*

sections follows different rules in the metron model, which is concerned with transformations of real solitons, than in quantum field theory, which computes transition probabilities between different states. An additional problem of the proposed metron picture is the experimental instability of the neutron, which contradicts the general stability result cited in Section 3.5.2. The interaction with the background field must enable in this case a "tunnelling" from the quasi-stable neutron soliton to the truly stable proton soliton. However, interactions leading to transitions between two or more soliton solutions, including also particle collision processes in which the solitons penetrate each other's core regions – although well defined mathematically – lie beyond the present overview.

We have also not discussed unstable mesons and the particles of the second and third families. However, the basic picture outlined above – leptons and quarks represented as four-spinor core-mode soliton fields – can be readily applied to the full elementary particle spectrum. Mesons, for example, can be described, in analogy with the Standard Model, as pairs of quarks and anti-quarks of the same colour, while the second and third particle family can be represented as soliton solutions corresponding to the second and third nonlinear eigenmodes of the trapped-mode equations (20) (as demonstrated for the case of the scalar analogue of the Einstein equations in H).

4. Quantum phenomena

The description of elementary particles as soliton solutions endowed with both local corpuscular and large-scale far-field properties represents a first step towards overcoming the wave-particle duality paradoxes of microphysics. The necessary second step is to explain the typical wave-like interference properties observed in particle phenomena at lower atomic-scale energy levels. These result, as mentioned earlier, from interactions of the periodic core mode with other matter.

The particle core modes are generally of too high frequency to directly affect the trajectory of another particle. However, if the core-mode field is scattered off other matter, the interaction of the scattered field with the primary core-mode field of the particle itself generates a low-frequency field through difference-frequency interactions. This is able to modify the particle's trajectory, producing the observed wave-like phenomena. We discuss in the following two cases: the interference patterns of a stream of particles passing through a single or double slit, and the discrete nature of atomic spectra. Similar interpretations can be given for other quantum phenomena, such as Compton scattering, the photo-electric effect and Planck's law.

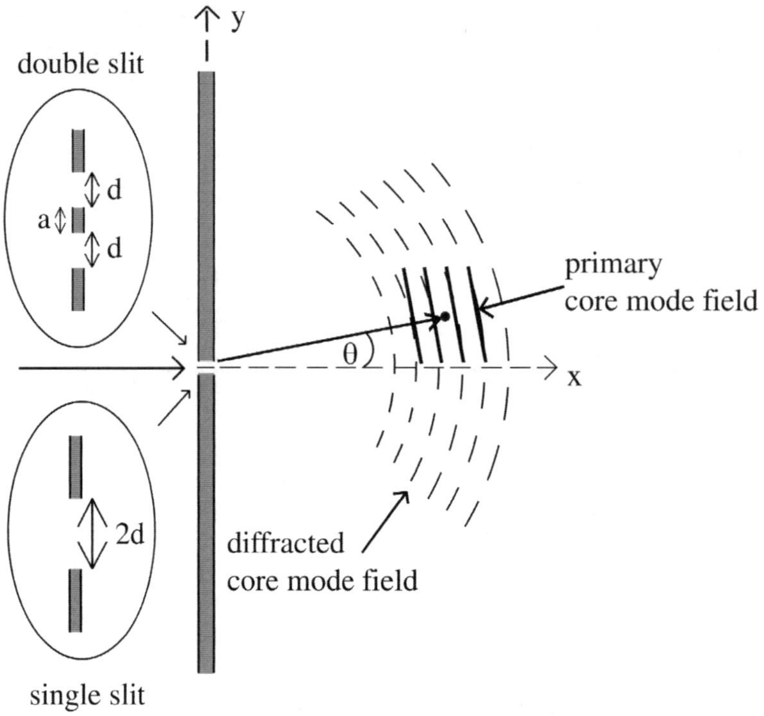

Figure 12. *Interacting primary and diffracted fields of the metron model.*

4.1. Single- and double-slit diffraction

In the metron picture of the single- and double-slit refraction experiment, each particle impinges on some particular position on the detector behind the screen after having passed through the single slit or one of the two slits on a well-defined, but in detail non-measurable, trajectory. The diffraction geometry is shown in Figure 12. The particle source and detector lie in the xy plane, with a screen in the yz plane orthogonal to the x axis at $x = 0$. The slits lie parallel to the z axis. The particles propagate with velocity u parallel to the x-axis, each particle passing through a slit at some random location $\mathbf{x} = (0, y_0, 0)$ at time $t = 0$.

The particle trajectories, after having passed through the diffraction screen, are determined by the force F produced by the interaction between the incident and diffracted core-mode fields. This is given by the negative gradient of the interaction potential V, $F = -\nabla V$, where V is proportional to the integral $< \ldots >$ over the particle core of the product of the (complex conjugate) diffracted secondary field $(\varphi')^*$ and the primary core-mode field φ:

$$V \sim < \Re((\varphi')^* \varphi) > \tag{64}$$

The diffracted secondary wave decays over a range of the order of κ^{-1} (eq.(22). This is large compared with the particle's Compton scale k_0^{-1} (eq.23), but nonetheless finite. The scale κ^{-1} is an important parameter determining the structure of the observed interference pattern (Figures 13, 14).

The interference pattern can be determined by Monte Carlo computations of the paths of individual particles. To compute the particle paths, one needs to: (i) determine the diffracted secondary core-mode field; (ii) evaluate the interaction between the primary and secondary core-

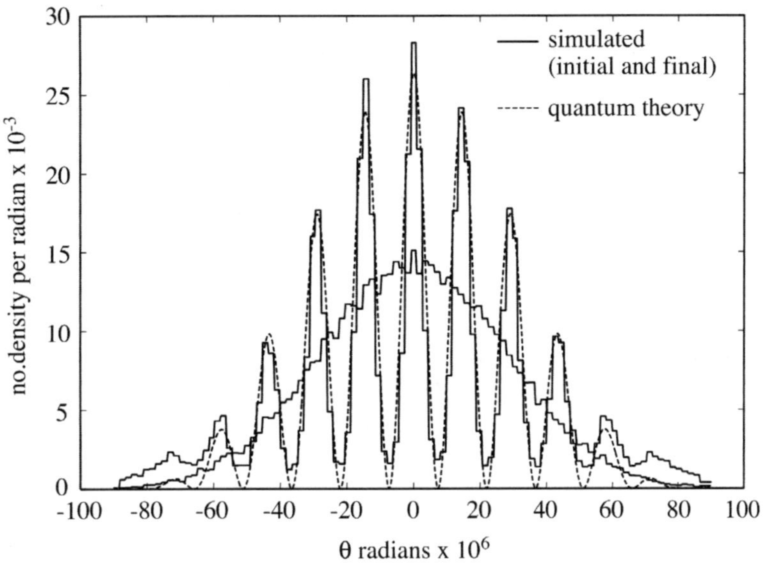

Figure 13. *Quantum theoretical and metron particle distributions for a monochromatic particle beam diffracted at a double slit.*

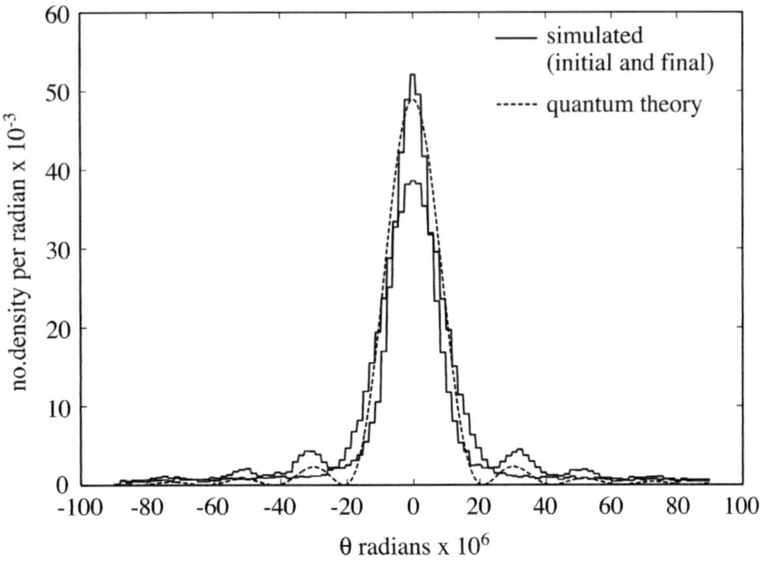

Figure 14. *Quantum theoretical and metron particle distributions for a monochromatic particle beam diffracted at a single slit.*

mode fields and the resultant force F acting on the diffracted particle; and (iii) integrate the trajectory of the diffracted particle underlying the force F.

The first step is similar to the standard computation of the diffracted state function in quantum theory. The only difference is that, whereas in quantum theory the incident wave is a monochromatic plane wave $\varphi_i \sim \exp[i(kx - \omega t)]$, in the metron model the amplitude of the incident core-mode field of the moving particle is dependent on space and time in the laboratory frame. Thus, both the primary core-mode field and the diffracted field need to be Fourier decomposed.

In the standard quantum theoretical case of an incident plane wave $\exp[i(k\,x - \omega t)]$ and small diffraction angle θ, the diffracted wave φ' at a location $\mathbf{x}' = (x', y' = \theta x', z' = 0)$ and time t' behind the screen is given by

$$\varphi'(x', \theta, t') \sim f_{1,2}(\zeta) \exp i(\mathbf{k}'.\mathbf{x}' - \omega t')(x')^{-1/2} \tag{65}$$

with

$$f_1(\zeta) = (2d/\zeta) \sin\zeta \quad \text{(single slit)}, \tag{66}$$

$$f_2(\zeta) = (2d/\zeta) \left[\sin\left\{\left(1 + \frac{a}{2d}\right)\zeta\right\} - \sin\left\{\frac{a}{2d}\zeta\right\}\right] \quad \text{(double slit)}, \tag{67}$$

where $\zeta = \theta k d$, d is the slit width (double-slit case) or slit half-width (single slit case), a is the double-slit separation, and $\mathbf{k}' = k(\mathbf{x}'/|\mathbf{x}'|)$ is the diffracted wavenumber.

If, instead of a plane wave, the incident field is the core-mode field of a particle travelling through a slit on the path $x = ut, y = y_0, z = 0$, the relations (65)-(67) remain basically unchanged, except that the incident field is multiplied by a modulation factor propagating with the particle velocity u. The diffracted wave field is then similarly modulated, the modulation factor also propagating with group velocity at the speed of the particle, in the slightly modified direction of the diffracted wavenumber. Thus, the incident and diffracted wave field are both transported in a frame moving with the particle.

In addition to propagating at the speed of the particle, the diffracted wave pattern experiences dispersion. This results in a spreading of the wave pattern in the x and z directions and a reduction of the diffracted wave amplitude by an additional factor $1/x'$. The net diffracted field at the position $(\mathbf{x}', t')$ then becomes

$$\varphi'(x', \theta, t') \sim f_{1,2}(\zeta) \exp i(\mathbf{k}'.\mathbf{x}' - \omega t') \exp(-\kappa x')(x')^{-3/2} \tag{68}$$

The probability distribution of the diffracted particles is governed by the tendency of the mean force $\bar{F} = -\nabla\bar{V}$ in the far field to drive the particles into the regions of high $\bar{V} \sim f_{1,2}(\zeta)^2$. The resulting probability distribution resembles, but is not identical to, the quantum theoretical pattern $f_{1,2}(\zeta)^2$.

Figures 13 and 14 show the distributions computed for an ensemble of 40,400 particle trajectories, together with the predictions of quantum theory, for the double- and single-slit cases, respectively. The classical and quantum theoretical distributions agree rather closely for the double-slit case. However, the single-slit distributions, although agreeing in general structure and in the positions of the diffraction maxima, differ in detail. The fit could presumably have been improved by suitable modification of the initial conditions and the parameters of the metron model. A systematic least-square fit was not attempted. However, the point of the exercise is not to reproduce the quantum theoretical predictions, but rather to demonstrate that both quantum theory and a classical model are able to explain, qualitatively, the observed wave-like interference features of particle experiments. Which model yields a better quantitative description of reality must be decided by experiment.

The detailed quantitative verification of de Broglie's 1924 prediction [11] of the wave-like nature of particles in particle diffraction experiments has proved notoriously difficult. Although the positions of the interference maxima predicted by quantum theory were verified in many experiments, quantitative verification of the complete quantum theoretical diffraction patterns was achieved only after almost seventy years using very low energy neutrons by Zeilinger *et al* [48] and Tschernitz *et al* [45]. In the first of these experiments, excellent agreement with quantum theory was shown for the case of double-slit diffraction, but in the single-slit case, small but statistically significant deviations were found between the predicted and measured diffraction patterns (as well as in a further unpublished experiment mentioned by the authors). Good agreement for the single-slit case was finally achieved by Tschernitz *et al*, but without clearly identifying the origin of the discrepancies in the first experiment. However, the good agreement shown by Zeilinger *et al* in the double-slit case also needs to be revisited: our repeat of the quantum theoretical computations revealed unresolved statistically significant discrepancies (paper in preparation). At this time, the detailed experimental verification or falsification of both the quantal and metron representation of the single- and double-slit diffraction experiments must be regarded as open.

4.2. Atomic spectra

The metron atomic model represents an amalgam of the original Bohr orbital model and the modern wave-dynamics approach of quantum electrodynamics. We consider individual electrons orbiting an atomic nucleus, with associated core-mode fields that are scattered at the atomic nucleus (Figure 15). The scattered core-mode fields ψ satisfy essentially the same field equations

$$(i\gamma^{\lambda}D_{\lambda} - \hat{\omega})\psi = F \tag{69}$$

as in QED, where $\hat{\omega} = (-k_l k^l)^{1/2} = (k_0^2 + \kappa^2)^{1/2}$ (eq.(57)) represents the electron mass, to within a small correction term of order $(\kappa/k_0)^2$ (cf. eq.(23)),

$$D_{\lambda} = \partial_{\lambda} - ieA_{\lambda}. \tag{70}$$

is the covariant derivative in the presence of the electromagnetic protential A_{λ} of the nucleus, and e is the electron charge,.

The only significant difference is that, in contrast to the homogeneous eigenmode equations of QED, eq. (69) includes a source term F representing the forcing by the orbiting electron.

The interaction of the scattered core-mode field with the electron's primary core-mode field produces a low-frequency field which modifies the electron orbit, in the same way as in the particle diffraction case discussed previously. For an arbitrary orbit, the interaction fields and the resultant orbital perturbations are normally small. The electrons therefore gradually drift into orbits of smaller radii through radiative damping. However, for certain discrete orbital parameters, resonances arise in the interacting fields. The perturbations can then no longer be treated as a small stationary response, but grow with time. The solutions stabilize in this case to a stationary state in which the radiative damping of the electron is balanced by the forcing due to the resonant interaction between the electron and the nucleus. The associated net radiative damping of the complete coupled electron-nucleus system vanishes.

For the case of circular orbits, the resonant interaction conditions for the primary and scattered core modes are found to be identical to lowest order to the QED eigenmode conditions, while the associated resonant orbital conditions correspond to Bohr's original orbital conditions (H). Detailed computations at the extremely high level of accuracy as achieved in the QED computations of the hydrogen atom, for example, have still to be performed. Although it may appear intrinsically unlikely that the same results as QED, based on an array of higher order corrections, including renormalization, can be achieved with an alternative classical theory, it is perhaps encouraging that Barut [3] has claimed that a non-second-quantized theory, without invoking renormalization, yields still higher agreement with experiment than QED.

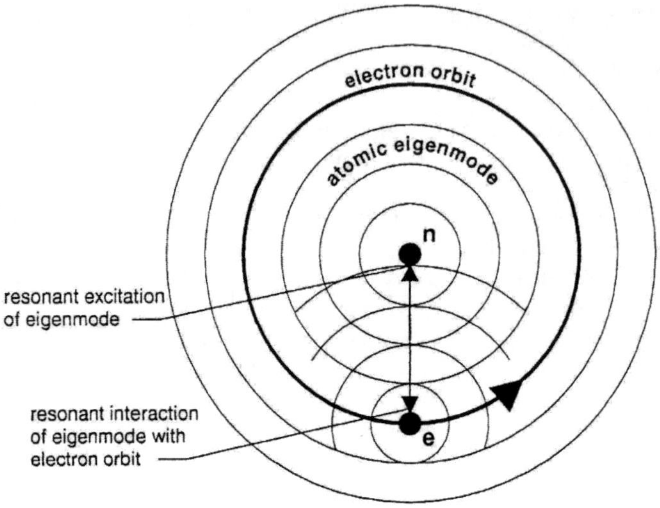

Figure 15. *Metron model of electron orbiting an atomic nucleus. The low-frequency field produced by the difference interaction between the primary core-mode field of the electron and the secondary core-mode field scattered at the atomic nucleus modifies the electron orbit. Resonance between the electron orbit and the difference-frequency field produces trapping of the electron in discrete orbits in accordance with Bohr's original discrete-orbit model. The resonant scattered core-mode fields correspond to the QED eigenmodes (reproduced from [27]).*

The attraction of the present ansatz is that it explains the discrete nature of atomic spectra as a resonance phenomenon of real existing particles supporting real continuous fields, without the need to switch pictures between fields and particles. Thus the photo-electric effect, for example, in which discrete electron particles are released from atomic states through incident electromagnetic radiation is readily explained without rejecting the concept of discrete localized particles in the description of the atomic states that release the observed discrete particles.

5. Summary

As precondition for the development of a classical unified theory, we have shown first that Bell's theorem, which has been widely interpreted as implying that classical theories are incapable of explaining the phenomenon of entanglement observed in microphysical phenomena, does not apply to classical models satisfying time-reversal symmetry. Entanglement is a universal property of all theories exhibiting time-reversal symmetry, whether classical or quantal. It follows directly from the relativistic implications of Newton's third law *actio* = *reactio*.

The basic concept of our classical model is very simple: the wave-particle duality paradoxes that led to the creation of quantum theory and its later special relativistic generalization, quantum field theory, can be explained in terms of the periodic soliton solutions of the Einstein vacuum equations $E_{LM} = 0$ in a twelve dimensional space. The finite extent and periodicity of the solitons provide a straightforward classical interpretation in terms of real existing objects of the various abstract features of quantum theory and quantum field theory.

The metron core modes correspond in many respects to the state function of quantum theory. From the metron viewpoint, the success of quantum theory, for example in atomic physics, is attributed to the theory reproducing the field properties of the core mode, while its conceptual difficulties arise from its inability to capture also the associated corpuscular features. This is elevated to a quantum theoretical theorem in Heisenberg's uncertainty principle. This is not in conflict with the metron model. Quantum theory is regarded as a theory of incomplete information. The uncertainty principle is justified empirically for both quantum theory and the metron model by the usual argument that initially limited information cannot be augmented subsequently by measurements using some (similarly imperfectly known) system that modifies the state to be measured.

Other basic quantal concepts, such as Fermi-Dirac and Bose-Einstein statistics, or vacuum state fluctuations, can be similarly translated in the metron model into classical processes involving real existing objects – avoiding thereby also divergent integrals and the need for renormalization.

In contrast to the mathematical formulation of quantum field theory based on the Standard Model Lagrangian and its symmetries, we have made no reference in the metron model as summarized here to the Einstein-Hilbert Lagrangian or the invariance of the Einstein equations with respect to coordinate transformations. We have simply exploited the fact that the Einstein equations are nonlinear, and then applied standard perturbation theory to study the interactions between the periodic core modes and higher-order components of the soliton solutions. The symmetries of the Standard Model correspond in the metron picture to the invariance with respect to translations in the fibre coordinates (i.e. phase shifts) and, in the case of strong interactions, the geometrical symmetries of the wavenumber configurations. Whether the Higgs mechanism and the internal breaking of parity symmetry can be replaced by the proposed simpler interaction with an external left-polarized high-mass background wave field remains to be investigated. In principle, however, the mathematics of the metron model is basically simple and can be summarized in terms of elementary interaction diagrams. This applies both to the structure of elementary particles and the explanation of the wave-particle duality paradoxes observed at lower energies.

Our overview of a proposed classical interpretation of elementary particles and quantum phenomena is necessarily of wide sweep and leaves open many questions. However, an alternative to a technically highly successful theory that has evolved over nearly a century can clearly not be developed in only a few papers. A comparison of the present presentation with its predecessors [23] - [28] reveals that, although the basic concepts have withstood the test of time, the metron model is still very much in flux. Nevertheless, a number of open questions listed in the conclusions of [26], such as the stability and uniqueness of the metron solitons and the origin of the basic physical constants, have now been successfully addressed.

Our motivation has not been to explain experimental results in conflict with current physical theory - often regarded as the litmus test of a new paradigm - but rather to offer an alternative view of physical reality that unifies the general relativistic theory of gravity with our understanding of elementary particles and quantum phenomena, avoiding in the process the conceptual difficulties of quantum field theory. Despite the undisputed successes of quantum field theory, its underlying conceptual fuzzines and its incompatability with general relativity have concerned not only Einstein, but many leading physicists. We believe that the path we have explored offers a new perspective for overcoming these inconsistencies and developing a conceptually better founded unified theory.

Acknowledgements
The extensive computer simulations of metron solitons were carried out by Susanne Hasselmann. The paper has profited from discussions with Gabriele Venanziano, Andrew Bennett, Armin Bunde, Wolfang Kundt and many other colleagues. Wolfgang Kundt, in particular, has followed the evolution of the metron model over many years and provided numerous constructive

comments.

References

[1] H. Abraham. *Ann.d.Physik*, 10:105, 1903.
[2] A. Aspect, P. Grangier, and G. Roge. Experimental Realization of Einstein-Podolsky-Rosen-Bohm Gedankenexperiment: A New Violation of Bell's Inequality. *Phys. Rev. Letters*, 49:91–94, 1982.
[3] A. O. Barut. Quantum-electrodynamics based on self-energy. *Physica Scripta*, 21:18–21, 1988.
[4] G. Basini, S. Capozziello, and G. Longo. The Einstein-Podolsky-Rosen Effect: Paradox of Gate? *General Relativity and Gravitation*, 35:189–200, 2003.
[5] J. S. Bell. On the Einstein-Podolsky-Rosen paradox. *Physics*, 1:195–200, 1964.
[6] D. Bohm. A suggested interpretation of the quantum theory in terms of "hidden variables", Parts 1 and 3. *Phys.Rev.*, 85:165–179 and 180–193, 1952.
[7] F. Cianfrani and G. Montani. Towards loop quantum gravity without the time gauge. *Phys.Rev.Lett*, 102:91301–91305, 2009.
[8] O. Costa de Beauregard. Time symmetry and the interpretation of quantum mechanics. *Foundations of Physics*, 6:539–559, 1976.
[9] J. G. Cramer. The transactional interpretation of quantum mechanics. *Rev. Mod. Phys.*, 58:647–687, 1986.
[10] W. C. Davidon. *Nuovo Cimento*, B 36:34, 1976.
[11] L. de Broglie. Recherches sur la thorie des quanta (researches on the quantum theory), thesis, paris 1924. *Ann. de Physique*, 3:22, 1925.
[12] L. de Broglie. *Compt. Rend. Acad. Sci.,Paris*, 183:447–448, 1926.
[13] L. de Broglie. *Nature*, 118:441–442, 1926.
[14] A. Einstein. Zum gegenwärtigen Stand des Strahlungsproblems. *Phys.Z.*, 10:185–193, 1909.
[15] A. Einstein. Quantentheorie der strahlung. *Phys.Z*, 18:121–128, 1917.
[16] A. Einstein, B. Podolsky, and N. Rosen. Can quantum-theoretical description of physical reality be considered complete? *Phys. Rev.*, 47:777–780, 1935.
[17] A.D. Fokker. An invariant variation principle for the motion of many electrical mass particles. *Z.Phys*, 58:386–393, 1929.
[18] S. J. Freedman and J. F. Clauser. Experimental test of local hidden-variable theories. *Phys. Rev. Letters*, 28:938–941, 1972.
[19] J. Frenkel. Zur Elektrodynamik punktörmiger Elektronen. *Z.Phys.*, 32:518–534, 1925.
[20] L. Gilder. *The Age of Entanglement: When Quantum Physics was Reborn.* Knopf, 2008.
[21] M. Göckeler and T. Schückeler. In *Differential Geometry, Gauge Theories and Gravity.* Camb. Univ. Press, Cambridge, 1987.
[22] B.R. Greene. *The Elegant Universe.* Vintage, London, 2000.
[23] K. Hasselmann. The metron model: Elements of a unified deterministic theory of fields and particles. part 1: The metron concept. *Physics Essays*, 9:311 – 325, 1996.
[24] K. Hasselmann. The metron model: Elements of a unified deterministic theory of fields and particles. part 2: The maxwell dirac-einstein system. *Physics Essays*, 9:460–475, 1996.
[25] K. Hasselmann. The metron model: Elements of a unified deterministic theory of fields and particles. part 3: Quantum phenomena. *Physics Essays*, 10:64 – 86, 1997.
[26] K. Hasselmann. The metron model: Elements of a unified deterministic theory of fields and particles. part 4: The standard model. *Physics Essays*, 10:269 – 286, 1997.
[27] K. Hasselmann. The metron model. towards a unified deterministic theory of fields and particles. In A.K. Richter, editor, *Understanding Physics*, pages 154–186, Copernicus Gesellschaft e.V. Kathlenburg-Lindau, Germany, 1998.
[28] K. Hasselmann and S. Hasselmann. The metron model. a unified deterministic theory of fields and particles - a progress report. In *Proc. 5th Intern.Conf., Symmetry in Nonlinear Mathematical Physics*, pages 788–795, Kyiv, 23-29 June 2004.
[29] P.R. Holland. *The Quantum Theory of Motion. An Account of the de Broglie-Bohm Causal Interpretation of Quantum Mechanics.* Cambridge Univ. Press, 1993.
[30] F. Hoyle and J.V. Narlikar. A new theory of gravitation. *Proc. Roy. Soc. A*, 282:191–207, 1964.
[31] C.J. Isham. *Modern Differential Geometry for Physicists.* World Scientific, 1999.
[32] T. Kaluza. Zum unitätsproblem der physik. *Sitzungsber.d.Preuss.Akad.d.Wiss*, pages 966–972, 1921.
[33] O. Klein. Quantentheorie und fünfdimensionale relativitätstheorie. *Z.Phys.*, 37:895–906, 1926.
[34] W. Kundt. Fundamental physics. *Foundations of Physics*, 37:1317–1369, 2007.
[35] H.A. Lorentz. republished in collected papers. volume II, page 281 and 343. 1892.
[36] H.A. Lorentz. The theory of electrons. Leizig, 1909.
[37] C.W. Misner, K.S.Thorne, and J.A.Wheeler. *Gravitation.* W.H.Freeman & Co., New York, 1973.

[38] R. Penrose. *The Road to Reality.* Vintage Books, New Yord, 2007.
[39] J. Polchinksi. *String Theory.* Camb.Univ.Press, Cambridge, 1998.
[40] W.M. Ritz. Recherches critiques sur l'électrodynamique géneral. *Ann. Chim.Phys.*, 13:145–275, 1908.
[41] E Schrödinger and M. Born. Discussion of probability relations between separated systems. *Math. Proc. Cambr. Phil. Soc.*, 31:555–563, 1935.
[42] L. Smolin. *Three Roads to Quantum Gravity.* Phoenix, Phoenix, 2001.
[43] N. Steenrod. *The Topology of Fibre Bundels.* Princeton University Press, Princeton, 1951.
[44] H. Tetrode. Über den Wirkungszusammenhang der Welt. Eine Erweiterung der klassischen Dynamik. *Zeitschrift für Physik*, 10:317–328, 1922.
[45] M. Tschernitz, R. Gähler, W. Mampe, B. Schillinger, and A. Zeilinger. Precision measurements of single slit diffraction with very cold neutrons. *Phys.Lett.*, A 164:365–368, 1992.
[46] J.A. Wheeler and R.P. Feynman. Interaction with the absorber as the mechanism of radiation. *Rev.Mod.Phys.*, 17:157–181, 1945.
[47] J.A. Wheeler and R.P. Feynman. Classical electrodynamics in terms of direct particle interactions. *Rev.Mod.Phys.*, 21:425–433, 1949.
[48] A. Zeilinger, R. Gähler, C.G. Shull, W. Treimer, and W. Mampe. Single- and doppel-slit diffraction of neutrons. *Rev.Mod.Phys.*, 60:1067–1073, 1988.
[49] A. Zeilinger, G. Weihs, T. Jennewein, and M. Aspelmeyer. Happy centenary, photon. *Nature*, 433:230–238, 2005.

Klein bottle logophysics: a unified principle for non-linear systems, cosmology, geophysics, biology, biomechanics and perception

Diego Lucio Rapoport

Department of Sciences and Technology, Universidad Nacional de Quilmes, Bernal, Buenos Aires, ARGENTINA

E-mail: diego.rapoport@gmail.com

Abstract. We present a unified principle for science that surmounts dualism, in terms of torsion fields and the non-orientable surfaces, notably the Klein Bottle and its logic, the Möbius strip and the projective plane. We apply it to the complex numbers and cosmology, to non-linear systems integrating the issue of hyperbolic divergences with the change of orientability, to the biomechanics of vision and the mammal heart, to the morphogenesis of crustal shapes on Earth in connection to the wavefronts of gravitation, elasticity and electromagnetism, to pattern recognition of artificial images and visual recognition, to neurology and the topographic maps of the sensorium,, to perception, in particular of music. We develop it in terms of the fundamental 2:1 resonance inherent to the Möbius strip and the Klein Bottle, the minimal surfaces representation of the wavefronts, and the non-dual Klein Bottle logic inherent to pattern recognition, to the harmonic functions and vector fields that lay at the basis of geophysics and physics at large. We discuss the relation between the topographic maps of the sensorium, and the issue of turning inside-out of the visual world as a general principle for cognition, topological chemistry, cell biology and biological morphogenesis in particular in embryology

1. Introduction

The objective of this article is to present a conceptual elaboration of fundamental issues of physics, namely the relation between divergences in hyperbolic geometry and non-linear science, to torsion fields and non-orientable topologies (the Möbius strip, the Klein Bottle and the projective plane) in several fields: cosmology, non-linear evolution equations and geophysics, to further apply them to biomechanics, perception, biology and chemistry. It is the nature of this paradigm and its current presentation in this article, that it can be read (presented) starting anywhere in its extension, alike to James Joyce *Finnegans Wake*, to return to this departing point finding the foundational elements once and again in the percourse of the presentation. Author Samuel Beckett has traced Joyce's recursive wake to the cyclical structure of Giambattista Vico's seminal text *Scienza Nuova (New Science)*, Newton's contemporary, pioneer investigator of complexity and systemics, critical of Cartesian dualism, whose surmountal is very much the issue of this article.

Physics has developed since Newton associated to a dualistic framework; the Third Law already is framed in terms of action and reaction; particles and antiparticles and CPT symmetry, are basic in quantum field theory. Einstein developed General Relativity with the conception of an absolute space, independent of self-reference and of the structure and processes of subjective experience, to which time, originally absolute in Newton's physics, became a variable related to space through the light-cone and the Lorentz invariance, while relativistic quantum mechanics indicates that it is the double covering of the Lorentz group the symmetry in question. Thus, two dimensions –though complex- are at stake; yet, this is also the case of the paradigm to be presented here, in which space does no longer hold the autonomous of experience character ascribed to it in physics. Time will turn out to be associated to depth, a phenomenological –and still, physiological- primal variable unseparable from non-dual logic by which, through self-penetration, the world manifests itself [14,134,135].

Biology has taken a similar dualistic stance, on facing the problem of what is life, first raised by Schrödinger [1] . The organization of the cell, the development of the embryo and morphogenesis, when elaborated from a Newtonian perspective of physics which is very much resisted by biologists of the molecular biology persuasion, has also taken dualism as its basic cornerstone [2,3,65]. Biophysicists dealing with the buildup of organisms from the spherical mother cell, will start by claiming that they are spherical [2,3,4], or still, to others those organisms having bilateral symmetry are claimed to be toroidal [5,6]. Both are topologies which in a static conception the division of the world is already established in terms of the outside and inside, as is the case of the embryonic sphere. This division is done in disregard that algae Volvox Carteri, which developmental biologists have come to consider the case study for morphogenesis, turns inside-out, a fact [7] that mathematicians have already established in the so-called eversion of the 2-dimensional sphere [8]. Yet, in a dynamic conception, it is the case that through smooth transformations, the sphere can be turned inside-out (i.e. everted). This is the Smale Paradox of topology; it is not restricted to a sphere, a torus can also be everted. Yet, it is important to remark, that although the sphere and the torus, are orientable surfaces, the eversion requires that the turning inside-out is carried through *half-way* surfaces which are non-orientable; otherwise, it is conceptually impossible, though a detailed proof is lacking, to evert a surface. This is the case of eversion of the sphere proposed by mathematician Morin, which has Möbius strips in the half-way transformation of Inside and Outside, and also is the case of a half-way eversion through the Boy surface, which is the embedding in Euclidean 3D space of the non-orientable projective plane. Yet, a remarkable precursor to evertions turning inside-out a sphere or a torus, is the Klein Bottle surface, a two-dimensional self-contained manifold for which there is no global inside nor global outside but a dynamic holonomic structure which processually turns inside-out and outside-inside. For this surface there is no half-way surface to produce the transformation; it is its very nature to do so, in its construction and in its form; neither it is a matter of statics nor of dynamics to produce this transformation, but rather an issue of holonomy, in a sense which had for precursor the theory of quantum mechanics due to David Bohm [94], yet which integrates his implicate and manifest realms [14].

In a serious of articles, it has been demonstrated that the Klein Bottle has a crucial role in biology (embryological development and differentiation [9,10], cellular biology [11], the genetic code and its relation with the Dirac algebra of Quantum Mechanics [12], the problem of what is life [9], biochemical recognition and a quantum theory of chemistry in terms of Bohm's theory of Quantum Mechanics [13], perception and anatomy-physiology [9,14,15,27-32], cybernetics and systemics [9,14], cognition and perception [9,15]. This appeared to be the case without positing a claim for totality as a petition of principle, but the derivation of seemingly unconnected scientific disciplines from a common unifying principle: The Klein Bottle as the embodiment of self-referentiality, rather than the petition of principle of dualism. Instead of an objective world as the usual take of theoretical physics would have it, the issue at stake is that of the experienced world and its joint constitution with the subject [140], enacted by active participation unrestricted to observation, thus superseding the conceptual framework of the Copenhagen interpretation of quantum physics.

Thus, whether this principle embraces *all* of reality or fails to do so, is by its very self-referential nature tied to this enaction rather than being an a priori manifesto: this is a physical (unrestricted to the reductive sense of the science of physics), sensorial, logical and cognitive bootstrapping which does not allow for an absolute framework other than self-reference. It is communicated through language, which whatever its quality may be, it rests on signs and utterances either generic (including formal languages as is the case of mathematics and its signs) or particular as applied to measurements which are designed by language themselves, thus having the same inherent self-referential being of this principle.

Allness is unseparable of self-referentiality, as Bertrand Russell would discover in the paradox later named after him, which lead him to pursue the provision of a dualistic basis for mathematics and metamathematics in his *Principia Mathematica* written with Whitehead, attempting to banish paradox of naive set theory and to banish paradox altogether. In this effort he was joined (Frege also played a central role) by mathematician and philosopher, founder of phenomenology, Edmund Husserl; (see [132,131], in particular his discussion on wholes and parts). Set theory is defined in dualistic terms, to belong or not to belong, and thus its blending with self-reference as in Russell's Paradox (namely, let R be the set of all sets that are not members of themselves; is R a member of R?) would manifest the paradoxical character of this blending. Set theory is non-processive but structurally static, and thus unappropiate to a conception of science in which self-reference is its foundational principle, and unsuitable for phenomenology as a philosophy of the lived world. Remarkably, this appears to have been altogether unconsidered by Husserl [131,132] and fully acknowledged by Spencer-Brown in his *Laws of Form* [18], to which we shall return below; see also the dualistic theory of autopoiesis in system theory, cognition science and biology [125] following Spencer-Brown, and the theory of ontopoiesis and the problem of what is life [9] elaborating the current paradigm.

To conclude, the universality of the paradigm hereby to be presented is unseparable from the universality of self-reference; it is independent of any statement based on quantifiers and independent as well of the notion of an elusive completeness which is beyond verification, if not through the lived world [9].

The paradigm hereby presented operates providing a non-dual multistate logic - and thus paradoxical states are inherent to this logic (as precursors of quantum superposition states): the Klein Bottle logic [9]. It has a Matrix Logic representation [14-16] which has quantum, fuzzy and Boolean logics for particular cases, and allows to demonstrate that dualism is nothing but a projection of the superposed topological states (produced by non-orientability of surfaces) to the Boolean states, and still to recover from them the superposed states; in other words, dualism is not a primeval independent logic but a projection of the Klein Bottle logic. This may provide a unifying principle for science and further integrate it with the analytical philosophical *tradition* (of which Russell and Frege were paramount exponents) that claimed to the possibility of establishing a relation between philosophy and science based on the latter, yet which its formalization was proved to be obsolete and misguided, through its *self-referential* reformulation in the Calculus of forms due to Spencer-Brown [18]. Thus, paradox and self-referentiality that the analytical traditional wished to banish (and Gödel Theorem remarked that it was impossible to do so for the natural numbers) was found to be the very roots of a logic with imaginary values, hitherto absent in logic, crucial to this paradigm.

The fundamental idea is that a sole sign of distinction, provided with two simple rules of generation (one of them being formal nilpotence, say of the exterior differential of calculus of functions of many variables real or complex) in which the sign is both operator and operand, allows not only to generate a formal system (free of logical quantifiers), but also time waves that stand for the imaginary logical values realizing the reentrance of the system on itself. This sign was further associated to the torsion geometry of spacetime [14,15], which is already crucial to the current paradigm of physics in terms of symmetries, since in the theory of Lie groups their structural coefficients are the Christoffel coefficients of a torsion geometry. In the Matrix Logic formulation of the Klein Bottle logic, this torsion appears through the lack of commutativity of the True and False operators embodied in a cognition operator through which cognitive statements are translated into quantum statements for quantum states, and in the photon field, in particular associated to singularities that sustain quantum jumps ([19,20,15] and the decomposition of the cognitive vacuum operator given by the null operator in Matrix Logic into twistors, thus unifying further the cognitive realm with rotational structures, which are further embodied as superposition states of the logic. In other words, the quantum and cognitive realms become integrated through vortical structures, whose topology is that of the Möbius strip and the Klein Bottle. So, torsion fields unseparably of non-orientable surfaces play a crucial role in this nascent paradigm of unification, providing the self-referential geometry of spacetime, cognition and logic, and still to the forms of life [21,22] and crucial to biology at large [9,10,11,12].

In the case of perception and the physiology of organisms, the division of Outside and Inside is also the case. In particular, the division of the world into the Exterior and Interior of a system, is very much ingrained in the sciences. Philosophers like Kant have tried to surmount this division by proposing noumenic and phenomenic worlds, which would somehow account for what he conceived to be a division between what there is and what can be known, in this case, of the world at large; in philosophical parlance, a division between ontology (what there is) and epistemics (how do we come to know about the world). Thus, for Kant, perception is the representation by our senses of the world out there, which is impossible to be fully cognized as it is, so we are constrained to represent a fleeting world, as from inside our senses and our cultural background as well. In fact, it appears to be untenable to prove that what we perceive is an outer world or an interior one, for that matter; cognition scientists have come to propose that instead of seeing an outer world, what we actually see is our brains [23], and thus it becomes impossible to establish a clear cut between the individual brain and that of the others. Indeed, unless we follow infinite regress, there is no absolute cognitive framework other than self-reference. This latter notion is widely unacknowledged –if not resisted- in science and in particular, in relation to consciousness as an epiphenomenon, an emergence that is postulated to originate in non-computability originating in quantum processes on the brain, which is formulated in terms of self-reference, yet denying it an ontological or epistemological principle [111].

We have argued that the photon is the ultimate physical process of self-referential nature and the eikonal equations of propagation of wavefronts of electromagnetic signals, to be the embodiment of self-reference [14,15,20]. This is rather general and even has a wide material manifestation. Indeed, Fock argued that the theory of relativity was conceptually wrong and that the ultimate reference frame to construct a theory of gravitation is provided by the solutions to the eikonal equation of the wavefront propagation of electromagnetic signals, i.e. fields that are discontinuous [111]. These wavefront propagation of electromagnetic fields are null ahead of the wavefront and are discontinuous in the wavefront, which depends on the higher order of the derivatives, two, and thus will present first order discontinuities; a spinor topological model of the photon was built as an electromagnetic signal –in the sense of Fock- by Kiehn [136]. They define, according to Fock, a universal reference frame, in terms of which physical reality comes to be [111]. Yet, due to the limitations of speed of light, for *any* field that has a limiting velocity of propagation, *indistinctly of the propagation being in the vacuum or in a matter field,* the equation for the wavefront propagation (the characteristics of the hyperbolic equations) is the eikonal equation, which rather than defining an objective geometry, they are related to torsion, which is essentially self-referential as the photon is –the photon is not seen but the seeing process [14,15,20]. So, *whether it is a propagation of elasticity, electromagnetic or gravitational fields, either on the vacuum or in matter*, say the surface of Earth, or the interior of a differentiating cell [11,2], they all have the same form of signalling their existence and manifestation in terms of the eikonal equations for light rays and for a physical manifestation of the singularities, the appearance of quantum jumps [19]. In fact, these signals are invariant by the Lorentz group transformations, so they are universal in their self-referentiality. This Lorentz group invariance, was argued by Fock to be the basis for a theory of gravitation; this group can be derived from the algebra of forms in Spencer-Brown's calculus [133]. These fields will all develop discontinuities on their propagating wavefronts which we shall see play a fundamental morphogenetic role, in particular on Earth's crust, as we proved it was the case of biological cells under differentiation. In this conception, this morphogenesis which manifests the diversity of reality as forms, has a self-referential origin, which is unconceivable for dualism, and still originates the physical world. In other words, rather than the emergence of consciousness and morphology, it is self-referentiality their source and that of the material world [9,14,15].

The physics and mathematics of perception has already a tradition, established in the 19th century by Fechner and von Helmholtz [24], that has been developed until present times, in particular in music [25], in vision [25,26,31,32] and pattern recognition [26,27], and in more general terms, in the search by neurologists of topographical maps of the sensorium, i.e. the existence of mathematical maps of the exterior sensorial input on our brains [27,28,29,30,31], mediated by the body [9,14,15,29]. In terms of the development of an early embryonic cell, the brain is by no means an interior to it, but a turning inwards of the *ectoderm* (exterior tissue), so this qualification of topographical maps being representations of an outer world in our brains as if the locus of our interiority, appears to be nothing but more of a the imposition of dichotomic thinking, which in this case, it turns out to be unwarranted, if not overtly wrong. It becomes patent that this conception is based on statics, and not in developmental considerations, the dynamics of forms and the problem of morphogenesis. They are the subject of non-linear science [33], and in particular, of the self-organization of systems, in particular, chemical, as dishpans convection experiments clearly shows, not to mention the generation of patterns from sound as empirically confirmed by Jenny's Cymatics [35] (with Galileo and Chaldni for predecessors), discussed in Rayleigh's treaty on sound [33].

In terms of number theory, the extension of the natural numbers to the integers, assume a duality and a non-compactification of them, which in placing the number system of measurements in terms of the real numbers (a practical impossibility, since cut-offs are inexorable) which Quantum Mechanics establishes to be the realm of what is being measured due to the use of self-adjoint operators whose spectra are real numbers. This is the case despite the fact that the complex numbers are essential to physics as the phase functions, yet appearing already in the classical harmonic operators, that linear combinations of the complex solutions of the characteristic polynomials of the differential equations allows to obliterate the square root of minus 1 in the solutions while keeping the oscillatory motions.

Yet, most surprisingly ignored at large, is the fact that the linear superposition of waves, not only modulates the amplitude of the resultant wave, but also produces a modulation of the phase, that is manifested as a time shift associated to the interference patterns. This allows to make away with the complex numbers for the purpose of writing the wave created by the superposition, which can be thus written as the product of sinusoidal functions, yet introducing anticipatory behaviour by this phase modulation [36]. We have discussed already the relation between the Klein Bottle and anticipatory phenomenae [9,10,11,12,14,15,130] yet in this article, the issue will be presented in another way which has been very much overlooked, apparently even by astronomer Möbius himself, a fundamental implicit harmonic of the holonomity of these non-orientable topologies. Yet, this modulation of phase in the case of linear superposition of waves, will turn out to be of crucial importance with regards to visual perception and pattern recognition, to the crustal morphogenesis of Earth, to anatomy-physiology, again related to the Klein Bottle logic.

2. The projective plane, the Möbius strip and the gnomonic projection

Complex analysis does away with the dual division between the positive and negative numbers, by representing the extended complex plane in the compact Riemann sphere, and thus the complex plane and the real numbers, by construction, become compactified, and in the associated stereographic projection, infinity is no allegory but the North Pole, and thus infinity and minus infinity become a single point. This point is both positive and negative, and neither one, and thus in terms of the Matrix Logic form of the Klein Bottle logic, it corresponds to the non-orientability Möbius strip states corresponding to topological and logical superposition [14,15,16]. Hence, by recourse of transforming the flat complex plane to the constantly curved sphere, the duality between negative and positive numbers becomes surmounted. Thus, the imposition of duality that is basic to contemporary approaches to quantum mechanics, is left wanting, since the number system which is basic to it, is not dual, when represented in the Riemann sphere. Yet, we can use this sphere, by further identifying the antipodal points through the sphere's origin, thus yielding a compact representation of the projective space. In this case, infinity is again no allegory for divergences, but the Equator. In fact, now we have embedded in the sphere, with antipodal points identified, the Möbius band, so non-orientability is the case, *locally.*

Fig. 1

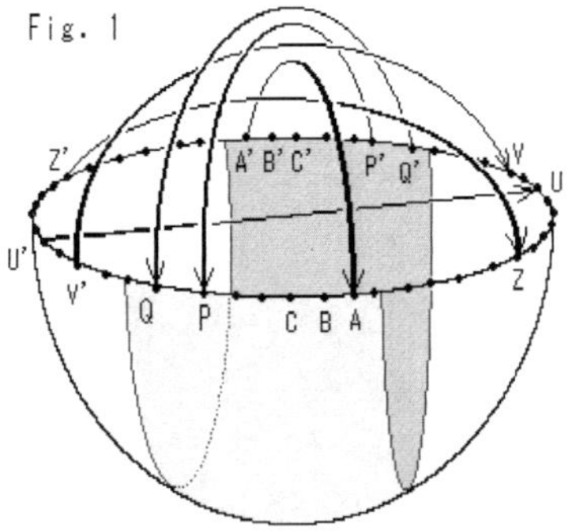

Figure 1.Two points at opposite ends of a sphere are called **antipodal points**. On the edge of a hemisphere there are countless anti-podal points. All antipodal points on the edge of the hemisphere made by central projection from the ground must be identified. That is, the ground that is spreaded infinitely has something like antipodal points. Such plane is projective plane. Though a visual model of the entire projective plane can not be made, we can get a part of it by cutting the hemisphere like the yellow belt. It is a well-known **Möbius strip**

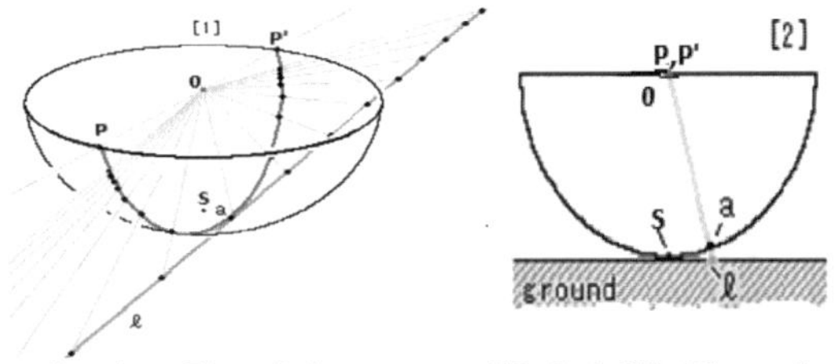

Fig. 2

Look at Figs. 2 (courtesy of T. Ito) [1]: The red spot is the center of hemisphere **O**. The black point *S* is its bottom (South Pole) where the ground plane touches. The green line ℓ is a straight line drawn on the ground. The edge of the hemisphere **O** is the line at infinity. The sky blue lines are projection rays from the green straight line ℓ to the surface of the semisphere. The red line **PP'** is the projected line that is a great circle. Point **a** is the middle point on line **PP'**. This red line **PP'** should be regarded as a straight line that stretches infinitely. We are going to see what happens if we put it onto Möbius strip. [2]: It is the cross section passing three points **O**, **a** and **S** for clarifing their positional relation. They are always in the positional relation as shown. At this moment, point **a** is apart from point **S** and it is not on green line ℓ. Naturally, in a case, point **a** is on point **S** and so is on green line ℓ.

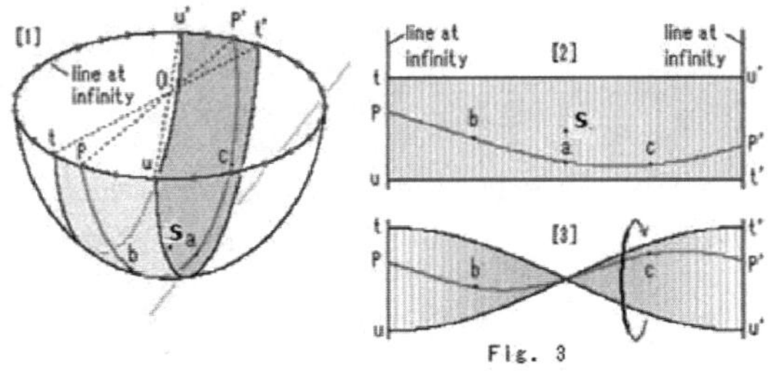

Fig. 3

Look at Fig. 3. [1]: The red spots on the edge (line at infinity) of hemisphere **O** show countless antipodal points. The dotted lines indicate that the end points of red line **PP'** (= straight line) are antipodal and points **t**, **t'** and points **u**, **u'** are also antipodal. We cut hemisphere **O** as shown. It is convenient for construction to cut hemisphere **O** in perpendicular to the equatorial plane. But you may freely cut hemisphere **O** as long as the red line **PP'** is in between lines **tt'** and **uu'**. Marks **a, b, c** and **G** are to trace points on hemisphere **O** to Möbius strip.[2]: It is a rectangular strip taken from hemisphere **O**. We treated the cut-out as if it is rubber made, and smoothed it flat. Since both end lines, right and left are at infinity, we may regard the length of rectangular strip is infinity. But the lines at infinity is that we pulled in our hands [3]: The right end of [2] is twisted by 180 ° while the left is fixed.

We continue to Figure 4. [4]: It is a view from top of [3].[5]: We jointed the end lines of [3] and got Möbius strip. The line at infinity is the joint line of Möbius strip, and point **a** is on the opposite side. [6]: We newly cut Möbius strip at point **a** and flat it. This time, the red line is smoothly connected. Thus we caught how a straight line that extends to infinity makes a loop at infinity. And the red "straight line" on Möbius strip that is a curved surface keeps its character as a straight line. No one knows whether the straight line acts like this at infinity or not. However, we feel nothing illogical. Naturally, It is based on an idea that there actually exists what is called infinity and a straight line can be stretched to that infinity.

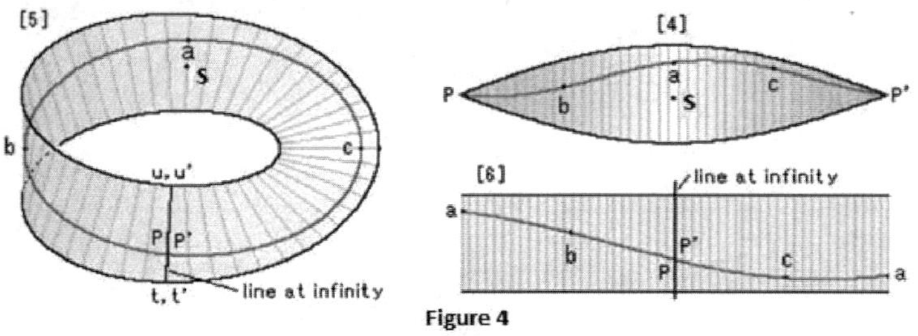

Figure 4

So far we have seen how the non-orientable Möbius strip is locally embedded in the projective plane and that infinity has a well defined meaning. Yet, the projective geometry is associated to hyperbolic geometry. Indeed, consider Figure 5 (courtesy of T. Ito):

Fig. 5

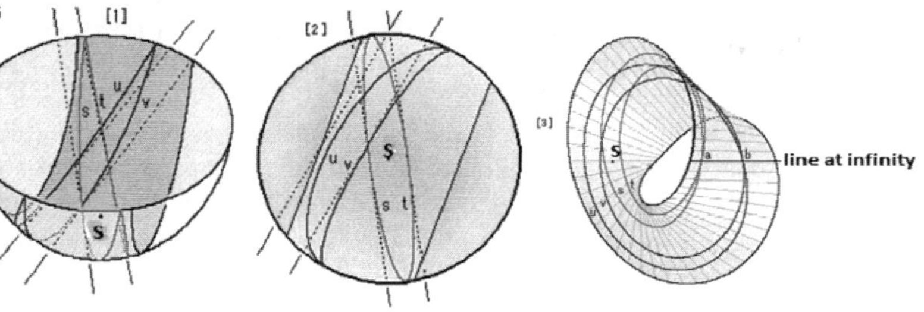

Two pairs of parallel lines now.[1]: We have to cut the hemisphere so widely that two pairs of parallel lines are contained. On the ground the red parallel lines are just under the hemisphere but the blue parallel lines are off to the side.[2]: It is the bird's-eye view of [1].[3]: It is Möbius strip made from [2]. It is drawn to show both of the line at infinity and the area where the two pairs cross each other are facing to us at a time. On the line at infinity, the red parallel lines meet at point **a** and the blue parallel lines meet at point **b**. The distance between points **a** and **b** is infinity though it is visible. Therefore we have to regard not ony the length but also the width of Möbius strip as infinity. There is no ordinary distance on the line at infinity.

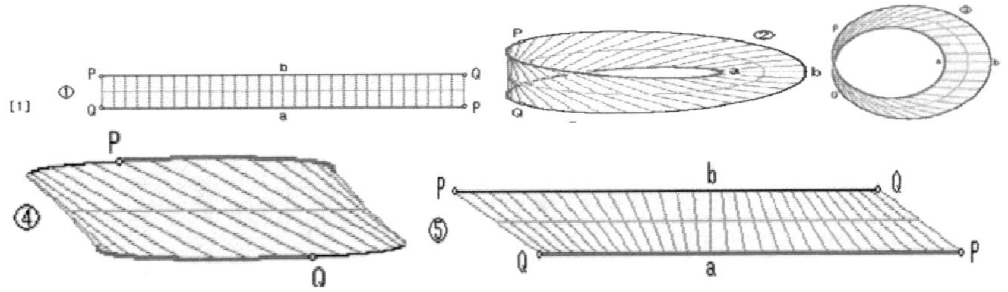

Figure 6: [1]: It is in case that the length of edges of the rectangular strip is identical to that of Möbius strip. ①We mark the rectangular strip into **n** equal parts, and draw the green center line. Points **P**, **Q** are to be connected for Möbius strip so that they are marked with °for attracting attention. One side of the edge of rectangular strip is colored in red to see which is which part on Möbius strip. ②The marked lines are broken at the center line. They shall be smoothly curved if we draw them accurately. But we omitted on purpose for attract attention. Compared with the rectangular strip. ①, the figure is enlarged. ③ It is a bird's-eye view. ④ It is the view around points **P**, **Q** from the left. Marks are leaning to one side. Only mark **ab** is perpendicular to the edge. It is quite different from paper-made Möbius strip. ⑤ Points on the edges and center line of the rectangular strip ①and Möbius strip correspond each other. But the length of their marked lines are quite different except line **ab**. So we shifted edges of the rectangular strip to right and left as shown. Now we possibly say that the length and angles of marked lines on the rectangular strip ④and Möbius strip are identical. However, if we make a paper-made Möbius strip with this kind of parallelogram, it will have no smooth edge but a cusp at the joint. So to actually construct a Möbius band a parallelogram will not do. Instead we need a trapezoid, as originally proposed by Möbius himself.

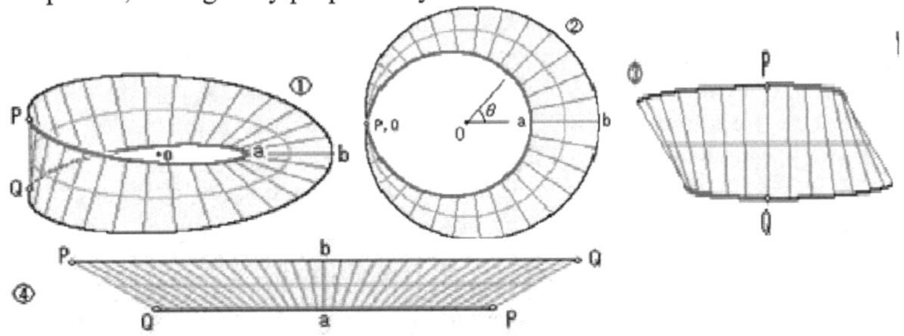

Figure 7.

This time, first we draw Möbius strip, and then we cut and open it as a rectangular strip. ①The green center line is a circle. Its center is **O**. Every notch line is perpendicular to the green center line at regular intervals of **n** division which is similar to [1]. Notches may look coarse but the number of them is the same as that of [1]. ②It is a bird's-eye view. θ is the angle of notches. See In [1] lengths of the red edge and black edge are the same as well as the rectangular strip. But in [2] it isn't. In the figure the length of black line is one and a half times as long as the red line. ③ It is the view around points **P**, **Q** from the left. Nick **PQ** is perpendicular to the edge. So is **ab**. ④ It is what we cut and opened Möbius strip while keeping the length of edge with notches. It is formed into a trapezoid and the length and slope of notches lines are distorted.This distortion is of great physical import. Indeed, while the fundamental rule of the sum of vectors, say forces, on Euclidean space is valid in the case of the parallelogram, in the case of Möbius strip, i.e. the trapezoid, this is no longer the case. In fact, this is shared by spaces with torsion, in which parallelograms do not close so that the sum of two vectors does not yield the diagonal, because the diagonal itself is ill-defined. Thus, instead in the non-closure of the parallelogram formed by the vectors u_P and $v_P$at P and their parallel transport, $u_R^{\parallel}$ at R and v_Q at Q, a fifth side T(u,v) closes the infinitesimal vectors.

Prolonging T(u,v) to the line joining P and Q, we have the trapezoid whose completion is depicted in red.

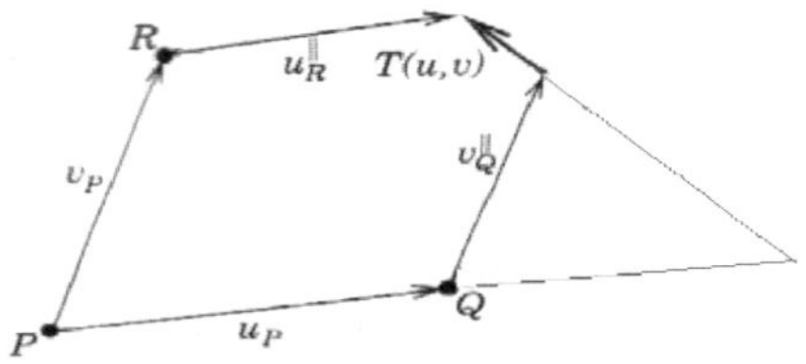

Figure 8. The infinitesimal parallelogram of vectors drawn in black does not close, the torsion being the fifth side closing the otherwise open parallelogram. We prolong it to draw in red the trapezoid naturally introduced. by torsion.

3. The self-referential torsion geometry

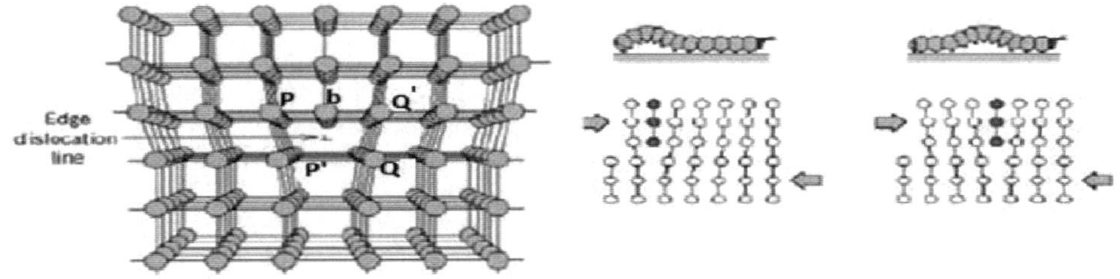

Observe figure 9 above. We introduce torsion by shear in a lattice; in the rhs by a caterpillar moving in the surface of a perfect crystal. In the lhs we have a perfect lattice, as is the case of a discrete space-time rendering of the homogeneous spacetime of General Relativity, but in the central area in which a dislocation is the case. We see then the transition from parallelograms that close (null torsion as in General Relativity), to non-closing and the formation of a fifth side, the torsion field. Figure 9 shows the meaning of torsion. In an otherwise perfect crystal (whose vertices are described in red), i.e. one free of inhomogeneities (in practice, difficult to achieve) an edge dislocation is produced, either by removal (i.e., introduction of singularities) of atoms of the crystal, as the figure shows, or by introducing extra material, in short, both inhomogeneities. Torsion can be introduced by shear (i.e. the relative motion of two planes) as figure 9 shows. Think of a caterpillar which moves a lattice a step at a time, and the shear produces the torsion of the crystal; this shear produces a vortical motion on the vertical plane to the shearing plane under the .mixing of layers for small that in can be. Another analogy is that of a rug, which moving in the perfect background of the homogeneous crystal; local changes affect the whole structure; the analogy strikingly applies to the crease (the folded rug) formation in the gastropore invagination in Embryology [9,10]. It can also be produced by a hole in the surface, producing an embryological expansion wave. Thus, it is an action-dependent participative geometry introduced in terms of inhomogeneities by the subject, say the caterpillar, or more basically the photon. This stands in stark contrast with the homogeneous situation of a Cartesian ideal geometry exterior to the subject, which corresponds in the continuum limit in which the atoms of the crystal approach indefinitively, i.e. the continuum hypothesis of Einstein, which due to the lack of a singularity, it corresponds to the zero torsion metric-based geometry of General Relativity. In short, to have loci, self-referentially dislocations are needed; these are the inhomogeneities that make a geometrical locus. The most basic dislocation is produced by a photon, a quantum particle, a singularity of an electromagnetic field. The parallelograms where inhomogeneities are present do not close ,while in the perfect crystal do close, indicating a self-referential trivial loop: by default is indeed a loop, i.e. closed without mediations. Instead, in the former case a pentagon is produced. The fifth newborn side at the upper right side of the center, joining Q' with b, is the torsion –the self-referential mediator; it is necessary and sufficient to the effect of completing the self-referential closure of the otherwise closed parallelogram; we shall explain this further below. As figure 9 shows in the center, the torsion appears producing a trapezoid with vertices P, Q, P' and Q', since the upper side is now longer than the lower one. Yet trapezoids are themselves Möbius strips cut transversally, say along the line on the Möbius strip below joining P (P') and Q (Q') below, and laid open on a plane, which by identifying the opposite lateral sides PP', QQ', on a previously 180° twist of the trapezoid, re-establishes thus the Möbius strip, as shown below in figure 10.

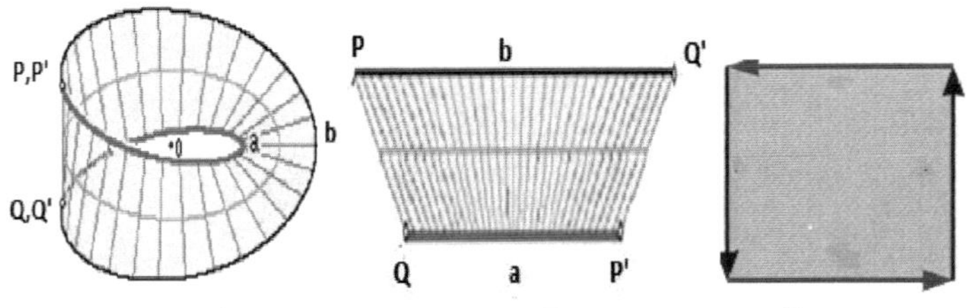

A B C

Figure 10: The Möbius band cut along PQ unfolds to the trapezoid in the middle with **P** on the Möbius strip connected to **Q' =Q** (the upper side), and Q to P' =P, the lower side, respectively. On the rhs, C depicts the identifications that produce the projective plane, a non-orientable surface.. This completion (actually a self-referential mediator), the torsion, establishes a loop, say starting on Q goes to P', continues to Q', further to P via b (changing thus the direction of this side contrarily to the direction of the shear) to finally reach back to Q; we depicted these directions on Figure 10C.Thus, it is a non-trivial self-referential action, to return to the identity, either on the sheared plane or out of it by the vortical twist installing the Möbius strip, depicted by the blue vectors in Figure 10C. As a loop yet with its upper and lower sides (drawn with red) now being considered as the edges of the Möbius strip, it establishes the characteristic uniqueness of edge of the Möbius strip, since the seemingly distinct edges PbQ' and QaP', which are transversed with opposite directions relative to the shearing motion, become parts of a single cycle; otherwise stated, this is the 2:1 harmonic. Thus, would we finally identify them topologically with their opposite directions as in Figure 10C with sides drawn on red, we get the projective plane, rather than the Möbius strip. In the case of Einstein's General Relativity, the closedness of the loop is by default, since zero shear is assumed and thus, the torsion Q'b does not appear if not the null vector, and thus the trapezoid collapses to a parallelogram; this is the case of the metric derived Levi-Civita symmetric connection of General Relativity, and the essential case is geodesic motion; yet, it is the spin coupling to linear momentum what produces a modification of this inertial motion on considering a connection with non-null torsion [61]. Would the fifth side Q'b, be equal in length to the other four, we would have the five-fold symmetry typical of quasicrystals and thus the Fibonacci sequence is embedded in this symmetry and a fortiori, the Golden Mean, Φ. As a final comment, while Fock noticed that in its dependence on metrics and covariant transformations of mathematical expressions on terms of which Einstein formulated his Theory of Gravitation had left unconsidered the problem of spacetime being uniform or not (see page 6 [111]), we clearly see that its implicit claim was that of uniformity.

4. Hyperbolicity and nonorientability of the Möbius strip

Hyperbolic geometry was introduced by Lobachevsky to lift the Euclidean axiom that parallel lines do not meet, as we already saw in the definition of torsion, the parallel transport of two vectors will not produce a parallelogram, so without any need of taking Lobachevski stance, parallel lines will intersect if the rule for parallel transport, i.e. the connection with torsion, is the case. Yet, it is related to projective geometry, in which appears related to lines at infinity. We want to dispel the ambiguity and elusiveness of this notion of infinity, relating it with the Möbius strip and the fact that infinity, as currently used by mathematicians and physicists alike, and no less to philosophers is a figure of speech, that can be found to have a more specific and yet universal sense than that of the figure of an allegory of an unbounded growth. In mathematics and physics in which appears in various guises, as divergences, the idea of a transcendental unknowable imaginary world somewhat related to the material world of physics and everyday, in the continuity of real numbers, in particular in the non-denumerability and non-compactness of them, and still in the issue of non-computability that has arisen in such diverse fields as complexity theory, consciousness studies, etc. [109].

To do this, we start with projective geometry and the so-called gnomonic projection which is the fundamental operation for its construction.

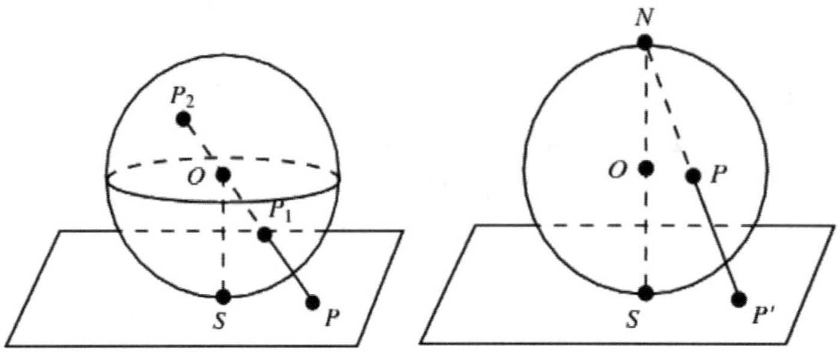

Figure 11: The gnomonic and stereographic projection of projective geometry and complex analysis, respectively.

We already introduced the gnomonic projection in figure 1. The gnomonic projection is a nonconformal map projection obtained by projecting point P_1 or P_2 on the surface of sphere from a sphere's center O to point P in a plane that is tangent to a point S. In the above figures, S is the south pole, but for the gnomonic projection can be any point on the sphere. Since this projection obviously sends antipodal points P_1 or P_2 to the same point P in the plane, it can only be used to project one hemisphere at a time. This is the why in Figures 1 and 2, we have also used the lower hemisphere. In a gnomonic projection, great circles are mapped to straight lines as shown in figure 1. The gnomonic projection represents the image formed by a spherical lens, and is sometimes known as the rectilinear projection. It is to be distinguished from the (conformal) stereographic projection of the Riemann sphere, in which the point of projection is the North Pole replacing the center 0, and the contact point with the surface can no longer be arbitrary but the South Pole, identified in the complex plane S as the origin of the complex numbers. The naming of the Gnomonic projection comes from the Greek gnomon, the pedestal of a sundial (solar clock), which indeed it is not placed on the centre of Earth but on any arbitrary point of its surface, and by which the shade of the Sun projecting on a surface would be marked to identify the time of the day. It was this projection which was used not only to keep track of the time of the day, but the tracking of the seasons due to the precession of the equinoxes derived from the wobbling of Earth's inclined axis further moving along the circumsolar elliptic orbit. The overall graph of this annual projection is the Analemma, the Möbius strip projected on the surface of Earth.

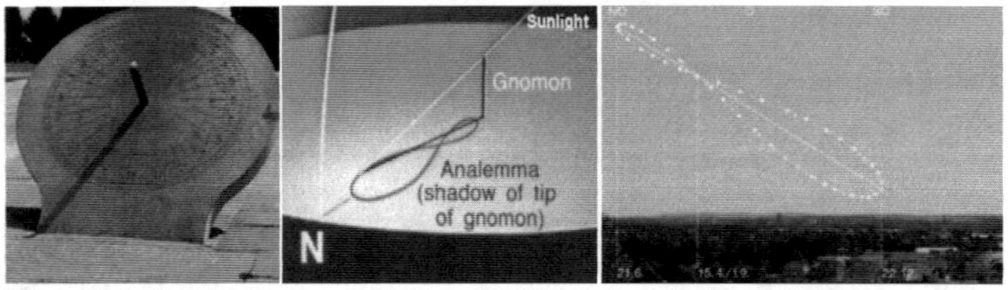

Figure 12: On the lhs a sundial produced by the shade of the Sun's projection on the gnomon, placed at the center of the device. In the middle, we see the figure 8 of infinity (the Analemma), produced by

the circumsolar rotation of Earth due to the wobbling tilted axis of rotation, on the elliptic orbit around the Sun. In the rhs the superposed annual photographs of this effect as captured by a camera lens placed in a fixed point and in a fixed direction, as in the rhs of Figure 12.

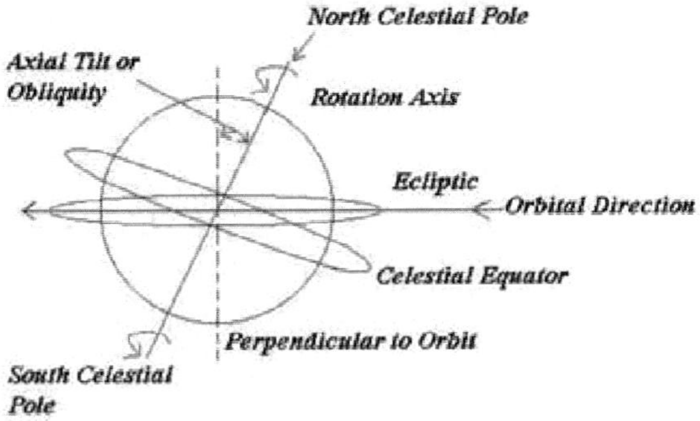

Fig. 13: The precession of the equinoxes produced by the circumsolar tilted wobble of the Earth's axis along the elliptic path of the centre of Earth; the angle of tilt and the eccentricity of the orbit are periodic variables, crucial to geophysics.

We return to the gnomonic projection to introduce the issue of infinity in projective geometry and the Möbius strip.

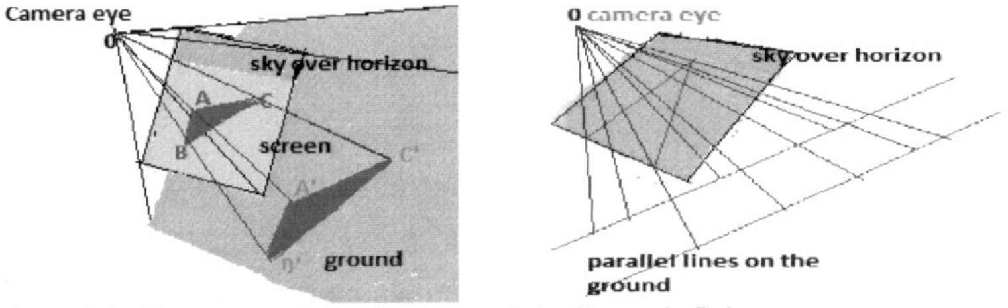

Figure 14: The gnomonic projection and the line at infinity.

So we imagine the projection of a point **0**, which now places the role of the centre of the sphere which we remove and still we take for the surface on which we practice the projection, to be a section of the tangent plane to Earth's surface.. The camera is the projection center and takes the picture of triangle **A'B'C'** on the ground. This ground is supposed to be the tangent plane. The light green area is the view from the camera. Black rays from the camera indicate stereo area (pyramid) of the viewfinder. The square board fitted to black rays is to show what we can look at through the camera. Let us call the square board the -screen. The photo of triangle **A'B'C'** on the ground is taken as triangle **ABC** on the-screen. Blue rays are projective lines of vertices of triangles. Consider now a pair of parallel lines on the surface and we lift them as if derived from the gnomonic projection of two lines drawn in red. These lines do intersect, in fact they do so on the horizon, on which we place the vertex/intersection point, with the horizon drawn in light green and the sky in light blue. Thus, the horizon which in projective geometry is called the line at infinity is not such line in the actual world in which the tangent plane ceases to provide a good approximation to the surface of the sphere. It is neither an imaginary line, but in the definition of mathematicians, as we already saw in identifying the line at infinity in the gnomonic projection as the equator of the sphere. Instead of the elusive imaginary line at infinity, we can identify it in the panel screen as the boundary between Earth and sky. It is no line at infinity at all. It can be thought as the manifestation of the non-local embedding of the Möbius strip on the projective plane, as we discussed at the beginning.

So again, what physicists would name a divergence, has a very concrete manifestation, and the very notion of infiniteness is called into question. Of course, we can take the camera's eye for the North Pole instead of its substitution for the center of the sphere, and the conclusions still apply to the stereographic projection; thus the conclusions do not require we have a conformal (stereographic) nor a non-conformal (gnomonic) projection. In fact, for visual perception, in which the centre of the retina plays the role of the point of organization of incoming light, it is the conformal mapping that applies, as we shall discuss below. We have already seen that the antipodal identification that is the case of the gnomonic projection, produces the Möbius strip, in which the line at infinity is precisely the line on which the Möbius strip is produced by the twisted identification of antipodals along a section of the equator/line-at-infinity. Let us see how is this further related to the actual asymptotic hyperbolic discontinuities on the real plane, which as we shall see, become continuous in the Riemann sphere by introducing the one-point compactification, and it is related to the non-orientable Möbius strip structure of the complex plane itself, which in the case of visual perception, it is related to the complex logarithmic map as defined on a Möbius strip instead of its usual orientable domain.

Consider the gnomonic projection now applied to an hyperbola placed on either a screen or on the projective plane as in figure 15.2 with the screen placed on top of the equator, which is the line at infinity; for the latter note that since the asymptotes all end on the line at the equator, this produces a compactification in which instead of discontinuity at the origin, we have a continuous curve so that the real positive and negative infinites are identified together with the positive and negative imaginary infinites.

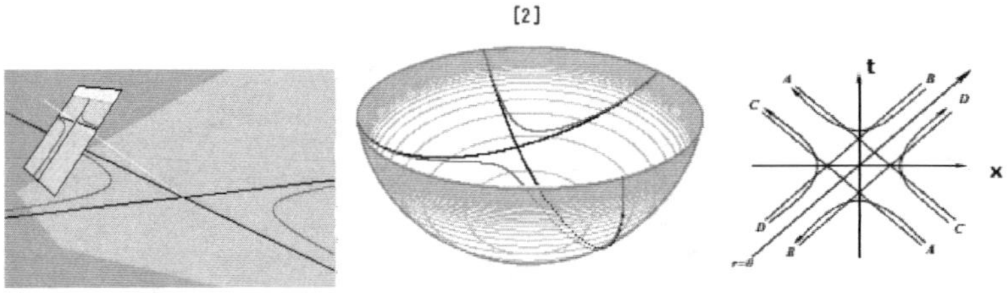

Figure 15. The projection of an hyperbola on the ground which in the case of the figure on the middle, rather than a discontinuity on both the real and imaginary axis, they are continuous via the line at infinity, which is the equator of the sphere, still identified by antipodal points. Right: we have the hyperbolic trajectories associated to the closed time loops in the Gödel solution to Einstein's equation (modified from [58]).

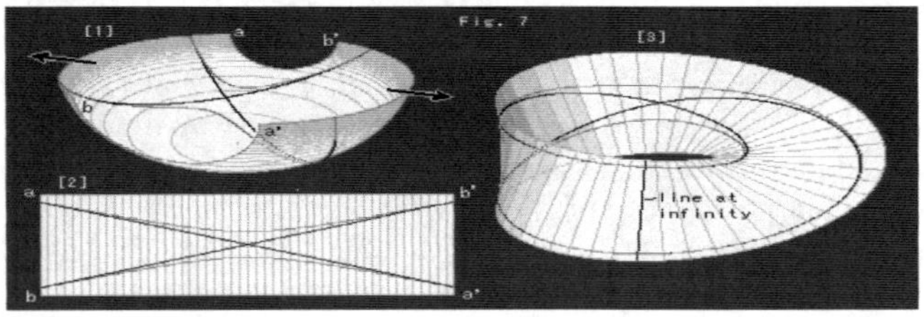

Figure 16 above is to see a full range of a hyperbola placed on Möbius strip. A part of projective plane is the Möbius strip. [1]: Cut the hemisphere in Fig. 16 straightly down as shown. [2]: Stretch it in the arrow direction by force, and make a belt. [3]: Twist one end (right or left) of it half, and glue it to the other, a to a' and b to b'. Then we get a Möbius strip with the hyperbola that was in [1]. The jointed line is the straight line at infinity. The red hyperbola is a double-turned loop now. Each asymptote forms a loop and meets with the hyperbola on the line at infinity.

Journal of Physics: Conference Series **437** (2013) 012024 doi:10.1088/1742-6596/437/1/012024

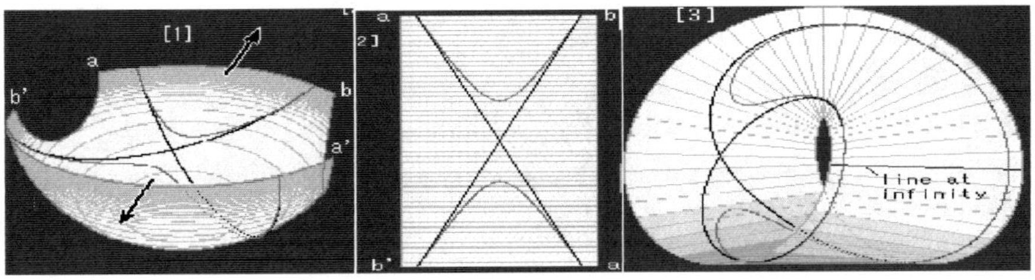

Fig. 17 above, is also to see full range of the hyperbola on the Möbius strip. The hyperbola is the same as that of Fig. 16 but we cut the hemisphere in another way. [1]: This time, we cut the hemisphere in another direction as shown. [2]: Stretch it in the arrow direction by force, and make a belt. [3]: Twist one end (top or bottom) of it half, just as we did before, and we get a Möbius strip. This time the red hyperbola is not a double-turned loop but a single loop; we shall return to this issue with regards to closed geodesics in General Relativity. It is because that Möbius strip is not the entire projective plane.

5. On turning inside out, the Klein bottle surface and the 2:1 resonance

We introduce the Klein bottle by producing the identifications depicted in the next figure:

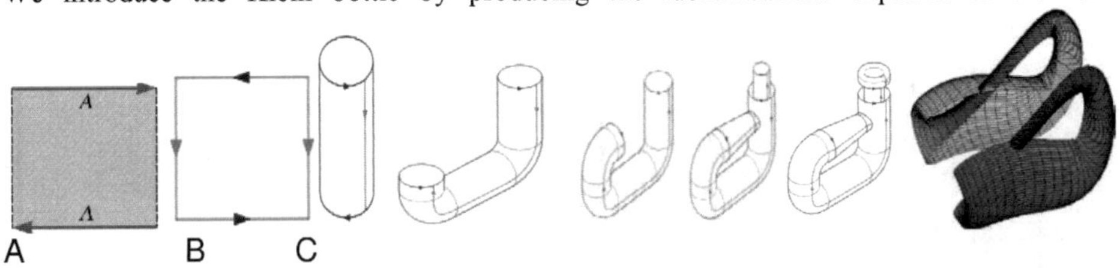

Figure 18: The identifications as they unfold to produce in A the Möbius strip and in B the Klein Bottle surface; courtesy of Inductiveload ©. C depicts the folding sequence to immerse the Klein Bottle in 3D. In the final picture –courtesy of Theon ©. we see two oppositely twisted MBs produced by cutting the Klein Bottle along the longitudinal section; conversely, zipping them we obtain the KB. Yet a single MB can be obtained from the KB

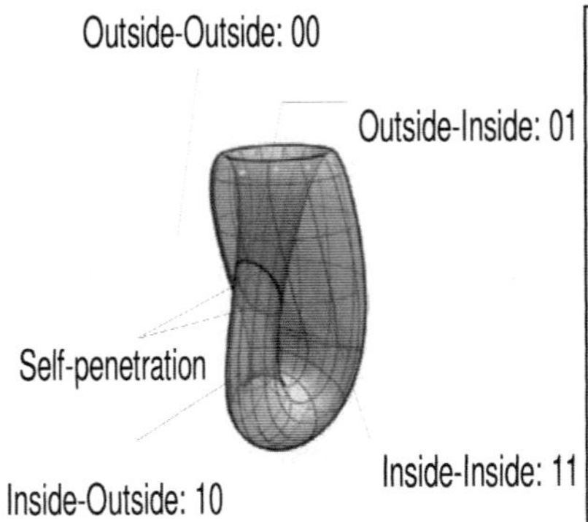

Figure 19: A natural logic of 4 states (the Klein Bottle Logic, KBL)which has two digits representation: Inside-Inside (represented as 11), Inside-Outside (represented by 10), Outside-Inside (represented by 01) and Outside-Outside (represented by 00). The mediation states arise from self-penetration, absent in the mechanical-dual-membrane and in the Cartesian conception of object-in-space-before-subject, are paradoxical states associated to time waves. From the KBL appears the genetic code with its 64 elementary codons, and furthermore the genome and its association to the mathematical structure of the Dirac algebra of Quantum Mechanics [12].

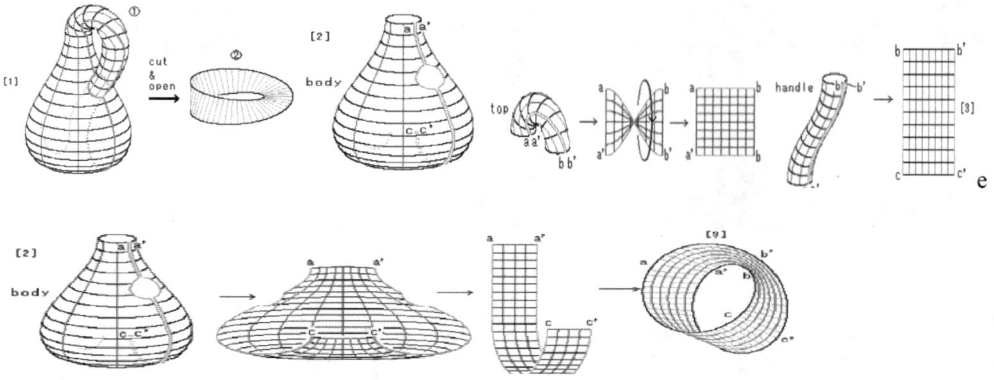

Figure 20: Surgery sequence to extract a single Möbius strip from the Klein Bottle.

Yet, by appropriate surgery on manifolds, we can find that a single Möbius strip lies on the surface of the Klein bottle. Indeed, consider [1]: We cut Klein Bottle ①along the green line. Then we get Möbius strip [2]: In order to see how the single Möbius strip comes out, we separate Klein Bottle with green cut line into "body", "top" and "handle". And we shape them into rectangles. Pay attention to that the cut line on the top is twisted by 180 °. So we untwist it on the way as shown with the arrow.[3]: Three rectangles are rejointed, and we get Möbius strip. The rectangle of the top was untwisted but what rejointed is automaticaly twisted again .[4]: It is the series of squares with arrows that represent the cutting above. The sequence of surgeries is represented by the figure below:

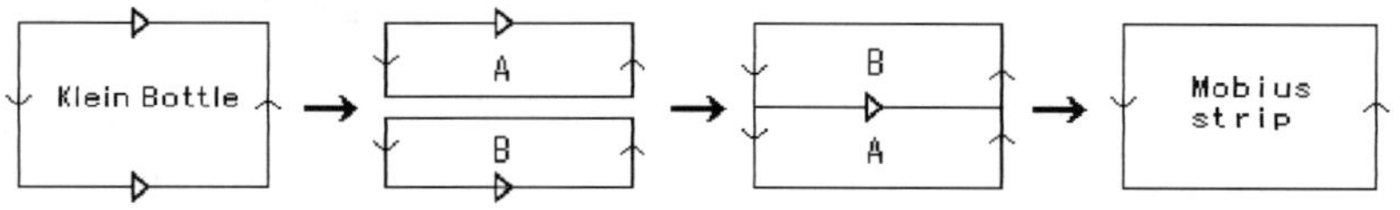

Figure 21: Schematic surgery of the Klein Bottle for extraction of a single Möbius strip.

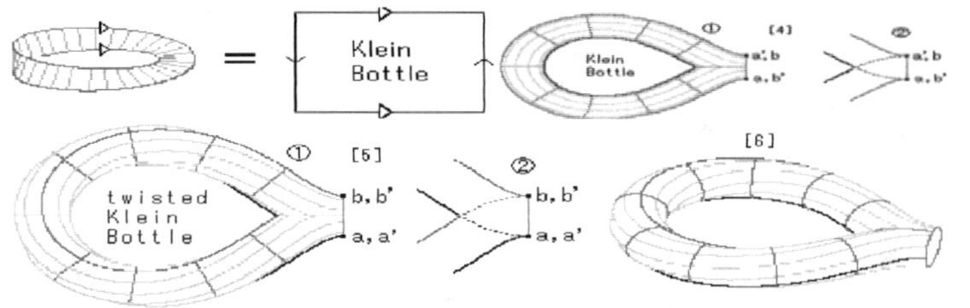

Figure 22. Constructing the Klein Bottle from the Möbius strip. [1]:The square with arrows ①represents Klein bottle. We begin with it. First we make a cylinder ②by gluing delta arrows. The red lines are just for illustrations.[2]: We try to make a torus-like, but the arrows never be in the same direction whatever we twist the tube. It is a well known matter. [3]: But bend the tube in U-shape, and the arrows are automatically arranged in order without twisting. [4]: Suppose the surface of U-tube is free to go through, and we joint both ends of U-tube while keeping their arrow directions unchanged as shown ①. ②is a transparent view around the joint.The shape is completely symmetric. [5]: The Klein bottle is made for an appearance. But it is not ready to correspond to Möbius strip of [1]. So we must twist the tube by

180 ° though arrows are already identified. ① is twisted one. Compare the transparent view ② with that of [4].[6]: It is a stereoscopic picture of [5]. Observe the black curve in [6]: which if continued another 360° turn it turns to be the red line of the top of the figure to return to an original point along any such green, red, or black curve. Here we find the topological genesis of the 720° rotation of the neutron submitted to electromagnetic field, or still, of the double covering of the Lorentz group which is the symmetry of the Dirac equation

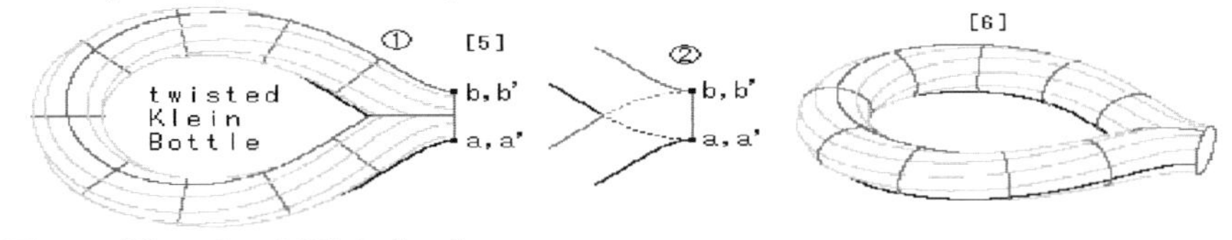

Figure 23: twisted Klein bottle

This also is the fundamental 2:1 resonance of celestial mechanics and of music, to be discussed below. Its importance to physics and astronomy, seems to have escaped the attention of Möbius , an outstanding astronomer. Indeed, consider the figure

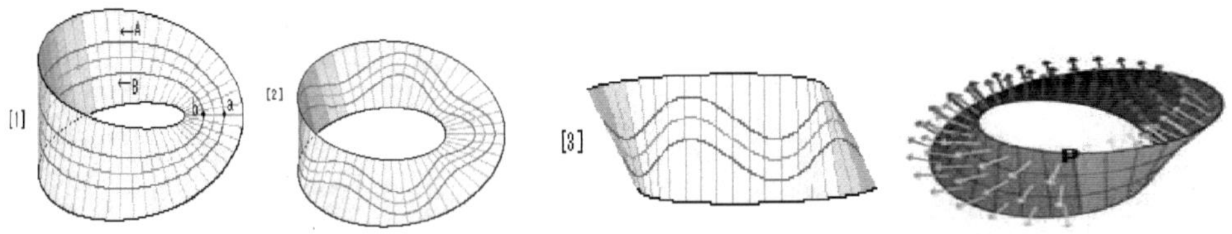

Figure 24: Left: the 2:1 harmonic of the Möbius strip. Right: the protoform of Newton's Third Law. Point **P'** lies ontop **P**, on the other side, modulo 360° turn.

[1]: The green line is a center line. Suppose the red line expends counterclockwise from point **a**. And it turns back and arrives at point **b** which is the opposite side of point, so to return to b again another complete 360° turn is required. The same is the case of the red line extends one more round, it gets back to the started point **a**. The red line never went across the center line. Nevertheless the red line runs both side of the center line. The red line looks like parallel lines when we see them partially. But it is not true. Route **A** and route **B** are actually a single route. The green line looks a median or a central reservation on a road but it does not divide Möbius strip into two. If we cut Möbius strip along the green line, its length becomes double. There is no opposite lane on Möbius strip. This is the topological origin of the 2:1 harmonic. It is also the origin of a protoform of Newton's Third Law, albeit it does not require, in distinction with Newton's formulation, a dualistic assumption, nor the instantaneous causality that this law implicitly assumes; causality, if any, is embodied by the 2:1 resonance produced by the 180° twist. Indeed, consider a normal vector to Möbius strip; if we move it along any curve in a 360° turn as before, we would find it pointing in the opposite direction, and equal in its length, without the need of an assumption as in the Third Law. Another 360° turn will return the vector to the same point and to coincide with its original configuration. Rather than having an action and a reaction, in the Möbius strip and in the Klein Bottle, the opposite and equal modulus of normal vectors is a resultant of the 2:1 harmonic, not an hypothesis for the foundations of physics at large. So it appears that this harmonic is more fundamental to physics than Newton's Third Law, which invokes an instantaneous symmetric causality. As we said before, this is the case of the origin of the 4π symmetry of relativistic quantum mechanics, and shows up in the precession of the neutron submitted to an electromagnetic field [60].

Interestingly, resonances –say in celestial mechanics and chaotic systems- are discussed in terms of the 2-torus [110], so we want to relate the torus to the Möbius strip and the Klein Bottle to the 2:1

resonance; the latter is already the case, say, of the Kirkwood gaps for Jupiter's periods in relation with the asteroid gaps. The relation is that the 2-torus is the double covering manifold of both surfaces. Yet we want to illustrate the meaning of this in terms of the 2:1 harmonic. This jumps to the eye on viewing either Figure 24 or 49 . It can also be seen in examining a 2-torus, which we assume to be solid, yet the 2:1 harmonic is already apparent in Figure 25 [1] below (courtesy of T. Ito):

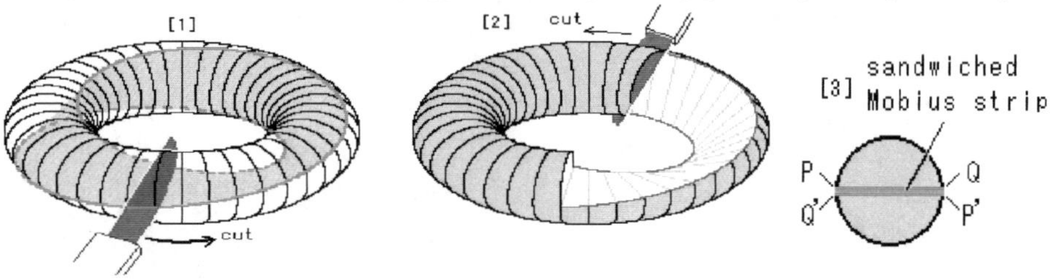

Figure 25: Cutting out the Möbius strip from the 2-torus by following the 2:1 harmonic

[1]: Möbius strip is lurking in the solid torus. We cut the solid torus along the green curve by the red knife as shown with an arrow. But the solid torus is not be separated into two. To get Möbius strip physically we have to scrape or melt the solid torus away. [2]: It is on the way of cutting. We are doing it while throwing unwanted parts off. We will be able to get Möbius strip if we continuously cut the torus by two circuits. [3]: We cut the solid torus with red knife. It is completed by two circuits. If we start to cut from the upper part **PQ**, the red knife will come to lower side **P'Q'** at a circuit, so that we have to continue cutting one more turn. That is why the length of new torus is double. During full cutting, the red knife screws 360 °. This 360 ° is what we observed on the surface of Möbius strip. The observation direction changes every moment while traveling. And the red knife itself turns around the hole of Möbius strip two times. It is similar to that of Earth circumsolar year rotation on an elliptic path, as it rotates precessing along its axis; this is crucial rotation around its axis, when submitted to an electromagnetic field.

Returning to the surgery of extracting Möbius strip from the 2-torus,

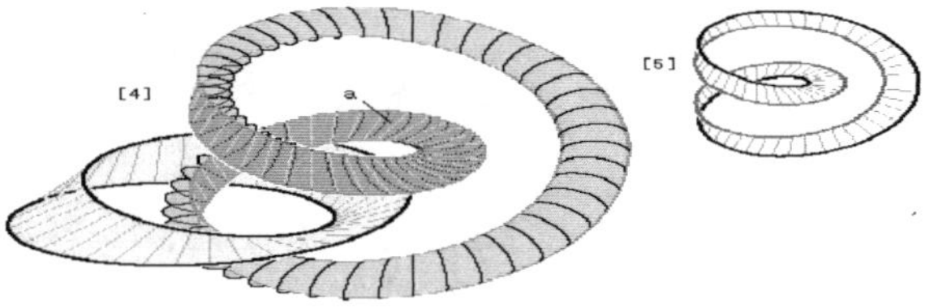

Figure 26: separation of the cutting of the Möbius strip from the 2-torus by following the 2:1 resonance.

[4]: Suppose the solid torus is rubber made. And we cut it without throwing anything off. We get the yellow Möbius strip and a long loop. The long loop is chained to Möbius strip and its length is double of Möbius strip. The dark surface **a** of the long loop was in contact with Möbius strip. We cannot get Möbius strip alone unless we cut off the long loop depicted shortened in [3]: Suppose the solid torus is rubber made. And we cut it without throwing anything off. We get the yellow Möbius strip and a long loop. The long loop is chained to Möbius strip and its length is double of that of the Möbius strip. The dark surface **a** of the long loop was in contact with Möbius strip. We can not get Möbius strip alone unless we cut off the long loop. [3]: Suppose the solid torus is rubber made. And we cut it without throwing anything off. We get the yellow Möbius strip and a long loop. The long loop is chained to Möbius strip and its length is double of Möbius strip. The dark surface **a** of the long loop was in

contact with Möbius strip. We can not get Möbius strip alone unless we cut off the long loop, depicted shortened in [5].

Therefore, to extract the Möbius strip from the 2-torus, we have to cut out the double length loop that is produced by the 4π rotation. In other words, the 2:1 harmonic is revealed as if embedded in the 2-torus by erasing its trace as an inherent rotation, that is neither an internal nor an external symmetry, the latter issue being another of the dualities that physics is used to claim to be fundamental. Of course, any astronomical body is extended, so the elliptic circumsolar motion is that of the centre, whose extended motion is indeed that of a solid 2-torus with the 2:1 built-in resonance.

6. Symmetries on the Möbius Strip and non-orientability in chemistry and biomembranes

We are lead to examine what are the inherent symmetries of the Möbius strip and the Klein Bottle, which we shall do next

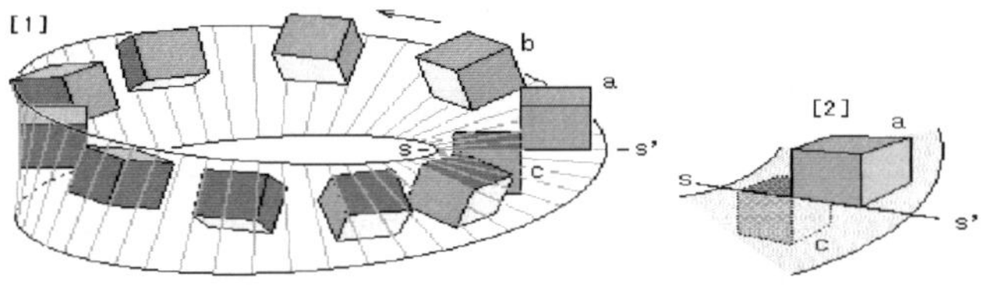

Figure 27: Representation of differentiated structures on a Möbius strip (courtesy of T. Ito).

Consider Figure 27: [1]: We slide a cube on the Möbius strip. It is coloured in light and shade of red, blue and yellow. If one side is light, its opposite side is shade. Cube **a** is at the start line **ss'**. We move it while keeping the shade red face is always on Möbius strip. Cube **b** on thw way is showing the three of light colors. [2]: Cube **c** that has traveled a circuit and now at the start line **ss'** again. Compared with cube **a**, it is upside down, right side left, and shifted to **s** under Möbius strip. In chemistry the issue appears, as we already said related to enantiomerism, the existence of chemical isomers, which have a symmetry which is non-specular but implies a change of chirality.

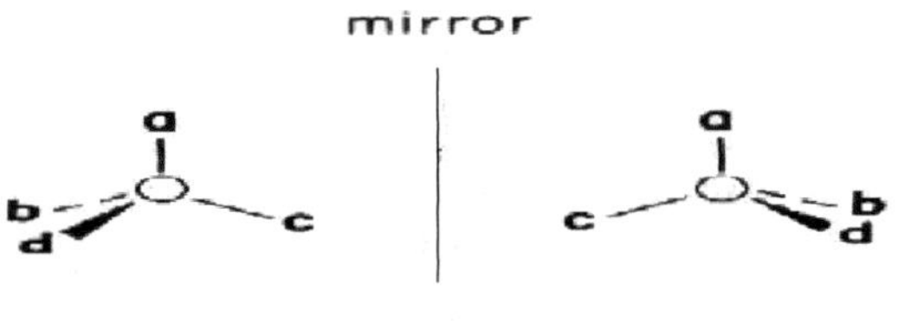

Figure 28: 2D representation of enantiomers in organic chemistry, after Emil Fischer.

The notation in Fig. 28 is the Emil Fischer' representation that chemists use for describing three-dimensional structures on 2D: a solid line in in the plane of the paper; a dashed line points in back of that plane; a wedge to the front. The molecule at left is not identical to the one at right; if we superpose **a** and **d**, **b** and **c** won't fit; thus, they are called chiral molecules, since the hand descriptor of a thumb, pinkie, palm and back, play the same role as **a**, **b**, **c** and **d**., the chemical groups that differentiate enantiomers. The rotation of a three dimensional structure on the Möbius strip describes precisely enantiomers. This is crucial to the biochemical activity of molecules, and in particular, of proteins [37]; we recall that Pasteur discovered the chirality of molecules, which lead him to propose the complementarity of form and function in biology, which is known today as the Pasteur-Curie Principle, in recognition of Pierre Curie's work on the subject. Yet, what this examples teach us is that the issue of chirality, in general, can be associated with the non-orientability of the Möbius strip, or still of the Klein Bottle.. There has been developed in the last years, a new paradigm of chemistry, and in particular of organic chemistry, in which non-orientable topologies are the very basis of the subject [38]. We here recall the cyclical structure of benzene.

Yet, the fundamental issue of turning inside out the charge configuration which is basic to the stereochemistry of molecules, is the basis for understanding this paradigm. In an early review to this topological paradigm for chemistry [38], it was claimed that ``...The examination of topological structures is not only intrinsically of great cognitive interest, in the sense of the opportunities for realising mathematical entities in a material, molecular form. It is no less significant that the actual properties of molecular systems may cause behaviour which is topologically indeterminate but nevertheless almost the same as for a topologically ideal system.′′ We notice that this is essentially the same thesis we have claimed for biology. and for systems in general [9,10,11,12,14,15]. One of the forms of topological isomerism deals with the same issue that cells do, the Outside and Inside which arises from the real non-pointlike character of atoms, namely, the set of spheres formed by the van der Waals radii of chemically non-bonded atoms in a molecule effectively create a closed surface. As a consequence, an atom or a molecule in the inner region is unable to pass into the outer region, since it is of finite size. Here we find at an atomic level the issue of penetration of biomembranes. The proposed topological hypothesis makes no demands on the size of the species or molecules to be transported (the migration of large molecules surpassing membranes are known. The main idea is that, at least in some cases, no real penetration takes place through the membrane (or the charge configuration in the case of a molecule), which is a one-sided non-orientable surface. The suggestion in this review is that if a particle Z is firmly attached to a site on the Outside of the membrane (or the charge configuration), induces a change in the conformation of the surface due to its bonding or some other cause, the site may move 'inside' in the sense of a Klein Bottle invagination, to wit: Z moves from the Outside-Outside through Outside-Inside and Inside-Outside to rest in Inside-Inside, as in Fig. 19. Thus the apparent result is the penetration of Z through the membrane, which is actually a consequence of the conformational change in the surface which becomes non-orientable—a special kind of the allosteric effect, which is conceptually predominant in biology. Biophotons emmission due to change of conformation by re-organization of chemical bonding is known in biology as chemiluminescence [42,48] ; it is to be noted that chemiluminscence is recognized to be the basic process for syntropic (in distinction of entropic) organization of organisms using stored energy of chemical machines, and that is the resonant energy communication of quantum entanglement in organisms as argued in [39,40,42]. Most remarkable is the fact that the Klein Bottle logic which includes quantum, fuzzy and Boolean logics as subcases, is essentially an entanglement topologic due to the non-orientability of the Klein Bottle, so that the quantum coherence of living organisms has in this logic a logophysical operator and in biochemiluminiscence a manifestation of quantum (topo)logical entanglement associated to stereochemical changes, which we have associated with the tensegrity structure of the cell's matrix associate to the tensegrity structure of the cell's matrix [10].Remarkably, a molecule that can switch both topologies, orientable and Möbius strip, without breaking any chemical bonds, has been deviced [41]. The Klein bottle topology has been revealed in Inorganic Chemistry [43], while the analysis of 'simple' molecular conformational data reveals that cyclo-octane rings have a Klein Bottle topology. Furthermore, the issue of non-orientable stereochemistries is the core of aromatics [44]. Sokolov's prophecy appears to be in the path of realization; also, the genetic code is based on the Klein Bottle Logic [12], the codons being the 3-fold tensor product of the four states of figure 19 resulting in the codons and anticodons lying one-to-one in either local side of a Klein Bottle which is further recurrent to a HyperKlein Bottle [9,12]. Therefore, we find already in stereochemistry, the logophysics we are claiming for biology. Thus, we suggest that it may be possible that bindings of molecules at the membrane can produce conformational changes without chemical bonds breakings, that transform its topology from two-sided orientable to one-sided non-orientable, as well as that of the molecule, with the production of bioluminiscence. Yet, the interior of the membrane as a vacuous interior to a non-orientable bilayer lipid would be identical to a massless quantum surface interface that appears in the Klein Bottle structure of the Periodic Table of elements [13], and thus associated itself to a boson field which we associated to the torsion photon field [15,19]; we shall retrieve this issue further below in examining the wavefronts of Earth's crustal tidal waves.

Alternatively, we could think of the bilayered cell membrane undergoing a stereochemical transformation induced by adhesion to its boundaries of integrine molecules [46]. Thus, would a stereochemical transformation through binding at the surface of the membrane occur, it would be accompanied with photon emmision of both the membrane and the adhered molecule, which would produce an electromagnetic signal that would lead to non-local effects both Exterior and Interiorwise, which under the topological change from orientability to non-orientability, would lose the dual distinction. The extracellular matrix as a tensegrity (i.e. in which tension is continuously distributed to the whole –think of the spider's web) structure. This matrix is connected with the cytoskeleton - through molecules known as integrines in adhesion cites of the membrane [46] which we proposed to modify the membrane's topology ; this holonomic system: connective tissue, cytoskeleton, nuclear matrix, has been termed the living matrix, and it is known that by the piezoelectric effect, electromagnetic fields are produced leading to instant signalling across the organism [47]. Hence, it is natural to conceive, unless we introduce the Exterior & Interior duality, that the tensegrity structure for the living matrix has the Klein bottle logic. This matrix does not only carry energy, but also information, and quantum phenomenae are at its roots. This radiation field might be associated with the mitogenetic field discovered by A.Gurwitsch [48], which lead to biophotonics as well as to the studies of the torsion phantom effect discovered by Gariaev [49]. In our conception, the photon is a universal objective-surjective gestalt, which cells have proved to be able to self-organize in terms of their 'perception'[50]; we can venture –following [50]- that the cell's tensegrity living matrix is the gestalt in question and that the signaling to which it responds are electromagnetic waves carriers of biophotons. We have identified in the bioresonance through biophotons, the source of selfhood [15] , which is identified as the quantum coherence of living organisms in syntropic self-organization [39,40]. The tensegrity structure of cells under differentiation was associated to torsion electromagnetic waves responsible for the contraction and expansion waves that determine the tissue to result in the differentiation process [10]. In fact, due to the identity of wave-fronts of elasticity and electromagnetic felds, this is a tensegrity which is both ruled by classical and quantum physics, which we shall further encounter, in geophysics, further below.

7. The Klein bottle, two-dimensional signals and the embedding in four-dimensional space

The Klein bottle appears as a form of forms for matter, as is the hidden symmetry-topology of the Mendeleev table of periodic elements when the stable nuclides are included [13]. The Klein Bottle as a 2D surface requires self-penetration for its construction (for the relation of this with homotopy see [31]); thus the classical conception of space as a container for objects,which is the imperating paradigm in physics and in the sciences at large, does not apply to this surface. Instead, the classical (in the sense of space as a container for forms) response to this impossibility of constructing the Klein bottle if not by self-penetration, is claimed to be surmounted by embedding it in four dimensions, where the self-penetration can be avoided, as an abstract construction, albeit one that can not in practice be carried out!

In pattern recognition through artificial means, analyzing digital photographs, the Klein Bottle appears as a form of forms, and its identification does not require an actual 4D construction, but in fact, in the artificial case, the construction of a seven-dimensional space, actually a sphere, from which the barcodes of the images are revealed to hold the Klein Bottle as a universal pattern [27,28]. The topological structure of simple signals in 2D with arbitrary phase and orientation is described by a Klein Bottle [30,31,51]; here the notion of a signal is the one in use in the theory of communication and image processing, any function (to be encoded/decoded, transmitted, filtered, enhanced, decomposed in Fourier harmonics or wavelets, etc.); yet for what we have just stated, most remarkably they appear as signals in the sense of wavefront discontinuities already discussed which we shall later elaborate.

Indeed, the issue is that *fringe* signals have a definite orientation (so here is where the boundary conditions come in) as opposed to a 1D signal, thus introducing a 2D phase, and thus the codification of such a signal turns out to be mathematically feasible in higher than 2 dimensions. Already given an image of dimension D (D = 2 for ordinary images), a mathematical representation of a local *intrinsically 1D* region (i.e. the number of dimensions necessary to describe the signal is one), say $f(\boldsymbol{x}) = g(<\boldsymbol{x},\boldsymbol{n}>)$, where f is the image intensity function which varies over a local image coordinate **x** in $\mathbb{R}^{\boldsymbol{D}}$, g is a one-variable function, $\boldsymbol{n}$ is a unit vector and $<\boldsymbol{x},\boldsymbol{n}>$ is the inner product of $\boldsymbol{x}$ and $\boldsymbol{n}$. The intensity function f is constant in all directions which are perpendicular to $\boldsymbol{n}$. Intuitively, the orientation of an D- dimensional region is therefore represented by the vector $\boldsymbol{n}$. However, for a given f, $\boldsymbol{n}$ is not uniquely determined. If $\boldsymbol{n}' = -\boldsymbol{n}$, $g'(x) = g(-x)$, then f can be written as $f(\boldsymbol{x}) = \mathbf{g'}(<x, \boldsymbol{n}'>)$, which implies that for antipodals, $\boldsymbol{n}' = -\boldsymbol{n}$, it is a valid representation for the local orientation.

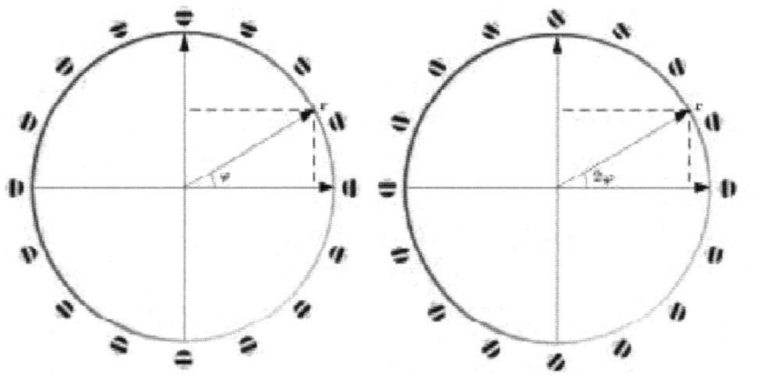

Figure 29. On the left, a non-unique representation for orientation of images on 2D for antipodals, except for a phase shift On the right, the double angle representation of orientation (reproduced modified from [51]). This is similar to the case of music perception of the half-octave (tritone); see Fig. 57. This duplication of angle is nothing else than using the 2:1 harmonic inherent to the Möbius strip and the Klein bottle, in this case to be able to assign a unique local orientation to a 2D image.

The conditions for representation of higher dimensional signals is as follows. Three requirements for a phase representationin higher dimension must be fulfilled. Firstly, the signal neighborhood must be intrinsically 1-dimensional. Secondly, the phase representation must be continuous. As a consequence of the two first requirements, the direction of the phase must be included in the representation.

Consider a signal $f(\mathbf{x}) = a(\mathbf{x}) \cos(\theta(\mathbf{x}))$ depending on a 2D phase $\theta(\mathbf{x})$. Locally, the phase can be described by the linear model as a function of the offset Δx from x,

$$\theta(\Delta\mathbf{x}) \approx \theta + <\nabla\theta, \Delta\mathbf{x}> = \theta + \rho<u, \Delta\mathbf{x}>, \quad (1)$$

where the local orientation has a direction, $\nabla\theta = \rho\mathbf{u}$. Let us here and henceforth assume that f is a sinusoid in 2D, i.e. the amplitude a(**x**) is assumed to be one, and $\mathbf{u} = (\boldsymbol{\cos\varphi}, \boldsymbol{\sin\varphi})$. The analytic signal for such a sinuosoid takes the form (via the Riesz transform of f),

$$\boldsymbol{r} = \boldsymbol{\sin\theta}\,(\cos\varphi, -\sin\varphi), \quad (2)$$

where we have fixed the amplitude a to be equal to one [51]. This vector is a sufficient representation of orientation for a fixed $\theta = 0$, but if θ varies problems occur. The amplitude of the signal, here equal is mixed up with the phase and for $\theta = 0$ the information about orientation is lost. Furthermore, the representation is not unique since identical signals can appear as antipodal points as seen in fig. 29. An elegant remedy to the latter problem is, in 2D, to compute the non-linear combination of the sinusoids, such that

$$r(\phi) = \boldsymbol{sin^2}\,\theta\ (\boldsymbol{cos^2\varphi} - \boldsymbol{sin^2}\varphi,\ 2\sin\varphi\cos\varphi), \tag{3}$$

where $r(\varphi)$ rotates with twice the speed; this is known as the double angle representation. This is, of course, equivalent to the 2: 1 resonance.

A representation of the Klein Bottle must satisfy the following identities

$$r(\varphi, \theta) = r(\varphi, \theta + 2\pi),$$
$$r(\varphi, \theta) = r(\varphi + \pi, 2\pi - \theta). \tag{4}$$

A Klein Bottle cannot be embedded without self-intersections in $\mathbb{R}^3$ so that the representation of fringe pattern requires to transport the self-penetration to a higher dimension in which a definite signal representation for orientation can be achieved: one adds dimensions to represent *as if* different, the superposition of singularities of the phase orientation as occurs in the cortical visual process (see figure no. which explains why the information of orientation vanishes in the phase representation and that simultaneous representation of orientation and phase could not be accomplished with only 3 elements. One possible representation of the Klein bottle in $\mathbb{R}^4$ is

$$\mathbf{r} = (\sin\theta\cos\varphi, \sin\theta\sin\varphi, \cos\theta\,(\cos^2\varphi - \sin^2\varphi), 2\cos\theta\cos\varphi\sin\varphi). \tag{5}$$

Consider the intrinsically 1D signal $f(\mathbf{x}) = \cos(\rho <\mathbf{u},\mathbf{x}>)$ where $\mathbf{u}$, as previously, is defined by φ, and the vectors

$$\mathbf{r}^1 = (\sin\theta\cos\varphi, \sin\theta\sin\varphi),\ \mathbf{r}^2 = (\cos\theta\, cos^2\varphi, \cos\theta\, sin^2\varphi, 2\cos\theta\cos\varphi\sin\varphi). \tag{6}$$

These vectors are the first and second Riesz transform of the 1D signal f [51]. Consider next the composed vector $\mathbf{r} = (\mathbf{r}^1, \mathbf{r}^2)$; it satisfies eq. (4). Thus, $\mathbf{r}$ gives a simultaneous representation in $\mathbb{R}^5$ for both amplitude and orientation; the effect of going to a higher-dimensional representation is to do away the singularities that may appear in low dimensional, concretely in 2D as the paradigmatic case; we have already discussed this issue in connecting hyperbolic geometry to the re-entrance of the Möbius strip and we shall later discuss it in relation of the divergences that arise with geophysics. In fact the Klein Bottle representation in $\mathbb{R}^4$ is the projection onto the spherical harmonics of order 1 and 2 of the 1D signal $f(\boldsymbol{x}) = \cos(\rho <\boldsymbol{u},\boldsymbol{x}>)$ [51].

This is quite remarkable: The form of forms is nothing else than the projection to order 1 and 2 in the space of spherical harmonics of the sinusoidal 1D signal with vector orientation that arises from the fringes of intrinsically 1D images. Of course, this is not exclusive to images, but to any such sinusoidal signal, and specifically, that arises from the fact that the orientation can not be uniquely defined on the boundary of the domains under examination, so it is about multivaluedness and superposition, which is precisely the essence of the Klein Bottle Logic and its Matrix Logic representation, and one of its physical manifestation is the lack of well defined tangential directions of wavefronts of hyperbolic partial differential equations, to which we turn further below.

8. Hyperbolic spacetime, time loops and nonorientability

The relation of non-orientability to the hyperbolic geometry of General Relativity can be approached as follows. It is widely believed that spacetime must be both orientable and time-orientable. Supporting arguments are that there is no evidence of a lack of orientability and that a non-orientable spacetime would be incompatible with the observed violations of P (parity) and T (time reversal) invariance. Yet, these stringent conditions can be relaxed by noting that spacetime (modeled as a Lorentzian manifold, and thus hyperbolic metrics are the case) are locally orientable (as is the case of the Möbius strip,the Klein Bottle – away from the self-penetration – and the projective plane) while on a non-orientable spacetime, there exists a closed path along which a consistent orientation cannot be defined; this follows from the definition of orientability and topological argument. Furthermore, if spacetime is not orientable, then through every point there exists a path on which an orientation cannot be defined.

These issues are the core of what has come to be considered a revolution in cosmology, i.e. the possibility that rather than an uni (single)-verse, the case is that of a multiverse. The connection in the multiverse is produced by ringholes or wormholes, namely holes in spacetime connecting the inter-universe tunneling . This hypothesis requires the existence of other universes should actually require choosing spacetime holes which be (i) non orientable and (ii) convertible into a time machine with completely unspecified relative speed between its mouths. One such space-time construct has in fact been already studied, in the shape of a Klein Bottle hole [111]. In this case, non orientability is guaranteed by the existence of a throat with the topology of a Klein Bottle and it was also shown that this space-time can be stable to vacuum fluctuations and is also convertible into a time machine with fully arbitrary inter mouths speed. Lensing effects appear associated to the non-orientable nature of the connecting Klein Bottle throat. The metrics that support these connections are solutions of Einstein's equations.

We recall that Gödel introduced a solution of the Einstein equations with cosmological constant under the following conditions, as stated by Boeyens [58]: 1) Spacetime is homogeneous and has rotational symmetry. 2) (Neighbouring) world lines of matter are equidistant. 3) A positive direction of time can be introduced in the whole system. 4) The direction of the time flow is not uniquely defined for each spacetime point. 5) Every worldline of matter is infinitely long and never returns to the same point; but there also exist closed time-like lines. 6) There are no space-like three-spaces. 7) There is no absolute time. 8) Matter everywhere rotates relative to the compass of inertia. Gödel's solution has been widely ignored because it fails to predict a Doppler redshift, does not give a clear definition of the compass of inertia, proposed as rotation axis, and allows closed time loops, which as already observed, is the case of non-orientable manifolds.It was shown by Boeyens that these objections are raised if we drop the condition 5), by adding a point at infinity, which turns, as already discussed, spacetime to be non-orientable and identifiable as the projective space. In Gödel's theory, in terms of this simple alternative geometry, geodesic transplantation fixes points on the line element to occur along the double cover of a narrow Möbius strip, as we showed in figures 6 and 7 above. In those figures we showed that for the Möbius strip, a point, which moves along the double cover, close to one edge, rotates around the central line without intersecting, as depicted in fig. 24. This is what Gödel describes as rotation with respect to a compass of inertia. Any geodesic in projective Gödel space therefore rotates around an inertial compass in the same way. The constant separation between symmetry related points determines the number of turns before returning to the starting point. To define a non-orientable plane this number must be odd. All essential properties are recovered, without loss of generality, by assuming a single Möbius strip. On adding points at infinity to all coordinates of Minkowski space, *spacetime* transforms into the projective space, the conformal compactification of Minkowski space-time. More important than the precise topology of the modified space-time is that conversion to projective space, with five homogeneous coordinates, leaves the gravitational potentials unchanged, while the extra degrees of freedom represent the electromagnetic field [59]. The Gödel solution therefore remains valid. In the geocentric mode, the Universe appears to rotate, as observed in radio astronomy, which would imply that Mach's Principle is invalid, as well as the isotropy of the cosmos appears not to be the case. As observed by Boeyens, the time paradox is also resolved in projective space. The closed time loops now connect any point, such as *P*, through involution, to its antipode P' in conjugate space-time. *P* and *P'* are related by CPT symmetry and separated in time. Yet, this separation in time is related to another most crucial law, in fact, a topological protoform of Newton's Third Law. Indeed, consider a normal vector placed at the point P on the Möbius band, and we rotate it 360° to reach for P', its conjugate lying in the other side, localwise; now the vector points in the opposite direction and certainly is equal to the original vector, but now points out from P', which is the local other side point of P; see figure 24, right .Yet, this identity does not require, in distinction to the Third Law, no such assumption on the identity of action and reaction, which are not connected by duality no more; this is also the case for the periodic table of elements which on inclusion of the stable nuclides appears to have the Klein Bottle topology: In one side we have what the dualistic approach would consider an atom of matter, connected under a 360° path on the Klein Bottle to an atom of antimatter. We remark that the Klein bottle, can be thought as a section of the projective plane of disk type, as we shall discuss further below, while we recall that we already showed that the Möbius strip is such a section (see figure 1).

9. The non-orientable topology of the complex plane and its extension to the Riemann sphere

With regards to hyperbolicity in Relativity, consider a pseudocircle on Minkowski space $x^2 - (ct)^2 = \pm r^2$*; it consists of two hyperbolas drawn in figure 15 (right).* By following the asymptotes of the hyperbolas as an involuted closed curve *e.g.* $(A \rightarrow B \rightarrow A)$ and $C \rightarrow D \rightarrow C$, two Möbius strips occur as sections through the projective plane. This path is analogous to the one that lies outside of the graph of the tangent function depending we approach, say, $\pi/2$ from the left or the right, which requires to identify ∞ and $-\infty$. In terms of the complex plane, this transformation occurs as the opening of the Möbius strip into its double covering with the following identifications

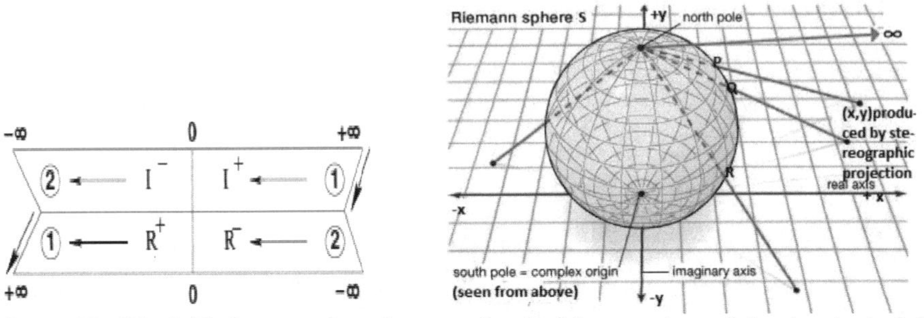

Figure 30: We fold the complex plane as the double covering of the topological identifications depicted above, proposed in [58]: . The folding of the Möbius strip from the extended (by the inclusion of ∞) complex plane as the double covering of the former; this non-orientable topology of the complex can be represented in the Riemann sphere in Figure 31, and is also the case of the visual topographic map on the brain's cortex.

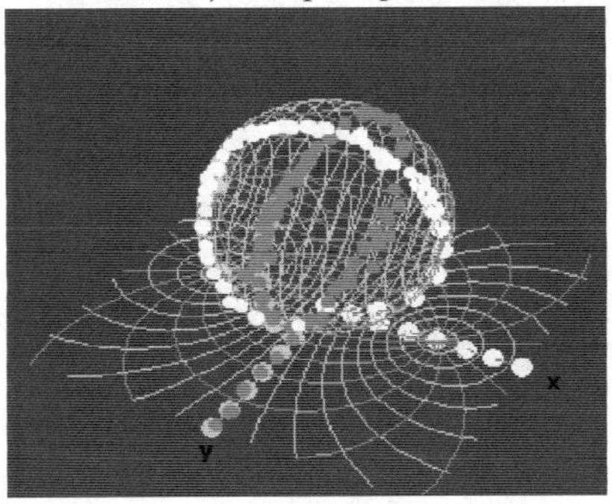

Figure 31: We represent the non-orientable topology of the complex plane presented in fig. 30, as the 2:1 fundamental octave on the Riemann sphere, lifting the discontinuity of hyperbolic singularities.

Journal of Physics: Conference Series **437** (2013) 012024 doi:10.1088/1742-6596/437/1/012024

In figure 30, we have depicted the unfolded double cover of a Möbius strip to model the relationship between real and imaginary numbers. Continuation of ℝ+ beyond +∞ at ① proceeds through ① at the positive imaginary semi-axis I = +∞, with a twist as indicated by the arrows. The conversion from ℝ to I happens gradually in terra incognito near infinity. This characterization of the complex plane in terms of the Möbius strip invokes a discontinuity and a transformation of the real numbers on the imaginary numbers at infinity, which is filled by the action of the Time operator of the Klein Bottle Logic, which is the analytical continuation of the reals by a 90º rotation of the real axis to the imaginary one, as is the case of the analytical continuation on the time variable. Yet, in the Riemann sphere, S, given by the one-point compactification of the complex plane, the continuity is intrinsic to the development of the real axis on the imaginary axis. It is produced by moving (along a band on the meridian drawn as pointed white dots) from the South Pole/origin along the East meridian which corresponds to the positive real numbers until reaching x= 1, y= 0, and up to the North Pole/infinity; on reaching the North Pole we give a 180º turn (for which we have two choices, an East-ward or a West-ward pointing twist, i.e. a choice of chirality) to this strip, to continue with a different orientation (drawn as green dots) to the imaginary axis along the corresponding meridian, which now continues to the point x= 0, y= 1, corresponding to $\sqrt{-1}$ to further return to the South Pole/origin, thus completing the ① percourse. This is the first half-octave of the 2:1 harmonic of the Möbius strip and the Klein Bottle; the motion further rises following the meridian corresponding to the negative imaginary axis, up to the North-Pole/infinity, thus starting the percourse ② ; on reaching the North Pole, we give a second 180º twist to the band, which returns to its original orientation/coloured surface, to follow now the West meridian corresponding to the negative imaginaries to complete ② . This yields the second-half octave and the completion of the 2:1 resonance. Since an even number of twists on the band have been produced, this indeed corresponds to the double covering of the Möbius strip, since the latter requires an odd number of turns. It is important to remark, that the 90º rotation on the complex plane on S that transforms +1 to $\sqrt{-1}$, corresponds to a 180º rotation from the South Pole to the North Pole followed by a motion (another 90º) to the point corresponding to $\sqrt{-1}$, and thus the 360º rotation on the complex plane corresponds to the 720º rotation and the motion South Pole-real axis-East-real axis-North-Pole-imaginary axis-South Pole-imaginary axis-North Pole-real axis-West-real axis-South Pole. We clearly see in this geometrical representation the 4π rotation of the double covering group of the Lorentz group, yet furthermore associated to the 2:1 resonance intrinsic to the Möbius strip and the Klein Bottle, as the transformation of the non-orientable topology of the complex plane, indicated by the figure , to the two sphere, whose local orientation is inverted twice. It is rather remarkable, that these characterizations have avoided recognition before. The change of orientation at ∞ along the meridians that allow to establish the continuity of the transformation between the real and the imaginary numbers is similar to that of the tidal forces on Earth's crust, in which the side pointing, say, to the Moon, with the most proximate moving point playing the role of ∞, has an orientation opposite to its antipodals. Again, we have a 2:1 harmonic associated with this geophysical displacement of the crust, associated with the change of orientation of the gravitational attraction on the side of Earth exposed to the Moon, turning to repulsion in the antipodal.

10. The non-orientable topology and geometry of geophysics: pattern formation

Earth rotates precessing about its North and South poles axis, with a variable angle, presently about 23.5^{a}; the origin of this precessional motion can be found in the precessional motion of a spinning test particle (say, Earth) submitted to an exterior gravitational field with torsion [61]. This is the origin for the precession of the equinoxes and its signature on Earth is the Analemma; see fig. 12. In his *Philosophie Naturalis Principia Mathematica* Newton modelled Earth's dynamics in terms of its deformation as an homogeneous fluid body following the hydrostatic principle, and submitted to the Sun's and the Moon's gravitational attraction. In the 18th century, Clairaut raised the homogeneity condition to one of a liquid with fluctuating density along its radius. Later, other researchers presumed the body's liquid to be an elastic and viscous substance. But the conditions of its hydrostatic equilibrium and inertial rotation of the planet have lingered up to recently reappear now as fundamentals of its dynamics as well as of the other celestial bodies.

It is currently understood that Newton's model is faulty. The study of oscillations of Earth's dynamics following earthquakes in 1960, lead to conclude that from the spectral decomposition of thousands of modes, only two modes appeared as the main ones: a spherical mode with direction of the radius, and a torsion mode normal to the radius (page 45, [76]). Of course, these two modes are well known in elasticity theory, yet, there is a most remarkable aspect that has been completely unconsidered in the geophysics literature, to the best knowledge of this author. The wave-fronts (or characteristics, in the theory of hyperbolic partial differential equations) of the equations of elasticity theory, in the linear regime, coincide with the wave-fronts of the Maxwell equations of electromagnetism [89], which themselves coincide with the wave-fronts of the Einstein's equations of gravitation, not only for a purely metrical connection [90], but one which also includes torsion, since in principle, they contribute in first-order of derivatives to the connection, while the characteristics involve second-order derivatives; in this setting, the gravitational field are given by harmonic equation. As we discussed in the Introduction, they are the fundamental signals/dislocations for the constitution of the physical world, yet they are of a self-referential nature.

These rotational elastic geophysical modes empirically verified are most visible on several places, and also appear as cases of the so-called unconformities of geology, in which newer stratae of depositions have surfaced ontop of older stratae, showing a circular time operator. These modes are very pervasive on Earth, visible through the naked eye (with the help of Google Earth) accounting for a design, and in several instances appear superposed to vortical motions on the atmosphere [75]. They appear to be related to harmonics of the fundamental 2:1 resonance, the very essence of the dynamics of the Möbius strip and the Klein Bottle. Their discoverer, Harold Overton, a retired engineering academic, calls it the Binary Theorem and relates them to the Croll-Milankovic cycles of the many-body problem posed by the Sun, Moon, Earth, and the other planets, the so-called *orbital forcing,* of crucial importance to the geophysical cycles [52,77,78,81].

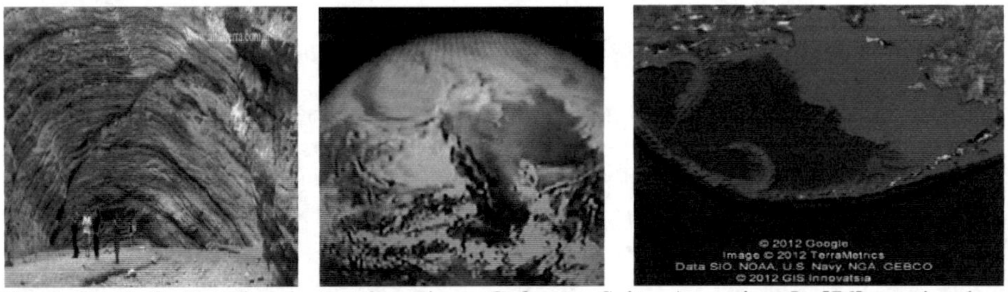

Figure 32. Left: Rotational crustal motions, Cafayate, Salta, Argentina. In [76] rotational modes are claimed to result from radial density inhomogeneities. Notice the dislocation from the ground climbing towards the right, and thus the lack of a well defined tangential direction of the tidal wavefront. Centre: Vortex over the Aleutians. Right: Partial view of the Aleutians Islands, a vortical crustal configuration, one of many in all scales (courtesy of Google Earth), indicating thus the existence of a time transformation of astronomical cycles forming Earth's crustal shapes, related to the 2:1 harmonic, and daily-scale climatological cycles, in which the ratio of the viscosity of the crust and that of the atmosphere play a crucial role.

A proposal is that the Earth is a self-gravitating body. Its matter moves in its own force field which is generated by mass particle interaction. The mass density distribution, rotation and oscillation of the bodies' shells result from the inner force field. And the orbital motion of the planet is controlled by interaction of the outer force fields of the planet and the Sun in accordance with Newton's theory. We remark two aspects of this proposal: firstly, it is about pointlike interactions of material constituents –

although it is recognized that the concept of point particles is untenable for it produces divergences in the mathematical computations so it is surmounted by considering charge densities (see page 52, [76].) and volumetric forces. Secondly, these volumetric forces cannot be reduced to applications to points but produce interior and exterior forces, so it is a dualistic approach, to address the formidable problem of the generation of Earth's shape, meaning by this the non-spherical shape, not the morphogenesis of crustal's shapes, which is left unconsidered. In this theory, the gravitational ''anomalies'' measured by satellites follow from the torsion oscillatory modes produced by the radial mass density distribution. as in fig. 32; see section 6.2 [76]. Other studies have found that the deviation of the hydrostatic gravitational potential due to the inhomogeneous density is an harmonic function (see chap. X, [53]); we recall that Fock showed that the equations for gravitation were given by harmonic fields [111]. Following these studies of modelisation of the interior of Earth, further extended to the crust modelled as a viscous fluid with displacement vector d satisfying the Navier-Stokes equations (which is fundamental example of torsion geometry with torsion vector given by the velocity of the fluid [62]), we find that the displacement vector field is bihamarmonic and its gradient and divergence are harmonic vector fields. This shows shows a deep relation between elasticity theory and potential theory (page 499, [53]) further extend to Maxwell equations and the Navier-Stokes equations and turbulence [79,80]; it further shows that the torsion geometry of elasticity, potential fluid-dynamics, and the eikonal equations is of relevance to geophysics. For a thorough discussion of harmonic vector fields and the gravitational field we refer to chap. X [53].

In this model, we can think of the crust's morphology, as a frozen (in the sense of kinetic energy having transformed to heat) to reach a temporary stationary distribution of this fluid displacement, which is associated to harmonic vector fields. Yet, the latter are keenly associated to minimal surfaces in classical differential geometry, which thus would be an embodiment for the morphology of Earth's crust.

Kiehn has developed a topological theory of physics which unfolds with a common foundational setting than the present one in which the electromagnetic signals with wavefront discontinuities play a fundamental role. The evolution of physical systems is related to these singularities [95], one of which is the photon [19,136]; furthermore the relation between minimal surfaces, electromagnetic signals, the so-called topological quantization program of physics (and its relations with quantum mechanics as a theory of ensembles [138]), spinors –even macroscopic-, irreversible thermodynamical processes and turbulence, and a chiral vacuum indistinguishable from the Lorentz vacuum except from the impedance of free space were discussed [136,137]. (It will turn out that one particular application of this setting is crustal morphogenesis, to be dealt with below). Kiehn discussed in this setting the relation between a solution of Maxwell's equations of electromagnetism for a complex electromagnetic field, **M**, due to Bateman [93] and minimal surfaces in 4D [108], i.e. mappings on 4D Euclidean space whose position vector, **X**, is harmonic. These vector fields can be represented as $\mathbb{R}^4$-valued maps F defined on $\mathbb{R}^4$, given by the expression:

$$F(x,y,z,t) = (a(x,y,z,t), b(x,y,z,t), v(a(x,y,z,t) + ib(x,y,z,t))) , \qquad (7)$$

where v= ϕ(a+ ib) + i ψ(a+ ib) is an arbitrary complex analytic function of a+ ib. Bateman proposed a complex 3-dimensional vector,$\mathbf{M} = \mathbf{B} \pm i\mathbf{E}$, with complex electric and magnetic field **E**, **B** could be used to express both the Maxwell Faraday and the Maxwell Ampere equations for the Lorentz vacuum as one combined set of complex vector equations. Consider the map be defined by

$$(a,b)\rightarrow(X^1(F(a+ib)), X^2(F(a+ib)), X^3(F(a+ib)),X^4(F(a+ib)), \qquad (8)$$

with F analytic, and X^k (k=1,2,3,4) harmonic. Define $\boldsymbol{M^k} = \partial X^k /\partial a - i\, \partial X^k /\partial b$. If **M** is isotropic, i.e. $\mathbf{M}^2 = 0$, and their complex square is not zero on the domain of definition of **M**, then **M** is defined in terms of a minimal surface which is regular (without self intersections or pinch points). Being **M** complex isotropic, it is a spinor [94,103]. If **E** and **B** are real, this is the case except at points where **E** and **B** are identically zero. If both **E** and **B** are complex, then the associated minimal surfaces will have 3D images that are not always regular. In general, two dimensional non-regular surfaces may have singularities consisting of curves of double points created by intersections of two local surface patches, or of triple points consisting of intersections of three local surface patches, or of curves of double points which terminate on "Pinch" points within the interior of the surface. These three types of self intersection singularities are the only three stable singularities in the sense of Whitney. Recall that Whitney proved that any N manifold can be embedded in 2N+1 euclidean space, and immersed in a 2N euclidean space. The induced surfaces may be orientable or non-orientable.

If the surface has no singularities, then the surface is said to be regular or embedded. The constraint of regularity implies that the surface normal vector never goes to zero over the surface, or the induced metric on the surface is always invertible. This implies that are always two linearly independent directions on a regular domain of the surface. If the lines of self intersection are divergence free on the domain (meaning that they stop or start only on boundary points, or are closed upon themselves, then the surface is said to be immersed in 3-Dimensions; the latter is the case of the Klein Bottle where the tangential discontinuity to the surface arises from the self-penetration and can still be embedded in 4D losing its singularities. The points where the divergence of the lines of intersection is not zero are defined as Pinch points. Such surfaces cannot be immersed in 3-D. The Pinch points are signatures of the fact the surface resides in 4-Dimensions (as an immersion), and cannot be immersed in 3-Dimensions. In this loss of regularity, which is very much the issue of the theory of hyperbolic partial differential equations and the constitution of the material world shapes, we find non-orientability as its signature.

A flow vector field may have domains where it is irrotational or solenoidal (we already saw that the displacement vector in the fluid crustal motion Navier-Stokes model is both), and these domains may be separated by a surface. If the surface of separation is a minimal surface, *then the flow on this surface is harmonic*. The minimal surface need not be regular, and may have lines of self-intersection. These lines of surface self-intersections (lines of singular double points) are not necessarily solenoidal. In fact, the Pinch points are points where the lines of self-intersection terminate not on themselves and not on a boundary, but in the surface interior. The Pinch points may be viewed as the "sources" of the divergence of the lines of self-intersection.

In terms of viscous fluids satisfying the Navier Stokes equations, if a velocity field V (alternatively the displacement vector in the geophysical model already discussed) started out with both harmonic and non-harmonic components, then after the passage of time the flow should reduce (by viscous dissipation) to the harmonic components alone. Such is the theory of wakes. This wakes correspond in the case of non-regularity of the minimal surfaces to an evolution in which dissipation has occurred, reaching to a stable evolution.Indeed, for incompressible fluid dynamics, with div V = 0, an initial (perhaps turbulent) fluid velocity distribution would decay by shear viscosity processes, such as those encoded by the Navier-Stokes term, $\nu\nabla^2 V = 0$. However, any components of the initial velocity field that are harmonic, $\nabla^2 V = 0$, will not decay, and it was argued that these components of the initial velocity distribution are those that form a residue or wake. The residue velocity fields admit a Hamiltonian representation, and therefore have a persistent existence: the initial domain with dimension four has been reduced by the evolution towards harmonicity to dimension two, the system organizes as a conservative claasical mechanics (symplectic) system. Returning to consider the geophysical fluid model, either by the celestial interactions as orbital forcing or by motions arising from the interior of Earth, these stable patterns may suffer dislocations which provoke earthquakes, tsunamis, and appears that climatological vortices are generated superposed to them [75], and yet they retain their stability due to the dissipation of vortical motion in the form of heat, until a new temporal cycle starts another evolutionary decay to the harmonic configuration, and so on.

Therefore, the observable patterns associated to the gravitational field, which as we have seen can be interpreted as an electromagnetic field or a viscous fluid, or an elastic medium, correspond to

minimal surfaces in which their tangential discontinuities, observable features of hydrodynamic wakes which can be put in correspondence with those characteristic surfaces of tangential discontinuities upon which the solutions to the evolutionary equations of hydrodynamics are not unique. Only the robust minimal surface subset, associated with a harmonic vector field, will be persistent and of minimal dissipation.
We consider the equations for space curves in the plane, which are generated by a pair of differential equations of the form,

$$dx/ds = \sin(Q(s)), \quad dy/ds = \cos(Q(s)). \tag{9}$$

Since the rhs of this system is a unit vector, **t**(s) it is possible to identify it with the system of Frenet in the classical differential geometry of curves, with s being arc length, and **n**(s) the normal vector. The system of derived equation is

$$d\mathbf{t}(s)/ds = \kappa\mathbf{n}(s) = dQ/ds\,(\cos(Q(s)), -\sin(Q(s)), \text{ with } \kappa = dQ/ds. \tag{10}$$

A simple set of solutions can be readily identified: a simple sequence is to be recognized : $\kappa = s^{-1}$, $\kappa = s^{0}$, κ = s, the Logarithmic spiral, circle, and the Cornu-Fresnel spiral (relevant to diffraction patterns and the Analemma), respectively, all of them pervasive in the Earth's crust, as examination with Google Earth will demonstrate (see also [75]). Would we consider the characteristic shapes of intrinsic space curves for which the curvature is proportional to an arbitrary power of the arc length, $\kappa = g(s) = s^{n}$, for all positive and negative integers (or rational fractions), of special importance are n=1 and n=2,

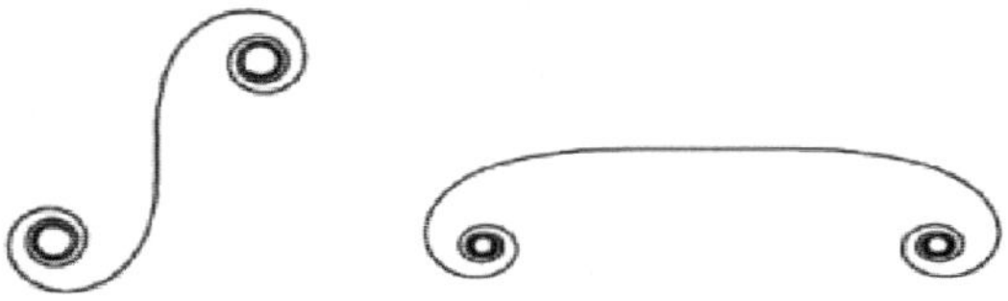

Figure 33. Left:The Cornu spiral, is the half-year trace of the Analemma. Right: The Mushroom wake. The asymptotic involutes in the Cornu spiral as a dipole system are actually reached at the time of the equinoxes, when the time seems to stop with the day and night length being almost constant during a three days lapse after which the Sun distinctly arises above the horizon initiating a new cycle of life.
Let us show some vivid examples of their crustal patterns:

Figure 34: Left: Aleutian Archipelago. Centre: Andaman Sea , Indonesia and Philippines (notice the subeddies in major eddies). Right: Caribean Sea. (the figures are deformed for lack of space; certainly their shape is kept).

Figure 35: Left to right, the Japanese and Sakhalin Archipelagos and the Kamchatka Peninsula, extending Northeastwise to the Aleutians, along a common dislocation curve that extends to connect to the East Pacific to run west of the whole American continent and southwise to the Northeast of Australia. The tangential discontinuities are visible.

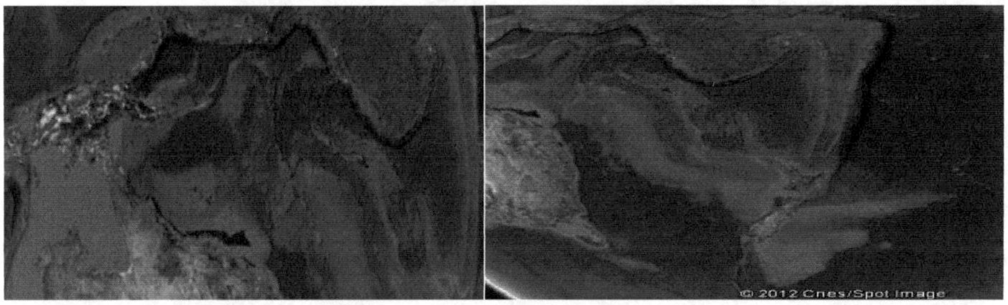

Figure 36: Northeast and Southeast Australian Sea; the tangential discontinuities are notorious.

Figure 37: We follow instead the wave front of Figure 35 in direction of the North to return to the North of Asia, or we turn our attention Northwise to next figure:

Figure 38: The Analemma formed by counter-rotating cells, on the left the Himalayas for its southern boundary, on the right China and Mongolia

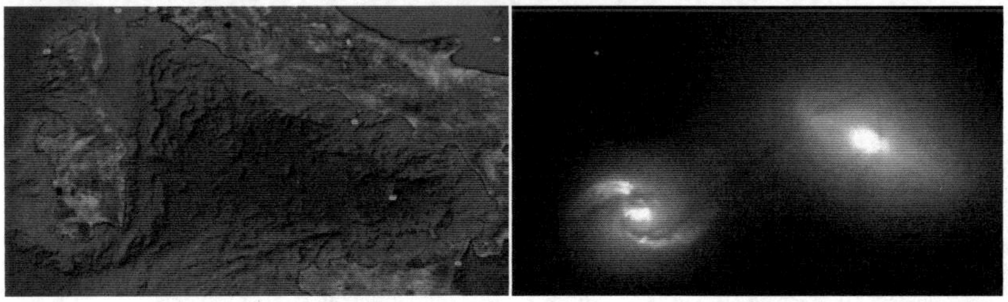

Figure 39: Left, the counter-rotating Analemma: Corsica and Sardegna, yet on a larger cell comprising the centre and south of Italy. Right: Galaxies NGC 1409 & NGC 1410 (courtesy of NASA).

Or we can change the scale to regard

Figure 40. Left and Centre, a tidal wave: The Wave, Paria Plateau Canyon, Vermillion Cliffs National Park, and Antylope Canyon, USA, families of Croll-Milankovitch rotational cells.

The present conception as applied for crustal formation has also a theoretical link with what what is claimed to be a new paradigm for systemics that stresses the fundamental character of torsion and vortical motions and structures (even in economics). In distinction of the linear attraction (with spherical symmetry) on points promoted to densities, the gravitational force is an eddy motion produced by the gradient of the gravitational potential with constant masses replaced by inhomogeneous densities [55,106,107]; more of this below.

The gravitational force on Earth produced by Sun and Moon, causes its orbital motion; in fact, it is this motion the largest effect of that force, which it further produces the deformations on oceans and Earth's crustal surface, Earth's tides. The main gravitational factor is the Moon and secondarily the Sun which accounts for about half the lunar attraction due to the distances; the other celestial bodies play a lesser yet still observable role; yet, other very important factor playing in the crustal morphogenesis and dynamics, is the amount of heating provided by the Sun, according to the Croll-

Milankovitch harmonic cycles [77,78]. While on the sea the gravitational attraction produces tides that may be as large as 10 m, the crust may move about 10-50 cm, and 10-20 cm in coastal areas.

The lunar-solar gravitational force can be decomposed into a part that is constant over Earth and causes the orbital motion, and still a variable on time and space force given by the remainder of this decomposition of the gravitational force. Thus we have the total force directed to Moon, the orbital force being an average of these forces, which in an approximation equals to the force on the origin. Would we substract this net force then we get another form of the gravitational force described in figure 41. Now the tidal force is towards and away of Moon and is radially inwards on the other sides. This inversion of the direction can be thought as being a change of orientation of the tidal force along Möbius strips covering Earth's crust along the meridians. This pattern remains fixed with respect to Moon, and Earth rotates around Sun. This causes the tidal force at each point and surface on Earth to be changing on time. So the mass distribution on Earth changes with time. Say, for the tidal force produced by Moon placed nearer to the Equator, every point in and on Earth,would travel over two bulges and two depressions of the tidal force and thus the period of this tide is $\Omega/2$, two cycles per day, i.e. 12 hours. Consider instead that the Moon is placed forming 45ª degrees with the equator; thus a point starting in a bulge at A moves after 6 hours it has moved to a inward depression at A' and returns to the original configuration in 24 hours; this is the case for any point in and on Earth.The period is now 1 cycle per day, the diurnal tides. If instead we consider the Moon to be over the North Pole, then any point on and in Earth would remain in ∞. Since rather than being fixed Moon is orbiting Earth with a period of 24 hours, we have a combination of the three cycles, diurnal, semidiurnal and infinite periods. Thus the tidal force can be represented as a Möbius strip inside a toroidal configuration, the latter representing the diurnal tide, and thus we have the 2:1 fundamental harmonic. We also have an infinite number of strips wrapped on the original one, which account for the unending period. It is quite remarkable that this is identical to the topology of the mammal heart, as characterized by the Ventricular Myocardial Band proposed by Torrent-Guasp, the similarity stopping short of being an identity on the grounds that the superposition of bands to the original 180ª twist that produces the heart, is not infinite. But then we recall that neither is the case for Earth on which the infinite cycle is asymptotic not actual, by the very nature of the North Pole being a point, i.e. unextended, and thus virtual, while what is physically manifest, say in the Riemann sphere, is the change of orientability in the neighbourhood of the North Pole., depicted in fig. 31. Thus we find that the gravitational (dominant) tidal forces, on the ocean and crust can be thought as distributed along Möbius strips (which we shall later propose to be better described by the Cross-cap) and still we find approximately the 2:1 harmonic characteristic of the Möbius strip and the Klein Bottle. They are visible on Earth's crust (and appear to have atmospheric associates) with the form of Analemmas of all sizes [75] established in what we can identify as the Shepard harmonics $f_n = f_0 2^n$, where f_0 is a fundamental frequency, that appear in music perception and its Klein Bottle representation to be presented further below.

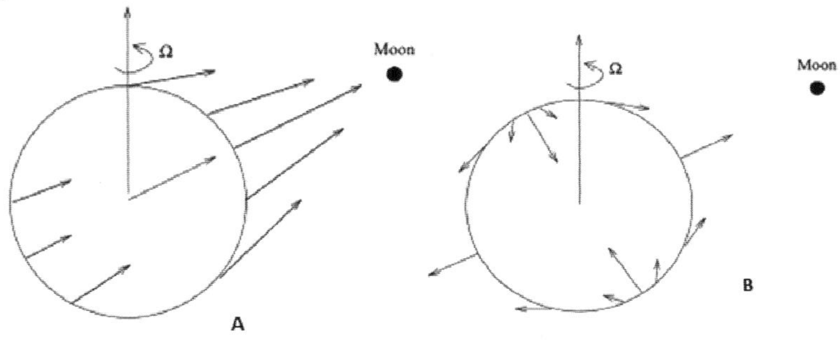

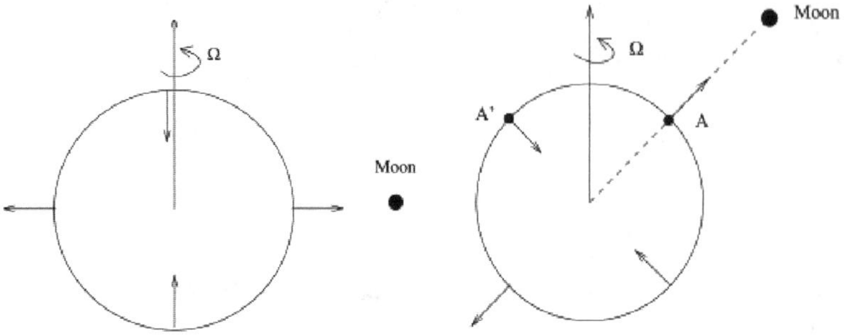

Figure 41 (reproduced from [59). Fig. A represents the attraction of the Moon (or the Sun) on Earth's crust; Ω represents the earth's rotation, which we remark that it is variable, which is of crucial importance to crustal configurations due to tides, storms, hurricanes,earthquakes [55,105,106], and to climatological Croll-Milankovic cycles [64], of crucial importance to paleogeology and paleobiology, and to pattern developments and growth of molluscs, as first observed by Leonardo da Vinci [65], . Note that the total force at every point is directed toward the moon, and that the arrows closest to the moon are the largest (the difference in arrow length is greatly exaggerated in the figure. Fig. B represents the attraction-repulsive tidal forces when the constant attraction forces are substracted from the former representation. We remark the change of orientation of the tidal force on the crust due to the depressions countering the bulges yet unified along meridians in which now the protoform of Newton's Third Law is the case. Note also that rather than an homogeneous force and thus gravitational acceleration, we have inhomogeneous acceleration is the case. This is the basis for the geoid, the non-spherical non-homogeneous gravitational equipotential surface; variations in the height of the geoidal surface are related to density anomalous distributions within the Earth. This validates Newton's Second Law for a non-constant mass and acceleration and that gravitation, rather than being a linear force, since the gradient for the gravitational potential with densities replacing point masses, generates a stirring motion and eddies, as discussed in [55,105,106] coinciding with the volumetric account of [76]. These stirring motion and the associated eddies are further related to the 2:1 fundamental and higher harmonics, and show up on Earth's crust and on the atmosphere as Analemmas of different scales corresponding to these harmonics.

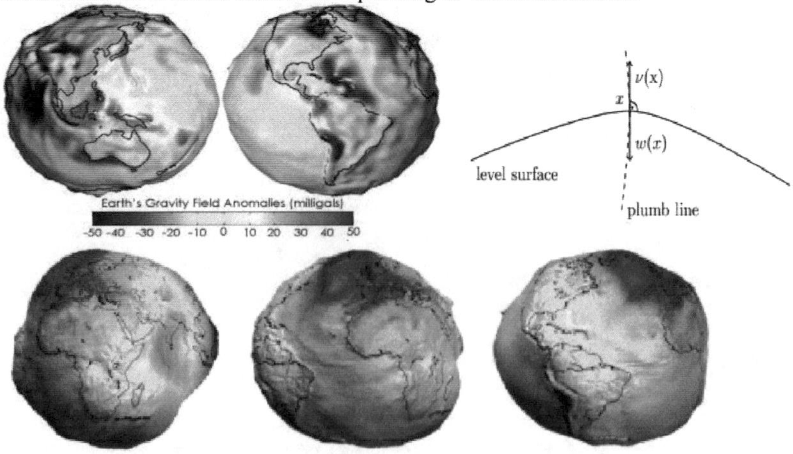

Figure 42. The geoid (courtesy of NASA). This shows that the gravitational potential is better described by its actions on extended areas, rather than virtual unextended points, the latter being the case of current geodesy. Alternatively, the gravitational constant becomes local (non-constant). Gravity, as observed on the Earth's surface, is the combined effect of the gravitational mass attraction

and the centrifugal force due to the Earth's rotation. The force of gravity provides a directional structure to the space above the Earth's surface. It is tangential to the vertical plumb lines and perpendicular to all level surfaces.
Consider now Fig. 43:

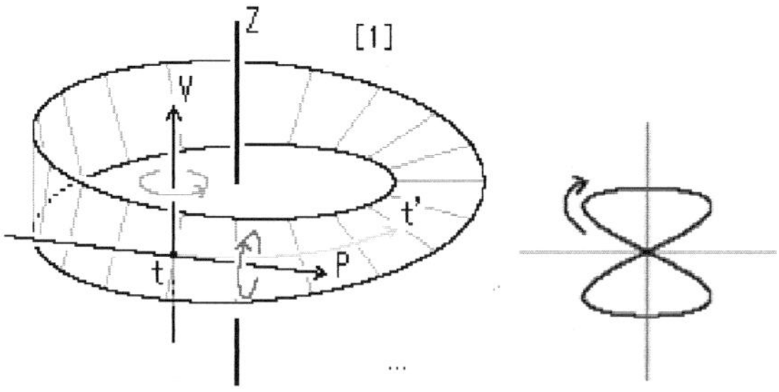

Figure 43: Tracing areal elements and motions on the Möbius strip. Right: the Analemma formed by tracing the motion of the surface elements on the 2:1 harmonic.

[1]: **Z** is the vertical axis that stands at the center of Möbius strip; it stands for the plumb line. The point **t** is moving to **t'** now under the rotational motion about **Z** of Earth. A small area at **t** somewhat turns around forward direction **tP** as shown the red arrow. Simultaneously the small area somewhat turns around **tV** that is in parallel with the axis **Z**. We can obtain the real twist of Möbius strip if we gather these two kind of twists. What we say "real" is the case we see Möbius strip from outside. Yet, Z if considered to be the vertical axis of the torus from which we cut the Möbius strip, we can further tilt it, as is the case of the precession of Earth's wobble around the North-South axis whose variable tilt is currently about 23.5°. As for the vector P we can think of it as a vector pointing in the direction of Earth's daily rotation or still the circumsolar rotation. Of course, we recall that it is the center of Earth that performs an elliptic orbit, yet the rotation of any element of area rather traces a 2-torus from which we can extract an embedded Möbius strip, as is the case above. Thus, rather than considering single points on Möbius strip for the origin of 2:1 harmonic it is elements of areas, i.e. densities, which are subjected to the gravitational pull of Sun and Moon, thus producing the crustal, ocean and atmospheric tides. This is consistent with the notion that the gravitational potential's action is not on ideal mathematical points but on actual physical extended areas, densities, so that the Newtonian expression for the gravitational force between two masses is substituted by two densities, and thus the gravitational potential given by the gradient creates a stirring vortical motion, rather than a linear motion as usually conceived. Whatever the angle of tilt of Z is with regards to the orbital rotation of area elements, the view from the left of these 2:1 resonances is a figure 8, i.e. the Analemma, yet not necessarily solely corresponding to a 2:1 harmonic produced by an annual rotation, but rather of cyclic phenomenae on Earth's crust and in growth of natural patterns [56,57]. This captures the two main oscillatory modes previously discussed that were found from measurements, yet combined in the two superposed motions described in Fig. 41. Let us give now a possible alternative topological description, that still captures the non-orientability, and still the fact that the gravitational field may, in this theory, be related to a projective spacetime.

11. The crosscap and non-orientability in geophysics

Cross-cap or projective plane can be divided into Möbius strip and a disk. It will be easier for us to observe relations among the representations of the non-orientable projective plane, Cross-cap and the Möbius strip.

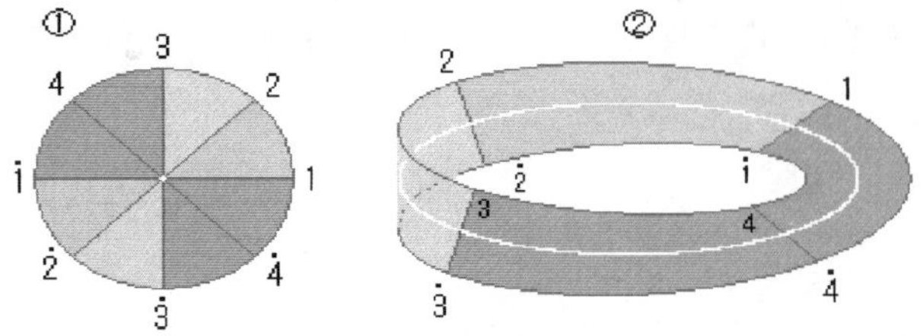

Figure 44: The cross-cap on the left and its representation on the Möbius strip on the right. Would we identify the antipodal points, we get the Klein Bottle, as shown in Fig. 10C.

①: We divide the disk into eight sectors equally by 4 diameters and color it as shown. The center, however, can not be divided by colouring. The end points of every diameter are numbered. ②: Disk ①became Möbius strip. The center of the disk is a single point, and now it is the center line of Möbius strip. Every coloured area corresponds to that of the disk except the center of disk. One to one correspondence stands up in the coloured area.

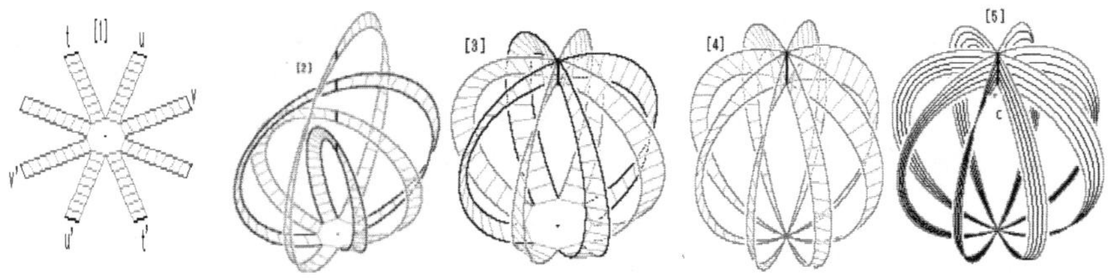

Figure 45 (courtesy of T. Ito): Surgery on the cross-cap.
[1]: Disk-type projective plane is a circle, which we cut the disk into rectangular strips arranged around the octagon that surrounds the center of disk, which is excluded. [2]: We join every pair of antipodal edges after twist. The twisting directions are made evenly. The black bridge on each thin strip is the joint line. Thin strips look like octopus legs. They are Möbius strips. Colours of the legs are changed alternately for easy observation. [3]: We make black joint lines free to pass through, and we let Möbius strips intersect each other as shown [4]: The octogon is shrunk to a single point. It is what the identification of antipodal points of [1] is done. The edges are coloured red in order to show that they are diameters of Disk-type projective plane. [5]: We split Möbius strips thinly. We assume that they become the diameters.[6]: It looks like an apple, it is made of the diameters. The joints of Möbius strips is the segment **c** now. The Cross-cap representation of the disk-type projective was seen to contain along the meridians of the sphere produced by antipodal identifications being glued, make of the projective plane a sphere whose meridians are Möbius strip, which we assume arbitrarily thin. Yet, on the case of the gravitational stirring force that produces the change of orientability associated to the tides, this is no longer the case. Thus, we are lead to propose the notion that they correspond to the natural topology of Earth, with regards to tidal forces, particularly of its crust, which are crucial to its geophysics. The zonal manifestations of this local changes of orientability is the appearance of Analemmas on all scales on Earth's crust and the atmosphere [75].

12. Blow-ups, divergences and the non-orientability of non-linear evolutions

The existence of time loops as reentrances of systems alike to the Gödel solution is by no means something exclusive of the hyperbolic geometry of Special or General Relativity, since it is further related to the existence of divergences in arbitrary non-linear systems. These discontinuities that involve the numerical notion of unbounded growth, is associated with the notion of blow-up, and they reflect transitional changes (the blow-ups) of systems described by, say partial or ordinary *non-linear* evolution equations. In the science of well posed linear systems, the assumption of initial conditions is the backbone for determinism and prediction of future states from the fixed initial state; yet linear systems do not have the ability to predict forthcoming transitional changes and the establishment of new systemic regimes posterior to them, which are the province of non-linear equations. In terms of non-evolution equations, blow-ups embody the destruction of old structures and their replacement by new ones. Prigogine already claimed that Einstein's theory of general relativity with its non-linear partial differential equations, lacked the historicity of evolutionary processes for the reason that time in the contemporary conception going back in science to Newton, is a mere parameter[105]. So the appearance of new regimes is not the exclusive province of non-linearity, but also of a conception of time which supersedes the mechanical notion of a mere additional variable to be compounded with space. Of course, in the theory of non-linear equations, the assumptions of initial and boundary conditions limits the ability of the equations themselves to account for interesting physics, climatology, mathematics, etc. [55,105,106].

Thus the concept of blow-up introduced by Yi Lin and Soucheng OuYang, reflects a very general phenomenology. For a given mathematical model, that we assume to describe a system's evolution, which we characterize by $u = u(t;t_0, u_0)$, with t the time variable, that satisfies $\lim_{t\to t_0} |u(t;t_0,u_0)| = +\infty$, and also, when $t \to t_0$, the system in question goes through a transitional change, then the solution $u = u(t;t_0, u_0)$ is called a blow-up. For non-linear systems in independent time and space variables, the blow-up is defined similarly.The question of this transition is what the meaning of ∞ is, rather than the mathematical indeterminacy that the real numbers unboundedness poses. The striking identity of this with the mapping of the double covering torus topology of the complex plane of its non-orientable topology as characterized by figures 30 and 31 is that the blowup rather than a divergence lifted by the one-point compactification of the complex plane represented in the Riemann sphere, it expresses the local change of orientability; already Yi Lin and OuYang claimed that the blow-up was equivalent to the passage of the system through the North Pole = ∞, showing a reversal of $+\infty$ to $-\infty$, yet they did not identify the relation of this with the topological issue of change of orientability, which we have found to be pervasive to science at large. Following these works, it is apparent that the issue of non-orientability is also crucial to the so called chaotic systems, as already the 2:1 resonance immersed in the 2-torus shows to be the case. Remarkably, ∞ corresponds to the Mö bius twist superposition cognitive state in the Matrix Logic representation of the Klein Bottle Logic, since it is both positive and negative, and nor positive nor negative, so that the change of orientability on N is the precise manifestation of this superposition. Boolean logic true and false states are the projection by the Hadamard operator which in this logic is the matrix representation of the Klein Bottle, which by further application of this operator, we retrieve the superposition states [14]. We see that ∞ is not associated to Boolean logic, but to the Klein Bottle Logic. This is of crucial importance on regards to hyperbolic divergences that arise in non-linear systems, which have been noted that are indeed outside of the cognitive realm established by Aristotelian-Boolean logic, as claimed in [55,105,106] Thus, rather than problems of indeterminacy of non-linear systems, blowups of non-linear systems *undergo a change of orientability as the topological signature of the change of regime with the appearance of new structures and processes.* This amounts to the manifestation of chirality following the explanation of figure 31 on the significance of transversing through the North Pole from the real to the imaginary axis with two different modes: Would the blown-up system reenter through the origin with imaginary values (green dotted path), then the chirality that has manifested with the first twist at infinity is preserved. Instead, would the system reenter through the negative real values (on the white dotted path), then the previous chirality is flipped by the second twist at infinity, and thus we have a complex orientable plane.

This reveals a deep meaning of singularities which has been overlooked. Yet, non-linearities is the case of vortical motions, as transpires already in the fact that the velocity field of the non-linear Navier-Stokes equations of fluid-dynamics is the torsion field, whose differential is the vorticity, and still the magnetic field of the kinematic dynamo equations of magnetofluid-dynamics. In this case, the vortices appear as a gradient of the velocity field of the inhomogeneous shearing motion of proximate sectors of the viscous fluid, and is also the case of magnetohydrodynamics [62], where the non-orientability of plasma flows as already manifested in solar coronal ejections.

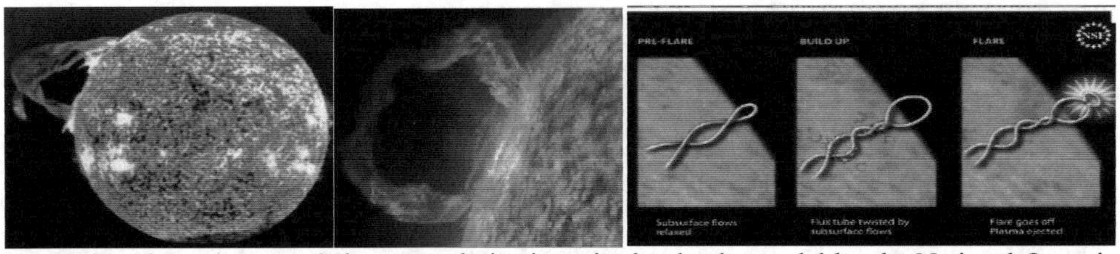

Figure 46: Lhs and centre: Solar coronal ejections; in the rhs the model by the National Oceanic and Atmospheric Administration, underdescribes the change of orientability that is manifested in the ejections (courtesy of NASA & NOAA, respectively)

In this conception, time is unseparable from material inhomogeneities as argued in the theory of blow-ups [55], yet this is so because the logic of the physical organization of matter and fields is the Klein Bottle logic of reentrances, and this extends to thinking and cognitive processes [9-16]. Eddy motions arise from the unevenness of space and materials [55], and the time cycles as occurs in Earth's crust and atmosphere, and the unevenness of time as argued by Kozyrev [67,68,70] (and empirically tested in very different situations, such as rotating tops [69], physiological cycles of plants, torsion balance anomalies and other phenomenae) in which efficient transformations or kinetic energy to heat occurs, with the consumption of kinetic energies establishing an equilibrium between them. But even more crucial, this shows that the notion of forces as exterior to systems, in the case of unevenness, is the physical manifestation of inhomogeneities. Yet, more fundamental to them, is the protoform of Newton's second law, embodied in the non-orientability of the Möbius band and the Klein Bottle, or still, the projective plane. Thus, with respect to Newton's third law, F= ma, as already the geoid shows, the acceleration a is no longer constant (but actually an average if considered as acting on an ideal point), nor is the mass m, and we rather have than a force applied to a non-extended virtual point of space or spacetime, a force that arises from the uneven distribution of a material or space. We can express this by saying that Newton's third law is contextual, and its absolute character resides precisely in that the physical constants are not such objective values but contextual variables which are, in usual practice, averaged out, losing thus the irregularities that rest at the physics of eddy currents and motions. Thus, instead of the gravitational potential, say acting on unmaterial virtual points of Earth's crust as assumed in the more standard geophysics [54], we have the gravitational attraction acting on the densities of real non-ideal extended matter distributions. Otherwise, it would be impossible to understand how morphogenesis of Earth's crustal structures and processes occurs, which is the case as Analemmas at several scales, and is also the case of the atmosphere with the occurrence of vertical structures both crustal as well as atmospheric. Furthermore, as much as the Time operator of the Klein Bottle Logic which geometrically is the $\pi/2$ rotation of the complex plane (identified with the cognitive plane) from the real to the imaginary axis, which represented on the Riemann sphere is a $3\pi/2$ percourse can also be seen through its action on two cognitive states of the Klein bottle logic, which is tantamount to compute its algebraic *difference*, systemic time can be understood associated to the dynamic *difference* that is embodied in the *unevenness* of materials. Thus, it is possible to understand the systemic time cycles that permeate Nature, and in particular the Croll-Milankovic climatological and geophysical cycles that we have argued that they arise from the 2:1 resonance of the Möbius strip as Analemmas on the Earth's crust produced by tidal action, or in the atmosphere, as the time operator that is intrinsic to this fundamental change of orientability that produces this harmonic resonance. Thus, generically the blow-ups are spinning currents which only appear with time, in which non-orientability supersedes the discontinuity proper of the divergences. With respect to the well-posedness of the initial conditions with their accessory uniqueness, existence and stability of the solutions of the non-linear evolution equations, they are not satisfied by nonlinear problems, so that blow-ups will only be avoidable under special conditions which are very different to meet in practice. While the curvature of the Riemann sphere and in general of spacetime allows for the hyperbolic divergences to be mapped to continuous transformations as self-reentrances of the systems in question, as already the Klein Bottle Logic that sustains this self-reentrance it is the torsion fields and their topological operations that fold and unfold the reentrances of the systems as new regimes that in turn will fade away, numerically diverge and transmutate to a new regime. The torsion fields are the vortical motions that embody the blow-ups.

With regards to the thermodynamics, it is very important to remark that on the case of non-linear systems described by the relations between entropy S, thermodynamic forces X_i and thermodynamical flows J_i, related by the non-linear equation $\frac{dS}{dt} = f(X_i J_i)$ that extend the linear case, and are the usual physical case, it follows that irreversible processes experience a blow-up, which signals the self-reentrance of the system following the increase of entropy with a finite time, reordering the system with the decrease of entropy [55]. It was argued that the cycles of re-entrance of systems is related to the Time Operator of the Matrix Logic formulation of the Klein Bottle logic, associated to the derivative of acceleration, control, neglected in physics but which shows up very pervasively even in quantum mechanics [55], and still the second-derivative of acceleration, self-control [15], which in the case of the Earth's crustal dynamics appears as the upthrust motion that inverts the geological layers, producing the older layers to rest upon the older ones [75]. This establishes a non-linear morphology of Earth's crust and atmosphere, as revealed in the figures we have presented above, and the notion of geological time. The latter was the subject of a much followed confrontation between Lord Kelvin and Charles Darwin, the latter claiming a uniform crust to avail his theory of evolution of species that required small uniform changes of the crust [107]. Leonardo claimed that biological patterns of molluscs on Earth were related to the orbital forcing due to the Sun and the Moon, which we know to determine the shapes of the crust, and paleobiology appears to support Leonardo's claim [56,57].

Therefore, rather than the universe expanding towards a heat death, we have a universe that reenters itself or to another universe through a Klein Bottle wormhole. The universe or the multiverse establishes an equilibrium through the cycle of destruction and recreation. This is the logophysics of all systems, a claim that begs further qualifications. We shall be brief, out of necessity.

In a lived world in which the anatomy-physiology of the cognizing-sensing-thinking-enacting subject, the physics and chemistry of systems, the logic and the fused perception and cognition of the subject, all co-conform the logophysics of self-reentrances of the Klein Bottle, we are still left to deal with the issue of complexity of interactions of monads conformed thus as well as to the constitution of each monad; this complexity transcends the reductive quantitative notion of algorithmic complexity, being of logophysical character. This requires the consideration of the HyperKlein Bottle in the framework of a hypersystemic theory of systems of self and mutual reentrances and closed/open containments, selfpenetrations and interpenetrations of Klein Bottles, which in the lived world are processed as a single integrated one [9,12]. Would we wish to recourse to a dualization (which as already mentioned, is a projection of the Klein Bottle Logic [9-12,14,15]), even in apparent contradiction with the principles of this lived world, claiming a specificity of a particular system due to complexity, it is this *hypersystemic integration and still relational specificity* that the HyperKlein Bottle embodies in a logophysics which the lived world through the body integrates *singulary*, and still *universally*, as a *single monad*, while preserving diversity, both perceptually and cognitively. In other words, cognitive complexity is maintained while integration is universal.

This monad whose reification waxes and wanes, is eternal; would a nihilification posture otherwise, it would stand in the previously mentioned projection of the Klein Bottle Logic to dualism, from which the superposition states (topological precursors of *Schrödinger's cats* states), by iterated self-referential action of the Klein Bottle in the Matrix Logic form of the Klein Bottle Logic, are recovered [14,15]. Thus, instead of the Nil/Void which quantum physics has long ago vanquished (yet physicists may associate with 0 [116]), the undifferentiated emptiness which logicians and cognition scientists appear to impose on the Plenum [124,125,127], it is this logophysical periodic oscillations of the Plenum what constitutes its manifestations, eternal as self-referential self-reentrant action is. The reentrance occurs through logophysical time waves, imaginary logical values [127] first introduced by Spencer Brown in his *Laws of Form*, which are also basic to the genetic code [12]. When a system comes to manifestation in the physical world, then its eigenstates are real, and divergences, either by initial conditions or constraints will make it blow-up, and then the reentrance will occur as previously discussed. They are the signature of the eternal return.

13. Nonorientability, the human heart and vortical dynamics

The mainstream science model of the human heart claims that the heart (weighing about 300 grams) is a pump , which as a vessel its walls have a fixed width, and is capable of `pumping' some eight thousand liters of blood per day at rest and much more during activity, without fatigue. In terms of mechanical work this represents the lifting of approximately 100 pounds one mile high [88]. In terms of capillary flow, the heart performs the task of `forcing' the blood with a viscosity five times greater than that of water through millions of capillaries with diameters often smaller than the red blood cells themselves! Due to the complexity of the variables involved, it has been impossible to calculate the true peripheral resistance even of a single organ, let alone of the entire peripheral circulation. One of the key concepts implicit to this model is that blood is naturally inert and therefore must be forced to circulate. Remarkably, this conception was expounded by Galileo and Leonardo da Vinci, who in his Notebooks depicts the left ventricle wall as having uniform width, as

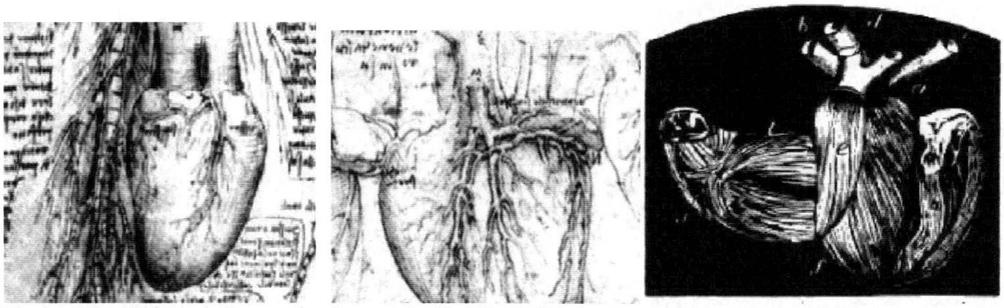

Figure 47. Left and Centre: by Leonardo da Vinci. Right: In stark contrast with Leonardo, John Bell Pettigrew showed in 1874, after mammal hearts were subjected to boiling and thus the fiber structures were revealed, the descending and ascending helical fiber structures surging from the apex on the bottom of a turkey's heart –anterior view-, revealing what Torrent-Guasp rediscovered almost one hundred years later. Pettigrew further gave a thorough discussion of the relation between the twisting motion of the fibers and the physiology of the heart [22,23], anticipating Torrent-Guasp [71,72]. He noticed the variable width of the fibres and their importance to its physiology. Currently, they are visible in vivo, through magnetic resonance imaging, and studied by the Biomechanics Group at Caltech [73]. We expect to find in a pressure chamber; quite the contrary, the left ventricle wall thickness varies by about 1800%, as was found by dissecting bovine hearts [71,72,88]. Furthermore, in the pumping conception the blood flow is passively shaped by the heart; this is the Cartesian concept of space as a container for objects, which was retaken by Newton and fully developed by Einstein in his theory of relativity, which excludes spin which requires torsion for its representation in a gauge theory of gravitation of the Poincaré group. To question the Cartesian model of the heart, it has been observed that the thin, flexible, cone shaped apex and suspension from the aorta suggest the accommodation of a twisting function especially, when taking into account the spiral orientation of the myocardial muscle layers. We cannot refrain on commenting the strangeness that this elicits with regards to Leonardo's description of the physiology of the heart, since he above all, was the master in the study of eddies in fluid-dynamics, designer and builder of canals and hydraulic systems, a unique pioneer in observing and representing Nature. In fairness, the limitations appear to rest on the

preparation of the dissected heart by boiling them, to be able to separate the layers of the tissue in dissolving thus the connective tissue.

The circulatory system can be seen as a transformation from a 2-torus to a Möbius strip at the root of the pulmonary artery and back to a 2-torus tube that becomes at the root of the aorta, as depicted in figure 48, left.

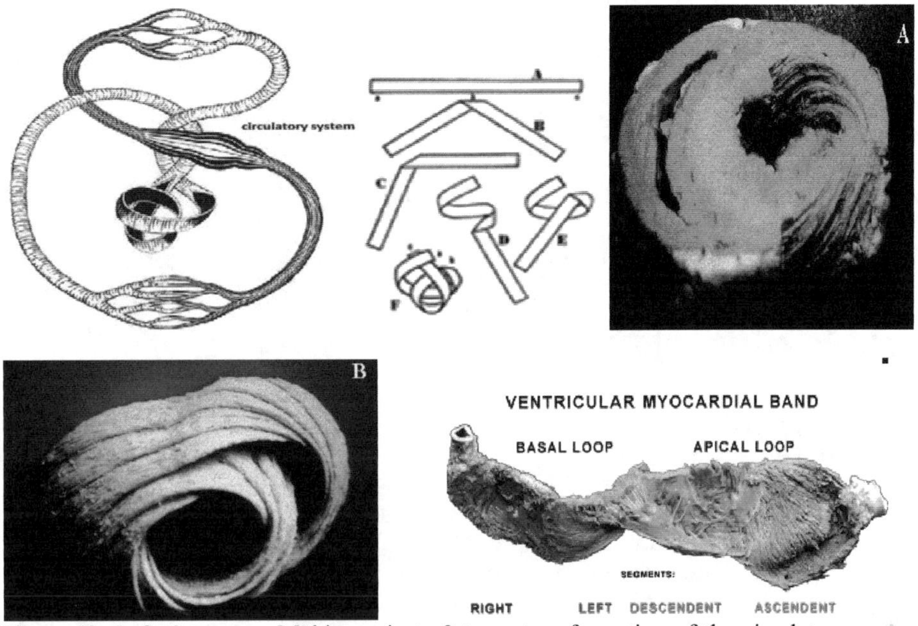

Figure 48: Left: 2-torus to Möbius strip to 2-torus transformation of the circulatory system. Center: We introduce the topological model on a paper strip of the Ventricular Myocardial Band by Torrent-Guasp. Start with a ribbon A (the pulmonary artery) on which we introduce a 180° turn, as in B, and C is the aorta, to form the apical and basal loops. Right and below: Unprecedented dissections by Torrent-Guasp A, transversal cut of the ventricles. B, posterior aspect of the ventricles seen from below. The left ventricular cavity is circular whereas the right one is semilunar. The right ventricular cavity cannot be considered as a true ventricle but , merely, as a fissure, a cleft opened in the thickness of the ventricular wall, alike to the separation of the Möbius strip from the 2-torus in figure 50 below. We note the iteration of the process covering the original single twist with a multilayered organ, the heart. The origin of the loop can be seen in the twist, in from the basal loop (in red) to the apical loop (in green) unfolding of the heart without any cuttings (photographs and drawings by Torrent –Guasp, courtesy of the Torrent-Guasp Foundation and its director, Dr Mladen Kocica).

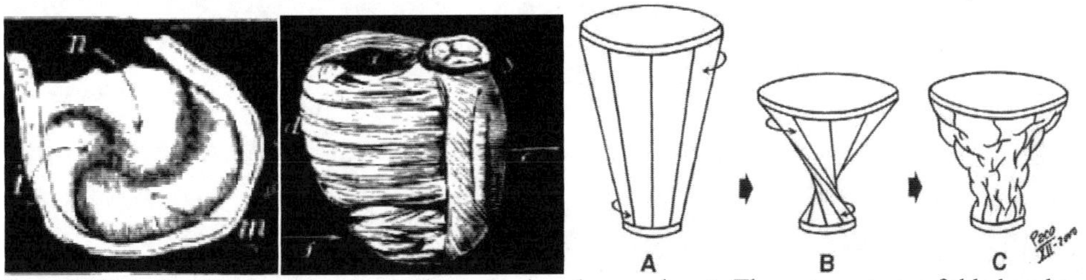

Figure 49: Left: ''Tricuspid valve action seen in a human heart. The segments are folded and twisted into each other by a screwing motion (vide arrows *m,i,n*) occasioned by the spiral movements of the

ventricles and the spiral impulse communicated to the blood contained within the right and left ventricle'' (Fig. 85, page 174, [22].Centre: The fourth layer of helical fibres, showing the point in which the external fibers turn to the inside (Fig. 99, [22]). Pettigrew referring to the layers: "an arrangement so unusual and perplexing, that it has long been considered as forming a kind of Gordian knot in Anatomy. Of the complexity of the arrangement I need not speak further than to say that Vesalius, Albinus, Haller and De Blainville, all confessed their inability to unravel it", and consequently it was disregarded until Torrent-Guasp's resurfaced it and elaborated his model. In A,B, and C we show the lateral view of the torsion, lengthening and shrinking of the layers of the heart (by Torrent-Guasp, courtesy of the Torrent-Guasp Foundation).

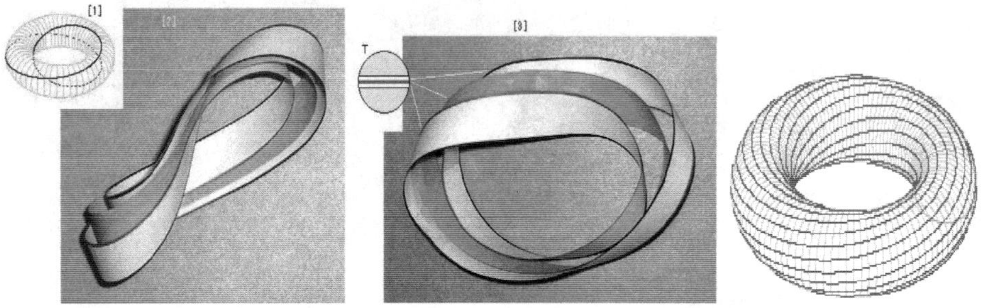

D

Figure 50: extracting the first layer of the heart in the multilayered heart that arises from the 2-torus cut of an iterated Möbius strip

Consider figure 50. [1]: We present the 2:1 resonance in the elastic 2-torus from which we have a Möbius strip in Figure 25. The black curved line is the track of cutting. [2]: The photo is a paper craft to show the situation at that time. The red belt is the Möbius strip and the white belt is the surface that contacted to it. They are loosened a little for easy observation. The white belt is two-sided. So we coloured edges brown and black separately. [3]: The picture is taken in different direction. **T** is the section of torus. Finally in the right we present a stack of n paths parallel to the 2.1 resonance which we can also cut to produce a multitier figure instead of the three-stack figure, yet it has one single twist so it is a long Möbius strip, yet with superposed red layers while the total angle of turns of the overlayed twisted red layers is $4\pi n$. This will be the topological model of the Ventricular Myocardial Band which indeed has a single 180º twist yet with superposed layers that make the heart's wall.

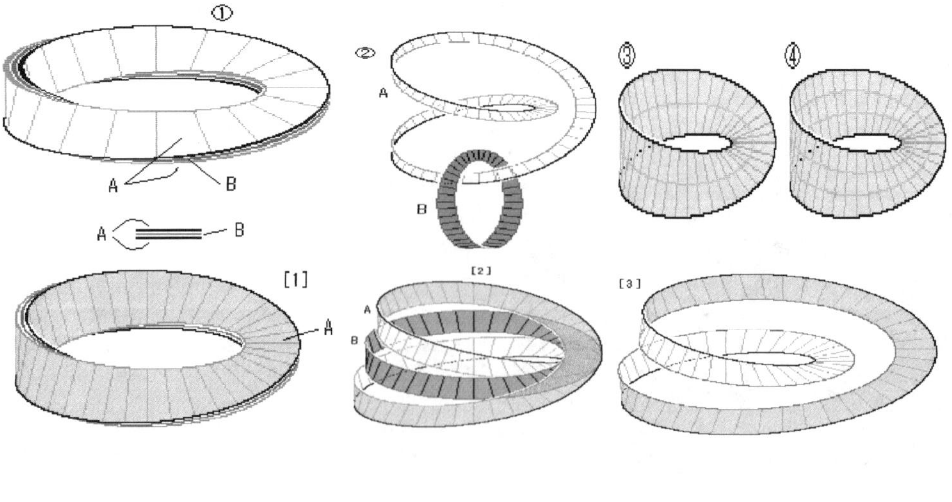

Figure 51. [1]: The belt painted sky blue is loop **A** in Fig.2. It is two-sided, and so we can distinguish its surfaces by color. [2]: The left parts of loop **A** are vertically pulled apart from Möbius strip **B** a little, alike the right ventricule in Figure 39. The white surface, rear side, of loop **A** appears now.[3]: Möbius strip **B** is omitted. The right parts of loop **A** are horizontally separated a little each other. The brown edge is coloured red now for contrast. The twist is 360° as long as we see this diagram. Roughly speaking, [3] is a coil of duplex winding. Iterating this by taking cuts along paths as in fig. 49D will form the Ventricular Myocardial Band. This topological model in which instead of paper we consider real elastic myocardial fibers may serve as a basis for a mathematical model of the heart in terms of the elasticity associated to the superposed harmonics of the iteration. Remarkably enough, the two-fold superposition of a Möbius strip, is a simple model of a continuous transformation of spin-one half into spin-2 unification states [95].

14. The non-orientable psychophysics of music cognition on, visual perception and pattern recognition.

Physicists and philosophers have studied visual perception and the geometry of visual space [97, 98]. The latter was a central issue to the Renaissance (Leonardo and Dürer its pioneers), which through projective geometry abandoned a trivial plane in which a fusion of object with subject operated, aiming to represent three dimensions on the plane. The identification on this plane, of object with subject, was the signature of the Middle Ages, evidenced in the lack of depth, the primal dimension of the lived world which phenomenologist Maurice Merleau Ponty associated with time [140]. Thus, the Renaissance introduced depth accompanied by a detachment of the subject to observe the world as an outsider, setting the stage and the scenery for Descartes. Lüneburg discovered that hyperbolic metrics (as in Einstein's Relativity) for which parallels converge, may describe the geometry of distant-object-vision instead of the near-vision Euclidean geometry, and that a psycho-metric function dependent on the observer is crucial to visual perception. Thus, a plural visual representation geometry (there is a third zone with a distinct geometry) was proposed. Thus the Principle of Non-Contradiction, the backbone of the Aristotelian Boolean Logic, is no longer valid for visual perception, which requires an interpretative contextualization by the subject, who enacts space , rather than the detached Renaissance lived world and its posterior Cartesian formula of object-in-space-before-subject. This ambiguity which the dual logic can not account for, is more notable in auditory space [99]. Yet the sensory modes have a topographic map representation in the body whose topology is the Klein Bottle, which will be presented below and thus its ontology is the Klein Bottle Logic, rather than a dualistic logic.

The term topographic maps refer to the existence of maps of the sensorial stimuli on the brain. In particular, a visual map refers to the existence of a non-random relationship between the positions of neurons in the visual centers of the brain (e.g. in the visual cortex) and the values of one or more of the receptive field (RF) properties of those neurons. The term is usually qualified by reference to the property concerned: thus the term retinotopic map refers to the orderly mapping of RF position in retinotopic coordinates in a brain region; orientation map refers to the orderly mapping of orientation preference and ocular dominance map refers to the orderly variation in relative preference for stimuli delivered to the contralateral vs. the ipsilateral eye. These quantities can be assigned to neurons in primary visual cortex on the basis of appropriate tests with physiological stimuli such as moving light or dark bars, or sine wave gratings, presented to one or both eyes while the activities of cells are monitored with techniques such as extracellular single unit recording, optical imaging, two-photon confocal imaging or high resolution fMRI. These are not the only quantities that may be mapped: in principle any well-characterised receptive field property is a candidate for mapping in any of the visual areas [82].

The issue of topographic maps lead Penfield to discover by stimulation of areas of the cortex to identify the surface areas of the body associated to them. This lead to identify the existence of contralateral (i.e. on the opposite side to another structure) maps of the cortex, one being somatosensory and the other being motoric; thus, sensation and action are integrated and seemingly lateralized in a feedback loop, yet as if functionally separate in each surface of the hemilateral hemispheres.

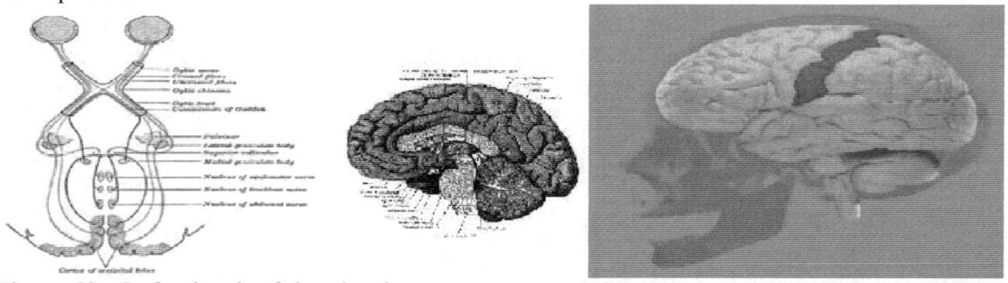

Figure 52. Left: sketch of the visual system anatomy (after Gray's Anatomy); Center and Right: the sensorymotor cortex; right figure licensed by eBodyParts3D, © The Database Center for Life Science licensed under CC Attribution-Share Alike 2.1 Japan".

Thus the cortical homunculus was discovered, a topographic 2D map associated to the primary motor cortex and the primary somatosensory cortex, i.e., the portion of the human brain directly responsible for the movement and exchange of sensory (in the left brain) and motor (in the right brain) information of the body. It can be thought as an anatomical-physiological section of the integrated sensorial modes on the sensorymotor cortex, whose topographic map has the topology of the Klein bottle [29].

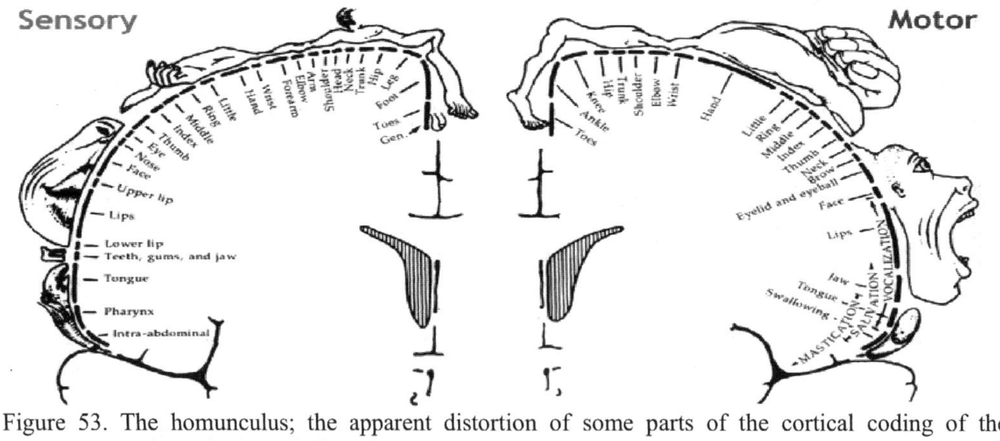

Figure 53. The homunculus; the apparent distortion of some parts of the cortical coding of the sematosensory input is the real distortion.

One striking aspect of the homunculus, is that on the left and right hemispheres, is the superposition of the motoric and sensorial locus. Alike the integrated crosslateral anatomy-physiology of vision, which is functionally representable by a Möbius-Klein bottle, with a hole playing the role of the X-cross of fig. 51:

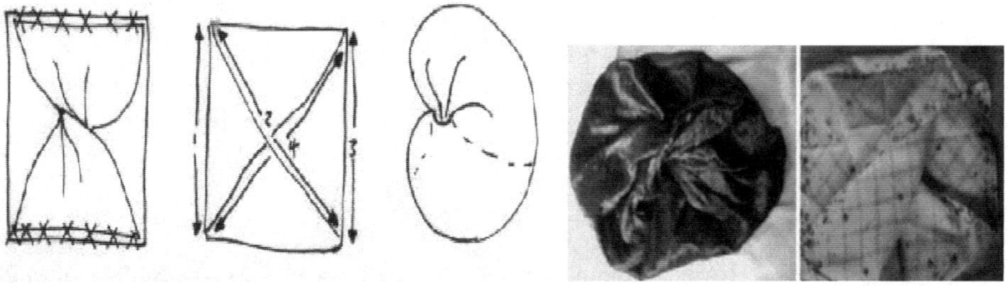

Figure 54. The Möbius Kleinbottle, a posture of the Klein Bottle, is easily constructed using three squares. This is the model described in Lewis Carol's *Mein Herr.* Two squares are joined together to make the twisted Möbius strip to finally stitch a third square as in B: Möbius Klein Bottle in C and D. The model in C made with three different coloured squares shows the twist in the purple square. In D, a model is of the same design and is offered for the reader to sew the third handkerchief onto the continuous edge of the Möbius strip. Being a fabric with a grid printed on it, the orientation of the diagonal and orthogonal weave is revealed. Reproduced with the kind permit of the author, Dr Melanie Purcell [100].

The integration of this superposed somatosensory and motoric topographic representations is produced by the corpus callosum, and that indeed we have a topological identification of the retina with the cortex, is given by the homunculus, while the cross-lateral identification is very much the design of the body. But then it natural to hypothesize that the hemilateral superposition of the motoric and somatosensory maps have precisely the non-orientable topology of the Möbius -Klein Bottle in which we can see each surface as being the local side of the globally one-sided Klein Bottle surface. Thus, the integrated feedback of the homunculus is that of the fundamental 2:1 resonance. For the physiological integration to occur, two half-octaves are playing it, each of them for the motoric and somatosensory cognitive modes, so that low and high frequencies signals are processed *as if* separate to each cognitive mode, and further integrated in the 2:1 resonance and superposed in higher octaves. Of course, as well known, both modes occur in quantized time, since elementary time frames both for perception and for action are the case. .As we said before, it is the Weber-Fechner law that rules the distinctions in the scales above the elementary thresholds [101]. Furthermore, the hemilateral brain asymmetries can be, in this setting, explained by the 2:1 resonance, since the information has to cross twice the corpus callosum for establishing a cognitive frame for integrated sensation and action. Already in the established paradigm for hemilateral functional asymmetries, it is postulated that they involve small qualitative differences (as shown in the setting of the Weber-Fechner law for just noticeable differences) rather than all or nothing differences.

So the logic of the cortical integration is non-Boolean. In fact, it is postulated that the brain asymmetries arise from fundamental physical asymmetries [102], of which chirality is most basic; already, chemical chirality can be linked to the non-orientability of the Möbius strip through enantiomerism; furthermore, the current paradigm of chemistry stresses the topology of chemical configurations; we have discussed this issues above. Returning to the biological development that may establish asymmetries, it is postulated that they correspond to different rates of maturation (so biological clocks are the case) and the changing embodiment of environmental information, and thus, the left brain has an evolutionary delay in the life of an organism to be able to learn high spatial frequencies, while the right hemisphere is earlier capable of processing low spatial frequencies [102]. Therefore, it seems plausible that an interpretation of evolutionary asymmetry can be set in terms of harmonics, and in fact, it was postulated that biological evolution complies with the Klein Bottle Logic [9,11] rather than with Boolean logic assumed to be the case by embryologists. But as much as the 2:1 resonance of the non-orientability of the Möbius Band and the Klein bottle might be the case of the evolutionary history towards establishing functional asymmetries, it can also be postulated that the quantized time frames that are related to the actual integration of both hemispheres, can also be related to the time lags that occur with respect to the functional response to stimuli, and still to sustain the anticipative response to stimuli as in the Libet's experiments, so that anticipation is undisociable from the holonomic character of the 2:1 resonance that is the case of the non-orientable Möbius and Klein Bottle surfaces. Already in aural interpretation of sequentially played half-octaves (the tritone paradox), the holonomity of the sensorial response is associated to the non-orientability of the aural map; we shall discuss this issue below. The relation of anticipation, non-orientability, hemispheric integration and Kozyrev's theory of time was discussed in [130].

Returning to the general discussion of topographic maps, the consequences of having multiple overlaid maps of different properties were realised by Hubel and Wiesel who speculated that maps ought to be superimposed in ways that would allow combinations of features such as eye and orientation preference to be equally well represented over all positions in visual space [83]. They also speculated that continuity in the layout of columns for different features would minimise connection lengths by bringing functionally related groups of cells close together in the cortex. These two ideas form the basis for a theoretical understanding of cortical maps that has been extremely fruitful [92].

In this approach, cortical maps are regarded as a dimension-reducing projection from a set of points or manifold in a multi-dimensional feature space onto the 2D surface of the cortex. It is noteworthy, that phenomenologically, these topographic maps of the sensorium produce a mapping of highly dimensional configuration (actually 6D for visual space, two axes are needed to represent receptive field position, the horizontal and vertical coordinates in visual space, two are needed for orientation (which is usually represented as a point on a circle), plus another for eye preference and another for spatial frequency) to 2D surfaces, so that their analytical structure of and the topologies associated to them are keenly related to the way that the notion of a lived world is structured and developed.

Rather than the usual approach in theoretical physics that follows Einstein inception of time as a fourth parameter to space, in adding dimensions to account for a phenomenology to later propose another related phenomenology to explain why these dimensions can not be empirically tested or even not manifest at all, the conception is that although many degrees of freedom may be considered, the world as is experienced, notably indicates dimension two. But then an abstraction of a primeval distinction or an abstract boundary -which we identified with the torsion field- on a 2D plane, allows not only to generate the Lorentz group, but also time waves as imaginary (i.e. related to $\sqrt{-1}$) logical states that correspond to the self-penetration of the Klein Bottle which occurs in 3D but does not occur in 4D, would we immerse the Klein Bottle in $\mathbb{R}^4$ erasing thus its self-penetration. We recall that it was this keeping of the singularities of self-penetration what allowed us to represent the complex electromagnetic field as a minimal surface with discontinuities and thus represent the shape of Earth's crust produced by tidal waves, later identified with harmonic wavefronts above [14].

The axes of the feature space are the parameters which vary in the maps, e.g. retinal position, orientation angle, spatial frequency etc., yet, neural networks modeling with Hebbian learning have repeatedly shown that reduction to 2D is achieved, though different parameter dimensions are brought into consideration and that singularities –which correspond to the change of orientability of the topographic map- naturally appear in the 2D representation, an issue of the upmost importance, to which we shall return The projection is assumed to be made subject to the two constraints identified by Hubel and Wiesel: completeness and continuity [82]. *Completeness* specifies that all functionally relevant points in the space should be mapped; this constraint is sometimes equivalently referred to as *coverage uniformity. Continuity* specifies that neighbouring points in the cortex should be mapped, as far as possible, to neighbouring points in the feature space. The issue of continuity is a most crucial issue to discuss next

In the case of orientation maps, each neuron in the primary visual cortex respond selectively to the orientation of edges and their direction of motion of stimuli presented by luminance gratings; this response is a Receptive Field (RF) of optimal orientation. Orientation preference is mapped in a systematic fashion across the cortical surface, such that neurons in adjacent columns have similar but slightly shifted preferred orientations; each ''column'' being a flat 2D slab codes for a single orientation .The mapping is continuous, and characterized by point singularities, the orientation centres at which all orientations from 0° to 180° converge in a pinwheel (i.e. vortical) singularity. These maps produce RF profiles of individuals neurons and cortical arrangements of parameters relevant to the RFs which are extracted from the RFs profile obtained from the self-organization of afferent inputs. Thus, the basis for the maps lies in the self-organization of inputs. The analytical basis for these maps lies in the Gabor function $R(\mathbf{x}) = A \exp(-\mathbf{x}^2/2\sigma^2) \cos(2\pi f\, e(\theta).\mathbf{x} - \phi)$ of the 2D position vector **x**; A, σ , F and e represent the normalized factor of response strength, width of the RF, spatial frequency and unit vector vertical to the orientation θ, respectively. We note here the presence of cosine term, which for a 2D signal already produces the Klein Bottle as the spherical harmonics expansion, discussed already. In particular, focusing on optimal orientationϕ and phase θ, the best fit of the Gabor function with the RFs is obtained by taking the values of the parameters with $\theta \in [0, \pi]$, $\phi \in [0,2\pi]$, as shown in the figure 56. The white and black ovals in figure 56 show the ON and OFF, of excitatory and inhibitory response of the neuron, respectively, in the RFs derived from the Gabor function which occur along the edges of the domain. Remarkably, identical configurations of ON and OFF appear along the opposite edges of the domain. Also OFF-flank and ON-Centre RFs rotate clockwise along both horizontal edges in the domain from 0° to 180° degrees of orientation, while the direction of rotation of the RFs change oppositely in the vertical directions, establishing thus the Klein Bottle. Yet, all in and all, it is, in first glance, a Boolean response ON or OFF of the cortical cells which supplemented with the states CENTRE and PERIPHERY, all together four states, produce the four state Klein Bottle logic of the representation of the RFs. Indeed, we must remark the paradoxical logic of the behaviour of the response with relation of PERIPHERY and CENTRE already noticed by Hubel and Wiesel, in the sense that the antagonism of CENTRE and PERIPHERY is non-dual, in that the effect of a stimulus on the CENTRE was countered by stimulating a cell in the surround –''as if you were telling the cell to fire faster and slower at the same time'', i.e. a *superposition* of inhibition and activation; see [83], page 41. Already the boundaries embodied in PERIPHERY have an orientation which will lead to the Klein Bottle Logic. The topology of visual space is identical if other degrees of freedom of visual space are considered, say, spatial frequency, retinotopic position and ocular dominance are introduced [92].

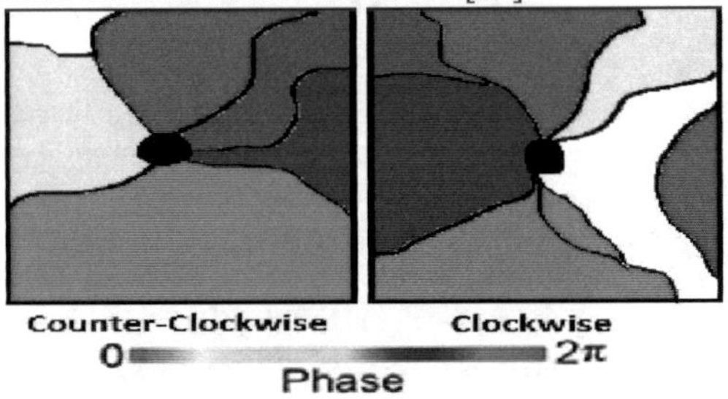

Figure 55: Representation of point singularities in the visual cortex. Each colour represents a different radial phase corresponding to an orientation column; Courtesy of Rtang 3 ©.

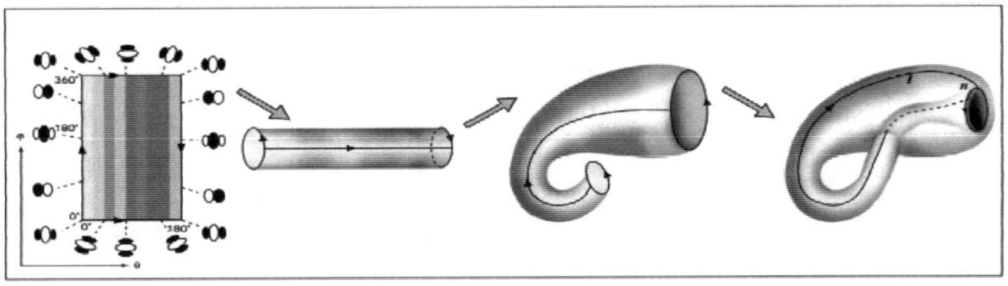

Figure 56: Demonstration of the Klein Bottle from the symmetry properties of simple cell receptive fields [30,31]. On the far left, receptive-field orientation is plotted on the horizontal axis, and receptive-field phase, which corresponds to the layout of excitatory (white) and inhibitory (black) subregions within the receptive field, is plotted on the vertical axis. Any combination of orientation and phase can be represented as a position within the rectangle. A continuous surface across which these parameters vary smoothly can (only) be constructed by joining together opposite edges of the rectangle to form a Klein bottle, as shown on the far right (reproduced, courtesy of Nick Swindale). The symmetry of the representation on the far left is identical to the representation of high-contrast pixel patches statistical analysis of natural images, which are thus found to have the Klein Bottle topology as the hidden universal form of those images [27,28].

A retinotopic map refers to the orderly mapping of receptive field position in retinotopic coordinates in a brain region. Retinotopic maps can be characterized quantitatively by the rate of change of position in the cortex as a function of position in visual space, expressed as mm/degree. This quantity is known as magnification factor (M). A high value of M means that a large area of cortex is devoted to analysing a small region of visual space. In primates, M is highest in the fovea and decreases reciprocally with visual field eccentricity. The overall relationship between visual field position and position in the cortex can be described approximately by a complex logarithmic function $\mathbf{w}=a\log(b+\mathbf{z})$ where $\mathbf{w}=(x+iy)$ gives the position in the cortex in mm, $\mathbf{z}=rei\theta$, where r is the visual field eccentricity in degrees, θ is the meridional angle in radians and a and b are experimentally determined constants. This function, first proposed by Schwartz [26,32], describes the type of mapping observed in primary visual cortex (V1) of primates, including humans, including the over-representation of the fovea (the centre of gaze) and the tendency for lines of constant meridional angle and constant eccentricity to be mapped as orthogonally intersecting straight lines on the surface of the cortex. The characterization of neuronal maps via analytic (conformal) functions, has a developmental interpretation as potential flows on surfaces, subject to boundary conditions imposed by the shape of the surface. Thus, Schwartz applied the Dirichlet Principle to demonstrate that the minimization of the average magnitude of the anatomical magnification factor (by unit area) is sufficient to encode the map based on the shape of the boundary conditions. In particular, the cortical domain is clearly a rectangular strip, since the logarithmic map conformally transforms the retinal annulus to the latter cortical strip. Yet, the domain of any branch of the complex logarithm function, implicitly states that due to the domain periodicity of the imaginary axis, the orientable 2-torus, is considered to be the case, where the implicit assumption is that the complex plane is considered with the usual orientability, rather than the non-orientable topology hereby presented in our previous discussions, which is crucial to non-linear dynamics. A consequence of this is that the center of the image point of the visual map is not conserved but pushed to the boundary of the hypercolumn, so to speak, the geometrical-physiological-anatomical center is not kept as an organizing center on the anatomical-perceptual topographic map on the cortex which thus becomes asymmetric. But as proposed by Schwartz, would we choose for the topology of the domain of definition of the complex logarithm, whose principal branch has for domain the non-orientable Möbius strip, then this conservation of the centre is the case and the symmetry of the image on the cortex corresponds to the symmetry of the input on the retina [32],. We recall that the principal branch of the complex logarithm, defined on $z= re^{i\theta}$, with $\theta \in (-\pi,\pi)$, so the imaginary axis establishes a 2π-periodicity, i.e. is torus-like on gluing along the boundary of the imaginary axis domain. While the multivaluedness of the complex logarithm is dealt by defining it on a Riemann surface, which by default is orientable (either locally a 2-torus or a 2-sphere), this construction of the domain as non-orientable, places the orientable Riemann surface of visual representation as the global 2-torus on which the non-orientable Möbius strip is embedded as the 2:1 resonance that unifies both hemispheres, as is the case already of the homunculus seen as the trans-corpus callosum integration of the Möbius strip of this homunculus to give the seemingly 2-torus topology of the body. Would surgery cut across the corpus callosum which integrates both hemispheres, the integration is no longer the case, separating the RFs into symmetric parts, as empirically verified [83].

15. The somatosensory system and the Klein Bottle

The somatosensory system embodies the cutaneous surface and pressure receptors enclosed by the body surface; thus, the perceptual modalities of it are touch and kinestesics. The anatomical constituents are the afferent nerve fibers which originate from the peripheral receptors, ascend in the posterior funicule of the spinal cord. It forms synaptic relays in the dorsal column and the ventrobasal nuclei of the thalamus, to finally project to a topographic representation map of the body in the postneural gyrus of the cerebral cortex, known as the somatosensory area 1. In the projections of afferents from the body to the cortex, the nervous receiving system receives signals from tactile (cutaneous) and kinesthesic receptors arranged in separate cell columnsss oriented perpendicular to the cortical surface. For the cells that compose each modality pure cell column, it was ascertained that the peripheral RF of a neuron situated at one depth of a cortical cell column is representative of the approximate RF location of all neurons in the same cell column. Thus, the same cortical columns that are the cortical arrangement for the topographic maps of the visual mode, have a somatosensory correlate. The neurons of each cell column contribute as one unit of which the cortical maps are composed. In fact, the cell columns of the two somatosensory modalities, touch and depth) intermingle in close proximity as a mosaic made of cell columns with RFs interior to the body surface and those representing the outer skin. Thus, they represent the functional-anatomical perceptual units in which the exterior world is interiorized to the cortical system, which being of the embryonic exterior cells that by torsion folding are brought inside, is thus in terms of developmental evolution of the embryo, units that sustain a topographic map from the outer world in an inner world which is historically and determinatively also exterior. This interiorization which is physiologically reversed as a sensory system, is the embodiment of the Klein Bottle (discovered by Werner [29) and its logic, fusing the exterior and interior worlds by self-penetration. This is a mapping and realization of the ectoderm 2D surface and depth RFS which due to the columnar organization of the sensorium as anatomical-physiological units on what is thus a 2D cortex, produces the notion of a body as if 3D on what is in its historical embryological development of turning inside-out and outside-in of the embryonic egg, is a 2D phenomenology which sustains our idea of a 3D body, and still a 3D world at large.

16. The topographic visual map, the conformal inversion and non-orientability

It is quite remarkable the possibility that the identification of the complex logarithm with the visual topographic map opens to the characterization of vision also in regards to the anatomy of the eye. Indeed, the eye points inward: the photoreceptors point not to the exterior light but inwards. We recall that the retina of the human eye is the returning migration to the exterior of the body of ectodermal (i.e. exterior) tissue that by torsion folding the development of the embryo constitutes the brain. Also, the modular anatomical-physiological unit of the cortex first proposed by Mountcastle and retaken by Hubel and Wiesel, the hypercolumn of cortical neurons piled as vertical columns (i.e. normal to the orientation of the cortex, which topologically is a sphere) all responding to sensorial input as a unit, have for precursors the migration in an early developmental stage of radial glial cells forming minicolumns so turning inside out is not only the phenomenae of formation of the eye, but also of the cortex [82]. Thus, it closes a cycle in which the exterior of the developing organism has infolded to outfold for completion of the mature organism.

The fact is that the derivative of the complex logarithmic map is the function 1/z, a particular case of a Möbius *transformation* in complex analysis: $z \rightarrow (az + b)/cz + d$, of one complex variable ; here the coefficients a, b, c, d are complex numbers satisfying $ad - bc \neq 0$. These transformations of the extended complex plane, the Riemann sphere Σ, form a Lie group, isomorphic to the Lorentz group of Special Relativity. There is a one-to-one mapping between Möbius transformations and Lorentz transformation on Minkowski spacetime [66]; in fact, this is the basis for the theory of spinors as elaborated by Penrose [103]. If we take a= 0, b=1, c=1 and d= 0, then we get the complex derivative of the logarithmic map.Yet, what is truly remarkable is that the Möbius transformations acting on a rectangle on the complex plane turns it inside out. In particular, the transformation $z \rightarrow 1/z$ produces the inversion and a reflection with respect to the real axis. Indeed, points outside and inside arbitrary circles on the complex plane are exchanged by the transformation $z \rightarrow 1/\bar{z}$, where $z = r\ \exp(i\theta)$, and the inversion is completed by a rotation transforming θ to $-\theta$, a reflection with respect to the real axis. It is easy to see that a directed segment of a line is mapped onto another directed segment with the opposite orientation [66]. If we now translate the results on $C$to Σ, we find that the inversion $z \rightarrow 1/z$ on C induces a rotation of Σ about the real axis through an angle of π. The inversion on C which transforms 0 to ∞ and ∞ to 0, swaps S= 0 with N= ∞ through the rotation of π about the x axis. It further transforms a neighbourhood of ∞ to a neighbourhood of 0. Thus to study the behaviour of a map on a neighbourhood of ∞ we transform to a neighbourhood of 0. That means that to study f(z) on a neighbourhood of ∞ we need to study the behaviour of F(z) = f(1/z) on a neighbourhood of 0. This can be used to extend the concept of a branch of a complex function at 0 to a branch at ∞. For example, log(z) has a branch at 0, so that F(z) = -log(z) has a branch at ∞.

Hence, the issue of divergences in real analysis, can, in principle, be surmounted by considering a complex extension in which divergences are considered analytically in the neighborhood of the origin; would this be the case, the failure of real analysis criticized in [55] would be surmounted by considering instead complex analysis, which is already the case of visual perception.

This swapping of the South Pole 0 with the North Pole ∞, would we associate the former to the centre of the iris and the latter to the fovea (the retinal point of the projection centre of gaze), characterizes the gist of the fundamental operation which transforms the Outside world to the Interior world, the transformation of visual perception. Furthermore, vectors on the complex plane that have a certain direction, under inversion they have the opposite direction, and on a map inversion acts as depicted below.

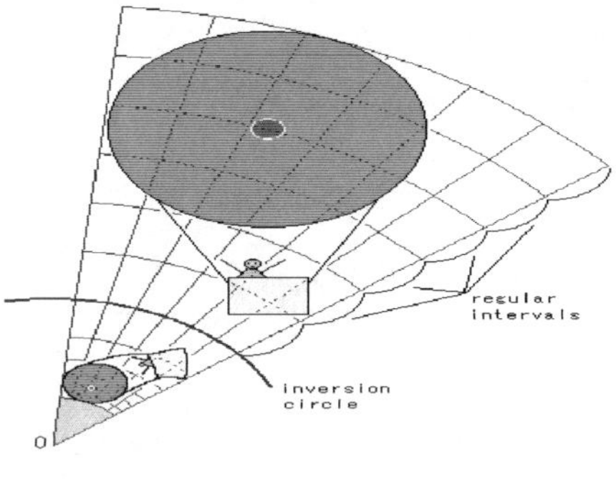

Figure 57: Action of the inversion on a map.

Thus, the inverted orientation of the photoreceptors of the retina, rather than being an anomaly, a defect of design, would there be any design in biology, a notion very strongly contested by Darwinists [139], are the very proof of the logic of the eversion by which organisms develop: they functionally operate assuring that the Outside world is mapped inside and viceversa. Their logic is the Klein Bottle. This eversion has for physiological manifestation the so called multistable perception, as is the case of the continuous transformation of the perception that arises from focusing on a 2D representation say of a cube, yielding two alternative visions, and established depth (actually the self-penetration of the Klein Bottle retinotopic map) as the primal time dimension [134,135,140] creating thus 3D space, in fact a double 3D space as we shall briefly comment further below [12,14].

Yet, as a Möbius transformation it swaps S and N, and we claimed to the be the gist of the turning outside in of the world through vision. It is by introducing the Riemann sphere, the one-point compactification $\infty = -\infty =$ N, allows to continuously connect N with the imaginary axis by introducing further a π twist at ∞. In this we see the transformation of Time, that in the complex plane rotates the real and imaginary axis, on the Riemann sphere through inversion it swaps S and N by a π rotation about the x axis but further requires a twist on N to continue to link S with $i=\sqrt{-1}$ as depicted in figure 31).

17. On the Klein bottle psychophysics of music perception

In contrast with visual space which is "out there'', the omni-directional nature of the auditory field places the listener still more firmly in space: auditory space is all around – and even inside - the listener (depth is thus crucial to audition), and the body is much more a participant in space. Thus, whereas visual space can give us the illusion of detachment, the auditory mode involves distinctly the body of the listener, whose sensorial modes have the Klein Bottle for the topology of its topographic maps. In this section we shall deal with music perception, which also has a non-linear topology This non-linear structure of aural space was empirically verified by psychologist Roger Shepard [86], which he further associated with the double helix, as in the standard model of DNA, and became a point of departure in studies in experimental psychology and music. To resume the gist of the phenomenon: 1) "Subjects interpret circularity'', i.e. the re-entrance of pitch space on itself, "in the frequency doubling at the octaves'' this is the fundamental 2:1 resonance, and 2) "Subjects interpret tones in an interval having a tendency or tension to move up or down on whether it is less than or greater than a half octave or tritone'' ([91], page 39 and 40). This is the so-called Tritone Paradox [84,85], "which indicates that we perceive pitch intervals as ascending when in an interval between 0° and 180°, and descending when in an interval between 180° and 360°'', the 180° being the twist that produces the Möbius strip, as in the Figure 49. This is the case for the perception of two pitches of an octave, presented either sequentially or simultaneously. Topologically, this means that the interpretation of pitch as ascending or descending, for an octave, can be identified with the Möbius strip, in one local side we interpret as ascending, in the other local side, as descending. Already, musicologist Dimitri Tymoczko through mathematical arguments in terms of orbifolds, proposed that ordered tones lie on the Möbius strip [87], yet stopped short of providing a connection with the physiology of hearing. Roughly speaking, the connection is that aural perception (not necessarily restricted to music), as is the case of the sensorium's codification, has the topology of the Klein Bottle , while anatomically, the outer ear has the topology of the Möbius strip (which is built-in in the Klein Bottle) and the cochlea has the shape of a Golden Number spiral, the generic architecture for harmonics.

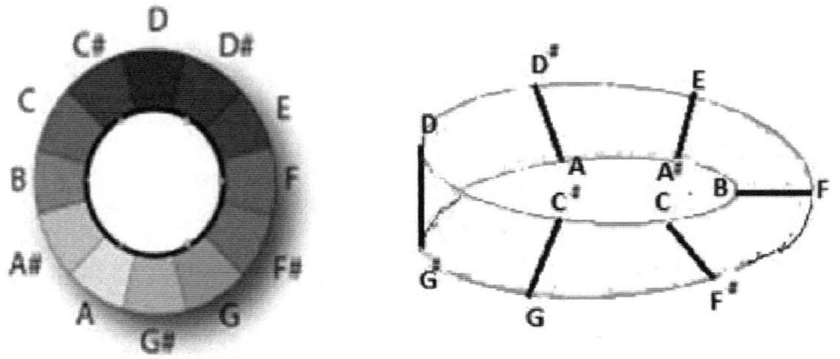

Figure 57. Left: topological representation of the chromatic aural space of an octave given by the disk-type real projective plane, represented as disk without a centre. The colours represent the synesthesic nature of perception in which the pitches are each associated with a colour according to the frequency [87]. Right: we depict without the colours the octave on the single edge of a Möbius with the opposite points joined by lines representing the tritone (half octave) perceptual identification of the lhs figure of a circular octave. This represents the fundamental 2:1 resonance: A complete rotation on the lhs circular pitch space of an octave perceptually translates into <u>two</u> complete rotations

in the Möbius strip, say D-D#-E-F-F#-G-G# followed by G#-A-A#-B-C-C#-D, completing the single edge of the Möbius strip The perceptual space turns to be the Klein Bottle surface, on identifying the antipodal points –depicted by lines on joining the antipodal pitches on the edge of the Meobius strip- as perceptual identities; the Klein Bottle thus arises as the identification of the equally oriented sides of Figure 18 B, since already the identification of opposite orientations of the other two sides produce the Möbius strip. These lines represent the perceptual identification of the tritone (half-octave). Therefore, perceptual space -according to the Tritone Paradox- is a Klein Bottle surface, which has not been acknowledged before. Music perception assents to the 2:1 resonance being a lived world. We recall that Kepler, in his *Harmonices Mundis*, created an intricate model of the solar system in terms of music harmonics, further related to the Platonic solids, yet this notion of the universal and foundational character of music has been intensively patronized as deeply wrong and misleading [115].

18. Forewords

The theory we have presented has rested in topological considerations and in terms of self-referential torsion geometries; we have chosen to leave equations aside, as they appeared to be unnecessary. The universal relevance of rotational vis-à-vis linear motions has been revealed, and spinor fields rather than exclusive to microscopic physics, are universal, as claimed by Kiehn [95]. The 2:1 resonance has appeared as a fundamental organizing principle. We have shown that the evolutionary generation of stable shapes in terms of wavefronts is related to dissipation process, as a form of attenuation of compressibility by dissipating kinetic energy through heat. Harmonic fields as proposed in Fock's theory of gravitation for its foundation have shown to be indeed essential even for geophysical morphogenesis. Yet, in terms of the organization of structures in terms of the 2:1 and higher resonances, the issue of dampening is crucial. Since without these dampening processes, the persistence of the processes and structures thus generated would not be the case. Interference must be at the roots of the persistence of these processes and structures, since on the contrary, superposition of resonances would lead to their destruction.

Richard Merrick proposed in his mathematical theory of music and its cognition/ perception, that the tones in aural space are produced by a standing wave [93].The Periodic Table of Elements of Chemistry of Mendeleev, which following its inception was claimed to be related to harmonics, has been proved to be related to Golden Mean spiral dispositions of the natural numbers (representing atomic numbers) which can be mapped, to a Klein Bottle topology for the atomic (and their stable isotopes) elements of the Periodic Table [58]. The non-orientable topology of the Periodic Table including the stable nuclides, which reflects the exclusion principle of fermions, appears to originate in the Fibonacci spiral geometry of the natural numbers that stand for the fundamental parameter of the periodic table, namely, atomic number, i.e. number of protons (electrons), and the ensuing proton number over neutron number ratio, and its relation with the Golden mean Φ; in this theory, the Aufbau orbitals filling rules that conform to Fermi's exclusion principle are related to the periodicity properties of this spiral geometry of atomic number, and particularly to those of prime numbers [129].

Furthermore, this topology of the Periodic Table implies a unification –rather than the duality purported to be a fundamental principle for Physics- of atoms of matter and anti-matter as lying in either side of the Möbius strip or the Klein Bottle, which can be mutually inter-transformed, alike the unification of pitch as ascending or descending, in the same form. Any perturbation to an atom producing radioactive decay returns to the stable Φ (the Golden Mean) structuration on the Klein Bottle of the Periodic Table. Furthermore, the atoms and stable isotopes can be generated from a sinusoidal standing wave, as in Merrick's theory of music. Therefore, the torsioned/twisted geometry of harmonics and the Golden Mean, Φ, lie at the very basis of bodyplans (as shown by Leonardo da Vinci in his *Man of Vitruvius*), the musical, perceptual and physical/chemical world, and further to the genetic code [12], and to the Matrix Logic derived from the Klein Bottle Logic [14,15,20]. Furthermore, the natural numbers have been proved to have a rotational planar structure [112] that allows for the potential generation of all the natural numbers and still to identify ad negatio the prime numbers. It is plausible to hypothesize that the natural numbers might be generated in toto by standing waves, at least already mathematically proved in the case of the Periodic Table and music; notably, the stratigraphic Möbius-Klein Bottle Analemma formations of geology, are produced by the harmonics of the Sun, Moon, Earth and other planets interactions [75], and structures crystalize/sediment with respect to the teleo-logical action of Φ. Indeed, it has been found that the diameters of the Analemmic counter-rotating cells on Earth's crust and in the atmosphere, have for quotient of their diameters the approximate value $\Phi \approx 1.618$ or $1/\Phi$. It is remarkable that the Cornu spiral, whose duplication by connecting both ways the dipole singularities in terms of which the Analemma is organized, were used by Huygens for evaluating the Fresnel integrals which show up in the evaluation of the diffraction intensities for the Fresnel diffraction of the light from a slit.

The essential role of standing waves is here evidenced: they provide for the generation, of structures and processes, while the Fibonacci sequence provides the meta-algorithm for assimilation of disturbances, whatever their domain may be. Indeed, already the photon which is a standing wave transaction between emitter and receptor of electromagnetic waves, manifesting as a singularity of the wave, biological, musical, geological, etc., yet in the present phenomenology, they appear as time waves from which the physical (in the ample sense) world comes to be organic (in the sense of wholeness), unseparably from the Klein Bottle, as the logophysics sustaining this organicity. In both the aural space and the Mendeleev table, the Klein Bottle embodies the organicity (in the sense of holonomy) of aural and atomic/chemical material space and particularly of massless particles (photon and neutrino), while Φ appears as the asymptotics teleo-logical result, of the most fundamental meta-algorithm claimed by Johansen [113]. The Fibonacci sequence as an algorithm, embodies the physically mensurable relations, of the unfolding fusion of the world of thought with the physical world, to reach for stable manifestations; this teleo-logical process is reached by interaction with the enviroment, so it is an issue of contextualization, embodied by the Klein Bottle Logic. Thus, it is universal, applying to all domains, whether biological, cognitive, economical, physical, in the natural numbers, music [91], astro/geo-logical (through the Croll-Milankovitch cycles and the Möbius strips stratigraphy,, astronomical (the distribution of planets in the Solar System), and chemical/cosmological, as related to the Klein Bottle topology of Mendeleev Table . In the latter case, we see that the conformation of periodicity in a graph with atomic number N as independent variable and the ratio of number of neutrons to the number of protons as dependent variable, where the periodicity of the table appears when the latter ratio converges to $1/\Phi$ following the Fibonacci sequence from which a ratio equal to 1 can be extrapolated. The latter value corresponds to the formation of more complex atoms starting from hydrogen, by the addition of α-particles, under very extreme pressures, say in the interior of a star, with a notable structure, which amounts to the turning inside-out of the electron orbital ordering, which later, in less extreme pressure conditions, turn to their usual electronic configuration proposed early in Quantum Mechanics, while the high-pressure conditions produce a neutron which itself has a Klein Bottle topology. A figurative representation of this would be that a spherical plasma from which the Solar system was thought to have been born, would have turned inside-out to constitute the planets which were further ordered into their present orbits following the Golden mean relations through the recurrence of the Fibonacci sequence, while the Sun retracted to become the center of the Solar System. Indeed, the inversion of orbitals occurring in the Periodic Table as a signature of its non orientable topology, points out to an unacknowledged eversion of the spherical O(4) symmetry of the 1/r potential in atomic physics. We owe to an anonymous reviewer's query the motivation for the present observation.

So, on the one hand we have that the naissance of atoms from hydrogen atoms, the latter being the building blocks of the atoms in the periodic table, by inversion of their electronic orbital-like configurations on reaching for their stable organization achieved by fusion of hydrogen atoms; this is similar to the case of aural perception in which the sequence of tones perception is inverted in aural space in its Klein Bottle topology. So the dual logic is not merely the logic of the evolutionary formation of the material world, but rather a logic for generic morphology in which Inside and Outside are fused, the Klein Bottle Logic, which we proved to be the logic of image recognition both in the body and in the recognition of natural images taken by digital devices. We have found that the turning-inside-out of an ovum, is also the logophysics by which a fertilized egg unfolds to become an embryo [11]. Furthermore, in the unfolding of the periodic structure of the material atomic/chemical world, basic also to the biological realm, the Fibonacci sequence acts as the generative meta-cognitive meta-algorithm which constrains standing waves, by dampening; as we already discussed in perturbation of atoms, this self-referential sequence produces their radioactive decay to their stable harmonic configurations. This meta-algorithm drives the Plenum to irreductible singularities; in the case of the natural numbers to evidence the prime numbers [112,113]; for a discussion on the ontology of this meta-algorithm and its role in number theory we refer to [120], who already claimed in [121] its primeval role as a universal meta-algorithm The relation between Φ, proportions of bodyplans of animals, the genetic code and music was elaborated in (Petoukhov & He). But still more important than Φ's guidance of body development to the proportions of lengths of bones (musical pitches, atomic weights, etc.) is the notion that it might be more fundamentally related to the gaps between them (more generically, of singularities) as the result of the contextualization with the environment; already the photon as the primeval domain of articulation of the physical and thinking realms, generates the Fibonacci sequence. In other words, Φ may be related to the domain of articulation as a teleo-logical feedback guide directing through recurrence of the Fibonacci meta-algorithm the unfoldment of the process towards stability and holonomic coherence, of music, of the material world as appears to be the case of the Periodic Table) and the freedom to move, create and recreate. (At this point we urge the reader to return to read the last two paragraphs of Section 12). This is also already the case of the Matrix Logic derived from the Klein Bottle Logic, in which the Golden Mean is associated to OR, the logical disjunction operator. This conception of the role of harmonics in the case of the gaps of bones, is related to the total null torsion of bodyplans, postulated in [64], following work in the surmountal of the Cartesian Cut and human gait [63]. Finally, inasmuch the Klein Bottle logic arises from the notion of a boundary or distinction (after Spencer Brown) which is subverted by time waves motions exchanging the dual states Inside-Inside and Outside-Outside, we have seen that whenever a fringe image is perceived, this requires that the multivaluednes of the orientation of the boundary, is dealt with the multistate Klein Bottle logic. Therefore, the issue of the discontinuity of the boundary value for the fundamental electromagnetic field propagation, has for its visual manifestation, the indefiniteness of the orientation of the phase of the signal on the boundary, producing thus the Klein Bottle , and for its geophysical manifestation, the Analemmas on the crust or still on the atmosphere, revealed in [75].

One of the foundational issues for mathematics and physics that we have dealt with in the framework of this paradigm is the non-orientable topology of the complex plane and the Riemann sphere, which has been unacknowledged before at large but by Boeyens. While physical measurements requires real Hilbert spaces and it is the spectral theory of Hilbert spaces that extends this to complex Hilbert spaces, it is the geometry of complex projective spaces (the Fubini-Study metric) of importance to quantum mechanics. There is still a torsion potential which drives two-state quantum *open* systems to manifest a single spectral value [118,119]; the setting for this is the random Brownian motions associated to the random Schrödinger equation for open systems, while the case of the linear and non-linear torsion geometry of the usual Schrödinger equation were presented in [118]; thus torsion geometries which are already intrinsic to fluid and magnetized fluids are also the case of quantum systems. Stueckelberg, on stressing the real Hilbert spaces noticed that charge required complex Hilbert spaces [122,123];. This leads to consider whether charge (following a query of an anonymous reviewer to this article) could be related to the Klein Bottle non-orientability of the complex plane.

In the last years a self-referential construction for physics and mathematics has been formulated in terms of the Dirac algebra of quantum mechanics in the framework of the Nilpotent Universal Rewrite System (NURS); it was further mapped to the double-strand model of DNA in the works by Rowlands and Hill [116,117]. While this theory was proposed to be based on fermion-boson dualism and dualism as a universal principle, it has been argued that this alleged dualism is actually based on the Klein Bottle logophysics [12]; we already saw in this article that there is a continuity between fermions and bosons, as provided in the two first windings that generate the self-referential torsion geometry and dynamics of the mammal heart, or still the tidal morphogenesis of Earth's crustal dynamical configurations. Thus the self-referential NURS was argued to be based in the same non-orientable topology basic to our current theory. The Dirac algebra representation in NURS requires complex quaternions, and thus a double 3D space appears: The elements with real coefficients will generate a 3D position space while the imaginary ones generate charge or still angular momentum space; this stems from the fact that the only truly fundamental parameters are space, time, mass(-energy) and charge. A double 3D space is necessary for the unfolding spiral growth of molluscs without breaking [128]. For further discussions on this construction of the double 3D space, the genesis of space and its relation with the multistable visual perception which is the functionality of the Klein Bottle Logic of visual perception discussed in this article, and the standard double helix model of DNA superseded by the Klein Bottle representation fractal structure of genomes, we refer to [12]. Remarkably, the 2:1 resonance characteristic of the non-orientability of the Möbius strip and the Klein Bottle, which we have evidenced in very diverse fields and demonstrated to be a precursor to Newton's second law, appears to be the case of the human and chimpanzee already sequenced genomes, an issue that we shall discuss elsewhere and further in relation with the quantum coherence that is established in genomes due to this resonance.

Acknowledgments
It is a pleasure to express my gratitude to Prof Martin Land and Prof Larry Horwitz for their kind support, to Craig Nevin and to Mikhail Bazanov for discussions on biomechanics (with the latter in 2009). To Sonia, without her love and support this work would not have come to be but a figment of my imagination.

References

[1] Schrödinger E 1991 *What is Life?* (Cambridge University Press Cambridge,UK)
[2] Gordon R 2002 *The Hierarchical Genome and Differentiation Waves: Novel Unification of Development, Genetics and Evolution* (World Scientific, Singapore)
[3] Gordon NK and Gordon R 2013 *Embryogenesis Explained* (World Scientific, Singapore)
[4] Forgacs G and Newman S 2000 *Biological Physics of the Developing Embryo (*Cambridge Univ.Press, Cambridge, UK)
[5] Maresin V M and Presnov E 1985 *J. Theor. Biol.* 114: 387–398
[6] Jockusch H and Dress A 2003, From Sphere to Torus: A Topological View of the Metazoan Body Plan. *Bull. Math. Biol.* 65: 57–65
[7] Nishii I and Miller SM 2010, Volvox: simple steps to developmental complexity? *Curr Opin Plant Biol.* Dec;13(6):646-53
[8] Levy S 1995, *A brief history of sphere eversions. In: Making Waves* (A K Peters Ltd, MA).
[9] Rapoport D L 2013, HyperKlein Bottle Ontopoiesis of Life and the Cosmos, 130 pages, to appear, Analecta Husserliana Volume CXVII..
[10] Rapoport DL 2013. Klein Bottle Logophysics, Cell Biology, Chemistry and Body Integration, In: Proceedings of CASYS'11, Conference on Computing Anticipatory Systems (Liege, August 2011).
[11] Rapoport D L 2011 On the Fusion of Physics and Klein Bottle Logic in Embryogenesis, Cell Biology and Evolution, *NeuroQuantology* 9 (4); 842-861
[12] Rapoport D L 2011 Surmounting the Cartesian Cut: Klein Bottle Logophysics, The Dirac Algebra & the Genetic Code *Neuroquantoloy* 9 (4): 862-881
[13] Boeyens J C A 2005 New Theories for Chemistry (Elsevier, Holland, 2005).

[14] Rapoport D L 2011 Surmounting the Cartesian with Philosophy, Physics, Logic, Cybernetics and Geometry: Self-reference, torsion, the Klein bottle, the time operator, multivalued logics and quantum mechanics, Found. Phys. 41(1): 33-76.
[15] Rapoport D L 2011, Torsion Fields, the Extended Photon, Quantum Jumps, The Klein Bottle, Multivalued Logic, the Time Operator, Chronomes,Perception, Semiosis, Neurology and Cognition. In: *Focus in Quantum Mechanics*, D. Hathaway and E. Randolph (eds), (Nova Science, NY).
[16] Stern A 2001 *Quantum Theoretic Machines* (Elsevier, Amsterdam).
[17] Russell B and Whitehead A 2011 *Principia Mathematica*(Rough Draft Printing)
[18] Spencer-Brown G 2010 *Laws of Form* (Bohmeier Verlag,Leipzig).
[19] Rapoport D L 2010 Torsion, propagating singularities, nilpotence, quantum jumps and the eikonal equations. In: Computing Anticipatory Systems, Proceedings CASYS'09 , Daniel M. Dubois (ed.), AIP Conf. Series 1303. (Springer, Berlin).
[20] Rapoport DL 2009, Torsion Fields, the Extended Photon, Quantum Jumps, the Eikonal Equations, the Twistor Geometry of Cognitive Space and the Laws of Thought. In: *Ether, Spacetime and Cosmology vol. 3: Physical Vacuum, Relativity and Quantum Mechanics.* Duffy M and Levy J (editors): 389-457, (Apeiron Press, Quebec, Canada).
[21] Pettigrew J B 1908 *Design in Nature* (Longmans, London).
[22] Petigrew, J B 1874 *The physiology of circulation in plants, in the lower animals, in man*(MacMillan, London).
[23] Lehar S 2003 *The World in our Heads: A Gestalt View of the Conscious Experience*, (Lawrence Erlbaum Associates, London)..
[24] Von Helmholtz H 1971, *Selected Writings of Hermann von Helmholtz* Kahl, Russell, ed. (Wesleyan Uni. Press, USA).
[25] Hartmann W 1997 *Signals, Sound and Sensation* (Springer, Berlin).
[26] Schwartz E L 1980 Computational Anatomy and Functional Architecture of Striate Cortex: A spatial Mapping Approach to Perceptual Coding, *Vision Research* 20. pp. 6.85.
[27] Carlsson G, Tigran I, de Silva·V and Zomorodian A 2008, On the Local Behavior of Spaces of Natural Images *Int J Comput Vis* 76: 1–12
[28] Carlsson G 2009, Topology and data, *Bulletin AMS Math.* Soc. *46*, No 2: 255 308.
[29] Werner G 1968, The Topology of the Body Representation in the Somatic Afferent Pathway. In: The Neurosciences Second Study Program, 605-617, Schmitt FO (editor), Rockefeller University.
[30] Swindale N 1996, Visual Cortex: Looking into a Klein Bottle, *Current Biology* 6 No 7:776–779
[31] Tanaka S 1997, Topology of Cortex Visual Maps, *Forma* 12:101-108
[32] Schwartz E L 1977 Afferent maps in the primate visual cortex and the generation of neuronal trigger features, *Biological Cybernetics* 28:1-14
[33] Scott A 2002 *Encyclopeadia of NonLinear Science* (Routledge, Taylor and Francis, London)
[34] Rayleigh J 1984 *Theory of Sound* (MacMillan, London)
[35] Jenny H 2001 *Cymatics: The Study of Wave Phenomena, (*B& W Publishers, London)
[36] Arévalo J, Aguilar L M, Robledo-Sánchez C, Arroyo Carrasco M L, Mendez Otero M, The principle of superposition for waves, (preprint, arXiv:1005.3237)
[37] Hoffman R 1995 *The Same and Not the Same* (Columbia University Press, New York)
[38] Sokolov I 1973 Topological Methods in Stereochemistry. *Russ Chem Rev* 42 (6).
[39] Mae-Won Ho 1994, What is (Schrödinger's) Negentropy? Modern Trends in BioThermoKinetics 1994,
[40] McClare C W F 1971 Chemical machines, Maxwell's demon and living organisms , J Theor Biol 30: 1-34
[41] Stapien M et al 2007 Angewandte Chemie 119, Issue 41, pgs. 7859-8047
[42] R. Van Wijk (2001), Journal of Scientific Exploration, 15, No. 2, pp. 183–197.
[43] http://www.cs.sandia.gov/newsnotes/2009newsnotes.html#klein

[44] Herges R 2006, *Chem. Rev.* 106, 4820-4842
[45] Rzepa R (2005) Chem. Rev.105, 3697-3715
[46] Ingber D. The architecture of life. *Scientific American*, January, 1998: 44-57
[47] Oschman J L 2000, *Energy Medicine: The Scientific Basis* (Harcourt Brace, Edinburgh)
[48] Gurwitsch AG.1932 and 1959 *Die mitogenetische Strahlung* (Fischer Verlag. Jena)
[49] Gariaev P, Gariaev P.P., Birshtein B.I., Iarochenko A.M, Marcer P.J.,Tertishny G G, Leonova E A, Kaempf U 2001, The DNA-wave biocomputer. «CASYS» – International Journal of Computing Anticipatory Systems (ed. D.M.Dubois), Liege, Belgium, 10, 290-310
[50] Albrecht-Buehler, G 1992 *Proc Nat. Academy of Sciences USA* 89, 8288–8292
[51] Svensson B 2008 A Multidimensional Filtering Framework with Applications to Local Structure and Image Enhancement, (Linkoping Studies in Science and Technology no. 1171, Sweden)
[52] House M R 1995,Orbital forcing time scales: an introduction, *Geological Society, London, Special Publications* 1995; v. 85; p. 1-18
[53] Freeden W and Schreiner M 2009 *Spherical Functions of Mathematical Geosciences:A Scalar, Vectorial, and Tensorial Setup*, (Springer, Berlin)
[54] Wahr J 1996, *Geodesy and Gravity: Class Notes*, (Samizdat Press, Denver)
[55] Yi Lin Y and Yong Wu 2002, *Beyond Nonstructural Quantitative Analysis - Blown-Ups, Spinning Currents and Modern Science*, (World Scientific, Singapore)
[56] Bennett K D 1990 Milankovitch cycles and their effects on species in ecological and evolutionary time, *Paleobiology*, 16 (1), 1-21
[57] Clark II G 1974 Growth Lines in Invertebrate Skeletons *Annual Review of Earth and Planetary Sciences* 2: 77-99
[58] Boeyens J C 2010, *Chemical Cosmology* (Springer, Berlin).
[59] Veblen O & Hoffmann H 1930, Projective Relativity,*Phys.Rev.* (36) 810–822.
[60] Fischer P, Iofe A, Jacobson D L, Arif F, Meze F 2000 4π-Periodicity of the spinorwave function under space rotation, *Nuclear Instruments and Methods in Physics Research A* 440; 575--578.
[61] Rapoport D L and Sternberg S 1984 On the interactions of spin with torsion *Annals of Physics* vol. 158, 447-475 .
[62] Rapoport DL 2005 On the unification of geometric and random structures through torsion fields: Brownian motions, viscous and magnetic fluid-dynamic, *Found. Phys.* 35, no.7, 1205-1244
[63] Nevin C 2010 Initiation and Control of Gait Principles: A Mathematically Animated Model of the Foot, Dissertation.com, Boca Raton, 2010.
[64] Nevin C and Rapoport D L. The self-referential bodyplan; Klein Bottle bauplans, to appear.
[65] Ling G. *A* 1982 *Revolution in the Physiology of the Living Cell.* (Krieger Publs, Malabar, Fl)
[66] Needham T 2009, *Visual Complex Analysis* (Clarendon Press).
[67] Kozyrev N A 1991 *Selected Works*, published by Leningrad State Univ., 1991. 488 p.
[68] Kozyrev N A and Nasonov VV 1980,. On some properties of time discovered by astronomical observations. In: Proyavleniye Kosmicheskikh Faktorov na Zemle i v Zvezdakh (Manifestation of Cosmic Factors on the Earth and in Stars). Problemy Issledovaniya Vselennoy (The Problems of Studying the Universe). 9th issue, pp.76-84. Moscow-Leningrad (in Russian)
[69] Hayasaka H and Takeuchi S 1989 Anomalous weight reduction on a gyroscope's right rotations around the vertical axis on the earth. *Physical Review Letters* 63, N.25, pp.2701-270
[70] Levich A P 1996 *A Substantial Interpretation of N.A. Kozyrev's Conception of Time.* On the Way to Understanding the Time Phenomenon: the Constructions of Time in Natural Science. Part 2. The "Active" Properties of Time According to N. A. Kozyrev . Singapore, New Jersey , London, Hong Kong: World Scientific. 1996. Pp. 109-134 (World Scientific, Singapore) p. 1-42

[71] Torrent-Guasp F, Kocica M Corno M, Komeda M, Carrera-Costa M, Flotats F, Cosin-Aguillar A, Wen Han J 2005, Towards new understanding of the human heart, *Eur. J. Cardiothorac. Surg.* 27, 191-201, and references therein;
[72] Nasiraei-Moghaddam, A., Gharib, M., "Evidence for the existence of a functional helical myocardial band",Am J Physiol Heart Circ Physiol 296: H127H131, 2009;
[73] Grosberg A 2008 A Bio-Inspired Computational Model of Cardiac Mechanics: Pathology and Development, Ph.D. thesis, Caltech
[74] Rapoport D L 2007 Torsion Fields, Cartan-Weyl Space-Time and State-Space Quantum Geometries, their Brownian Motions, and the Time Variables, Foundations of Physics 37, nos. 4-5, 813-854, 2007
[75] Overton H, www.PorOgle.*blogspot*.com
[76] Ferronsky V I and Ferronsky S V 2010 *Dynamics of the Earth Theory of the Planet's Motion Based on Dynamic Equilibrium* (Springer, Berlin)
[77] Milankovic M 1941, *Canon of insolation and the ice-age problem: (Kanon der Erdbestrahlung und seine Anwendung auf das Eiszeitenproblem) Belgrade* (Israel Program for Scientific Translations, Jerusalem, 1969)
[78] Schwarzacher W 1993 *Cyclostratigraphy and the Milankovitch Theory* (Elsevier, Amsterdam)
[79] Marmanis H 2000, Analogy between the electromagnetic and hydrodynamic equations: Application to turbulence, Ph.D. thesis, Brown University
[80] Guyone E, Hulin P and Petit L 2001, *Hydrodynamique physique.* Deuxième édition, (InterEditions / CNRS Éditions,Paris)
[81] de Boer P and Trabucho J 2012, Orbitally forced sedimentary rythms in the stratigraphic records: is there place for tidal forcing? *Sedimentology* 59 (2):379-392
[82] Smith C M U, B 2008, *Biology of Sensory Systems* (Wiley,Blackwell, UK)
[83] Hubel D 1988, *Eye, Brain and Vision*, (Scientific American Library, NY)
[84] Deutsch D, 1987, The tritone paradox, *Percept and Psychology* 42 :563-567
[85] Deutsch D 1988, The semitone paradox, *Music Perception* 6: 115-134
[86] Shepard RN 1964. Circularity in judgments of relative pitch. *Journal of the Acoustical Society of America*, 36, 2346-2353
[87] Tymozcko D 2006, The geometry of musical space, *Science* 7 July 313 (5783): 72-74
[88] Marinelli R, Branko F, van der Zee H, McGinn A, Marinelli A 1995, The heart is not a pump: A refutation of the pressure propulsion premise of heart function, Frontier Perspectives (Journal of the Center for Frontier Sciences at Temple University in Philadelphia, Pa.)
[89] Bóna A and Slawinski MA 2011, *Wavefronts and rays as characteristics and asymptotics.*(World Scientific, Singapore)
[90] Fock V 1964, *The Theory of Space, ime and Gravitation* (Macmillan, London)
[91] Merrick R 2009, *Interference Theoy: A grand scientific Musical Theory* (USA)
[92] Durbin R and Mitchson G 1990, A reductional dimension framework for understanding cortical maps, Nature 343, July 15, 644-547
[93] Bateman H 1995, *Electrical and Optical Wave Motion*,(Dover, N.Y.)
[94] Cartan E 1966, *The Theory of Spinors* (Dover, New York, p. 39)
[95] Kiehn R *2009 Non-Equilibrium Systems and Irreversible Processes;* Adventures in Applied Topology Vol. 5 Topological Torsion and Macroscopic Spinors from a Perspective of Continuous Topological Evolution (Lulu, USA)
[96] Bohm D 2002 Wholeness and the Implicate Order (Routledge and Kegan, London).
[97] Heelan P 1989 *Space Perception and the Philosophy of Science* (University of California Press, Berkeley).
[98] Luneburg R 1948 *Mathematical Analysis of Binocular Vision*, (Edward Brothers,USA)
[99] Varèse E 1959 Spatial Music, in Schwartz & Childs (eds.): *Contemporary Composerson Contemporary Music*, p205.
[100] Purcell M C 2006 *Towards a new era: epistemological resolution analysis by through and from*

the Klein bottle wholeness, Ph.D. Thesis, Department of Philosophy, University of Newcastle, Australia.

[101] Stevens S S 1975 *Psychophysics: Introduction to its Perceptual, Neural and Social Prospects*, ed. G. Stevens (Wiley,New York)

[102] Hellige J B 1993 *Hemispheric Asymmetry: What's Right and What's Left* (Harvard University Press, Cambridge, USA)

[103] Penrose R and Rindler W 1996 *Spinors and Spacetime vol. I* (Cambridge University Press, UK)

[104] Prigogine I 1991 *From Being to Becoming* (Freeman, San Francisco)

[105] Lin Y 2008 *Systemic Yoyos: Some Impacts of the Second Dimension* (CRC Press, Boca Raton, Fl, USA)

[106] Lin Y and OuYang S 2010 *Irregularities and Prediction of Major Disasters* (CRC Press, Boca Raton, FL USA)

[107] Burchfield J 1976 *Lord Kelvin and the Age of Earth* (University of Chicago Press, Chicago)

[108] Kiehn R 012/29/89 to 04/12/95 Notes on Minimal Surfaces and Surfaces with Pinch Points:Applications to Electromagnetism Following Bateman.

[109] Penrose R 2002 *Shadows of the Mind* (Oxford University Press, UK)

[110] Gutzwiller M 1990 *Chaos in Classical and Quantum Mechanics* (Springer, Berlin)

[111] González-Díaz P and Alonso-Serrano A Observing other universe through ringholes and Klein Bottle Holes, In: Towards New Paradigms, Proc of the Spanish Relativity Meetingm (29 August–2 September) AIP Conf. Proc. 1458, pp. 311-314 (Springer, Berlin)

[112] Johansen S E 2010 Complete Exposition of Non-Primes Generated from a Geometric Revolving Approach by 8x8 Sets of Related Series, and thereby *ad negativo* Exposition of a Systematic Pattern for the Totality of Prime Numbers'', *Journal of Dynamical Systems and Geometric Theories* 8 (2): 101-171

[113] Stein Johansen 2013 *Outline of a Differential Epistemology,*. English translation of the Norwegian, published by University of Trondheim (1991). To appear

[114] Petoukhov S and He M 2011 *Symmetrical Analysis Techniques for Genetic Systems and Bioinformatics: Advanced Patterns and Applications* (IGI Global Publs, USA).

[115] Stephenson B 1994 *Kepler's Harmonics: the Music of the Spheres* (Princeton Univ. Press, Princeton, USA)

[116] Rowlands P 2009 *From Zero to Infinity: the Foundations of Physics* (World Scientific, Singapore)

[117] Hill V and Rowlands P 2008 Nature's Code, in Computing Anticipatory Systems. In: CASYS'07—Eighth International Conference; Dubois D M (editor) American Institute of Physics Conference Proceedings. Vol. 1051, pp. 117-126 (Springer, Berlin)

[118] Rapoport D L 2007 Torsion Fields, Cartan-Weyl Space-Time and State-Space Quantum Geometries, their Brownian Motions, and the Time Variables, Foundations of Physics vol. 37, nos. 4-5, 813-854

[119] Rapoport D L 2007 On the state-space and spacetime geometries of geometric quantum mechanics, In: *Foundations of Probability and Physics IV*, Proceedings of the Conference, Center for Interdisciplinary Mathematics and its applications, Univ . of Vaxho, Sweden, Khrennikov A and Adenier G, American Institute of Physics Conference Series (Springer, Berlin)

[120] Johansen S 2010 Fibonacci Generation of Natural Numbers and Prime Numbers In: Proceedings of the Third International Conference on Lie-admissible Treatment of Irreversible Processes (ICLATIP - 3) Santilli R M (editor), Kathmandu University, Nepal, April (2011)

[121] Johansen S 2006 Initiation of 'Hadronic Philosophy', the Philosophy Underlying Hadronic Mechanics and Chemistry. *Hadronic Journal* 29: 111-135

[122] Stueckelberg E C G 1960 Quantum theory in real Hilbert space, *Helv. Phys. Acta* 33, 727

[123] Adler S L 1995 *Quaternionic quantum mechanics and quantum fields (*Oxford, New York)

[124] Kauffman L 2012 *Laws of Form - An Exploration in Mathematics and Foundations*, In: http://homepages.math.uic.edu/~kauffman/Laws.pdf
[125] Varela F 1979 *Principles of Biological Autonomy* (North Holland, Amsterdam)
[126] Kauffman L 1978 De Morgan Algebras, Completeness and Recursion, In: Proceedings of the Eighth International Symposium in Multiple Valued Logics 82-86
[127] Rosch E.Varela,F and Thompson E 1991 *The Embodied Mind* (MIT Press, Cambridge)
[128] Illert C 1992 *Foundations of Theoretical Conchology* (Hadronic Press, Palm Harbor, Fl)
[129] Boeyens J C A and Levendi D 2008 *Number Theory and the Periodicity of Matter* (Springer, Berlin)
[130] Rapoport D L 2011 Self-reference, the Möbius and Klein bottle surfaces,multivalued logic and cognition, *International Journal of Computing Anticipatory Systems* 23, 103-113
[131] Husserl E 1970 Logical Investigations Part II (Routledge and Kegan, London)
[132] Ortiz C H 2003 Incomplete Symbols, Dependent Meanings, and Paradox, In: *Husserl's Logical Investigations*, Daniel O. Dahlstrom (ed.) pp. 69-93 (Dordrecht: Kluwer)
[133] Kaufffman L 2013 *Knots and Physics* (World Scientific, Singapore)
[134] Rosen S 2004 *Dimensions of Apeiron: A Topological Phenomenology of Space, Time and Individuation*, (Rodopi,New York)
[135] Rosen S 2006 *Topologies of the Flesh: A Multidimensional Exploration of the Lifeworld* (Ohio University Press, Ohio)
[136] Kiehn R 2003 Propagating Topological Singularities in the lightcone: the Photon http://www.findthatpdf.com/search-3518144-hPDF/download-documents-spie2ext.pdf.htm
[137] Kiehn R 2011 *Topological Torsion and Macroscopic Spinors Vol. 5 Non-equilibrium Systems and Irreversible Processes* (Lulu, USA)
[138] Post J 1995 *Quantum Reprogramming: ensembles and single systems, a two-tier approach to quantum mechanics* (Kluwer, Amsterdam)
[139] Dawkins R 1996 *The Blind Watchmaker: Why the Evidence of Evolution Reveals a Universe without Design* (W W Norton, UK)
[140] Merleau Ponty M 1962 Phenomenology of Perception (Routledge & Kegan Paul, London)
[141] Durbin R and Mitchinson (1990) *A dimension reduction framework for understanding cortical maps Nature* **343**, 644 - 647 (15 February 1990)

AUTHOR INDEX